World Sustainability Series

Series Editor

Walter Leal Filho, European School of Sustainability Science
and Research, Research and Transfer Centre "Sustainable Development
and Climate Change Management", Hamburg University of Applied Sciences,
Hamburg, Germany

Due to its scope and nature, sustainable development is a matter which is very interdisciplinary, and draws from knowledge and inputs from the social sciences and environmental sciences on the one hand, but also from physical sciences and arts on the other. As such, there is a perceived need to foster integrative approaches, whereby the combination of inputs from various fields may contribute to a better understanding of what sustainability is, and means to people. But despite the need for and the relevance of integrative approaches towards sustainable development, there is a paucity of literature which address matters related to sustainability in an integrated way.

Notes on the quality assurance and peer review of this publication

Prior to publication, the quality of the works published in this series is double blind reviewed by external referees appointed by the editor. The referees are not aware of the author's name when performing the review; the referees' names are not disclosed.

More information about this series at http://www.springer.com/series/13384

Walter Leal Filho · Rudi Pretorius ·
Luiza Olim de Sousa
Editors

Sustainable Development in Africa

Fostering Sustainability in one of the World's Most Promising Continents

Springer

Editors
Walter Leal Filho
FTZ-NK
HAW Hamburg
Hamburg, Germany

Rudi Pretorius
Department of Geography
University of South Africa
Johannesburg, South Africa

Luiza Olim de Sousa
Community-Based Educational Research
(COMBER), Geography Education
and Environmental Education,
School of Mathematics, Science
and Technology Education,
Faculty of Education
North-West University
Potchefstroom, South Africa

ISSN 2199-7373 ISSN 2199-7381 (electronic)
World Sustainability Series
ISBN 978-3-030-74692-6 ISBN 978-3-030-74693-3 (eBook)
https://doi.org/10.1007/978-3-030-74693-3

This Springer imprint is published by the registered company Springer Nature Switzerland AG
The registered company address is: Gewerbestrasse 11, 6330 Cham, Switzerland

Preface

The scope of sustainable development is not only about education but also about agriculture, transport, energy, climate change or consumption.

Therefore, if we want efforts to foster sustainable development to succeed, we need to take these wide-ranging issues into account.

Sustainability as a process or a way of thinking can be practised every day, be it in making one aware of how much water one consumes when opening the water tap to brush one's teeth in the morning, or as part of a decision to buy a product, or choose the use of private transport where it is widely available. It can be applied in various sectors as part of our lives.

In Africa, a continent rich in culture and resources, its inhabitants are challenged by various social, economic and political issues, as well as by problems such as climate change. A wide spread of a sense of sustainability is important as part of efforts to promote human development on the one hand and the conservation of natural resources on the other. The book "Sustainable Development in Africa: Fostering Sustainability in one the World's Most Promising Continents" is one attempt to support these efforts.

This book serves the purpose of documenting and promoting a wide range of African experiences on sustainable development, which encompasses both formal and non-formal education. As African nations strive to pursue the UN Sustainable Development Goals, it is imperative to cater for the information needs seen across the continent and to foster the dissemination of experiences and case studies, which may support economic diversification, address inequalities and contain environmental degradation. Consistent with this aim, the scope of the book is deliberately kept wide, and it is structured around four parts:

Part 1- Indigenous Knowledge Systems and Sustainable Development
Part 2- Educational Institutions, Approaches and Innovations
Part 3- Resource Use, Environment and Agriculture
Part 4- Sustainable development: Innovation, Industry and Cross-Cutting Themes

A special feature of this book is that it not only presents a wide range of philosophies, approaches, methods and analyses about sustainable development across Africa but also documents and disseminates concrete case studies, which show sustainable development may be realised in practice.

We would like to thank all authors and reviewers for making available their experience in their chapters and the willingness to share their ideas. Much can be gained by offering a platform for the debate on sustainable development in Africa, in a pragmatic way. By providing their inputs, the authors have made a positive contribution towards a debate which needs to be continued and reach a depth far beyond what conferences, workshops or seminars may be able to offer.

Summer 2021 The Editors

Contents

Resources, Environment and Agriculture

Indigenous Knowledge Systems and Sustainable Development

The Role of Indigenous Communities of Practice in Fostering Community Engagement and Partnerships for Sustainable Development in Africa

Mothusiotsile Edwin Maditsi and Simeon Albert Materechera

Abstract Partnerships between government, the private sector and civil society are critical for any sustainable development agenda. However, experience in Africa has shown that participation by local communities in development projects is rarely interactive or empowering. The crux of community engagement and partnership is that it should be inclusive, responsive, equitable and ethical. This chapter argues that indigenous communities of practice (iCoPs) provide a platform where the voices, rights, knowledge and experiences of local communities could be heard, respected and negotiated in development agendas. An indigenous community of practice is developed, acquired and owned by a close group of people who are usually ethnically or culturally linked and use indigenous knowledge to address problems in order to survive. iCoPs are largely products of local indigenous knowledge and governance systems. The chapter uses a thematic analysis of two South African iCoPs to illustrate how the principles, values, shared vision and goals of sustainable community engagement and partnerships can be built among stakeholders. It is concluded that indigenous knowledge is a social capital of any community and development agencies in Africa should use iCoPs along with their knowledge producers, propagators and practitioners as pillars and building blocks in fostering sustainable engagement and partnership pathways.

Keywords Communities of practice · Sustainable development · Co-production and Co-management of knowledge · Indigenous knowledge systems · Civil society organizations

---

M. E. Maditsi (✉) · S. A. Materechera
Indigenous Knowledge Systems Centre, Faculty of Natural and Agricultural Sciences, North West University, Mmabatho, South Africa

S. A. Materechera
e-mail: Albert.Materechera@nwu.ac.za

W. Leal Filho et al. (eds.), *Sustainable Development in Africa*, World Sustainability Series,
https://doi.org/10.1007/978-3-030-74693-3_1

1 Introduction

Sustainable development is defined as development that meets the needs of the present generation without compromising the ability of future generations to meet their own needs (Emas 2015). It is linked to the ability to maintain and improve upon the outcomes and goals achieved with external support. Furthermore, sustainable development is based on socio-economic development, redistribution of resources and the possibility of long-term usage of resources to ensure an acceptable quality of life for future generations (Klarin 2018). It is within the definition and the importance of sustainable development that the 2030 Agenda for Sustainable Development was set in September 2015 and was adopted and made more actionable on the 6th of July 2017 by the United Nations (UN) General Assembly (UN 2017). The Sustainable Development Goals (SDGs) or Global Goals are a collection of 17 interlinked goals designed as a blueprint to achieve a better and more sustainable future for all (UN 2017). The SDGs came into action in January 2016 and aim at offering a broad and inclusive framework for ending poverty worldwide. It is widely recognised that successful achievement of this agenda, which is comprised of 17 SDGs and 169 targets, requires national, regional and local efforts across all sectors of society (Nunes et al. 2016).

One goal which is in line with the approach proposed in this chapter is Goal 17 of the SDGs, which is to "strengthen the means of implementation and revitalise the global partnership for sustainable development". This goal supports community engagement (CE), forming of partnerships with indigenous communities of practice (iCoPs) and the realisation of the value and importance of indigenous knowledge. iCoPs are communities of practice that are formed by a close group of people bound by geographical location, culture and a shared domain. They use indigenous knowledge to address various problems they come across in their communities in order to survive. While the principles of CE have been well documented and publicised, there is a lack of comparable guidance for applying it to foster partnerships for development in Africa. CE is viewed as a goal for achieving sustainability thereby making it a much needed and important approach to guide and foster the development agenda of any project (Hes 2017).

This chapter aims to describe and provide an in-depth understanding of iCoPs and how they can assist various stakeholders and organisations in fostering sustainable development in Africa. An analysis of two iCoPs within the North West Province of South Africa is applied in order to illustrate the relevance and value of these organisations as sustainable development drivers. Furthermore, this chapter puts in perspective the lessons learned from these iCoPs, which can guide organisations on issues of sustained engagements and partnerships. These lessons, together with the iCoPs perspective, can further assist developmental agencies, organisations and governments to sustain the type of engagement which will facilitate meaningful partnerships and collaboration in Africa, thereby enabling a sustainable future for all Africans.

2 Communities of Practice (CoPs) and Their Role in Community Engagement and Partnerships

The term community of practice (CoP) was introduced in the early 1990s by Eugene Lave and Etienne Wenger. They defined a CoP as a group of people who share a concern or a passion for something they do and learn how to do it better as they interact regularly (Wenger and Wenger-Trayner 2015). For many years, CoPs have been at the core of CE. These CoPs have served to provide and create strategic advice and knowledge. CoPs provide an approach to form partnerships and collaborate with various stakeholders which in turn provides a mutual understanding and participative decision making (Kapucu 2012). A CoP is an important theoretical construct that brings forth a learning experience whereby people join in legitimate peripheral participation which serves as a home of shared practices (Hoadley 2012). CoPs are also vital organisations as they share experience and tacit knowledge which is embedded in indigenous knowledge in free flow, thereby improving abilities and skills, and fostering learning (Agrifoglio 2015).

Furthermore, within a CoP, community members develop common sets of codes and language, share norms and values, conduct critical reflection, and engage in dialogue with each other at a professional level. This generates an environment characterised by high levels of trust, shared behavioural norms, mutual respect and reciprocity (Sharratt and Usoro 2003; Agrifoglio 2015). It is within this context that the relevance and importance of CoPs are realized. These are organisations that value mutual respect and reciprocity. Therefore, these two aspects are in line with fostering sustainable CE and partnerships with various communities in Africa that can drive sustainable development initiatives. Wenger and Wenger-Trayner (2015) point out that CoPs have been around for as long as human beings have learned together. Therefore, CoPs have been catalysts of CE and have always fostered partnerships for as long as human beings have been in existence. Most of the people who are within a CoP might not even be aware that they belong to one, as these organisations are usually informal in nature (Wenger et al. 2002).

CoPs are essential organisations which enable learning jointly for purposes of collaborative development and research. These organisations have one important characteristic, which is to have a clear and relevant domain with shared interest that is demand driven rather than abstractly focused (Urama et al. 2014). Therefore, application of the CoP concept to sustainable development facilitates networks and links that enable identification of peers to work with. Forming partnerships with CoPs further provides an idea of potential partners and stakeholders who are vital to drive sustainable development, also in Africa. CoPs can also serve as links to government advisory services and ensure continued interaction through various procedures.

Johnson and Khalidi (2005) point out that CoPs, as informal networks of profession and practice, are key drivers in achieving sustainable development. This is because these organisations contribute to a more informed dialogue with decision-makers. They also facilitate problem-solving among individual members, stimulate learning and promote professional development in their daily work, which positions them as relevant stakeholders within sustainable development networks.

3 The Link Between Indigenous Knowledge Systems (IKS) and Sustainable Development

Indigenous knowledge systems (IKS) refer to the complex set of knowledge, skills and technologies existing and developing around specific conditions of populations and communities indigenous to a particular geographic area (Noyoo 2007). This is a knowledge form that is deeply rooted in community practices, institutions, relationships and rituals (Materer et al 2001). IKS is informed by and relates to all domains of life and the environment. It can be summarised as a total of knowledge and practices, whether explicit or implicit, used in the management of socioeconomic, ecological and spiritual facets of life (Osman 2010).

A huge interest in IKS by the international community has recently developed. This interest has made IKS to be regarded as a remedy for many problems faced by development agencies regarding CE strategies (Magni 2017). The realisation of many stakeholders and development agencies towards the value of IKS has increased, thereby making IKS a key catalyst for sustainable development, especially in Africa. This is mainly because Africa remains a gold mine of indigenous knowledge (IK) for sustainable development (Akinwale 2012). Recently, almost all the developmental actors have recognised the value of participatory approaches in decision-making for sustainable development. This is mainly because IK provides the basis for grassroots decision-making (Senanayake 2006).

It is a well-known fact that development activities that work with and through IK have several advantages over projects that operate outside them (Breidlid 2009). This is evidenced by Odora Hoppers (2002) whereby she asserts that the inspiration by the African Renaissance has been a driver for the development of IKS culture and worldviews which in turn drives the sustainability of indigenous communities. Various people resort to calling IKS a reductionist science and knowledge system, but this has only led to further exploration of the role of social sciences and natural sciences in supporting the development of IKS (Odora Hoppers 2002) thereby making it a fountain of solutions to Africa's development issues. The lack of respect for local or indigenous knowledge and the assumption by many Western scientists about the superiority of Western epistemology and scientific discourse is a serious obstacle to sustainable development in Africa in the light of their apparent failure to meet some human development needs (Breidlid 2009).

The importance of IKS for sustainable development in Africa is a key aspect to achieve all the SDGs. SDG 17 is focused specifically on global partnerships which is very much in line with the development and management of IKS within the global sphere. Partnering and engaging with various indigenous communities and applying their IK is of great importance as it can drive the sustainability momentum in Africa. IK is dynamic and changing and it is also cumulative and adaptive which makes it a non-stagnant knowledge form. Also, the international community has accepted IK as both a driver and enabler of sustainability (Bedini 2020). It is therefore important to explore the linkages between sustainable development and IK, intended here as local knowledge of indigenous communities having its own epistemology and scientific validity and not as opposite to Western knowledge (Magni 2017). This therefore requires forging meaningful partnerships with the wider society including other indigenous communities as well as iCoPs. This partnership approach therefore places IKS in a more advantageous position to achieve sustainability.

Due to cultural imperialism and colonization, IK has been subjected to stigmatisation of it being labelled and compartmentalised to ancestor worship (Dube and Hlalele 2018). However, various foregoing observations by indigenous scholars have dispelled this stigmatisation of IK (Amoah and Bennett 2008). IKS as a local knowledge is complimented by other forms of knowledge (Dube and Hlalele 2018), which better positions it as a knowledge that can contribute to provide a sustainable future for the people of Africa. Additionally, IKS provides a culturally sensitive development paradigm as it has shown concretely how economic and human rights dimensions can be addressed simultaneously, while providing solutions to complex development issues in an innovative and multisectoral manner (Bedini 2020).

Indigenous knowledge is intricately interwoven with the tenets of sustainable development (Priyadarshini and Abhilash 2019). It is therefore imperative to incorporate IKS within scientific systems as this can support socio-economic development that could bear positive effects on various targets of the SDGs. Furthermore, relevant indigenous practices can be part of broad-based approaches and processes aimed at meeting the need for sustainable development (Mbah and Fonchingong 2019). Its therefore imperative to note that IKS and the practice thereof provide a societal and environmentally responsive approach to attain sustainability. This knowledge benefits the community and local stakeholders, thereby enabling them to achieve sustainable solutions that are well aligned to the SDGs.

4 Indigenous Communities of Practice (iCoPs) as a Catalyst for Meaningful Community Engagement and Partnerships

An iCoP is a community of practice located within a local community that uses their own IKS to address problems and survive. This is whereby the CoP originating from a local community organise themselves to address issues affecting them, or simply to learn from one another by sharing knowledge and interacting regularly. An iCoP is a CoP that is developed, acquired and owned by a close group of people who are usually ethnically or culturally linked (Mapara 2009; Tirivangasi and Tayengwa 2017). They are also largely products of indigenous knowledge and governance systems (Luutu 2014).

iCoPs are locally established organisations which are in touch with the reality of local indigenous communities. They exist within these communities mostly because they want to assist in overcoming various issues facing their communities. These iCoPs have different domains and they focus on such domains. They provide a link between the community and development agencies, governments and a host of other stakeholders.

In addition, iCoPs are known as grassroots organisations, managed by members on behalf of members (Datta 2005). These organisations are the only ones in which people in indigenous communities feel that they own, trust and can rely upon. The success of these iCoPs is largely dependent on the people who initiate it, i.e. the indigenous peoples themselves or local leaders (Datta 2005). iCoPs foster greater decision and action on the topic of participatory networks thus, contributing positively to sustainable development in Africa (Osuchukwu and Edewor 2016). iCoPs are recognised and acknowledged organisations that play vital roles in community development and upliftment, with each iCoP having primary functions in protecting the interest of its members and the wider community in which they exist.

iCoPs are organisations that are viewed as sustainable development drivers because they use local resources embedded in IKS to drive sustainable community development programmes and projects (Ceptureanu et al 2018). These are organisations that improve transparency and equitable distribution of benefits of development while, on long term basis tend to improve the living standards and quality of life within the involved communities. Datta (2005) alludes that the sustainability of iCoPs can be viewed as a stage of empowering these organisations, where members are capable of planning and implementing their development initiatives independently.

5 Methodological Approach

A qualitative exploratory approach was used in order to gain in-depth insights into iCoPs and how various sectors can draw lessons from them. A purposive sampling technique was followed in order to select the two iCoPs used as case studies in this chapter. Purposive sampling is best described by Etikan et al (2016) and Taherdoost (2016). Firstly, all CoPs within the North West Province of South Africa were identified. Self-developed criteria in a form of a checklist focusing on aspects of IKS by iCoPs and cultural protocols such as the use of local languages, appropriate dress code and taking into consideration the traditional authority of communities were used as indicators to determine which amongst all the CoPs in the province are iCoPs. After applying the criteria, two iCoPs were identified as they were consistent with the criteria mentioned above. The two iCoPs that are used as case studies in this chapter are classified as knowledge form iCoPs. This means that most of the organisational learning is focused on developing new knowledge and sharing it amongst other iCoPs and also with the wider society. These two iCoPs are mostly concerned with the development and sharing of African indigenous knowledge with the aim of decolonising how people perceive the knowledge and to address and redress social ills and pertinent problems that exist in society. Data was collected using in-depth face to face interviews with iCoP leaders. The interviews were guided by an interview guide which covered aspects of knowledge development, sharing, dissemination and community engagement as applied by these two iCoPs.

The interview data was analysed using thematic analysis (Maguire and Delahunt 2017; Braun and Clarke 2006). The steps followed in the thematic analysis are briefly summarised in the rest of this paragraph. The data collected during the interviews was first transcribed verbatim from the voice recordings and translated from Setswana into English. Using Creswell (2007), Kovach (2010), Braun and Clarke (2006) and Lincoln and Guba (1985) the following procedure for manual analysis was followed. In stage 1, the researchers read all the transcribed interview data thoroughly. In stage 2, recurrent phrases and concepts that were considered important were colour-coded in all the interview data. In stage 3, concepts and phrases from the different data sets were matched and categorised into themes. In stage 4, themes were reviewed, and similar themes were merged. In stage 5, the themes were reported as the results of the study. The themes that emerged from the analysis and which dealt with fostering of community engagement are discussed in this chapter to illustrate the lessons that could be learnt in fostering community engagement and partnerships for sustainable development in Africa.

6 Background and Context of the Two Identified iCoPs

The two iCoPs identified for the analysis reported on in this chapter are based in the North West Province of South Africa. These iCoPs are Bafokeng iCoP that is implemented through Ikitse Cultural Solutions in a village called Phokeng and the Black Leadership Association Conveying Knowledge (BLACK).

The Bafokeng iCoP consists of academics, arts and crafts experts and it also partners with some universities in South Africa, namely: University of Johannesburg (UJ) and North-West University (NWU). This iCoP's domain is on indigenous life skills education and its practice is developing and sharing their knowledge with the youth, academics and other research institutions or government departments. The Bafokeng iCoP is motivated by continuously creating a dialogue where people can ask questions and talk about African centred problems with the intention of finding indigenous solutions to them.

For BLACK, the mandate is to incorporate indigenous knowledge systems and practices within communities in order to promote and appreciate the role of education. This iCoP is based in Mmabatho close to the Mahikeng Campus of NWU. BLACK's domain is on education transformation within poorer communities. Its community consists of academics and indigenous knowledge holders who are passionate about learning and IKS. This iCoP is motivated by innovation, transformation and excellence. These two iCoPs are classified for this chapter as knowledge form iCoPs as most of their focus is on developing new knowledge and sharing or disseminating it within their communities and the wider society.

7 Emerging Themes from Interviews

Through the thematic analysis, four themes emerged. These themes are from the two iCoPs in the North West Province of South Africa, as they are viewed as examples of key drivers for fostering CE and partnerships. The results first present the definition or meaning of CE as per the provision of the two iCoPs and then proceeds to report the interview results in terms of the themes. The four themes that emerged from the analysis are collaboration, dialogue, cultural protocols and language.

7.1 Meaning of Community Engagement (CE)

CE is defined differently by various institutions to suit their mission and vision (Center for Disease Control and Prevention 2011). CE differs from organisation to organisation and from institution to institution. This is because CE means different things to different people and it is viewed as such (National Collaborating Centre for Determinants of Health, 2013). It is within the definition of this term that various misunderstandings and misinterpretations arise. As much as all these institutions and organisations follow a certain definition that blends with their goal, iCoPs also have their own perspective to what CE is to them.

The Bafokeng iCoP leader defined CE as "*working collaboratively with the community with the intent to find solutions to an existing problem and to help each other*". Furthermore, CE as defined by the Bafokeng iCoP leader is viewed as "*a learning curve and working together or to work with other people. It is to love yourself as a person and also to love others. It is to remember others. It's also to work collaboratively with other people. Lastly, it is to help other people who need help*". In essence CE is a social endeavour to address issues affecting the community.

The second iCoP (BLACK) defined CE differently from the Bafokeng iCoP. The BLACK iCoP leader revealed that "*CE is about going to the people with the intention to share an idea that you might have. It is to share the knowledge with the intent to further develop it*".

It is within the definition of CE that various points are focussed on to show how these iCoPs excel in fostering partnerships with local communities. Issues as collaboration, knowledge sharing, and development are key aspects that ensures sustainability. It is therefore imperative to understand further how these iCoPs operate in order to draw relevant lessons applicable to the continent.

7.2 *Sustaining Community Engagement and Partnerships Through Collaboration*

The notion of collaboration between iCoPs and other stakeholders is essential for iCoPs. This is one aspect that can be applied by developmental agencies to foster sustainable development in Africa. The two iCoP leaders both revealed that collaboration is of great importance to them. CE is a collaborative approach that can be used to generate meaningful knowledge to address social problems. Participation by public in every key aspect that directly involves them is a very important aspect. This is because iCoPs, just as local communities who apply their IKS, prefer to make decisions that impact on their communities together. The participation ranges from general education that involves a small degree of direct influence on decision making, to public collaborations in the final decision-making processes, which involves a significant degree of influence in decision making and can imply empowerment for the broader community. Furthermore, both iCoP leaders revealed that collaboration is necessary as it can provide the development of alternatives and the identification of the preferred solution for both the community and the collaborators alike.

The Bafokeng iCoP leader expressed her sentiments on collaboration as follows: "*collaborating with other iCoPs affords our iCoP the opportunity to develop other new interventions and strategies that are much needed in society*". The iCoP leaders have further shown that collaboration with other iCoPs within the same domain enable them to develop a fruitful engagement. The BLACK iCoP leader indicated that "*the collaborations are not only with other iCoPs but also with IK holders within the various communities we engage with*". In this way, the iCoPs are able to develop new knowledge and further grow their domains. This knowledge that is created and enhanced allows them to tackle problems or issues that they face within their domains. Both iCoP leaders share the same sentiments that through collaboration, they are able to come up with new and different approaches to drive their various initiatives in their domains and apply them to develop their communities and the iCoP itself.

7.3 *Dialogue as a Way to Foster Community Engagement and Partnerships for Sustainable Development*

Creation of a dialogue between the iCoPs and the society is of great importance. It is within the dialogue that possible solutions to challenging situations can be established. iCoPs provide a platform that creates an environment that enables dialoguing with the community and various other stakeholders. Thus, collaboration that enables fruitful dialoguing between iCoPs and the society is key. The two iCoP leaders, asserted that creation of a dialogue (known in Setswana as *puisano*) between the iCoPs and the society is of great importance. It's within the dialogue that possible solutions to situations can be addressed.

The Bafokeng iCoP leader expressed this sentiment as follows: *"puisano ke mo batho ba botsang dipotso, ba buisana ka mathata a ba kopanang le one, mme gape le go ithuta go go tseneletseng ka sengwe kgotsa dilo dingwe"*. The above direct quote translates in English as follows: *"a dialogue is where people ask questions, discuss various problems or issues that they come across and also learn in-depth about something/s"*. The Bafokeng iCoP leader further explained that her iCoP is centred around dialogue. She pointed out that a dialogue helps uncover knowledge that sometimes is not apparent even to the iCoP members (tacit knowledge).

The BLACK iCoP leader shared a different view on dialogue. He argued that dialogue is important, but for their iCoP dialogue takes place only within the iCoP, with IK knowledge holders and academics within the IKS spectrum. He believes that within dialoguing, questions can be raised and an in-depth learning about a certain phenomenon can be achieved. "*When dialoguing, a lot of academic information is uncovered, and such knowledge can be used to influence various policies*" he explained. Also, dialogue enables people to talk about a shared problem or area of interest (domain) which is key when trying to understand various things. In a way, dialoguing can be asserted to be a platform that enables social justice. iCoPs can be viewed as advocates of social justice initiatives in their respective disciplines. Several organisations, including universities have dedicated themselves to knowledge building and evaluation of place-based community engagement toward aspirations of social justice and reciprocity.

7.4 Cultural Protocols as a Vital Aspect of CE and Partnerships

It is well known that all indigenous and ethnic communities have their own set of cultural protocols that need to be adhered to. These cultural protocols are the cornerstone of each and every community and are held dear by the respective communities. The Bafokeng iCoP leader explained that cultural protocols are key in every indigenous organisation and society stating that "*tshwanetse re gate mo tlhako ya pele e gatileng teng*". This statement by the iCoP leader translates into English as follows: *"we need to follow in the footsteps of the elders"*. This is because the cultural protocols, when followed, enable any other procedure or approach to be successful. iCoPs, since they are embedded within IK and indigenous ways of doing things, are organisations who apply and follow these cultural protocols. iCoPs are well vested in culture and indigenous knowledge that they are well suited to promote these customary protocols of their indigenous communities. These protocols are upheld at all times.

Furthermore, these protocols encompass taboos and traditions that should not be violated in any way. The BLACK iCoP leader asserts that these community protocols are vital as they are used as a form of protection by communities. iCoPs have therefore adopted most of these cultural protocols. "*BLACK engages with various indigenous communities in the province and therefore we need to always know how to present ourselves in these communities in order to show that we respect their norms and values*". Therefore, it is imperative to follow these cultural protocols in order to foster meaningful development and engagements with various indigenous communities. Cultural protocols encompass a way of life for many societies as they inculcate the norms and values of such communities. The iCoPs are aware of these since they originate and exist in the indigenous society within which these protocols are upheld. The iCoPs also follow and implement such protocols when engaging and sharing their expertise and knowledge.

7.5 Language as a Key Element for Fostering Sustainable Development

Language is an important element and a priority of every community. All four of the themes above emanate from the use of language. There can never be a fruitful collaboration and dialoguing if those involved are not conversant with the local language of the people being engaged. The two iCoP leaders revealed that language plays a fundamental role in the engagement process. Firstly, it serves as a medium of communication amongst iCoP members and the wider society and secondly, it is a means by which indigenous knowledge can be shared and promoted. The BLACK leader expressed their sentiment in the following excerpt: "*language is a key aspect of communication and therefore, the use of local indigenous language to communicate as an iCoP and with the wider community is vital as those communicating can make informed decisions*". Everyone can participate fully when a language that they are familiar with is used to communicate.

Within the language aspect, the community making up the iCoP is able to share experiences and thoughts regarding the issue at hand. Their lived realities are reflected in language and they are also able to conceptualise issues within their indigenous languages. The Bafokeng iCoP leader indicated that "*language is how you get to understand oneself. You will know who exactly you are, where you are from and where you want to be in future*". Within the norms and values of every community and iCoP is the aspect of language which is vital for engagement. Use of local languages such as Setswana for the iCoPs investigated in this chapter is vital. Communities and iCoPs alike prefer being engaged with their local languages as this create mutual understanding and encompasses a collective engagement. No one feels left out when engaging using their local language. Therefore, a lot of knowledge that can help to foster development and forge partnerships is shared within a local language parameter.

8 Discussion and Implications for Sustainable Development

Collaboration, as one of the key aspects promoted by iCoPs, is a suitable way in which development agencies can drive and foster sustainable development, especially within local indigenous communities who apply their IK. Some note a paradigmatic shift away from professional experts who either seek to do things to and for communities, toward those who seek to work with community members as equally valued collaborators (Stout 2018). Additionally, collaborations create a more innovative relationships and stronger communication between the organisations and communities. Such relationships, communications, and collaboration can ultimately lead to more participatory methods that can enhance partnerships and CE thereby informing how development can be fostered.

The aspect of dialoguing is also of importance when trying to foster sustainable development. Dialoguing opens channels of engagement that can be beneficial to development agencies, governments, iCoPs and also the wider society. Analysis of the data from the interviews with the two iCoP leaders showed that no development can be sustained without dialoguing. Engaging in dialogue with iCoPs and learning from them can help foster sustainable development and partnerships. This is because when dialoguing, a lot of tacit knowledge that is embedded in indigenous knowledge is shared. The norms and values of a community that needs to be engaged arise from this dialoguing. In this manner, all the necessary information will be shared when dialoguing with the communities and also with iCoPs.

The two iCoP leaders indicated that dialoguing is an important factor that can drive the engagement and partnerships between developmental agencies and society. Dialoguing with indigenous communities therefore allows the uncovering of values, customs and traditions that can be applied to drive the development agenda. A review by the International Labour Organization (ILO) (ILO 2012) supports the notion of dialoguing for sustainable development. The ILO refers to these dialogues as "social dialogue" and they argue that this is whereby successful experiences in partnerships and engagement can be identified. The iCoPs, as organisations which engage and partner with the society in which they exist, are powerful organisations that can be applied to drive the dialogue which can be used to push and drive sustainable development.

For effective fostering of sustainable development in Africa, governments and other stakeholders need to engage and learn from iCoPs in order to drive the development in a proper manner. A UNESCO report on post-2015 dialogues on culture and development specifically (UNESCO 2015) positions dialoguing in a positive manner and sets out how the cultural protocols which are upheld by iCoPs can contribute to the achievement of sustainable development. Sustainability in rural development depends largely on the way stakeholders perceive the proposed change and the way they are involved in assessing and deciding about how that change should be achieved.

Thus, one of the main roles of dialoguing is to facilitate participation by people, thereby acquiring a rapidly increasing relevance in sustainable development (Mefalopulos 2005). The interviewed iCoP leaders viewed cultural protocols as both enablers and drivers of sustainable development by iCoPs. These protocols represent an important aspect that is always vital to uphold. Cultural protocols as recognized by the Nagoya Protocol serve as a guide to communities. These protocols guide the interactions within communities, between communities, with external actors, and with the environment (UNEP and EDO NSW 2013). In fostering development, it is vital to understand and follow these cultural protocols. Developmental agencies should therefore engage and partner with iCoPs so that they can be better informed about such protocols. iCoPs as earlier stated, are well vested in their communities and they are aware of any community protocols that needs to be followed.

Using indigenous languages in initiatives to foster sustainable development in Africa is key in any development agenda. In this way, stakeholders, governments and development agencies are able to observe a reflection of the thinking of iCoPs and the wider community. This can lead to the development of a mutually beneficial engagement and collaboration, with various resulting developments. Language, as indicated by the two iCoP leaders, is vital since a mutual understanding is required. Development goals will be easier to achieve if using an indigenous language, since the language allows everyone to express their sentiments about any matter fully. Community cultural protocols are communicated using the local languages of indigenous communities. An understanding an indigenous language is therefore vital for fostering sustainable development through partnerships with communities.

9 Concluding Remarks

It is concluded that iCoPs are a much-needed bridge that can assist various departments, organisations and universities to implement and design fruitful CE projects, activities and partnerships that can foster and sustain development in Africa. It is vital to note that these iCoPs apply various visions, values and knowledge that are different from those found in mainstream public and private institutions. This therefore positions them as relevant organisations to assist in driving sustainable development activities and projects across Africa. iCoPs consider IK as a vital tool that can assist in implementing sustainable development. This is because IK is dynamic, adaptive and changes with time thereby making it a flexible resource to foster sustainable development. Development agencies can obtain valuable experience about engagement and communication strategies from iCoPs that will improve the way they relate and engage with the wider society. iCoPs bring about a paradigm shift in defining and implementing development and the way it is approached. This paradigm shift therefore needs to be fully realised by development agencies and governments. It has been revealed that collaboration, dialogue, language and cultural protocols are key aspects that have to be considered by development agencies in Africa to foster meaningful partnerships and development.

Analysis of the interview data obtained from two iCoP leaders point towards the fundamental role of iCoPs within society, with specific reference to the key aspects of collaboration, dialogue, language and cultural protocols. In this way iCoPs is positioned as a perfect steppingstone to assist with fostering sustainable development and partnerships in Africa. The evolution of iCoPs as a social indicator will experience an increase in participation and engagement between stakeholders, development agencies and governments. Applying an iCoPs approach to drive and sustain development could turn a process which is often deemed as too reactive, and in some cases non-existent, into a proactive one. This approach can therefore produce a distinctive locally grounded framework that can benefit all.

However, some limitations should be noted. The use of iCoPs as a new term coined by the researchers may be subjected to bias as it is only focusing on local organisations. In addition, the lack of previous research studies on the topic of iCoPs implies a lack of ontological evidence. Bias rooted within cultural protocols also needs consideration and can point towards other issues. For example, an aspect still in need of research is to establish whether women in African society are not actually disadvantaged by these protocols. Lastly, the use of case studies of iCoPs only found in one province of the country provided access to and use of a limited data set. Consulting various iCoPs across the country is advisable and could have provided data on a variety of issues and developmental aspects that are vital for sustainable development in Africa but would imply a much larger scale research project and supporting logistics and funding.

References

Agrifoglio R (2015) Knowledge preservation through community of practice: theoretical issues and empirical evidence. Springer, Naples

Akinwale AA (2012) Digitisation of indigenous knowledge for natural resources management in Africa. Paper presented at the 20th anniversary summit of the African Education Research Network at North Carolina State University, Raleigh

Bedini G (2020) Indigenous Knowledge as a Driver of Sustainable Development: a Study on Māori Resource Management. Course of Contemporary History

Braun V, Clarke V (2006) Using thematic analysis in psychology. Qual Res Psychol 3(2):77–101

Breidlid A (2009) Culture, indigenous knowledge systems and sustainable development: a critical view of education in an African context. Int J Educ Dev 29:140–148

Center for Disease Control and Prevention (CDC) (2011) Principles of community engagement Second edition https://www.atsdr.cdc.gov/communityengagement/pdf/PCE_Report_508_FINAL.pdf

Ceptureanu SI, Ceptureanu EG, Luchian CE, Luchian I (2018) Community based programs sustainability. a multidimensional analysis of sustainability factors. Sustainability 10:870

Creswell JW (2007) Qualitative inquiry and research design: choosing among five approaches. Sage, Thousand Oaks

Datta R (2005) Sustainability of community-based organizations of the rural poor: learning from Concern's rural development projects Bangladesh. . Commun Dev J 42(1):47–62

Dube B, Hlalele D (2018) Engaging critical emancipatory research as an alternative to mitigate school violence in South Africa. Educ Res Soc Change 7(2):74–86

Emas R (2015) The concept of sustainable development: definition and defining principles. Brief for GSDR 2015: 1–3. https://sustainabledevelopment.unorg/content/documents/5839GSDR%202015_SD_concept_definition_rev.pdf

Etikan I, Musa SA, Alkassim RS (2016) Comparison of convenience sampling and purposive sampling. Am J Theor Appl Stat 5(1):1–4
Hes D (2017) Impact of community engagement on sustainability outcomes – there is no sustainability without community engagement, Melbourne, Australia
Hoadley C (2012) What is a community of practice and how can we support it? In: Jonassen D, Lund S (eds) Theoretical foundations of learning environments, 2nd edn. Routledge, New York, pp 286–300
International Labour Organization (2012) Social dialogue for sustainable development: a review of national and regional experiences, Geneva
Johnson EC, Khalidi R (2005) Communities of practice for development in the Middle East and North Africa. Knowl Manag Dev J 1(1):96–109
Kapucu N (2012) Classrooms as communities of practice: designing and facilitating learning in a networked environment. J Pub Affairs Educ 18(3):585–610
Klarin T (2018) The concept of sustainable development: from its beginning to contemporary issues. Zagreb Int Rev Econ Bus 21(1):67–94
Kovach M (2010) Conversational method in Indigenous research. First Peoples Child Family Rev 5:40–48
Lincoln YS, Guba E (1985) Naturalistic inquiry. Sage, New York
Luutu BM (2014) Beyond cognitive captivity some Afrikological reflections on university community sites of knowledge as employment pathways. In Alvares C (ed). Multicultural knowledge and the university. Penang, Malaysia: Multiversity and Citizens International: 34–39
Magni G (2017) Indigenous knowledge and the implications for the sustainable development agenda. Eur J Educ 52:437–447
Maguire M, Delahunt B (2017) Doing thematic analysis: a practical, step by step guide for learning and teaching scholars. All Ireland J Teach Learn High Educ 9(3):3351–33514
Mapara J (2009) Indigenous knowledge systems in Zimbabwe: Juxtaposing postcolonial theory. J Pan African Stud 3(1):139–155
Materer S, Valdivia C, Gilles (2001) Indigenous knowledge systems: characteristics and importance to climatic uncertainty. Department of Agricultural Economics working paper, AEWP 2001–2003. University of Missouri, Columbia: USA
Mbah M, Fonchingong C (2019) Curating indigenous knowledge and practices for sustainable development: possibilities for a socio-ecologically-minded university. Sustainability 11:4244
Mefalopulos P (2005) Communication for sustainable development: applications and challenges. Media and Global Change: Rethinking Communication for Development, Latin American Council of Social Sciences (CLACSO), Buenos Aires
National Collaborating Centre for Determinants of Health (2013) A Guide to Community Engagement Frameworks for Action on the Social Determinants of Health and Health Equity. Antigonish, NS: National Collaborating Centre for Determinants of Health. St Francis Xavier University. ISBN 987-1-926823-58-4
Noyoo N (2007) Indigenous knowledge systems and their relevance for sustainable development: a case of southern Africa. Tribes Tribals 1:167–172
Nunes AR, Lee K, O'Riordan T (2016) The importance of an integrating framework for achieving the Sustainable Development Goals: the example of health and well-being. BMJ Global Health 1(3)
Odora Hoppers CA (ed) (2002) Indigenous Knowledge and the Integration Of Knowledge Systems. New Africa Books, Claremont
Osuchukwu NP, Edewor N (2016) Stimulating and enriching partnership with community based organizations: inclusive participatory platform with libraries in Nigeria. Paper presented at: IFLA WLIC 2016 – Columbus, OH
Connections. Collaboration. Community in Session S27 - Africa. In: Building Cross Cultural Capacities for Universal Access to Information and Knowledge in Africa, 11–12 August 2016, Athens, Ohio, USA

Osman A (2010) Indigenous Knowledge in Africa: Challenges and Opportunities. University of the Free State, Free State, An inaugural lecture

Priyadarshin P, Fonchingong PC (2019) Promoting tribal communities and indigenous knowledge as potential solutions for the sustainable developmentof India. Environ Dev 32:100459

Priyadarshini P, Abhilash PC (2019) Promoting tribal communities and indigenous knowledge as potential solutions for the sustainable development of India. Environ Dev 32:100459

Senanayake SGJN (2006) Indigenous knowledge as a key to sustainable development. J Agric Sci 2(1):87–94

Sharratt M, Usoro A (2003) Understanding knowledge-sharing in online communities of practice. Electron J Knowl Manag 1(2):187–196

Stout M (2018) Pursuing community change through radically democratic practice. In Reframing non-profit management: Democracy, inclusion and social change. Edited by Angela M Eikenberry, Billie Sandberg & Roseanne Mirabella. Melvin & Leigh Publishers, Irvine, CA

Taherdoost H (2016) Sampling methods in research methodology; How to choose a sampling technique for research. Int J Acad Res Manag 5 (2):18–27, ISSN 2296-1747

Tirivangasi HM, Tayengwa D (2017) Indigenous knowledge systems and food security in South Africa: Is land reform a prerequisite? J Hum Ecol 57(3):118–124

UNEP and EDO NSW (2013) Community Protocols for Environmental sustainability: A Guide for Policymakers. UNEP, Nairobi and EDO NSW, Sydney

United Nations (2017) Resolution adopted by the General Assembly on 6 July 2017, Work of the Statistical Commission pertaining to the 2030 Agenda for Sustainable Development. https://UN.resolution.generalassembly. Accessed 5 Nov 2020

United Nations Education, Scientific and Cultural Organization [UNESCO] (2015). Post-2015 dialogues on culture and development, UNESCO, Paris. ISBN 978-92-3-100086-7

Urama K, Ozor N, Acheampong K (2014) Achieving sustainable development goals (SDGs) through transformative governance practices and vertical alignment at the national and subnational levels in Africa. Regional practice paper, SDplanNet

Wenger-Trayner E, Wenger-Trayner B (2015) Communities of practice. A brief introduction. Internet: https://wenger-traynercom/wp-content/uploads/2015/04/07-Brief-introduction-tocommunities-of-practice.pdf. Accessed 12 Dec 2016

Wenger E, McDermott R, Snyder WM (2002) Cultivating communities of practice, 1st edn. Harvard Business School Press, Watertown

Mr Mothusiotsile Edwin Maditsi received his master's degree in Indigenous Knowledge Systems from the North-West University in 2018. He has been actively participating in various community engagement activities with various communities and stakeholders that are part of the IKS Centre. He has also lectured modules such as introduction to Indigenous Knowledge Systems and Innovation, Comparative culture and Biodiversity, African Indigenous Signs and Symbols for community development, the Rights of Indigenous Peoples and Indigenous Peace Building and Conflict Resolution in Africa. He is currently a PhD candidate in Indigenous Knowledge Systems at the North-West University.

Professor Simeon Albert Materechera is a Professor and researcher in the Indigenous Knowledge Systems (IKS) Centre at the North-West University. His research is focused on sustainable agriculture, food security and the application of indigenous methodologies in research and community engagement.

Fostering University Contribution Towards Sustainable Development: The Instrumentality of Indigenous Knowledge System

Marcellus Forh Mbah and Ane Turner Johnson

Abstract Universities are often seen as drivers of change in their regions of operation through research, teaching, engagement and enterprise activities. This is significant in Africa where several universities are state owned, rely on government subsidies, and possess a mission to promote the sustainable development of the nation. Whilest the UN sustainable development goals provide an opportunity for Africa to achieve its development targets, we examined the role of the university. More specifically, we considered the significance of Indigenous knowledge to a university's mission towards development. This is pivotal, as Africans and their institutions provide leadership to the transformation of their nations, not just in terms of knowledge production but also integration. We adopted a multiple case-study design that recruited participants from Zambia (N = 50) and The Gambia (N = 40) comprising academics, university managers and community members. Participants took part in relational dialogues that address the intersection between Indigenous knowledge, the university's mission and sustainable development. Findings from a comprehensive data analysis posit the need for the university in Africa to re-envision its teaching and research architectures for sustainable development. The chapter underscores that Indigenous knowledge holders should be provided space to contribute to the curricula if the teaching mission of the university would result in graduates who are suited to contribute to the continent's development with sustainable outcomes. Similarly, it is argued that Indigenous people can be co-researchers, who can identify and provide indigenised methodological insights into the investigation of complex development challenges faced by their communities.

M. F. Mbah (✉)
Nottingham Institute of Education, Nottingham Trent University, Clifton Lane, Nottingham NG11 8NS, UK
e-mail: marcellus.mbah@ntu.ac.uk

A. T. Johnson
College of Education, Rowan University, 201 Mullica Hill Rd, Glassboro, NJ 08028 USA
e-mail: johnsona@rowan.edu

W. Leal Filho et al. (eds.), *Sustainable Development in Africa*, World Sustainability Series,
https://doi.org/10.1007/978-3-030-74693-3_2

1 Introduction

Africa as a continent is endowed with great wealth, characterised by the abundance of natural resources (Lebert 2015), with an estimated 30% of the world's recognised mineral reserves (Sharaky 2014). Whilest the continent could be considered the envy of the world in terms of natural resources, this has also resulted in the new scramble for Africa (Pakenham 2015; Carmody 2017). However, the extreme poverty in the continent, evident by many of its nations amongst the poorest on the planet (Christiaenson and Hill 2019; Beegle and Christiaensen 2019) does not demonstrate the reality of the abundance of natural wealth. Income inequality; protracted conflicts; the scourging impact of climate change on different sectors such as agriculture, transport and energy; corruption; political instability; poor infrastructures; lack of good support systems; hunger and health crises are some of the challenges facing the continent with abundance of mineral deposits (Dude et al. 2016; Aryeetey et al. 2012; Coates 2020).

Regardless of the many challenges facing the continent, it can be argued that the wealth of Africa can be leveraged to promote sustainable development. Although sustainable development can be defined as development that meets the challenges of the present without compromising the ability of future generations to meet their own needs (WCED 1987), it can also be confined to achieving relevant goals such as the UN Millennium Development Goals or the ensuing Sustainable Development Goals. The attendant effect of realising these goals has the potential to transform the living conditions of Africa's masses and instead of its natural resources being seen as a liability, evident via protracted conflicts, natural hazards, corruption and slow economic growth (Van der Ploeg 2011; Scheffran et al. 2019; Adams et al. 2019; Tiba and Frikha 2020), they can be reliable assets supporting national unity, nurturing collective self-reliance, and reducing social inequalities. Although Africa made significant progress in achieving the Millennium Development Goals (MDGs), demonstrated through a galvanised effort to meet the needs of those living in extreme poverty, halting the spread of HIV/AIDS and providing access to primary education, many development targets are yet to be realised (UN 2015a).

The UN sustainable development goals provide yet an opportunity for the continent to meet many of its development priorities and given the strategic part it played in negotiating the post-2015 development agenda. The development aspiration of the continent has been captured in its Agenda 2063, adopted by the Heads of State and Governments of the African Union. This Agenda highlights inclusive growth and sustainable development that touch on poverty eradication, human capital development, improved agricultural activities and prioritising adaptation to climate change amongst others (UN 2015b). To realise its development targets, many authors have opined the instrumentality of Africa's home-based solutions (Shaw 2019; Ani 2019; Mngomezulu 2019). Similarly, African institutions such as their universities have been considered over the years to possess the potential and orientation to catalyse the development of the continent's economy and society (McCowan 2016; Tikly 2019; Mbah 2016). Whereas the role of higher education in fostering sustainable development is foregrounded in literature (Hallinger and Chatpinyakoop 2019; Leal Filho et al. 2019; Leal Filho 2011), including the operational idea of the ecological university (Barnet 2011), there is need to continue to interrogate novel ways African universities can consider in their attempt to engender development in their region and country. This is particularly significant within the remit of home orchestrated solutions to Africa's challenges, that seek to engage relevant knowledge holders and their knowledge-bases.

2 Engaging Indigenous Knowledge (IK) for Sustainable Development

The idea of engaging IK for sustainable development can be nested in the conceptual notion of the ecological university propounded by Ronald Barnet. Barnett (2011) suggested that an ecologically driven university, unlike the entrepreneurial and managerial university models proliferating the world, is characterised as one 'for others' (452), reflective of Ubuntu philosophies, often associated with shared responsibilities and communal solidarity (Ramose 2004; Waghid 2014). According to Barnett (2011), the ecological university takes the form of "an engaged university, a critical and an enquiring university and a university-for-development, acting to put its resources to good effect in promoting world well-being" p. 452. Whilest it is not clear how the ecological university can engage for development within an African context, it is worth noting that most African universities capture a development agenda for their regions in their mission statements. This overarching development mandate of African Universities was equally captured in the mission statement of the Association for African Universities: "to enhance the quality and relevance of higher education in Africa and strengthen its contribution to Africa's development". This statement is consistent with many considerations of African universities' potential to foster the continent's development (Coleman 1994; Teferra and Altbach 2004; Sawyerr 2004; Preece et al. 2012). A notable voice is that of the former UN Secretary General Kofi Annan, who asserted that "the university must be a primary tool for Africa's development in the new century" (Annan in United Nations Information Service (2000), cited by Bloom et al. 2006, p. 4). Similarly, different authors have pointed to the relationship between universities and communities in Africa via increasingly concrete engagement policy orientation (Mtawa et al. 2015); as partners in peace (Johnson 2019); through an indigenized curriculum (Dei 2014); Africanisation (Preece 2013); and through an institutionalized Ubuntu framework (Mbah 2016).

Dei (2014) conceived of the university as an African Academy in which academic excellence is context-specific, anti-colonial (see also Dei 2013), community engaged through reflexivity, and, most importantly, epistemically indigenous. Engagement with Indigenous communities within the structure of higher education, or what some refer to as the Africanisation of higher education, 'emphasizes the importance of rethinking development and changing educational policies in order to take account of local curriculum with local values, local culture and local languages' (Babaci-Wilhite 2015, 18). An Indigenous, co-generative epistemology is critical to the reconsideration of higher education as a development actor in Africa. So called modern, western, northern, minority world, or developed notions of development and the theory and discourse backing such notions have been fundamentally insolvent. It can be argued that

> "the lack of respect for local or Indigenous knowledge and the assumption by many Western scientists about the superiority of Western epistemology and scientific discourse is a serious obstacle to sustainable development in light of their apparent failure to meet human development needs" (Breidlid 2009, 142).

The ecological premise of an African university cannot overlook the role of Indigenous knowledge holders and their knowledge-base as it positioned itself to promote sustainable development in the continent. It can also be maintained that interventions designed to address the development needs of a given people or community should adopt an engagement approach that would capture and respect the views of those living within that environment, if such interventions are not be considered an intrusion by the locals, irrespective of the good intensions. This justifies the United Nations' emphasis on inclusion of Indigenous peoples and local communities in achieving different sustainable development goals (Kaya 2014; Magni 2017). In this light, the university can institutionalize mechanisms to overturn bankrupt development discourses and schemes via authentic engagement with Indigenous knowledge and align its mission toward sustainability. Whilest notions of sustainable development have been present in African Indigenous knowledge and understanding for generations, connecting balanced use of natural resources to social harmony (Mawere and Awuah-Nyamekye 2015) for instance, the university can also leverage on this form of knowledge to foster its outreach mission.

3 Contextual Background

The study that underpins this chapter focused on two countries not commonly addressed in the African education literature, namely: Zambia and The Gambia. Two public universities were examined, because they are often guided by and evaluated on their contribution to sustainable development. Additionally, each country claims many Indigenous communities and cultures.

Zambia
Located in Southern Africa (see Fig. 1) and a member of the Southern Africa Development Community (SADC), Zambia is a country of over 16 million people, claiming diverse ethnicities. The country is a Lower Middle-Income Country (LMIC), whose economy relies heavily on copper mining exports. The brunt of climate change was felt in the country recently, because of an increase in temperature and more extreme weather events. These events negatively impacted the management of natural resources and the agricultural sector, which are key aspects of the economy.

Fig. 1 Map of Africa, showing the location of Zambia

The public university of interest in the research that underpins this chapter was the University of Zambia (UNZA), established in 1966 to respond to the shortage of skilled manpower and development needs of the country (Chipindi and Vavrus 2018). The student body at UNZA has grown tremendously ever since the founding of the institution, with an estimated number of around 30,000 as at the 2018/19 academic year. There is also an estimated academic workforce of 802 and 2,000 administrative staff members. Although the country boasts over 70 languages and dialects, the official language of instruction is English.

The Gambia

The Gambia is located in the Economic Community of West African States (ECOWAS) (see Fig. 2), with a dense population of 2 million people (176 people per square kilometre), belonging to the Manidaka (34%), the Fula (24%), Wolof (15%), and Jola (10%) ethnic groups, among others, 57% of which live in urban or peri-urban centres (World Bank 2020). According to the United Nations Human Development Index, The Gambia ranks below most African countries and sits at 174 out of 189 countries in the world in terms of human development (UNDP 2019). The economy relies heavily on agricultural exports (predominantly peanuts), employing more than 68% of the workforce. Due to climate change, The Gambia has experienced changing weather patterns, drier conditions, excessive salinity in the river, coastal erosion, and increased temperatures.

Fig. 2 Map of Africa, showing the location of The Gambia

The public university of interest in The Gambia at the nexus of this paper is The University of The Gambia, established by an Act of the National Assembly of The Gambia in March 1999. In the country's National Development Plan (2018–21), the university has a mission to promote equitable and sustainable socio-economic development of communities through relevant, high-quality gender-sensitive teaching, research and outreach programmes. It has also been mandated to develop its information and communication technologies infrastructure as a driving force for the education of more people rapidly and for the improvement of the efficiency and academic quality associated with the goals of poverty alleviation and national development.

Methods and Cross-Case Analysis

The research employed a collaborative, exploratory multiple case study design (Stake 2013). After receiving ethical clearance from each university, semi-structured interviews and talking circles were the main instruments for data collection. These entailed responsive and relational dialogues (Chilisa 2012) with academics engaged in community-based research, community counterparts and university managers that support the research mission of the institution (Table 1). These dialogues were flexible enough to allow for serendipitous moments (Simons 2009).

Table 1 Project participants, total

University	Academics	Managers	Community members
University of Zambia	22	12	16
University of the Gambia	17	11	12

Conversation with academics and university managers centred on their understanding of Indigenous knowledge, community-based research, sustainable development, and structures that support the university's engagement with Indigenous knowledge and practices for sustainable development. Essentially the dialogues which lasted between 30 min—1 h, were structured to place the participants' expertise and experience at the centre of the engagement. Community counterparts who participated in the research were recruited via relevant university contacts due to their engagement in community-based research projects. They were asked about their roles in the community, participation in research, and what they have shared about their practices with researchers. The implementation of talking circles with community members was appropriate as "in African contexts and among [I]ndigenous peoples, there are many occasions when people form a circle…and given a chance to speak uninterrupted" (Chilisa 2012, p. 213). Such a context encourages idea sharing, togetherness, respect, and equality of members (2012). Not unsimilar to focus groups, talking circles replicated the Bantaba found in The Gambia, where men in the community meet in a public place (often around a large tree) to discuss community business. In our methods, however, we were sensitive to gender issues in each country and systematic in creating circles that were gender segregated and representative. Given that a majority of community participants could not speak English, the data collection process was aided by an interpreter. Interviews and talking circles were recorded and later transcribed.

The process of data analysis started with coding within and between cases. The initial phase involves using coding to parse the data or taking the data corpus apart in order to make sense of the whole (Stake 1995). Initially, we captured conceptual phrases and participant-driven examples consistent with the research questions (Saldaña 2016). This phase of coding focused on defining concepts such as Indigenous knowledge, community-based research, and sustainability, and generating examples, connecting to institutional support frameworks, and matching with participant experiences. The next phase of analysis entailed pattern coding that grouped the structural codes into a smaller number of categories, effectively reducing the data into analytic units, to facilitate emerging explanations of the case (2016).

It was essential that we also isolated participant stories and examples from the corpus to exemplify the analysis. Simpson asserts that 'stories formed in everyday conversation, which may include those generated in research processes, are directly linked to the experience of organisational members and their desire to account for and make sense of their lives' (p. 94–95). Finally, we focused on constructing the cases by using categorical aggregation to put the parts of the corpus deconstructed during coding back together to create a whole (Stake 1995) and provide an opportunity for an in-depth interpretation of the phenomenon under investigation. In the ensuing findings, we represent the cases in the form of naturalistic generalisations (1995). Furthermore, we supported the findings with the use of participant stories and thick descriptions (Ponterotto 2006) to illustrate aspects of the cases in our work. We present a description of our major findings based on this analysis.

4 Findings

Data analysis resulted in two important findings. These capture the significance of Indigenous knowledge (IK) to sustainable development and legitimisation of IK for sustainable development by the university. Insights into these findings have been presented below:

The Significance of Indigenous Knowledge (IK) to Sustainable Development
To understand our participants' perception of the significance of IK to sustainable development, it was needful to capture their definition of the key concept. Their responses subsequently reveal a great deal of variety of insights on what indigenous knowledge represents to them. An academic at the University of The Gambia noted:

> Well, Indigenous knowledge by my understanding simply means knowledge that is native to a particular environment or locality, and not just any knowledge. It is a functional knowledge because that is what is keeping the community... So, certainly, those people have one form of knowledge or the other, which they have been using to move their enterprise over the years. Basically, they have some knowledge which we can always fall back on and we can also use as a baseline for any other thing we want to do

Given this assertion, a few insights come to the fore. These include Indigenous knowledge being localized and the knowledge holders reside in a specific geographical community. It is also functional, in that, it forms part of a peoples' cultural capital and informs their way of life, that has been passed from one generation to the other. A few participants also noted its distinctiveness from other forms of knowledge. Some contrasted and disengaged IK from "Western" knowledge, while others identified a complementary evolution between scientific knowledge and IK, and many focused on its context, temporality, history and dissemination of that knowledge. The following participants posited:

> You know the problem with the Indigenous knowledge is, it cannot be used to explain some things to you. But, if you really think scientifically with your scientific background you can easily see the reason why it happened. For example, in The Gambia, it is a tradition, common in all the communities, that do not graze your animals early in the morning. They believe if you do it, the animals will be sick. Scientifically, some worms or larvae, early in the morning are on the leaves and are active. When animals graze early in the morning, they ingest the larva and the worms develop in their systems. The farmers wait until when the sun is out; that sun desiccates those larvae and they die or fall off, and that's the time it's safer for the

> animals to graze. But they will not be able explain all these parasitology stuff to you. But you have to use your own knowledge and relate to whatever they do. (An academic in The Gambia)
>
> Okay now, in the first place these terms are terms that you also need to very careful with because during the colonial system, the local knowledge was given very negative representation. When you call it native education, native was taken in a very negative connotation and so is indigenous. So quite a number of times when you are using these terms, you are already [devaluing] that knowledge because of that history. (An academic in Zambia)
>
> I think this is knowledge that people have which has nothing to do with maybe foreign knowledge for example; knowledge that people have used from time immemorial is what Indigenous knowledge is. It has nothing to do with western knowledge. (An Academic in Zambia)

What often pervaded these definitions was intangibility: IK was also difficult to *identify* due to dissemination practices, fatigue, trust, secrecy, and inhibited access by community members. Community members underscored these issues, suggesting that researchers engaged in community-based projects that furthered distrust:

> The farmers would tell the researcher to, say, you know we are facing this type of problem, but when they go back when they go to their centres after doing the research usually they don't come back and report to say you can solve these problems by these and these, they don't come back. (A community member in Zambia)

There was an awareness of this problem among university administrators as well:

> We need to have better trust developed between universities and the communities where this information is coming from because the minute those people will know that you are there to milk them of what they know then disappear, you are cutting the very source of that, so we need to develop better mechanisms of ensuring that the information is flowing but also the community is benefiting from what they had. (A university manager in Zambia)

Clearly, these participants pointed to the need for transparency, predicated on effective communication and sensitization of Indigenous knowledge holders if their knowledge systems are to be accessed for sustainable development. The insights from participants denote that Indigenous knowledge holders may be hesitant to communicate with someone they consider a stranger, and should the researcher wants to access their knowledge, the individual may consider spending extended time with the indigenes to build rapport and breach any cultural gap. An academic at the University of Gambia noted:

> I do a lot of sensitization, you have to know their culture, you have to make sure that they accept you, you have to accept them, then you will be able to have those. And, you must go and stay with them, get involved in their social activities and things like that, then we'll be able to know some of the things. Because, the moment they trust you, they invite you to their places, saying "come and see this, this is what I do".

This participant's account posits that as relationship is established and trust is built, access to Indigenous knowledge and practices can be given. Potentially, the resultant effect on any development endeavour can be maximised. Mostly, participants noted that IK is the gateway to sustainable development and provides the baseline of any project in terms of identifying what community needs are, ongoing solutions/interventions and gaps that need to be addressed. The following participants asserted:

> To be honest, if we are truly serious about sustainable development, if we want development to be sustainable (yeah), obviously, it has to make use of Indigenous knowledge, it has to make use of Indigenous links and networks, only then it will succeed. (An academic in Zambia)

> I think is very important because you can't develop a people if you don't understand the knowledge they have about development. Indigenous people have a knowledge regarding development. They may not be comfortable wearing clothes; they may be comfortable wearing their lace animal skin and all that stuff, and you may ask yourself but why? You go there and carry out research, and they may tell you that the weather here is so hot that they can only feel comfortable in patches of animal skin, so that when the wind blows, it can keep them cold and fresh all the time. (An academic in The Gambia)

Given these assertions, it can be argued that Indigenous knowledge holders and their knowledge-base are crucial to any development endeavours, especially if the outcomes are to be sustainable. The role of IK in any development architecture does not lie in Indigenous knowledge holders providing legitimacy to their knowledge systems as it has served them over the years and therefore its validity can be established. However, it can be argued that the onus lies with the university to give IK legitimacy or the pride of place it deserves in its development mission.

Legitimization of Indigenous Knowledge for Sustainable Development by the University

During the enquiry process, many participants noted that although Indigenous knowledge is instrumental for sustainable development, it has not been given a prime place. According to participants, it is considered inferior to other forms of knowledge. A participant put it this way: "So you are looked down upon and you are not taken seriously and that's another challenge if you are into African Indigenous knowledge systems and even you publishing that… I don't think people will take you serious" (An academic in Zambia). Whilest definable and significant, participants expressed concern over the legitimacy of IK in academia and noted that it should be recognised in every facet of a university mission, notably, teaching and research.

Teaching Indigenous Knowledge for Sustainable Development

Despite the challenges associated with engaging IK, efforts can be made to incorporate it into the teaching and learning mission of the university, increasing its relevance to students and the community. A participant highlighted the fact that students being trained at the University are going to increase the work force of the nation and will be serving different communities, as such it is expedient that they understand the context of Indigenous knowledge for easy integration:

> Let me start by talking about the university structure and the education structure. You see, we are training our students, our graduates to have an impact on the community, alright. So we are not training them for export to Europe, we are training to have an impact on the community… I think its strongly important for our students to understand the contexts of Indigenous knowledge, we should highlight this kind of knowledge as we teach. (An academic in Zambia)

Another participant noted that in order to understand abstract concepts or western forms of knowledge, there is need to start with a deconstruction and create space for local knowledge which can be used to cast meaning on other forms of knowledge or render them relevant to local contexts:

> So, here, we have the advantage of tapping or buying into the Indigenous knowledge available and seeing how we can use it to adapt to the environment we find ourselves. There was a time I proposed the idea of bringing in the study of Indigenous culture into our curriculum, especially in my own field of sociology. We have a lot to read, we have a lot to learn from that area. Basing everything on Western system of knowledge hardly contribute to our understanding. What we can only do is borrow the knowledge, contextualize it to our own situation. (A university manager in The Gambia)

Whilest so little precedent exists for steerage on how to institutionalize indigenous knowledge in a university teaching mission. A participant pointed to a context at the University of Zambia.

> I think we need to train a new cream of students to appreciate their own Indigenous knowledge, similar to the way we have done it here at the University of Zambia. We have established a new program on Zambian Cultures and Ceremonies (ZCC). So, the intention of that program is to train a new cream of youths to appreciate their own cultures. (An Academic in Zambia)

As the university institutionalized IK in its curriculum, it can be considered an act of legitimisation and cooperation with different knowledge holders in the community can flourish for sustainable development.

Researching with Indigenous Knowledge Holders for Sustainable Development

It was noted during fieldwork that a few academics engage in different forms of research amongst Indigenous people. Whileit may be a laudable initiative to research Indigenous people and different issues in their contexts, this can also result in some forms of resentment against researchers. Many participants posit that there was distrust between the researched and researchers, and this was because local people are often not part of framing the research protocols. Some of the participants considered themselves being used and abandoned without due access to research findings. In order to address this unease between both parties, a participant noted that local people should also be considered as researchers:

> We know that Indigenous people, they do their own research through observation, experimentation, trial and error. There are different types of research methods that are being used by Indigenous people to come up with their own form of knowledge. For instance, you see the Indigenous people, they can predict the time as to when it is going to rain. Over time and through observation, they may say that, "okay, when a particular bird cries, it shows that rain will soon come". Over time, they've observed, and they know that this is true. When it comes to farming, they know that the intercropping of certain crops can improve the yield. They have known this through experience. (An academic in The Gambia)

Insight into the data reveals that as co-researchers or investigators, Indigenous people can articulate community problems that need attention, as well as the appropriate methodology that will be best suited in each context, and in addressing issues of concern to them. Once they are part of an entire research process including the data analysis, they will be privy to research findings. A participant revealed:

> Indigenous knowledge in my research has been a source of problem identification because often Indigenous knowledge is not reflected in conventional knowledge and in policies. The second part is that Indigenous knowledge has helped to define my research methodology. For instance, you can read a book that states that a focus group discussion should have no more than eight persons and then in the field, people wondered and say "but this guy, does he really know that actually in this rural area you cannot chase someone away from a group discussion, because we are one and live as a community?". (An academic in Zambia)

Given this account, it can be asserted that for the research mission of the university to yield sustainable outcomes, Indigenous knowledge holders should be involved in every stage of the research process, that is, from conceptualization, to design and to implementation. Another participant exhorted:

> Most development interventions in Africa are not sustainable for the fact that somebody will sit in DC and think that they know what the local people want... they know what needs to be done for local people. That is why they're not sustainable. Now for development to be sustainable especially the SDGs, for example, education, health; for them to be sustainable, we need to involve local communities not only in the design phase, but also the implementation phase. Their knowledge is key, because they understand more the issues that are important to them. They understand the historical and cultural contexts. (An Academic in Zambia)

By including Indigenous people in every process of a university's research mission aimed towards sustainable development, participants also argued that this can be empowering to local people, as it gives them a sense of agency and ownership of their development.

5 Discussion

Similar to participants' understanding, Indigenous knowledge has been defined in varying ways—with no consensus of a universal thought, especially given its interdisciplinary nature (Battiste 2005; Daes 1993). Notwithstanding, it has been associated to a native way of knowing (Semali et al. 2002) and an embodiment of the historical experiences of a people over time and which has informed their way of life, enhance community cohesion and provided solution or coping mechanisms toward societal and environmental challenges (Katerere et al. 2019). However, some participants were of the opinion that IK being referred to in a particular way such as being native, draws a negative connotation of primitivity and rejection. Other authors maintain a similar line of thought by asserting that some of the challenges associated with Indigenous knowledge are the continuing marginalisation, devaluation, primitivization, domination, rejection, subjugation, invalidation and exclusion by Eurocentric scholars and western oriented actors (Dei 2002; Shava 2013). However, it can be argued that one cannot disentangle the Indigenous knowledge systems of a people from what they may consider to be sustainable development, disease prevention, and food provision. It forms part of their individual and collective identity and memory, can cement their social, as well as cultural capital, and provide communities with a sense of resilience in the face of adverse conditions. It could also be argued that this form of knowledge underlies the South African philosophy of Ubuntu, one of solidarity and shared humanity (Mbah and Fonchingong 2019; Muwanga-Zake 2009).

Given that IK is a knowledge system that is being passed down from one generation to another (Banda and Banda 2018, Daes 1993; Semali and Kincheloe 2002), it is not static but dynamic (Dei et al. 2002; Shava 2013, Masuku Van Damme and Neluvhalani 2004). It is adaptable (Dei 2008) and renewable (Katerere et al. 2019). Research participants were emphatic in their view that development cannot be sustainable without a significant hold on IK. Similarly, scholars on IK have pointed out the issues with Eurocentric or western epistemologies as unable to generate sustainable solution to global crises without engaging other forms of knowledge (Shiva 1993; Shava 2013). It has been recognised that development initiatives that capture Indigenous inputs and ways of life are more prompted to generate sustainable solutions and meet people's needs (Sillitoe 1998). Notwithstanding, there is need for a conscientized empowerment of Indigenous knowledge holders as they present alternative ways of looking at and understanding development. Katerere et al. (2019) assert that they present alternative epistemologies and therefore, IK needs to be move from the margins of sustainable development architecture to its core. Notably, IK can meaningfully inform the fundamental elements of teaching and research architectures of the university towards achieving sustainable development in Africa.

Re-envisioning the Teaching Architecture of the University for Sustainable Development

A key facet of the marginalization of Indigenous knowledge in Africa is that it is not recognised in dominant curricula across different disciplines as they are heavily Eurocentric or westernized. The attendant outcome of such marginalization by fuelling the curriculum with what is not relevant to the people of Africa is the impediment of development. As African universities envision their role to educate development agents of the future, a scrutiny of the content of the curriculum is needed. Chilisa (2017) affirms that 'African scholars, and academia as a whole, need to contest the role that colonization, imperialism, and its new form of globalization continue to play in suppressing and silencing knowledge systems of formerly colonized, historically marginalized, and oppressed groups' (p. 814). By indigenizing the curriculum or reclaiming space for Indigenous knowledge in the teaching element of the academy, university students, who are potential agents of change can be impacted with relevant knowledge for the sustainable development of their communities. Therefore, the curriculum can be realized with inputs from Indigenous knowledge holders, with no inferior consideration of their knowledge system.

Whilest it can be argued that Africa's Indigenous knowledge system is not homogenous and therefore one form that pertains to a certain culture or community cannot be generalized (Ndofirepi and Gwaravanda 2019), the academy can work with dominant forms to provide a steerage or exemplar in teaching IK for sustainable development. Although there are growing voices calling for the deconstruction of westernized curriculum for sustainable development in Africa, with some traces of ongoing practices, more needs to be done (Ndofirepi and Gwaravanda 2018; Mawere 2015; Emeagwali and Dei 2014). An awareness of the significance of teaching Indigenous knowledge for sustainable development needs to be intensified. As existing courses are being enriched with content on IK, universities can also create short courses or professional modules on Indigenous knowledge that address specific sustainable development goals as these can appeal to different stakeholders or sectors of the wider community. Support opportunities for module, course or programme developers on the intersection between Indigenous knowledge systems and sustainable development in Africa can also be established to share best practices and boost continuity.

Re-envisioning the Research Architecture of the University for Sustainable Development

As African universities exercise their mission as catalysts of sustainable development, they must rethink how they conduct themselves in Indigenous communities. Any sustainable outcome of development must be predicated on trust and respect of Indigenous knowledge holders and their knowledge system. Some research participants decried how university researchers often use them to gain vital knowledge during fieldwork, without returning to the local community to communicate their findings, thereby fuelling distrust. Similarly, other authors have opined that Indigenous knowledge have been engulfed by western knowledge without due credit to the knowledge provider (Hountondji 2002; Akena 2012; Shava 2013). Given that the value system in communities where Indigenous knowledge thrive is encapsulated by relational existence and shared respect, a form of legitimacy is needed for African universities to maximize IK in their contribution towards sustainable development.

It is important for Indigenous people to own the research aimed at their development. By owning the research, the reasoning is not on single handed conceptualization of research ideas, implementation of research methodologies and diffusion of research findings but in the context of coevolution or cocreation of knowledge. As research participants noted, Indigenous knowledge holders are researchers themselves and primarily through observation and experimentation over the years, they have been able to validate or evidence certain occurrence in agriculture, medicine, climate change, food security, governance and conflict resolution amongst others. Therefore, it is fitting for Indigenous people not to be restricted to the periphery of research or deciding what challenges in their context need addressing for optimal sustainable development but should also be part of the entire research process and contributing indigenized methodologies and approaches (Chilisa 2012; Banda and Banda 2018). This involvement as coresearchers can be empowering to Indigenous knowledge holders by unleashing their potential to be owners and drivers of their own destiny. It can rightly be observed that people cannot be developed; they can only develop themselves by exercising a sense of ownership over processes or projects aimed toward their development. As university research becomes increasingly pivotal for sustainable development in Africa, new ways of conducting research are needed (Waas et al. 2010). Consequently, we argue that new forms of research to aid sustainable development in Africa must be rooted in Indigenous leadership and epistemologies.

6 Conclusion

In this chapter, we have drawn attention to the fact that African universities have been considered as engines of change to pilot the continent's development in the new millennium. This position has been echoed in the mission statements of most African universities, including the Association of African Universities (AAU). As the UN sustainable development goals provide an opportunity for Africa to realize its development targets, we brought to the fore, the idea of the ecological university propounded by Barnett (2011). Drawing on inputs from research participants, we examined the instrumentality of a University's teaching and research missions to sustainable development and what difference Indigenous knowledge can bring to bear. Following a cross case analysis, a number of key lessons can be leant. Firstly, it can be ascertained that development cannot be sustainable without incorporating and acting on the inputs from the intended beneficiaries of development. Therefore, it is needful to legitimize Indigenous knowledge systems in the academy via a re-envisioning of the university's teaching and research architectures. Secondly, it can be posited that space must be reclaimed in the curricula for inputs from Indigenous knowledge holders if the teaching mission of the university must produce the needed human capital to stimulate the continent's development drives. Thirdly, Indigenous people are researchers, who can participate as co-investigators to identify and provide indigenized methodological insights into the study of complex issues faced by their communities. Finally, it can be affirmed that Indigenous knowledge systems and sustainable development in Africa are inseparable and therefore, any future prospect to frame and firm-up development architectures in Africa must be predicated on mutual respect and trust. These virtues should be underscored in any joined venture between Indigenous knowledge holders and the academy in Africa for sustainable development.

References

Adams D, Adams K, Ullah S, Ullah F (2019) Globalisation, governance, accountability and the natural resource 'curse': implications for socio-economic growth of oil-rich developing countries. Resour Policy 61:128–140

Akena FA (2012) Critical analysis of the production of Western knowledge and its implications for Indigenous knowledge and decolonization. J Black Stud 43(6):599–619

Ani NC (2019) Three schools of thought on "African Solutions to African Problems." J Black Stud 50(2):135–155

Aryeetey E, Devarajan S, Kanbur R, Kasekende L (eds) (2012) The Oxford companion to the economics of Africa. Oxford University Press, New York

Babaci-Wilhite Z (2015) Local languages as a human right in education: comparative cases from Africa. Sense Publishers, Leiden
Banda F, Banda D (2018) Framing theoretical/conceptual frameworks and research processes in African indigenous knowledge systems and everyday experiences. Excellence in Higher Education
Barnett R (2011) The coming of the ecological university. Oxf Rev Educ 37(4):439–455
Battiste M (2005) Indigenous knowledge: foundations for first nations. WINHEC: Int J Indig Educ Scholarsh (1), 1–17
Beegle K, Christiaensen L (eds) (2019) Accelerating poverty reduction in Africa. The World Bank, Washington D.C.
Bloom DE, Canning D, Chan K (2006) Higher education and economic development in Africa, vol 102. World Bank, Washington, D.C.
Breidlid A (2009) Culture, Indigenous knowledge systems and sustainable development: a critical view of education in an African context. Int J Educ Dev 29(2):140–148
Carmody P (2017) The new scramble for Africa. Polity Press, Cambridge
Chilisa B (2012) Indigenous research methodologies. SAGE, Thousand Oaks
Chilisa B (2017) Decolonising transdisciplinary research approaches: an African perspective for enhancing knowledge integration in sustainability science. Sustain Sci 12(5):813–827
Chipindi FM, Vavrus F (2018) The ontology of mention: contexts, contests, and constructions of identity among University of Zambia faculty. FIRE: Forum Int Res Educ 4(3):135–150
Coates SJ, Enbiale W, Davis MD, Andersen LK (2020) The effects of climate change on human health in Africa, a dermatologic perspective: a report from the international society of dermatology climate change committee. Int J Dermatol 59(3):265–278
Coleman JS (1994) Nationalism and development in Africa: selected essays. Univ of California Press
Christiaensen L, Hill R (eds) (2019) Poverty in Africa. The World Bank, Washington D.C.
Daes EIA (1993) Study on the protection of the cultural and intellectual property of indigenous peoples. Special Rapporteur of the UN Sub-Commission on Prevention of Discrimination and Protection of Minorities and Chairperson of the Working Group on Indigenous Populations. https://www.refworld.org/docid/3b00f4380.html. Accessed 21 Sep 2020
Dei G (2002) Rethinking the role of Indigenous knowledge in the academy. Lecture paper presented at the studies in national and international development. Queen's University, Kingston, Ontario
Dei SGJ, Hall BL, Rosenberg DG (2002) Indigenous knowledges in global contexts: multiple readings of our world. University of Toronto Press, Toronto
Dei GJS (2008) Indigenous knowledge studies and the next generation: pedagogical possibilites for anti-colonial education. Aust J Indig Educ 37(1):5
Dei GJS (2014) Indigenizing the curriculum: the case of the African university. African indigenous knowledge and the disciplines. Sense Publishers, Leiden, pp 165–180
Dei GS (2013) Critical perspectives on indigenous research. Socialist Studies/Etudes socialistes
Dube T, Moyo P, Ncube M, Nyathi D (2016) The impact of climate change on agro-ecological based livelihoods in Africa: a review. J Sustain Dev 9(1):256–267
Emeagwali G, Dei GJS (2014) African indigenous knowledge and the disciplines. Sense Publishers, Rotterdam
Hallinger P, Chatpinyakoop C (2019) A bibliometric review of research on higher education for sustainable development, 1998–2018. Sustainability 11(8):2401
Hountondji PJ (2002) Knowledge appropriation in a post-colonial context. Indigenous knowledge and the integration of knowledge systems: towards a philosophy of articulation, pp 23–38
Johnson AT (2019) University infrastructures for peace in Africa: the transformative potential of higher education in conflict contexts. J Transform Educ 17(2):173–194
Katerere DR, Applequist W, Aboyade OM, Togo C (eds) (2019) Traditional and indigenous knowledge for the modern era: a natural and applied science perspective. CRC Press, Boca Raton

Kaya HO (2014) Revitalizing African indigenous ways of knowing and knowledge production. Restoring indigenous self-determination, p 105

Leal Filho W (2011) About the role of universities and their contribution to sustainable development. High Educ Policy 24(4):427–438

Leal Filho W, Salvia AL, Pretorius RW, Brandli LL, Manolas E, Alves F, Azeiteiro U, Rogers J, Shiel C, Do Paco A (2019) Universities as living labs for sustainable development. World Sustainability Series

Lebert T (2015) Africa: a continent of wealth, a continent of poverty. New Int 24

Lockwood Commisison (1963) The report on the development of a university in Northern Rhodesia (Zambia). Lusaka, Government Printer

Magni G (2017) Indigenous knowledge and implications for the sustainable development agenda. Eur J Educ 52(4):437–447

Mawere M (2015) Indigenous knowledge and public education in sub-Saharan Africa. Afr Spectr 50(2):57–71

Mawere M, Awuah-Nyamekye S (eds) (2015) Between rhetoric and reality: the state and use of indigenous knowledge in post-colonial Africa. Bamenda, Langaa RPCIG

Mbah MF (2016) Towards the idea of the interconnected university for sustainable community development. High Educ Res Dev 35(6):1228–1241

Mbah M, Fonchingong C (2019) Curating indigenous knowledge and practices for sustainable development: possibilities for a socio-ecologically-minded university. Sustainability 11(15):4244

McCowan T (2016) Universities and the post-2015 development agenda: an analytical framework. High Educ 72(4):505–523

Mngomezulu BR (2019) Revisiting the notion of 'African solution to African problems': successes and challenges. J Afr Union Stud 8(2):9–23

Mtawa NN, Fongwa SN, Wangenge-Ouma G (2016) The scholarship of university-community engagement: interrogating Boyer's model. Int J Educ Dev 49:126–133

Muwanga-Zake JW (2009) Building bridges across knowledge systems: Ubuntu and participative research paradigms in Bantu communities. Discourse: Stud Cult Polit Educ 30(4):413–426

Ndofirepi AP, Gwaravanda ET (2019) Epistemic (in) justice in African universities: a perspective of the politics of knowledge. Educ Rev 71(5):581–594

Pakenham T (2015) The scramble for Africa. Hachette UK, London

Ponterotto JG (2006) Brief note on the origins, evolution, and meaning of the qualitative research concept thick description. Qual Rep 11(3):538–549

Preece J, Ntseane P, MmaB O (eds) (2012) Community engagement in African Universities: perspectives, prospects and challenges. Leicester, NIACE

Preece J (2013) Towards an Africanisation of community engagement and service learning. Perspect Educ 31(2):114–122

Ramose MB (2004) In search of an African philosophy of education: perspectives on higher education. South Afr J High Educ 18(3):138–160

Saldaña J (2016) The coding manual for qualitative researchers, 3rd edn. SAGE, Thousand Oaks

Sawyerr A (2004) Challenges facing African universities: selected issues. Afr Stud Rev 47:1–59

Scheffran J, Link PM, Schilling J (2019) Climate and conflict in Africa. In: Oxford research encyclopedia of climate science. Oxford University Press, New York

SDGC (2019) Reflection on a three year journey. https://sdgcafrica.org/wp-content/uploads/2019/09/SDGCs-Implementationin-Africa-Reflection-on-3Y-Journey_Report.pdf. Accessed 18 Aug 2020

Semali LM, Kincheloe JL (2002) What is indigenous knowledge?: voices from the academy. Routledge, New York

Sharaky AM (2014) Mineral resources and exploration in Africa. Department of Natural Resources, Institute of African Research and Studies, Cairo University, Egypt

Shava S (2013) The representation of indigenous knowledges. In: International handbook of research on environmental education, pp 384–393

Shaw TM (ed) (2019) Alternative futures for Africa. Routledge, New York

Shiva V (1993) Monocultures of the mind. Perspectives on biodiversity and biotechnology. Zed Books, London
Sillitoe P (1998) The development of indigenous knowledge: a new applied anthropology. Curr Anthropol 39(2):223–252
Simons H (2009) Listen, look, document: methods in case study research. In: Case study research in practice, pp 43–66. Sage, London
Simpson P (2008) Stories. In: Broussine M (ed) Creative methods in organizational research. Sage, London, pp 93–107
Stake RE (2013) Multiple case study analysis. Guilford Press, New York
Stake R (1995) The art of case study research. Sage, Thousand Oaks
Teferra D, Altbachl PG (2004) African higher education: challenges for the 21st century. High Educ 47(1):21–50
Tiba S, Frikha M (2020) Africa is rich, Africans are poor! A blessing or curse: an application of cointegration techniques. J Knowl Econ 11(1):114–139
Tikly L (2019) Education for sustainable development in Africa: a critique of regional agendas. Asia Pac Educ Rev 20(2):223–237
UN (2015a) MDGs Report. https://sustainabledevelopment.un.org/content/documents/2036MDG%202015%20rev%20(July%201).pdf. Accessed 18 Aug 2020
UN (2015b) Agenda 2063 – The Africa We Want. https://www.un.org/en/africa/osaa/pdf/au/agenda2063.pdf. Accessed 18 Aug 2020
Van der Ploeg F (2011) Natural resources: curse or blessing? J Econ Lit 49(2):366–420
Van Damme LSM, Neluvhalani EF (2004) Indigenous knowledge in environmental education processes: perspectives on a growing research arena. Environ Educ Res 10(3):353–370
Waas T, Verbruggen A, Wright T (2010) University research for sustainable development: definition and characteristics explored. J Clean Prod 18(7):629–636
Waghid Y (2014) African philosophy of education as a response to human rights violations: cultivating Ubuntu as a virtue in religious education. J Study Relig 27(1):267–282
WCED (1987) Our common future: the world commission on environment and development. Oxford University Press, Oxford

Relevance of Indigenous Knowledge Systems in Sustaining Agriculture Under Smallholder Farming in Sub-Saharan Africa

Simeon Albert Materechera

Abstract The majority of rural population in Sub-Saharan Africa (SSA) rely on smallholder agriculture to wade off hunger, food insecurity and undernourishment. Sustainable agriculture focusses on producing crops and livestock on a long-term basis while having minimal effects on the environment. This chapter argues that African Indigenous Knowledge systems (AIKS) affords the knowledge, practices, innovation and technologies that can contribute to the sustainability of agriculture. It is generally acknowledged that all indigenous knowledges have an inherent organic sustainability norms as they have sustained communities over many years. This chapter highlights some features of AIKS which gives it salient relevance to sustainable agriculture. It examines cases and experiences from AIKS to elucidate that components of sustainability in food production such as resilience, genetic diversity and yield stability are present in the traditional agricultural systems. Examples of indigenous ways of knowing and practices in areas such as land use planning, soil water conservation and fertility management, crop and livestock husbandry, pest, weed and disease control are used to draw guidance on principles and indicators of agricultural sustainability. The chapter concludes by providing pointers by which AIKS can be tapped to accelerate sustainable agricultural production and agribusiness among poor households in SSA.

Keywords African indigenous knowledge systems · Smallholder farmers · Sub-Saharan Africa · Sustainable agriculture

1 Introduction

According to the World Bank (1989), the Sub-Saharan Africa (SSA) consists of all African countries and territories that are fully or partially geographically located south of the Sahara. The population of SSA was estimated to be over one billion in 2018 and with a growth rate of 2.3%, the UN predicts it to rise between 2 and 2.5

S. A. Materechera (✉)
Indigenous Knowledge Systems Centre, North-West University, P/B x2046, Mmabatho 2735, South Africa
e-mail: albert.materechera@nwu.ac.za

W. Leal Filho et al. (eds.), *Sustainable Development in Africa*, World Sustainability Series,
https://doi.org/10.1007/978-3-030-74693-3_3

billion by 2050 (UN 2019). Ajani Mgbenka and Okeke (2013) have suggested that more than 60% of the population in SSA live in rural areas and depend on rain-fed smallholder (used synonymously with small-scale) agriculture for subsistence and livelihood. Agriculture contributes to development not only as an economic activity and as a source of livelihoods, but is also an important provider of environmental services (Saunderson 1995). Consequently, development of the agricultural sector in SSA is central to combating hunger, reducing poverty and generating economic growth especially for women and the youth (Buthelezi and Hughes 2014).

Although some quarters have been critical of the value of smallholder agriculture due to issues of scale (e.g. Briggs 2005), the sector remains important and indigenous knowledge is its relevant component (Safakish 2015). It cannot be disputed that agriculture has a central role in SSA's march towards achieving the UN Sustainable Development Goals (SDGs) especially goals 1, 2, 3 and 13 (UN 2015). The African Union (AU) recognises that smallholder agricultural producers are central to this goal as reflected by the adoption of the Competitive African Agriculture Development Programme (CAAP) in 2003 (NEPAD 2003). The CAADP recognises that sustainable rural livelihoods on the continent will be supported by sustainable growth in the agricultural sector. With a wide variety of climate zones and cultures, agriculture in SSA is highly complex and diverse, the majority of farmers in the region rely on local knowledge and experience gained over the years thereby affirming the potential role of Indigenous Knowledge Systems (IKS) in development (Lwoga, Ngulube and Stilwell 2010).

This chapter is aimed at projecting a conceptual understanding of the relevance and role of IKS in sustaining agriculture under smallholder farming in SSA. The chapter employed document analysis qualitative research method of Bowen (2009) to review existing literature around IKS, sustainable agriculture, and smallholder farming in SSA. The analytic procedure entailed finding, selecting, appraising and synthesising data contained in documents. This approach is rooted in a critical paradigm which enables a researcher to critique conventional and dominant discourses of knowledge on the subject and develop own appraisal and judgement of the issues (Asghar 2013).

The chapter starts by exploring the concepts of smallholder farming and the challenges posed by rain-fed farming including climate change. It reviews relevant issues for sustainable agricultural development and demonstrates the practical applications of AIKS by smallholder farmers in countries of SSA. The chapter concludes by drawing lessons from the discourse and outlines some future prospects on the role of IKS in sustaining agriculture under smallholder farming in SSA.

2 Indigenous Knowledge and Its Role in Sustainable Agriculture and Development

Indigenous Knowledge (IK), also referred to as traditional knowledge, farmers' knowledge or local knowledge, is used to make reference to knowledge and know-how unique to every culture or society and forms the basis for local-level decision making in diverse areas of human endeavour including agriculture (Nwonwu 2011).

According to Warren (1991:1), indigenous knowledge provides problem-solving strategies for communities and remains tacit and embedded in community practices, technologies, institutions, relationships, and rituals. Even though the connotation of the term is sometimes contested (e.g. Briggs and Moyo 2012), they share sufficiently consistent meanings that IKS is generated by local people through a systematic process of observing local conditions, experimenting with solutions and re-adapting previously identified solutions to modified environmental, socio-economic and technological situations (Hart and Mouton 2005: 251).

The World Bank (2004), has acknowledged that it is knowledge not financial capital, which is key to sustainable social and economic development. It is not surprising that Ghorbani and Bakhtiar (2011: 183) have claimed that 'building on local knowledge as the basic component of any region's knowledge system, is the first step to mobilize such capital' Studies have shown that some local communities in SSA have successfully achieved a level of sustainable livelihoods by adapting continuously in their farming and other income-earning activities (Mortimore 2000; Eyong 2007; Warren 1991; World Bank 2004). In recognition of the importance and relevance of IK, some development agencies have called for its inclusion in the research and development agenda of SSA countries (World Bank 1998b; 2004). As Briggs and Moyo (2012: 67) have observed, "the drivers of food production under smallholder farming in SSA are embedded within the socio-economic and cultural realities of the community, and so are the knowledges which underpin this production". In this case, indigenous knowledge provides knowledge diversity which is one of the principles of sustainability (Sanusi 2014). This illustrates why AIKS is intricately embedded and linked to smallholder agriculture and how it underpins their resilience and sustainability.

3 Indigenous Knowledge Systems as a Key Driver of Resilience and Sustainability in Smallholder Agriculture of SSA

Smallholder farmers in SSA, have for centuries, been drawing from IK to inform their production systems and practices based on own accumulated experiences, experimentation, circumstances and reality (Briggs and Moyo 2012).

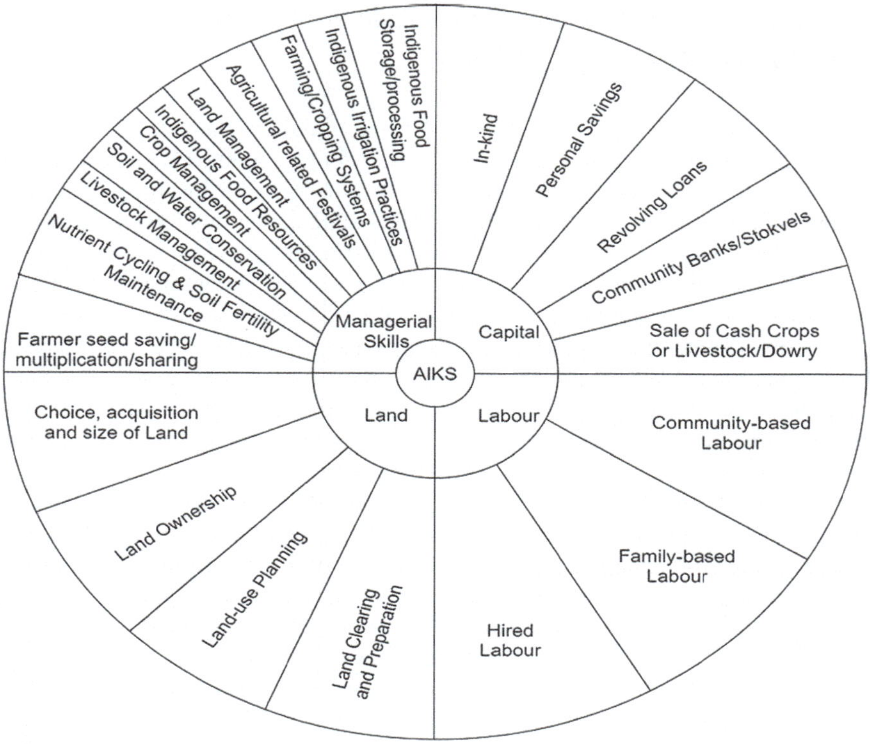

Fig. 1 The embeddedness of African Indigenous Knowledge Systems (AIKS) in the drivers of food production under smallholder farming in Sub-Saharan Africa. *Source* the author

Figure 1 illustrates how AIKS in SSA is intricately embedded and linked to the four drivers of production (land, labour, capital and management) and how it underpins the resilience and sustainability of smallholder agriculture.

3.1 Labour and Capital

Smallholder farmers in SSA rely on pooling labour resources between and among relations (community) as well as 'the prudent use of family labour to help control and limit cash out flows thereby reducing the overall cost of production' (Briggs and Moyo 2012: 69). Capital is perhaps the most limiting factor in smallholder agriculture of SSA as most farmers produce for subsistence with minimal earnings and savings. The average smallholder farmers' access to credit is also extremely limited. Nwonwu (2011) has indicated that the main sources of capital among small-scale farmers include in-kind (transfer of capital goods), personal savings, revolving loans, community banks (*Stokvels*), sale of cash crops or livestock and dowry.

3.2 Land

3.2.1 Choice, Acquisition and Ownership of Land

The philosophy of *Ubuntu* (humanity to others) and sacredness of land constitutes a main principle guiding all forms of land acquisition and ownership for agricultural purposes among most smallholder farmers in SSA. According to Nwonwu (2011: 16), in the olden days, land was not sold for money but could be in a form of exchange between willing owners with the purpose of capturing mutual benefits between users. He further observes that, "part of the agreement in such agreements often included a provision that the tenant would engage in land improving practices and investments on the land such as planting leguminous trees and shrubs and undertake erosion control measures (Nwonwu 2011: 17). It could also include mutual and symbiotic benefits through sharing of the land and labour resources.

3.2.2 Indigenous Land Use Planning

The existence of indigenous knowledge of land suitability classification based on the understanding of biophysical and social elements was recently indicated in Uganda by Nuwategeka and Nyeko (2019) who have shown that modeling land suitability based on IK and Geographic Information System (GIS) tools produces locally-generated agricultural land management models that are more effective in achieving sustainable agricultural production and environmental management. This is because they operate within the eco-cultural context of the farmers, therefore are easily useable. In another study, Niemeijer and Muzzucato (2003) shows that key indigenous indicators such as steepness of land (slope), colour (fertility), relative location of land from village (topographic position of land and location) were successfully used to allocate land into categories of forestland, cropland and grass and.

Farmlands for smallholder farmers in SSA typically comprises of small (between 0.5 and 2.0 ha) and deliberately fragmented land holdings in order to spread the risk of crop failure (Baiyegunhi and Makwangudze 2013). There are different categories of landholdings named according to their proximity to the homestead, types of crop cultivated, soil fertility and use (Nwonwu 2011). The farm closest to the homestead, commonly called a home garden, is of social importance to the household as this is where food security investments in terms of enterprises diversity and soil management efforts of the farmer are concentrated (Moyo 2010).

Home gardens contain mixtures of all-season and early maturing crops such as vegetables, maize and fruits planted in mixtures in a multi-storey structure (Watson and Eyzaguirre 2001). The diversity in variety and stage of maturity of the crops in the home garden represents a sound insurance against crop failure, loss of produce from pest and diseases and damage of crops from extreme weather vents (Musotsi et al. 2008). In so doing, the garden guarantees the family food supply all the year round. The proximity of this garden to the homestead makes it the biggest recipient of managerial land and soil improvement inputs such as organic waste from household and livestock, ash from fire places, fallen leaves from shade trees, all which are incorporated into the soil to provide organic matter and nutrients for crops (Nwonwu 2011). Apart from providing easy access to food for children, old and people with disability, the proximity of food gardens provides the farmer opportunity to experiment on new crops (Moyo 2010).

Main farms, are farmlands located in the outskirts of the home or village and stretches into natural boundaries such as rivers, streams, valleys or hills, are larger than home gardens. They are typically grown with food crops under mixed cropping pattern or grazing livestock (Moyo 2010). The crops grown here are those that do not require constant attention, are not harvested on daily basis and do not require much external inputs (Nwonwu 2011: 68). The range of crops provide the household with a balanced diet and some commercial crops for cash income. Gana (2009) has observed that smallholder farmers in SSA prefer to use indigenous plant species because, although they may yield lower, they are more tolerant to drought, and reduced available soil water, resistant to diseases and low soil fertility which ensures production under conditions of climate change. Other goods and services provided by the field crops to the household include firewood, fodder for small stock, fruits, vegetables and mushrooms.

Another type of landholding is that located in a wetland ecosystem, if stream(s) happen to dissect the area. This farmland, known locally as wetland, *dambo* or *vlei* is one of the most important as it has the attribute of having continuous water supply and being wet all year round (Adams 1993). In this way, a wetland is of strategic importance in bridging the food supply gap between time of planting and harvesting in the other farmlands. The wetland is productive from the combined effect of its continuous wetness and the deposition of alluvial material and organic matter when the stream or river overflows its banks during the rainy season (Nwonwu 2011). Consequently, it is planted early and provides farmers with early harvest in the year and thus safeguards the family from hunger at the stage of the cropping season when there is extreme food scarcity (Adams 1993). Furthermore, staggered planting across the different fields offers more food security for the household. The wetland also provides green grass for feeding livestock.

3.2.3 Clearing and Tillage of Land

Smallholder farmers in traditional farming systems of SSA use simple tools such hoes, *matchete* and animal drawn implements to prepare land for planting. Ridges (with or without imbedded organic residues), are widely used in the region to grow vegetables, maize, groundnuts, sweet potatoes and many other crops. On plateaus, some farmers make flat top ridges and on hilly areas round-top ridge systems are used. Contoured ridges prevent soil erosion and retain some moisture that is good for plant growth (Ellis-Jones and Tengberg 2000).

3.3 Indigenous Managerial Skills in Crop and Livestock Production

3.3.1 Farmer Seed Saving, Multiplication and Sharing Systems

The importance of farmers, especially smallholders in SSA to have control over their seed in order to sustain agriculture cannot be overemphasized. This requires a knowledge and appreciation of alternative technologies for crop and seed production, genetic conservation, and crop improvement (Eyong 2007). Smallholder farmers have employed various methods in this practice, including growing from own saved seed, collection of seedlings and vegetative propagation. Smallholder farmers grow and use traditional food crops because the practice is familiar. They understand traditional food plants and select varieties to meet the needs and constraints of their environment and use trusted seed from their own land races to produce reliable crops (Shava et al. 2009). Indigenous systems, including seeds and associated practices, are a rich source of alternatives.

3.3.2 Smallholder Farming and Cropping Systems

The prevalent farming and cropping systems among smallholder farmers in SSA include mixed/inter cropping, relay/sequential cropping, the *machobane* system, shifting cultivation, fallowing, agroforestry and crop rotation. Diversity is the norm in these systems due to its diverse cultures and climates (Eyong 2007). Crop diversification guarantees the farmers production and supply of food for the household and some surplus for sale. The different crop maturing times of the crops during the growing season minimises the risk of total crop failure (Scoones et al. 1996). Leguminous plants are intercropped with other crops or are planted during the fallow period in a bid to enrich the soil through biological nitrogen fixation (Kang et al. 1984). Other benefits of these traditional farming systems include: use of organic fertilizer which is locally produced; ensuring perennial vegetation cover; natural pest control and preservation of natural pest predators (Abate et al. 2000).

3.4 Indigenous Nutrient Cycling and Soil Fertility Maintenance Strategies

A number of indigenous practices contribute to improved nutrient levels and physical soil properties that provide better aeration, infiltration, water holding capacity, and soil moisture (Table 1). This, in the long run, contributes to improving soil ecological functions (productive, supportive, regulation and recreational).

Table 1 Indigenous practices and mechanisms of cycling nutrients and maintaining soil fertility by smallholder farmers in SSA

Practice	Mechanism
In-situ burning of vegetation	The burning of trees, branches and grass produces heat that kill weed seeds and soil borne pests and microorganisms e.g. the *Guie* or soil burning practice in Ethiopia (Abebe 1981)
Ash from burning biomass	The ash produced from burnt biomass provides nutrients such as Calcium and Magnesium (Edje et al. 1988)
Biomass transfer	Biomass such as crop residues and leaves are heaped and left to decompose on/in the soil to produce organic matter or the biomass is fed to livestock which produces manure that is transferred to the field (Nwonwu 2011)
Fallowing	Farmers leave the land to fallow in order to allow natural fertility to regenerate by plant species, bushes, grasses and trees that colonise the agricultural land (Nye and Greenland 1960). Due to reducing farm sizes, this system has largely been replaced by 'agroforestry'
Agroforestry	A system where fast growing leguminous trees such as *Gricidia sepium* are planted to replenish nutrients and allow more intensive and continuous cropping. This system recycles nutrients, fixes nitrogen, reduces soil erosion, lowers soil surface temperature and provides organic mulch and humus to the soil (Kang et al. 1984). The trees and shrubs in agroforestry land use systems are recognised to play an important role in sequestering and storing carbon in the soil (Ajani et al. 2013)
Termitaria (Termite mound soil)	Farmers in many parts of SSA plant crops which require good water supply and high nutrient levels, especially calcium and potassium, on anti-hills (Nyamapfene 1986). Some farmers apply *termitaria* in "dollop" form close to the base of the plant in order to improve its nutrients (Edje et al. 1988)
Organic amendments	The use of organic materials on and in the soil as amendments adds organic matter and soil cover to the soil, thus making it more resilient to flood, leaching and drought (Bationo and Mokunye 1991). Nyong et al. (2007) have suggested that the dependence and wide use of organic materials under smallholder farming in SSA systems is capable of reducing GHG emissions
Manure	Smallholder farmers in SSA have developed local knowledge on the use and management of animal (kraal) manure for improving and maintaining soil fertility (Harris 2002; Materechera 2010). Green manuring is practiced by some farmers when space allows, however, because of small land-holdings, most farmers practice green manuring only inadvertently when an early crop fails and has to be ploughed in before replanting (Scoones et al. 1996)

(continued)

Table 1 (continued)

Practice	Mechanism
Composting	Compost made from household waste, sometimes combined with livestock dung are applied mostly to home gardens and small plots around the homestead. The resultant high content of incorporated organic matter increases soil fertility and promotes rainfall infiltration (Harris 2002)
Nitrogen fixation	In many dry areas, the indigenous agrosilvo-pastoral system takes advantage of the multiple benefits provided by *Faidherbia* (formerly *Acacia) albida* (Weil and Mughogho 1993). The tree sheds its leaves at the onset of the wet season, permitting enough light to penetrate for the growth of sorghum and millet, yet still providing enough shade to reduce the effects of intense heat. The tree also fixes nitrogen for improving crop yield (Kang, Wilson and Lawson 1984)
Mulching	Application of mulch is used to conserve soil and water, maintain soil fertility and reduce weed growth (Nwonwu 2011)

3.5 Indigenous Soil and Water Conservation (ISWC) as a Strategy of Sustainable Land Management

There exists a wide array of ISWC strategies such as: water harvesting, hand-dug wells, run-off farming, low earth bunds, man-made terraces, traditional planting pits (*Zai*), stone and trash lines, micro-catchments, mulching, grass strips, among others (Reij et al. 1996). These systems enable production where purely rain-fed crops would fail. Smallholder farmers in the Sahel zone and other semi-arid areas of the SSA have employed a wide variety of innovative traditional rainwater harvesting techniques such as pits to produce water for crops including millet and sorghum (Critchley et al. 1994). The water harvesting systems involves collecting rainwater or runoff from a surface and storing it for crop production, livestock and domestic use (Pike 1938). Other indigenous practices used such as trash lines formed by placing crop residues in lines, across the farm impede runoff and enhance infiltration if carefully and effectively done. The trash lines are constructed from materials that are slow to decompose (such as sorghum and millet clovers) and are of lower palatability to livestock (Reij 1991). These technologies improve water infiltration and increase nutrient availability to crops leading to significant increases in yields, soil cover and reduced flooding (Ellis-Jones and Tengberg 2000).

3.6 *Indigenous Irrigation Practices in Sub-Saharan Africa Agriculture*

Irrigation is broadly defined in this chapter to include the application of water to crops by means of technology, and its removal by drainage (Adams and Anderson 2016). It also includes the management of water through selection of crops with known water requirements and tolerances, and the adaptation of cropping patterns and strategies to suit harsh environmental conditions (Adams et al. 1994). The main sources of water available to smallholder farmers include surface sources, streams and rivers but also springs. There is a wide range of traditional small-scale irrigation practices, mostly along small rivers in mountainous terrain, riverbeds, using residual moisture in valley bottoms, or tapping shallow groundwater (Watson et al. 1998). Adams and Anderson (2016) have reported flood plain cultivation of crops based on the exploitation of residual soil moisture left in a flood plain soils following inundation of rain caused depressions in East Africa. These flooded areas leave deep layers of deposited silt into which crops such as cotton, maize or rice are planted in deep holes dug with special tools (Adams and Anderson 2016: 523).

Another widely used practice involve the diversion of streams and transporting the water by canals (also called furrows) into fields where crops such as vegetables are grown. These furrows represents an example of engineering skill that is applied in extensive and complex indigenous water management system (Tafesse 2003). This brief description of indigenous irrigation practices, although far from being exhaustive, serves to indicate the richness of cultural practices in agricultural water control and management among smallholder farmers in SSA. As Martin et al. (1986) have suggested, the relevance and potential significance of this knowledge is in the development of modern technological improvement of irrigated agriculture in SSA.

3.7 *Indigenous Pest Management Approaches*

The majority of African farmers still rely on indigenous pest management approaches to manage pest problems, although many government extension programs encourage the use of pesticides (Abate et al. 2000). Traditional pest management practices consist mainly of cultural control methods such as crop associations, planting and harvesting time, closed season, mechanical control, use of herbal products, and, sometimes, dealing with pests in a supernatural way. Other approaches include reducing pest attack with crop rotations and intercropping, use of plants with repellent or insecticide action, traps, scarecrows and smoke (Nwonwu 2011). These control practices are based features that are built into the cropping systems, such as the location of the field.

4 Contribution of Indigenous Plant and Animal Genetic Resources to Food Security and Sustainable Agriculture in Smallholder Farming Systems of SSA

4.1 Indigenous Food Crops as Components of Food Systems and Sustainable Smallholder Farming in SSA

The practice of growing indigenous plants has been in existence among SSA communities over a long time and has provided one of the best mechanisms for addressing many agricultural challenges related to sustainability (Modi 2003). According to Shava et al. (2009), indigenous crops such as millet, sorghum, cassava, and other root crops are more important in smallholder farming because they are hardy and resistant to pest and disease; require fewer fungicide and insecticide applications; compete better with weeds and so require less herbicide; use water more efficiently and so can produce higher yields with less water. These species are known to use nitrogen more efficiently require less nitrogenous fertilizer thereby saving fossil fuel (Eyong 2007). These crops with their high genetic variability have a huge untapped potential for cultivation in many marginal environments of Africa that are threatened by climate change in order to provide greater stability and ecological resilience of cropping systems under extreme climatic conditions (Ajani et al. 2013: 32).

Many traditional plants make an important contribution to crop productivity in smallholder farmer systems in many ways. When grown as rotation crops between successive main crops, they may have the advantage of reducing the building up of pests and diseases (Abate et al. 2000). On the other hand, when grown as interplants, indigenous plants can act as an ecological barrier to diseases and ground cover to prevent soil erosion, reduce evaporation and suppresses weed growth. In some cases these plants are grown as green manure and ploughed in to increase soil organic matter which improves soil structure (Eyong 2007). Indigenous legumes fix atmospheric nitrogen, enriching the soil for the following crop or far near the plants (Weil and Mughogho 1993).

4.2 Role of African Indigenous Vegetables (AIV) in Small Scale Urban and Peri-Urban Agriculture

According to the UN, the Sub-Saharan Africa is experiencing the highest rates (about 4.6 per cent per annum) of migration from rural to urban areas (UN-Habitat 2007). Although cities will continue to depend largely on rural agriculture, substantial contribution (up to 20%) is now increasingly coming from within the urban and the peri-urban areas (Shackleton et al. 2009). Those who engage in small-scale urban farming activities mainly engage in horticulture, mushrooms, herbs, including keeping small stock such as poultry, pigs goats/sheep and fish for meat, eggs and milk to provide substantial family food requirements and income (Shackleton et al. 2009).

In many cities of SSA, the horticultural sector, particularly fruit and vegetables, occupies a significant niche in urban food systems that contributes to food security through increased food availability (Aphane et al. 2003). Indigenous (or traditional) African vegetables are an important part of farming and consumption systems of smallholder farmers in SSA (Jansen van Rensburg et al. 2004). The use of these vegetables is part of cultural heritage, playing a significant role in customs and traditions and in maintaining equity within the family structure of African societies (Mnzava 1997).

Even though the diversity of consumed AIVs varies significantly from one community group to another, and between groups living in similar conditions due to cultural reasons, one of the characteristic features of traditional African vegetables is that they contain high levels of vitamins, proteins and minerals (especially calcium, iron and phosphorus) than exotic vegetables (Schippers 2000). Thus, indigenous African leafy vegetables make significant contributions to the diets of many urban populations in the provision of elements, particularly iron, zinc and vitamin B complex and vitamins C and A, which have been noted to be deficient in many parts of SSA (Aphane et al. 2003).

Apart from food and nutrition security, most traditional vegetables in Africa have other uses such as medicines for curing a range of different illnesses (World Bank 2002). Some AIVs have been shown to contain phyto-chemicals or Nutraceuticals which are biologically active, non-nutrient compounds that provide health benefits by lowering risk of occurrence of chronic diseases like cancer (Shackleton et al. 2009). African indigenous vegetables have considerable potential as cash income earners, enabling the poorest people, particularly women, to earn a living (Schippers 2000). They provide employment opportunities and generate income for a rural population (World Bank 2002).

Ecologically, indigenous vegetables are well adapted to harsh climatic conditions and disease infestations of the SSA and are easier to grow in comparison to their exotic counterparts (Shava et al. 2009). Furthermore, African indigenous vegetables can produce seed under tropical conditions unlike exotic vegetables. They have a short growth period with most of them being ready for harvesting within 3–4 weeks, and respond very well to organic fertilizers (Uusiku et al. 2010). Most AIVs have an in built ability to withstand and tolerate some biotic and abiotic stresses (Schippers 2000).

4.3 *Indigenous Livestock and Their Role on the Sustainability of Smallholder Farming*

In many African societies, the number and quality of livestock which a farmer and his household own, is a measure of wealth. Livestock can be sold to purchase food, fibre, fertilizers, fuel and it provides draught power and transportation as well as being used as collateral for credit and as buffer against crop failure and other risks (Thornton and Herrero 2015).

Pastoralism is the most suitable livelihood and land-use system in arid and semi-arid areas as it constitutes an extensive production system with low emission of phosphates, no accumulation of dung and gases and so does not pollute land or water resources. This land-use system enables people to adapt by moving livestock according to the shifting availability of water and pasture. Pastoralists have developed grazing practices and patterns according to the available fodder and water (Msangi 2014).

Smallholder farmers in SSA have learnt mechanisms to preserve their livestock during times of drought, notably through devising cattle grazing and drinking schedules (Nwonwu 2011). The farmers prepare cattle fodder banks, which they use to feed the animals during the dry spell. This practice has a long history in rural areas and is considered as an additional adaptation strategy (Enete et al. 2011). Fodder selection requires good knowledge of the species composition, palatability and non-poisonousness, all which is based on the application of indigenous knowledge. Water scarcity caused by drought also demands that the locals devise appropriate coping mechanisms by planning cattle drinking schedules.

5 The Power of Indigenous Knowledge Systems in Strengthening the Resilience of Smallholder Farming to Climate Change and Variability

The SSA region experiences one of the highest climate variability in the world and this negatively impacts on food security particularly at the level of smallholder agricultural producers (Msangi 2014).

Smallholder farmers in SSA have, through IKS, developed intricate systems of gathering, predicting and interpreting weather and climate which are used for: forecasting weather and climate patterns; adaptation-responding to anticipated disasters such droughts and floods; predicting the seasonal climate; decisions about crop and livestock practices; decisions about farming activities e.g. choice of crops, planting dates etc. (Jiri et al. 2016; Mafongoya and Ajayi 2017; Nyong et al. 2007). Indigenous early warning systems for predicting or forecasting weather or climatic events have been an important contributor to reducing the vulnerability of such hazards (Ajibade and Shokemi 2003).

There is an abundance of literature which shows that SSA farmers have been using various local weather indicators such as plants, animals, insects, the solar system and wind in predicting the seasonal climate (Enete et al. 2011; Ziergovel and Opere 2010). This knowledge has been gained through many decades of experience and has been passed down through generations. The knowledge is also adapted to local conditions and needs. Indigenous knowledge is increasingly recognised as an additional important source of climate knowledge for adaptation strategies in the face of climate change and variability (Mapfumo et al. 2015).

6 African Indigenous Practices and Technologies of Post-Harvest Food Storage and Processing Enhance Household Food and Nutrition Security

In African societies, food is an integral component of people's culture and important knowledge, skills and technologies favourable to the food supply, quality and safety aspects of food security has been accumulated by the local people over time (Nwonwu 2011: 53).

There are numerous indigenous knowledge, practices and simple technologies which farmers across the SSA region use for storage and processing agricultural produce and keeping food products longer (Oniang'o et al. 2004: 88). These practices and techniques, which are predominantly used by women and youth, provide the majority of households and communities a basis of their food security by providing affordable adequate and nutritionally balanced diets over extended periods of time (Wane 2014).

The implementation of indigenous knowledge in post-harvest food storage and processing is an important asset upon which sustainable food security interventions can be built as IK provides for mutual learning and adaptation which contributes to the empowerment of local African communities in shaping their own food security agenda (FAO 1997). Apart from food security, indigenous practices and technologies of food storage and processing create employment among its practitioners and thus provide income to the neediest (Oniang'o et al. 2004). Due to its dynamic nature, indigenous knowledge provides the potential for improving food and livelihood security of SSA through innovation, adaptation and experimentation (Warren 1991).

7 Concluding Remarks

This chapter has explored the value of smallholder farming in sub-Saharan Africa and highlighted the challenges posed by rain-fed agriculture.

The purpose of the chapter was to show that African Indigenous Knowledge Systems (AIKS) have the potential of being used with success to achieve sustainability. It has identified and illustrated the embeddedness of AIKS as drivers of smallholder agriculture and highlighted the contribution of indigenous knowledge, technologies, plant and animal resources to the resilience and sustainability of smallholder farming.

The lessons learnt from the chapter are that smallholder agriculture remains a critical component of meeting the UN SDGs in SSA especially as it relates to household food security. Indeed, the global disruption of traditional food production and supply chains caused by the COVID-19 pandemic has necessitated the need to consider and emphasize on strengthening local food production systems and enhancing their resilience against any future disruptions. It is acknowledged that Indigenous Knowledge Systems (IKS) are key drivers of traditional ways of identifying and solving agricultural challenges especially those related to sustainable use of natural resources.

Some features of the smallholder farming systems which gives them relevance to agricultural sustainability include: high species diversity of indigenous plants and animals in time and space; recycling of plant nutrients from organic materials; reliant on locally available resources; use of indigenous land use planning and ISWC; reliance on indigenous ways of weather forecasting to adapt and mitigate against vulgarises induced by climate change and variability.

It is recommended that more research should be invested into AIKS and used in the formulation of Science and Technology-based agricultural practices that will work sustainably in SSA. There is also need to integrate AIKS in the teaching and learning curricula of educational institutions in the region in order to enhance its role in sustainable development. AIKS should be used to develop climate change adaptation strategies that are effective, participatory and sustainable.

Acknowledgements The author acknowledges support from the North-West University Library staff and the Centre for Teaching and Learning (CTL) for checking similarity index.

References

Abate T, van Huis A, Ampofo JKO (2000) Pest management strategies in traditional agriculture: an African perspective. Annu Rev Entomol 45:631–659

Abebe M (1981) Soil burning in Ethiopia. Ethiopian J Agric Sci 3:57–74

Adams WM, Anderson DM (2016) Irrigation before development: indigenous and induced change, in agricultural water management in east Africa. Afr Affairs 10: 519–535. https://afraf.oxfordjournals.org/. Accessed 13 Oct 2020

Adams WM (1993) Indigenous use of wetlands and sustainable development in West Africa. Geograph J 159(2):209–218. https://doi.org/10.2307/3451412 https://www.jstor.org/stable/3451412. Accessed 06 Sept 2020

Adams WM, Potkanski T, Sutton JEG (1994) Indigenous farmer-managed irrigation in Sonjo, Tanzania. Geograph J 160(1):17–32. https://doi.org/10.2307/3060138, https://www.jstor.org/stable/306013. Accessed 11 Dec 2020

Ajani EN, Mgbenka RN, Okeke MN (2013) Use of indigenous knowledge as a strategy for climate change adaptation among farmers in sub-Saharan Africa: implications for policy. Asian J Agric Extension Econ Sociol 2(1):23–40

Ajibade LT, Shokemi OO (2003) Indigenous approaches to weather forecasting in Asa LGA, Kwara State, Nigeria. INDILINGA Afr J Indigenous Knowl Syst 2:37–44

Aphane J, Chadha ML, Oluoch MO (2003) Increasing the consumption of micronutrient-rich foods through production and promotion of indigenous foods. In: AVRDC—the World Vegetable Center

Publication No. 03-561. International workshop proceedings, Arusha, Tanzania. FAO-AVRDC AVRDC, Rome, 5–8 March 2002, pp 77–92
Asghar J (2013) Critical paradigm: a preamble for novice researchers. Life Sci J 10(4):3121–3127
Baiyegunh LJS, Makwangudze KE (2013) Home gardening and food security status of HIV/AIDS affected households in Mpophomeni, KwaZulu-Natal Province, South Africa. J Hum Ecol 44(1):1–8
Bationo A, Mokwunye A (1991) Role of manures and crop residue in alleviating soil fertility constraints to crop production: with special reference to the Sahelian and Sudanian zones of West Africa. Fert Res 29(1):117–125
Bowen GA (2009) Document analysis as a qualitative research method. Qual Res J 9(2):27–40
Briggs J, Moyo B (2012) The resilience of indigenous knowledge in small-scale African agriculture: key drivers. Scott Geogr J 128(1):64–80
Briggs J (2005) The use of indigenous knowledge in development: problems and challenges. Progress Dev Stud 5(2):99–114
Buthelezi NN, Hughes JC (2014) Indigenous knowledge systems and agricultural rural development in South Africa: past and present perspectives. INDILINGA Afr J Indigenous Knowl Syst 13(2):231–250
Critchley W, Reij C, Wilcocks T (1994) Indigenous soil and water conservation: a review of the state of knowledge and prospects for building on traditions. Land Degr Rehabil 5:293–314
Edje OT, Semoka JMR, Haule KL (1988) Traditional forms of soil fertility maintenance. In: Wortman C (ed) Proceedings of a workshop on soil fertility research for bean cropping systems in Africa. CIAT African workashop series No. 3, pp 7–29
Ellis-Jones J, Tengberg A (2000) The impact of indigenous soil and water conservation practices on soil productivity: examples from Kenya, Tanzania and Uganda. Land Degrad Dev 11:19–36
Enete AA, Madu II, Mojekwu JC, Onyekuru AN, Onwubuya EA Eze F (2011) Indigenous agricultural adaptation to climate change: study of southeast Nigeria. The African technology policy studies network, research paper No. 6 Nairobi, Kenya
Eyong CT (2007) Indigenous knowledge and sustainable development in Africa: case study on Central Africa. Tribes Tribals Spec 1:121–139
FAO (1997) Agriculture, food and nutrition for Africa. Food & Agriculture Organisation, Rome
Gana F (2009) The usage of indigenous plant materials among small-scale farmers in Niger State agricultural development project-Nigeria. INDILINGA-Afr J Indigenous Knowl Syst 2(1):43–51
Ghorbani E, Bakhtiar F (2011) The importance of indigenous knowledge in agricultural development. J Am Sci 7(6):180–184
Harris F (2002) Management of manure in farming systems in semi-arid West Africa. Exp Agric 38(2):131–148
Hart T, Mouton J (2005) Indigenous knowledge and its relevance for agriculture: a case study in Uganda. INDILINGA-Afr J Indigenous Knowl Syst 4(1):249–263
Jansen van Rensburg WS, Venter SL, Netshiluvhi TR, van den Heever E, Vorster HJ, de Ronde JA (2004) Role of indigenous leafy vegetables in combating hunger and malnutrition. South African Journal of Botany 70(1):52–59
Jiri O, Mafongoya PL, Mubaya C, Mafongoya O (2016) Seasonal climate prediction and adaptation using indigenous knowledge systems in agriculture systems in Southern Africa: a review. J Agric Sci 8(5):156–172
Kang BT, Wilson GF, Lawson TL (1984) Alley cropping, a stable alternative to shifting cultivation. International Institute of Tropical Agriculture, Ibadan, Nigeria
Lwoga ET, Ngulube P, Stilwell C (2010) Managing indigenous knowledge for sustainable agricultural development in developing countries: knowledge management approaches in the social Context. Int Inf Libr Rev 42:174–185
Mafongoya PL, Ajayi OC (eds) (2017) Indigenous knowledge systems and climate change management in Africa. Wageningen, The Netherlands, CTA

Mapfumo P, Mtambanengwe F, Chikowo R (2015) Building on indigenous knowledge to strengthen the capacity of smallholder farming communities to adapt to climate change and variability in southern Africa. Climate Dev 8(1):77–82
Martin E, Yoder R, Groenfeldt D (1986) Farmer-managed irrigation: Research issues. Irrigation Management Network Paper 86/3c, Overseas Development Institute (ODA), London
Materechera SA (2010) Utilization and management practices of animal manure for replenishing soil fertility among smallscale crop framers in semi-arid farming districts of the North West province, South Africa. Nutr Cycl Agroecosyst 87:415–428
Mnzava NA (1997) Vegetable crop diversification and the place of traditional species. In: Guarino L (ed) Traditional African vegetables. Promoting the conservation and use of Underutilized and neglected crops: Proceedings of the IPGRI international workshop on genetic resources of traditional vegetables in Africa. 29–31 August 1995, Nairobi, Kenya. Institute of Plant Genetic and Crop Plant Research, Gatersleben. Plant Genetic Resources Institute, Rome, Italy, pp 1–15
Modi AT (2003) What do subsistence farmers know about indigenous crops and organic farming? Preliminary experience in KwaZulu-Natal. Dev South Afr 20:675–684
Mortimore M (2000) Profile of rainfall change and variability in the Kano-Maradi region: working Paper 25, Somerset, UK
Moyo BHZ (2010) The use and role of indigenous knowledge in small-scale agricultural systems in Africa: the case of farmers in northern Malawi. Ph.D. thesis. University of Glasgow, Glasgow. https://theses.gla.ac.uk/2022/. Accessed 14 Dec 2020
Msangi J (2014) Food security among small-scale agricultural producers in Southern Africa. Springer, Cham
Musotsi AA, Sigot AJ, Onyango MOA (2008) The role of home gardens in household food security in Butter division of Western Kenya. Afr J Food Agric Nutr Dev 8(4):375–390
NEPAD (2003) Comprehensive Africa agriculture development programme. AU/NEPAD, Pretoria
Niemeijer D, Muzzucato V (2003) Moving beyond indigenous soil taxonomies: local theories of soil for sustainable development. Geoderma 111:403–424
Nuwategeka E, Nyeko M (2019) Using indigenous knowledge to model land suitability for crops in Amuru district, Northern Uganda. Afr J Food Agric Nutr Dev 19(04):14805–14828
Indigenous knowledge systems in Igbo traditional agriculture. AuthorHouse, London
Nyamapfene KW (1986) The use of termite mounds in Zimbabwe peasant agriculture. Tropical Agric 63:191–192
Nye PH, Greenland DJ (1960) The soil under shifting cultivation. Technical communication no. 51, Commonwealth Bureau of Soils, Harpenden
Nyong A, Adesina F, Elasha BO (2007) The value of indigenous knowledge in climate change mitigation and adaptation strategies in the African Sahel. Mitig Adapt Strat Glob Change 12:787–797
Oniang'o R, Allotey J, Malaba S (2004) Contribution of indigenous knowledge and practices in food technology to the attainment of food security in Africa. J Food Sci 69(3):87–91
Pike AH (1938) Soil conservation among the Matengo tribe. Tanganyika Nous Rec 6:79–81
Reij C (1991) Indigenous soil and water conservation in Africa. Gatekeeper series no. 27. Centre for Development of Cooperative Services, Amsterdam. Vrije University, Van der Boechorstraat 7, The Netherlands
Reij C, Scoones I, Toulmin C (eds) (1996) Sustaining the soil: indigenous soil and water conservation in Africa. Earthscan Publications Ltd., London
Safakish M (2015) The role of indigenous knowledge in sustainable rural economic development. J Appl Environ Biol Sci 5(9S):285–289
Sanusi ZA (2014) Integrating knowledge diversity for sustainability. In: Alvares C (ed) Multicultural knowledge and the university. Multiversity & Citizens International, Penang, Malaysia, pp 274–277
Saunderson M (1995) Sustainable agriculture and rural development in South Africa. Outlook Agric 24(3):163–165

Schippers RR (2000) African indigenous vegetables: an overview of the cultivated species. Universty of Greenwich, Natural Resources Institute

Scoones I, Chibudu C, Chikura S, Jeranyama P, Machaka D, Machanja W, Mavedzenge B, Zirereza B (1996) Hazards and opportunities- farming livelihoods in dryland Africa: lessons from Zimbabwe. Zed Books Ltd., London

Shackleton CM, Pasquini MW, Drescher AW (eds) (2009) African indigenous vegetables in urban agriculture. Earthscan Publications Ltd., London

Shava S, O'Donoghue R, Krasny ME, Zazu C (2009) Traditional food crops as a source of community resilience in Sub-Saharan Africa. Natural Resources Institute/ACP-EU Technical Centre, Chatham

Tafesse M (2003) Small-scale irrigation for food security in sub-Saharan Africa. Report and recommendations of a CTA study visit to Ethiopia, 20–29 January 2003. CTA Working Document Number 8031. The ACP-EU Technical Centre for Agricultural and Rural Cooperation (CTA), Netherlands

Thornton P, Herrero M (2015) Adapting to climate change in the mixed crop and livestock farming systems in sub-Saharan Africa. Nature Climate Change 5(9):830–836

UN (2015) Transforming our world: the 2030 agenda for sustainable development... Sustainable development knowledge platform. https://sustainabledevelopment.un.org/post2015/transforming ourworld/. Accessed 29 Nov 2020

UN (2019) World population prospects: the 2019 revision. United Nations, Department of Economic and Social Affairs, Population Division, Washington

UN-Habitat (2007) United Nations human settlements programme (UN-HABITAT). The challenge of slums: global report on human settlements 2003. Earthscan Ltd, London

Uusiku NP, Oelofse A, Kwaku GD, Megan JB, Mieke F (2010) Nutritional value of leafy vegetables of sub-Saharan Africa and their potential contribution to human health: a review. J Food Compos Anal 23:499–409

Wane NN (2014) Indigenous African knowledge production: food-processing practices among Kenyan Rural Women. University of Toronto Press, Toronto

Warren DM (1991) Using indigenous knowledge in agricultural development. Word Bank Discussion Paper No. 127. The World Bank, Washington

Watson EE, Adams WM, Mutiso SK (1998) Indigenous irrigation, agriculture and development, Marakwet, Kenya. Geogr J 164:67–84

Watson JW, Eyzaguirre PB (2001) Home gardens and in-situ conservation of plant genetic resources in farming systems. Proceedings of the 2nd international home gardens workshop, Witzen-Hausen, Germany, 17–19 July 2001

Weil RR, Mughogho SK (1993) Nutrient cycling by Acacia albida (syn. Faidherbia albida). In: Ragland J, Lal R (eds) Technologies for sustainable agriculture in the tropics. ASA special publication number 56. SAS, CSSA, SSSA, Madison, Wisconsin, pp 97–108

World Bank (1989) Sub-Saharan Africa: from crisis to sustainable growth. The World Bank, Washington

World Bank (1998b) Indigenous knowledge for development: a framework for action. Knowledge and Learning Centre, Africa Region, The World Bank, Washington. https://worldbank.org/afr/ik/ikrept.pdf. Accessed 07 Nov 2020

World Bank (2002) The contribution of indigenous vegetables to household food security. IK Notes. No. 44, The World Bank, Washington

World Bank (2004) Indigenous knowledge: Local pathways to global development – marking five years of World Bank Indigenous Knowledge for Development Programme. The World Bank, Washington

Ziervogel G, Opere A (ed.) (2010) Integrating meteorological and indigenous knowledge-based seasonal climate forecasts in the agricultural sector. International Development Research Centre (DRC), Ottawa, Canada

Professor Simeon Albert Materechera holds a PhD in Soil Science from the University of Adelaide (1994) and a Postgraduate Diploma in Knowledge and Information Systems (PGDip Knowl & InfoSys) from the University of Stellenbosch, South Africa (2015). He is currently a Professor in the Indigenous Knowledge Systems (IKS) Centre at the North-West University. His research is focused on sustainable agriculture, food security and the application of indigenous methodologies in research and community engagement.

Women's Economic Empowerment for Sustainable Development through Indigenous Basket and Mat Weaving in Zimbabwe

Sylvia Madusise

Abstract Investing in women's economic empowerment sets a direct path towards gender equality, poverty eradication and inclusive economic growth. This chapter reflects on how women's economic empowerment could serve as a prerequisite for sustainable development. Zimbabwe is still experiencing a generally low status for women with respect to access to economic resources and economic opportunities. Women's economic empowerment may be achieved through education. This chapter considers the significance of non-formal education in Zimbabwe. It analyses how holistic community-based non-formal education organisations might offer education to women that could lead to sustainable development through training women on how to weave baskets and mats using indigenous materials. Three indigenous basket and mat weavers based at Great Zimbabwe Monuments were observed and interviewed whilst weaving. It was established that the weaving skills being used could be passed on to other community women through non-formal training. Production costs could be minimised through using locally available materials. Markets for the products are available at most tourist centres in Zimbabwe, which is a famous tourist destination. Entrepreneurial skills could be developed through non-formal training. It is argued that the weaving project, if well-established, could be a source for the much-needed foreign currency in Zimbabwe.

Keywords Economic empowerment · Sustainable development · Non-formal education · Indigenous basket and mat weaving

1 Introduction

Eliot (2008) posits that the economic empowerment of women can be achieved if women engage in income generating projects. Economic empowerment of women can help women stave off poverty and become self-reliant. A woman can be economically empowered by affording her economic success, economic advancement and

S. Madusise (✉)
Department of Teacher Development, Great Zimbabwe University, P.O Box 1235, Masvingo, Zimbabwe
e-mail: smadusise@gzu.ac.zw

W. Leal Filho et al. (eds.), *Sustainable Development in Africa*, World Sustainability Series,
https://doi.org/10.1007/978-3-030-74693-3_4

decision making on economic matters (Golla et al. 2011). Women need necessary skills which can enable them to economically succeed. For women to exert influence in society their economic empowerment and participation are critical in as much as bolstering their rights enabling them to control their lives. DAC Network on Gender Equality (2011) accedes to these ideas when it says women' s economic empowerment affords them access to resources, job opportunities, property ownership and a plethora of other services that are necessary in life. In every society, matters about women empowerment are important. The same matters speak to human rights by drawing the attention of leaders in business and the making of policies. González (2016) reiterates that for global inclusive and sustainable development, empowering women economically is an inevitable reality. Gender inequality can also be eradicated through economically empowering women since women are key contributors to economies. Women participate in various facets of economies such as food production, management of natural resources and entrepreneurship.

Many International bodies support women's economic empowerment for sustainable development, these include the Beijing Platform for Action, United Nations Development Programme (UNDP), International Trade Centre (ITC), the Convention of All Forms of Discrimination against Women and a series of International Labour Organisation (ILO) conventions on gender equality. United Nations Women supports women's economic empowerment to earn higher incomes and have better access to and control of resources. Economic equality benefits the individual, the family, the nation and the world. United Nations global development agenda champions the goal to economically empower women so as to eradicate poverty (González 2016). Such a fit validates equitable and sustainable growth. Poverty is a significant challenge to human well-being, and its eradication is a high priority goal for development. The economic empowerment of women is fundamental to women's ability to move and stay out of poverty.

There are a variety of interventions in the attempt to alleviate poverty, as highlighted by González (2016) in her presentation at the Business and Professional Women conference in Auckland. In Mongolia, a joint development agency of United Nations and the World Trade Organisation, ICT, economically empowered women through assisting them to secure markets for their products. Mongolian women managed to export cashmere, wool and yak hair and raised up to $50 000 in 2016. In Cambodia rural women weavers and business women were trained to develop business plans and prepare new products. The training enabled the women to increase their exposure to markets and raised their income by 30%. Vibrant platforms to sell the products were also provided by the training. In 2016, an ICT programme in Nairobi economically empowered women through skills development to produce ethnic fashion clothes. The transformational engagement enabled the empowered women to generate income, thus positively impacting on the society. In Papua New Guinea, ITC economically empowered women coffee growers and in 2016 they received export licences for the first time to export coffee. In Angola UNDP trained women in charcoal production. The training which helped women to preserve their incomes and forests focussed on methods the workers could use to pass onto others:

increasing environmental awareness and the knowledge of the benefits of sustainably maintaining forests (UNDP 2019). Poverty eradication and inclusive economic growth may be achieved through investing in women's economic empowerment. Women therefore, enormously contribute to economies, be it in farms, in businesses, as employees or entrepreneurs.

Eradication of poverty is a high-priority goal for development. To achieve this goal, UN Women offers a helping hand to most needy grassroots women. The group of grassroots women comprises low-skilled women. In Errachidia, a province in Morocco, associations of farming women were helped to adapt to climate by UN Women. Un Women economically empowered them through training them how to grow aromatic and medicinal plants that require little water but producing higher yields as compared to traditional crops. Over 1500 women entrepreneurs were trained by Un Women in Kenya since 2013. The training provided women with comprehensive business skills. UN Women developed Women's Empowerment Principles on corporate social responsibility for gender equality. The Women's Empowerment Principles were signed by over one thousand (1000) CEOs around the world (UN Women 2013). Promoting economic empowerment of women and gender equality benefits the economy and the society at large. The government of Zimbabwe has been collaborating with the United Nations Country Team (UNCT) to promote the economic empowerment of women and support their inclusion in all spheres of life (Tarinda 2019). UNCT has a cohort of UN agencies such as UNICEF, UN Women, UNDP and others. In support of economic empowerment of women, Bill Clinton the former president of USA reiterated that in issues of improving education in developing countries, or fighting global change or addressing any other experienced challenge, the critical part of the equation is based on empowerment of women. In his presentation at the Clinton Global Initiative (September, 2009), bill Clinton argued that 66% of the world's work is performed by women and they produce 50% of the food. However, the same women earn 10% of the income and own only 1% of the property (DAC Network on Gender Equality 2011). This analysis depicts a scenario where women are unfavourably treated in the world of work. Therefore, women's economic empowerment is to be considered as a prerequisite for sustainable development. In Eastern and Southern Africa the Federation of National Associations of Women in Business (FEMCOM) promotes women's economic empowerment across various fields like mining, agriculture, fisheries and construction (Tarinda 2019). Behind all these stories there are hundreds and sometimes thousands of women who positively impact on future generations as change agents. González (2016) argued that trade and women empowerment are intertwined. Women's empowerment is supported by trade. The empowered women's contribution to the economy and society is monumental.

Considering the economic empowerment of women so as to reduce poverty, this chapter proposes non-formal training of Zimbabwean women in indigenous basket and mat weaving to produce artefacts for sale. This could then lead to development of micro and small-scale enterprises in which women are highly represented, thus, economically empowering women for sustainable development.

2 Women's Economic Empowerment and Sustainable Development

Governments and non-governmental organisations the world over have adopted sustainable development as a panacea for poverty. Sultana and Haque (2019) argue that sustainable development is development that meets the needs of the present without compromising the ability of future generations to meet their needs. Sustainable development refers to growth that aims at reducing poverty and building shared prosperity of today's generation and to continue meeting the needs and expectations of future generations. For individuals and institutions to function in a manner that supports and meets challenges of both the present and the future, innovative ways must be applied to sustainable development. According to the UN Secretary General women's economic empowerment and the closure of gender gaps in the world of work are key to achieving sustainable development goals and also achieving the 2030 Agenda for Sustainable Development. Particular goals to be achieved are Goal 8 which relates to the promotion of full and productive employment and decent work for all, Goal 1 on ending poverty and Goal 5 on achievement of gender equality. Other goals which may be achieved by economic empowerment of women are: Goal 2 on food security, Goal 3 on ensuring health and Goal 10 on reducing inequalities (UN Women Headquarters 2018). McKinsey & Company (2018) when arguing for women's economic equality estimated that companies with three or more women occupying senior management posts score very high in all dimensions of organisational performance. Increasing employment and leadership opportunities for women impacts positively on companies by way of:

- Organisational effectiveness
- Economic equality for women which is good for many a business (UN Women Headquarters 2018).

Organisations that empower women economically realise a win-win scenario which benefits not women only but society at large for sustainable development. Women who acquire the right skills can influence the growth of businesses and markets (Golla et al. 2011)

Research proffers good reasons why women empowerment is critical to sustainable development programmes. A big number of the world's poor are women, so argue Golla et al. (2011). Poverty reduction can be achieved by empowering women, economically so that they achieve their potential and advance their rights. Women are also known to be good at investing extra income on their children.

3 Addressing Women's Economic Empowerment

The economic empowerment of women is a prerequisite for sustainable development and requires sound public policies, a holistic approach and a long-term commitment.

Kampilipili's (2017) analysis of the research carried out in Southern Africa shows that the phenomenon of women with multiple responsibilities of wife, mother and paid worker is significantly being entrenched in societies. Kampilipili (ibid), in the same vein, presents that Africa struggles to realise its potential of full growth because a marked portion of its full growth reserve, the women, is not utilised fully. To achieve women's economic empowerment, organisations must address the underlying factors that contribute to it, that is, individual and community resources. For women to succeed economically they need resources such as:

- Education to cater for impartation of skills and training
- Financial resources
- Networking
- Land and implements.

Resources available to women can be used to economically empower women through engaging in projects thereby emboldening sustainable development. The projects build women's power base as agents of improved quality of life since they would engage in economically empowering activities at organisational, household and individual level (Golla et al. 2011).

Understanding the wider opportunities and challenges women face in the specific context of the project is important in order to: determine where resources are best spent to meet the goals; design the best intervention; and identify areas outside of the project focus that may help or hinder its success.

This chapter proposes non-formal training of women on indigenous basket and mat weaving at community level for informal sector employment.

4 Non-formal Education and Informal Sector Employment

Islam and Mia (2007) have hinted that the only strategic intervention for poverty eradication or alleviation is non-formal education, a multi-faceted education programme. Some of its programmes are equipping the poorer and disadvantaged sections of society with survival skills necessary for sustainable development. Lives of almost half of the world's population would be saved from poverty if it is equipped with the survival skills. La Belle (1986) concurs with Islam and Mia's (ibid) ideas by positing that non-formal education is a regular educational activity which exists outside formal education system. It aims at imparting varying kinds of learning to specific groups of society, normally the disadvantaged. According to Coombs and Ahmed (1974), quoted by Mok (2011), non-formal education is normally linked with community groups and other groups, unlike formal education which is linked with schools and training institutions.

Non-formal education affords a second chance to education to those who could not fit in the formal system. No matter how informal the economy is, society cannot afford to forgo skills training. Members of society can access non-formal education to acquire practical skills needed in real life for sustainable development in a context where the informal economy is a source of livelihood. According to Sultana and Haque (2019) non-formal education is considered as an educational and training avenue outside formal education.

Many non-governmental organisations in Zimbabwe provide non-formal education to the poor and disadvantaged groups. Products of non-formal education usually end up in informal employment. The informal economy employs a larger share of the economically active population in Africa. According to the International Labour Organisation (ILO)'s 2018 report on the informal economy, 85.8% of employment in Africa is informal. The proportion is 68.2% in Asia and the Pacific; 68.6% in Arab States; 10.0% in the Americans and 25.1% in Europe and Central Asia. In Zimbabwe, according to the Zimbabwe National Statistics Agency (ZIMSTAT) report, 94.5% of the 6.3 million people defined as employed are working in the informal economy (ZIMSTAT 2015). In Tanzania, out of 700 000 people who enter into the labour market annually, only 40 000 can expect to be employed in the formal economy (Vanek et al. 2014). In Kenya, the business writer Amadala (2019) reiterates that 6% of Kenya's total employment is in the formal sector's non-agricultural employment and almost 49% is in informal non-agricultural employment, with 45% employed in agriculture. According to some survey (Chen 2017), informal employment accounts for two-thirds (66%) of non-agricultural employment in Sub-Saharan Africa. It accounts for 82% in Mali and 76% in Tanzania. However, variation within the regions is significant. In South Africa, informal non-agricultural employment accounts for 33% and 44% in Namibia. Therefore, in most African countries the informal sector continues to contribute enormously to employment creation while the growth of employment in private and public sectors has almost stagnated.

Non-formal education is then directly linked to income generating programmes for poverty alleviation. This chapter is significant because it explores the contribution of non-formal education to income generation for sustainable development and meeting sustainable development goal 1.

5 Theoretical Framework

The current chapter is anchored on Vygotsky's sociocultural theory of human learning. Vygotsky views learning as occurring through social interactions with more skilled tutors (McLeod 2018). Vygotsky referred to the more skilled tutor as the more knowledgeable other (MKO). To Vygotsky, anyone who has a better understanding or a higher ability than the learner with respect to the task is the more knowledgeable other. Another principle of Vygotsky' sociocultural theory is culture. According to Vygotsky, mental development occurs in a unique way in each culture. For example, people in a farming community learn how to farm as a way of functioning in their culture. In this chapter the more knowledgeable ones are the experienced indigenous basket and mat weavers who are expected to non-formally train community women to weave indigenous baskets and mats through social interactions. The involved cooperative or collaborative dialogue in the social interactions is expected to promote cognitive development. The tutors are expected to demonstrate some basic weaving strategies to the trainees. The trainees then seek to understand the actions or instructions and internalise the information using it to guide their own performance. The community women learn to weave indigenous baskets and mats as a means of production to boost their income for sustainable development.

6 Sample and Sampling Procedure

The sample in this qualitative study consisted of three indigenous basket and mat weavers (one male and two females) who were purposively chosen. The permission to carry out the research with the weavers was sought from the Great Zimbabwe Monuments Marketing Director. In selecting the weavers, the researcher was guided by the Marketing Director at Great Zimbabwe Monuments. The Director introduced the researcher to the three weavers whom he described as the best weavers among the cohort of weavers who were participants at the Monuments' cultural village. The weavers completed the consent form which sought their permission to photograph them and their products. According to the Marketing Director, the weavers were chosen for their exceptional expertise and they supplied artefacts to the curio shop at the Monuments. Their products were highly rated. The weavers also displayed their products for sale at the Monuments' entrance and they were also supplying artefacts for international market through the Monuments' marketing department.

7 Methodology

Driven by the need to economically empower women for sustainable development, the study set out to explore the possibility of equipping community women with indigenous basket and mat weaving skills through non-formal training. This view is premised on Sultana and Haque's (2019) argument that non-formal education is considered as a wider spectrum of educational and training activities. The focus of this study was on non-formal training of women on indigenous basket and mat weaving. The research adopted a qualitative research approach in which the descriptive survey was used as the research design. The descriptive survey was deemed appropriate because it accorded the researcher ample time and means to explore the contribution of non-formal education to income generation for sustainable development. This was explored through direct observation and relatively unstructured one-to-one interviews and a group interview in a natural field setting. The observation checklist considered the following guidelines:

- The weaving knowledge being applied and the possibility of acquiring that knowledge through non-formal training
- The type of weaving material required
- Where and how the materials were obtained/gathered.

Observations and one-to-one interviews were simultaneously carried out. Open-ended questions were asked, and the interviews flew like everyday conversations. The respondents could use their own indigenous language, and this allowed the respondents to talk in-depth, using their own words. It also helped the researcher to develop a real sense of the weavers' knowledge of basket and mat weaving. Although the interviews were unstructured, the researcher had certain topics in mind which were to be covered in all the interviews. Main areas of information sought from weavers included:

- Level of attained education required
- Availability and cost of raw materials
- Possible market
- Preparedness of the weavers to offer non-formal training to the community women
- How closely non-formal education was linked to income generating projects.

The collected data were in the form of words and pictures. The weavers and some of their artefacts were photographed. Since the researcher had told the weavers that the pictures were to be shown to the UN Women Organisation when campaigning for the sponsorship of the Weaving Project, the weavers were enthusiastic to be photographed as they saw it as a way of marketing their wares. According to Bogdan and Biklen (1992), qualitative articles and reports are described as being 'anecdotal'. They often contain quotations and try to describe what a particular situation or view of the world is like in a narrative form. After one-to-one interviews with the three weavers, a group interview was then carried out as a way of triangulating what came out from one-to-one interviews and observations. Use of different sources helped the researcher to validate data, as data from one source could be checked against another source. The whole process of data collection, therefore, included the following activities:

- List of topics to be covered by the interviews
- Checklist and observation guidelines.

Quotations from interview transcripts were used as narrative stories told by weavers evaluating the possibility of training community women to weave indigenous baskets and mats for sale. The narrative stories were thematically analysed. The interviewed weavers were referred to as Weaver 1, Weaver 2 and Weaver 3.

8 Findings and Discussion

Weaving Skills Development
The weavers responded in their indigenous language (Shona). However, the researcher transcribed both the questions and answers to English for presentation purposes.

The three interviewed weavers had this to say about the development of weaving skills:

Researcher: Who taught you how to weave?
Weaver 1: I was trained by my father who was also trained by my grandfather.
Researcher: Was your grandfather formally trained how to weave at school?
Weaver 1: Among weavers in my family no-one was formally trained how to weave. Weaving skills do develop through practice. Through observing someone weaving, you can also learn how to weave. My father used to ask me to simultaneously weave a similar basket as his. He could use his basket to demonstrate to me from the beginning until the basket was complete.
Researcher: After what age and level of education did you start weaving?
Weaver 1: I joined my father in weaving after completing Grade Seven. I was about fifteen years old then. Weaving comes with passion; it can be learnt at any age. You do not need bookish education except a little bit of mathematics for counting and measuring.

Weaver 1 was in his early fifties.

Similar unstructured questions were asked to Weaver 2 and 3. Weaver 2 said that she had been taught how to weave by her mother. She then started producing artefacts for sale at the age of fifteen. Some of her artefacts are displayed in Fig. 2. By the time of the interview she was in her mid-sixties. She only completed Standard Two which is equivalent to Grade Four. She posited that to be a good weaver, one needs to be committed to one's work.

Weaver 3 said that she was a first born in a family of six girls and her father was a weaver. Due to financial problems, she had to drop out of school at a tender age of twelve after completing Grade Five. As a result, she spent most of her time helping her father prepare raw materials for weaving. She learnt about weaving circular trays *(tsero)*, triangular trays and fruit bowls (see Fig. 1 and Fig. 3). She described her work as unusual for women as most women normally make mats using reeds (*nhokwe*). Weaver 3 was forty-eight years old at the time of the interview.

The discussion with the weavers revealed that weaving skills could be developed through non-formal training. Therefore, community women who were not gainfully employed could be organised and be informally trained to weave indigenous baskets and mats for sale. It was also discovered that even someone with minimum bookish education could be trained to weave indigenous baskets and mats. Weaving comes with a passion and commitment as reiterated by the weavers. Due to their expertise, the interviewed weavers produce quite beautiful artefacts (see photographed artefacts).

Figure 1 is a fruit bowl produced by Weaver 3.

Fig. 1 A fruit bowl

Fig. 2 A sugar basin and a fruit bowl produced by Weaver 2

Fig. 3 Weaver 3 proudly displays her completed circular tray

9 Non-formal Training of Indigenous Basket and Mat Weavers

Upon discussing with weavers, it was gathered that they were prepared to share their weaving knowledge with other community members. The weavers asserted that from their weaving experience, they were competent enough to train others.

> I have been in this weaving business for more than forty years. I can train others on the skills involved. However, the trainees should be committed and should have a lot of patience, understanding the motive for training (Weaver 2).

Effective facilitation of non-formal education entails knowing that adult learners are self-directed and internally motivated. They need to be respected. They are practical, relevancy and goal oriented since they apply real life experiences as they seek knowledge when learning (Knowles 1984). Hague and Logan (2009) suggest that encouraging and supporting self-directed non-formal education is important to development.

> Weaver 1: In our community there are no income generating projects especially for women. Women only engage in vegetable gardening. Every household has its own garden now.

Weaver 1 posited that people in their community especially women were living in poverty. The only project they were engaged in was vegetable gardening which was rather flooded as every household had its own garden. Therefore, there was a limited market for the vegetables. This then showed the need for women's income generating projects. Badu-Nyarko and Zumapkeh (2014) state that non-formal education can change people's incomes. Boosting these incomes can be achieved through engaging in various projects, for example, market gardening. Jayaweera (1979) is in agreement with Badu-Nyarko and Zumapkeh (ibid) ideas by stating that women can be development agents as they embrace the teaching and training of non-formal education. Women can also be pivotal in poverty eradication in their communities since they will be empowered to participate in economic development projects earning them money for socio-economic growth (UNDP 2016). Their confidence in taking up opportunities is also emboldened by playing positive roles and openly speaking before others about the economic projects there are to engage in.

Fig. 4 Weaver 1 demonstrating mat weaving

Fig. 5 Weaver 1's completed mat

The possibilities of starting an intervention programme which trains community women weaving skills were discussed. The weavers argued that they were going to charge a nominal training fee, affordable by most community women. When asked about the availability of raw materials, the weavers indicated that the raw materials most needed were locally available in their community. For mats, they used reeds which were available along riverbanks (see Fig 4 and Fig 5). For basins and fruit bowls (see Fig 1, Fig 2 and Fig 3), they used what they referred to as *mbebwa* which is said to be awash in the mountains, and for stronger baskets, they used material from bamboo trees. Materials from bamboo trees were sold locally by fellow villagers who grow the trees in their backyard plantations. Most of the materials used were indigenous.

When it is not the ploughing season, weavers spend almost all weekdays working at the cultural village housed at the Great Zimbabwe Monuments. They suggested that they could use the cultural village as their training venue. During the planting season, they do not work in the fields on Wednesdays, so they could dedicate this day to training. Out-of-school programmes should be considered as vital to the community's development. The programmes could meet the immediate needs of the community (Latchem 2014). Discussions with the weavers revealed a high feasibility of a successful weaving training programme. Weavers seemed to have acquired a satisfactory level of skills and were prepared to develop fellow community members. Islam and Mia (2007) argue that the trainers are required to have a certain technical competence level. The competence level can be measured against the ability to translate theory into practice for alleviation of poverty and generation of income.

10 Income Generation and Poverty Alleviation through Basket and Mat Weaving

Responses from the weavers indicate that the market facility was very rich since Great Zimbabwe Monuments is an attractive tourist centre. Most of the products they displayed at the monuments were bought by both local and international tourists. The international customers bought using foreign currency. The sale of indigenous baskets and mats brought the much-needed foreign currency to the country. Weaver 3 indicated that business was very lucrative during festive holidays and they sometimes ran out of the products. This then suggests the need for a weaving project which could lead to the production of enough artefacts for the market. If that happens it means they could even take their wares to other tourist centres such as Victoria Falls and Chinhoyi Caves.

Weavers reiterated that they were having problems in accessing loans. They envisioned that if a well-established weaving project was to be registered, this could facilitate their loan applications. Owing to lack of property ownership, women find it very difficult to access loans where collateral is a prerequisite. Only 5% of female respondents in a World Bank survey of female micro-entrepreneurs in Zimbabwe had received formal credit (Dejene 2007). One of the weavers reiterated that for future sustainability of the project, there was need for loan facilities in case the project ran out of raw materials. Many enterprises owned and managed by women experience low growth due to lack of credit facilities.

One good thing about the proposed project, according to the weavers, was that after training, the project could be managed from home. Due to the multiplicity of home responsibilities for women, they preferred engaging in income generating projects which they could manage from home. The advantage of managing business from home is being able, at the same time, to get time for other pressing domestic chores. A review of African micro and small enterprises (MSEs) indicated that 45% of female-headed micro and small enterprises were home-based as compared to 19% of male-headed MSEs (DAI 2005).

Discussions on how the weavers were keeping accounts of their income and expenditure revealed that they had inadequate business and financial management skills. The advantage of running a community weaving project is accessing non-formal education on entrepreneurial skills, costing and pricing, record keeping and quality control, among other key related areas. The weavers were made aware that being in a project would enable them to invite UN Women Zimbabwe whose responsibility includes strengthening the capacity of rural women, providing them with skills in entrepreneurship and business management and giving them greater opportunity to access financial resources and markets. It became clearer to the weavers that working in a community project would eventually increase their individual financial benefits and improve their standard of living, thereby making them economically empowered for sustainable development.

11 Conclusion

Considering the role of women in a developing nation like Zimbabwe, it has been proposed in this chapter that non-formal training of women on indigenous basket and mat weaving could provide employment for the informal sector. Empowering women through non-formal education could lead to women's economic empowerment. Non-formal education is a key player in addressing the problems of rural poverty and it is also an important instrument in boosting the image, self-esteem and confidence among rural people, especially women, thus enabling them to participate more actively in the nation's development efforts. Non-formal education may ultimately lead to achieving the sustainable development goals by 2030, especially poverty alleviation through informal sector employment.

By engaging in indigenous basket and mat weaving, women's participation in the development of community economy is enhanced. If women's weaving skills are well-developed and they participate in a community weaving project, producing artefacts for the local and international market, not only will their statuses change, but also the buoyancy of their community's economy will stabilise. A buoyant and stabilised economy minimises poverty. The project could also be a source of the much-needed foreign currency in Zimbabwe for the development of its economy. It is no gainsaying that women are catalysts for the economic development of a nation. Empower a woman and you empower the nation.

12 Recommendations

It is recommended that:

- to be economically empowered women should engage in community projects;
- if women are economically empowered, they would be able to participate in sustainable development endeavours.
- the buoyancy of the community projects could be attained if women are accorded access to loan facilities that disregard collateral demands; and
- experienced indigenous basket and mat weavers should be encouraged to train other community members. Empowering members of the society benefits society at large.

References

Amadala V (2019) Kenya's unemployment rate to rise. https://www.the-star.co.ke/business/kenya/2019-07-30-kenya-unemployment-rate-to-rise-world-bank/

Badu-Nyarko SK, Zumapkeh ESK (2014) Effects of non-formal education on the socio-economic development of women in Nadowli District, Ghana. Asian J Humanit Soc Sci (AJHSS) 2(2):118–132

Bogdan RC, Biklen SK (1992) Qualitative research for education: an introduction to theory and methods, 2nd edn. Allyn & Bacon, Needham Heights

Chen MA (2017) The informal economy in african cities: key to inclusive and sustainable urban development. https://oecd-development-matters.org/2017/04/04/theinformaleconomy-in-africa-cities-key-to-inclusiv

DAC Network on Gender Equality (2011) Women's economic empowerment: issues paper. www.oced.org/dac/gender-development/47561694.pdf

Dejene Y (2007) Promoting women's economic empowerment in Africa. https://www.afdb.org/fileadmin/uploads/afdb/Document/Knowledge/25040341-EN-DRAFT-DEJENE.9.15.07DOC.pdf

DAI Development Alternatives, Inc. (2005) Gender and pro-poor growth, USAID Publication

Elliot CM (2008) Global empowerment of women: responses to globalization and politicized religions. Routledge, New York

Golla AM, Malhotra A, Nanda P, Mehra R (2011) Understanding and measuring women's economic empowerment: definition, framework and indicators. International Centre for Research on Women (ICRW). https://www.icrw.org/wp-content/uploads/2016/10/Understanding-measuring-women-economic-empowerment,pdf

González A (2016) Economic empowerment of women. https://www.intracen.org/news/Economic-Empowerment-of-Women/

Hague C, Logan A (2009) A review of the current landscape of adult informal learning using digital technologies. Future Lab, Bristol

International Labour Organisation (2018) Informal economy. https://www.ilo.org/global/about-the-ilo/newsroom/news/WCMS_627189

Islam M, Mia A (2007) The innovative elements in non-formal education of Bangladesh: perspective of income generating programmes for poverty alleviation. Int J Educ Dev Using Inf Commun Technol (IJEDICT) 3(3):89–104

Jaraweera S (1979) Programmes of non-formal education for women. Convergence 12(3):21–31

Kampilipili N (2017) Women empowerment vital to sustainable development. https://zimbabwe-today.com/women-empowerment-vital-sustainable-development/

Knowles M (1984) The adult learner: a neglected species, 3rd edn. Gulf Publishing, Houston

La Belle T (1986) Non-formal education in Latin American and the Caribbean: stability, reform or revolution? Praeger, New York

Latchem C (2014) Informal learning and non-formal education for development. J Learn Dev 1(1):1–13

McKinsey C (2018) Women matter: time to accelerate. Ten years of insights into gender diversity. https://www.empowerment.org/-/media/files/un%20women/resources/hlp%20briefs/unhip%20report.pdf

McLeod SA (05 August 2018) Lev Vygotsky. Simply psychology. https://www.simplypsychology.org/vygotsky.html

Mok ONA (2011) Non-formal learning: clarification of the concept and its application in music learning. https://files.eric.ed.gov/fulltext/EJ952003.pdf

Sultana M, Haque S (2019) Non-formal education for sustainable development in Bangladesh: a study. N Am Acad Res 2(1):31–44

Tarinda S (2019) Gender, women's economic empowerment and financial inclusion in Zimbabwe. https://www.afi-global.org/sites/default/files/pulication/2019-08/AFI_BTG_Zim_CS19_AW_DIGITAL.pdf

UN Secretary General's High Level Panel on Women's Economic empowerment. (2016). Leave no one behind: a call to action for gender equality and women's economic empowerment. https://www.peacewomen.org/sites/default/files/UNWomen%20Full%20Report.pdf

UN Women Headquarters (2018) Facts and figures: economic empowerment. https://www.unwomen.org/en/what-we-do/economic-empowerment/facts-and-figures

UNDP (2016) Gender and poverty reduction. https://www.undp.org/content/undp/en/home/ourwork/povertyreduction/focus_areas/focus_gender_and_poverty.html
UNDP (2019) Sustainable charcoal leads the way in Angola. https://medium.com/@UNDP/sustainable-charcoal-leads-the-way-in-angola-fdb161a338d0
Vanek J, Chen M, Carre F, Heintz J, Hussmanns R (2014) Statistics on the informal economy: definitions, regional estimates and challenges. WIEGO Working Paper No 2. Cambridge, USA. https://www.wiego.org/sites/wiego.org/files/publications/files/Vanek-statistics-WIEGO-WP2.pdf
UN Women (2013) Economic empowerment of women. https://www.unwomen.org/-media/headquarters/attachments/sections/library/publications/2013/12/un%20women_e_th
ZIMSTAT (2015) Zimbabwe national statistics agency: facts and figures. www.co.zw/wp-content/uploads/publications/Miscellenious/Facts-and-Figures/Facts-Figures-2015.pdf

The Role of Indigenous Social Protection Systems in the Sustainable Development of Contemporary African Communities

Zitha Mokomane, Oko Chima Enworo, and Thato Setambule

Abstract Indigenous social protection systems are the main source of risk management social security for many Africans. Despite this, these systems are often overlooked in the sustainable development efforts of many governments in the region. However, taking cognisant of the systems' long history and rich practices, the past decade has seen increasing calls for greater recognition and support of indigenous social protection at all levels. Against this background, this chapter adopts a literature review approach to illustrate how indigenous social protection systems can build the resilience of individuals and communities to various shocks and thus contribute to sustainable development. To do so, the key dimensions of three selected indigenous social protection systems are reviewed within the Social Risk Management framework. The overall conclusion is that these systems play a major role in the realising of the social and economic pillars of sustainable development in Africa.

1 Introduction

Poverty is a persistent global challenge and calls to reduce it have been a recurring theme in international commitments to achieve sustainable development. Perhaps the most notable instrument in this regard was the Millennium Development Goals set by the United Nations in 2000 which had, as their first goal, the eradication of poverty and hunger by 2015. Although this goal was attained by 2010 (five years ahead of schedule), the aim to reduce poverty still tops the 2030 Agenda for Sustainable Development; the first Sustainable Development Goal (SDG 1) calls on United Nations Member States to "End poverty in all its forms everywhere" by 2030. This concerted focus on global poverty eradication can be attributed to the multidimensional nature of poverty and it affects the very core of being and existence for individuals, households, families, and communities. In essence, poverty is:

Z. Mokomane (✉) · O. C. Enworo · T. Setambule
Department of Sociology, University of Pretoria, Private Bag X20, Hatfield 0028, Pretoria, Republic of South Africa
e-mail: zitha.mokomane@up.ac.za

W. Leal Filho et al. (eds.), *Sustainable Development in Africa*, World Sustainability Series,
https://doi.org/10.1007/978-3-030-74693-3_5

> … a denial of choices and opportunities, a violation of human dignity. It means lack of basic capacity to participate effectively in society. It means not having enough to feed and clothe a family, not having a school or clinic to go to, not having the land on which to grow one's food or a job to earn one's living, not having access to credit. It means insecurity, powerlessness and exclusion of individuals, households and communities. It means susceptibility to violence, and it often implies living in marginal or fragile environments, without access to clean water or sanitation[1]

It is in recognition of the foregoing, and other negative effects of poverty, that the global discourse on sustainable development has seen increasing calls for the design, implementation, and expansion of social protection measures. Target 1.3 of the SGD 1, for example, calls on all United Nations Member States to "implement nationally appropriate social protection systems and measures for all, including floors, and by 2030 achieve substantial coverage of the poor and the vulnerable". Broadly construed, social protection refers to the "set of public and private policies and programmes aimed at preventing, reducing and eliminating economic and social vulnerabilities to poverty and deprivation" (UNICEF 2016: 1).

It can be argued, therefore, that social protection plays a key role in particularly achieving the social equality pillar of sustainable development. The latter is defined as "development that meets the needs of the current generation without compromising the ability of future generations to meet their own needs' (Mensah 2019:6). Its social equality pillar, on the other hand, connotes a system of social organisation that fosters the social and economic empowerment of individuals and communities by providing enabling conditions for achieving meaningful life (Kolk 2016; Mensah 2019). These conditions, which also form the core of social protection, include access to basic services such as healthcare; education. It also includes improved gender equality; social insurance to protect people against the risks and consequences of livelihood shocks; social equity to protect people against social risks such as discrimination and abuse; as well as peace and stability (Holmes and Lwanga-Ntale 2012; ILO 2017; Tirivayi and Rodriguez 2017; Mensah 2019). All in all, through the realisation of the foregoing, social protection can also enhance the achievement of the other two pillars of sustainable development: economic sustainability (a system of production that satisfies present consumption levels without compromising future needs) and environmental sustainability, which relates to the sustainable use of natural capital as a source of economic inputs (Mensah 2019: 10).

In Africa, the potential role of social protection is acknowledged in various continental instruments. These include the Ouagadougou Declaration and Plan of Action (2004), the Livingstone Call for Action on Social Protection (2006) and the 2008 Social Policy Framework for Africa (Mokomane 2013). In line with the calls made by these instruments as well as by the 2030 Agenda, many African governments have, over the last decade, increasingly developed and implemented social protection programmes as part of their overall strategies to achieve economic growth, poverty

[1]The Statement for Action to Eradicate Poverty adopted by the Administrative Committee on Coordination in May 1998, quoted in the Report of the Independent Expert on Human Rights and Extreme Poverty (E/CN.4/1999/48) See Indicators of Poverty and Hunger: www.un.org.esa/socdev/unyin/documents/YdiDavidGordon_poverty.pdf.

reduction, and social development. Despite this, social protection coverage in Africa remains limited. According to the most recent World Social Protection Report (ILO 2017), only 17.8% of Africans are covered by at least one social protection benefit. This is less than half of the global average of 45.2%, and much lower than the proportion covered in other developing regions such as Asia and the Pacific (38.9%) as well as Latin America and the Caribbean at 61.4% (ILO 2017).

It is noteworthy, however, that the limited coverage of social protection programmes in Africa does not mean that individuals and communities in the continent are devoid of means to cope with risk and vulnerabilities. As was the case in the traditional setting, the majority of Africans continue to rely on indigenous social protection as the main source of risk management and social security in times of adversity (Patel et al. 2012; Mokomane 2013). Also known as informal social protection, indigenous social protection has been described as "bundle of measures taken at household and community level to protect against risks or to combat shocks, in the absence or presence of public or market based arrangements" (Balgah and Buchenrieder 2010: 357). This type of social protection is based on widely agreed norms or obligations and reciprocity.

Despite their long history and demonstrated contribution to social welfare (see, for example Holmes and Lwanga-Ntale 2012; Arrey 2018; Dafuleya 2018; Mokomane 2018; Noyoo and Boon 2018), indigenous social protection systems are not a visible aspect of contemporary African social protection programming. This is largely due to notions that the systems are fragile, rapidly declining, or not as effective as before (Dafuleya 2018). Other arguments are that informal systems only provide short-term solutions and fail to build communities' resilience to future shocks (Balgah and Buchenrieder 2010). To this end, only a handful of African countries such as Chad, Mauritania, Ethiopia, Kenya and Tanzania explicitly acknowledge the existence and importance of indigenous systems in their national social protection policies.

A plausible source of the foregoing criticisms is the paucity of research and evidence on the dynamic processes of indigenous social protection systems and the extent to which they can be used to effectively manage shocks (Balgah and Buchenrieder 2010). Thus, as Ugiagbe and Ugiagbe (2015: 366) assert, "the quest for sustainable grassroots' development by [African] governments calls for the need to look inwards and re-examine and evaluate our indigenous cultural practices in an attempt to attain the desired development goals". In light of calls such as these, this chapter adopts a literature review approach to illustrate how indigenous social protection systems build the resilience of individuals and communities to various shocks and thus contribute to sustainable development in Africa. To do so, the chapter reviews the key dimensions of three selected indigenous social protection systems – regarded herein as case studies – that are common in different parts of the continent. Using the Social Risk Management (SRM) framework as a conceptual lens, particular focus is placed on understanding the operations of the selected systems, the extent to which they provide social protection in contemporary society, and how this can contribute to sustainable development.

In the next section, a brief explanation of the SRM framework is presented. Thereafter the key dimensions of the three selected indigenous social protection systems are presented before their potential contribution to sustainable development is discussed in the concluding section.

2 Social Risk Management Framework

The SRM framework conceptually extends the traditional framework of social protection (which focused on labour market interventions, social insurance, and social assistance) to include strategies that can protect the poor and the most vulnerable from various types of shocks and/or vulnerability risks. The framework's strategies fall in three broad categories that have been described by Holzmann et al. (2003) as follows:

(i) **Prevention strategies.** These are introduced before a risk occurs and they aim to reduce the probability of losing expected income or increasing income variance. Examples include policies regarding sound macroeconomics, public health, the environment, and education and training.

(ii) **Mitigation strategies.** These are also employed before a risk occurs but their aim is to decrease the potential impact of a future downside risk. Mitigation strategies take several forms including *portfolio diversification* (acquisition of different physical, financial, human and social capital assets to reduce the variability of income); and *informal and formal insurance mechanisms* characterised by risk sharing or pooling by a number of participants whose risks are not (very) correlated;

(iii) **Coping strategies**. The aim of these strategies is to relieve the impact of a risk once it has occurred. Examples of coping strategies include individual dis-saving/borrowing, migration, selling of labour, reduction of food intake, and/or the reliance on public or private transfers.

3 Indigenous Social Protection and Risk Management

3.1 Preventive Strategies: Age Grades

Age grades are a social organisation of community members born within two to three years of the same period into age cohorts, often by gender (Erim et al. 2011; Ugiagbe and Ugiagbe 2015; Ujumadu 2017; Thacker et al. 2019). Although the number of age grades in a community, the range of years in each grade, and the terminology used vary significantly between traditions and ethnic groups, a common mark of entry into an age grade is the participation in an initiation rite (Ugiagbe and Ugiagbe 2015). The rites typically entail extensive set of ceremonies including education and training on various socio-cultural aspects, physical conditioning, and in some cases, male circumcision (Benardi 1985; Malisha et al. 2008; Morton 2012; Ugiagbe and Ugiagbe 2015; Timbs 2018).

The age-grade system can be traced back to pre-colonial times where, in articulation of societal unity and identity as well as a vision for continuity (Raymond 2006: 223), they protected their communities from other warring ethnic groups or mercenaries of other communities (Ujumadu 2017). In Botswana for example, age grades (known as *mephato*), were after initiation, transformed into important political and military units to carry out the instructions of senior royals in service or defence of their authority (Morton 2012). According to Morton, as years went by, the *mephato* took up tasks beyond the military. These included herding and defending community cattle from predators and raiders, undertaking policing duties as well as raiding and subjugating servile groups. In the absence of modern-day governments and/or non-governmental organisations, age grades were also important agents of community development as well as societal sustenance and cohesion. They, for example, built or maintained community infrastructure such as roads and water points; cleared village squares and market places; and helped rebuild victims' houses in times of natural disasters such as floods or fires (Iroh 2019: 3).

Despite socio-economic transformations that have brought modernisation, industrialisation and urbanisation, the roles of age grades remain significant and impactful in many contemporary African societies including in East Africa (Lienard 2016; Kung'u, and Waweru 2016), and Southern Africa (Deacon and Thomson 2012; Morton 2012). However, much of the available literature on the contemporary dimensions of this age long socio-cultural institution is based on evidence from West Africa, especially Nigeria (for example, Erim et al. 2011; Dike 2012; Ada 2014; Ndukwe 2015; Adeyemi-Suenu 2016), where their roles have been shown to fall broadly into three categories: socio-religious, socio-political, and socio-economic (Dike 2012).

According to Dike, through their socio-religious role, age grades enhance societal equality, create a sense of social cohesion, and afford members a type of social security insurance in times of adversity such as the death of a member. They also uphold the morality of their members through a number of rules that guide social conduct. Promoting communities' cultural heritage through, for example, participating in national festivals is another aspect of the system's socio-religious role (Dike 2012). Through their socio-political role, on the other hand, age grades are represented in local governing bodies and are typically consulted in matters involving duties in the community (Dike 2012: 86). The socio-economic role of age grades entails the performance of various economic activities for the benefit of the members and their communities at large. Examples, according to Dike, include providing economic aid to their members in times of need, generating economic opportunities for members, and embarking on economic ventures for the development of the community.

In addition to the foregoing, age grades facilitate the achievement of economic, social and political wellbeing of communities through the development and execution of community initiatives and projects such as crime control and the enforcement of law and order (Theophilus 2012; Ujumadu 2017). In Botswana, for example, the last decade has witnessed increasing instances where *mephato* partner with state institutions in villages to maintain law and order as part of rural crime prevention strategies (Modikwa 2009; Molomo and Maundeni 2015). With crime prevention recognised as a sustainability indicator due to its association with stable conditions, safety in localities, good governance, and overall improved quality of life (see for example, Du Plessis 2000; Kitchen 2009; Marzbali et al. 2011; van Dijk 2014), age grades are playing a major and essential contributory role in securing sustainable development. The voluntary nature of age grades also makes them particularly relevant in this regard as a shortage of both human and financial resources is often blamed for the lack of effective crime prevention and sustainable development initiatives in Africa (Du Plessis 2000: 24).

Age grades are also widely recognised for participating in projects aimed at ensuring access to basic social services such as health services education, as well as water and sanitation. For example, commenting on the activities of age grade associations in Nigeria's Cross River State, Erim et al. (2011: 168) reported that they assisted in the implementation of projects such as the provision of electricity, pipe-borne water and recreational centres in rural areas. They also assisted with preparatory work – such as carrying out stones, clearing sites, and digging foundations – for community projects. Given that "infrastructure either directly or indirectly influences the attainment of all of the Sustainable Development Goals (SDGs), including 72% of the targets" (Thacker et al. 2019), the role of age grades in this regard is evident. In essence, as UNOPS (2019: 3) posits, infrastructure development holds great potential to deliver all three pillars of sustainable development. In terms of economic sustainability, for example, dividends of infrastructure development such as transportation and telecommunications range from jobs created during construction and maintenance to the ability for infrastructure to generate economic activity; for environmental sustainability, infrastructure assets play a key role in conserving natural resources and reducing the impact of climate change. Finally, infrastructure development ensures social sustainability by, *inter alia*, facilitating the delivery of basic services such as health care, clean water and sanitation, and education.

Indeed, in addition to building community halls, providing boreholes, and building bridges, age grades in many Nigerian communities are widely recognised for renovating schools and instituting national and international scholarship schemes to support gifted but indigent children in the community (Ndukwe 2015; Ujumadu 2017). In other regions age grades also encourage older community members who have never been to school to register for adult literacy classes as a way of assisting them to "embrace new visons, and participate in community development projects" (Erim et al. 2011: 170). According to Erim and colleagues, some age grades also make in-kind donations such as books, school furniture as well stationery to some schools. Thus, given the broad recognition of the role of education in breaking the intergenerational cycle of poverty, increasing earning potential, improving health outcomes and reproductive choice, as well in strengthening environmental resilience (Bangay 2016: 6), age grades' role in improving quality of life and sustainable development through education is evident.

3.2 Mitigation Strategies - Rotating Savings and Credit Associations

Rotating Savings and Credit Associations (ROSCAs) are a long-standing tradition of community members forming groups to help each other in times of difficulties (Mbamaonyeukwu 2013; Mushuku and Mayisa 2014; Kabuya 2015). In Africa, these associations can be historically traced back to the early 19th Century before the introduction of monetary economy. Then many Africans used the barter system to make periodical contribution of grain such as wheat, rice, corn, and other paddy (Bouman 1995 cited in Kabuya 2015). These were fairly shared in times of food insecurity brought about by shocks such as droughts and floods.

ROSCAs continue to exist in many parts of Africa where they are known by various local names such as *Tontines* in parts of West Africa, *Muzikis* or *Likelambas* in Democratic Republic of the Congo, *Ekub* in Ethiopia, *Stokvel* in South Africa, *Mukando* in Zimbabwe, *Isusu* in Nigeria, *Susu* in Ghana, and *Metshelo* in Botswana (Kabuya 2015; Mokomane 2018). In contemporary society, their existence is largely due to the many obstacles that make it difficult for low income and poor households to access formal financial services. Similar to what Salahuddin and Rashmi (2015) noted in India, these obstacles include: the existence of a large informal sector which is financially excluded from the formal financial sectors; difficulties in obtaining institutional credit and collateral for many people; strict documentation requirements and often inflexible terms and conditions, and poor functioning of the formal sectors in many countries. The majority of members also join either because they are not covered by any social protection scheme, or are covered but the level of protection is not adequate (Matuku and Kaseke 2014).

While models of operation vary, ROSCAs typically comprise of at least five people who contribute an agreed periodic amount to a central 'collective savings vehicle' or 'central pot' (Mulaudzi 2017). At the very basic level, the collected funds are disbursed to individual members in line with an agreed order or cycle of receiving the pooled funds until every member has received their share (Matuku and Kaseke 2014; Kabuya 2015). Other models of operation include where each member receives a lump sum of the combined contributions only at the end or beginning of the year; or where members contribute an agreed amount and at the end of the agreed period, collectively buy items that are shared among them (Radipotsane 2007; Mulaudzi 2017). Regardless of model, savings from the ROSCAs are used to build up reserves that can be used to mitigate the effects of unforeseen contingencies by providing easy access to credit and mini-loans for members; enhancing the ability to meet basic needs; promoting saving and investment; providing moral support and mutual assistance; and creating social capital (Radipotsane 2007; Matuku and Kaseke 2014). Recent years have seen the emergence of other types of ROSCAs that aim to provide other types of assets, beyond financial capital. As an example, the box below draws on previous studies such as Irving (2005), African Response (2012), Semenya (2013) and Van Wyk (2017) to summarize the key features of some of the new forms of ROSCAs found in South Africa.

- *Investment stokvels*, also known as cooperative buying societies. The aim is to accumulate capital by investing in business ventures, promoting savings through bulk purchases of goods etc.
- *High-budget stokvels. These* operate as financial institutions and are established for the purpose of promoting savings and investments. Members contribute substantial amounts that eventually enable them to receive large lump-sum payments which can be used to purchase household goods or meet other needs that require an outlay of considerable resources.
- *Event stokvels.* Members contribute towards a hosted event. An entrance fee is charge and beverages and food are sold. Profits from the event are shared among the members.
- *Grocery stokvels.* Members contribute a fixed a monthly amount for a given period. Thereafter the money is used to buy groceries and/or shopping vouchers or coupons for all members.
- *Purchasing stokvels.* The aim is to buy income-generating assets using members' contributions. The assets are then rented out to generate additional income for the stockvel. Proceeds are shared between members at a specified interval.
- *Property stokvels*[2] aimed at property acquisition, ownership and investment are a reality and accessible option for those who would otherwise not qualify for property loans. In these types of ROSCAs, prospective property owners or investors pool their financial resources which, after an agreed period, are used for purchasing homes or investment properties for members. The funds can also be variously used for investing in rental properties where each member receives a share of the rental income; buying land to develop and build on; purchasing shares in a property investment portfolio; acquiring commercial and industrial properties; creating annuity income for members; putting down a deposit on particular properties; and securing homes or land for each member in a cascading fashion.

What the foregoing demonstrates in part is that contemporary ROSCAs in many African countries are transitioning from being consumer to investor-based, and have become more formal undertakings that employ sophisticated financial models despite not always having the financial education to back up their ideas. Other intriguing facts about these associations are that, consistent with the global picture, there is a remarkably low default rate among members. That is, despite ROSCAs not having any external regulation, members tend to keep their obligations to contribute even after their turn to receive funds. ROSCAs have also been widely noted for being relatively more inclusive when compared with more formal credit schemes as they often include the most marginalised and vulnerable socio-economic categories such as rural residents and women. This not only gives vulnerable populations a sense of agency and control over decisions, but it also leads to a relatively more equitable society (Okurut and Botlhole 2010; Storchi 2018).

While ROSCAs have historically operated outside the realm of traditional financial and banking systems, these sectors are increasingly seeing the potential and

---

[2]A new type of stokvelhttps://www.privateproperty.co.za/advice/property/articles/how-does-a-property-stokvel-work/7086.

collective strength of these associations. For example, current estimates in South Africa are that there are over 820 000 stokvels in the country with a combined membership of 11.4 million people (out of an estimated total population of about 59 million), handling over R44billion (US$260 billion) per annum (Mulaudz 2017). This suggests that the observation made by van den Brink and Chaves almost three decades ago may still be at play. That is ROSCAs typically outperform all other credit institutions "in terms of efficiency and equity ... [and] by reaching virtually every household in [need] handling thousands of credit transactions every year, and doing so at a relatively low level of transactions costs" (van den Brink and Chava 1991: 29). Indeed, recognition of the efficiency of this indigenous system, and in an attempt to secure the various advantages of its market share virtually all the major banks in South Africa as well as neighbouring countries such as Botswana have developed stokvel and *metshelo* accounts, respectively.

All in all, it can be concluded that, at their core, the activities of ROSCAs have strong empowerment and sustainability aspects (Mbamaonyeukw 2013). In terms of the former, the members draw on group funds to access resources for productivity and advancement. In essence, as Mushuku and Mayisa (2014: 131) assert, regular and sustainable access to microfinance through these associations enables members "to attain human capital through educating their children, physical capital through accumulation of assets and financial capital through starting up businesses or expanding existing ones". The rotational nature of accessing funds, on the other hand, suggests a focus on expansion and sustainability of the livelihood projects that the members embark on with the funds from the associations. As Mbamaonyeukwu (2013: 17), points out, the members continuously work toward "building on their future, and are not interested in halting growth once they have achieved just one objective".

3.3 Coping Strategies - Burial Societies

Burial societies are informal insurance schemes made up of friends, relatives, workmates or community residents who come together to insure themselves and their extended families against death and funeral-related expenses. The societies are particularly common in Southern Africa. In South Africa, for example it is reported that 19% of the population belongs to a burial society (Semenya 2013). Burial societies are also popular among many regional migrants who depend on them for purchasing of coffins, repatriation of bodies, psychosocial support, and the provision of social networks that enhance coping with death and bereavement in a foreign country (Wheeler 2011).

As with ROSCAs, members of burial societies share risk by pooling their collective resources. This entails regularly paying an agreed subscription fee that is used to build a collective investment fund with a set return to be used towards a burial and/or a repatriation upon a member or their nominated beneficiaries' death (Ngwenya 2003; Wheeler 2011; Semenya 2013). Each burial society generally specifies those who are eligible to receive funeral support. Typically, the core group of recipients includes parents (whether begetting or fictive parents and or affines), parents-in-law, the members' children and spouse. Other societies allow a specified number of additional members, such as extended family, at an additional but often minimal fee. Each burial society also decides how much funeral support eligible recipients get (Ngwenya 2003). These societies typically all have member-nominated management committees who oversee the financial assets. For example, contributions from each member are recorded by the treasurer and in many cases are deposited with a financial institution. The committees also ensure good governance and smooth functioning of the societies.

In addition to financial assistance, burial societies also provide support to members and their relatives in the form of physical labour for cooking and serving guests during the bereavement period, relieving the bereaved family of any domestic work, running funeral-related errands, and lending the bereaved family any catering utilises and equipment (Ngwenya 2003; Semenya 2013; Dafuleya 2018). Psychosocial support such as attending and/or holding prayer meetings and attending traditional night vigils in honour of the deceased (Thomson and Posel 2002) is another common form of support provided by these societies. Overall, therefore, burial societies facilitate the removal of anxiety from where it is most intense (at household level) to a domain that is relatively free from anxiety (the burial society itself).

Ngwenya (2003) argues that the popularity of burial societies should be seen in the context of the ways in which in Africa death, as a sociocultural event and process, demands massive mobilization of labour and resources (both monetary and nonmonetary). She argues:

Family groups are expected to access and redistribute the resources needed to give the deceased person a socially defined "dignified funeral". Any failure to do this offends the fundamental [African] humanistic ethos and ethic. Death and funeral events thus in and by themselves create social arenas for collective social action and the regeneration of community life (Ngwenya (2003: 89).

Furthermore, for many Africans, funerals are seen as "the bridge from the living into death", a transition that is often celebrated in the form of a big ceremony (Mathe, 2019). Having sufficient funds to enable this celebration one way or the other somehow bestows dignity on everyone and hence reduces social inequality to some extent. The societal importance of 'socially acceptable' funerals can be partly seen in the fact that older and destitute members of societies often use their cash transfers, such as pensions and destitute cash allowance, to pay for burial society subscriptions (Seleka et al. 2007).

Similar to the case with ROSCAs, formal financial service providers such as banks, insurance companies down to retailers like mobile service providers, clothing stores and even soccer teams attempt to have a share of the large burial society market by offering funeral plans or policies. In advertising their funeral products, these providers often highlight the unregulated nature of burial societies, and the extent to which these societies are susceptible to fraud, mismanagement and the possibility of insufficient funds when the time for claiming comes. Counter arguments, however, include that like many indigenous systems, burial societies have stood the test of time and remained more cost effective than formal providers.[3] Others have also argued that with their focus of 'core fixed costs' such as buying coffins, providing hearses etc., funeral policies are "not structured for the way that black people's funerals are conducted".[4] Yet others argue that the sense of community you get to experience with a burial society cannot be matched by a funeral plan.[5]

Of particular note is also that, in both their makeup and in the execution of their activities, burial societies are entrenched in traditional values of community support and have a strong focus on solidarity (Mbamaonyeukwu 2013). Thus, to the extent that people-centeredness and a strong sense of community are some of the essential components of sustainable development (Mbamaonyeukwu 2013; Kolk 2016; Mensah 2019), this subsection has highlighted the extent to which burial societies can be an important aspect of realising sustainable development in the parts of Africa where they exist.

4 Conclusion

The last decade has seen increasing calls for greater recognition and support of indigenous social protection in Africa at the academic, policy practice and service delivery levels (see for example, Dafuleya 2015; Holmes and Lwanga-Ntale 2012; Patel et al. 2012; Noyoo and Boon 2018). The basic thesis in this regard is that formal social protection systems are not only limited in coverage but they are also too narrowly focused on income security, the need to respond to risk and vulnerability, and the need to respond to particular lifecycle needs (Holmes and Lwanga-Ntale 2012). Indigenous social protection systems, on the other hand, are recognised for "placing a strong cultural obligation on communities to promote cooperation, build solidarity, and help vulnerable families and individuals" (Patel et al. 2012: 13). It is noteworthy, however, that the two systems have the same goal – that of providing various forms of support to a variety of vulnerable social groups and ensuring their socio-economic inclusion and empowerment.

This was affirmed in this chapter through the presentation of the main characteristics and operations of three selected indigenous social protection systems common in different parts of Africa. Overall the chapter showed that despite the various criticisms and scepticism around them, indigenous social protection systems are not only sustainable in themselves, but they are also great enablers of the socio equality pillars of sustainability and, albeit to a lesser extent, the economic and environment pillars. It was shown, for example, that where they exist, age grade associations are what Erim et al. (2011: 167) describe as "the bedrock of community socio-economic development". The safety net provided by ROSCAs and burial societies were, on the

[3]https://www.news24.com/citypress/personal-finance/burial-society-vs-funeral-insurance-which-is-best-20180730.

[4]https://mg.co.za/article/2019-07-11-00-the-unholy-cost-of-a-sacred-death/.

[5]https://www.oldmutual.co.za/articles/funeral-cover-and-burial-societies/.

other hand, shown to not only be a critical component for a cohesive society but also an important and financial and social capital that can lead to sustainable development through avenues such as support for micro-entrepreneurs, funding of education, and home ownership, among other things (African Response 2012: 4).

In terms of future prospects, this chapter therefore suggests that in the quest for sustainable development in Africa, and in line with the *Leave No One Behind* principle, it will be worthwhile to build synergies between the two systems rather than seeing them as alternatives to each other. Given the paucity of research and literature on indigenous social protection systems, pathways to achieving this are not precisely clear. This chapter therefore underscores the need for African countries to invest resources to build a comprehensive evidence base on the overall dimensions and mechanisms of indigenous social protection systems in the continent and to use this information to strengthen and foster linkages with formal systems for improved social protection coverage and overall sustainable development.

The chapter also clearly revealed the role of indigenous social protection systems in enhancing the attainment of the social pillar of sustainable development. What remains relatively unclear is the role of these systems in realising the economic and environmental sustainability pillars. It is thus critical that efforts to understand this become an important component of the future research agenda in the broad area of social protection and sustainable development in Africa.

All in all, as Patel et al. (2012: 28) posit, in view of their demonstrated potential to inform sustainable development activities in different African countries, greater evidence-based recognition and support of indigenous social protection at the regional, and indeed international level is needed. In this way, mutually beneficial reciprocal exchanges that enhance people's welfare and promote sustainable development can be fostered at all levels.

References

Ada IM (2014) Role of Nnannebuife Age Grade in promoting rural development in Nsugbe community Anambra East L.G.A in Anambra State. Unpublished BSc Dissertation, Caritas University, Amorji- Nike Enugu State

Adeyemi-Suenu A (2016) Age grade system and the politics of development and social cohesion in Ikoroduland. Int J Rese Econ Soc Sci 6(7):166–172

African Response (2012) Stokvels- A hidden economy: Unpacking the potential of South Africa's traditional savings schemes. https://base.socioeco.org/docs/2012stokvelhiddeneconomy.pdf

Arrey AE (2018) The challenge of HIV/AIDS on the indigenous social security systems: the case of Cameroon. In: Noyoo N, Boon E (eds) Indigenous social security systems in Southern and West Africa. Sun Press, pp 197–2015

Balgah RA, Buchenrieder G (2010) The dynamics of informal responses to covariate shocks. J Natural Resour Policy Res 2(4):357–370

Bangay C (2016) Protecting the future: the role of school education in sustainable development–an Indian case study. Int J Dev Educ Global Learn 8(1):5–19

Benardi B (1985) Age class systems: Social institutions and polities based on age. Cambridge University Press, Cambridge

Bouman FJA (1995) Rotating and accumulating savings and credit associations: a development perspective. World Dev 23(3):371–384

Dafuleya G (2018) (Non)state and (in)formal social protection in Africa: focusing on burial societies. Int Soc Work 61(1):156–168

Deacon H, Thomson K (2012) The Social Penis. Traditional Male Circumcision and Initiation in Southern Africa, 1800-2000: A Literature Review. CSSR Working Paper No 304. Centre for Social Science Research, University of Cape Town

Dike U (2012) The role of age grade in Ogba Land. J Relig Hum Relat 1:83–89

Du Plessis C (2000) The links between crime prevention and sustainable development. In: Lawrence R (ed) Sustaining human settlement: a challenge for the new millennium. Urban International Press, North Shields, UK, pp 239–270

Erim CM, Akpama SI, Asor LJ (2011) Age grade associations' activities in enhancing community development projects in Odukpani and Biase Local Government Areas of Cross River State Nigeria. LWATI: J Contemp Res 8(2), 167–176

Holmes R, Lwanga-Natale C (2012) Social protection in Africa: a review of social protection issues in research. PASGR, Nairobi

Holzmann R, Sherburne-Benz L, Tesliuc E (2003) Social risk management: The World Bank approach to social protection in a globalizing world. http://documents.worldbank.org/curated/en/494981468762925392/Social-risk-managementThe-World-Banks-approach-to-social-protection-in-a-globalizing-world

ILO (2017) World Social Protection Report 2017–19: Universal social protection to achieve the Sustainable Development Goals. International Labour Organisation, Geneva

Iroh WO (2019) Age grade system and community development in Ohafia local government area. MSc Research project. https://www.researchgate.net/publication/330213492_AGE_GRADE_SYSTEM_AND_COMMUNITY_DEVELOPMENT_IN_OHAFIA_LOCAL_GOVERNMENT_AREA

Irving M (2005) Informal savings groups in South Africa: Investing in social capital. Working Paper Number 112. Centre for Social Science Research, University of Cape Town

Kabuya FI (2015) The rotating savings and credit associations (ROSCAs): Unregistered sources of credit in local communities. J Hum Soc Sci 20(8):95–98

Kitchen T (2009) Crime prevention, the planning system and sustainable development: addressing policy challenges in English practice. Built Environ 35(3):328–345

Kolk A (2016) The social responsibility of international business: from ethics and the environment to CSR and sustainable development. J World Bus 51(1):23–34

Kung'u JN, Waweru P (2016) Generation sets as institutions of cohesion and mutual existence in pre-colonial kenya. Scholars J Arts, Hum Soc Sci 4(9):1189–1197

Lienard P (2016) Age grouping and social complexity. Curr Anthropol 57(13):S105–S117

Malisha L, Maharaj P, Rogan M (2008) Rites of passage to adulthood: traditional initiation schools in the context of HIV/AIDS in the Limpopo province, South Africa. J Health Risk Soc 10(6):585–598

Marzbali MH, Abdullah A, Razak N, Tilaki MJM (2011) A review of the effectiveness of crime prevention by design approaches towards sustainable development. J Sustain Dev 4(1):160–172

Mathe T (2019) The unholy cost of a sacred death. Mail & Guardian Newspaper, 11 July. https://mg.co.za/article/2019-07-11-00-the-unholy-cost-of-a-sacred-death/

Matuku S, Kaseke E (2014) The role of stokvels in improving people's lives: the case in Orange Farm, Johannesburg. South Africa Soc Work/MaatskaplikeWerk 50(4):504–515

Mbamaonyeukwu VS (2013) ROSCAs: An instrument for the sustainable development of the solidarity economy. Paper presented at the 5th RIPESS International Meeting of SSE, Manila, Philippines, October 15–18. https://base.socioeco.org/docs/valerie-mbamaonyeukwu-long-paper.pdf

Mensah J (2019) Sustainable development: Meaning, history, principles, pillars, and implications for human action: Literature review. Cogent Soc Sci 5:1. https://doi.org/10.1080/23311886.2019.1653531

Modikwa O (2009) Botswana: Mephato arrest crime in Tswapong. https://allafrica.com/stories/200903310793.html
Mokomane Z (2013) Social protection as a mechanism for family protection in sub-Saharan Africa. Int J Soc Welfare 22(3):248–259
Mokomane Z (2018) Indigenous social security systems in Botswana. In: Noyoo N, Boon E (eds) Indigenous Social Security Systems in Southern and West Africa. Sun Press, pp 13–28
Molomo M, Maundeni Z (2015) Botswana. In: Gumedze S (ed) Promoting partnerships for crime prevention between state and private security providers in Southern Africa. Private Secur Ind Regul Authority, Pretoria, pp 45–94
Morton F (2012) Mephato: the rise of the Tswana militia in the pre-colonial period. J South Southern an Stud 38(2):385–397
Mulaudzi R (2017) From consumers to investors: an investigation into the character and nature of stokvels in South Africa's urban, peri-urban and rural centres using a phenomenological approach. Unpublished MPhil Dissertation, University of Cape Town
Mushuku A, Mayisa J (2014) Unlocking the door from poverty through rotating savings and credit associations: a study of group based savings in Ward 14 of Gutu District. Zimbabwe. Bangladesh e-Journal Sociol 11(2):118–131
Ndukwe C (2015) Changes and continuity in age grade practices in Nigeria: a study of the Igbo traditional and modern administrative system. J Policy Dev Stud 9(3):176–184
Ngwenya BN (2003) Redefining kin and family social relations: burial societies and emergency relief in Botswana. J Soc Dev Africa 18(1):85–110
Noyoo N, Boon E (2018) Introduction and background. In: Noyoo N, Boon E (eds) Indigenous social security systems in Southern and West Africa. Sun Press, pp 1–13
Okurut FN, Botlhole T (2009) Informal financial markets in Botswana: a case study of Gaborone City. Dev South Africa 26(2):255–270
Patel L, Kaseke E, Midgley J (2012) Indigenous welfare and community-based social development: lessons from African innovations. J Commun Pract 20(1–2):12–31
Radipotsane MO (2007) Determinants of household saving and borrowing in Botswana. IFC Bull 25:284–296
Raymond E (2006) The age politics: concept and meanings. Ellorf Crest Inc., Princetown, Ghana
Salahuddin M, Rashmi C (2015) Rotating savings and credit association: an alternate of financing among low income households. Abhinav National Monthly Refereed J Res Commer Manage 4(3):16–20
Seleka TB, Siphambe H, Ntseane D, Mbere N, Kerapeletswe C, Sharp C (2007) Social safety nets in Botswana: Administration, targeting and sustainability. Lightbooks, Gaborone
Semenya DK (2013) Burial society versus the Church in the Black society of South Africa: a pastoral response. Verbum et Ecclesia **34**(1), Art. #698
Storchi S (2018) Impact evaluation of savings groups and stokvels in South Africa: The economic and social value of group-based financial inclusion. Pietermaritzburg, SaveAct
Thacker S, Adshead D, Fay M et al (2019) Infrastructure for sustainable development. Nature Sustainability 2:324–331
Theophilus AT (2012) The role of age grade in crime control in TIV society of central Nigeria. Journal of Social Science and Public Policy 4(1):46–53
Thomson RJ, Posel DB (2002) The management of risk by burial societies in South Africa. South Afr Actuarial J 2:83–127
Timbs L (2018) An In(ter)vention of tradition: medical male circumcision in KwaZulu-Natal, 2009–2016. J Natal Zulu History 32(1):55–77
Tirivayi N, Rodriguez O (2017) Social protection for forest-dependent communities. FAO Policy Brief. http://www.fao.org/3/a-i7008e.pdf
Ugiagbe EO, Ugiagbe I (2015) Transcending the Eurocentric development paradigms in Nigeria: the traditional age grades in discourse. J Siberian Federal Univ. Humanit Soc Sci **3**(8):366–376
Ujumadu V (2017) Age grades as facilitators of development in Igboland. Vanguard Metrol. https://www.vanguardngr.com

UNICEF (2019) UNICEF's Approach to Social Protection. https://www.unicef.org/easterncaribbean/media/851/file/Social-Inclusion-Summaries-%20UNICEF's-Approach-to-Social-Protection-2016.pdf
UNOPS (2019). The critical role of infrastructure for the Sustainable Development Goals. https://unops.economist.com/wp-content/uploads/2019/01/Thecriticalroleofinfrastructurefor theSustainableDevelopmentGoals.pdf
van den Brink R, Chavas J (1991) The microeconomics of an indigenous African institution: the rotating savings and credit associations. Cornell Food and Nutrition Policy Program. http://www.cfnpp.cornell.edu/images/wp15.pdf
van Dijk JJM (2014) Crime. In: Rowe D (ed) Achieving sustainability: visions, principles, and practices. Detroit, Macmillan Reference
Van Wyk M (2017) Stokvels as a community-based savings club aimed at eradicating poverty: a case of South African rural woman. Int J Commun Divers 17(2):14–26
Wheeler BL (2011) (Un)Rest in peace: the agents of human remains repatriation and the lives of living migrants: A study of agency, process and effect in repatriating bodies from South Africa and the U.S.A. Unpublished MA Research report, University of the Witwatersrand

Sustainable Quality Education Through Self-directed Multimodal Learning, for Open Indigenous Knowledge Co-creation

Jako Olivier

Abstract A requirement for sustainable development is quality education, which – according to the United Nations Sustainable Development Goal 4 – relates to education that is inclusive, equitable and promotes lifelong learning opportunities. Such aims require embracing self-directed learning, multimodality, open education and indigenous knowledge. In this conceptual chapter these four foci are explored through a critical analysis of current discourses with a specific emphasis on the context and needs of Africa. For indigenous communities, the concept of self-directedness is key to effective knowledge production, access and governance. Self-directed learning implies that students take charge of their learning individually or communally by setting goals and choosing resources, and learning strategies in order to support lifelong learning. Knowledge functions multimodally and on different levels. Interaction, instruction and delivery can be done through different modes or blending of modes: either face-to-face or via technologically-supported manners. Furthermore, knowledge used or created should adhere to the principles of open education. In this context, open licensing is embraced and access to resources is more equitable in terms of both access and localization, in order to account for indigenous knowledges and languages. This chapter concludes with recommendations of practical steps towards fostering sustainable development in education, through self-directed multimodal learning, for open indigenous knowledge co-creation.

Keywords Self-directed multimodal learning · Multimodality · Open educational resources · Indigenous knowledge

1 Introduction

It is clear that education can be regarded as a requirement for wider sustainable development (cf. Roeder et al. 2017; Uvalić-Trumbić and Daniel 2016; Urbančič et al. 2019). The focus of this chapter is on higher education and it is evident from

J. Olivier (✉)
Research Unit Self-directed Learning, Faculty of Education, North-West University, Mahikeng, South Africa
e-mail: jako.olivier@nwu.ac.za

W. Leal Filho et al. (eds.), *Sustainable Development in Africa*, World Sustainability Series,
https://doi.org/10.1007/978-3-030-74693-3_6

literature that higher education has an important role to play in the realisation of the Sustainable Development Goals (SDGs) (Aarts et al. 2020).

From the inception of the concept of *sustainable development* in the 1987 Bruntland Commission report, *Our Common Future*, the focus has been on meeting the current needs without impacting future needs negatively (Emas 2015). Furthermore, for Emas (2015) "… [t]he overall goal of sustainable development (SD) is the long-term stability of the economy and environment; this is only achievable through the integration and acknowledgement of economic, environmental, and social concerns throughout the decision making process." (p. 2). With a specific focus on higher education, these issues all converge in a complex context.

According to MacKinnon et al. (2016) the link between sustainability and education is generally made in terms of education for sustainability, where the focus is on teaching students about environmental challenges or otherwise in terms of economic or ecological issues around sustainability. Consequently, this research also does not relate to *Education for Sustainability* which "[e]ncourages ecologically sustainable behaviour via education …", but rather *Sustainability of Education* which involves being "[c]oncerned with the ongoing ability of an educational endeavour to meet its goals" (MacKinnon et al. 2016). The focus in this chapter is on the latter. However, both aspects influence the way higher education functions - and will function in future.

This chapter should be interpreted in terms of Goal 4 of the SDGs as adopted by the United Nations General Assembly in 2015 (cf. Lane 2017). This goal relates to the intention to "[e]nsure inclusive and equitable quality education and promote lifelong learning opportunities for all" (United Nations 2020, p. 4). From this goal, the four main aspects addressed in this chapter are derived: self-directed learning, multimodality, open education and indigenous knowledge.

Indigenous knowledge is, therefore, central to the context of this chapter. The concept of indigenous knowledge is defined by Keane et al. (2016) as "the understandings and skills developed by communities and passed from generation to generation over long periods of time" (p. 164). Consequently, this community-driven source of knowledge has historically shown a sustainability of its own, through its transference over centuries through the rich oral histories all over the world. Importantly, this body of knowledge is not static and is, in fact, quite dynamic and adaptable (Keane et al. 2016). Through the definition of indigenous peoples of the then United Nations Human Rights Council's Sub-Commission on Prevention of Discrimination and Protection of Minorities, the emphasis was placed on self-identification, recognition and acceptance, and specifically, empowerment (Magni 2017).

In a context where education increasingly needs to prepare students for lifelong learning, self-directedness has become essential. Furthermore, the learning landscape has become totally multimodal with learning spaces becoming increasingly reliant on technology and on different modes of communication and learning. Inclusive and equitable education relies on resources being openly available; hence open educational resources (OERs), delivered in a multimodal manner, are pertinent to this discussion. The importance of OERs is also evident in the wider drive towards opening access to higher education (cf. Casserly and Smith 2008; Ossiannilsson et al. 2016). The benefits of OERs for widening general participation in higher education (Lane 2012), access to education for excluded communities (Lane 2008) and the empowerment of women (Perryman and de Los Arcos 2016) is clear from the literature. Finally, all these aspects are considered within the African context in this chapter and this introduces complexity in terms of existing diversity, but also the burden of colonialism and a constant need to embrace local epistemologies which are locked up in indigenous knowledge.

Within this setting, this chapter aims to explore how sustainable quality education can be achieved through self-directed multimodal learning, for open indigenous knowledge co-creation. This will be achieved in this conceptual chapter through critical engagement with the relevant scholarship in this context.

2 Sustainable Development and Education in Africa

Globally – but especially in Africa – the SDGs are of importance in ensuring sustainable development for the continent. However, according to Ramutsindela and Mickler (2020), there is some conflict between the SDGs from the United Nations 2030 Agenda (cf. Agbedahin 2019) and the African Union (AU) Agenda 2063, in terms of what should be the driving agenda. However, despite an ideological disconnect there is some overlap between the two. Since the focus in this chapter is on education, it is relevant to indicate the similarity between Goal 4 of the SDGs noted at the start of the chapter, and the second goal noted under Aspiration 1 of the AU Agenda 2063, where it is aspired to reach "[w]ell educated and skilled citizens underpinned by science, technology and innovation for a knowledge society is the norm and no child misses school due to poverty or any form of discrimination" (African Union Commission 2015, p. 2).

The importance of higher education in terms of the SDGs is also evident (Avelar et al. 2019; Elfert 2019; Franco et al. 2019). In this regard, Ramutsindela and Mickler (2020) note that there is agreement among academics and governmental structures that institutions of higher education would be strategically well-suited to address the SDGs. According to these authors, this aspect is actualised through including SDGs in curricula as well as research. Consequently, the SDGs are also of great importance for higher education in Africa.

Within the context of the SDGs, some challenges have been identified for Africa. These include the different levels of diversity between – and inside of – African countries (Majgaard and Mingat 2012; Ramutsindela and Mickler 2020). In this chapter, diversity is recognised not just in terms of countries or inside of countries; but in terms of education this extends far beyond, to the level of culture and language. As such, higher education should consider such issues especially in terms of decolonisation of the curriculum (Uleanya et al. 2019). In this regard, Uleanya et al. (2019) make the following observation "African nations are free from colonization, yet continue to experience colonization in new forms via various means, such as education, economy, amongst others" and furthermore, "[t]he educational system of a society must first be re-imaged before such society can experience new and sustainable form of development".

Finally, SDGs are also impacted by increases in the use of technology (Walsh et al. 2020) by people in Africa and the growing need for students in Africa to be prepared for ever-changing circumstances (Asongu and Odhiambo 2019) as envisaged within the context of the Fourth Industrial Revolution (cf. Schwab 2016). Importantly, Asongu and Odhiambo (2019, p. 653) observe that "reducing inequality by means of policies designed to enhance ICT will also tackle SDGs that are related to inequality". As such, students must become lifelong learners (Elfert 2019) who are self-directed (Brockett and Hiemstra 2019) and who can function effectively in dynamic multimodal (Olivier 2020a) contexts. These two aspects are explored in the next section.

3 Self-directed Multimodal Learning

3.1 Self-directed Learning and Lifelong Learning

The importance of self-directed learning (Brockett and Hiemstra 2019; Gibbons 2002) as required for sustainable development is evident, as this is a requirement for lifelong learning (Mbagwu et al. 2020; Salleh et al. 2019). Karani and Preece (2020) acknowledge that lifelong learning is essential for reaching several of the targets of the SDGs; but they also highlight that lifelong learning is not currently adequately reflected in the SDGs. Ossiannilsson et al. (2016) also show how lifelong learning in addition to opening access has become an important aim for universities around the world. Consequently, self-directed learning within an open educational context is highly relevant and self-directedness can be interpreted as self-directed multimodal learning in technology-enhanced contexts (Olivier 2020a).

Despite the close association made in this chapter between lifelong learning and self-directed learning, it is important to delineate the latter concept clearly. Knowles (1975) defines self-directed learning as "… a process in which individuals take the initiative, with or without the help of others, in diagnosing their learning needs, formulating learning goals, identifying human and material resources for learning, choosing and implementing appropriate learning strategies and evaluating learning outcomes" (p. 18). This process and student characteristics inform how higher education should be approached. Towards addressing the needs of an African student population students should also be self-directed in an open education context, where increased opportunities exist.

The link between self-directed learning and open education is evident. Ossiannilsson et al. (2016) aptly state that "[a]n open education pedagogical approach, towards self-directed learning is also essential to foster openness in both praxis and culture" (p. 161). Moreover, according to Blessinger and Bliss (2016), open education relates to the cultivation of personal agency as well as life-long self-determination and self-regulation. In addition, these aspects of self-determination and self-regulated learning are prerequisites for self-directed learning.

Consequently, in order for students in an African context to be able to become self-directed learners, quality and appropriate resources should be available in order for them to take charge of their learning. To this end, OERs can be an option to provide appropriate resource choices. However, the distribution and access of OERs require certain technologies and, as such, the relevance of multimodal environments is clear. In the African context, the issue of the digital divide (Aikins 2019; Mutsvairo and Ragnedda 2019; Warf 2019) is highly relevant as this has an impact on access to OER. The digital divide relates to differences in access to computers and the internet (Aikins 2019). Peroni and Bartolo (2018, p. 101) describes the digital divide as "the gap between those who can access digital information and use it and those who are excluded, whether partially or totally". This divide is ascribed to lack of access to infrastructure and challenges in terms of digital literacy (Chetty et al. 2018) as well as economic reasons (Peroni and Bartolo 2018). The issue of access is especially relevant in Africa where only 39.3% of the population has internet access (Internet World Stats 2020). Despite persistent challenges in terms of the digital divide in the African context, there has been some positive developments in terms of access and use of mobile devices (Samarakoon et al. 2017), which also makes multimodal learning more relevant for this context.

3.2 From Multimodality to Multimodal Learning

The issue of multimodal learning is appropriate to this discussion not only because technology is being used increasingly in education, but also because of open education being associated with technology.

Multimodality and multimodal learning are both relevant in this discussion and these terms are described by Olivier (2020a) as follows: "… multimodality refers to the dynamic application of different modes, while multimodal learning refers to individual modal preferences, communicating through different modes, learning and teaching by means of different modes, and education taking place through different modes of delivery." (p. 119). Consequently, when considering multimodal learning, the four levels of multimodality for multimodal learning should be kept in mind (Olivier 2020a) and these involve:

- individual multimodality, which pertains to individual modal preferences in learning;
- interactional multimodality, which specifically relates to modes of communication;
- instructional multimodality, which pertains to the manner in which learning takes place, whether face-to-face, online or through blended learning; and

- institutional multimodality, which refers to the mode of delivery, which could be through contact, distance or hybrid delivery.

Within this context, sustainable quality education acknowledges personal preferences in the way knowledge is conveyed but also, importantly, the manner through which communication and learning as communication take place. Finally, for sustainable education, the affordances of blended learning technologies and distance education should also be considered.

According to Lane (2009) and Ossiannilsson et al. (2016), digital technologies have created many opportunities for open education and the use of OERs. Yet, as stated before in terms of the digital divide (Aikins 2019; Mutsvairo and Ragnedda 2019), there are still challenges regarding access to technology in the African context. However, the flexibility of OERs – having the potential of being print-based or not requiring constant internet access might emphasise its relevance over traditional commercial resources. The affordances of OER in contexts with little resource and as a way to bridge the digital divide is evident (Nashipudi 2017). Blessinger and Bliss (2016) add that through open education, physical and geographical challenges can be bridged and that mobile information technologies allow for increased and more effective access. These issues are discussed in the next section.

4 Open Education

4.1 Open Educational Resources

The idea of open education extends beyond the mere use of OERs and has a long history within the context of open universities (Blessinger and Bliss 2016; Lane 2009). Historically, the description of education being 'open' implied that education is flexible, available to all people in different places - and even in terms of methods and ideas - and consequently this aspect has also been closely associated with distance education (Blessinger and Bliss 2016; Lane 2009). Furthermore, open education contributes towards democratizing education and providing opportunities to students taking up the fundamental human right to education (Blessinger and Bliss 2016). However, this democratization of education cannot merely rely on openness and would also, in the African context, imply decolonization initiatives (cf. Waghid and Manthalu 2019). OERs can play a role in such initiatives (Olivier 2020b). Moreover, McGreal (2017) emphasizes the importance of OERs within the context where "... more and more nations and institutions adopt the view that publicly-funded research and educational content belongs to the people and should therefore be open and accessible to them" (p. 292). Hence, an open approach makes access to such content more easily available and also retains it for posterity.

In terms of open education, the concept of OERs is relevant to this chapter. UNESCO (2019) defined OERs at the General Conference meeting in Paris in November 2019 as "... learning, teaching and research materials in any format and medium that reside in the public domain or are under copyright that have been released under an open license, that permit no-cost access, re-use, re-purpose, adaptation and redistribution by others" (p. 5). Open licences, such as Creative Commons licensing can be used to indicate how OERs can be used. For Lane (2009) "… openness can be equated with freedoms, but the degrees of freedom available within a particular openness can vary (as seen in the spectrum of Creative Commons licences themselves) and can be influenced by many other factors beyond the licence and particularly how potential users perceive their openness" (p. 3). The use of such licences for indigenous knowledge is quite important in the sharing and curation of such knowledge.

OERs presents a number of affordances for education. In this context, Casserly and Smith (2008) state that OERs "… alter(s) the balance of power around the accessibility to knowledge throughout the world, and they provide the opportunity for people to become empowered by taking control of their educational content and other materials," while in addition "… much of the better known OER material is of higher quality than even the best private sector or otherwise proprietary content" (p. 263). For this chapter, the role of OERs for sustainability is also pertinent.

4.2 *Open Educational Resources and Sustainability*

It is clear from the literature that open education is associated with sustainability (MacKinnon et al. 2016; McGreal 2017; Urbančič et al. 2019). Essentially, OERs also allow to counter certain barriers to learning in terms of access and cost and thereby opens up opportunities for learning for individuals who might have been excluded otherwise (Lane 2008; Lane 2009; Lane 2012). Despite the costs and accessibility issues alluded to here, OERs are more cost effective (Buchanan 2019; Wiley et al. 2012) and flexible in terms of adaptation and copy permissions than commercial resources that would have been used in classrooms. The work done by Urbančič et al. (2019) through an online mentoring programme called *Open Education for a Better World* (OE4BW), also showed promise in using OERs in support of achieving the SDGs.

McGreal (2017) describes how OERs are essential to achieving, specifically, Sustainable Development Goal 4: Quality Education, and that this aspect was also acknowledged by UNESCO from the inception of OERs, as well as the OER Paris Declaration. Roeder et al. (2017) show the affordances of OERs for Education for Sustainable Development (ESD), while Lane (2017) also highlights the advantages of open education in terms of contributing to the SDGs. The *UNESCO Recommendation on OER* highlights the importance of OERs being contextualized and localized. In this regard, one of the objectives in this document is "[f]ostering and facilitating international cooperation: supporting international cooperation between stakeholders to minimize unnecessary duplication in OERs development investments and to develop a global pool of culturally diverse, locally relevant, gender-sensitive, accessible, educational materials in multiple languages and formats." (UNESCO 2019, p. 6).

This localization relates to what Ossiannilsson et al. (2016) call *contextualisation* as according to them, in terms of OERs, "... contextualization and interconnections affect how learners construct and relate to knowledge, and sharing knowledge through social interaction and negotiation has become crucial" (p. 163). So for sustainability, the diversity of student populations in Africa should be considered and localized OERs can play an important role towards self-directed open educational practices (OEPs) (Olivier 2020b).

However, the disconnect between OERs and the practices and demands of curriculum is evident. Roeder et al. (2017) observe that German OERs and the curation thereof sometimes lack sufficient connections with educational frameworks and that the resources do not necessarily inform teachers how they can be used, or whether they would be effective without the teachers going through the whole resource. This aspect should also be considered in an African context where the curriculum should allow for the use of OERs, open practices and effective localization.

Similarly, a persisting issue with OERs is the hegemony of English and a bias to the Global North or West when it comes to content, as the majority of such resources are only available in English and created within this context (Cobo 2013; Krelja Kurelovic 2016; Mays 2020; McGreal 2017;). However, OERs allow for translation and localization (Olivier 2020b). Within the African context there is evidence of positivity towards contributing to OERs. In this regard, Nkuyubwatsi (2017) found that in Rwanda academics were willing to contribute to OERs and Mays (2020) also describe various very successful OER initiatives within the South African context. Consequently, the localization and contextualization of OERs should be considered for the African continent and part of such an approach is including indigenous knowledge in OERs.

5 Indigenous Knowledge in Education

The inclusion of indigenous knowledge in education relates to open education's striving towards localization and contextualization. But this aspect should also be considered within a wider move towards decolonizing higher education. Waghid and Manthalu (2019) explain the need for decolonization in African higher education institutions where, historically, there has been a hegemony of Eurocentrism. This dynamic source of knowledge should also be carefully considered within its original context, as it is socio-culturally situated (Magni 2017).

This chapter relates to multimodal learning and, as such, the affordances of technology for teaching and learning in relation to indigenous knowledge have been noted in the literature (Olivier et al. 2019). Yet, the affordances should be regarded within a diverse context where resources are not equally distributed and where access to technology is a challenge (Calandro et al. 2019). In addition, the increased inclusion of indigenous knowledge in education is essential in establishing what the African Union Commission (2015) phrases as "... an African knowledge society through transformation and investments in universities, science, technology, research and innovation; and through the harmonization of education standards and mutual recognition of academic and professional qualifications" (p. 15).

Open education is especially relevant for indigenous knowledge as, traditionally, within the context of ubuntu, knowledge has historically been shared in an open manner (Olivier 2020b). Letseka (2012) describes *ubuntu* as an African approach where "… a human being is a human being because of other human beings" (p. 57). For McGreal (2017) "… OER can be used to support the 'participatory principle' that is common in many indigenous communities, as well as preserving and distributing indigenous knowledge that has traditionally been open" (p. 293). So similarly to OER, indigenous knowledge is regarded as being owned by the community.

Consequently, in terms of indigenous knowledge in the wider educational context the important role of OERs should also be considered and this is evident from the literature (McGreal 2017; Olivier 2020b; Olivier et al. 2019). In this regard, McGreal (2017) notes that OERs can employed in order to ensure the preservation and distribution of indigenous knowledge in many different countries. This aspect can also be explored further within the context of sustainable development in education on the African continent. Furthermore, this issue also raises the importance about efficient licensing towards opening up access and the use of open indigenous knowledge while also providing appropriate recognition of indigenous knowledge-holders.

Embracing indigenous knowledge is part of a discourse about the need for wider Africanisation of education. For Ramutsindela and Mickler (2020), this means that universities should follow and specifically share the African experience as an impetus for scientific inquiry. Waghid and Manthalu (2019) are of the opinion that universities should be "… meaningfully and not tokenistically open to indigenous epistemologies and pedagogies without firstly transforming such epistemologies and pedagogies and benchmarking them with 'intelligible' Eurocentric paradigms" (p. 55). Hence it is suggested that embracing indigenous knowledge and the African experiences should be driven by higher educational institutions that are already embedded within an African context and which can draw on the experiences of students, staff and communities as resources.

The next section involves a synthesis of the main aspects: self-directed multimodal learning and open indigenous knowledge co-creation, in terms of a way to foster sustainable development in education.

6 Fostering Sustainable Development in Education Through Self-directed Multimodal Learning for Open Indigenous Knowledge Co-creation

Firstly, the use of OERs is essential as, according to McGreal (2017), OERs can be beneficial for developing countries as it can open up access to learning where there might be constraints in terms of cost, quality and even equity in terms of knowledge distribution. In terms of content, UNESCO (2019) recommends that states encourage

"... OER stakeholders to develop gender-sensitive, culturally and linguistically relevant OER, and to create local language OER, particularly in indigenous languages which are less used, under-resourced and endangered" (p. 8). In this regard, the fact that OERs can be translated and localized for different contexts is affirmed in the literature (McGreal, 2017). Through the adaptability and localization process, OERs afford the opportunity for content to be changed and expanded in order to accommodate indigenous knowledge. However, the use of OERs does not ensure effective learning; hence a self-directed multimodal approach is recommended in this chapter.

For the fostering of sustainable development through self-directed multimodal learning for open indigenous knowledge co-creation, some focus on OEPs would be necessary. Such OEPs involves promoting the freedom for teachers and students to be able to reuse, remix and redistribute OERs (cf. MacKinnon et al. 2016). Consequently, it implies renewability but also involvement of both teachers and students. From the policy research by MacKinnon et al. (2016) in Australia, it is evident that supporting OEPs in terms of sustainable development means not only bottom-up processes by individuals, but also top-down policy support and incentives, in order to create a space for wider participation. Consequently, the source of knowledge is not outside or decontextualised, but rather a hybrid approach of drawing from existing open licensed content; and then adding and creating new contextualised content. In this regard, teachers can utilise not only the expertise students have but also draw from the communities around them.

OEPs also imply effectively embracing open licenses like Creative Commons. Licenses such as these not only provide freedom in terms of sharing indigenous knowledge, but they also empower knowledge-holders to determine how such knowledge can be used or reused. Therefore, not only should teachers and students be empowered to understand and use licensing, but they should also contribute to the use thereof within indigenous communities, for the curation of knowledge. OEPs is but one aspect of the learning process and as stated before, this chapter recommends a self-directed multimodal learning approach.

However, a key requirement for effective self-directed multimodal learning in an open context is the fact that issues around the digital divide need to be addressed. This digital divide implies that certain groups and societies have different levels of access to digital technologies (Lane 2008; Lane 2009). In terms of wider use of OERs, the issue of infrastructure is also evident (McGreal 2017). In this regard, Lane (2009) also observes that "[t]he digital educational divide can mean that some learners are much more sophisticated users of digital technology for learning than their (subject-focused) teachers, while such fluency (or not) with the technology can exacerbate the educational divide as modes of communication, collaboration, and computation multiply or become more sophisticated" (p. 8). In terms of open education, it is noted that, despite the fact that OERs imply opening up access to knowledge, it may actually also accentuate existing inequalities in terms of access to technology, due to the digital divide (Lane 2009). However, with the spread of the mobile technologies and through increased access to the internet, OERs can still be regarded as a more viable option. In addition, not all OERs need to be accessed through the internet and a more hybrid approach is also recommended here, where certain content can either be accessed online or be downloaded with the option of it being printed in hard copy – at a cost, but potentially cheaper than commercial textbooks.

Co-creation is central to the proposed approach in this chapter. The selection of resources in education settings may be teacher-centred, but an open co-creation approach would imply increased inputs by students. Lane (2009) acknowledges that in open education teachers may be responsible for selecting OERs but acknowledges the potential role of students in also making inputs. Casserly and Smith (2008) highlight the learning possibilities where students create content within such a context. The generation of knowledge, therefore, not only serves a practical purpose in expanding the resources, but - in itself - becomes a learning activity.

Blessinger and Bliss (2016) furthermore declare that open education not only implies allowing access to resources, it can actually be a vehicle towards supporting the agency of students and teachers, as they take control of the content and the technologies they want to use in a co-creation process. The importance of co-creation as collective activity relates to the statement by Ossiannilsson et al. (2016) that, "… collective interactive processes between individuals within frameworks of contextualization shape the process of learning, that is, how to learn, change, construct, and relate to enormous amounts of information in order to meet learners' expectations and needs in a meaningful way" (p. 163). Thus not only the needs of students which relate to the identified learning needs in terms of self-directed learning, are addressed: this process also allows for contextualising content and learning for students' own specific contexts.

The focus in this chapter is on formal education, but the importance of lifelong learning in terms of non-formal and informal learning in terms of the SDGs in the African context is stated by Karani and Preece (2020). McGreal (2017) also notes how OERs can be supportive of lifelong learning. This issue brings the importance of self-directed learning to the foreground, as through engagement with OERs, students can be exposed to resource repositories and sources of knowledge outside of the classroom and textbook.

As stated earlier, central to fostering self-directed learning is the ability to identify relevant human, as well as material, resources in the learning process. To this end, teachers should infuse the use of OERs in strategies that can be supportive of self-directed learning and this can include problem-based and collaborative learning strategies. Consequently, problems can be identified within a localized context, which would imply using - or even creating - OERs relevant to addressing the issue. Furthermore, the localization of OERs can become an effective collaborative exercise. A further option would be to embrace renewable assignments which would be a way to generate OERs. These assignments can be defined as "assignments which both support an individual student's learning and result in new or improved open educational resources that provide a lasting benefit to the broader community of learners" (Wiley and Hilton 2018, p. 137). Renewable assignments counter the idea of assignments being done just to be read by the teacher and not having any purpose after assessment as they can be reused. The use of such assignments would not only support students' learning, but may result in the creation of contextualised OERs that can be to the benefit of the broader student community (Wiley and Hilton 2018).

Consequently, renewable assignments can be regarded as one vehicle through which to infuse the classroom with indigenous knowledge.

7 Recommendations

The proposed focus on self-directed learning, multimodality, open education and indigenous knowledge provides a conducive environment in which the SDGs and specifically Goal 4 can be reached. This also supports Ramutsindela and Mickler's (2020) view that the SDGs should be adapted to the African context and realities, whilst not negatively affecting existing developmental activities

Localization of OERs is an essential step towards sustainable quality education. For Pereira (2007) localization pertains to the process in which local partners work with OER creators in order to make content appropriate for a specific context. To this end, collaboration is key to any localization effort (Wolfenden and Adinolfi 2019). Localization efforts should also address the needs in terms of what Waghid and Manthalu (2019) call 'local epistemologies and languages' (p. 49), which have traditionally been ignored; or regarded as less important than Eurocentric equivalents in the African context.

Casserly and Smith (2008) observe that OERs create opportunities for localizing materials in terms of language and culture, while also allowing for further potential individualization and personalization. Therefore, the scope of localization could even extend form mere adaptation for the needs of a specific culture or language community and also be considered at the level of school - and ultimately – the individual. The work done by Urbančič et al. (2019) in terms of OERs through online mentoring has also emphasised the importance of personalization and learning by doing.

Central to any open education initiative would be supporting research and collaboration. For Ossiannilsson et al. (2016) this can be phrased as follows: "Opening up education requires the adoption of practices and cultures that foster academic research and collaboration to enhance learning and teaching" (p. 162). This prompts the need for further research in terms of open education within the African context as well as policies and practices supportive of collaboration within the learning and teaching environment and beyond.

The actions discussed in this chapter require adequate access to technology especially in the African context. In this regard, the opportunities provided by open education might not be as open or available to all. Lane (2008) notes that, "… availability, accessibility and acceptability of this mode of teaching and learning is extremely variable, with socially excluded groups or communities being those who do not have much access to such technologies, may find few opportunities available to them in their circumstances and are worried that they cannot cope with these new technologies and ways of learning" (p. 1). Worldwide digital exclusion of women is also part of this discourse (cf. Perryman and de Los Arcos 2016). Consequently, an increased and ongoing commitment to providing access to technology and to providing adequate support and training, would be essential for any form of sustainable quality education in Africa.

8 Conclusion

The aim of making self-directed multimodal learning open for indigenous knowledge co-creation is aligned with the African Union Commission's (2015) aim to create, open up and strengthen access to higher education and supporting scientific reforms that are central to the transformation of Africa.

An important implication regarding sustainable quality education through self-directed multimodal learning for open indigenous knowledge co-creation, is the origin of the word *sustain*. This word can be traced back to the Latin *sustinēre* which pertains to holding (*tenēre*) something from below (*sub-*) (AHD 2000). Consequently, despite the importance of policies and governmental support educational initiatives - within a context of self-directed multimodal learning for open indigenous knowledge co-creation - would benefit from bottom-up initiatives.

This bottom-up approach starts with students. In this regard, this chapter aligns itself with the following statement by Blessinger and Bliss (2016):

> Open education is designed for participation when it is well-designed for access, agency, and ownership. In other words, these aspects lead to participation by learners and educators. As open education promotes these fundamental principles, students and teachers are more likely to collaborate and participatory in inclusive activities. Indeed, one of the goals of open education is to move learners closer to the center of a community of practice specifically through providing opportunities and infrastructure for participation and collaboration. (p. 14)

Therefore, the need for establishing communities of practice around open education, where students are at the centre of such activities, is emphasised. However, this requires appropriate physical and human resources for those students to effectively take ownership of their learning.

Certain limitations are relevant for this chapter. In this regard, the discussion and findings are limited to the selected sources and consequently certain views might have been excluded. However, an attempt was made to present a balanced discussion based on the critical review of the selected sources.

A number potential areas of future research have been identified. Empirical research on using localized OERs infused with indigenous knowledge can be explored within different classroom settings. In addition, students' roles in creating OERs at different educational levels can also be explored in terms of their engagement with local communities towards curating and infusing indigenous knowledge within the curriculum.

In conclusion, sustainable quality education can be supported through self-directed multimodal learning for open indigenous knowledge co-creation. The needs in terms of sustainable development and the SDGs, prompt practices that foster self-directed learning among students in Africa. In this context, access to multimodal learning in terms of all levels of multimodality is essential in order to address the changing technological nature of education globally. However, for sustainable quality education, content needs to build on OERs and allow for localization and contextualization in order to make content appropriate for students throughout the diverse learning environment. To this end, teachers and students should become co-creators of knowledge by engaging with OERs. Through such a process even indigenous knowledge can be infused in curricula throughout the continent.

References

Aarts H, Greijn H, Mohamedbhai G, Jowi JO (2020) The SDGs and African higher education. In: Ramutsindela M, Mickler D (eds) Africa and the sustainable development goals. Springer, Cham, pp 231–241

African Union Commission (2015) Agenda 2063: The Africa we want Addis Ababa: African Union Commission https://au.int/sites/default/files/documents/36204-doc-agenda2063_popular_version_en.pdf. Accessed 24 Aug 2020

Agbedahin AV (2019) Sustainable development, Education for Sustainable Development, and the 2030 Agenda for Sustainable Development: Emergence, efficacy, eminence, and future. Sustain Dev **27**(4):669–680

AHD (American Heritage Dictionary) 2000 The American Heritage Dictionary of the English Language Fourth Edition. Houghton Mifflin Harcourt, Boston, MA

Aikins SK (2019) Determinants of digital divide in Africa and policy implications. Int J Public Adm Digit Age (IJPADA) 6(1):64–79

Asongu SA, Odhiambo NM (2019) How enhancing information and communication technology has affected inequality in Africa for sustainable development: An empirical investigation. Sustainable Dev 27(4):647–656

Avelar ABA, da Silva-Oliveira KD, da Silva Pereira R (2019) Education for advancing the implementation of the Sustainable Development Goals: a systematic approach. Int J Manage Educ 17(3):

Blessinger P, Bliss TJ (2016) Introduction to open education: towards a human rights theory. In: Blessinger P, Bliss TJ (eds) Open education: international perspectives in high education. Openbook Publishers, Cambridge, pp 11–30

Brockett RG, Hiemstra R (2019) Self–direction in adult learning: perspectives on theory, research, and practice. Routledge, London

Buchanan S (2019) Implementing a Framework for Reducing Textbook Costs by Utilizing OER and Other Textbook Alternatives in Online Course Development. Doctoral dissertation, University of Delaware

Calandro E, Chavula J, Phokeer A (2019) Internet development in Africa: a content use, hosting and distribution perspective. In: Mendy G, Ouya S, Dioum I, Thiaré O (eds) e–Infrastructure and e–Services for Developing Countries AFRICOMM 2018 Lecture Notes of the Institute for Computer Sciences, Social Informatics and Telecommunications Engineering, 275. Springer, Cham

Casserly CM, Smith MS (2008) Revolutionizing education through innovation: can openness transform teaching and learning? In: Iiyoshi T, Kumar MSV (eds) Opening up education: The collective advancement of education through open technology, open content, and open knowledge. MIT Press, Cambridge, pp 261–276

Chetty K, Qigui L, Gcora N, Josie J, Wenwei L, Fang C (2018) Bridging the digital divide: measuring digital literacy. Econ Open-Access, Open-Assess E-J **12**(2018–23):1–20

Cobo C (2013) Exploration of open educational resources in non–English speaking communities. Int Rev Res Open Distrib Learn **14**(2):106–128

Elfert M (2019) Lifelong learning in Sustainable Development Goal 4: What does it mean for UNESCO's rights–based approach to adult learning and education? Int Rev Educ 65(4):537–556

Emas R (2015) The concept of sustainable development: definition and defining principles Brief for GSDR. https://sustainabledevelopment.un.org/content/documents/5839GSDR%202015_SD_concept_definiton_rev.pdf. Accessed 24 Aug 2020

Franco I, Saito O, Vaughter P, Whereat J, Kanie N, Takemoto K (2019) Higher education for sustainable development: actioning the global goals in policy, curriculum and practice. Sustain Sci 14(6):1621–1642

Gibbons M (2002) The self-directed learning handbook: challenging adolescent students to excel. Jossey-Bass, San Francisco

Internet World Stats (2020) Africa Internet Users, 2020 Population and Facebook Statistics. https://www.internetworldstats.com/stats1.htm. Accessed 19 Nov 2020

Karani FA, Preece J (2020) Lifelong Learning and the SDGs In: Ramutsindela M, Mickler D (eds) Africa and the Sustainable Development Goals Springer, Cham, pp 23–31

Keane M, Khupe C, Muza B (2016) It matters who you are: Indigenous knowledge research and researchers. Educ Change 20(2):163–183

Knowles MS (1975) Self-directed learning: a guide for learners and teachers. Follett, Chicago, IL

Krelja Kurelovic E (2016) Advantages and limitations of usage of open educational resources in small countries. Int J Res Educ Sci 2(1):136–142

Lane A (2008) Am I good enough? The mediated use of open educational resources to empower learners in excluded communities In: Fifth pan–commonwealth forum on open learning, 13–17 Jul 2008, London, UK

Lane A (2009) The impact of openness on bridging educational digital divides. Int Rev Res Open Distrib Learn 10(5):1–12

Lane A (2012) Widening participation in higher education through open educational resources. In: Okada A, Connolly T, Scott PJ (eds) Collaborative learning 20: open educational resources. IGI Global, Hershey, PA, pp 1–15

Lane A (2017) Open education and the sustainable development goals: mking change happen. J Learn Dev 4(3):275–286

Letseka M (2012) In defence of Ubuntu. Stud Philos Educ 31(1):47–60

MacKinnon T, Pasfield–Neofitou S, Manns H, Grant S (2016) A meta–analysis of open educational communities of practice and sustainability in higher educational policy. Apprentissage des Langues et Systèmes d'Information et de Communication 19(1). https://journals.openedition.org/alsic/2908. Accessed 23 Aug 2020

Mays T (2020) Open Educational Resources in South Africa. In: Huang R, Liu D Tlili A, Gao Y, Koper R (eds) Current state of open educational resources in the "belt and road" countries. Lecture Notes in Educational Technology. Springer, Singapore

Magni G (2017) Indigenous knowledge and implications for the sustainable development agenda. Eur J Educ 52(4):437–447

Majgaard K, Mingat A (2012) Education in sub–Saharan Africa: a comparative analysis. The World Bank, Washington, DC

Mbagwu FO, Chukwuedo SO, Ogbuanya TC (2020) Promoting lifelong learning propensity and intentions for vocational training among adult and vocational educational undergraduates. Vocations Learn 13:419–437

McGreal R (2017) Special report on the role of open educational resources in supporting the sustainable development goal 4: quality education challenges and opportunities. Int Rev Res Open Distrib Learn 18(7):292–305

Mutsvairo B, Ragnedda M (eds) (2019) Mapping the digital divide in Africa: a mediated analysis. Amsterdam University Press, Amsterdam

Nashipudi M (2017) OER footprints bridging the educational digital divide in India pearl. J Libr Inf Sci 11(2):124–133

Nkuyubwatsi B (2017) Willingness to engage in open educational practices among academics in Rwandan public higher education and responsive actions. J Learn Dev 4(3):322–337
Olivier J (2020a) Self-directed multimodal learning to support demiurgic access. In: Burgos D (ed) Radical Solutions and eLearning. Springer Nature, Singapore, pp 117–130. https://doi.org/10.1007/978-981-15-4952-6_8
Olivier J (2020b) Self-directed open educational practices for a decolonized South African curriculum: a process of localization for learning. J e-Learning Knowl Soc **16**(4):20–28. https://doi.org/10.20368/1971-8829/113533
Olivier J, Van der Westhuizen C, Laubscher D, Bailey R (2019) The affordances of technology for teaching indigenous knowledge. In: de Beer J (ed) The decolonisation of the curriculum project: The affordances of indigenous knowledge for self-directed learning. AOSIS, Cape Town, pp 277–317
Ossiannilsson E, Altinay Z, Altinay F (2016) Transformation of teaching and learning in higher education towards open learning arenas: A question of quality. In: Blessinger P, Bliss TJ (eds) Open education: international perspectives in high education. Openbook Publishers, Cambridge, pp 159–177
Pereira M (2007) Open educational resources: A bridge to education in the developing world. Educational Technology 41–43
Peroni M, Bartolo M (2018) The Digital Divide. In: Bartolo M, Ferrari F (eds) Multidisciplinary Teleconsultation in Developing Countries TELe–Health. Springer, Cham, pp 101–109
Perryman LA, de Los Arcos B (2016) Women's empowerment through openness: OER, OEP and the Sustainable Development Goals. Open Praxis 8(2):163–180
Ramutsindela M, Mickler D (2020) Global Goals and African Development. In: Ramutsindela M, Mickler D (eds) Africa and the sustainable development goals. Springer, Cham, pp 231–241
Roeder I, Severengiz M, Stark R, Seliger G (2017) Open educational resources as a driver for manufacturing–related education for learning of sustainable development. Procedia Manuf 8:81–88
Salleh UKM, Zulnaidi H, Rahim SSA, Bin Zakaria AR, Hidayat R (2019) Roles of self-directed learning and social networking sites in lifelong learning. Int J Instr 12(4):167–182
Samarakoon S, Christiansen A, Munro PG (2017) Equitable and quality education for all of Africa? The challenges of using ICT in education Perspectives on Global Development and Technology 16(6):645–665
Schwab K (2016) The Fourth Industrial Revolution. World Economic Forum
UNESCO (2019) Recommendation on Open Educational Resources (OER).: UNESCO, Paris https://unesdoc.unesco.org/ark:/48223/pf0000373755/PDF/373755eng.pdf.multi.page=3. Accessed 24 Aug 2020
Uleanya C, Rugbeer Y, Olaniran SO (2019) Decolonization of education: exploring a new praxis for sustainable development. African Identities 17(2):94–107
United Nations (2020) Global indicator framework for the Sustainable Development Goals and targets of the 2030 Agenda for Sustainable Development. https://unstats.un.org/sdgs/indicators/Global%20Indicator%20Framework%20after%202020%20review_Eng.pdf. Accessed 24 Aug 2020
Urbančič T, Polajnar A, Jermol M (2019) Open education for a better world: a mentoring programme fostering design and reuse of open educational resources for sustainable development goals. Open Praxis 11(4):409–426
Uvalić-Trumbić S, Daniel J (2016) Sustainable development begins with education. J Learn Dev 3(3):3–8
Waghid Y, Manthalu CH (2019) Decoloniality as democratic change within higher education. CH Manthalu. Y Waghid Education for Decoloniality and Decolonisation in Africa. Palgrave Macmillan, Cham, pp 47–68
Walsh PP, Murphy E, Horan D (2020) The role of science, technology and innovation in the UN 2030 agenda. Technol Forecasting Soc Change 154:
Warf B (2019) Teaching digital divides. J Geogr 118(2):77–87

Wiley D, Green C, Soares L (2012) Dramatically Bringing down the Cost of Education with OER: How Open Education Resources Unlock the Door to Free Learning. Center for American Progress
Wiley D, Hilton JL (2018) Defining OER–Enabled Pedagogy. The International Review of Research in Open and Distributed Learning 19(4):133–147
Wolfenden F, Adinolfi L (2019) An exploration of agency in the localisation of open educational resources for teacher development. Learn Media Technol 44(3):327–344

Jako Olivier is the holder of the UNESCO Chair in Multimodal Learning and Open Educational Resources and is a professor of Multimodal Learning in the Faculty of Education at North-West University (NWU), South Africa. He obtained his PhD in 2011 in which he researched the accommodation and promotion of multilingualism in schools through blended learning. Before he joined the NWU as lecturer in 2010, he was involved in teaching information technology and languages in schools in the United Kingdom and in South Africa. His research, within the NWU's Research Unit Self-Directed Learning, focuses on self-directed multimodal learning, open educational resources, multiliteracies, blended and e-learning in language classrooms as well as multilingualism in education. He currently holds a Y rating from the South African National Research Foundation (NRF) and was awarded the Education Association of South Africa (EASA) Emerging Researcher Medal in 2018.

Integrating Indigenous Knowledge on Environmental Management Practices in the Teaching of Geography in Secondary Schools: A Case of Chimanimani Community in Zimbabwe

Joshua Risiro

Abstract The rapid population and industrial growth in Zimbabwe has resulted in land degradation and negative impacts on the human and physical environment. The primary purpose of this chapter was to find out the various indigenous environmental management practices available in Chimanimani that can be adopted by other communities in the conservation and sustainable use of resources. Secondly, the study explored Indigenous Knowledge (IK) and environmental management practices that can be integrated in the teaching of Geography in secondary schools, in order to ensure there is sustainable use of resources for the current and future generations. The study was guided by Edward Said (1978) post-colonial theory. An interpretive research paradigm was adopted in the community case study. Data were generated through a documentary analysis of the Form 1–4 (2015–2022) Geography syllabus, individual and group WhatsApp chats with Geography teachers and face-to-face interviews with community indigenous environment experts. The population sample consisted of ten (10) Geography teachers drawn from three (3) secondary schools and (6) community elders who were purposively sampled from the Chimanimani community. Data were analysed by generating thick descriptions and themes. The study revealed that the Chimanimani community has cultural practices that can be adopted in the conservation of resources. The research further exposed that IK related to environmental management can be integrated in the teaching of Geography in secondary schools in order to enhance sustainable use and environmental management of resources. The study recommends the integration of Indigenous and Eurocentric models in the conservation and sustainable management of resources. The incorporation of more indigenous environmental management practices in the Geography curriculum in secondary schools can help to inculcate a culture of environmental conservation among learners and communities.

Keywords Indigenous Knowledge (IK) · Integration · Environmental Management · Chimanimani community · Geography · Sustainable

---

J. Risiro (✉)
Great Zimbabwe University, Masvingo, Zimbabwe

W. Leal Filho et al. (eds.), *Sustainable Development in Africa*, World Sustainability Series,
https://doi.org/10.1007/978-3-030-74693-3_7

1 Introduction

The argument in this chapter is that the British education system in Zimbabwe disregarded Indigenous Knowledge (IK) which is important in the environmental management and sustainable use of resources within indigenous communities. The recent decades have been characterised by massive destruction of biodiversity due to rapid industrialisation, urban expansion and population pressure on land. The scramble for resources by colonialists destroyed indigenous preserved forests and animal species (Sarma 2016). Foreign invasions desecrated existing sacred places, disregarded taboos and cultural beliefs on environmental management within indigenous communities (Risiro et al. 2013). Worse still, the colonial British government in Zimbabwe did little to integrate IK on environmental practices in the school curriculum as observed by Masaka (2016), Ogunniyi (2016) and Pedzisai (2013).

The primary purpose of this study is therefore, to explore the indigenous environmental management practices that could be integrated in the teaching of Geography in secondary schools in Zimbabwe in order to achieve Sustainable Development Goals (SDGs) on the conservation of resources. A case study of Chimanimani community in the Eastern Highlands of Zimbabwe was adopted. The study was guided by Edward Said (1978) post-colonial theory. The interpretive paradigm was adopted in the study. Data were generated using qualitative methods namely, document analysis of the Form 1–4 Geography syllabus (2015–2022), individual and group WhatsApp chats and face-to-face interviews with community indigenous environment experts. Participants in the research were purposively sampled from Geography teachers and community elders. The thematic approach was used to present and analyse the data.

2 Background to the Study

2.1 Colonial Influences on Education and Environmental Sustainability

The study is undertaken within the context that Western education has disregarded and despised Indigenous Knowledge (IK) as a form of knowledge worth being taught in schools (Masaka 2016). The indigenous people ended up believing that their indigenous practices were pagan and backward (Hwami 2014; Masaka 2016). Several studies have shown that the Western curriculum imposed among the indigenous people was inappropriate and retrogressive resulting in poor performance and a high dropout of the indigenous learners in schools (Barnhardt and Kawagley 2005; Cushner et al. 2012). Public education has tended to ignore the knowledge that indigenous people bring into the classroom.

Zimbabwe was a colony of Britain until the attainment of its independence in 1980. Zimbabwe therefore inherited the British education system which promoted the interests of the colonisers (Masaka 2016). The British curriculum implemented prepared the Europeans for future leadership in industry and Government offices. On the contrary, the education for the Africans was meant to provide a source of cheap labour in the colonies (Mavhunga 2006). Furthermore, colonial education disempowered Africans such that they lost control of their lives (Masaka 2016). Some of the Africans who acquired British education used it to oppress fellow Africans (Hwami 2014; Shizha and Kariwo 2011).

There was preferential treatment and segregation between the European and African schools. European schools in Zimbabwe were government-funded while African schools were donor-funded by Missionaries (Government of Zimbabwe 2005). Further, there was a bottleneck system to restrict progression of Africans through a rigorous grading system. In addition, schools were classified as academic schools (F1 schools) for the brightest students and F2 schools for the less gifted learners (Government of Zimbabwe 2005). The system created many dropouts for Africans who would then provide a cheap labour force in industries (Mavhunga 2006; Shizha 2006).

In relation to sustainable use of the environment, Obiora and Emeka (2015) believe that, Indigenous people in the Igbo community in Nigeria lived in sync with their natural environment. The community sustainably conserved their resources through respect of their beliefs, taboos, totems and myths. However, the colonialists' mechanistic view disregarded indigenous cultures resulting in massive land degradation. Niesenbaum (2019) observed that the disregarded indigenous knowledge on biodiversity conservation as the primary aim of the colonisers was to exploit resources from colonies. In the Himalaya forests and Central India, Sharma (2016) found that the indigenous community sustainably used their forest resources which, however, were destroyed by colonial greed that ignored the indigenous people's harmonious relationship with nature.

The views given by different scholars in the ongoing discussion suggest a need to review the colonial education system and incorporate indigenous environmental management practices in order to achieve SDG 14 (conservation of life under water) and SDG15 (conservation of life on land).

2.2 *International Conventions and Sustainable Development Goals (SDGs)*

The global population expansion has resulted in massive land degradation from agricultural practices, urbanisation and industrialization thereby impacting negatively on biodiversity. Forests have declined while some animal species are going into extinction. The decline in forests has resulted in increased carbon dioxide emissions into the atmosphere thereby causing global warming and acidification of sea waters (Risiro et al. 2013).

In view of the extensive destruction of biodiversity, various international conventions have been held in order to put in place measures to mitigate the rampant loss of biodiversity. The Stockholm Convention held in 1972 guided people in the preservation of the environment. In 1977, the Tbilisi Convention environmental education was convened. Further, The Brundtland Commission was constituted in 1983 in order to find a link between environmental issues and economic development. The concept of sustainable development was formalized. This is the development that allows the present population to meet their needs without disadvantaging future generations to meet their own wants (World Commission on Environment and Development 1987). In 1992, the Earth Summit in Rio de Janeiro produced the Rio Principles Agenda 21. Agenda 21 gives an outline of actions to be carried out in order to achieve sustainable development at various scales (Chenje et al. 1998). Other international conventions held in support of environmental education included the Convention on Climatic Change, the Biodiversity Convention, the Desertification Convention, the Montreal Protocol on ozone layer and the Convention on the International Trade of Endangered Species (CITES) which seeks to protect wildlife (Chenje et al. 1998).

Linked to International conventions are SDGs which were adopted in New York at the United Nations Summit. There are seventeen (17) goals to be achieved by 2030 (Filho et al. 2018). Of particular note to the study are SDG 14, which aim at the conservation and sustainable use of life under water; SDG 15 (conservation and sustainable use of terrestrial ecosystems); SDG 2 (food security and sustainable Agriculture); SDG 6 (sustainable use of water and sanitation) and SDG 13 (climate change) (Filho et al. 2018). The study aims to explore how Chimanimani community indigenous environmental management practices can achieve the stated SDGs.

2.3 Indigenous Methods on Sustainable Resource Management

From time immemorial, indigenous people have practised their culture in order to manage resources. Emmanuel (2020) found that taboos have been used to protect endangered species. Killing of some animals was said to cause bad omen to the community such as drought (Diawuo and Issifu 2015). Each clan in Zimbabwe has got a totem belonging to an animal which the clan would not kill and eat. The belief is that eating an animal belonging to one's totem would result in one losing his or her teeth (Bourdillon 1976). Taboos were upheld to protect fruit trees and those trees associated with ancestral spirits (Anoliefo et al. 2015) Further, sacred places have been used to preserve flora and fauna in the Cape Coast of Ghana (Emmanuel 2020). The indigenous people also had ways for sustainable use of soils through shifting cultivation and nomadic herding which allowed vegetation to regenerate and for the soil to regain its fertility (Burger 1990). In the Zaka area of Masvingo in Zimbabwe, the local community used stone ridging to reduce surface run-off and soil erosion (Risiro et al. 2013).

The indigenous people cherish Ubuntu values characterised by caring, sharing, community engagement and respect for nature and the environment. Ubuntu is holistic in nature and involves the whole community (Mahlatsi 2017). Its attributes displayed by the indigenous people are essential in the study as their integration in the Geography curriculum can contribute to environmental management and sustainable use of resources.

2.4 *Edward Said's Post-Colonial Theory*

The research was guided by Said's (1978) post-colonial theory which argues that the colonisers used education to perpetuate their dominance over the indigenous people. The theory explains how the colonised people resisted the supremacy of Eurocentric education after attaining their independence (Subedi and Daza 2008). The effect of colonial education on the local people was to make them despise their own indigenous practices. In the British colonies, Christianity was used as a tool to suppress the local people (Jaware 2012). The colonial education system in Zimbabwe was tailored to serve the interest of the colonial power through provision of cheap labour in industries (Mavhunga 2006). The post-colonial theory is suitable for the study as the research seeks to integrate IK on environmental management practices that has received little attention from the inherited Geography curriculum in the secondary schools. Despite various curriculum changes in Zimbabwe since independence in 1980, very little IK in environmental management practices have been integrated into the school curriculum (Ogunniyi 2016; Pedzisai 2013).

3 Statement of the Problem

The colonial education system in Zimbabwe disregarded indigenous education as a form of knowledge resulting in the colonisers imposing their education that ignored indigenous environmental practices on the Africans. The inappropriate education received by the indigenous people coupled with rapid population growth and industrialisation resulted in massive land degradation and loss of biodiversity. The study further argues that the current Form 1–4 Geography syllabus (2015—20,220 is deficient on IK practices related to environmental management. Scholars among them Masaka (2016), Ogunniyi (2016) and Pedzisai (2013) have all acknowledged that there is minimum application of IK in the classroom teaching in Zimbabwe. This is despite the United Nations Declaration on the Rights of Indigenous People [UNDRIP] (United Nations 2007) that gave the green light and institutional support for the integration of IK in the school curriculum. Article 31-1 of the UNDRIP articulates the rights by indigenous people to pass on to their future generations, their traditions and philosophy. Further, Article 14-1 of the same UNDRIP gave legal power to the indigenous people to have a stake in the establishment of educational institutions which meet the expectations of their culture and language. In

Zimbabwe, the Nziramasanga Commission of Inquiry into Education and Training (CIET) recommended the incorporation of indigenous practices in the Zimbabwean education system (Government of Zimbabwe 1999). It is against this background that the study explores the integration of indigenous practices on environmental management into the teaching of Geography in secondary schools as a strategy to achieve the Sustainable Development Goals (SDGs) on the preservation and conservation of resources.

4 Research Questions

1. What indigenous environmental management practices are used in the Chimanimani community in managing their environment?
2. Which topics in the Form 1–4 Geography syllabus (2015–2022) can be integrated with indigenous environmental practices used to manage the environment in Chimanimani District?

5 Methodology

The section discusses methodological issues comprising of the research design, data generation methods, population sample, data presentation and analysis as well as ethical considerations.

5.1 Research Design

An interpretive paradigm was adopted to carry out a case study of the Chimanimani community situated in the Eastern Highlands of Zimbabwe. The paradigm was selected as it accommodates multiple perspectives from various participants on a phenomenon (Pham 2018). A case study approach was adopted in the study in order to allow in-depth data generation as suggested by Creswell (2013) and Patton (2015).

5.2 Data Generation and Sampling Methods

Data was generated from 26 May to 25 June 2020 through interviews and WhatsApp chat group discussions. The sample population consisted of six (6) community elders and ten (10) Geography teachers purposively sampled from around the community of three (3) secondary schools. Data was generated during the Covid-19 lockdown restriction period through a WhatsApp chat group which was appropriate to gather data from ten (10) Geography teachers. The researcher is a member of a WhatsApp chat group consisting of Geography Examiners in Chimanimani District. WhatsApp group chats with Geography teachers were useful in generating a large amount of information within a short space of time as alluded to by Houser (2015). The WhatsApp group chats allowed participants to either confirm or disconfirm facts raised by other colleagues.

Individual interviews were used to generate data from the community elders. Patton (2015) observed that interviews yield in-depth responses about a phenomenon under study. Community elders were chosen since they are knowledgeable on IK practised in the study area. This was in line with Khupe (2014) who viewed the elderly people as custodians of local knowledge. An interview guide was used to generate data from the participants as recommended by Patton (2015). The interview questions were translated from English into the local language, *Ndau* to allow community elders to explain fluently their indigenous practices. Kaya and Lyana (2014) and Khupe (2014) found that the use of a local language by participants allowed them to freely express their views.

Document analysis of the Form 1–4 Geography syllabus (2015–2022) was conducted. (Form 1–4 refers to the first four years of secondary education in Zimbabwe). The analysis of the syllabus was important in order to identify topics in the current Geography syllabus which can be integrated with IK on environmental management practices.

5.3 *Data Presentation and Analysis*

Data analysis was done according to Creswell (2014) guidelines. After recording the data, it was transcribed, coded and grouped into themes. Data was then presented in tabular form using main themes and sub-themes as well as quotes from participants. This was followed by brief explanation and discussion of the findings.

5.4 *Ethical Issues*

Permission to carry out the research was sought from and granted by the local leadership consisting of the Councilor and Headman as well as from the research participants. The procedures that guided the study to ensure the participant's rights were protected were largely drawn from Cohen et al. (2018) and Creswell (2014). Informed consent forms were availed to the participants for signing. In addition to written permission, oral permission was requested from the same people.

6 Findings and Discussion

For anonymity and ethical considerations, participants were identified using codes. Teachers were identified as T1, T2 in that numerical order. Similarly, community elders were identified as C1, C2 and so on. Data presented in Tables 1, 2, 3, and 4 addresses research question one (1) that reads: What indigenous environmental management practices are used in Chimanimani community in managing their environment? Table 5 addresses research question 2: Which topics in the Form 1–4 Geography syllabus (2015–2022) can be integrated with indigenous environmental practices used to manage the environment in Chimanimani District?

6.1 Indigenous Environmental Management Practices on Plant Species

Indigenous ways of environmental management and sustainable use of trees and herbs are presented and explained in this section (Table 1).

Indigenous people of Chimanimani used taboos, totems and believed in the spiritual world that help them to preserve and conserve biodiversity including water

Table 1 Indigenous environmental management practices on plant species

Main theme	Sub-theme	Quotes
Indigenous environmental management practices on plant species	-Sustainable utilisation of fruit trees	*-No-one is allowed to harvest unripe fruits such as mazhanje. Otherwise wild animals like baboons will eat our crops.* **T3** *-We are only allowed to pick wild fruits in the mountains after the 'mutoro' ceremony is performed. The ancestors have to be thanked before harvesting the fruits. The chief can impose a fine of a goat if you are caught harvesting fruits before the ceremony.* **C4**
	-Sustainable use of medicinal plants	*- It is not allowed to cut the whole tree of a medicinal plant. Instead either the roots, leaves or bark are used. In addition, the knowledge on the use of herbs is limited to a few people* **C6** *-Traditional medicine can be re-used, for example an ill person can boil plant roots and drink the concoction several times.* **T5**

(continued)

Table 1 (continued)

Main theme	Sub-theme	Quotes
	-Sacred forests	*- In this area, we have got sacred places where Chiefs are buried. Shara, a place used to bury the Chiefs of [illegible] clan and Nyamwazha is used to bury chiefs of the Muusha clan. No one is allowed to collect firewood or hunt in this forest.* **C1** *- Each Village head has a sacred place which he uses to perform ceremonies such as doro remusoso (rainmaking ceremony), for example in Nyeredzawanda village we have a place called Ma[illegible]uwe, which has got some stone enclosures. People are not allowed to cut trees or collect fire wood* **C3**
	Use of taboos in plant species conservation	*- Branches of trees such as 'mutarara' and 'murungu' are placed on graves and it's a taboo to cut the tree for firewood or bring it to your home. It will give bad omen like death in the family. Trees like 'mukonde' are associated with the ancestors. Ceremonies like 'musoso' (rain making ceremony) are often done under this tree.* **T7**

and land resources. Taboos are used to conserve and sustainably exploit fruit trees such as *muzhanje* and *muchakata* medicinal plants. The taboos ensure that there is no wastage of resources. Only ripe fruits are harvested after performing *mutoro* (harvesting ceremony) in order to ensure the fruits are harvested when they are ripe to reduce wastage. A thanksgiving ceremony held is to ensure that the ancestors continue to provide such fruits in future.

The selective extraction of herbs for medicinal purposes is another strategy used by the Chimanimani community in the sustainable use and management of natural ecosystems. The herbalist either selected the roots, bark or leaves for treatment of patients. This ensured the plant continues to survive and used in the future. The medicinal plant parts can be consumed several times before being disposed of. Few selected individuals in a clan are knowledgeable in the use of medicinal plants.

In the Cape Coast of Ghana, Anoliefo et al. (2015) observed that it was a taboo to cut down fruit trees. In Zimbabwe, power shortages and the high cost of electricity has caused rampant deforestation of forests for firewood in both urban and rural areas. However, the belief by the indigenous people that cutting down fruit trees can cause calamity such as smoke and invasion of field crops by baboons has promoted the conservation and sustainable use of fruit trees in Chimanimani.

In the study area, Village heads also perform *musoso* (rainmaking) ceremony under the *muonde* or *muhacha* tree, hence the trees are not cut for domestic use. Other trees associated with the dead such as *mutarara* survive from deforestation because the community believes that cutting the tree for use in the home can cause the death of a family member. *Mutarara* is placed at the top of the grave after burying the deceased in order to confuse the witches who may want to come and collect the buried body.

Sacred places are believed to be inhabited by ancestral spirits. The sacred places include burial places for Chiefs, which are usually mountain tops; forests where ceremonies such as *musoso* (rainmaking ceremony) are performed uner individual trees such as *muonde* and *muhacha.* In Chimanimani, the sacred places include Maruwe, Shara and Nyamwazha forest areas. The sacred places have developed thick forests that are inhabited by a great diversity of flora and fauna. People are not allowed to cut down trees, collect firewood or hunt wild animals in the forest lest the community is befallen by a calamity such as drought.

Ngara and Mangwizvo (2013) found that the Shangwe community in Gokwe performed rainmaking ceremonies under trees such as *muchakata* and *mubvumira* in the Chidoma hills. The places are regarded as rain spirit shrines where rain spirits inhabit. The Chidoma hills have thick forests that have become home to a range of flora and fauna since it is taboo to interfere with the biome. Similarly, Risiro et al. (2013) observed that the Chikona hills in the Zaka District of Masvingo is regarded as sacred since the hills have got some caves where Chiefs are buried. Further, rainmaking and *mutoro* (fruit and insect harvesting) ceremonies are performed in the hills. There is a strong belief that the ancestral spirits inhabit in the mountains, hence no one destroys the forests for fear of reprisals by the spirits in the form of drought or reduced fruit and insect production if the ancestors get angry. Elsewhere, in North East India, Khan et al. (2008) found that plants and animals have been conserved in sacred grooves.

The findings are in sync with the United Nations Declaration on the Rights of Indigenous Peoples (UNDRIP 2007) Article 12 that gave rights to the indigenous people to practise their spiritual traditions and ceremonies. Further, it is observed that indigenous environmental management practices strive to achieve SDG 14 and 15 whose thrusts are to conserve and promote the sustainable use of resources under water and on land surfaces respectively. The inclusion of these indigenous practices in the teaching of Geography will therefore inculcate a culture of sustainable use of resources among the learners.

6.2 Indigenous Environmental Practices on Animal Species

The section presents how totems and taboos are used in environmental management and conservation of animal species in Chimanimani community (Table 2).

Table 2 Indigenous environmental practices on animal species

Main theme	Sub-themes	Quotes
Indigenous environmental practices for animal species	-Use of totems	*Every clan has a totem in form of an animal or part of it. My totem is 'Zhou', I don't eat an elephant, otherwise I will lose my teeth.* **T9**
	-Use of taboos	*-Animals like python and 'harakabvuka' are not permitted to be killed. It is only the Chief who should eat meat from these animals.* **C5** *-Killing of frogs in water can result in one developing some pimples in the face and cause drying of water in the well or river* **T4**

Most animal species are a totem to a certain clan. A clan has got a totem they identify with and relate to themselves. People do not kill, harm or eat an animal belonging to their totem. This ensures that in an area where, for example, the Zhou clan is dominant, elephants flourish in numbers. In Kenya, Ayaal and Waswa (2016) found that people respected animals that belonged to their totem such that they would not harm or kill the animal.

Animals including endangered species like the python and small defenceless creatures have also been conserved by observing taboos against killing these animals. It is believed that, killing of small and defenceless animals could cause some incurable pimples to an individual offender or drought in the whole community. Diawuo and Issifu (2015) showed that observing taboos preserved endangered species and small and defenseless animals from extinction. Further, Risiro et al. (2013) in a study in Masvingo found that aquatic life such as frogs were not supposed to be killed since this could cause drying of water sources.

6.3 *Indigenous Environmental Practices on Land Resources*

The indigenous environmental practices used into conserve land resources in the study are presented in Table 3. The land resources presented include water, soil, landforms and land use planning.

Water related taboos have kept water unpolluted in some pools in Chikwakwa and Mhakwe rivers. People desist from bathing and washing clothes with detergents like soap in the pools. Further, it is a taboo to fetch water from the wells using smoke-coated utensils other than traditional clean gourds locally known as *tsani/mukombe*. Non-observance of the taboos can lead to calamities such as drying of the well or the offenders may disappear in the pools.

The Teso community in Kenya believed that, rivers, marshes and swamps are habitats for ancestral spirits and are not supposed to be tempered with (Ayaal and

Table 3 Indigenous environmental practices on land resources

Main theme	Sub-theme	Quotes
Indigenous environmental practices on land resources	- Sustainable use of water resources	*-There are some pools in Chikwakwa and Mhakwe rivers where mermaids are said to reside. there are records of people who have disappeared and later found dead in these pools after bathing or swimming in the pools.* **T6** *-People are not allowed to fetch watch using modern utensils coated with suit in some of the wells. If that is done the water may dry. Only traditional gourds(tsani) are supposed to be used to fetch water in the wells.* **T3**
	- Sustainable use of soil	*- This area has got a lot of stones so we practice stone ridging (marozhi) to reduce our soil being washed away when it rains.* **C1** *-We use organic manure (murakwani) and animal manure which we spread in the field.* **C4** *-Here in Rusitu valley, the community protect the soil from erosion by planting banana trees along the ridges.* **T1** *-Under Chief Satiya there is chisi (resting day) once in every month. No one is supposed to do any form of work in the fields or collecting firewood in the forest. One can be fined a goat if seen working in the field on 'chisi day.* **T5**
	-Conservation of landforms	*-In some parts of Runhenga and Murwangemhanga mountain, there are sacred places where Chiefs are buried in caves. People in the area are not allowed to cut down trees, practice any land use or practising immoral behaviour. If one disobeys the taboos, he/she can disappear***T8**

(continued)

Table 3 (continued)

Main theme	Sub-theme	Quotes
	-Land-use planning and food security	*-We practise land fallowing and mixed farming to allow the soil to gain its fertility The crops mature at different times thus the soil is protected from erosion.* **T6** *- We have dura (crop storage hut) that is smeared with cow dung to prevent termites from eating harvested crops, we also apply 'mushani' (lipica javanica) to prevent pests from eating our crops. Selected seeds are smoked in the kitchen to protect them against termites and maize stock borers* **C3** *-Food stuff is preserved through sun drying of vegetables, 'watinoti mufushwa'. This enables us to have food up to the next season.* **T4**

Waswa 2016). Emmanuel (2020) concurs that the respect of taboos has kept water unpolluted in Ghana. The use of *tsani* (gourd) ensured the sustainability of water resources to the community. *Tsani* is a small gourd used to collect water such that it is cumbersome to collect large amounts of water. Water is therefore collected sparingly for other consumers to collect the water. Further, the use of *tsani* ensures no wastage of the water resource as there is limited water that can spill off from it. The use of smoke-coated gourds pollutes the water unlike using tsani which is always clean to collect water for domestic use.

Water related taboos across the indigenous communities are important in ensuring that there is sustainable use of water resources and sanitation (SDG6). The indigenous community of Chimanimani ensured that water resources are protected from pollution by respecting taboos related to water bodies. The researcher argues that the inclusion of indigenous practices in the Geography curriculum can be an environmental management strategy of instilling values of water conservation within learners and communities thereby achieving SDG6 by 2030.

The Chimanimani community promotes sustainable use of soil resource through observing *chisi* (resting day), construction of *marozhi* (stone ridging) and application of *murakwani* (humus) and animal manure to the soil. *Chisi* reduces over exploitation of natural resources since no one is allowed to work on land on *chisi* day. Offenders are made to pay a goat or an ox by the local Village Head or Chief. *Marozhi* are constructed across the slope. This reduces surface runoff and soil erosion. Further, in areas like Rusitu, Ndima and Mutsvangwa, the local people plant banana crops and pineapples to anchor the stone ridges and to allow the crop residues (humus) to add fertility to the soil. Similar findings have been recorded by Mapara (2009) among the Tangwena people of Nyanga who practise *marozhi* in order to reduce soil erosion.

The local people of Chimanimani conserved biodiversity through land use planning and crop preservation strategies. The community practise fallowing and mixed farming to allow the soil to regain soil fertility. Land fallowing encourages grass and plants to regenerate and provide humus to the soil while mixed farming ensures that there is continuous soil cover that reduces surface runoff and soil erosion. Reiss and Chapungu (2000) argue that inter-cropping is believed to reduce pest damage by 80%. Similarly, Magni (2017) observed that indigenous people reduced damage to crops and animals through modifying farming practices. In Thailand, the Karen community practise shifting cultivation such that the soil is given enough time to regain fertility. The community of Chimanimani adopted land use practices (*chisi*; use of *murakwani*; *marozhi*; preservation strategies) that ensured that there is protection and sustainable use of land resources (SDG15).

Mountains such as Runhenga and Murwangemhanga, where Chiefs are buried are sacred places believed to be inhabited by the ancestors. Such mountains are protected from deforestation, farming and hunting of wild animals. The mountains have become habitats for a diversity of flora and fauna. People who have behaved immorally, demeaned the ancestors or cut down trees in the sacred mountains have been reported to have disappeared in those mountains. Indigenous people believe in spiritual practices to ensure biological diversity and sustenance of ecosystems (Luzar and Fragoso 2013). The belief in the ancestral spirits inhabiting the sacred places has conserved biodiversity in burial places in the mountains such as Chidoma hills in Gokwe (Ngara and Mangwizvo 2013) and burial groves in Kenya (Ayaal and Waswa 2016).

In order to ensure food security, the Chimanimani community preserved their crops in *dura*. Vegetables are also boiled and sun-dried. Seeds for the next season are smoked in the kitchen to prevent termites and other pests from eating them. These indigenous practices conform very well to SDG2 that strives to achieve food security. Mashoko (2018) found that in the Chivi community in Masvingo the local people used indigenous ways such as drying and salting to preserve foodstuffs.

6.4 *Indigenous Environmental Practices on Weather and Climate Changes*

In this section, indigenous ways of weather forecasting, rainmaking ceremonies and growing of small grains are presented to show how they are used in environmental management.

The Chimanimani community relied largely on subsistence farming for their survival. Accurate weather forecasting is important for them since it influences land-use planning, crops to grow and planning for animal grazing. The sacred forests of Shara and Nyamwazha have been conserved and have got flora and fauna useful for indigenous weather forecasting. The sacred forests have become home for biodiversity since tempering with the forests is prohibited by the spirit mediums and local Chiefs.

The forests are also preserved as places to perform traditional ceremonies such as *musoso* (rainmaking ceremony). During the ceremony, the spirit medium would request the ancestors to grant them adequate rains and protection from pests, diseases and weather hazards that may destroy the land, crops and animals. In Western Kenya, the Nganyi shrine has a large biodiversity of flora and fauna preserved for weather forecasting by the Nganyi clan (Gumo 2017). The Borana community in South Ethiopia practise weather forecasting so that the cattle herders can scout for areas that are not affected by drought. Similarly, in Mozambique small farmers depend on indigenous weather forecasts to plan farming activities (Salite 2019). The use of indigenous forecasting by Chimanimani community is in line with SDG13 that aims at reducing the impact of climate change.

6.5 Integrating IK on Environmental Management Practices with the Form 1–4 Geography Syllabus (2015–2022)

The Form 1–4 Geography syllabus (2015–2022) was analysed in order to identify topics that could be integrated with IK on environmental management practices. The section answers research question 2. The topics (main concepts) and content (sub-concepts) are presented in Table 5. The indigenous environmental management practices (Table 5) that could be integrated in teaching the Geography topics are based on the discussions with the participants whose findings are presented in Tables 1, 2, 3 and 4.

Table 5 shows that the Form 1–4 Geography syllabus (2015–2022) had topics that could be integrated with IK on environmental management in the teaching of Geography. On the topic on weather and climate, indigenous ways of forecasting can be integrated to produce more accurate results. In view of global climate change and the need for food security, indigenous farming practices such as growing small grains, use of *murakwani* (organic manure), *marozhi* (stone ridging), fallowing and food preservation methods such as smoking, sun drying should be promoted through integrating them in the Geography syllabus.

The topics on ecosystems, natural resources and environmental management in the Form 1–4 Geography syllabus can be effectively dove-tailed with indigenous environmental practices such as use of taboos and totems in the conservation of biodiversity (Table 5). The indigenous people have effectively managed to conserve forests such as Shara, Nyamwazha and Muruwe through their belief in the spiritual world. Taboos related to water resources have ensured reduced pollution in rivers such as Chikwakwa and Mhakwe. The participants believed that observance of taboos,

Table 4 Indigenous environmental practices on weather and climate changes

Main theme	Sub-theme	Quotes
Weather, climate and climate change and mitigation	-Indigenous weather forecasting Rainmaking ceremonies	*-Indigenous weather forecasting guides the farmers on land preparation and types of crops to grow. If suitable crops are grown, they provide cover to the soil thereby protecting it from degradation.* **T7** *-The Satiya clan has a sacred forest called Shara with types of trees useful in weather forecasting. Such sacred forests are protected by the Chief from degradation.* ***T5*** *-Our grandfathers used to go to Chief Satiya with some few coins to request for some rains during the drought period. A team from Satiya will then go to Rain Spirit medium who is Chief Mutema. On their way back from Mutema it will definitely rain to rub off footprints for the rain beggars This ensured we had adequate crop yields and grass and water for our livestock and animals.***C2** *-Chiefs and headman have sacred places where they offer musoso/mukwerere (rain making ceremony beer) to the ancestors as they request for rains. During the ceremony the Spirit medium will request the ancestors to provide some rains, good health, good crop yields and livestock as well as protection from environmental hazards.* **T10**

chisi and performance of rainmaking ceremonies promoted environmental management practices among the indigenous people of Chimanimani, hence they should be integrated in the Geography curriculum.

Despite a range of topics which can be integrated in the Geography curriculum, there is very little IK integration in the Geography syllabus. There are twelve (12) topics in the Form 1–4 Geography syllabus (2015–2022). However, there are only two (2) topics where IK is mentioned in the Geography syllabus namely, weather and climate and environmental management. Ogunniyi (2016) and Pedzisai (2013) have acknowledged that there is minimum application of IK in the classroom teaching. UNDRIP (2007) emphasise the right of the indigenous people to revitalise their cultural traditions (Article 11) and the right to teach their spiritual traditions (Article

Table 5 Integrating IK on environmental management with the geography syllabus

Form 1–4 geography syllabus (2015–2022) topic(s)	Content	IK on environmental management to integrate
Weather and climate	-Weather forecasting -Weather hazards	-Indigenous weather forecasting -Rainmaking ceremonies
Landforms	-Benefits of landforms	-Sacred places in mountains
-Ecosystems	-Biomes -Sustainable management of ecosystems -Conservation of ecosystems	-Taboos on fruit trees -Taboos on medicinal trees -Sacred forests -Myths on use of plant species
-Natural resources	-Conservation of resources -Wildlife management	-Taboos on endangered plant species
-Environmental management	-Climate change and adaptation -Environmental legislation -Environmental management legislation -Land use planning	-Indigenous ways of weather forecasting -Rainmaking ceremonies -Role of Chiefs in resource conservation -'Chisi' (resting day) -Fallowing -Mixed farming -*Marozhi* (stone ridging)
-Agriculture and Land use	-Food security -Climate change and Agriculture	-Indigenous ways of preserving food (eg. Smoking, sun drying, salting) -Growing of small grains

12) as well as the right to conserve and protect the environment (Article 29). It is prudent therefore to integrate taboos, totems, ceremonies, spiritual and cultural practices (Table 5) in the Form 1–4 Geography syllabus (2015–2022) for the sustainable management of the environment.

Learners bring to the classroom knowledge from their communities on indigenous environmental management practices. According to Vygotsky (1978), a child acquires knowledge through interactions with his or her environment. The integration of the learner's background and beliefs helps the learner to comprehend new concepts taught in the classroom (Castagno and Braboy 2008; Shizha 2014). The teaching of IK on environmental management practices in the Geography class helps the learners to understand issues on environmental management.

The discussion has revealed that IK environmental management practices can be integrated with many topics in the Geography curriculum. The inclusion of indigenous environmental practices in the Geography curriculum could promote values and attitudes of environmental conservation of biodiversity thereby possibly achieve SDGs on under water life, land resources, food security and adaptation to climate change by 2030.

7 Conclusions

The contribution of this paper to new knowledge is that, it has established indigenous environmental management practices implemented in Chimanimani which can be adopted by other communities in order to achieve sustainable development goals on the preservation and conservation of resources. The indigenous people of Chimanimani believe in the spiritual world, totems, taboos and indigenous land-use planning and governance that promote sustainable use of resources. Secondly, the study has contributed new knowledge on how indigenous environmental practices can be integrated with the teaching of Geography curriculum in order to enhance sustainable utilisation and conservation of resources. More IK on environmental practices can be integrated with the largely Eurocentric Geography syllabus in order to promote values, attitudes and cultural practices within learners and communities that enhance environmental conservation and sustainable use of resources. The paper therefore argues for the integration of Indigenous and Eurocentric models in the sustainable use and conservation of resources in order to achieve the 2030 SDGs on the preservation, restoration and conservation of resources.

8 Recommendations

Based on the findings of the study the following is recommended:

- There is need for more awareness of indigenous practices on environmental management so that indigenous practices complement modern measures on environment management. This ensures that there is sustainable use of resources in the communities. Awareness can be done through IK integration in the Geography syllabus or holding workshops in the communities on indigenous environmental management practices.
- There is little IK on environmental management practices in the Form 1–4 Geography syllabus (2015–2022). There is need to review the curriculum in order to incorporate indigenous environmental management practices.
- More research can be carried out on IK integration in the classroom so that indigenous knowledge on environmental practices can be documented and integrated in the teaching of Geography in secondary schools.
- In future, more research could be carried out on how Indigenous models on environmental conservation could be integrated with Eurocentric conservation models in order to enhance environmental conservation and sustainable use of resources.

References

Anoliefo GO, Nwokeji PA, Ikhajiagbe E (2015) Influence of traditional taboo practice on natural resource conservation in Uli, Ihiala local government area of Anambra State Nigeria; sustainable community development. J Environ Sustain 4(4):1–14. JES.007.02

Ayaal DD, Waswa F (2016) Role of indigenous knowledge systems in the conservation of the bio-physical environment among the Tesc community in Busia County-Kenya. Afr J Environ Sci Technol 10(12):467–475. JES.007.02

Barnhardt R, Kawagley AO (2005) Indigenous knowledge systems and Alaska native ways of knowing. Anthropol Educ 36(1):8–23

Bourdillon MF (1976) The shona peoples: an ethnography of the contemporary Shona with special reference to their religion. Mambo Press, Gweru

Burger J (1990) The gaia atlas of first peoples: a future for the indigenous world. Penguin Books, pp 40–44

Castagno AE, Braboy BMJ (2008) Culturally responsive schooling for indigenous youth: a review of the literature. Rev Educ Res 78:941–993

Chenje M, Sola L, Paleczny D (1998) The state of Zimbabwe. Ministry of Mines, Environment and Tourism, Harare

Cohen L, Manion L, Morrison K (2018) Research methods in education, 8th edn. Routledge, London and New York

Creswell JW (2014) Research design: qualitative, quantitative and mixed approaches. Sage Publications Ltd., California

Creswell JW (2013) Qualitative inquiry and research design: choosing among five approaches. SAGE, Los Angeles

Cushner K, McClelland A, Safford P (2012) Human diversity in education: an intellectual approach, 7th edn. McGraw Hill, New York

Diawuo F, Issifu AK (2015) Exploring the African traditional belief syastems in natural resource management in Ghana. J Pan Afr Stud 8(9):115–131

Emmanuel AE (2020) The role of taboos in solving contemporary environmental degradation in Ghana: the case of cape coast metro. Soc Sci 9(4):89–97. https://doi.org/10.11648/j.ss.20200904.11

Filho WL, Azeiteiro U, Alves F, Pace P, Mifsud M, Luciana Brandli L et al (2018) Reinvigorating the sustainable development research agenda: the role of the sustainable development goals (SDG). Int J Sustain Dev World Ecol 25(2):131–142. https://doi.org/10.1080/13504509.2017.1342103

Government of Zimbabwe (1999) Report of the presidential commission of inquiry into education and training. Government Printers, Harare

Government of Zimbabwe (2005) National action plan of Zimbabwe: education for all towards 2015. Government Printers, Harare

Gumo S (2017) Praying for rain indigenous systems of rainmaking in Kenya. The ecumenical review world council of churches. Wiley, pp 386–397. https://doi.org/10.1111/erev.12301

Houser RA (2015) Counselling and educational research: evaluation and application. 3rd edn. Sage Publications, California

Hwami M (2014) Education of the peasantry in Zimbabwe as internal colonialism. SAGE 1–10. https://doi.org/10.1177/2158244014536743

Jaware PR (2012) Post-colonial analysis of 'riot'. In: Proceedings of national seminar on postmodern literary theory and literature, 27–28 January 2012

Kaya HO, Lyana A (2014) Knowledge and perceptions of rural communities on world food resources consumption in Tanzania. J Hum Ecol 48(1):53–60

Khan ML, Khumbongmayum AD, Triphathi RS (2008) The sacred grooves and their significance in conserving biodiversity: an overview. Int J Ecol Environ Sci 34(3):277–291

Khupe C (2014) Indigenous knowledge and school science: possibilities for integration. A Ph.D. thesis. Faculty of Science, University of Witwatersrand, Johannesburg

Luzar JB, Fragoso JM (2013) Shamanism, christianity and culture change in amazonia. Hum Ecol 41:299–311

Magni G (2017) (2017) Indigenous knowledge and implications for the sustainable development agenda. Eur J Educ 52:437–447

Mahlatsi M (2017) Botho/Ubunhu philosophy: education from childhood to adulthood in Africa. Int J Sci Technol Res 6:94–98

Mapara J (2009) Indigenous knowledge systems in Zimbabwe: juxtaposing post-colonial theory. J Pan African Stud 3(1):139–155

Masaka D (2016) The impact of western colonial education on Zimbabwe's traditional and post-colonial educational systems. Published thesis. University of South Africa

Mashoko D (2018) Integrating indigenous knowledge of food preservation with school science. Ph.D. thesis, University of Witwatersrand, Johannesburg

Mavhunga F (2006) Africanising the school curriculum. A case for Zimbabwe. In: Zindi F (ed) The Zimbabwe Journal of Educational Research, vol 18, no 3

Ngara R, Mangwizvo RV (2013) Indigenous knowledge systems and the conservation of natural resources in the shangwe community in Gokwe District, Zimbabwe. Int J Asian Soc Sci 3(1):20–28

Niesenbaum RA (2019) The integration of conservation, biodiversity and sustainability. Sustainability 11:4676. https://doi.org/10.3390/su11174676

Obiora AC, Emeka EE (2015) African indigenous knowledge system and environmental sustainability. Int J Environ Prot Policy 3(4):88–96. https://doi.org/10.11648/j.ijepp.20150304.12

Ogunniyi MB (2016) Explicating the philosophy of ubuntu into science education: a project experience. In: The proceedings of SAARMSTE 24th annual conference at Tshwane University of Technology, South Africa, 12–15 January 2016, pp 417–431

Patton M (2015) Qualitative research and evaluation methods, 4th edn. Sage Publications, Thousand Oaks

Pedzisai C (2013) Teachers' perceptions on inclusion of agricultural indigenous knowledge systems in crop production: a case study of Zimbabwe's ordinary level agriculture syllabus (5035). J Biol Agric Health Care 3(16):37–44

Pham L (2018) Qualitative approach to research, a review of advantages and disadvantages of three paradigms: positivism, interpretivism and critical inquiry. University of Adelaide: ResearchGate. https://doi.org/10.13140/RG.2.2.13995.54569

Reiss M, Chapungu J (2000) Environmental biology. Cambridge University Press, New York

Risiro J, Tshuma D, Basikiti A (2013) Indigenous knowledge systems and environmental management: a case study of Zaka District, Masvingo Province, Zimbabwe. Int J Acad Res Prog Educ Dev 2(1):19–39

Said E (1978) Orientalism. Pantheon, New York

Salite D (2019) (2019) Traditional prediction of drought under weather and climate uncertainty: analyzing the challenges and opportunities for small-scale farmers in Gaza Province Southern Region of Mozambique. Nat Hazards 96:1289–1309. https://doi.org/10.1007/s11069-019-03613-4

Sarma C (2016) Indigenous practices and sustainability. Indian J Sustain Dev 2(2):55–63

Shizha E, Kariwo T (2011) Education and development in zimbabwe: a social, political and economic analysis. Sense Publishers, Rotterdam

Shizha E (2006) Legitimising indigenous knowledge in Zimbabwe: a theoretical analysis of post-colonial school knowledge and its colonial legacy. Youth and Children's Study, Paper 2. https://works.bepress.com/e_shizha/11

Shizha E (2014) Rethinking contemporary Sub-Saharan african school knowledge: restoring the indigenous African cultures. Int J Cross Disciplinary Subjects Educ (IJCDSE) Special Issue 4(1):1870–1878

Subedi B, Daza SL (2008) The possibilities of post-colonial praxis in education. Race Ethn Educ 11(1):1–10. https://doi.org/10.1080/13613320701845731

United Nations (2007) Declaration on the Rights of Indigenous Peoples (UNDRIP). https://www.un.org/esa/socdev/unpfii/documents/DRIPS_en.pdf en.pdf
Vygotsky LS (1978) Mind in society: the development of higher psychological processes. Harvard University Press, Cambridge
World Commission on Environment and Development (1987) Our Common Future. Report Published as Annex to Document A/42/427: Development and International Co-operation: Environment Transmitted to General Assembly. https://www.un-documents.net/wced-ocf.htm

Sustainable Future for Early Childhood: Applying the African Ubuntu Philosophy to Contribute to the Holistic Development of Young Children

Mariëtte Koen

> *Africans believe in something that is difficult to render in English. We call it ubuntu, botho. It means the essence of being human. You know it when it is there and when it is absent. It speaks about humaneness, gentleness, hospitality, putting yourself out on behalf of others, being vulnerable. It embraces compassion and toughness. It recognizes that my humanity is bound up in yours, for we can only be human together.*
> Archbishop Desmond Tutu (Tutu and Tutu 1989: 71)

Abstract The world is under enormous pressure as it faces major environmental, health, social, political and economic threats. Although disasters and challenges can, at any time, pose a threat to the sustainability of any nation or continent, it seems as if children happen to be hardest hit. Using a theoretical approach, this chapter systematically reviews research studies on early childhood education to explore the way the African Ubuntu philosophy could contribute to the holistic development of young children. In this chapter, the roles of *education for sustainability* (EfS) and *early childhood education and care for sustainability* (ECECfS) are recognised as essential means of creating sustainable building blocks for adult lives. EfS comprises the holistic development of the inseparable social, emotional, cognitive and physical facets of child development in order to form behaviours towards preserving the environment and towards promoting fairness and social justice for life. Given that the holistic development comprises both on the whole child and the environment, this chapter explores how an awareness of the principles of the African Ubuntu philosophy, constituting principles such as sharing, respect and interaction may be used to young children with values, skills and knowledge that form the basis of sustainable development. The aim is not to see Ubuntu as an ideal agenda to the exclusion of Western knowledge systems, but rather to integrate ideas from both

M. Koen (✉)
School of Psycho-Social Education (Learner Support), North West University Potchefstroom, Building B11: Office G09, Private Bag x6001, Internal Box 539, Corner Malherbe and Esselsen Streets, Potchefstroom 9321, South Africa
e-mail: Mariette.Koen@nwu.ac.za

W. Leal Filho et al. (eds.), *Sustainable Development in Africa*, World Sustainability Series,
https://doi.org/10.1007/978-3-030-74693-3_8

worlds in the process of seeking ways to construct a just and sustainable world in early childhood.

Keywords Early childhood · Sustainable · Ubuntu · Holistic · Quality · Education

1 Introduction

The world is under enormous pressure as it faces major environmental, health, social, political and economic threats. Following a year of civil unrests, virus outbreaks and climate crises like high temperatures, floods, severe droughts and raging bush fires it is evident that Planet Earth is in a deplorable state. In fact, the United Nations environment chief recently declared that the natural world is sending us a wake-up call in the form of the coronavirus pandemic (Carrington 2020). The media has moreover provoked an increased awareness of human activity that should be held responsible for the environmental destruction and the damaged relationship between humans and nature. As a consequence, nations are being required to rethink ways of building sustainable futures for young children (Lehtonen et al. 2018).

Georgeson (2018) points out that a sustainable future requires focusing the attention on early childhood and care for sustainability towards a holistic approach to the environment and society. At the heart of sustainable development lies quality education (MacDonald 2015). The contested nature of the terms *education for sustainability* (EfS), *environmental education* (EE) and *early childhood education and care for sustainability* (ECECfS) are roundly acknowledged within literature (Hill et al. 2014). In this chapter, the roles of EfS and ECECfS are recognised as essential means of creating sustainable building blocks for adult lives (Warwick et al. 2018). EfS comprises the holistic development of the inseparable social, emotional, cognitive and physical facets of child development in order to provide a solid foundation for healthy and sustainable futures (Zhou et al. 2016). Young children need to be provided with experiences to develop their basic values, attitudes and skills and to form behaviours towards preserving the environment and towards promoting fairness and social justice for life (Pramling Samuelsson and Siraj-Blatchford 2015; Siraj-Blatchford 2018). The World Health Organization [WHO] (2018:2) supports this idea by stating that "if we change the beginning of the story we change the whole story". Researchers believe that in order to realise quality education it will be essential to sensitise children to their lifestyle choices, to develop autonomous thinking, to foster both cultural transformation and tolerance between people, and to develop children into responsible citizens who will help to preserve their own futures (Taylor et al. 2015). A transformative mindset of this kind includes a broader context, namely a holistic developmental view that is externally supported by the community, school, teachers, mentors, parents and guardians (Morgan and Waite 2018).

By embedding the holistic development of children in an integrated collective Ubuntu approach, the aim of the chapter is to explore how respect and values for the good and benefit of society may be embraced and sustained. Ubuntu has gained recognition through the writings of Archbishop Desmond Tutu (1999 154) who believes that "no one can be human alone". Oppenheim (2012) maintains that Ubuntu provides much scope for people to realise their full potential both as individual or as part of the society to which they belong. Although Ubuntu is often overlocked in Western thinking in that it is considered to be an ancient philosophy, it has significance in education and can contribute to sustainable development today (Shumba 2011). Dolamo (2013) points out that, no matter whether one is African, Western or Asian, everyone is capable of living more compassionately with others. He moreover states that all should join hands to sustain our world. In this process, Ubuntu could be employed as the organising principle.

Using a theoretical approach, this chapter systematically reviews research studies on early childhood education to explore the way the African Ubuntu philosophy could contribute to the holistic development of young children. First, an overview on education for sustainability in early childhood is explored. A research based discussion is then given to explore how the holistic development of young children can be embedded in the African Ubuntu philosophy. Lastly a relationship model is used to investigate how the different layers of relationships could contribute to the all the facets of holistic development in the early years. The aim of this chapter is not intended to advocate the Ubuntu philosophy as an ideal agenda to the exclusion of Western knowledge systems. The goal is rather to integrate ideas from both worlds in the process of seeking ways to construct a just and sustainable world in early childhood. It is acknowledged that Ubuntu is unable to solve all the risk factors encountered in early childhood development—be they challenges of a socioeconomic, political, social or cultural nature. The discussion aims to acknowledge elasticity with which Ubuntu cuts through the spheres of quality education in early childhood by focusing on an integrated way of supporting the holistic development of young children, one that is able to contribute to building a sustainable society. The goal is to present a philosophy that would be applicable in different cultural contexts and life situations by focusing on the holistic development of young children and by empowering young children with values, skills and knowledge that form the basis of sustainable development.

2 Education for Sustainability in Early Childhood

Sustainability is a concept first introduced in the Brundtland Report in 1987. It begged for a development plan that would satisfy the needs of the present without compromising the capacity of future generations (Sen 2013; World Commission on Environment and Development 1987). Although the concept *sustainability* is open to different interpretations in various contexts, in the report it is portrayed as a need for awareness around the way nations use natural resources. In the report, the interconnected role of social progress and economic growth for both current and future generations is also emphasised (Boyd and Hirst 2018). In finding possible solutions to the negative environmental consequences of economic growth and globalisation, ideas were particularised on 25 September 2015 when the United Nations approved 17 sustainable development goals (SDGs) with 169 specific targets to be achieved by 2030 (Siraj-Blatchford et al. 2016). The set of 17 interlinked, universal goals, targets and indicators recognises sustainability as a holistic concept covering both environmental dimensions and social, economic, political and cultural agendas for radical reform (Warwick et al. 2018). The adoption of the SDGs is considered to be a turning point in the understanding of what a transformed world would look like. Pramling Samuelsson and Park (2017) point out that Goal 4 on quality education focuses our attention on the importance of using education at every level towards sustainable development.

Over the past decade, EfS has been articulated as the most persuasive way to address human and environmental threats that can affect the holistic well-being in children's early years. According to Wertlieb (2019), ECECfS should be understood as an integration between early childhood education and environmental education, the so-called 'hole' or research gap as identified by Davis (2009:229). Scholars, however, focus our attention on the fact that while educating young children about sustainability happens to be a step in the right direction, a sustainable future requires more than mere content-related knowledge (Ärlemalm-Hagsér and Pramling-Samuelsson 2018). Sustainability requires critical thinking and taking moral responsibility that provoke thinking and acting in relation to a family, school and community (MacDonald 2015). This means that sustainability also includes a socially transformative approach that empowers young children to cope with the changing needs of society in order to create a sustainable world (Alghamdi et al. 2017; Borg et al. 2017; Li et al. 2019; Pramling Samuelsson and Park 2017; Wertlieb 2019).

It is a fact that environmental and social aspects are vital towards accomplishing transformation of the world. Instead of simply seeing young children as sufferers of existing sustainable challenges, they should be engaged, supported, protected and educated to make them resilient and active in terms of pursuing healthy, fair and sustainable futures (Davis 2015). Warwick et al. (2018:28) define the abovementioned tasks in the early childhood education and care years as constituting creative "pedagogies of hope" to safeguard a good life for all. These pedagogies of hope are aimed at the development of the whole learner so as to engage them socially, physically, emotionally, cognitively and spiritually to pursue more sustainable futures.

The holistic development of young children is shaped by the experiences afforded them (Peckham 2017). It means that providing children with opportunities to conceptualise, plan, do and reflect will motivate them to pursue sustainable futures (Warwick et al. 2018). Huggins and Evans (2018) consider the abovementioned actions to be essential in ECECfS with a view to instilling the principles involved in conserving the environment, in recycling and in social justice. Given that the holistic development of children focuses both on the whole child and on the environment, it seems worthwhile to investigate how an awareness of the principles of the Ubuntu philosophy—constituting characteristics such as sharing, social responsibility, compassion, respect, trust, problem-solving, co-operation and interaction—may be used in the ECECfS environment.

3 Embedding Holistic Development in Ubuntu

The rationale behind holistic development lies in the idea that, in ECECfS, young children should be educated to think critically; to construct moral thinking; to be sensitive both to the environment and to other people; to make value judgements; to participate with others; and to act as agents of change within a society—this with a view to creating a more equitable and sustainable environmental, economic and social world (Huggins and Evans 2018). By teaching these life skills to young children, recognition is accorded to the importance of developing not only pro-environmental but also pro-sociocultural behaviour (Warwick et al. 2018). Bondai and Kaputa (2016) add that holistic education aimed at the development of the whole child is arguably enshrined in the Ubuntu philosophy. According to Ngwaru (2014:62), 'wholeness'—a stamp of the Ubuntu philosophy—integrates various facets of life and bundles it up in an indivisible whole. Ubuntu is founded in this indivisible whole and it is debated that through this common tie that binds humans together, people can be empowered to discover their own human strengths.

3.1 Human Worth in Communal Relationships

The family serves to act as the first school for children. It lays the ground for developing attitudes, values, behaviour, habits and skills by promoting awareness for sustainable development (Agut et al. 2014). Although research demonstrates that the parent–child dyad (including primary caregivers) is fundamental for children's well-being, family disintegration is fast becoming a trending issue, especially in poor socioeconomic environments (Gumbi 2019). As a result, an overwhelming majority of young children around the world are deprived of a nurturing and caring family environment (The Organisation for Economic Cooperation and Development [OECD] 2011). Ngwaru (2014) proposes the use of extended involvement beyond the family circle, arguing that it could be the missing link in the holistic development of young children in Africa. The significance of close relationships within families and communities are, however, not limited only to Africa. Davis (2010) reiterates that partnerships in communities provide valuable opportunities to engage children in aspects of sustainability where they can practice what they learn It appears as if the Ubuntu philosophy increasingly qualifies as a viable tool for recognising the human worth in communal relationships while also respecting values, the natural environment and the resources it provides (Shumba 2011).

The Ubuntu philosophy forms part of how African communities raise their children, the key message being that people should care for others as they do for themselves (Mugumbate and Chereni 2019). Ubuntu, translated as 'humanness' in English, dates back to precolonial ways and is seen as a cultural strength that constitutes the African way of life (Le Grange 2012; Msengana, 2006; Van Breda 2019) This philosophy derives from proverbial expressions found in most African cultures. So, for instance, the isiZulu phrase 'umuntu ngumuntu ngabantu' translates as a 'person is a person through other people' (Shumba 2011:90). Phrases such as these and other common sayings like 'I am because we are' or 'it takes a village to raise a child' acknowledge the human worth in communal, culturally connected African societies (Chikunda and Ngcoza 2017:84).

Harmonious interactions between the child, family members and those outside the family circle inculcate respect for others and are indispensable in terms of realising human excellence (Metz 2016; Mugumbate and Chereni 2019). Although it is recognised that it may be problematic to apply Ubuntu as an ideal framework in education for sustainability (Sheya 2014), embracing an Ubuntu approach in early childhood development may provide opportunities to explore how African values and practices may be able to extend the sustainability agenda.

3.2 Relationship Circle

The idea of a united relationship is a core element of Ubuntu and it is claimed that when young children develop an understanding of the interconnectedness between people and the environment they may feel empowered to create sustainable change. Davis (2008) maintains that sustainable change is only possible if interactions are extended beyond the individual child. In agreement with the idea of an extended relationship, Mugumbatu and Chereni (2019) have identified five relationship levels, which include the individual, the family, the community, the environment and the spiritual level as components of an integrated framework of Ubuntu. These authors explain how young children are decentred as the prime unit of analysis. The focus of attention should rather be on the development of the child in a community of shared life. As depicted in Fig. 1, the focus in this nested relationship model is on the *we* rather on the *I*. Museka and Madondo (2012) add that although the focus of Ubuntu is on a shared approach, it also recognises the holistic approach to educating children. In fact, they argue that holism is seen as a fundamental aspect in the 'fine-tuning' of development for a sustainable future.

Figure 1 demonstrates the relationship between the child and the other levels as one of interdependence. The different layers influence the holistic development of the young child and what happens in a particular layer also affects the other layers. The young child depends on others and they, in their turn, depend on the child. In this model, the young child is part of a larger system that values the well-being of all people while promoting change for a sustainable future. In maintaining that environmental and sustainable challenges are often too complicated for one

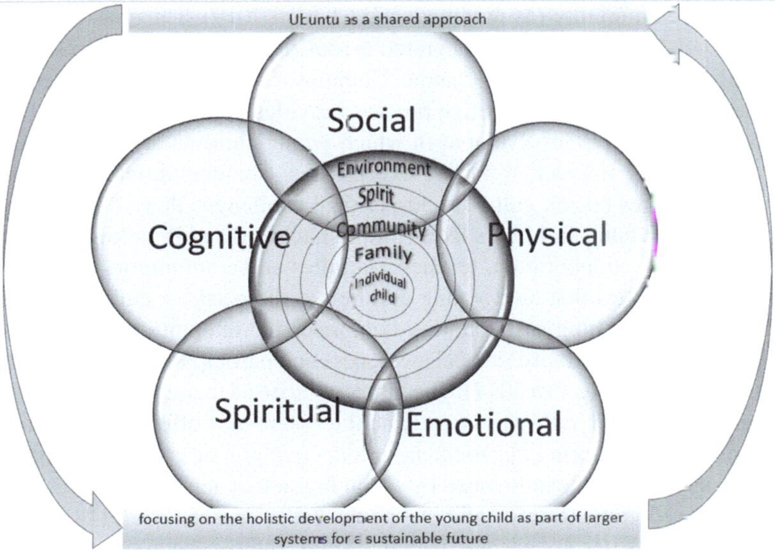

Fig. 1 Integrated framework focusing on developing young children as agents of change for a sustainable future (Adapted from Mugumbate and Chereni 2019: 31)

individual to solve and input from others is required, Shumba (2011) supports this view. In his opinion, the term *ecological intelligence* suggests that it is essential for young children to share knowledge and collaborate with others. Applying ecological knowledge denotes exhausting inputs from all the facets of holistic development, which means that if one of the dimensions is out of balance, it may cause physical, social, emotional, spiritual or cognitive imbalances (Mcunu 2004).

3.2.1 Social-Emotional Development

Social-emotional development denotes an integrative process in which young children start to understand who they are, how to express and manage emotions and establish meaningful relationships with others (Bennett and Palaiologou 2016). According to Zhou et al. (2016), social-emotional skills include being able to utlise a range of intrapersonal and interpersonal skills. These skills include getting along with others, cooperation, regulating and demonstrating emotions, learning to share and to demonstrate empathy, and being able to understand and to follow basic social behavioural rules. The skills further include showing commitment, becoming independent while also being able to work in a group, respecting others, caring for peers, and understanding the local community, the culture and the country. Finally, it is about understanding diversity and equity, and respecting all cultures.

Figure 1 indicates that children form part of a family, families are part of a community and communities are part of the environment, which, in its turn, forms part of a larger spiritual community (Mugumbate and Chereni 2019). The question arises as to how the social qualities in this integrated model can be used to act as a socioemotional catalyst for broader societal change. Ubuntu sociocultural thinking stems from sharing and caring. This participative process provides opportunities for collaborative and communicative involvement in which young children are able to talk and to learn while they interact in shared experiences. Socioemotional sustainability in ECECfS includes social, cultural and political challenges that affect the quality and continuity of young children's lives (Siraj-Blatchford and Pramling Samuelsson 2016). Apart from collaboration, teachers, families and communities are obligated to create opportunities that alert young children to the need for cultural sensitivity (Follari 2019). Embracing sustainability comprises the integration of principles of inclusivity, cultural specificity, critical awareness and participation in a multicultural world (Chikunda and Ngcoza 2017). The aim is not to use Ubuntu as a language game (Van Niekerk 2013) but rather to investigate the potential it offers for education for social transformation during engagement in reflective practices (Waghid 2018).

Social understanding is influenced by social literacies such as prejudice, discrimination and stereotyping (Sensoy and DiAngelo 2017). Jansen (2011:114) terms one of these literacies *bitter knowledge*, which, to him, refers—both verbally and nonverbally—to the way parents (specifically white Afrikaner parents in South Africa) have raised their young children by sending them skewed world views of their privileged upbringing. These views are reinforced by the messages of cultural insensitivity that young children may have received from different contexts like the school, the church, friends or organisations. In addition, cultural sensitivity involves a process whereby young children acquire, model and then apply the necessary knowledge, attitudes and skills to promote social understanding (Folque 2018). If an ethos of compassion, empathy and respect is not modelled to young children, this may result in bitter knowledge perceptions. Globally, this bitter knowledge notion is mirrored in tragic racial-inequity incidents such as the recent demonstrations and unrest following the death of George Floyd in the United States of America. According to Losavio (2020), these protests are not the result of one incident but are rather a manifestation of frustration resulting from a history of racial inequities or, as Pilon (2013:3) cautions, the real glitches "lie deep inside the pot … not in the bubbles". Losavio (2020) agrees by stating that until racial inequity is addressed, other sustainable problems will remain secondary and perpetual. Msengana (2006) sees the role of Ubuntu as one of tolerance and compassion, a philosophy that embraces forgiveness. In his opinion, Ubuntu is able to facilitate a healing process to build nations plagued with problems of cultural diversity. Because he regards it to be a theory that focuses on the recognition of all human beings regardless of their skin colour, all societies therefore have the potential to become skilled in utilising the assets of Ubuntu.

The aim of the above discussion is not to portray Ubuntu as a magic wand that will resolve all sociocultural issues. The aim is to rather debate that, in the light of the holistic development of young children, Ubuntu may develop prosocial knowledge not only to help young children meet current social needs but also to support future generations to maintain a socially healthy community (Ampuero et al. 2015). Msengana (2006) maintains that if societies were to look through the Ubuntu lens and see all people as equal, and not feel threatened by cultures that differ from theirs, this may play a significant role in rebuilding relationships in racially and socially divided societies. Ubuntu offers young children the opportunity to become critical thinkers who embrace diverse cultures (Shepherd and Mhlanga 2014) and who consider consequences before engaging in detrimental behaviours towards others and to the environment.

3.2.2 Spiritual Development

In this chapter, the concept *spirituality* is confined neither to religious beliefs nor to a particular faith. Rather, the spiritual dimension of holistic development in the early years is interpreted as the maintenance of moral values cherished by the community (Shepherd and Mhlanga 2014). A holistic approach for sustainable living includes the ability to make value judgements about destructive behaviour, to change harmful patterns of thinking and to apply positive action skills (Hedefalk et al. 2015). Walker (2017) is indeed concerned that, despite substantial work on sustainability education, considerably less scholarly work has to date been devoted to children's responses to moral and educational messages featuring the environment as an object of concern.

It is believed that morals and values can be instilled through shared communal moral teaching at all levels of society (Masango 2006; Shepherd and Mhlanga 2014). Values and respect for sustainability promote qualities that teach children to be creative, to think and to reflect (Pramling Samuelsson and Park 2017). Pascal (1993) adds that children should be given responsibility in order for them to learn how to act responsibly. She further maintains that teaching young children responsible behaviour will only be effective in a community setting in which there are boundaries for acceptable behaviour and where children learn the consequences of behaviour that does not honour these.

For Africans, spiritual development is holistic and involves an awareness of the whole self and the environment (Seehawer 2018). Mugumbate and Chereni (2019) consider the spiritual development of children to include a representation of the inner self of who the child is, the child's morals, including the values and the culture of the community. Children are considered to be part of the entire surrounding world, comprised of God, the ancestors, the elders, the family and the community (Mugumbate and Chereni 2019). In this holistic approach, Ubuntu is seen as the glue that unites not only present communities with ancestors and descendants but also preceding and future communities (Van Breda 2019). It does however not suggest an exclusive focus on ancestor veneration, but it should rather be interpreted as a means of respecting leaders who live a noble life (Masango 2006). In other words, it is a means of including living people and connecting with them in a respectable way by internalising values, norms and ethics (Masango 2006). Although this does not mean that individuality is forbidden (Van Niekerk 2013), it signifies that community members' moral principles are mostly the product of shared social experiences stemming from past times (Shepherd and Mhlanga 2014).

When reflecting on the consequences of unsustainable practices, the Ubuntu philosophy could offer present and future generations opportunities for learning that positive social behaviour could engender respect, sharing, caring and responsibility (Molefe 2019). In fact, Dolamo (2013) urges that if Africa as a continent is to be saved, Africans have the responsibility to embed themselves in a life based on the norms and values of Ubuntu. Some scholars believe that Ubuntu can assist the young child in his/her search for individual identity by nurturing respect and an ethical stance towards society, the environment and all forms of life in order to secure a sustainable future (Borg et al. 2017; Molefe 2019). This idea links up well with the 2030 agenda for sustainable development, which states that present actions have an impact on the future, implying that present behaviours should be checked or moderated by future possibilities (Van Breda 2019).

3.2.3 Cognitive Development

Spiritual-competence experiences can benefit young children by teaching them different ways of conduct in which strong foundations for sustainable development are formed. Morgan and Waite (2018) confirm that the intrinsic motivation process allows young children to be able to participate as active problem solvers. It is argued that young children should no longer be regarded as empty vessels waiting to be filled with information that will later be retrieved. They need to learn to ask critical questions, to solve problems, to make decisions and to envision more positive and sustainable futures (Tilbury 2011). Ndofirepi (2011) cautions that children cannot be truly critical if they do not know of what or of whom they should be critical. Sustainable development is rooted in teaching young children to be critical within everyday-life practices (Percy-Smith and Burns 2013).

This, however, does not mean that young children should be left on their own to develop critical thinking skills. Facilitation by and support from adults are fundamental towards promoting the role of children as agents of change for sustainability (Percy-Smith and Burns 2013). Walker (2017) points out that while researchers frequently refer to young children as agents of change in households, schools and communities, opportunities should be created that expose young children to first-hand experiences in which they are confronted with sustainable challenges. It is vital that young children learn how critically to address and solve these challenges faced by communities (Munkebye et al. 2020). This means that a community of inquiry can be effective in the sharpening of children's critical reasoning aptitudes to allow them to live in a sustainable way. While decision making in an individualist approach focuses on a voting process to steer everybody in the same direction, the Ubuntu collaborative approach prefers rather to deliberate until consensus is reached (Msengana 2006).

In the Ubuntu collaborative approach, individuality is neither ignored nor repudiated (Takyi-Amoako and Assié-Lumumba 2018). It is believed that the sharing, the togetherness, the focus on the community rather than on the individual should take precedence. Although Ubuntu will surely not resolve all sustainable challenges, it may however provide children with opportunities for participating in discussions and practical activities that concern their future lives. Collective processes allow for knowledge generation in decision making, which would contribute both to own learning processes and to learning in the community (Folque 2018). After all, the goal of holistic development is to develop competencies in young children that will motivate, inform and develop critical abilities to empower them as rescuers to patch up the sustainability errors of past generations (Munkebye et al. 2020).

3.2.4 Physical Development

Holistic development in the early years requires that young children be provided with opportunities to be physically active in their daily routine and that they have a supportive network to facilitate these. Beauchamp (2016) argues that physical well-being in the early years will help to sustain a healthy lifestyle that contributes to social-emotional, cognitive and moral development.

It is widely known that poor socioeconomic environments affect the early stimulation, which is fundamental to development in all dimensions (Takyi-Amoako and Assié-Lumumba 2018:10). Although the Ubuntu philosophy is not a miracle worker able to fulfil young children's basic needs instantly, it can transform empty homes into educational and stimulating environments for young children. Norren (2014) maintains that Unbuntu can be used as a nurturing tool for collaboration and for developing the full potential of each young child. An example of providing nurturing learning dispositions by stimulating young children is apparent in the work of the charity organisation Singawkenza in KwaZulu-Natal, South Africa. Julie Hay, one of the board members of Singakwenza (meaning "we can do it" in isiZulu), explains that although they cannot change the world or even an entire town, it is worthwhile seizing the opportunity to change the future of one child (Singakwenza 2020). They believe that by developing the potential of young children, they can make a difference in society. The focus of their programme is on the development of teachers, practitioners and caregivers at the grassroots level to enhance quality education in the early years. Perceptual stimulation through play is integrated into the process of teaching communities how to make their own toys and educational resources by using recycled material (Singakwenza 2020). According to News 24 (2019), the "Waste 2 Toys" project run by Singakwenza Workshops provided stimulating opportunities for nearly 49 800 children in preparing them for formal schooling while also teaching them the basic life skills necessary for building a sustainable society.

Siraj-Blatchford and Pramling Samuelsson (2016) refer to global projects in which the physical well-being of young children has been addressed in order to develop sustainable futures. It is therefore evident that acting for sustainability requires dynamic problem-solving citizens who act as a responsible society and who are willing to work together, to learn from others and to find creative solutions. Msengana (2006) already warned 20 years ago that Ubuntu education regards life as an indivisible whole emphasising the collective interdependence in all spheres of life. Oladokun (2020) summarises the important role of an Ubuntu plan by arguing that there is no better time for a worldwide, synchronised Ubuntu approach than today. More than ever before, it is time that societies around the world respond to calls for sustainability by strengthening global cooperation. The Ubuntu philosophy demands financial priorities to preserve human self-worth, health, and socioemotional well-being. Oladokun (2020) cautions that "[i]t demands nothing less".

4 Conclusion

In this chapter it is recognised that the world is under enormous pressure as it faces major environmental, cultural, health, and economic threats. As a consequence, countries are being required to rethink ways of building sustainable futures for young children. It is debated that working towards sustainability, require four important considerations. First, promoting and supporting quality early childhood education is essential to enable young children to reach their full human potential. Quality education for sustainable development is seen as a tool where adults could present young children with opportunities in which they can gain knowledge, understand problems, make decisions and act as agents of change in grasping how the environment relates to society. It is reasoned that ECECfS should no longer be regarded as knowledge transmission, but it should engage young children as agents of change to apply values, responsibility and lifestyles that are consistent with a sustainable future.

Second, it is debated that the holistic development in the early years is not only crucial for laying the foundation for well-being, but that it plays an important role in the pursuit of sustainability. The reciprocal, multifaceted and complex relationships in the early years have been highlighted (see Fig. 1). This model suggests that there are layers of interacting systems allowing opportunities for change, growth and development in the socioemotional, physical, cognitive and spiritual domains.

Third, the views of Davis (2010) are supported by arguing that partnerships in communities provide valuable opportunities to engage young children in aspects of sustainable futures where they can practice what they learn. It appears that the Ubuntu philosophy qualifies as a viable tool to engage young children in communal practices. This chapter acknowledges the way holistic education aimed at the development of the whole child is enshrined in the Ubuntu philosophy.

In this chapter it is sketched how an Ubuntu-inspired recognition of wholeness may contribute towards reaching quality education in the early years. It is argued that if the Ubuntu philosophy is embraced, it may result in educational change influencing the socioemotional, cognitive, spiritual and physical development of young children. Finally, this chapter resonates with the words of the former president of the United States of America, Barack Obama:

> "…there is an African word – Ubuntu – that labels the greatest human gift. A recognition that all societies are bound together in ways invisible to the eye, that there is a oneness to humanity, that we achieve ourselves by sharing ourselves with others and caring for those around us" (Williams 2018).

References

Agut MPM, Ull MA, Minguet PA (2014) Education for sustainable development in early childhood education in Spain. Evolution, trends and proposals. Eur Early Child Educ Res J 22(2):213–228

Alghamdi A, Ernest J, Hafiz F (2017) Teaching sustainable practices as part of a holistic education in the Saudi Context. Int J Whole Child 2(2):29–40

Ampuero C, Miranda CE, Delgado LE, Goyen S, Weaver S (2015) Empathy and critical thinking: primary students solving local environmental problems through outdoor learning. J Adventure Educ Outdoor Learn 15(1):64–78

Ärlemalm-Hagsér E, Pramling Samuelsson I (2018) Early childhood education and care for sustainability: historical context and current challenges. In: Huggins V, Evans D (eds) Early childhood education and care for sustainability. International perspectives. Routledge, London, pp 13–28

Beauchamp G (2016). Understanding the world. In: Palaiologou I (ed) The early years foundation stage. Theory and practice, 3rd edn. Sage, London, pp 423–439

Bennett J, Palaiologou I (2016) Personal, social and emotional development. In: Palaiologou I, (ed) The early years foundation stage. Theory and practice, 3rd edn. Sage, London, pp 345–365

Bondai B, Kaputa TM (2016) Reaffirming Ubuntu/unhu mainstreaming in the education curricula: a panacea for sustainable educational change in Southern Africa. Int J Acad Res Reflection 4(6):37–44

Borg F, Winberg M, Vinterek M (2017) Children's learning for a sustainable society: influences from home and pre-school. Educ Inquiry 8(2):151–172

Boyd D, Hirst N (2018). Recognising Beach Kindy as a pedagogical approach for critical agents of change within early childhood education. In: Leal Filho W (ed) Handbook of sustainability science and research. Springer, Cham, pp 877–892

Carrington D (2020) Coronavirus: nature is sending us a message, says UN environment chief. https://www.theguardian.com/world/2020/mar/25/coronavirus-nature-is-sending-us-a-message-says-un-environment-chief

Chikunda C, Ngcoza KM (2017) Integrating Afrocentric approaches for meaningful learning of science concepts. In: Lotz-Sisitka H, Shumba O, Lupele J, Willmot D (eds) Schooling for sustainable development in Africa. Springer, Cham, pp 79–92

Davis JM (2008) What might education for sustainability look like in early childhood? A case for participatory, whole-of-settings approaches. In: Pramling Samuelsson I, Kaga Y (eds) The contribution of early childhood education to a sustainable society. UNESCO Paris, pp 18–24

Davis JM (2009) Revealing the research "hole" of early childhood education for sustainability: a preliminary survey of the literature. Environ Educ Res 15(2):227–241

Davis JM (2010) Action research today in early childhood (ARTEC). Special Issue January 2010. Early childhood education for sustainability: why it matters, what it is and how whole centre action research and systems thinking can help, pp 1–10

Davis JM (2015) What is early childhood education for sustainability and why does it matter. In: Davis JM (ed) Young children and the environment. Cambridge University Press, Port Melbourne, pp 7–27

Dolamo R (2013) Botho/Ubuntu: the heart of African ethics. Scriptura 112(1):1–10

Follari F (2019) Foundations and best practices in early childhood education. History, theories and approaches to learning, 4th edn. London, Pearson

Folque AM (2018) Nestling into the world: the importance of place and mutuality in the early years. In: Huggins V, Evans D (eds) Early childhood education and care for sustainability. International perspectives. Routledge, London, pp 51–67

Georgeson J (2018) Sustainable leadership in the early years. In: Huggins V, Evans D (eds) Early childhood education and care for sustainability. International perspectives. Routledge, London, pp 124–137

Gumbi EB (2019) Indigenous fundamentals for early childhood development in a shared educational praxis for the 21st century: pedagogical perspective on a multi-cultural race. Gender Behav 17(3). https://www.ajol.info/index.php/gab/issue/view/18545

Hedefalk M, Almqvist J, Ostman L (2015) Education for sustainable development in early childhood education: a review of the research literature. Environ Educ Res 21(7):975–900
Hill A, McCrea N, Emery S, Nailon D, Davis JM, Dyment JE, Getenet S (2014) Exploring how adults who work with young children conceptualise sustainability and describe their practice initiatives. Australas J Early Child 39(3):14–22
Huggins V, Evans D (2018) Introduction. In: Huggins V, Evans D (eds) Early childhood education and care for sustainability. International perspectives. Routledge, London, pp 1–13
Jansen JD (2011) Knowledge in the blood: confronting race and the apartheid past. Stanford University Press, Stanford
Le Grange L (2012) Ubuntu, ukama, environment and moral education. J Moral Educ 4(3):329–340
Lehtonen A, Salonen AO, Cantell H (2018) Climate change education: a new approach for a world of wicked problems. In: Cook JW (ed) Sustainability, human well-being and the future of education. Springer, Cham, pp 339–374
Li M, Zhang Y, Yuan L, Birkeland A (2019) A critical analysis of education for sustainability in early childhood curriculum. Documents in China and Norway. EGNU Rev Educ 2(4):441–457
Losavio J (2020) George Floyd: these are the injustices that led to the protests in the United States. World Economic Forum. https://www.weforum.org/agenda/2020/06/this-is-what-has-led-to-george-floyd-protests-in-the-united-states/
MacDonald M (2015) Early childhood education and sustainability: a living curriculum. J Child Educ 91(5):332–342
Masango MJS (2006) African spirituality that shapes the concept of Ubuntu. Verbum Et Ecclesia 27(3):930–943
Mcunu TN (2004) The dignity of the human person: a contribution of the theology of Ubuntu to theological anthropology. Unpublished Master of Theology dissertation, University of South Africa, Pretoria, South Africa
Metz T (2016) Recent philosophical approaches to social protection: from capability to Ubuntu. Glob Soc Policy 16(2):132–150
Molefe M (2019) Ubuntu and development: an African conception of development. Africa Today 66(1):97–113
Morgan A, Waite S (2018) Nestling into the world: the importance of place and mutuality in the early years. In: Huggins V, Evans D (eds) Early childhood education and care for sustainability. International perspectives. Routledge, London, pp 51–67
Msengana NW (2006) The significance of the concept "Ubuntu" for educational management and leadership during democratic transformation in South Africa. Unpublished doctoral thesis, University of Stellenbosch, Stellenbosch
Mugumbate J, Chereni A (2019) Using African Ubuntu theory in social work with children in Zimbabwe. Afr J Soc Work 9(1):27–34
Munkebye E, Scheie E, Gabrielsen A, Jordet A, Misund S, Nergård A, Bergljot AB (2020) Interdisciplinary primary school curriculum units for sustainable development. Environ Educ Res 26(6):795–811
Museka G, Madondo MM (2012) The quest for a relevant environmental pedagogy in the African context: insights from Unhu/Ubuntu philosophy. J Ecol Nat Environ 4(10):258–265
Ndofirepi AP (2011) Philosophy for children: the quest for an African perspective. South African J Educ 31(2):246–256
News 24 (2019) Edutainment from recyclables. https://www.news24.com/news24/southafrica/local/maritzburg-fever/edutainment-from-recyclables-20191127
Ngwaru JM (2014) Promoting children's sustainable access to early schooling in Africa: reflections on the roles of parents in their children's early childhood care and education. New Dir Child Adolesc Dev 146:61–76
Norren DE (2014) The nexus between Ubuntu and global public goods: its relevance for the post 2015 development agenda. Dev Stud Res Open Access J 1(1):255–266
Oladokun V (2020) Covid-19: Africa urgently needs an Ubuntu plan. Health & Welfare News Africa. https://www.bizcommunity.africa/Article/410/629/202141.html

Oppenheim CE (2012) Nelson Mandela and the power of Ubuntu. Religions 3:369–388
Organisation for Economic Cooperation and Development (2011) Families are changing. https://www.oecd.org/els/soc/47701118.pdf
Pascal C (1993) Spiritual and moral development. A discussion paper. National Curriculum Council York. https://www.educationengland.org.uk/documents/ncc1993/smdev.html
Peckham K (2017) Developing school readiness. Creating lifelong learners. Sage, London
Percy-Smith B, Burns D (2013) Exploring the role of children and young people as agents of change in sustainable community development. Local Environ 18(3):323–339
Pilon AF (2013). Building a new world: an eco-systemic approach for global change & development design. Munich Personal RePEc Archive (MPRA) Paper 45360. University Library of Munich. file:///C:/Users/21656568/Downloads/MPRA_paper_62124.pdf
Pramling Samuelsson I, Siraj-Blatchford J (2015) Education for sustainable development in early childhood care and education: A UNESCO background paper. https://doi.org/10.13140/RG.2.1.3197.2564
Pramling Samuelsson I, Park E (2017) How to educate children for sustainable learning and for a sustainable world. Int J Early Child 49:273–285
Seehawer MK (2018) Decolonising research in a sub-Saharan African context: exploring Ubuntu as a foundation for research methodology, ethics and agenda. Int J Soc Res Methodol 21(4):1–14
Sen A (2013) The ends and the means of sustainability. J Hum Dev Capabilities Multi-Disciplinary J People-Centered Dev 14(1):6–20
Sensoy Ö, DiAngelo R (2017) Is everyone really equal? An introduction to key concepts in social justice education, 2nd edn. Teachers College Press, New York
Shepherd N, Mhlanga D (2014) Philosophy for children: a model for Unhu/Ubuntu philosophy. Int J Sci Res Publ 4(2):1–5
Sheya E (2014) Indigenous knowledge and environmental education: a case study of selected schools in Namibia. Master of Education in Curriculum Studies (Environmental Education). Stellenbosch University, Stellenbosch
Shumba O (2011) Commons thinking, ecological intelligence and the ethical and moral framework of Ubuntu: an imperative for sustainable development. J Media Commun Stud 3(3):84–96
Singakwenza (2020) We can do it. Early childhood education. https://www.singakwenza.co.za/our-story/
Siraj-Blatchford J (2018) Preface. In: Huggins V, Evans D (eds) Early childhood education and care for sustainability. International perspectives. Routledge, London, pp xv–xix
Siraj-Blatchford J, Mogharreban C, Park E (2016) Preface. In: Siraj-Blatchford J, Mogharreban C, Park E (eds) International perspectives on early childhood education and development. Springer, Cham, pp x–vi
Siraj-Blatchford J, Pramling Samuelsson I (2016) Education for sustainable development in early childhood care and education: an introduction. In: Siraj-Blatchford J, Mogharreban C, Park E (eds) International perspectives on early childhood education and development. Springer, Cham, pp 1–16
Takyi-Amoako EJ, Assié-Lumumba NT (2018) Re-visioning education in Africa. Ubuntu-inspired education for humanity. Palgrave, New York
Taylor N, Quinn F, Eames C (2015) Preface. In: Taylor N, Quinn F, Eames C (eds) Educating for sustainability in primary schools: teaching for the future. Sense, Rotterdam, pp ix–x
Tilbury D (2011) Education for sustainable development: an expert review of processes and learning. UNESCO, Paris
Tutu D (1999) No future without forgiveness. Cox & Wyman, Berkshire
Tutu D, Tutu N (1989) The words of Desmond Tutu selected by Naomi Tutu. Newmarket Press, New York
Van Breda AD (2019) Developing the notion of Ubuntu as African theory for social work practice. Soc Work/maatskaplike Werk 55(4):439–450
Van Niekerk J (2013) Ubuntu and moral value. Unpublished doctoral thesis, University of the Witwatersrand, Johannesburg

Waghid Y (2018) On the educational potential of Ubuntu. In: Takyi-Amoako EJ, Assié-Lumumba NT (eds) Re-visioning education in Africa. Ubuntu-inspired education for humanity. Springer, Cham, pp 55–65

Walker C (2017) Tomorrow's leaders and todays agents of change? Children, sustainability education and environmental governance. Child Soc 31:72–83

Warwick P, Warwick A, Nash K (2018) Towards a pedagogy of hope: sustainability education in the early years. In: Huggins V, Evans D (eds) Early childhood education and care for sustainability. International perspectives. Routledge, London, pp 28–40

Williams HS (2018) Obama said. https://www.globalcitizen.org/en/content/ubuntu-south-africa-together-nelson-mandela/

Wertlieb D (2019) Inclusive early childhood development (IECD): a twin-tracking approach to advancing behavioural health and social justice. Am J Orthopsychiatry 89(4):442–448

World Commission on Environment and Development (1987) Our common future. Oxford University Press, Oxford

World Health Organization (2018) Nurturing care for early childhood development. A framework for helping children survive and thrive to transform health and human potential. Geneva: World Health Organization

Zhou X, Liu Z, Han C, Wang G (2016) Early childhood education for sustainable development in China. In: Siraj-Blatchford J, Mogharreban C, Park E (eds) International research on education for sustainable development in early childhood. Springer, Cham, pp 43–59

Educational Institutions, Approaches and Innovations

Transdisciplinary Education: Enabling the Sustainable Development Goals Using the Fourth Industrial Revolution

Saurabh Sinha and Vishaal Lutchman

Abstract More than ever before, the COVID-19 period has amplified the inequality of society. This inequality is particularly evident in South Africa and other African countries where the consumption expenditure Gini coefficient remains high. In the case of South Africa, the Gini coefficient measures 0.63 (2015). Despite the timespan since 1996, when this measure was 0.61, the inequality situation has thus barely changed. In 2015, South Africa became a signatory to the UN Sustainable Development Goals (SDGs) and in Dec. 2018, there was a call to establish the Presidential Commission on the Fourth Industrial Revolution (4IR). Attaining the ambition of SDGs is more likely through an intertwined approach with 4IR. This is particularly true for countries in Africa where resources are constrained, infrastructure lags that of other countries. Still, there is an opportunity, through innovation and a skills revolution, to leapfrog the journey towards the SDGs. This, however, requires human capacity; this realisation is partly an undertaking of higher education institutions. Higher education progression entails disciplinary as well as inter-, multi-, and cross-disciplinary education. Education in forms other than disciplinary education has occurred sporadically and not always by design. This manuscript, therefore, makes two contributions. Firstly, it discusses the intertwined aspect of 4IR and SDGs. Secondly, it calls for an explicit and institutional approach—transdisciplinary thinking about higher education and in favour of the SDGs. The transdisciplinary thinking then brings together the multiple lenses of the sustainable development framework to achieve the delivery of graduates who enable the realisation of SDGs.

S. Sinha (✉)
Research, and Internationalisation, University of Johannesburg (UJ), Auckland Park Kingsway Campus, Johannesburg, South Africa
e-mail: ssinha@uj.ac.za

V. Lutchman
South African Institution of Civil Engineering (SAICE), Thornhill Office Park, Bekker Rd, Vorna Valley, Midrand 1685, South Africa
e-mail: vishaal@saice.org.za

W. Leal Filho et al. (eds.), *Sustainable Development in Africa*, World Sustainability Series,
https://doi.org/10.1007/978-3-030-74693-3_9

1 Concept of Inequality

This contribution explores inequality in South Africa and by extension regionally. Inequality is a phenomenon that dates back to prehistoric times before written history, classified by archaeologists and anthropologists as an attribute of a society being nonegalitarian (Price and Feinman 1995). This contribution explores the skills revolution in the provision of quality education, innovative industrialisation, and infrastructure concerning the Sustainable Development Goals (SDGs). Furthermore, the approach leverages the Fourth Industrial Revolution (4IR) and a pursuit for the SDGs to bring about social justice ultimately.

The initial findings in a nonegalitarian society indicated that inequality manifested in instances of political instability, societal distress, economic shortcomings, and ideologies involving disagreements. Inequality is relevant to South Africa because of its roots in significant social distress and injustice. Based on the Gini coefficient, South Africa is a nonegalitarian society, indexed at 0.63, making it the second most unequal country in the world (World Population 2019). The South African government has implemented several initiatives to combat inequality in a post-apartheid era. Addressing inequality remains an economic focus in the 'government's 2030 National Development Plan (NDP). Structured initiatives to address inequality are in place with a greater push year on year to ensure that targets are met (Gumede 2016). Upon review of the overall gains made in addressing inequality, a macro-level assessment (McLennan 2016) indicates that existing initiatives have not made sustainable contributions to inequality. Furthermore, the study recommends that further work should be done to integrate initiatives while considering additional views. It is predicted that inequality will worsen in response to the COVID-19 pandemic (Broadbent et al. 2020).

The opportunity to advance equality by innovation and a skills revolution lies in the appropriate usage of 4IR. The weaving together of solutions that achieve more than one SDG, such as quality education (SDG4) and innovative industrialisation (SDG9) in the journey to progress equality (SDG10), involves the adoption of transdisciplinary methodologies to construct thinking, ideas, and future opportunities.

2 Research Context

This study makes two contributions: exploring transdisciplinary approaches to quality education and to attaining SDGs through 4IR. A transdisciplinary approach adds breadth to the otherwise discipline-specific approach. Transdisciplinary in this concept refers to inquiry requiring the development of solutions while maintaining the focus on goals analogous to the SDGs (Todorova 2013). This contribution motivates a transdisciplinary methodology being adopted in the development of sustainable solutions (Sachs et al. 2019). The adoption of transdisciplinary methodologies in advancing existing education requires additional curricula or reconstructing existing disciplines for study while improving the quality of education (Hadar 2018). The role of 4IR supports the opportunity to advance SDGs rapidly through integrated solutions supported by digital technologies in addition to education (Sutherland 2020). This integrative culmination is referred to as the skills revolution (Sachs et al. 2019).

The NDP includes initiatives that suggest transdisciplinary methodologies are plausible considerations to achieve multiple outcomes across many sectors (Naude 2014). After review of the NDP, a synergy of initiatives could be of greater value. For example, investments in coal will increase coal export. In addition to an increase in mining activities, the increase in coal consumption will increase greenhouse gas emission, while climate change will have an impact on food security and water supply. The South African government remains at loggerheads with organised labour concerning the content, prioritisation, and implementation of recommended initiatives (Mosala, Venter and Bain 2017). The public sector infrastructure component of the NDP is gazetted in the National Infrastructure Act (Act 23 of 2014), which establishes a focussed national oversight and implementing agency called the Public Infrastructure Coordinating Commission (PICC) (National Planning Commission 2015). The economic and social focus of the 'commission's mandate correlates with the techno-economics and social dynamics disciplines, indicating that policy, aims, and objectives are in place. Effective implementation of the plan is underway.

The NDP's objectives as facilitated by the PICC seek to advance SDG 10 while addressing the effects of colonialism, racism, apartheid, sexism, and repressive labour policies (Naude 2014). Furthermore, there is a correlation between the NDP and the SDGs (World Bank Group 2019). Aside from the need to progress economic growth, the more recent urgency for implementing the plans has come about owing to COVID-19 (Channing et al. 2020). Herein lies the opportunity for a transdisciplinary approach in reviewing the feasibility of planned initiatives, obtaining buy-in, and giving confidence that goals will be achieved.

3 Exploring the Constructs of Inequality

SDG 10 promotes a socially just and inclusive world in which the needs of the most vulnerable are met (World Bank Group 2019). The entire world has experienced an average rise in the Gini coefficient, which implies a rise in inequality in the last 15 years (United Nations 2019). Gains made to date have been significantly undone by the COVID-19 pandemic (United Nations 2019). The most significant construct of inequality is income and wealth inequality (United Nations 2013). As high income and extreme wealth grows in the top 1% of the population, the world has experienced decreased access to food, healthcare, education, land, clean water, and other resources required to live life with dignity. The most marginalised social groups are people in rural areas, women, youth, people with disabilities, and indigenous people (Ostry et al. 2019). SDG 10, as defined by the UN, is split into local, regional, and global goals (World Bank Group 2019). Table 1 indicates the targets.

Table 1 SDG 10—reduced inequalities—targets, aims, and disciplines (World Bank Group 2019)

Geography	Target	Description (Aims)	Disciplines (Primary Field)
In-country	Income growth	Prosperity	Techno-economics
	Social, economic, and political	Inclusion of all people	Techno-economics, social dynamics
	Ensure equal opportunities	Social protection	Social dynamics
	Adopt policies that converge income and social protection policies	Greater equality	Social dynamics
Inter-country (regional)	Regulation and monitoring of markets and institutions	Strengthening regulations to control fairness	Techno-economics
	Ensure a voice for developing countries	Access to credible global institutions	Geopolitics
	Facilitating the safe migration of people	Well-planned migration policies	Social dynamics
Global	Implementing a policy of special treatment of developing countries in terms of the World Trade Organisation (WTO) agreements	Allow developing countries to progress economically	Geopolitics
	Offer development assistance to emerging economies	Allow development in countries that need it most	Techno-economics, Geopolitics
	Reduce costs to migrate	Allow free movement of migrants between countries	Geopolitics, techno-economics

The targets can be assembled into the disciplines. If one ignores the geographical aggregation (column 1) of the targets, aims may be arranged into broad disciplines.

The constructs of inequality relate to the disparity in income, inclusivity, access to opportunities, fairness, migration movement policies, and geopolitical support. These are broad constructs that in themselves have subconstruct specificities as applicable to countries, communities, and individuals (United Nations 2013).

The inequality disciplines from Table 1 give rise to four disciplines namely geopolitics, social dynamics, techno-economics, and earth systems. This contribution postulates that integration (transdisciplinary methodology) of these four disciplines will lead to the formulation of sustainable solutions that progress the achievement of SDG 10. Incorporating a transdisciplinary methodology will apply to the effective achievement of other SDGs (Sachs et al. 2019). As it pertains to Africa, it was agreed in the transition from the Millennium Development Goals to the UN SDGs in "Transforming our World: the 2030 Agenda for Sustainable Development" that Africa has not made significant progress towards achieving SDG 10. Specifically, little progress was made to achieve a quality education (SDG4) and innovation (SDG9). These two goals were postulated and accepted as crucial focus goals underpinning inequality. The weaving of the four mentioned disciplines into education and innovation will advance the achievement of SDG4 and SDG9 faster than has been the case in the past owing to the ability to unify knowledge (Jeder 2014).

4 Exploring the Opportunity for Quality Education (SDG 4) to Reduce Inequality (SDG 10)

This contribution explores the notion of diversifying existing curricula using transdisciplinary curricula, which incorporates the disciplines of geopolitics, social dynamics, techno-economics, and earth systems to form a new approach to education (SDG 4). Two of the aforementioned disciplines are relevant: (i) using a transdisciplinary approach in the development and resourcing of new curricula and (ii) using a transdisciplinary approach to satisfy the need for quality education system level rapidly. SDG 4 has the potential to develop new, effective methods of teaching and learning, supported by digital technologies that could mitigate current shortcomings in providing quality education. Education has at least three contributions to make to economic growth, helping to reduce income inequality. These are an increase in human capital productivity, an increase in innovation, and diffusion of knowledge to develop and implement new technologies (Hanushek and Wößmann 2010). It is therefore plausible to assume that where infrastructure development is most needed to afford quality and dignified life, education and infrastructure will assist in achieving SDG 10.

SDG 10, which relates to addressing inequality as generally defined by the UN, has specific intentions and targets (United Nations 2013). Concerning Table 1, the goal has three geographical boundaries that relate to the local, regional, and global context. Concerning South Africa (local country-level aims)–Table 2, for example, improvement in the quality of education will assist with developing overall human capacity, thereby increasing the possibility of finding employment and raising income, increasing prosperity. At this time 80% of the population in South Africa that remain unemployed have not completed secondary schooling having been excluded

Table 2 SDG 10—inequality—national aims (United Nations 2013)

National target description	Aim
Income growth	Prosperity
Social, economic, and political	Inclusion of all people
Ensure equal opportunities	Social protection
Adopt policies that converge income and social protection policies	Greater equality

due to socio-economic and political factors, while 8% of graduates remain unemployed (Stats 2020). This reality concurs with SDG 4 national aims in Table 2 that educating all people will allow for greater inclusivity and participation in economic activities as individuals, leaders, and entrepreneurs while allowing for an increase in income (United Nations 2013). SDG 10 as defined by the UN is split into national and global aims (World Bank Group 2019).

This section postulates that the constructs of SDG 10, when assembled into possible study disciplines, have the potential to give students a holistic understanding and appreciation of the overall SDGs. Such understanding may promote a transformed workplace, careers including postgraduate study, target setting, and goals. Such a study has the potential to encourage support for the SDGs (Sachs et al. 2019).

Concerning Table 3—SDG 4—quality education, the applicable disciplines are similar to inequality, comprising economic and social issues. This suggests that quality education has the potential to improve social standing while growing an economy supported by investment in the targets indicated.

In summary, Sachs et al. (2019) refer to education as a critical enabler of all SDGs that should be integrated with all SDG initiatives. To date, Africa has not made significant strides in the quality of primary and secondary education that feeds into the tertiary system (Hadar 2018). The investments made in education in South Africa have not yielded the economic growth forecast year on year, nor income equality. A percentage increase in income has been higher for lower-wage workers, but off a low base (Arestis and Sawyer 2018). The increase for the higher wage earners has been smaller, but of a very high base, which has had the effect of widening the

Table 3 SDG 4—quality education targets, aims, and disciplines (World Bank Group 2019)

Target (SDG 4)	Description (Aims)	Disciplines
Equitable primary and secondary education	Free	Techno-economics
Childhood education	Quality	Social dynamics
Technical and vocational education and training (TVET) and tertiary education	Quality	Social dynamics
Technical and vocational education	Skills	Techno-economics—based on available employment
Education for vulnerable	Equal access	Techno-economics, social dynamics
Youth and adult literacy and numeracy	Make gains to ensure basic adult and youth education	Techno-economics, social dynamics
Education facilities	Safe, good infrastructure, reduced bullying	Social dynamics, Techno-economic
Scholarships	Awarding of growing scholarships	Techno-economic
Teachers	Reduce the ratio of teachers/pupils	Techno-economics

inequality gap over time. The education system has not invested meaningfully in teacher training, reduction of student–teacher ratios, TVET colleges, and providing good quality education facilities (Eroer, Bai and Fonseca 2019). Therefore a disconnect persists between the budget, quality of education, and economic and income growth. A transdisciplinary approach has the potential to aggregate the disconnects in the education system to deliver holistic goal-orientated solutions (Oke and Fernandes 2020).

5 Transdisciplinary Methodologies to Accelerate Achieving SDGs

The transdisciplinary methodology is an inquiry-based method to explore, describe a problem, and develop solutions aligned to an overall goal. As it pertains to this study, the topic relates to exploring inequality with a focus on education and infrastructure as sub-goals to address inequality in the South African context. Because of the complexity of the topic and the integrated nature of the solutions required, a single-discipline approach may sacrifice the additional constructs of the solution, thus creating blind spots. It defeats the initial objectives; for example, private schools that provide education worsen the state of inequality. As it pertains to the narrative of inequality in South Africa, four disciplines have served to connect the topics, namely geopolitics, social dynamics, techno-economics, and earth systems. The solution lies in ensuring that the dynamics between the four are appreciated and fully understood before embarking on any initiative. Hence a transdisciplinary methodology will assist in exploring and scoping each discipline and its interconnectedness, relationships, interactions, and context with other disciplines, ultimately leading to sustainable solutions while new disciplines are created in the future.

More profound complexity comes to the fore in the relationships between the disciplines (Todorova 2013; Montuori 2014). Paradigms play out; for example, toll roads in South Africa have stalled in response to resistance from various stakeholders, which has led to a stop in much-needed spending on roads in the country. Stakeholders, unions, and the taxi industry opposed the toll-road concept, yet the same union members and taxi commuters are now unemployed because of economic contraction caused in part by lack of spending on roads.

The need to focus on efficient and effective initiatives cannot be overemphasised, especially given the COVID-19 pandemic. This study postulates that a focussed initiative to reduce inequality and promote inclusivity (SDG 10) in conjunction with quality education (SDG 4) and innovation in infrastructure provision (SDG 9) will lead to much-desired sustainable economic growth. Economic growth will provide the income lost to poor economic policy implementation pre-pandemic and economic contraction during and after the pandemic.

Concerning Table 4—Disciplines applicable to 4, this study seeks to examine the constructs of the three SDGs mentioned in terms of disciplines. This examination is

Table 4 Disciplines applicable to SDGs 4, 9, and 10 (World Bank Group 2019)

No	Target (SDG 10)	Description (Aims)	Disciplines
1	Income growth	Prosperity	Techno-economics
2	Social, economic, and political	Inclusion of all people	Techno-economics, social dynamics, Geopolitics
3	Ensure equal opportunities	Social protection	Social dynamics
4	Adopt policies that converge income and social protection policies	Greater equality	Social dynamics
5	Regulation and monitoring of markets and institutions	Strengthening regulations to control fairness	Techno-economics
6	Ensure a voice for developing countries	Access to credible global institutions	Geopolitics
7	Facilitating the safe migration of people	Well-planned migration policies	Social dynamics
8	Implementing a policy of special treatment of developing countries in terms of WTO agreements	Allow for developing countries to progress economically	Geopolitics
9	Offer development assistance to developing countries	Allow for development in countries that need it most	Techno-economics, Geopolitics
10	Reduce the costs to migrate	Allow for free movement of migrants between countries	Geopolitics, techno-economics
⇧	Target (SDG 4)	Description (Aims)	Disciplines
11	Equitable primary and secondary education	Free	Techno-economics
12	Childhood education	Quality	Social dynamics
13	TVET and tertiary	Quality	Social dynamics
14	Technical and vocational	Skills	Techno-economics—based on available employment
15	Education for vulnerable	Equal access	Techno-economics, social dynamics
16	Youth and adult literacy and numeracy	Make gains to ensure basic	Techno-economics, social dynamics
17	Education facilities	Safe, good infrastructure, reduced bullying	Social, techno-economics
18	Scholarships	Awarding of growing scholarships	Techno-economics
19	Teachers	Reduce the ratio of teachers/pupils	Techno-economics

(continued)

Table 4 (continued)

⇧	Target (SDG 9)	Description (Aims)	Disciplines
20	Reliable and resilient infrastructure	Affordable	Techno-economics
21	Industrialisation	Inclusive	Social dynamics, Techno-economics
22	Small-scale development	Increase	Techno-economics
23	Upgrade and retrofit industry	Climate conscious	Earth systems
24	Innovation and technology in the industry	Research and development workers	Techno-economics
25	Infrastructure development	Financial support in developing countries	Techno-economics
26	Domestic technology development	Value-adding industrial development	Social dynamics, Techno-economic
27	Access to information	Access to the internet	Techno-economics

done in conjunction with the respective targets and aims of the said SDG's. This is key to consolidating whilst integrating the targets and aims into a focussed matrix giving rise to the initial thinking of a conceptual model. The suggestion is that focus on the disciplines will give rise to integrated solutions that meet the aim and targets of the individual SDG's.

The discipline (last column) gives the first broad focus areas in integrating solutions to reach the SDGs. Each 'discipline's attributes contribute to promoting several SDGs, in part. The transdisciplinary approach allows for the integration of multiple knowledge areas, as it applies to the roles and responsibilities of each stakeholder such as government, society, academia, and business to be implemented as a joint initiative. If not executed in unison, the chances of realising the goals will be low and the objectives will not be achieved.

A further contribution is expediting the adoption of initiatives that embrace 4IR solutions intertwined with a transdisciplinary approach, thus providing the toolkits necessary to achieve the SDGs. Disruptive technologies can assist in unlocking the integration of disciplines where it remains difficult to reach sustainable solutions. For example, virtual meetings became a new way of working globally in a matter of weeks, forced by COVID-19 curfews and lockdown protocols. As a result, social dynamics have changed in the home space, requiring adaptations of families in tight spaces (Hiscott et al. 2020). This is a case of forced adoption of technology; however, this study recommends a proactive approach to develop other such uses of technologies to enable the implementation of transdisciplinary solutions.

6 Accelerating the Attainment of the SDGs in the Context of the Fourth Industrial Revolution

This chapter offers two contributions, namely incorporating transdisciplinary methodologies in education to achieve SDGs and exploring the role of 4IR in achieving SDGs. In the context of the 4IR, a short meta-analysis on how 4IR can be considered in the South African context shows that 4IR is seen as a threat to employment and economic growth in the short term, therefore a risk rather than an opportunity. A solution lies in this chapter allaying perceived 4IR threats through education while adopting a transformative approach to the SDGs that addresses inequality, education, and infrastructure development (Sutherland 2020).

South Africa has not been a leader in the development of technology solutions. Government, academia, and technology providers are proponents of the country embracing 4IR to progress SDG 10. The creation of the commission on the Fourth Industrial Revolution enables initiatives to achieve SDGs by improving competitiveness and assessing focussed sectors of the economy that must leverage technology.

Like its proponents, many sceptics are vigorously airing their views, with little or no understanding of what is tangible about 4IR. The constructs of 4IR comprise artificial intelligence using machines to learn, using complex mathematical calculations, robotics, automation, and electronics. This broadly entails technology to perform tasks, creating obsolescence, and threatening employment. However, (Ostry et al. 2019) mitigate the need for a fearful response by pointing out that technology has throughout recorded history assisted humanity to become more efficient. In this era of 4IR, the rate of change and advancement has been the fastest it has ever been, which may have caught people unprepared and continues to make them fear it, rather than getting to know and embrace it (Sutherland 2020).

Furthermore, 4IR needs to put a few things in place before its benefits can be comprehensively appreciated. An example is a digital readiness, which refers to the availability of good and fit for purpose internet (SDG 9) to all (World Bank Group 2019 and World Economic Forum 2020). Trust in technology is required for the 4IR to be exploited to its fullest potential. (Sachs et al. 2019) have found that preparing to embrace technology requires governments to scrap old and prohibitive legislation and policy to become pro-growth and embrace technology.

Traditional government bureaucracy tends to discourage progress and slow the pace at which technology is being developed. Many who are not involved in working with technology risk becoming ignorant. Adopting conservative paradigms will limit economic liberalisation. As a result, market shares are gravitating towards those who are willing to embrace industrialisation (Sachs et al. 2019). In the adoption of transdisciplinary methodologies, existing paradigms and assumptions need to be discussed and so does the role of government (Cockburn et al. 2013).

Geo-political agendas that do not incorporate the disciplines of social dynamics, techno-economics, and earth systems are fast falling out of favour. The disruption that COVID-19 has caused in each discipline is incomparable to any other event in terms of cost burdens to global economies in a global recession (Hiscott et al. 2020). This global economic reset will create the potential for new, nimble technology-knowledgeable entrants to redefine markets at the expense of dominant firms. The opportunity is supported by (Price and Feinman 1995), (Sachs et al. 2019), and (Sutherland 2020), who urge governments, businesses, and civil society to act with care and balance in respect of the postulated disciplines.

Fear and lack of a transformative economic policy hinder South Africa from moving past the propaganda of employing 4IR immediately in its processes in government and businesses; this will in time diminish opportunities to take the steps needed to grow economically (Bloom and Sachs 1998). As markets reorganise themselves, for example, countries that have managed to deal well with COVID-19 are in the process of entering into new trade deals, which are transforming international trade and geopolitical relationships. As the concept of fear and lack of trust in embracing 4IR in South Africa inhibits its potential, it is plausible that South Africa may not adopt a transformative economic policy, supported by a macrolevel long-term plan. Adoption of a transdisciplinary methodology can serve to enhance, adjust, and revise a liberal economic policy (Azeiteiro et al. 2018). To date, economic policy has been based on protectionist practices, largely entailing the state creating jobs and labour protecting jobs. Such practices, albeit necessary to address the ills of an apartheid path, should have been transformed over the last 25 years of democracy to combine initiatives with business and civil society to grow the economy jointly. To date, a disconnect remains, affecting geopolitics, techno-economics, society, and earth systems, which should be integrated to achieve the aims of the 2030 NDP and SDGs 4, 9, and 10 (Seta et al. 2017).

The disruption potential of fear of embracing 4IR in South Africa could be mitigated through policy reform adopting a long-term macro-economic view, including plans that are congruent, practical, and in support of the SDGs (Mosala, Venter and Bain 2017). Konratiev (2013) theorised that the world goes through economic cycles that last for 50 to 60 years between depression and economic growth in a sinusoidal wave profile for capitalist economies. Such an economic theory indicates the opportunity to adopt a macro-economic policy to be in phase with global cycles. Attainment of the SDGs is a medium- to long-term goal aligning with such cycles, giving insight into the prioritisation of initiatives for significant results.

7 Conclusion

This chapter has highlighted the opportunity to advance conventional approaches to the simultaneous attainment of the NDP and SDGs. South Africa has adopted various initiatives to address inequality, support education, and develop infrastructure (World Bank Group 2019). These initiatives have had varying levels of success over time. COVID-19 has accelerated the timeline for the meaningful and sustainable accomplishment of these initiatives, which could provide alternative and unconventional solutions. Technology development, including disruptive solutions to new ways of looking at opportunities, will result from revised approaches to education (Broer et al. 2019). While education has medium- to long-term results, innovative industrialisation infrastructure spending is required to yield financial results in the short term. Adopting social emphasis in the short term is likely to gain support because of contracted economic growth, unemployment, and poverty (Channing et al. 2020). Education and infrastructure are postulated to be considered in two dimensions, namely strategy and implementation. As education affects implementation, technology can assist in allowing more students to gain access to quality education. Education and the unification of the disciplines of geopolitics, social dynamics, earth systems, and techno-economics need to be expanded and evaluated in the South African context.

Relating to infrastructure implementation, this chapter has motivated that technology can be used in the urgent rolling out of infrastructure development, taking into consideration a revised look at the ratio of production splits between labour and capital, to make an immediate impact on income inequality. Infrastructure plans are in place and lean towards effective implementation of techno-economic policy and the need for congruent focus on aligning infrastructure planning while ensuring integration with geopolitics, social dynamics, and earth systems.

South Africa, and the broader continent of Africa's emerging economies are grappling with the stubborn effects of inequality. As it pertains to prospects, this chapter can lead to the development of a model that can be incorporated into methodologies for development planning at the national level. For example, a holistic approach to education requires infrastructure, that is supported by an economic plan that can fund education, but also providing employment. Initiatives cannot stand alone, and departments can no longer work in silos. The adoption of such a model can also be included in the criteria used by such African states to take investment decisions and perform pre-feasibility studies that need to integrate the mentioned disciplines. In addition, the model may be used in the approval and authorisation process for infrastructure development applications. At a global level, should developmental states achieve congruence across the disciplines, it has the potential to mitigate the current climate risks created by development by those not adopting such a methodology.

References

Arestis P, Sawyer M (2018) Inequality: trends, causes, consequences, and relevant policies. 3. Palgrave MacMillan, Cham

Azeiteiro UM, Akerman M, Filho WL, Setti AFF, Brandli LL (2018) Lifelong learning and education in healthy and sustainable cities. 1. Springer, Hamburg

Bloom DE, Sachs JD (1998) Geography, demography, and economic growth in Africa. Brookings Papera Econ Activity 2(1):207–295

Broadbent A, Kaziboni A, Smart B, Tella O, Walker D (2020) A framework for decisions in a post COVID world. Institute for the Future of Knowledge - University of Johannesburg, Johannesburg

Broer M, Bai Y, Fonseca F (2019) Socioeconomic inequality and educational outcomes. 1. Springer Open, Cham

Chang CC (2020) Economic inequality, war and pursuit of tax fairness. J Hum Values 26(2):114–132
Channing A, Davies R, Sherwin G, Laurence H, Konstanin M, Boipuso M, Sherman R, Witness S, Dirk S, Lillian A (2020) Impact of Covid-19 on the South African Economy: An Initial Analysis. Accessed 10 Aug 2020. https://sa-tied.wider.unu.edu/sites/default/files/pdf/SA-TIED-WP-111.pdf
Cockburn J, Dissou Y, Duclos J, Tiberti L (2013) Infrastructure and economic growth in Asia: economic studies in inequality, social exclusion, and well-being. 1. Springer, New York
Dipio D (2013) Traditional Leaderships Wisdoms and their Contemporary Parallels - The Madi of Uganda. Journal for African Culture and Society, pp 37–63
Gumede V (2016) Post Apartheid South Africa. 1. Cambria Press, New York
Hadar IB (2018) Education, finance, equality, and equity. 1. Springer, Cham
Hanushek EA, Wößmann L (2010) Education and economic growth. Int Encycl Educ 2(1):245–252
Hiscott J, Alexandridi M, Muscolini M, Tassone E, Palermo E (2020) The global impact of the coronavirus pandemic. Cytokine Growth Factor Rev 53(1):1–9
Jeder D (2014) Transdisciplinarity - the advantage of a holistic approach to life. Procedia Soc Behav Sci 137(2):127–131
Konratie ND (2013) Do business cycles shape the world. Today Zaman, Istanbul
McLennan DJ (2016) The lived experience of inequality in post-apartheid South Africa: measuring exposure to socio-economic inequality at a small area level. 1. University of Oxford, London
Montuori A (2014) The complexity of transdisciplinary literature reviews. Res Gate (university Alberta) 10(1):45–55
Mosala SJ, Venter JCM, Bain EG (2017) South Africa's economic transformation since 1994: What influence has the national democratic revolution had. Rev Black Polit Econ 44(1):327–340
National Planning Commission (2015) National Development Plan 2030. https://www.gov.za. Accessed August 25 2020
Naude P (2014) Economic policy and theological reflection in South Africa. Int J Public Theol 8(1):445–470
Oke A, Fernandes FAP (2020) Innovations in teaching and learning: exploring the perceptions of the education sector on the 4th industrial revolution. J Open Innov 6(31):1–22
Ostry JD, Loungani P, Berg A (2019) Confronting Inequality: How Societies can chcose Inclusive growth. 1. Columbia University Press, New York
Price DT, Feinman MG (1995) Foundations of prehistoric social inequality. 1. Plenum Press, New York
Sachs JD, Traub GS, Mazzucato M, Messner D, Nakicenovic N, Rockstrom J (2019) Six transformations to achieve the sustainable development goals. Nat Sustain 2(1):805–814
Seta F, Sen J, Biswas A, Khare A (2017) From poverty, inequality to smart city. 1. Springer, Singapore
Stats SA (2020) Labour Force Survey Quarter 2 2020 - Statistics South Africa. https://www.statssa.gov.za. Accessed 18 Aug 2020
Sutherland E (2020) The fourth industrial revolution: the case of South Africa. The South African Journal of Political Studies 2(47):233–255
Todorova P (2013) A contemporary transdisciplinary approach as a methodology to aid students of humanities and social sciences. Bulgarian Comparative Educ Soc (bulgarian Acad Sci) 11(1):187–191
World Population (2019) https://www.worldpopulationreview.com. Accessed 4 Aug 2020
World Bank Group (2019) United Nations: Department for Economic and Social Development. Accessed 9 Aug 2020. https://sdgs.un.org/sites/default/files/documents/21453SDG_10_EGM_2019_concept_note_30Jan_consolidated.pdf

Straddling the Divide: A Framework for Digital Skills Education in Support of Both the UN Sustainable Development Goals and the Fourth Industrial Revolution in South Africa

Hugo Lotriet and Hossana Twinomurinzi

Abstract South Africa has endorsed both the UN Sustainable Development Goals (UN SDG) which is to be aligned to its National Development Plan and the agenda of the Fourth Industrial Revolution (4IR), which will inform economic direction and is supported in the country's forthcoming Science and Technology White Paper. These two agendas are potentially divergent, and care is required to align the actions taken in support of these agendas. Both ICT and education are core drivers of both agendas and thus the development of human capital through digital skills education has the potential to play a significant role in enhancing both agendas, but if approached inappropriately may contribute to deepening the digital divide in the country. This paper proposes a conceptual framework for digital skills education that would allow human capital development and the development of digital skills that would enable digital support for both the SDG and 4IR agendas. The paper provides an example of a South African digital skills training programme undertaken on behalf of the South African government by the National Electronic Media Institute of South Africa (NEMISA) as an example of the rollout of digital skills to an African nation. The paper finally indicates the implications of the proposed conceptual educational framework for the provision of digital skills in South Africa.

1 Introduction

This paper focuses on digital skills education in the context of two divergent agendas. These are the United Nations Sustainable Development Goals (SDGs) (United Nations 2015) and the World Economic Forum's Fourth Industrial Revolution (4IR) (Schwab 2016).

The South African government explicitly supports both agendas. As a subscriber to the UN SDGs South Africa has the intention for alignment between the SDGs and the South African National Development Plan (Fourie 2018; Haywood et al. 2019). President Ramaphosa has included the principles of 4IR as part of the country's economic

H. Lotriet (✉)
Department of Information Systems, University of South Africa, Pretoria, South Africa

H. Twinomurinzi
Department of Computer Science, University of South Africa, Pretoria, South Africa

W. Leal Filho et al. (eds.), *Sustainable Development in Africa*, World Sustainability Series,
https://doi.org/10.1007/978-3-030-74693-3_10

vision and created a Presidential Commission on the Fourth Industrial Revolution (Sutherland 2020). The forthcoming White Paper on Science and Technology of the South African Department of Science and Technology focuses strongly on strategies and institutional measures that would support South Africa taking advantage of the 4IR (National Planning Commission 2020).

Both SDG and 4IR agendas require effective Information and Communication Technology support towards achieving their overall goals. In the case of 4IR, the technology agenda is explicit and the agenda is associated with various technologies such as Artificial Intelligence (AI), robotics and the Internet of Things (IOT) (Sutherland 2020). In contrast, the SDG agenda with its focus on more global humanitarian goals means that the role of ICT is less explicitly stated. Nevertheless it is implied that education, human collaboration and partnerships, and adequacy of infrastructures require the effective use of ICT to achieve for instance SDG's 4, 5, 9 and 17 (Pollitzer 2019; National Planning Commission 2020).

The SDG and 4IR agendas are divergent with different priorities and goals, and there is the potential that in the pursuit of technology support and digital skills for these agendas, inequalities such as the digital divide can increase rather than be erased (Pollitzer 2019). A harmonization of the agendas would imply that an equilibrium is found between humanitarian, societal, environmental and economic needs and demands (World Economic Forum 2016a; Pollitzer 2019). Poor outcomes for a country such as South Africa would be the growth of the digital divide and digital inequality, while a desired outcome would be "digital harmony" where there is alignment between society and its technology needs, and where there is universal access to appropriate technologies and skills (Pollitzer 2019). For this reason, care needs to be taken in managing the digital requirements of both agendas.

One of the main drivers that have a direct impact on the attainment of digital equality is high quality education and human capital development, and specifically ensuring that high quality educational opportunities are provided to ensure that citizens have opportunities to gain appropriate digital skills (Pollitzer 2019). In order to reap maximum benefit from both agendas it is therefore important that citizens have appropriate skills to both contribute meaningfully to the attainment of the goals of the agendas, but also to share in the benefits of sustainable development and economic opportunities afforded to those who are suitably skilled.

As stated in the introductory paragraph, this chapter focuses specifically on digital skills education in support of the two agendas. An important aspect of this is to make the digital skills associated with the SDG and 4IR skills explicit and to define these skills. A similar exercise has been undertaken by Van Laar et al. (2017) for explicitly defining the digital skills related to the so-called "21st century skills" set (Trilling and Fadel 2010). However, the key skills sets for SDG and 4IR differ from the "21st century skills set", both in terms of scope and in terms of priorities.

For SDG skills this chapter draws on the SDG skills proposed by UNESCO (2017) augmented by an extensive survey by Lozano et al. (2017), that identifies a set of skills that are appropriate in terms of the SDG agenda. Similarly, the WEF indicates a skills set for 4IR (World Economic Forum 2016a), which is used for the purposes of this chapter. This chapter extends the work of Lozano et al. (2017 op. cit.) by defining the digital skills associated with both SDG and 4IR key skills sets that will need to be addressed by digital skills education programmes. The chapter also discusses the educational implications of digital skills training in the context of the SDG and 4IR agendas.

Finally, the chapter provides an example of digital skills education in South Africa. The National Electronic Media Institute of South Africa (NEMISA) manages a digital skills training initiative on behalf of the South African government.

2 Defining a Set of Consolidated Digital Skills for SDG and 4IR

In this section, a set of digital skills associated with key SDG and 4IR skills is proposed in Table 1. The work is similar in format to the proposed digital skills set associated with 21st Century key skills proposed by Van Laar et al. (2017).

The proposed digital skills were compiled by means of a hermeneutic approach that combines and adapts the following elements:

1) SDG critical skills, based mainly on UNESCO (2017) guidelines, and augmented by an extensive systematic review by Lozano et al. (2017).
2) WEF critical skills for 4IR as listed in World Economic Forum (2016a), and Gray (2016).
3) The Occupational Information Network (O*NET) skills categories (Peterson et al. 2001), which provide a common reference framework for describing the broad categories of required skills. The World Economic Forum (2016b) uses this categorization in categorizing the key 4IR skills. The O*NET skills classification system allows for the broad categories of fundamental abilities, basic skills (enabling learning and knowledge acquisition) and cross-functional skills, that enable functioning across different contexts (Peterson et al. 2001 op. cit.).

The set of digital skills is shown in Table 1, with columns for key SDG skills, key WEF skills and key skills that are common to both SDG and WEF. For each key skill listed, the corresponding digital skills are defined in the final column of the table. Where appropriate, previous definitions of Van Laar et al. (2017) were used, but for a number of the key skills new definitions are provided.

The digital skills proposed in Table 1 could be considered complex skills, mainly related to higher order thinking dimensions, if for instance categorized in terms of Bloom's taxonomy. Churches (2010) argues that for learning complex digital skills it should be remembered that these follow on the learning of lower, more basic digital skills as part of a natural scaffolding process associated with learning. By presenting the skills in Table 1, the authors are therefore not arguing for the negation of basic digital skills education. It should however be clear that the SDG and 4IR agendas ultimately would require to be supported by more complex digital skills and that digital education programmes in South Africa would ultimately have to consider the implication of this in terms of intended outcomes.

Table 1 Digital skills related to SDG and 4IR key skills

O*Net Content model categories	4IR key skills	SDG key skills (UNESCO 2017)	Skills indicated by both SDG and 4IR as key	Digital skills associated with core skills description
Abilities: Cognitive	Cognitive flexibility			Skills to use technology to enable adaptation of thinking on the basis of changes in the environment (Adapted from (van Laar et al. 2017)
		Tolerance for ambiguity and uncertainty		Skills in using technology to improve tolerance for ambiguity and uncertainty; Conversely, understanding and dealing with the ambiguity involved in the adoption and use of technology (for an agricultural example see Barham et al. (2014)); also, understanding and dealing with the impact of technology ambiguity on human activity (see for example Hackett et al. (Hackett et al. 2017))
	Creativity (originality)			Skill to use technology for the generation of novel concepts and ideas (van Laar et al. 2017)

(continued)

Table 1 (continued)

O*Net Content model categories	4IR key skills	SDG key skills (UNESCO 2017)	Skills indicated by both SDG and 4IR as key	Digital skills associated with core skills description
		Anticipatory Thinking		Using technology to support and enable anticipatory thinking, e.g. technology mediated scenario planning (Piirainen and Lindqvist 2009)
Content (Basic)	Learning, writing, reading, comprehension			Skills to use technology for improved basic content skills. Conversely, the acquisition of basic content skills in support of more efficient use of technology and systems artefacts
Process (Basic)			Critical Thinking and analysis	Skills to employ technology in the process of critical thinking and analysis (van Laar et al. 2017)
Complex problem-solving skills		Integrated problem solving		Skills to employ technology in the process of integrated problem solving
Systems skills			Systems Analysis, judgment and decision making	Skills in the use of technology to analyze systems and make decisions
			Systems Assessment and evaluation	Skills in the use of technology to assess and/or evaluate systems

(continued)

Table 1 (continued)

O*Net Content model categories	4IR key skills	SDG key skills (UNESCO 2017)	Skills indicated by both SDG and 4IR as key	Digital skills associated with core skills description
Social skills			Interpersonal collaboration and communication	Skills to use technology in the development of networks, enabling of interactions between persons and communication (van Laar et al. 2017)
			Empathy and emotional intelligence (as part of self-awareness)	Skills to use technology as mediators for the development of empathy and emotional intelligence, e.g. through gaming (e.g. Wu et al., (2020)) or by using virtual environments (see for example the discussion by Shin (2018) on empathy and digital storytelling)
	Negotiation			Skills to employ technology in the process of negotiation
	Service orientation			Skills to employ technology to improve service and in the training of persons for service orientation

(continued)

Table 1 (continued)

O*Net Content model categories	4IR key skills	SDG key skills (UNESCO 2017)	Skills indicated by both SDG and 4IR as key	Digital skills associated with core skills description
		Strategic action		Understanding and ability to use technology to enable strategic action. Conversely, understanding the impact of strategic action on the shaping of the technology landscape (see for instance Fjeldstadt et al. (2004))
Resource management	Management of finances, materials, persons and time			Skills to use technology to support the efficient management of resources
Technical	Technical skills			The skills that enable understanding and use of technology artefacts in the execution of (online) tasks (van Laar et al. 2017)
			Media use	Skills in the effective use of digital media
Complex problem-solving skills		Complex problem solving		The skills to use technology to assist in developing solutions to complex problems, e.g. through simulations (Scanlon et al. 2005)
		Integrated problem solving		Skills to employ technology in the process of integrated problem solving

(continued)

Table 1 (continued)

O*Net Content model categories	4IR key skills	SDG key skills (UNESCO 2017)	Skills indicated by both SDG and 4IR as key	Digital skills associated with core skills description
			Interdisciplinary work	The skills to use technology for the support of interdisciplinary work, e.g. computer mediated virtual team work (Vroman and Kovacich 2002)
Non-O*NET categorized skills		Normativity and the application of values and norms		The skills to employ technology to enable justice, responsibility and ethics. Also, conversely the ethical use of technology (van Laar et al. 2017)
		Personal involvement		Using technology to support learning, participation in action, and self-motivation (Lozano et al. 2017)

3 Educational Implications of the Proposed Digital Skills Framework

It is important to highlight the educational and pedagogical implications related to a set of skills or competencies (Lozano et al. 2017). For the proposed set of skills in Table 1, the educational and pedagogical implications are presented in this section as the intersection of three sets of educational discourses. Firstly, there is the general scholarly discourse on digital skills education; secondly the discourse on education within the context of the SDG agenda; and thirdly the discourse on education within the context of the 4IR agenda. In this section each of these discourses are briefly discussed, after which the implications for digital skills education in South Africa are set out.

The digital skills set proposed in this chapter relates to SDGs and 4IR, and for both agenda's some of the educational implications are set out in the following sub-sections.

3.1 Considerations Related to Scholarly Discourses on Digital Skills Education

Although the intersection between ICT and education has been extensively researched (Ricaurte 2016), the pedagogy of digital skills remains complex. The pedagogy of teaching digital skills needs to consider both the fact that there is not a single umbrella conceptualization of digital skills and competencies, instead portions of it go under a large number of other terms, e.g. "digital literacy" or "ICT skills", and secondly that the notion of pedagogy is also widely defined (Lakkala et al. 2011). This makes research into the pedagogy of digital competences complex, and therefore relatively little literature exists (Lakkala et al. 2011).

From this study, it should also be clear that the digital skills identified in Table 1, could not simply be separated from the SDG and 4IR key skills that the digital skills are supporting (Lakkala et al. (2011) support this principle). Therefore, a simple adoption of a technology-centric approach in teaching the digital skills to operate technologies would not adequately equip citizens to be productive in either working effectively towards sustainability or economic growth and development. Instead, a meaningful approach should clearly include pedagogical approaches that allows a learner to not only for instance master the technology of scenario planning, but also acquire the skill of anticipatory thinking, in order to be able to make a more meaningful contribution in the area of SDGs.

Grand-Clement (2017) indicate a number of key issues that in principle needs consideration in the context of digital skills education:

1) Acquisition of digital skills is a life-long process (Grand-Clement et al. 2017).
2) Digital skills cannot be acquired in isolation, it should happen in conjunction with other (non-digital) skills, which in the context of this chapter would be the SDG and 4IR skills listed in previous sections.
3) The roles and responsibilities of stakeholders (educators, learners, others) will change from "traditional" education.
4) There is a strong focus on the use of technology as educational tools in this context.
5) Digital skills are dynamic, as the required skills are the equivalent of a moving target.

It may be seen that there is natural alignment between these issues and the issues indicated under the issues for 4IR skills training.

3.2 SDG Education Implications

The UN SDG 4 on education emphasizes various aspects of a sustainable educational agenda (Scott 2015).

For the purpose of digital education in South Africa, the following aspects of the SDG discourse on education (UNESCO 2017) need to be noted:

1) The vision of sustainability needs to be clear for education to contribute towards it.
2) The outcomes of education for sustainable development are reflective individuals who can assess their own actions within both local and global contexts (UNESCO 2017).
3) SDG education is holistic and engages the cognitive, affective and behavioural domains.
4) Learning methods are diverse and include inter alia collaboration (group and inter-group), action learning, case studies, internships and job-shadowing, enquiry-based learning, site visits, scenario development, storytelling, project-based learning.
5) A supportive context is required for sustainability education to be viable. This includes explicit educational policies, integrated curricula, training of teachers, institutional support.
6) In addition, the goals of SDG 4 (quality education) should be pursued. These include (1) universal access; (2) the importance of quality related to all aspects of education including teaching, the pedagogies employed, content and relevance; and equity and gender equality in education. Achievement of educational quality is complex (Boeren 2019) and needs a comprehensive systematic approach.

3.3 4IR Education Considerations

In support of the ideals of 4IR, the WEF proposes a system of "social and emotional learning" (SEL) (World Economic Forum 2016b). Essentially it argues for lifelong learning that will result in a set of foundational literacies, competences and character qualities (the so-called "21st century skills") (World Economic Forum 2016b). It emphasizes the tools of education and the educational possibilities offered by educational technologies.

From a 4IR education perspective, the following points are of importance for digital skills education:

1) Learning is expected to be distributed geographically (Aziz Hussin 2018)
2) Learning is expected to be more individualized with the learner having a stronger choice on the method, nature and content of learning and curricula (Aziz Hussin 2018). By implication, learners will also become more independent.
3) Teaching methods can include among others blended learning, project-based learning, field visits, internships, collaboration, knowledge application (Aziz Hussin 2018).
4) A supporting environment is required that involves policy, government, employers, educators.

One of the significant implications of the 4IR view is the short time span involved in the disruptions taking place (World Economic Forum, 2016a). Skills development programmes therefore also need to be agile to respond to demands. In addition, requirements may be very context and situation specific (World Economic Forum 2016b op. cit.).

3.4 Implications for Digital Skills Education South Africa

When the previous sub-sections are considered together, several important points arise that digital skills education in South Africa need to take into consideration, if digital skills education is intended to support both the SDG and 4IR agendas. These can be separated into broader considerations around the educational system at large and pedagogical considerations.

Broader considerations include the following:

1) It is clear from the UN, WEF and digital skills education agendas that success in digital skills education can only take place as a collaborative effort between government, education institutions, society, citizens and other role players such as businesses and employers.
2) The visions of the SDG and 4IR agendas should be clearly stated in the education policies that relate to digital skills education, and great care should be taken to negotiate the priorities and directions of digital skills education in order to meet the requirements of both agendas
3) The overall principles of the UN SDG4 should be explicitly applied to digital skills education. This would imply that a digital skills education agenda for South Africa should actively strive towards gender equality, equitable and high quality education for all, and creation of safe education spaces for digital skills education and prioritization of the allocation of resources for digital skills education (human, financial and other).
4) Another important aspect that SDG education emphasizes is the importance of ethical awareness of human action and the enabling of justice, responsibility and ethics. Other than the skills associated with this, normativity (both in digital and associated skills) should be a foundation of digital skills education.
5) One of the main potential tension lines that need to be considered by all role players is the essentially bottoms-up (democratic) nature of the 4IR mindset, in terms of direction of technology, jobs and skills. In contrast the implied mindset of the SDGs is that there are limited choices in terms of the direction that needs to be followed in order to achieve sustainable human condition on the planet and that this is underpinned by certain ethical considerations. Negotiating the balance between the essentially capitalist free-market mindset of the 4IR and the more directed SDG mindset may well be complex at times but will have to be dealt with responsibly by all role players.

Considerations regarding digital skills pedagogy:

1) Digital skills cannot be taught in isolation but should be closely integrated to the core SDG and 4IR skills that define the required digital skills set. As the core skills sets evolve rapidly, especially for 4IR, core skills sets can be expected to vary over time.
2) The integrated skills sets can be taught through a multitude of appropriate methods, which for the sake of the learners should include amongst others group work and collaboration, case studies, field work, action learning and internships.
3) Digital skills education is a lifelong process, probably composed of multiple short periods of intense acquisition of specific skills.
4) The expectation is that digital skills will increasingly be taught by means of digital technologies. The essence of good education still is that people teach people. The tools can enhance the educational experience but cannot guarantee quality digital skills education.

4 Example: Digital Skills Education in South Africa Through NEMISA

The National Electronic Media Institute of South Africa (NEMISA) is a state-owned entity (SOE) with a mandate to develop digital skills capacity across South Africa through collaborative partnerships with all societal stakeholders, that is, industry, government, academia, civil society, organized labour and international bodies. The focus is primarily on those who were historically disadvantaged (because of apartheid) and those who are highest at risk such as unemployed youth, women and those in marginal areas. The aim is to ensure that all South Africans have the necessary digital skills to participate in the 4IR and for sustainable development.

NEMISA began in 1998 with a specific mandate to train black women to participate in the broadcasting industry, which at the time was primarily male dominated. In 2008, NEMISA's mandate was expanded to include the E-Skills Institute (e-SI) and the Institute for Satellite and Software Application (ISSA). E-SI had been established in 2008 as a presidential think-tank to deliberate on how South Africa could participate in the fast advancing digital economy. ISSA was focused on satellite engineer development.

NEMISA currently offers digital skills training to government and in marginalized communities, especially rural areas. The digital skills training is mainly in digital literacy but extends to 4IR type courses such as programming and robotics. These partnerships are led through colab partnerships established at national universities. The Universities are regarded as impartial centres of excellence for skills areas considered as critical to South Africa. At the moment, there are eight established colabs as shown in Table 2.

Table 2 Digital skills education in South Africa through NEMISA

Digital skills focus	University	Province
Public sector and policy implications of a digital future	Durban University of Technology	KwaZulu Natal
Digital inclusion in society and SMEs	University of the Western Cape	Western Cape
Digital literacy	Vaal University of Technology	Northern Cape and Southern Gauteng
ICT for rural development	Walter Sisulu University	Eastern Cape
Digital connected health	University of Limpopo	Limpopo
E-Agriculture and e-tourism (e-Agro Tourism)	North West University	North West
Cybersecurity	University of Mpumalanga	Mpumalanga
Digital skills research	University of South Africa	Research

5 Conclusions

This chapter presents a set of digital skills that are required in support of the key skills for SDG and 4IR agendas and discusses the educational implications for the teaching of these skills in South Africa. The South African National Development plan, which is the primary vision for South Africa's development agenda has allowed alignment with the UN SDGs. As a result, an understanding of the implications of the UN SDGs for digital skills development and education is significantly important. The work presented in this chapter extends the work by Van Laar et al. (2017) on key 21st century skills. The digital skills requirements cannot be seen in isolation from digital skills education, and therefore the chapter has a reflection on the educational implications of the two agendas for digital skills education in South Africa.

Even though the 4IR agenda is more explicit about technology than the SDG agenda, digital skills education is significantly important to both. Digital skills cannot be taught in isolation but should be linked to the key skills sets associated with the SDGs and 4IR. The digital skills that are required to support SDG and 4IR are complex higher order skills some of which are not explicitly addressed in either agenda but are implied e.g. cognitive skills. Digital education programmes can therefore not only be limited to the provision of basic digital skills, as these would not adequately equip learners to participate fully in support of SDG or 4IR. Also significantly important for the digital skills agenda is the emphasis by the SDG agenda on the teaching of appropriate values and norms.

Digital skills education policies and frameworks in South Africa should explicitly incorporate the vision of SDG on education and ensure that this integrates as intended with the other SDGs to contribute towards a sustainable society.

Effective digital skills education relies on a holistic approach, including policy support, involvement of a variety of role players including government, educational institutions, educators, learners, industry and civil society.

Lessons learnt from digital skills development in South Africa show that the existence of enabling multi-stakeholder institutions within the public sector to stimulate such a dynamic digital skills programmes that meet both the SDG and 4IR agendas, without widening the digital divide any further is necessary. Such institutions need to remain responsive to the changing technology environment separate from inevitable government bureaucracy, which is often antithetical to the speed at which technology advancement occurs, especially regarding the 4IR.

The conversations between all stakeholders mean that all those affected by the intended and unintended consequences of the new holistic approach will be able to engage with these consequences (Habermas 1972). The available technologies today allow for the facilitation of such public discourse, potentially free from the domination of political interests (or groupthink). Such public discourse could enable South Africa to oscillate towards a shared and common understanding of "a better world", one that incorporates local interests as well as the ideals of the divergent 4IR and SDG agendas.

There is significant potential for future research work on digital skills education in complex developmental contexts, such as South Africa. Understanding how digital skills can contribute to sustainability would be of benefit to multiple stakeholders involved in the provision of these skills worldwide.

References

Aziz HA (2018) Education 4.0 Made Simple: Ideas For Teaching. Int J Educ Lit Stud 6:92. https://doi.org/10.7575/aiac.ijels.v.6n.3p.92

Barham BL, Chavas JP, Fitz D et al (2014) The roles of risk and ambiguity in technology adoption. J Econ Behav Organ 97:204–218. https://doi.org/10.1016/j.jebo.2013.06.014

Boeren E (2019) Understanding sustainable development goal (SDG) 4 on "quality education" from micro, meso and macro perspectives. Int Rev Educ 65:277–294. https://doi.org/10.1007/s11159-019-09772-7

Churches A (2010) Bloom's Digital Taxonomy. Aust Sch Libr Assoc 1–44

Fjeldstad ØD, Becerra M, Narayanan S (2004) Strategic action in network industries: an empirical analysis of the European mobile phone industry. Scand J Manag 20:173–196. https://doi.org/10.1016/j.scaman.2004.05.007

Fourie W (2018) Aligning South Africa's national development plan with the 2030 agenda's sustainable development goals: guidelines from the policy coherence for development movement. Sustain Dev 26:765–771. https://doi.org/10.1002/sd.1745

Grand-Clement S, Devaux A, Belanger J, Manville C (2017) Digital Learning: Education and skills in the digital age

Gray A (2016) The 10 skills you need to thrive in the Fourth Industrial Revolution. World Econ Forum, Davos 1–4

Habermas J (1972) Knowledge and Human Interests (Paperback). Heinemann Educational, London

Hackett PT, Lemoine PA, Richardson MD (2017) Impact of Technology Ambiguity on Leadership in Global Higher Education. In: Encyclopedia of Strategic Leadership and Management. pp 270–281

Haywood LK, Funke N, Audouin M et al (2019) The sustainable development goals in south africa: investigating the need for multi-stakeholder partnerships. Dev South Afr 36:555–569. https://doi.org/10.1080/0376835X.2018.1461611

Lakkala M, Ilomäki L, Kantosalo A (2011) Which pedagogical practices and methods best support learning digital competences? Linked portal Eur Sch (EUN)., Brussels

Lozano R, Merrill MY, Sammalisto K et al (2017) Connecting competences and pedagogical approaches for sustainable development in higher education: A literature review and framework proposal. Sustain 9:89. https://doi.org/10.3390/su9101889

National Planning Commission (2020) Digital Futures: South Africa's Digital Readiness for the Fourth Industrial Revolution (draft)

Peterson NG, Mumford MD, Borman WC et al (2001) Understanding work using the occupational information network (O*NET): implications for practice and research. Pers Psychol 54:451–492. https://doi.org/10.1111/j.1744-6570.2001.tb00100.x

Piirainen K, Lindqvist A (2009) Enhancing business and technology foresight with electronically mediated scenario process. PICMET Portl Int Cent Manag Eng Technol Proc 2397–2410:52. https://doi.org/10.1109/PICMET.2009.5261830

Pollitzer E (2019) Four scenarios for how digital technologies could change the world. J Int Aff 72:75–90. https://doi.org/10.2307/26588344

Ricaurte P (2016) Pedagogies for the open knowledge society. Int J Educ Technol High Educ 13:16. https://doi.org/10.1186/s41239-016-0033-y

Scanlon E, Blake C, Joiner R, O'Shea T (2005) Technologically mediated complex problem-solving on a statistics task. In: Learning, Media and Technology. pp 165–183

Schwab K (2016) The Fourth Industrial Revolution, 1st edn. Wiley, Geneva

Scott CL (2015) Preparing learners for 21st century competencies and skils overall vision of twenty-first century pedagogy conclusions. next steps and future issues, issues the futures of learning 3: what kind of pedagogies for the 21st century? UNESCO Educ Res Foresight 15:1–21

Shin D (2018) Empathy and embodied experience in virtual environment: to what extent can virtual reality stimulate empathy and embodied experience? Comput Human Behav 78:64–73. https://doi.org/10.1016/j.chb.2017.09.012

Sutherland E (2020) the fourth industrial revolution-the case of South Africa. Politikon 47:233–252. https://doi.org/10.1080/02589346.2019.1696003

Trilling B, Fadel C (2010) 21st century skills: learning for life in our times. Choice Rev Online 47:47-5788-47–5788. https://doi.org/10.5860/choice.47-5788

UNESCO (2017) Education for Sustainable Development: Learning objectives. UNESCO

United Nations (2015) Sustainable Development Goals. In: Sustain. Dev. Goals - 17 Goals To Transform Our World. http://www.un.org/sustainabledevelopment/sustainable-development-goals/

van Laar E, van Deursen AJAM, van Dijk JAGM, de Haan J (2017) The relation between 21st-century skills and digital skills: A systematic literature review. Comput Human Behav 72:577–588. https://doi.org/10.1016/j.chb.2017.03.010

Vroman K, Kovacich J (2002) Computer-mediated interdisciplinary teams: theory and reality. J Interprof Care 16:159–170. https://doi.org/10.1080/13561820220124175

World Economic Forum (2016a) The Future of Jobs Employment, Skills and Workforce Strategy for the Fourth Industrial Revolution

World Economic Forum (2016b) New Vision for Education : Fostering Social and Emotional Learning through Technology. World Econ Forum, 36

Wu L, Kim M, Markauskaite L (2020) Developing young children's empathic perception through digitally mediated interpersonal experience: Principles for a hybrid design of empathy games. Br J Educ Technol 51:1168–1187. https://doi.org/10.1111/bjet.12918

Case Study: Higher Education and the Education for Sustainable Development in Africa (ESDA) Program

Jackline Nyerere, Farai Kapfudzaruwa, Olushola Fadairo, Alice Odingo, James Manchisi, and Shogo Kudo

Abstract Using reflective assessment, this case study shares the experience and lessons of the Education for Sustainable Development in Africa (ESDA) initiative. ESDA is an inter-university collaborative program among eight partner African universities focusing on graduate training and field-based transdisciplinary research aiming to promoting sustainable development in Africa. The assessment shows how interdisciplinarity in curriculum content and design and the transdisciplinary field-oriented pedagogy of the ESDA programs could be a useful model for Education for Sustainable Development (ESD), especially when higher education institutions (HEIs) collaborate. Given the resource and capacity constraints of African HEIs, universities have to collectively re-imagine their societal role by working together and sharing their complementary strengths to attain the critical mass associated with developing the human capital required to sustainably transform Africa. Evidence presented through the ESDA master's programs in mining, shows how sharing learning materials and teaching approaches in an interdisciplinary setting enriches the learning experiences of students as they obtain a holistic understanding of the complex nature of sustainability. Transdisciplinary field-oriented internships and

J. Nyerere (✉)
Department of Educational Management, Policy and Curriculum Studies, Kenyatta University, Nairobi, Kenya
e-mail: nyerere.jackline@ku.ac.ke

F. Kapfudzaruwa
Centre for the Advancement of Scholarship (CAS), University of Pretoria, Pretoria, South Africa

O. Fadairo
Department of Agricultural Extension and Rural Development, University of Ibadan, Ibadan, Nigeria

A. Odingo
Department of Geography and Environmental Studies, University of Nairobi, Nairobi, Kenya

J. Manchisi
Department of Metallurgical Engineering, University of Zambia, Lusaka, Zambia

S. Kudo
Graduate Program in Sustainability Science – Global Leadership Initiatives, Graduate School of Frontier Sciences, The University of Tokyo, Tokyo, Japan

W. Leal Filho et al. (eds.), *Sustainable Development in Africa*, World Sustainability Series,
https://doi.org/10.1007/978-3-030-74693-3_11

research activities are shown to create "living labs for sustainability" which allow students to learn that working together in real-world settings with multiple actors (including non-academics) instead of individually is valuable. In co-creating knowledge and co-innovating sustainable solutions—the resulting societal impact is higher. Reflections on the experiences of the Next Generation Researchers (NGRs) of the ESDA initiative reveals that for ESDA to succeed, the creation of learning materials reflecting the African context and the capacity development of young faculty at African institutions need to be prioritized.

Keyword Inter-university collaborative programmes · Graduate training · Education for Sustainable Development (ESD) · Inter- and transdisciplinary pedagogy · African universities

1 Introduction

The increased role of higher education institutions (HEIs) in sustainable development involves two dimensions:

1. The importance of universities in creating, distributing and using knowledge and technological innovations to drive sustainable development, and
2. The strategic emphasis in higher education on human capacity development to produce researchers, educators, professionals and entrepreneurs who recognize the importance of sustainable development (Nagao et al. 2018).

The breakthrouh study by Kates and Dasgupta (2007) on sustainable development notes that Africa's poverty is a complex sustainability issue of global importance - similar to climate change and biodiversity reduction. Therefore, increasing research to describe and understand the causes and develop policies and solutions to reduce poverty in Africa is important. In 2002, the World Summit on Sustainable Development (Rio + 10) showed that African governments are interested in sustainable development. However, this interest was not mirrored by African researchers and universities. In 2010, an Association of African Universities (AAU) survey revealed that the limited involvement by African universities in sustainable development was due to financial and human resources capacity constraints (AAU 2011).

Responding to this and the adoption of the UN Decade of Education for Sustainable Development (UNDESD) initiative, a collaborative project was initiated in 2008 between the United Nations University (UNU) in Japan and the University of Tokyo (Tōdai) Graduate Program in Sustainability Science (GPSS). This was supported by the United Nations Educational, Scientific and Cultural Organization (UNESCO), the United Nations Environment Programme (UNEP), the United Nations Human Settlement Programme (UN-HABITAT) and the African Development Bank (AfDB). The idea was to develop an overview of the state of research and Education in Sustainable Development in Africa (ESDA). The ESDA project seeks to shed light on the readiness of African universities to engage in graduate-level capacity building in sustainable development. The study by Popoola (2014) indicates a significant gap between field-oriented, problem-solving approaches, stakeholder interaction emphasized in sustainability science, and the actual teaching and research practice in sustainable development-related courses and programs offered by African universities. Fuh (2014) indicates that there is an urgent need to strengthen the capacity of researchers at African universities and institutions to do research and publish on sustainable development and related issues. Nyerere et al. (2014) emphasize the need for consortiums involving national universities as well as science and technology institutions (in or outside the African region). This is because sustainable development efforts are trans-boundary in nature.

In 2013, three ESDA graduate training programs were launched by eight African partner universities to contribute to sustainable development by training professionals who could serve as planners, organizers, instructors, researchers and practitioners. These programs aim to support research and information sharing, particularly among emerging researchers (see Table 1). A unique feature of these programs is the collaboration between universities and governmental and non-governmental stakeholders outside universities. This was necessary due to the complexity of sustainable development and the requirement to share partnerships, resources and expertise. Table 1 shows the eight partner ESDA universities that implemented three masters programs, namely Sustainable Urban Development (SUD), Sustainable Integrated Rural Development in Africa (SIRDA) and Mining and Mineral Resources (MMR). They also collaborate in joint research projects through the Next Generation Researchers (NGRs) team.

Table 1 ESDA partner universities and programs

African partner university	Master's program	Next Generation Researchers (NGRs)
Kenya: • Kenyatta University (KU) • University of Nairobi (UoN)	Sustainable urban development (SUD)	NGRs program in support of master's program: • Rural–urban connections • Corruption, gender and governance • Out-migration of youth and the impact on rural communities (in collaboration with NGRs Asia)
Ghana: • Kwame Nkrumah University of Science and Technology (KNUST) • University for Development Studies (UDS) • University of Ghana (UG) Nigeria: • University of Ibadan (UI)	Sustainable integrated rural development in Africa (sirda)	
South Africa: • University of Cape Town (UCT) Zambia: • University of Zambia (UNZA)	Mining and mineral resources (MMR)	

The ESDA masters and NGRs programs are designed to respond to key sustainable development challenges in Africa. The SIRDA program targets rural development agents such as agricultural extension officers, forestry field workers, rural social community workers, mid-level policymakers and researchers who are involved in the eradication of poverty and food insecurities, and the provision of water, sanitation services and sustainable energy in rural areas. Africa's rural regions require dedicated attention because a high proportion of the sub-Saharan populace live and make their livelihoods there (Olaniyan and Fadairo 2019). The SUD program develops capacity to tackle sustainable development challenges in African urban areas associated with rapid urban population growth. Its design focuses on training innovative and action-oriented professionals capable of contributing to sustainable livelihoods of vulnerable urban communities. The MMR program addresses the complex and interrelated challenges of mining in Africa, including managing and interacting with stakeholders to safeguard and balance production targets with regulatory, environmental, safety, health and socio-economic targets in mining communities.

The aim of the case study presented in this chapter is to illustrate how collaboration between African universities through the ESDA program can provide effective teaching and learning and build research capacity in graduate programs through inter- and transdisciplinary approaches necessary for fostering sustainable development.

2 Education for Sustainable Development (ESD)

Since education, public awareness and training were recognized as crucial elements of sustainable development implementation in Agenda 21, increased attention was placed on universities' role in transforming society by educating decision-makers, leaders, entrepreneurs and academics (Lozano-García et al. 2008). Initial arguments for ESD were captured in Chapter 36 of Agenda 21 (UN 1992). Presently, ESD is at the core of the 2030 Agenda for Sustainable Development and the Sustainable Development Goals (SDGs) adopted by the UN General Assembly in 2015. To increase action and build on the SDGs, other interventions such as the UNDESD (2005–2014) and the Global Action Programme (GAP) on ESD (2015–2019) were implemented. In 2019, UNESCO's 40th General Conference adopted a new global framework on ESD. The framework, "ESD for 2030" seeks to:

> …build education systems that support learners of all ages to be active contributors to more peaceful and sustainable societies and develop a sense of responsibility for our planet in line with the 17 SDGs (UNESCO 2019).

"ESD" is defined by UNESCO (2015) as:

> … including key sustainable development issues into teaching and learning; … climate change, disaster risk reduction, poverty reduction, and sustainable consumption. It also requires participatory teaching and learning methods that motivate and empower learners to change their behavior and take action for sustainable development. [ESD] consequently promotes competences like critical thinking, imagining future scenarios and making decisions in a collaborative way …

The definition emphasizes two elements: (1) content and (2) pedagogy (Pauw et al. 2015). Content covers broad disciplines linked to "wicked problems" such as climate change and poverty (Pryshlakivsky and Searcy 2013). These problems are often complex and require "interdisciplinarty" to understand them (Munkebye et al. 2020; Ventakataraman 2009; Walshe 2017). In the literature, "interdisciplinarity" is labelled as "holism" or a holistic approach, since it includes multiple perspectives on the content (Pauw et al. 2015). This perspective emphasizes the need to include the environmental, socio-cultural and economic dimensions of sustainable development, and focuses on their interrelationships and interactions over space and time (Sinakou et al. 2017; Summers and Childs 2007). Thus, it is important to reflect on the ability of the ESDA program to heighten efforts to integrate an interdisciplinary dimension in the curriculum to train African professionals responsible for promoting sustainable development.

An essential aspect of ESD related to teaching and learning focuses on the development of skills and action competences for sustainability (Mogensen et al. 2010). This pedagogical approach labelled "pluralism" (Santoianni 2017) strives to acknowledge and engage varying perspectives, views and values of different stakeholders at multiple levels. The underlying idea of transdisciplinarity is to teach (and learn) that the complex issues dealt with in sustainable development make it difficult to operate in silos (Sinakou et al. 2017). Therefore, ESD focuses on multi-method approaches (Rafols and Meyer 2010), social learning (Jacobi et al. 2016), as well as participatory and "real-life" action-oriented learning (Joseph and Said 2020). Essentially, ESD facilitates learning so that students engage other actors to understand sustainability issues and develop competences to take action. By reflecting on participatory field-oriented student internships and the collaborations by the NGRs the value of transdisciplinarity in learning and knowledge generation is illustrated.

The context in which ESDA members operate raises questions about the effectiveness of African universities in providing ESD training. Evidence shows that African governments invest less than 1% of their gross domestic product (GDP) towards higher education (Gandhi 2020). The COVID-19 pandemic exacerbated these financial constraints of African HEIs. These dwindling resources undermine the need for collaborative efforts among African universities to ensure effective ESD. In using the ESDA master's programs as a case study this chapter demonstrates how African universities can collaborate to address sustainable development issues through the lens of the ESDA coordinators.

3 Methodology

Using critical reflection as a methodology—also referred to as "reflexivity"—this section reflects on the capacity of the ESDA initiative to develop professionals and researchers tasked with promoting sustainable development in Africa (Fook 2011; Mezirow 1998). Used in several disciplines, this methodology relates to a "process of identifying, questioning, and assessing our deeply held assumption[s] about our knowledge, the way we perceive events and issues, our beliefs, feelings, actions" (Bilson 2006). Critical reflection is based on the Maturana's concept (1988) that individuals do not experience a single stable reality. Instead, we all exist in and experience a "multiverse" with many realities experienced differently based on our varying assumptions, values, beliefs and actions. Thus, the experience with the ESDA programs could be perceived differently by the different people involved; hence, the process relies on multiple processes to collect the data used.

First, the critical reflection process relied on the annual ESDA consortium meeting. Attendees include the ESDA secretariat, vice chancellors, program and NGR coordinators, lecturers, students, sponsors, employers and other stakeholders (such as civil society and other policymakers) at local, national and regional level. Reflection on the progress of attaining set goals and priorities happens as well as planning strategically for the future. Therefore, the six meeting reports (2014–2019) and four meeting reports discussing the conceptualization of the ESDA initiative (Fuh 2014; Nagao 2009, 2010, 2011; UNU 2014, 2015, 2016, 2017, 2018a) were used. Second, the process relied on the field-oriented research activity reports of the NGRs and the ESDA Sustainable Development Workshop series which provides a platform for engagement between academic and non-academic stakeholders. Third, the process relied on external evaluation reports produced in October 2018. This evaluation of ESDA was carried out by two external experts. The first report focused on an accountability review to assess the effectiveness and impact of the programs. The second report focused on a strategic review to support the program's future design and execution (Mugabe 2018; Ofir 2018; UNU 2019). These reports relied on document reviews and ESDA consortium member interviews. Fourth, another source of information was the selection of program graduates from various institutions to represent and provide their views on the different programs.

4 Education for Sustainable Development in Africa (ESDA) Academic Programs

4.1 Sustainable Urban Development (SUD) Programme

The SUD program is jointly offered by the University of Nairobi (UoN) and Kenyatta University (KU) in Kenya. The program design considers the fact that Kenya's population is rapidly urbanizing. As of 2019, Kenya's urban population stood at 27.3%, meaning more than 14 million Kenyans live and work in urban centers (Worldometer 2020). The country's proportion of urban dwellers is projected to quadruple by 2045. Most of the migrants are escaping poverty in rural areas as they seek jobs to support their immediate families and dependents. Due to limited opportunities, many of the migrant youth end up unemployed or underemployed (Nyerere 2018), driving them to informal settlements. More than 60% of Nairobi's residents live in low-income neighborhoods, with poverty ranging between 60 to 78% in various city suburbs (Ayaga et al. 2005).

The SUD program targets community change agents (policymakers; non-governmental organizations [NGOs]; community health care and social workers; women and youth groups; and entrepreneurial, energy, housing, water, sanitation and manufacturing extension workers). Essentially, the two-year Master of Science in Sustainable Urban Development (MSc in SUD) trains and equips students with knowledge, skills and attitudes necessary for effective and sustainable urban development. The program's priority is to alleviate the poverty of the urban poor, typically residing in mushrooming urban slums, through imparting quality and market-responsive technical and entrepreneurial skills. The skills are expected to bring about an all-inclusive space in which a secure base is established to ensure self-employment opportunities for the disadvantaged and vulnerable youth and women in poor urban communities. The fundamental guiding principle is to empower the urban poor by enhancing their capacity to operate sustainable income-generating activities and accumulate surplus resources to deal with development challenges.

The program is offered part- and full-time at UoK and full-time at KU. It features coursework, fieldwork internships and research components. The curriculum's core subjects are designed to encourage entrepreneurship and innovation for sustainable development at community level through subjects such as "green economy and sustainable entrepeneurship" and "sustainability innovations". The elective courses (covered in Year 1) are organized into three critical areas identified for community-based innovation: (1) water and sanitation, (2) food security and (3) renewable energy. In Year 2 students must complete an internship, seminar paper and research project addressing a real-life community problem.

The program's multidisciplinary and interdisciplinary nature seeks to produce innovative and action-oriented professionals ready to contribute to sustainable livelihoods of vulnerable urban communities in line with the SDGs. Specifically, the program equips graduates with skills and competencies for:

- Innovative sustainable urban development,
- Technology-oriented research with a direct bearing on enhanced social-economic development,
- Entrepreneurial culture and practices for wealth creation and poverty alleviation, and
- The formulation and implementation of sustainability policies.

The course work applies a variety of methodologies: community-based conceptualization and development of innovative projects, mandatory fieldwork within target communities, exchange visits, and a demonstration of community-based innovations. This approach provides exciting experiences for students as captured in this quote from a graduate (class of 2018):

> [The] ESDA program remains my most enriching academic experience. I was attracted to the program because of its innovative pedagogical methods that focused on solving real sustainability challenges in Africa ... though we couldn't have a lot of fieldwork as was earlier envisaged (due to funding challenges), the assignments issued and case studies discussed during

> our coursework enabled us to develop critical and creative thinking to solve development challenges in real urban contexts …

The program enhances the entrepreneurial capacity of students and has led to the development of entrepreneurial ventures. An example of such a venture is Nico Green, started by one of the graduates (class of 2018):

> [The] ESDA program empowered me with the inspiration, competency and sense of urgency to lead Africa's sustainable transformation as a sustainability researcher and eco-entrepreneur. [I] co-founded Kenya's pioneering sustainability consultancy and green technology firm—NIKO GREEN LTD (https://institute.nikogreen.com)—which is already providing evidence-based policy advice to government, UN agencies, NGOs, businesses as well as deploying green technologies in the areas of clean energy, green building and sustainable health care …

The program's approach allows students to co-create a collegial scholarly environment, such as the sustainability knowledge café, that provides a platform for dissemination of ESDA knowledge with scholars and decision-makers from other programs. In addition, there have been activities involving research, publications and workshops funded by the AfDB to support knowledge production and dissemination during the implementation period. The NGRs in this program engaged in research both independently and with ESDA student involvement. The studies address a range of relevant topics including the energy-water-sanitation nexus as it relates to sustainable urban development, as well as social entrepreneurship and youth employment. Findings were presented and published as part of the reference materials for the course (Kapfudzarura 2018; Kapfudzarura et al. 2018; Nagao et al. 2019a). Engagement of the NGRs with the students serves to build the research capacity of students as the NGRs take part in teaching and supervision of these students. Other significant activities are the workshops jointly organized by the two universities. These provide a debate platform for sustainable development issues. Importantly, the workshops have generated literature published as case studies. These efforts by the NGRs have greatly enhanced the programs, especially in creating and designating local knowledge (UNU 2018b). Exposure to local knowledge and methods helps fill the gap of over reliance on materials produced in the Global North that sometimes do not fit the African context.

However, the program implementation has challenges. The first of these is the the inability of the program to secure internship tuition and funds, which is a course completion requirement. A second challenge relates to the low enrolment numbers across the universities (for example, only three per year at KU). This problem is partly attributed to the program's costs, and to the fact that ESD programs are new to African universities. The funding issue has affected the envisioned program implementation—because co-teaching and joint research activities are resource intensive. Since these programs are offered by specific departments (see Table 2) it makes it easy to resort to traditional methods of interdisciplinary teaching and research—thereby complicating matters. Where such programs are hosted in dedicated institutes or research centers they tend to do better in pooling expertise across disciplines than when hosted in single departments. Thus, the envisioned co-teaching of the programs with experts across faculties and disciplines was not achieved.

Table 2 SUD program structures at the UoN and KU (Kenya)

Institution	Program	Year 1	Year 2
Department of geography and environmental studies (UoN)	MSc in SUD	• Subjects: Core & electives • Assignments	• Internship • Seminar paper • Research project
Department of environmental sciences and education (KU)			

Table 3 SIRDA program structures at the UI (Nigeria) and KNUST, UDS and UG (all three in Ghana)

Institution	Program	Year 1	Year 2 (MPhil/Master's)
Centre for sustainable development (UI)	MSc in SIRDA (18 months)	• Subjects: Core & electives • Assignments	• Internship • Field-oriented research project
Bureau of integrated rural development and department of agricultural economics, agribusiness and extension (KNUST)	MPhil in SIRDA (18 months)		
Department of agricultural extension, rural development and gender studies (UDS)	Masters in SIRDA (18 months)	• Subjects: Core & electives • Assignments • Seminar paper	• Internship • Field-oriented research project
Department of geography and resource development (UG)			

4.2 *Sustainable Integrated Rural Development in Africa (SIRDA) Program*

The SIRDA program is part of a consortium of four West African universities: Nigeria's University of Ibadan (UI) and Ghana's University of Ghana (UG), University for Development Studies (UDS) and Kwame Nkrumah University of Science and Technology (KNUST) (see Table 3). SIRDA provides capacity to professionals to provide leadership in finding enduring socio-economic and environmental solutions for rural areas. The course structure, classroom and field orientation as well as the internship exposure is similar among all four institutions (see Table 3). Minor variations exist based on differences in local contexts (for example, admission criteria, program duration and tuition). The curriculum combines coursework, a workshop series and field practicum/internship within a minimum of three semesters.

As with all three ESDA programs, the SIRDA structure allows for inter-disciplinary and holistic learning reflective of West Africa's complex developmental issues. The conception of the program came with the realization that the region's sustainable development was hindered by the unacceptable paradox of West Africa's rich natural resource endowments, together with incidences of widespread poverty and poor living conditions. The challenges are pronounced in rural areas were nutritional deficiency, food insecurity, limited access to safe water, low agricultural production capacity, land degradation and abnormal weather and climate is present (Odusola 2017). These complex and interconnected sustainable development challenges in rural areas in Ghana and Nigeria are reflected in the curriculum.

The UI program is coordinated by interrelated departments and faculties offering multidisciplinary courses including energy, waste management, rural sanitation and water management, geographical information systems (GIS), integrated crop management and food security, integrated livestock management, entrepreneurship and management of rural enterprises. Furthermore, it is coordinated by the Centre for Sustainable Development (CESDEV)—a transdisciplinary platform coordinating the sustainable development activities of the university.

Most of the coursework taught at the four universities is done by faculty members; however, the program still depends on external technical and financial resources which at times undermine interdisciplinary learning. AfDB funding only supports research activities and capacity development of NGRs and does not cover other components of student learning. This limits the financial resources available to hire technical experts such as rural extension officers, policy makers and civil society leaders. Due to the financial constraints, the envisioned inter-university faculty and student mobility of the initiative did not materialize. This resulted in the universities coordinating their programs without any student and faculty exchanges to share complementary teaching and research resources (see Table 3).

Despite limited funding, the program implemented community resident practical training using integrated participatory rural development and practice-oriented learning through internships. These activities are often implemented together with the field-based research projects which enriches the learning process. The NGRs used some of their allocated AfDB research grants to support the supervision of student's research activities. The transdisciplinary approach to learning allows for the incorporation of theories, concepts, methods and data from diverse disciplines while co-creating knowledge as well as co-implementation of actions for sustainability by involving various non-academic stakeholders throughout the research—resulting in valuable experience for students and the community. This is captured by the following comment from a graduate (class of 2016):

> The internship component of the … program has transformed my initial vague perspective about rural/community development intervention from the supply-driven viewpoints to a new understanding of working with the people, identifying their resources and building on it. The practical exposure in the form of internship makes the programme very unique and useful for early career development practitioners …

From 2015–2020, the program trained over 50 students/professionals in Nigeria and Ghana through these transdisciplinary field-oriented initiatives. The field-action oriented activities carried out by the UI in Oyo State, Nigeria promoted water, sanitation and hygiene intervention in rural settlements. The project used a "living labs" approach to facilitate co-production of knowledge and collective learning of stakeholders to co-innovate solutions to the community's sanitation challenges. This illustrated the possible long-term social transformations achievable from transdisciplinary action-oriented field activities.

4.3 *Mining and Mineral Resources (MMR) Program*

The Management of Mineral Resources (MMR) program is a two-year master's program jointly offered by Zambia's University of Zambia (UNZA) and South Africa's University of Cape Town (UCT) with students graduating from the university where originally registered (see Table 4). The program aims to "educate and train future managers and leaders to have a high-level understanding of, and sensitivity to, the critical factors of sustainable development in the context of mining and mineral processing in Africa" (UNU 2018a/2018b). The first-year coursework for the joint program is offered in the form of blocks of four core subjects shared among three universities (UNZA, UCT and South Africa's Stellenbosch University [SU]). SU does not admit its own students but hosts those admitted to UNZA and UCT for some subjects. The four core subjects are delivered in blocks of approximately 10 days each in Year 1. During the first and second teaching blocks, the UNZA students travel to UCT and SU for approximately two weeks to participate in three subjects offered by these institutions. During the third block, the UCT students travel to UNZA to participate in classes.

The inter-university collaboration in the design and teaching of subjects as well as the sharing of common module materials, methods and faculty was facilitated by government guaranteed loans from the AfDB to both UCT and UNZA. The national governments of other universities were unable to guarantee AfDB loans for their institutions due to varying national debt and budgetary limitations. The program's financial resources allow for the mobility of students and staff between institutions, while the integrated approach creates synergies which enable new pathways to transformative learning. The multidisciplinary involvement of departments and units across three universities allowed transdisciplinary engagement with the concepts and approaches of sustainable development. Responses by graduates and employers

Table 4 MMR program structures at Zambia's UNZA and South Africa's UCT

Institution	Program	Year 1	Year 2
Department of metallurgy and mineral processing (UNZA)	MPhil in sustainable mineral resource development	• Subjects: • Core & electives* • Assignments	• Internship • Research project
Department of chemical engineering (UCT)			

*At UCT, Master of Science in Engineering students can take some of the subjects as electives

suggest that the program has delivered the desired human capital—especially the "scholars" and "leaders" —relevant to job market demands and research skills based on Africa's sustainable development demands. Quotes from two graduates and an employer attending one of the ESDA consortium meetings reflect this:

> The MMR was my second [master's] and the structure of the program with different interconnected strands motivated me to want to continue studying as I enjoyed the experience (2017 MMR student currently enrolled in USA PhD programme)

> Before I enrolled … I didn't know anything about systems thinking and sustainability and how I could apply it to my job. The learning experience from lectures and fieldwork in both South Africa and Zambia was really valuable. Now I can teach [a] thing or two to my engineering colleagues (2018 MMR graduate employed by a South African mining firm)

> As an employer, we usually have to retrain our graduates because they don't understand complex nature of doing business in Africa, especially the mining sector. This program is perfectly designed to nurture a professional who can deal with multiple and complex stakeholders at different levels (Director of Sustainability at a Zambian copper mine during 2018 ESDA-MMR Sustainable Development Workshop)

Year 2 involves internships and an analytical research project on key sustainability challenges in the mining sector. The modular format of the program attracts students who are employed full-time, many of whom hold senior management positions in mining companies in Zambia and South Africa. Consequently, students partake in the internship programs with their employers to immediately apply the principles taught in real-world scenarios. Students in this particular internship process do not explicitly experience a transdisciplinary approach to sustainability transformation in their organizations. This is because companies often impose boundaries which inhibit the engagement with other stakeholders in knowledge co-creation and co-implementation of sustainability actions. Many students report that mining conglomerates have brands to protect. Therefore, to reduce brand risk exposure, the integration of sustainability actions is often limited to corporate social responsibility (CSR) initiatives which are narrow organizational activities with little to no dynamic interaction with actors outside the mining company. This experience is opposite to student's experiences with non-profit organizations. The high tuition fees for the program, particularly at UNZA means that 80% of students study using bursaries from mining companies which employ them, while the other students are self-funded. Although an AfDB loan was awarded, the funding was not for scholarships but program activities such as faculty exchange, curriculum development and research activities.

Despite limited transdisciplinary engagement opportunities, the student's diverse experiences allow for peer-learning and classroom idea exchanges. Further, the inter-university collaboration in the program provides the opportunity that the second-year research projects be structured in such a way that students from both institutions are allocated to transdisciplinary research teams supervised by the NGRs, using their AfDB research grants. This approach was partly implemented in the program to compel employed students to commit to and complete their research projects.

Through this transdisciplinary team-based research approach, the research projects created "living labs" settings which allowed students and researchers to learn and implement applied research methods directly to co-create knowledge and co-innovate solutions to "real-world" problems, in partnership with non-academics (Almirall and Wareham 2011; Gasco 2017). One such "living labs" project is the Konkola Copper Mine (KCM) in Nampundwe, Zambia—a collaboration between MMR NGRs and students, together with their counterparts at Japan's University of Tokyo. The project seeks to develop a framework that will induce "autonomous" development of local communities who are over-dependent on mining activities/companies to support their livelihoods. Research shows that there are limited post-mining adaptation strategies in place in many communities once mining operations cease. The project seeks to collaboratively support these communities to develop pathways to sustainable livelihoods without being over-dependent on mining. Outputs from this research resulted in contributions towards the books and policy briefs produced by the ESDA NGRs (Nagao et al. 2019a; Nagao et al. 2019b).

These "living labs" projects illustrate how ESDA's work has the potential to produce research outputs and transform lives. Thus, ESDA's success in this regard is reflected in the the following comment by an external evaluator:

> …outcomes and impact is strongly reliant on its ability to (i) establish effective 'boundary-spanning' scholarship and leadership amongst its researchers and students, and (ii) engage more explicitly with the balance between demand for and supply of specific types of human capital and knowledge (ESDA External Evaluator) (Ofir 2018).

5 The Next Generation Researchers (NGRs) Project

During the conceptualization and initial implementation phase of the ESDA program, there were several concerns linked to the underdeveloped nature of ESDA. These included: (1) the limited ESDA research and learning material and (2) the limited research capacity of African researchers involved in ESD (UNU 2013b). Responding to these concerns, ESDA initiated the NGRs project to develop the research capacity of faculty members of participating universities to enable them to generate research outputs that would enrich the programs by providing research-informed instructional material and supporting fieldwork. Illustrative of the collaborative research by the NGRs, eight coordinators were nominated in 2016 by the vice chancellors of their respective universities to each receive a US$50,000 grant from the AfDB to conduct transdisciplinary research on the theme of rural–urban connections (Kapfudzarura et al. 2018). The theme was selected because it cuts across the three programs and the connection between rural and urban areas is an important process in the socio-ecological and economic sustainability of both spaces (Ingelaere et al. 2017; Tacoli and Satterthwaite 2014). The research outputs included two books used as instruction materials in the programs (Kapfudzarura et al. 2018; Kapfudzarura et al. 2018).

For the research component, each coordinator was requested to form a team of researchers from different disciplines within their universities. At KNUST, the NGRs team investigated uneven industrial growth and its environmental implications on rural and urban communities. The team comprised of agricultural economists, environmentalists, land surveyors and GIS experts. The team's composition, coupled with the diverse research techniques being used, emphasized the importance of transdisciplinarity to understand and address "wicked problems" linked to sustainable development challenges in Africa (Porter et al. 2006; Rafols and Meyer 2010). After the conclusion of the rural–urban connection project, the reflection and evaluation phase revealed the impact of the research—the need for more inter-university collaboration. The researchers realized that a comparative study across countries involving researchers at the universities would have allowed more peer-learning, cross-fertilization of ideas and generalizable data useful to increase the impact across Africa. Consequently, it was agreed that the next phase of research by the NGRs, including the 2021 fieldwork, would be collaborative among all the NGRs at all the ESDA universities (UNU 2018a).

The implementation of the rural–urban connection research project revealed pertinent research capacity limitations not identified at the start of the project. The increasing research activity and competitive research environment among the continent's researchers was found to sometimes result in unethical research practices due to low capacity in the area. The former ESDA chair noted:

> … we need to nurture a culture of research ethics amongst our young scholars if we are to grow a new crop of world class researchers from the continent. We had incidences at my institution which we should avoid amongst our [NGRs] (UNU 2017)

Furthermore, several NGR coordinators had limited research management and writing experience. This resulted in a five-day workshop held in 2018 in partnership with the research office of UCTs to provide research capacity development to the NGRs. In 2020, before the disruptions caused by the COVID-19 pandemic, ESDA in partnership with the University of Pretoria (UP) in South Africa received a US$250,000 grant from the Carnegie Cooperation of New York (CCNY) to coordinate another five-day African Early Career Researchers Meeting (AECRM), to assemble over 70 NGRs. The AECRM allows the NGRs to network, share experiences and acquire skills through transdisciplinary team-based activities and training with support for senior academics and professionals (Nepomuceno 2020).

6 ESDA Sustainable Development Workshops

After the 2013 launch and implementation of the ESDA programs the staff involved faced two immediate tasks:

1. How to continually improve the quality of programs offered using pedagogical approaches reflective of Africa's socio-cultural and economic context; and
2. How to proactively contribute to the socio-ecological and economic needs of society.

To address both tasks, ESDA decided to implement a Sustainable Development Workshop series involving all partner universities. Since 2017, each university is expected to organize two annual workshops with a common structure—the first, to discuss theories and methods for sustainable development studies; the second, to share case studies with stakeholders at national and local level in an effort to create a science-policy interface. In 2017 and 2018, the first workshops on theories and methods resulted in an edited volume designed to enrich the instruction through the program. This was considered a priority because most of the sustainable development curriculum used in Africa is imported from the Global North. It resulted in an increased emphasis within ESDA to develop course material and research approaches which are responsive to the African socio-cultural context. This desire within ESDA is linked to current debates on decolonizing the African higher education curriculum (Mahabeer 2020; Vandeyar 2020).

To address the second task, consideration was given to the need to nurture the convening power of ESDA universities in creating science-policy interfaces which support evidence-based decision-making. Consequently, the second ESDA workshop series focused on sharing the research outputs produced through interactions and dialogues with political, industrial and civil society leaders as well as regional and international organizations. The UCT program comprised of a series of collaborative workshops with external stakeholders during the annual Cape Town Mining Indaba—an international conference on mining investment in Africa for mining sector policymakers and professionals.

These workshops produced a set of policy briefs which have been useful in integrating sustainable development in mining activities. In addition to ensuring that knowledge generated by researchers has an impact on both corporate and national policies as well as mining sustainability community activities, they also seek to secure the understanding and support of external partners on specific follow-up activities the universities should carry out. Towards the end of ESDA's first phase in October 2018, the consortium meeting discussions recognized the value and impact of combining these workshops on specific themes aligned to the programs to enable the universities to share their complementary strengths. UNZA and UCT held one of these first "ESDA Open Days" in December 2019 in Lusaka to interact with political, industrial and civil society leaders from across the continent in support of Africa's Mining Vision.

7 Key Lessons

ESDA's mandate and focus highlight the relevance and importance of African universities collaborating through transdisciplinary approaches to co-create impactful knowledge and producing the type of human capital required to advance Africa's sustainable development. The reflection in this chapter on the ESDA programs, the NGRs project and the ESDA workshop series using a reflective methodology, provided an outline how the initiative attained this goal.

Despite the resource and capacity constraints in the SIRDA and SUD programs, the inter-university collaborative nature of the MMR program shows that African universities should not only seek excellence in research and education individually but also work collectively toward effective interaction with society and significant impact on sustainable development. The MMR program also shows that through its design, it allows student and faculty exchange in joint-program offerings (teaching and learning), transdisciplinary co-creation and sharing of knowledge between UCT and UNZA. ESD programs that are designed to foster cooperation based on sharing and co-creating information, insights and pedagogical approaches will contribute to sustainable development in Africa. Further, the responses by graduates of the MMR program indicate that the sharing of experiences and cross-fertilization of ideas nurtured during the interdisciplinary learning phase is ideal to produce human capital with creative and complex problem-solving skills required in tackling Africa's sustainability problems. The reflections by employers during ESDA consortium meetings point towards the need for this type of human capital (Ofir 2018).

The other components of the ESDA program, particularly, the field-based research activities and internships, highlighted the importance of applying transdisciplinary approaches in learning, co-creation and sharing of knowledge with non-academic stakeholders to ensure greater societal impact. This is evidenced in the "living labs" NGRs projects as part of the MMR program, the field-based community residential practical training supported by NGRs and the practice-oriented learning through internships in the SIRDA program. Research on ESD by O'Byrne et al. (2015), Kudo et al. (2018) and San Carlos et al. (2016) show how valuable transdisciplinary is in acquiring:

1. Systemic knowledge that enables the understanding of the challenges of the mechanisms and dynamics of sustainable development challenges;
2. Normative knowledge that incorporates multiple perspectives based on different values and social norms; and
3. Design-thinking knowledge that enables co-design and co-innovation to initiate concrete actions that will lead to social transformation for sustainability.

The value of transdisciplinary approaches in field-oriented activities combined with the interactions across universities and departments evidenced by the ESDA Sustainable Development Workshops, show how universities are re-imagined to co-create and share knowledge as well as co-innovate solutions to societal problems. The 2019 SUD Sustainable Development Workshop saw the participation of all the universities, several stakeholders and the UNZA and UCT "ESDA Open Days" workshops. This revealed that for universities to have a greater impact. their efforts should be collective since this will permit the creation of a stronger knowledge and technical base to mobilize resources from and experiences of partners in industry, government and civil society.

The capacity limitations of the NGRs in research ethics, project management and writing skills illustrate the importance of practical approaches to capacity development of African NGRs. Given the long time horizon set for sustainable development in Africa—AU Agenda 2063—it may be worthwhile to emphasize the role of the next generation in considering leadership and ownership of development efforts. Fundamentally, they require the necessary capacity development. Given the limited

ESD learning material suitable for African contexts, these NGRs will play a critical role in shaping pedagogies reflective of sustainable development concerns in Africa.

8 Conclusion

Though launched 2013, the ESDA programs started at different times at different universities. The reflective methodology used in this chapter to analyze these programs provided valuable insights concerning the importance of collaboration between African universities in human capacity development and research to create a more significant societal impact. For Africa to generate its own knowledge and foster future talents that will tackle the sheer scale of the sustainability challenges that the continent is facing, it is essential for African HEIs to collaborate beyond academic disciplines and across geographical borders. The inter-university nature of ESDA adopts such boundary-spanning perspectives in its curricular implementation that emphasizes fieldwork, integrative thinking, holistic approaches, problem-solving orientation and interdisciplinary research. Kudo et al. (2018) notes that sharing the experiences of program operations between partner universities and the Africa-Asia knowledge exchange scheme within ESDA, illustrated the value of peer-learning and facilitated co-generation of knowledge for sustainable development. The increased participation by the NGRs participation in collaborative research to produce ESDA learning materials and student supervision serves as a collective investment by ESDA in its younger faculty members who will be responsible for the promotion of ESDA in future.

However, the ESDA program implementation has its challenges. Though some participating universities have advanced and graduated a number of cohorts, others are yet to graduate their first students, while one—UG has not admitted its first cohort of students as yet. This lack of uniformity in implementation, coupled with low graduation rates has afforded a reflection based on only the limited experiences of a few ESDA graduates. Therefore, the low enrolment and graduation rate challenge created a significant limitation to the research. In contrast with the latter limitation (the corroboration of information from different sources), reflections of ESDA coordinators, program reports and external evaluation reports provided an in-depth assessment of the experience of implementing the ESDA programs. In future, there will be a need to increase investment in ESD programs at national, regional and continental level in Africa to attract more students and offer quality training covering the coursework, fieldwork and research components.

Acknowledgements This work was supported by the AfDB through the Japan Trust Fund as allocated by the Ministry of Finance.

References

AAU (2011) Mobilizing regional capacity initiative. Accra

Allmirall E, Wareham J (2011) Living labs: Arbiters of mid- and ground-level innovation. Technol Anal Strateg Manag 23(1):187–102

Ayaga G, Kibata G, Lee-Smith D, Njenga M, Rege R (2005) Policy prospects for urban and peri-urban agriculture in Kenya. KARI, Urban Harvest – CIP and ILRI

Bilson A (2006) Rationalities, reflection and research. In: Fook J, Gardner F, White S (eds) Critical reflection in health and social care. Open University, Berkshire, pp 135–143

Fook J (2011) Developing critical reflection as a research method. In: Bridges D, Higgs J, Titchen A (eds) Creative spaces for qualitative researching: Practice, education, work and society, Sense, Rotterdam, vol 5, pp 55–64

Fuh D (2014) An annotated bibliography of publication by scholars and researchers based in and affiliated to African universities and research institutions. Tōdai GPSS–GLI, Kashiwa

Gandhi D (2020) Figures of the week: public spending on education in Africa. Africa Focus. Brookings

Gasco M (2017) Living labs: Implementing open innovation in the public sector. Gov Inf Q 34(1):90–98

Ingelaere B, Christiaensen L, De Weerdt J, Kanbur R (2017) Why secondary towns can be important for poverty reduction – a migrant's perspective. The World Bank, Washington

Jacobi PR, Toledo R, Grandisoli E (2016) Education, sustainability and social learning. Braz J Sci Technol 3(3):19–39

Joseph C, Said R (2020) Community-based education: a participatory approach to achieve the SDGs. In: Azul AM, Brandli L, Filho WL, Ozuyar PG, Wall T (eds) Quality education. Springer, Cham, pp 123–145

Kapfudzaruwa F (2018) The future of youth entrepreneurship and sustainable industrialization. In: Kapfudzaruwa F, Mutisya E, Nagao M (eds) Youth entrepreneurship and Africa's sustainable industrialization. Spears Media, Denver, CO., pp 204–224

Kapfudzarura F, Kudo S, Mfune O, Hansen M, Nyerere J (2018) (eds) Rural-urban linkages and sustainable development in Africa. Spears Media, Denver, CO

Kates RW, Dasgupta P (2007) African poverty: a grand challenge for sustainability science. PNAS USA 104(43):16474–16750

Kudo S, Mursaleen H, Ness B, Nagao M (2018) Exercise on transdisciplinarity: lessons from a field-based course on rural sustainability in an aging society. Sustainability 10(4):12–28

Lozano-García FJ, Gándara G, Perrni O, Manzano M, Hernández DE, Huisingh D (2008) Capacity building: a course on sustainable development to educate the educators. Int J Sustain High Educ 9(3):257–281

Mahabeer P (2020) Decolonizing the school curriculum in South Africa: black women teachers' perspectives. Third World Thematics TWQ J 5(1):97–119

Maturana H (1988) Reality: the search for objectivity or the quest for a compelling argument. Ir J Psychol 9(1):25–82

Mezirow J (1998) On critical reflection. AEQ 48(3):185–198

Mogensen F, Schnack K (2010) The action competence approach and the 'new' discourse of education for sustainable development: competence and quality criteria. Environ Educ Res 16(1):59–74

Mugabe J (2018) Education for ESDA external evaluation: accountability component. UNU ESDA Programme, Tokyo

Munkebye E, Scheie E, Grabrielsen A, Jordet A, Misund S, Nergard T, Øyehaug B (2020) Interdisciplinary primary school curriculum units for Sustainable Development. Environ Educ Res 26(6):795–811

Nagao M, Masinja J, Alhassan A (2019a) Sustainable development in Africa. Spears Media Denver, CO

Nagao M, Broadhurst JL, Edusah S, Awere KG (2019b) Sustainable development in Africa: concepts and methodological approaches. Spears Media, Denver, CO

Nagao M, Mutisya E, Kudo S (2018) The role of higher education in developing sustainability. In: Bakar R, Ismail A (eds) Sustainability of higher education: a global perspective. Penerbit Universiti Sains, Malaysia, pp 135–159

Nagao M (2009) Sustainability development curriculum: insights for higher education in Africa. Tōdai GPSS–GLI, Kashiwa

Nagao M (2010) Report on sustainable development curriculum in Africa workshop. Tōdai GPSS–GLI, Kashiwa

Nagao M (2011) Concept note for the development of a master's program in education for sustainable development in Africa. UNU, Tokyo

Nepomuceno K (2020) AECRM. https://www.up.ac.za/peer-learning-for-emerging-researchers-knowledge-and-advancement-perka-program/article/2877361/about-aecrm. Accessed 28 Sept 2020

Nyerere J (2018) Youth unemployment in Kenya: incorporating entrepreneurial and transferable skills education. In: Kapfudzarura F, Mutisya E, Nagao M (eds) Youth entrepreneurship and Africa's sustainable industrialization. Spears Media, Denver, CO

Nyerere J, Mfune O, Fuh D, Sulemana N, Mutisya E, Yiran G, Fadairo O, Ameyaw J, Odingo A (2014) The role of higher education in building a sustainable African society. AJSD 4(3):17–38

O'Byrne D, Dripps W, Nicholas KA (2015) Teaching and learning sustainability: an assessment of the curriculum content and structure of sustainability degree programs in higher education. Sustain Sci 10(1):43–59

Odusola A (2017) Agriculture, rural poverty and income inequality in sub-Saharan Africa. In: Bhorat H, Conceiçao P, Cornia GA, Odusola A (eds) Income inequality trends in sub-Saharan Africa: divergence, determinants and consequences. UNDP, Regional Bureau for Africa, New York, NY

Ofir Z (2018) Education for Sustainable Development in Africa (ESDA) external evaluation: strategic review component. UNU ESDA Program, Tokyo

Olaniyan O, Fadairo OS (2019) Rural development and household livelihood. UI, Ibadan

Pauw J, Gericke N, Olsson D, Berglund T (2015) The effectiveness of education for sustainable development. Sustainability 7(11):15693–15717

Popoola L (2014) Survey of African universities' engagements in the area of Sustainable Development. An overview report on Sustainable Development in Africa. Tōdai GPSS–GLI, Kashiwa

Porter AL, Roessner JD, Cohen AS (2006) Interdisciplinary research: meaning, metrics and nature. Res Evol 15(3):195–197

Pryshlakivsky J, Searcy C (2013) Sustainable development as a wicked problem. In: Kovacic SF, Souza-Poza A (eds) Managing and engineering in complex situations. Springer, Dordrecht, pp 109–128

Rafols I, Meyer M (2010) Diversity and network coherence as indicators of interdisciplinarity: case studies in bionanoscience. Scientometrics 82(2):263–287

San Carlos RO, Tyunina O, Yoshida Y, Mori A, Sioen GB, Yang J (2016) Assessment of fieldwork methodologies for educational purposes in sustainability science: exercise on resilience, Tohoku Unit 2015. In: Akiyama T, Chiahsin C, Esteban M, Ikeda I (eds) Sustainability science: field methods and exercises. Springer, Cham, pp 67–91

Santoianni F (2017) Models in pedagogy and education. In: Bertolotti T, Magnani L (eds) Springer handbook of model-based science. Springer, Cham, pp 1033–1049

Sinakou E, Pauw JB, Van Petegem P (2017) Exploring the concept of sustainable development within education for sustainable development: implications for ESD research and practice. Environ Dev Sustain 21(3):1–10

Summers M, Childs A (2007) Student science teachers' conceptions of Sustainable Development: an empirical study of three postgraduate training cohorts. Res Sci Technol Educ 25 (3):3017–3027

Tacoli C, Satterthwaite D (2014) Seeking an understanding of poverty that recognizes rural-urban difference and rural-urban linkages. In: Loyd-Jones T, Rakodi C (eds) Urban livelihoods: a people-centered approach to reducing poverty. Routledge, London, pp 75–93
UN (1992) Report of the UN conference on environment and development. Chapter 36 of Agenda 21: Promoting education, public awareness and training. https://www.un.org/esa/documents/ecosoc/cn17/1997/ecn171997-2add26.htm, Last Accessed 15 July 2020
UNESCO (2015) ESD – Building a better, fairer world for the 21st century. https://unesdoc.unesco.org/ark:/48223/pf0000216673, Accessed 20 June 2020
UNESCO (2019) Framework for the implementation of ESD beyond 2019. 40 C/23. https://www.oneplanetnetwork.org/sites/default/files/40_c23_framework_for_the_implementation_of_esd_beyond_2019.pdf, Accessed 19 Apr 2020
UNU (2019) ESDA long-term strategic plan 2020–2025. Tokyo
UNU (2018a) ESDA 2018 consortium and symposium report. Tokyo, 29–30 October. Tokyo
UNU (2018b) ESDA master's programs: training leaders for development in Africa. Tokyo
UNU (2017) ESDA 2017 consortium meeting summary report. Cape Town, 16–18 October. Tokyo
UNU (2016) ESDA 2016 consortium meeting summary report. Lusaka, 17–19 October. Tokyo
UNU (2015) ESDA 2015 consortium meeting summary report. Nairobi, 19–21 October. Tokyo
UNU (2014) ESDA 2014 consortium meeting summary report. Ibadan, 20–22 October. Tokyo
Vandeyar S (2020) Why decolonizing the South African university curriculum will fail. Teach High Educ Crit Perspect 25(7):783–796
Ventakataraman B (2009) Education for sustainable development. Environ Sci Policy 51(2):8–12
Walshe N (2017) An interdisciplinary approach to environmental and sustainability education: developing geography students' understandings of sustainable development using poetry. Environ Educ Res 23(8):1130–1149
Worldometer (2020) Kenya Demographics. https://www.worldometers.info/demographics/kenya-demographics/, Accessed 24 Mar 2020

Jackline Nyerere is a Senior Lecturer of Educational Leadership and Policy at Kenyatta University, Kenya. Her focus is on internationalisation of education, open and flexible learning, and sustainability. Her recent studies include Academic student mobility in Africa, and transnational research partnerships in East Africa. She is currently a co-principal investigator in the transforming universities for a changing climate research project. She is a member of the African Network for Internationalisation of Education, and she co-chairs the Education for Sustainable Development in Africa – Next Generation Research (ESDA-NGR) Network.

Farai Kapfudzaruwa holds a doctoral degree in Business Administration from the Graduate School of Business at the University of Cape Town in South Africa. His research interests focus on the role and effectiveness of private governance in the provision of public goods, mostly in Africa and Asia. Kapfudzaruwa is currently based at the University of Pretoria within the Strategic International Partnerships office. He is also the project coordinator for the Education for Sustainable Development in Africa (ESDA) initiative – a collaborative graduate and research programme between eight leading African universities with support from the African Development Bank.

Olushola Fadairo is a senior faculty member in Agricultural Extension & Rural Development, University of Ibadan. Has research interests in rural-urban linkages, integrated rural development and sustainable rural livelihoods. He was a visiting fellow at the University of Cape Town, South Africa and a member of the United Nations Next Generation of Researchers team.

Alice A. Oluoko-Odingo is a Senior Lecturer in the Department of Geography and Environmental Studies, University of Nairobi. She holds Honours Degree in Geography and Master of Arts and PhD in Agricultural Geography from the University of Nairobi. She has research focus and publications in agriculture and food security, poverty and livelihoods, climate change, Energy, Gender and disaster risk reduction, Ecosystem management, agro-biodiversity, health cleaner production, and sustainability issues. She is immediate former Deputy Chairperson of the Education for Sustainable Development in Africa – Next Generation Research (ESDA-NGR) Network.

James Manchisi holds a PhD in Chemical Engineering from the University of Birmingham (United Kingdom). His research interests include the development of chemical processes for sustainable exploitation of mineral resources and environmental issues of mining projects. James currently teaches hydrometallurgical process engineering, chemical thermodynamics and environmental stewardship courses at the University of Zambia. He is also an academic staff member of the Education for Sustainable Development in Africa – Next Generation Research (ESDA-NGR) Network.

Shogo Kudo is an Assistant Professor at the Graduate Program in Sustainability Science – Global Leadership Initiative, Graduate School of Frontier Sciences, The University of Tokyo. His research proposes a translocal learning approach, which is a new form of collective learning across rural communities in different socio-economic, cultural and political settings. He serves as Co-coordinator in Education for Sustainable Development in Africa (ESDA)program and organizes Africa-Asia knowledge exchange activities for early career researchers.

The Challenges and Opportunities for Climate Change Education at Makerere University in Kampala Uganda

David Ssekamatte, Karsten Speck, and Bernd Siebenhüner

Abstract Societies across the globe are battling with the effects of climate change as seen in the scientific evidence produced periodically by the Intergovernmental Panel on Climate Change (IPCC). Africa has been identified as more vulnerable to climate change given its weak governance systems, low technology, dependency on ecological resources for livelihood as well as conflict and high poverty levels. Educational responses to climate change though recognized, are yet to be explored by various countries on the continent. The researchers conducted a qualitative study to examine the perspectives and views of university lecturers, administrators and students on climate change related courses and programmes regarding opportunities and challenges for climate change education at Makerere University as part of a larger multiple comparative study in East Africa. A case study design was adopted. Semi-structured in-depth interviews and focus group discussions were conducted to collect data from a total of 25 participants at the case university. Data was analysed using thematic analysis based on Braun and Clarke (2006) with the help of MAXQDA software. Findings revealed a wide range of perspectives and views regarding the challenges and opportunities for climate change education at this university. It is expected that the findings of this study may contribute to empirical scientific knowledge on climate change education at universities in the African context.

Keywords Challenges · Opportunities · Climate change education · Makerere University

D. Ssekamatte (✉)
Department of Management, Uganda Management Institute, P. O Box 20131, Kampala, Uganda

K. Speck
Faculty of Education and Social Sciences, Carl von Ossietzky University of Oldenburg, Ammerländer Heerstraβe 114-118, 26129 Oldenburg, Germany
e-mail: karsten.speck@uni-oldenburg.de

B. Siebenhüner
Faculty of Computer Science, Economics and Law, Carl von Ossietzky University of Oldenburg, Ammerländer Heerstraβe 114-118, 26129 Oldenburg, Germany
e-mail: bernd.siebenhuener@uni-oldenburg.de

W. Leal Filho et al. (eds.), *Sustainable Development in Africa*, World Sustainability Series,
https://doi.org/10.1007/978-3-030-74693-3_12

1 Introduction

Experts especially those of the Intergovernmental Panel on Climate Change (IPCC) continue to create awareness about the challenge of climate change and how it is/will continue impacting societies globally. Scholars such as Stern (2008); Farmer and Cook (2013); Filho (2015); Giugni et al. (2015) and Vale (2016) have written extensively on the same challenge warning that if actions are not taken to mitigate it, humanity will pay a huge price in all aspects of life. Kagawa and Selby (2010) in their introduction to education and climate change: living and learning in interesting times, contend that;

> ...there is widespread consensus among the international scientific community that human-induced climate change is happening and that, even with the most immediate and sustained mitigation efforts, there will be no escaping some time-lagged catastrophic consequences from past emissions of carbon dioxide and other gases (p. 1)

Therefore, all this evidence points to the fact that climate change is a serious and urgent issue that needs to be dealt with by everyone at all levels. Education is poised to play a key role in the battle against climate change. This role has been extensively written about by various scholars with a number of illustrative examples of how climate change issues are being integrated into education programmes (UNESCO (2011). Boyde and Hume (2015) aver that education at all levels can play a key role in addressing climate change through their curriculum and co-curricular activities. Higher education is particularly key given its level of influence on other sectors in terms of capacity building, scientific research, and community engagement functions. Universities could engage students on climate change issues through integration in their study courses and programmes. They can also conduct scientific research on climate change and based on the findings come up with innovative technologies and products for mitigation and adaptation purposes.

In addition, Kotecha (2010) presents the idea that African universities in particular can develop long term and short-term training interventions on climate change, offering short courses, academic and research programmes on the phenomenon. They could also effectively engage communities and governments to adopt climate friendly practices as well as policies. Thus, universities need to take lead in promoting behavioural change among actors (Virtenen 2010).

This paper presents findings of a larger study that investigated opportunities and challenges of climate change education at universities in African context at two (2) universities (Makerere University and University of Dar es salaam) in East Africa. The findings being presented in this particular paper emerged from the analysis of data collected at Makerere University in Uganda. The study sought perspectives, views and experiences of university lecturers, administrators, researchers and students on climate change related courses and programmes regarding opportunities and challenges for climate change education at Makerere University. The article begins with the methodology followed in undertaking the study and then presents the key findings on opportunities and challenges with illustrative excerpts from the participants.

The paper then provides a discussion of key findings and how they relate to existing literature and a short conclusion.

2 Methods

2.1 *Approach and Design*

Using a social constructivist lens, the researchers sought to examine the lived experiences and perspectives of study participants with respect to the opportunities and challenges for climate change education at the case university. The lens is known for enabling the researcher to construct reality by interacting with participants in their own context and therefore derive multiple meanings of the phenomenon under study as participants share their experiences and perspectives (Guba and Lincoln 2005). The qualitative approach was adopted to "describe and understand the meanings" that study participants attach to the phenomenon from different perspectives (Polkinghorne 2010, p. 425). Using this approach, the researchers were able to analyse both individual and group experiences as well as perspectives at the case university (Flick 2007). The research design adopted was the case study design taking Yin's model. Yin (2003) explained a case study as "empirical inquiry that investigates a contemporary phenomenon within its real-life context" (p.13). This explanation guided the researchers to undertake the study by interacting with lecturers, administrators and students on climate change related course and programmes within their university context. Given Yin's typology, the researcher adopted a holistic descriptive case study design in order to harvest "rich and revealing insights" regarding what the participants perceive to be the opportunities and challenges for climate change education at their university (Yin 2012, p. 49).

2.2 *Sample Size and Sampling Procedures*

At Makerere University, data was collected from 25 participants. These included three (03) administrators, four (04) lecturers, two (2) researchers and sixteen (16) undergraduate and postgraduate students in the college of Agriculture and Environmental sciences. These participants came from various units of the college including the Makerere University Centre for Climate Change and Innovations (MUCCRI), the department of geography and school of forestry. The selection of these units and the participants was purposively done because the units house courses and programmes related to climate change and environmental sustainability. The students were mainly those from bachelor of science (undergraduate) and Master of science in meteorology as well as PhD in Geography (postgraduate) programmes. The participants are knowledgeable and experienced on issues of climate change as researchers, lecturers, and administrators while the students undertake courses and programmes related to climate change in these units. The lecturers, researchers and administrators were selected to participate in semi-structured in-depth interviews while the students were selected to participate in Focus Group Discussions (FGDs).

2.3 Data Collection

Data was collected using document review, semi-structured in-depth interviews, and Focus Group Discussion (FGDs). These are explained below;

Document Review

At Makerere University data collection started with doing a document review. The researchers collected a variety of official and personal documents including annual reports, the university strategic plan documents, programme and course manuals, research agenda and publications, policy documents and other relevant internal and external documents related to climate change study, research, and policy programmes. These documents provided rich data that was useful to generate background information about the case university, the units selected for the study and the programmes as well as courses offered related to climate change (Berg and Lune 2012). The review also included collecting data from curriculum documents, university prospectuses, university research policy documents, national climate change policies and strategies as well as the state of environment reports for Uganda. The analysis of these documents greatly informed the study especially collecting contextual information about the case university and the units as well as participants under study.

Semi-structured In-Depth Interviews

Data from lecturers, researchers and administrators on climate change related courses and programmes was collected using semi-structured in-depth interviews. Kvale (2007) explains that an interview allows the researcher to ask questions and listen to participants as they tell their views, lived experiences, opinions as well as their social life regarding a phenomenon. During an interview, the researcher interacts with the participant or interviewee and together constructs knowledge about the phenomenon being studied. Through semi-structured interviews the researchers engaged lecturers, researchers, and administrators to discover the ways in which they experienced and understood their reality with respect to the opportunities and challenges for climate change education at their university (Kvale 2007).

The researchers adopted Kvale's process of undertaking interviews as illustrated below (Fig. 1);

The interviews started with thematizing that involved formulation of key aspects of the study and coming up with interview questions which were checked for validity

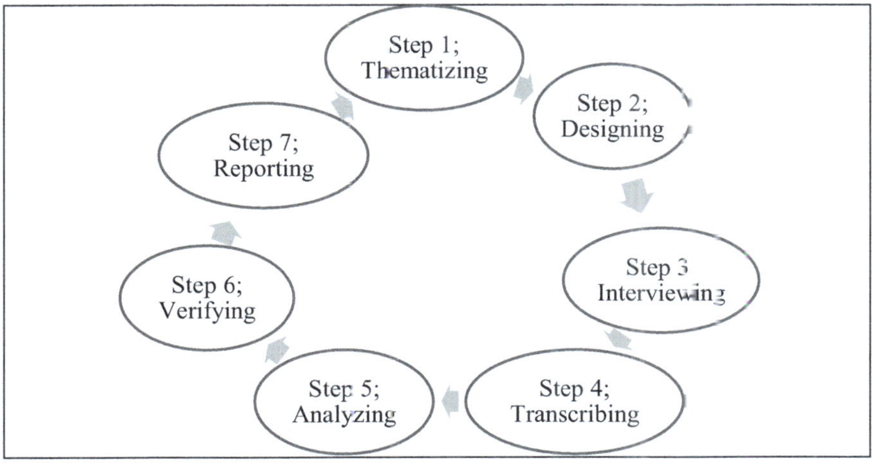

Fig. 1 Stages of an interview *Source* Adapted from Kvale (2007)

and ethical considerations. The next stage was to design the interview guide and procedures that would be followed for the interviews as well as a schedule which was agreed upon by the participants. Interviewing took place at different times depending on the schedule and availability of the participants at the university. The researchers and the note taker organized all the materials including notebooks, information for participants and consent forms, the audio recorders and interview guides before going to the field. Interviewing was done accordingly, and responses recorded using an audio recorder.

After interviewing all the nine (09) participants at the case university, recorded data was transcribed and prepared for analysis. Transcription was done by listening to the audio recording several times and typing the script, ensuring no transcription errors are made. The interview transcription process ensured proper sentence structure, quotation marks and that mistaken words as well as omissions were avoided (Braun and Clarke 2013). The transcripts were later verified and prepared for analysis and reporting.

Focus Group Discussion

Data from undergraduate and postgraduate students at the case university was collected using Focus Group Discussions (FGDs). Barbour (2007) argues that focus groups need to be homogenous to allow members of the group to interact freely within the group as the discussion progresses. Therefore, undergraduate students constituted a focus group since they could easily speak or talk freely with each other without any intimidation from the postgraduate students. The same applied to the postgraduate students comprising of masters and PhD students on courses and programmes related to climate change. Each of the focus groups were made up of eight (08) participants that were selected using simple random sampling based on the list of students provided by the heads of departments geography and meteorology at the university. The selected students were mobilized and assembled in different rooms where the discussions were held for an hour per group with the help of a focus group discussion guide and an audio recorder. All focus group data was recorded, and field notes compiled. The method provided an opportunity to examine the diverse ways in which individual participants (students) collectively make sense of the opportunities and challenges for climate change education at their university (Bryman 2008).

2.4 Data Analysis and Interpretation

Analysis of interview and focus group data was done using Braun & Clarke's thematic analysis model with the help of MAXDA software. The analyst began with transcription of all the recorded audio data to ensure it is ready for analysis. After transcription, errors in sentence structure, punctuation, mistaken words, or phrases as well as omissions were checked to ensure validity of the transcripts (Braun and Clarke 2013). In the process of preparing the transcripts for analysis, the analyst developed transcription notations, and participant identifiers to ensure anonymity. Then transcripts were imported into MAXQDA software. Thematic analysis was done following Braun and Clarke (2006) model as shown below (Fig. 2);

The process generated initial codes which were got through open coding of the entire data set and these were later categorized based on the research questions and theoretical framework. The initial codes gave rise to a well-designed code book that enable the analyst to search for possible groupings or categories. The analyst searched for related codes and kept on grouping them into conceptually and empirically grounded categories. This is what Merriam and Tisdell (2016) referred to as axial coding. The axial

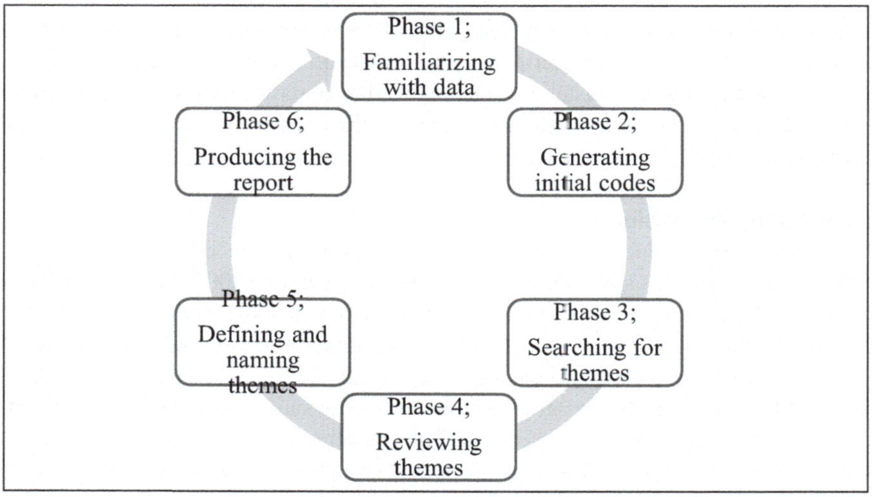

Fig. 2 Thematic analysis process used *Source* Adapted from Braun and Clarke (2006)

coding enabled the search for potential or candidate themes which were reviewed for their relevance, validity, and empirical soundness. The process led to identification of final themes and the corresponding categories and therefore a final thematic map was generated. The analyst then critically reviewed each of these themes and defined the scope and content for each. The last stage was to write down the analytical description of each theme including selection of illustrative excerpts or quotations from the transcripts (Braun and Clarke 2006).

2.5 *Ethical Considerations*

Every researcher has ethical obligations to the scientific community ensuring that ethical issues within their research endeavours are considered through out the research process (Berg and Lune 2012). To ensure ethical research practice the study was cleared by the ethics commission at Carl von Ossietzky University of Oldenburg, and Uganda National Council of Science and Technology before going to the field. All participants filled an informed consent form and their privacy and confidentially were ensured during data collection and analysis as well as writing of findings. The research team ensured a high level of professional research practice.

3 Key Findings

3.1 *The Study Context*

Makerere University is the oldest university in Uganda and was established in 1922 by the British colonial government. The university is known to have produced most of the political and professional elites in the country and East African region. It has been ranked among the first 10 universities in Africa especially in medical and engineering research. It had a total of 275 academic programmes by the time of the research ranging from ordinary diplomas to bachelor's degree, masters, and PhD programmes. At the time of the study, it had a total enrolment of 48,102 students (MAK 2016). The university has several research collaborations with other universities in Africa and beyond in areas of natural production technology, infectious disease management, engineering, and natural resource management.

The research agenda at the time of the study included research on: health and health systems agriculture transformation; food security and livelihoods; natural resources management and climate change; education and education systems; governance, culture, visual arts, social justice, communication and sustainable development; science and technology including foundational sciences, ICT, innovations, technology and engineering and; cross cutting activities like biotechnology, knowledge transfer, gender and human resources development.

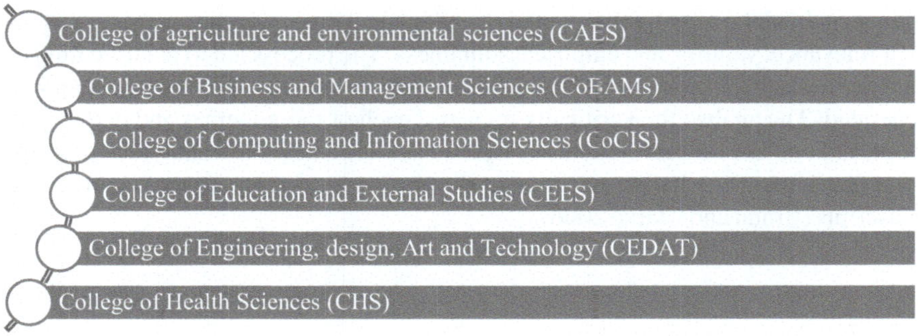

Fig. 3 Academic colleges at Makerere University *Source* MAK (2017)

The university has been engaged in various innovations from its research including: automobile design and production (Kiira Motors and Kayoola Solar Bus automobiles); design of a mobile fruit processor for value addition in fruit and vegetable processing; development of the ceramic organic filter used in decontaminating water without boiling; development of a traffic monitoring computer application by PhD researchers and; development of the whitefly computer application for automated mobile survey technology. The university is organized in various academic colleges that house undergraduate and postgraduate programmes. These include (Fig. 3);

The study was conducted in the college of Agriculture and Environmental Sciences. This college houses the school of agricultural sciences; the school of forestry, environmental and geographical sciences and; the school of food technology, nutrition, and bioengineering. From the document review, climate change related study programmes and research are housed in the school of environmental and geographical sciences as well as the school of forestry. This therefore informed our purposive selection of these two schools where our study participants (lecturers, researchers, administrators, and students) were purposively selected.

3.2 *Empirical Findings Regarding Challenges for Climate Change Education at Makerere University*

The participants were asked to share their experiences and perspectives regarding the key challenges that affect the university in addressing issues of climate change in its programmes. From the analysis of interview and focus group data, several challenges emerged and were categorized as shown in the box below;

Box 1: Participants Views on Challenges Faced By Implementing Units of Climate Change Education at Makerere University

Sub theme 1: Enemies of progress to university climate change interventions

- Institutional barriers
- Resource related impediments
- Organizational environment related obstacles

Institutional Barriers to Climate Change Education
In this category of challenges were those that arise from the institutional failures and weaknesses within the university. Some of these are related to funding political interference, government policies and organizational culture. The participants reported various challenges including limited access to literature and reference materials; limited staffing on climate change courses and programmes as well as high levels of bureaucracy within the institution.

Participants reported that limited access to literature and reference materials on climate change science was one of the key challenges facing the implementing units. This was due to the limited funding available to acquire such literature and reference materials to adequately support training and research on climate change. One of the meteorological student said;

> "The problems I see with research, may be the availability of research books in the university. As we said our course is new and books are not very many at the moment, so we resort to using internet, but there are other things that really require using hard copies for references. Because these things you need them each and every day. The second thing I would see is the data access. Data access like satellite data, there are some programmes for satellite data which are not available and you have to really go, it is bureaucratic to get them, yeah" [MUK GD21 UG MUK; Position: 44-44]

As noted above, the participant laments about the challenge of unavailability of research books on climate change science. Even when they are referred to access reference materials on the internet, they feel unsatisfied. She also decries of lack of access to up-to-date data on climate trends for example. Most of the satellite climate data is owned by the Uganda National Meteorological Authority (UNMA) which is not easy to access by the students due to the high bureaucracy since it is a government agency. This constrains student learning and research on climate related programmes at the university.

The other challenge mentioned was limited staffing on climate change related courses and programmes at the university. Participants reported that due to the limited number of academic staff on these courses and programmes, it is difficult to access them one-on-one for effective support and helping them to practically apply the knowledge they get in class. One of them noted;

> "Among the challenges we face is few and busy lecturers. They fail to give us ample time to effectively fulfil the course requirements. For instance, our department has a weather station. But at undergraduate as we were studying climatology either at introduction or an advanced level, we were never introduced into weather station and when you keep on reminding the person in charge, he or she tells you she has no time for that. What is taught in class many times is not related to what happens outside. Practical access to various equipment used in studying the phenomenon can be helpful in building our capability and advance our knowledge. But most of such equipment is not accessible to us." [MUK GD22 PGD MUK; Position: 28-28]

The participant decries of the few numbers of lecturers who are also busy most of the times. The students do not often have adequate time with their lecturers since they have to attend to many other students as well as other administrative engagements within the university. This makes students fail to be exposed to practical application of the knowledge they acquire in class and opportunity to access equipment they need to learn and do research on climate change. Another participant reported that some of the lecturers come from other schools and departments and therefore they only come for a class and will go back to their home departments which makes it difficult for students in this department accessing these lecturers. In his own words he reported "Now there is this issue I talked about, that we get staff from the different units, so you sometimes may need staff and they are occupied elsewhere. So we have those challenges." [MUK PA21; Position: 57-57].

High bureaucracy at the university was reported to be one of the challenges that greatly affect programme development and review. The participants reported that often staff put in a lot of effort in developing new programmes, but the process of approval is really long stretching to almost two (2) years. One of the interview participants said;

> "In terms of developing curricular and reviewing curricular, you know that universities are so bureaucratic. For you to be able to develop new curricular and have it accredited, it takes a number of years and some of these programmes have taken three to four years without being accredited. But for the new universities that don't have a lot of bureaucracy do it in months and have their new programmes accredited." [MUK PA21; Position: 55-55]

As seen above, the university being a public one, is so bureaucratic and therefore the process of accreditation of new programmes takes a long time. This hampers the efforts of academic staff in coming up with new courses and programmes as well as reviewing the old programmes on climate change.

During the focus group discussion, participants reported that failure to walk the talk was yet another challenge that impedes climate change education at their university. The failure to walk the talk arose from the fact that the university has study programmes and conducts research on climate change, but still encourages activities on campus that emit carbon dioxide to their atmosphere. One of the student participants claimed;

> "Makerere being no.1 university in Uganda, there is this thing that is making me sad, when you go in the halls of residence, because Makerere itself is a community, has more than 1000 people in the borders of the university, but it hurts me so much going around the university halls of residence, the kitchens and restaurants of the university are still using wood fuel. I don't know whether that is helping the communities in any way, because the more you use the wood, it's like encouraging people to cut down trees. You buying more, so if I analyze Makerere as a community, which is first of all governed and a centre of knowledge, I think we are not faring well in that aspect." [MUK GD22 PGD MUK; Position: 50-50]

Participants cite the use of wood fuel in the halls of residence, kitchens, and restaurants within the university to prepare meals by private service providers. The continued use of wood as a source of fuel within the university campus is not walking the talk and therefore betrays the essence of climate change education at the university.

Resource Related Impediments to Climate Change Education

Participants mentioned several challenges that relate to resources needed to facilitate the implementation of climate change education programmes at the university. These included: inadequate funding for research on climate change; high cost of training amidst high levels of poverty and inadequate support from government and local institutions.

Participants decried the challenge of inadequate funding for climate change education interventions at the university which seriously affects their operations. They reported that mostly the university provides limited funds due to the limited support from government. Government has most of the times only funded salaries and wages and therefore many of the operations and programmes are funded by external donors and private student through tuition fees they pay for the courses and study programmes. Some of them reported;

> "Funding has always been an issue ahmm, all over. But it also depends on how creative people are. So if you look into this university which is very big, there has been several projects on climate change from different donors, development agencies and things like that. What comes from the university itself is quite small, but it is also understandable that the subvention may be from government is small. And so they grapple with issues of re-allocating it, like most of the funds are going to salaries." [MUK PL22; Position: 33-33]

> "Yes, the first one is of course, the funding. This can be quite challenging. As I said, many times you don't drive your own programmes, the donors drive the programmes. So sometimes you propose you want to do this and they say that they don't see the use of this, simply because they have their own interests. So one of the issues is funding." [MUK PA21; Position: 73-73]

As indicated above, the participant talks about funding being an issue and that the university provides a small portion of what is required. The university being a public one gets limited funding from government and most of it goes into salaries. This affects innovativeness and high-level research relevant to the needs of the country since most of the research is donor driven and therefore in the interest of the donors rather than local needs.

The high cost of training especially at postgraduate level amidst high levels of poverty was one of the key challenges mentioned by participants at the university. Participants reported that postgraduate courses and programmes have low numbers of students due to the cost involved. In most universities in Africa, students privately pay for their education and the cost of postgraduate studies is never subsidized by government implying it is borne by the individual students and their parents. One of the participants who was a student noted;

> "The other thing I want to bring out is monesy issues, poverty in Uganda and Africa is killing every one. University education is very expensive and how does this affect the student? The student is already in university, and the hustles that the student goes through to raise the fees makes it tough. So whoever is teaching climate change courses may assure all is well when actually, students are either absent minded thinking of how to get fees and also cover other costs or are physically not present to attend classes. That's one of the biggest challenge we face on these courses." [MUK GD22 PGD MUK; Position: 31-31]

University education was seen as a very expensive service by this participant and that students really struggle to raise the tuition fees and therefore, given the high poverty levels many of the young people can't afford to pay adequately for all fees and other requirements as well as funding their climate change related research at the university.

The other key challenge for climate change education interventions at the university cited by participants was inadequate support from government and local institutions around the university. During the interaction with participants in a focus group discussion, many of them indicated that generally governments in Africa have not seriously provided resources for research by the universities. One said;

> "On top of that, though I will bring out the challenging bit of it, ahmm as you know, Africa predominantly involves developing countries that have limited access to finances. Little is invested in research, yet research is the driver of economies. Most research done in these institutions is funded by western world. So the west benefits from the research curried out here where as our governments put in little. I am calling upon governments to begin funding research projects. It would be nice to hear that a country like Uganda has come up with an innovation funded by the government, without or with limited external funding. That we will be proud of it and it will encourage young practitioners to do more research." [MUK GD22 PGD MUK; Position: 36-36]

The participant highlighted the limited investment by governments and local sources despite the fact that research drives economies in those countries. She re-echoes earlier submissions that most of the research is funded by donors from the developed world and therefore, the donors determine the purpose and nature of the research to be done and benefit more compared to the host countries. This possibly explains the few publications produced by academics in the country since they have limited funding for climate change research despite the fact that most of them have the capability to produce state of the art research.

Environmental Related Challenges for Climate Change Education at the University

Among the challenges reported included those that relate to the environment around the university that are outside the control of the university but seriously affects climate change education interventions at the university. These included: the limited expertise in climate change science; the negative attitudes of communities towards the new technologies and products introduced by the university through research; limited access to climate data and; political interference and instability that affects the activities of the university.

The challenge of limited expertise in climate change science around the country was cited as a key barrier to climate change education at the university by participants. According to them, competent and qualified human resource on climate change science are not readily available in the environment. One of them who was an administrator reported;

> "The the human resource, this has been a problem, but I think we are trying to come up. Because initially everyone was trying to be an expert in climate change, and you don't have people who have that expertize on climate change. But at the moment, I think they are trying to come up" [MUK PL23; Position: 67-67]

The university depends on local and international community to supply the required human resource and expertise on sophisticated areas like climate change science. However, often the university has failed to have adequate experts and scientists on climate change whenever it advertises for these posts. This limits the ability of the university to engage in more specialized teaching and research as well as innovations in this field.

> It was also reported that often the university develops new technologies and products that are climate friendly or can support adaptation to climate change within local communities. However, often some members of the communities have a negative attitude towards these new products and technologies which affects research and innovation. The attitudes are shaped by religious and cultural beliefs as well as political ideologies and inclinations of communities. For example, one of the participants noted that; 'communities are not so receptive to improved varieties and therefore it takes a lot of time and resources to have their buy in' [MUK PL21; Position: 71-71].

The other challenge was limited access to quality climate data. Study participants reported that even when one has all the resources to collect climate data, accessing quality data is a serious challenge. This is because of two factors. First, is that due to Uganda's turbulent political history, it is difficult to get good quality climate data spanning 40–50 years. Secondly, climate data is collected by government institutions (Uganda Bureau of Statistics and Uganda National Meteorogical Authority) that have high levels of bureaucracy for students and individual researchers to access data. They reported that the university had no effective arrangement with these institutions to support student access to climate data and therefore this affects student research on climate change.

Political interference and instability in Uganda over the years was one of the challenges that greately affect climate change education programmes at the case university. Makerere university being public and the biggest university, it has always been affected by political swings in the country. Sometimes politicians require that university leaders and staff subscribe to political ideologies and if this is not the case, politicians will frustrate the efforts of the university. In addition, participants reported that political instability has affected activities of the university especially during community outreaches and field research on climate change.

3.3 Empirical Findings Regarding Opportunities for Climate Change Education at Makerere University

During the study, the researchers asked participants to share their perspectives, views and experiences regarding the opportunities that could be exploited by the university to enable it effectively address climate change issues in their study and research programmes. Their views and perspectives were categorized as below:

Box 2: Participants perspectives and views regarding the opportunities for climate change education at Makerere University.

Sub theme 2: The current openings for our climate change work

- Legislation backup
- The nature of the phenomenon
- The conducive environment around us

Legislative Backup

The category includes those opportunities that exist for exploitation arising from legislations in place on climate change. Participants mentioned several opportunities that relate to legislation including: Uganda being a signatory to UNFCC; the national climate change policy for Uganda and other national support policies and legislations on climate change being in place. They also reported that climate change research is enshrined in the university strategic plan as one of the items on the research agenda. One of them noted;

> "….Uganda is a party to UNFCC, and it participates in the COPs. Uganda government is leading as far as addressing climate change is concerned. So with all these, and Makerere university is a premier university in Uganda." [MUK PA21; Position: 47-47]

According to the participant, Uganda is a signatory to the UNFCC and annually it participates in the COPs. This implies that the country has the mandate and commitment to come up with other national legislations and strategies to deal with climate change. Makerere University being a premier university in Uganda can capitalize on this as an opportunity for it to come up with climate change education interventions. The legislative backup would also make it easy for the university to solicit funding from both internal and international sources for a legitimate cause. Legislative back up also offers many opportunities for the university to be a centre of policy discussions, scientific discussions and information to address climate change in the country.

The fact that climate change is on the research agenda in the strategic plan of the university, offers a lot of opportunities for climate change education. This means that implementing units can legitimately come up with climate change related courses and programmes as well as research projects without any hindrance. The university will also ensure that the limited resources available are shared including some for climate change education but also support fundraising drives for climate change research and other interventions from external sources.

The Nature of the Phenomenon

There were opportunities identified by the participants relating to the nature of the climate change phenomenon. Participants recognized that climate change affects all aspects of life and therefore adds to other problems facing humanity. This comes with various opportunities including interventions by the university to support comunities and governments to address the problem. One participant stated;

> "Definitely climate change is a problem or a challenge to the development of Africa, because even prior to the spread of climate change phenomenon, Africa was already challenged with so many other problems and so having this added on to the so many we already have. I

think it is stressing the people, many are dying of starvation, people are separating in their families, government budgets have been constrained, governments are confused and people do not really understand the impacts of climate change. Yes, it is a reality but its something that has not been explained thoroughly and for instance people in government have not been equipped on how to deal with this challenge at national level and when it fails at national level, definitely it will fail at the local level. So I see it as a great challenge." MUK GD22 PGD MUK; Position: 11-11]

According to him, the havoc created by the climate change phenomenon, especially on the continent of Africa that already suffer other several challenges, in itself offers vast opportunities for the university. Every government is looking for ways to deal with the phenomenon and therefore would support the university to implement scientific research based interventions that can help in mitigation and adaptation. This is a huge opportunity for the university to come up with various interventions to address this phenomenon.

Participants also reported that Uganda being dependent on rainfed agriculture, makes it more vulnerable to climate change. The country like other developing nations lack adaptation capacity to impacts of climate change and therefore communities will suffer the more. The country finds itself accumulating debts and depending on donors or foreign funding to deal with effects of climate change. This situation presents vast opportunities for the university to invest more efforts in research and other interventions to support communities and government in dealing with this climate change phenomenon.

The Conducive Environment Around the University

Participants identified varous opportunities that exist from the conducive internal and external environment in which the university operates. These included: the existing partnerships with various donors, agencies and international organizations; the current stability and peace in the country; the willingness among several other universities in the country to collaborate on cliamte change programmes and; the fact that governments in the region have not yet built capacity to deal with the phenomenon.

Makerere university has established a vast number of partnerships with local and international universities in the region and beyond and these have climate change programmes in place. This offers an opportunity for the university to partner with these on a number of programmes. There are also many other universities that have realised the need to intervene but have not yet started interventions on climate change. The university could explore potential partners and reach out to collaborate on some programmes to address climate change.

It was also noted that the existing stability and peace in the country offers an opportunity for the university to effectively engage communities and government without any instability. Uganda has in the past, experienced serious political instabilities and therefore the current stability and peace offers vast opportunities for the university to intervene effectively. Participants also indicated that the fact that many countries in the region and on the continent have not yet bult the capacity to effectively deal with climate change, offers the university an opportunity to develop interventions to support these countries. Opportunities for capacity builiding programmes, policy advise and consultancies, provision of innovative solutions to governments and communities for adaptation as well as other interventions at community level can be explored by the university.

4 Discussion of Findings

Challenges for Climate Change Education at Makerere University
The social learning theory provides that social learning of individuals is determined by the environment in which they leave. This is the principle of reciprocal determinism. Bandura (1977) maintained that the behaviour of individuals is determined by their environment and context which is characterised by motivators (positive forces) and demotivators (negative forces). The motivators and demotivators often lead to a change of behaviour among individuals and groups of individuals in any context. Based on the reciprocal determinism principle of the social learning theory, the researchers explored participant's views and opinions as well as perspectives regarding the challenges that the university face in implementing climate change education programmes given their environment and context. The analysis of interview and focus group data revealed a number of challenges including; limited access to reference materials and facilities for students on climate change science; inadequate funding support by government and local institutions and the challenge of conceptualizing climate change among local people or communities. These findings were in support of the claims by Teferra and Albach (2004) that African universities face serious financial constraints that affect their delivery of quality study, research and community engagement programmes across disciplines. The findings also supported the claims by Higgins and Thomas (2016) that universities face various challenges including globalization, massification of students, diversification and financial pressures amidst changing economic times.

Findings also revealed that the implementors at the case university face: high bureaucratic tendencies in approving and accrediting new courses and programmes; lack proper coordination and conflict over which academic unit should host climate change programmes and; failure to adopt climate friendly management practices on campus despite having climate change programmes. These findings support the claims by Gale et al. (2015) who contended that the organizational structures of universities constrain efforts to embed sustainability education because of their rigidity and hierarchical operations. Sustainability education including climate change education calls for interdisciplinary and multi-disciplinary approaches which are not easy to implement in such institutions given their rigid structures and bureaucratic operations.

Opportunities for Climate Change Education at Makerere University
The social learning theory provides for learning that transcends an individual to groups or social units or entities. According to Barth (2015) social learning within an entity is seen as system-wide changes as a result of the individuals within the

social entity learning and applying the learning to change from the usual. Based on this theoretical lens, the researchers sought views and perspectives of study participants regarding the opportunities available for enhancing climate change education programmes at Makerere University. Analysis of data revealed that Africa's high vulnerability to climate change and its effects across sectors and the opportunity to collaborate with other universities on the continent and beyond were significant opportunities for climate change programmes at the case university. The participants reported that the legislative backup since Uganda is a signatory to UNFCC; the limited capacity of governments to deal with climate change which calls for the support of universities and; climate change research being part of the university research agenda in its 5 year strategic plan were serious opportunities for climate change education at Makerere University. These findings should be added to existing scholarly literature on climate change education and should be handy for university administrators to tap such opportunities to enhance climate change education interventions.

5 Conclusion

This study identified the challenges and opportunities for climate change education at Makerere University from the perspectives of university lecturers, administrators, and students on climate change related programmes. The challenges identified will require university administrators to come up with effective solutions so as to enhance effective climate change education interventions. The university needs to think about innovative ways to deal to engage government and local institutions to effectively support climate change interventions especially research to address local climate change adaptation and mitigation efforts. There is also a need to innovatively engage communities using indigenous knowledge to clarify their conceptualization of climate change and therefore enhance local adaptation and mitigation initiatives. The findings also revealed a vast number of opportunities that can be tapped or explored to enhance climate change education interventions as well as sharing with other universities that are yet to start such programmes at their campuses. With the knowledge of these opportunities, the university needs to find innovative ways to tap into prospects around supporting government through consultancies and capacity building on climate change adaptation and programming. They could also creatively develop effectively country wide initiatives to support local and regional climate adaptation programmes through effective collaborations with the private sector and other universities in the region. The study findings on challenges and opportunities may add to scholarly literature on climate change education in the African context and also contribute to improvement in the university interventions on climate change at the case university and other universities in Africa. Future research will be needed to explore perspectives on how implementing units on climate change education programmes could document and integrate indigenous knowledge into their climate change education programmes.

References

Bandura A (1977) Social learning theory. General Learning Press, New York

Barbour R (2007) Doing focus groups. Sage Publications Inc., Thousand Oaks

Barth M (2015) Implementing sustainability in higher education: learning in an age of transformation. Routledge Taylor and Francis Group, London

Berg BL, Lune H (2012) Qualitative research methods for the social sciences, 8th edn. Pearson Education Inc., Upper Saddle River

Boyde MC, Hume T (2015) Addressing the challenge of climate changes: the potential role of development education in the tertiary sector. In: McCloskey S (ed.) Policy and practice: a development education review, pp 63–86. Centre for Global Education

Braun V, Clarke V (2006) Using thematic analysis in psychology. Qual Res Psychol 3(2):1–41

Braun V, Clarke V (2013) Successful qualitative research: a practical guide for beginners. Sage Publications Ltd, London

Bryman A (2008) Social research methods. Oxford University Press, New York

Farmer T, Cook J (2013) Climate change science : a mordern synthesis, vol 1. Springer, Heidelberg

Filho WL (ed) (2015) Handbook of climate change adaptation, 2015th edn. Springer, Heidelberg

Flick U (2007) Designing qualitative research. Sage Publications Ltd., London

Gale F, Davison A, Wood G, Williams S, Towle, n. (2015) Four impediments to embedding education for sustainability in higher education. Aust J Environ Educ 31(2):248–263

Giugni M, Simonis I, Bucchignani E, Capuono P, Paola FD, Engelbrecht F, Topa ME (2015) The impacts of climate change on african cities. In: Pauleit S, Coly A, Fohlmeister S, Gasparini P, Jorgensen G, Kabisch S, Kombe WJ, Lindley S, Simmonis I, Yeshitela K (eds.) Urban vulnerability and climate change in Africa: a multidisciplinary approach, pp 37–75. Springer, New York

Guba EG, Lincoln YS (2005) Paradigmatic contraversies, contradictions and emerging confluences. In: Denzin NK, Lincoln YS (eds) The sage handbook of qualitative research, 3rd edn. Sage Publications Inc., Thousand Oaks, pp 191–216

Higgins B, Thomas I (2016) Education for sustainability in universities: challenges and opportunities for change. Aust J Environ Educ 32(1):91–108

Kagawa F, Selby D (2010) Introduction to "education and climate change: living and learning in interesting times" . In: Kagawa F, Selby D (eds) Education and climate change: living and learning in interesting times. Routledge Taylor & Francis Group, New York, pp 1–11

Kotecha P (2010) A framework for action on climate change and adaptation in Higher Education in SADC. South African Regional Universities Association (SARUA) Leadership dialogue series, pp 1–144

Kvale S (2007) Doing interviews. Sage Publications Inc., Thousand Oaks

MAK (2016) Fact book; special edition, tracking the performance of the Makerere University strategic plan 2008/09-2015/2016. Mkerere University Press, Kampala

MAK (2017) Makerere University Strategic Plan Review Report 2017. Retrieved from Kampala

Merriam SB, Tisdell EJ (2016) Qualitative research: a guide to design and implementation. Wiley, Hoboken

Polkinghorne DE (2010) Qualitative research. In: Thomas J, Hersen M (eds) Handbook of clinical pyschology competencies. Springer, Heidelberg, pp 425–455

Stern N (2008) The economics of climate change: the stern review. Cambridge University Press, New York

Teferra D, Albach PG (2004) African higher education; challenges for the 21st century. High Educ 47:21–50

UNESCO (2011) Education for Sustainable Development : An Expert Review of Processes and Learning. Retrieved from Paris France

Vale PM (2016) The changing climate of climate change economics. Ecol Econ 121(1):12–19
Virtenen A (2010) Learning for climate responsibility: via consciousness to action. In: Filho WL (ed) Universities and climate change. Springer, Heidelberg, pp 231–240
Yin RK (2003) Case study research design and methods. Sage Publications Inc., Thousand Oaks
Yin RK (2012) Application of case study research, 3rd edn. Sage Publications Inc., Los Angeles

Possibilities of Translanguaging Pedagogy for Sustainable Education in Africa

Nomathamsanqa Mpofu

Abstract African nations are working on modifying their educational curricula to fulfil the fourth Sustainable Development Goal on quality education. However, if the language of educational instruction remains foreign or secondary to the African child, accessing that adapted curriculum may remain impeded. This chapter, which is informed by a case study, proposes that Africa adopts translanguaging pedagogy towards ensuring that the African child effectually accesses the adapted curricular. Qualitative data were collected from fourteen final year students who were majoring in English and three lecturers of English from one tertiary college in Zimbabwe. Student participants were purposively chosen from a total of ninety-eight while lecturer participants were conveniently sampled out of five. Evidence was collected using two focus group discussions with students, an open-ended questionnaire with lecturers, and participant observations of five lessons. Collected data were analysed thematically. The findings were that the monolingual pedagogy currently used is inadequate. Secondly, although translanguaging proved to be time-intensive and ethnic-based such that it threatened assessment, it could be adopted for its benefits of improving content grasp, relaxing the classroom atmosphere and enhancing multimodality. The main conclusions from these findings are that translanguaging is beneficial in learning and teaching in a multilingual classroom and possible to adopt. Thus, the chapter recommends that Africa shifts from exclusively using foreign languages for educational instruction to engaging African languages through translanguaging instead.

Keywords Translanguaging · Language of learning and teaching · Sustainable education · Tertiary education · Home languages

N. Mpofu (✉)
Department of Teacher Development, Great Zimbabwe University, Box 1235, Mas-ingo, Zimbabwe
e-mail: nmpofu@gzu.ac.zw

W. Leal Filho et al. (eds.), *Sustainable Development in Africa*, World Sustainability Series,
https://doi.org/10.1007/978-3-030-74693-3_13

1 Introduction

African nations are working on modifying their curricula to effectively embrace sustainable education for the African child (Filho et al. 2016). This is a quality education that equips learners with the "knowledge, skills, values, and behaviours" needed by societies towards living their lives without destroying the future (United Nations Educational, Scientific and Cultural Organization [UNESCO] 2019a). Effectual access to the curricular is determined by several factors, chief among them is the language(s) of instruction used (Brock-Utne 2010 and Nyika 2015). While most African nations (from mid-primary education level upwards) use languages of instruction which are either foreign or secondary to the African child, they shun learners' elementary languages (Oluoch 2017). These languages of instruction are languages of former colonizers (Milligan and Tikly 2016) yet learning in a language other than one's home language thwarts "the child's cognitive and scholastic ability" (Ovando et al. 2003, cited in Baker 2008: 133). Thus, there is a need for Africa to engage learners' home languages if the "knowledge, skills, values and behaviours" packaged for sustainable education development (SED) are to be effectively accessed, sustained, and reached by all.

Translanguaging pedagogy is being practiced in many multilingual educational settings at all levels around the world. In early childhood education and primary levels translanguaging pedagogy is recorded to be practiced in New York City, United States of America (USA) (Cskillik and Golubenia 2020), Hong Kong (Leung 2019), and Nepal (Knee 2017) among other countries. In secondary and higher education it is recorded in Canada (Anwaruddin 2018), East London, Britain (Hassan and Ahmed 2015), Australia (Heugh et al. 2017), Qatar (Hillman, Graham and Eslami 2018), USA (Robinson et al. 2018) French Guiana (Solphie and Leglise 2018), and China (Wang and Curdt-Christiansen 2019) among other countries. Translanguaging is also preferred because it resonates with current educational trends of inclusion in education and singularization of plurality. Inclusion in education respects diversity among learners and ensures that all barriers limiting learner participation and achievement are removed (UNESCO 2019b). Singularization of plurality is a concept that also considers individual differences in a whole class (Makoni et al. 2012). Thus, in view of the stated African situation and global educational practices, three objectives guide this study. The study will discuss reasons necessitating the shift from using foreign or secondary languages in educational instruction, educational benefits of translanguaging pedagogy, and perceived limitations of translanguaging. The chapter draws its evidence from a case study conducted in one African country.

2 Current Practices of the Language of Instruction in Education

African nations have multilingual educational settings but their official educational policies are predominantly monolingual (Bakamba 2015). The language(s) of a country's former coloniser is used for teaching and learning purposes with learners' home languages only engaged in the lower primary grades (Oluoch 2017). Thus, the African child is forced to use a language s/he is also learning to access the curriculum. However, unofficially, some classroom practitioners engage learners' home languages in class (Mareva 2016) and some students do the same on their own to cope with the challenges of using a foreign language in learning (Marie 2013).

3 Translanguaging: The Concept

Many studies have been conducted on translanguaging worldwide including in Africa. Originally termed "trawsieithu" in the 1920s, the concept of translanguaging first started when bilingual Welsh/English students alternated languages in a bid to make meaning and produce information (Baker et al. (2012), cited in Mgijima and Makalela 2016). The concept is now viewed as the perspective that understands a multilingual speaker's use of features of his/her unitary idiolect or an internal language where s/he does not name those features as belonging to named languages (Garcia 2017). It asserts that (despite the many named languages possessed) the speaker's language system is one whole unit in horizontal continua (Garcia and Vogel 2017). Thus, translanguaging does not view the linguistic possessions of the bi/plurilingual speaker as separate languages but as one (Makalela 2013). The speaker then chooses certain features from what s/he calls his/her language (although these would be many languages) and manipulates those in his/her discourse. Thus, a bi/plurilingual speaker possesses a language system which is a unitary, integrated communication system and this should not be seen as comprising separate languages.

To the African child who is endowed with many languages, translanguaging becomes an invaluable resource in that s/he will employ all those languages (instead of one foreign or second language) to effectively understand the sustainable education curriculum.

4 The Study

The study adopted a phenomenological perspective that allowed investigation of participants' real learning experiences as they engaged with translanguaging (Kivunja and Kuyini 2017). It adopted the qualitative research approach and used a case study design to gain a deep, intense, and holistic picture of the impact of the

Table 1 Distribution of Student Participants according to home languages

	Ndebele	Shona	Kalanga	Tonga
Focus Group Discussion 1	3	2	1	1
Focus Group Discussion 2	5	0	0	2

lived phenomenon (Gray 2009). The results from the case study design, though not generalizable to the whole of the African continent (Best and Khan 2010), could be taken as indicative of possible trends (Gray 2009). The case was the United College of Education which is a primary teacher education institution in Zimbabwe, Africa. Although the study was delimited to exploring the feasibility of translanguaging in the teaching and learning of phonetics and English phonology at tertiary level in Zimbabwe, the findings are discussed in this chapter to advance the incontestable shift from using the colonial languages to translanguaging in accessing sustainable education curricula in Africa.

4.1 Participants

Out of a population of 98 students that were majoring in English studies at the college, 14 were purposively sampled. These were final year students. Most studies interrogating the performance of translanguaging at tertiary level worldwide such as Brown (2019), Marie (2013) and Ngcobo et al. (2016) among others, used first-year students as participants. This study sought opinions of students with theory and practice of evaluating teaching and learning methods. These participants were purposefully chosen because they had passed their practicum teaching practice stint and were best positioned to give a near-professional evaluation of pedagogic experiences. Although all the student participants were plurilingual, speaking English and one or two other African languages, they registered their mostly used home languages as represented in Table 1. Three out of 5 lecturers of English at the College were conveniently sampled.

4.2 Data Collection

The study collected evidence using participant observation, focus group discussions, and an open-ended questionnaire. Participant observation allowed formal and informal interviews to be carried out on site (Agar (1996), cited in DeWalt and DeWalt 2010). As the researcher was the students' lecturer for the linguistic component, participant observation provided a natural atmosphere and context throughout the lessons (DeLamater et al. 2018). Five lessons of one and a half hours long each were observed using a guide based on domains 2 and 3 of Charlotte Danielson's Framework (Danielson 2011).

Two focus groups comprising seven students each held two discussion sessions between them. Each session lasted about 35 min. The questions asked in the discussions were guided by the objectives of the study. Sessions were audio recorded with participants using pseudonyms. The recorded data were later transcribed and verified by the participants. Focus group discussions uncovered ranges of experiences while also refining views (Creswell 2014).

An open-ended questionnaire was also used with lecturers for "insightful...original and illuminating data" (Gray 2009: 137). It was accessed electronically. Lecturers were coded as Lecturer 1 to 3.

All collected data were analysed thematically and reported under the three research objectives using thick descriptions.

5 Findings

5.1 *Reasons for Language Shift*

5.1.1 Lack of Proficiency in English Language

The lack of proficiency in the English language gave rise to the following challenges:

Low Classroom Participation
The focus group discussions and the questionnaire both identified that participation in class is low if English Language is used in the lessons. Lecturer 1 complained that *"Content delivery... is mostly a one-way channel"* and *"the same students [participate] over and over again"*. These sentiments from the lecturer prove that very little learning was taking place in classes delivered in the English language. Students expressed the same challenges differently. Tatenda said *"ndandisingaunzvisise saka [I could not understand so] participation was very low"*, Nomagugu *"It [English] didn't make it easy for me to participate... I didn't do any talking"* and Shanaz said, *"[I would] just sit there, watch things happening and watch things being said"*. Apparent from the students is lack of understanding and passiveness which meant the lecturer could not gauge content grasp by students.

Low Confidence in Students
Using English in the lessons also lowered students' confidence in class. Lecturer 3 said *"Most students are... less confident in using it [English Language]"* and Kudzai, a student said *"I used to fear using English because I couldn't easily say out everything"*.

Restricting Atmosphere

Serious restrictions were registered by both sets of participants. Lecturer 1 said at times the atmosphere was so tense that he felt that if he encouraged individual answers he was "*victimising and exposing the student's incompetence*". This suggests that the use of English language was more inhibitive than facilitative to the teacher. Students, for example, Rass commented that the *"Atmosphere was tense", "[it was] hard to open up and say esasifisa ukukukhuluma [what we wanted/wished to say]"* and Masi simply put it as that the *"Atmosphere was not free"*.

Ineffective Content Delivery and Grasp

Both sets of participants perceived using English language in learning and teaching as resulting in ineffective content delivery and grasp. Lecturer 1 commented that written assignments "*reflect misunderstandings, half-truths and low performance*" and Lecturer 2 said, "*lecturer labours to explain concepts to them [students]*". Students expressed it differently. Zee said "*I would listen, try to cram and memorise*" while Raas said "*ngangingabambi konke [I wouldn't grasp everything]*". Thus, students used rote strategies most likely to pass the examination and not to comprehend content; hence another student said he could not grasp everything.

Disinterest

Lecturer 1 said more students were inactive in lessons and likened them to *"parcels"* and Lecturer 2 said "*Those less proficient in English disengage*". A graver and sadder expression of disinterest was articulated by Shanaz who said *"I wish I could bunk Main Study [English Main subject lessons]"*. Thus, both lecturers and students identified serious disinterest in the curriculum due to the use of exclusive English language in class.

5.1.2 Other Languages Are Already Used in Learning Outside Class

All students testified that whenever they are discussing curriculum outside class and in the absence of lecturers they translanguage. Rass said using all the languages at his disposal made it "*easier to express ourselves in our L1*" and Tatenda said *"We mostly mixed with Shona in concepts zvaunenge usinakunzvisisa, saka munhu akakuexplenera neL1 yako unokwanisa kunzvisisa. [We mostly used Shona in concepts that we would not have understood, if someone explains in your L1 you can understand. English helped us in some other terminologies]*. Thus, students showed they understood better when discussing curriculum mixing English with their African languages.

5.2 *Benefits of Translanguaging*

5.2.1 Relaxed Atmosphere

Lecturer 1 termed the atmosphere as "*relaxed*" whilst students, for example, Toboka said, *students [bengaselafiye] were no longer afraid"*, Mbali said lessons *"ayesec-haza" [became more interesting]*, and Jayden said *"ngakholisa" [I enjoyed]*. All these testify to a positive and conducive ambience. Observed lessons where the class used translanguaging showed a jovial, relaxed atmosphere with all students participating, including those who were usually quiet in class.

5.2.2 Improved Content Grasp

All three lecturers testified that taught concepts were better clarified in an African language. Lecturer 1 added that "*other students can explain concepts better, more so in a friendlier language"*. Thus, besides the content being better explained, learner to learner interaction is encouraged. Students agreed to this as well Mbali said she did not know what the uvula is even from the model of the human respiratory system used *"and then sengikwazi ngelangweji yami sikhuluma ngesilakalaka [I knew from my language that we are talking about the uvula]*. Raas said *"kwakusiba izi uku-expleyinelwa loku-expleyina [it was easy for someone to explain to you [a concept] and for you to explain [it] to someone as well"*. In the observed lessons concepts were clearly explained and formative assessments carried out proved accurate information was understood.

5.2.3 Strengthened Student-Centredness

Lecturers identified student-centredness to be created by translanguaging. Lecturer 1 expressed that *"those [students] with poor English are not closed out"*. In observed lessons, all students participated and interacted together for the greater part of the lessons.

5.2.4 Enhanced Cross-Referencing

Lecturers and students identified translanguaging as enhancing cross-referencing to known experiences. Lecturer 3 said, *"it may help or enable students to relate what they know in L1 with what they are learning in L2"* and Lecturer 1 commented that *"one can easily compare and contrast the concepts with the same concepts in their language"*. Zee, a student, also said, *"we were doing the production of sounds practically"*, *"we were able to link it[phonetics"] to uhlelo esalufunda kuSiNde-bele [phonetic concepts learnt in Ndebele [reference to 'Ordinary' or 'Advanced' level Ndebele syllabuses].* These quotes show that translanguaging enabled both lecturers and students to compare and contrast concepts from their experiences toward curriculum understanding.

5.2.5 Versatility

Lecturers and students viewed translanguaging as versatile and perceived that it can be employed in other English subject concepts such as "*literary appreciation, composition writing and literature*" and Jayden, a student, said possibly "*even in some other subjects*". Thus, lecturers and students were agreeing that translanguaging can be used in the teaching and learning of many other skills in the curriculum.

5.3 Challenges with Translanguaging

5.3.1 Time Intensive

Translanguaging was found to require a lot of time in group work and report back sessions. Lecturer 2 advanced that there could be a need for "*someone to translate from my language to my students*" and Ruth observed that "*it takes time for feedback*". Being time-intense may be against the scheduled time slots allotted to a particular subject.

5.3.2 Ethnic Based

Observation found translanguaging grouping students who understand each other's main language. Thus, it seemingly divided the class along ethnic lines and paused the following limitations:

Assessment Unfriendly

Both lecturers and students registered that translanguaging proved unfriendly in accessing content grasp in students. In Lecturer 1's words, LC1: "*non-standard language negatively affects their expression in essay writing [usual assessment mode]*", and "*[if the teacher is unfamiliar with the language] gauging [students'] concept explanation and grasp could be a challenge.* The lecturer identified that since formal assessment is in written form, students may not develop adequate formal and technical language to use and the teacher may not confidently judge students' concept understanding. Students, for example, Ruth and Tatenda observed that at times the lecturer would not know if a word used by a student in one language is the correct one.

Numbers-Determined
One lecturer and one student identified that translanguaging is limited especially if there is only one student in the class speaking a language that is different from the other languages spoken in class. Kudzai, a student, voiced the challenge that translanguaging is dependent on someone else explaining in one's language. The dependency habit associated with translanguaging means that it is impracticable in some cases.

Threatened Content Grasp
While content grasp was a benefit of translanguaging, it also surfaced as a challenge. Mbali, a student, explained that at times students did not understand other languages at feedback stage while Shanaz and CK were concerned that *"some of the terms can't be substituted to mother language"*. This meant that African languages are seemingly not adequately developed as languages of instructions.

Facilitated Labelling
Lecturer 2 observed that, since translanguaging thrives on grouping students, he perceived that translanguaging might be a conduit that could facilitate labelling of some ethnic groups.

6 Discussion

6.1 Reasons for Language Shift

The first reason for shifting from the use of the former colonizer's language to students' home languages is that some of the students are not proficient in the colonizer's language that is used for instruction. This means that the English language was inadequate for content delivery and grasp to those students. The effect of lack of proficiency resonates with what is underlined by Krashen (1985) who explains that if one poorly performs in the language of instruction, one would perform equally poorly in the subject content learnt or taught in that language. The findings also back up sentiments by UNESCO (2018). These are that learning in a language that is not one's own has the potential of impacting negatively on the learning processes and outcomes. Thus, using the English language in this case opposed the concepts of inclusive education and singularization of plurality.

Low classroom participation is another prompt for changing exclusive foreign or secondary language instruction. Participation in class is integral in learning. In their action research with first year students, Debele and Kelbisa (2017) found that participation is affected by students' feelings of being inadequate and poor teaching approaches and methods used. Thus, the findings of the current study echo the discoveries by Debele and Kelbisa (2017). Poor approaches in terms of the inadequately developed language of instruction in some students discourage them from participating in class, thereby affecting the learning process. Therefore, as the use of English language lowers participation which is crucial in learning, there is a need for it to be changed.

The other finding identified in the case study was of low confidence. Self-confidence or self-regard is an essential element in learning. Briggs (2014) opines that it enables one to overcome everything else. This means that confidence overcomes challenges of non-participation and disinterest in learning. Thus, being a determinant of the learning processes and outcomes, confidence is a crucial factor whose lack solicits a change in the language of instruction.

The atmosphere in lessons that were exclusively taught in English was described as being *"tense"*, *"not conducive"*, and *"not free"* thereby revealing a restrictive ambience which stifled learner participation. These findings echo sentiments by Debele and Kelbisa (2017) who argue that any atmosphere created in the classroom should cultivate self-confidence and determine participation. Thus, while cementing this symbiotic relationship between atmosphere and learning, the findings add the element of the restricting atmosphere as being directly created by the use of a language with which one is not very familiar. These findings further echo the sentiments by Qorro (2006) who says that language difficulties can be a barrier to academic processes. Since a language of learning and teaching can impede course intake, changing that language is most logical.

Lecturers and students also viewed the English language as negatively affecting content delivery and grasp typified by *"misunderstandings"*, *"half-truths"* and *"low performance"* emerging from written formative assessments. These comments suggest perennial poor content grasp. One could argue that the language could also yield *"misunderstandings and half-truths"* of the sustainable education curriculum. These findings are in agreement with the observation made by Baker (2008) namely that often a second language is insufficiently developed in a student to be efficiently used to access abstract curricular. Thus, the studies blame ineffective grasp of curricula on the lack of competence in the foreign or second language with which the student accesses curricular. Shohamy (2012) also comments that low academic achievement is often registered in people who acquire curricula via a language that they barely understand. The findings are also a reflection of those made by Ovando et al. (2003, cited in Baker 2008:13). The authors found out that any other language which is other than one's own thwarts a student's "cognitive and scholastic ability" if used to interface curriculum. Thus, the case study findings, together with many other previous studies, blame poor grasp of knowledge, skills, attitudes, and values on the foreign language which students use to understand the contents.

The research also found that students discussed phonetics and English phonology concepts using translanguaging and not English language only when they were outside class. This bears witness that students had been engaging all languages at their disposal strategically every time they were learning phonetics and English phonology on their own. To the students, English language was restricting effective communication and compromising their understanding of the concepts discussed hence the shift. The findings are congruous to those by many other studies done across educational levels particularly in Africa. These include studies by Bagwasi (2016) in Botswana, Banda (2017) in South Africa, Charamba (2019) in Zimbabwe,

Marie (2013) in Rwanda, Mwinda and van der Walt (2015), and Nghifingiwanga (2014) in Namibia. Therefore, as home languages are already being successfully used for learning, logic would suggest that they should be co-opted officially into the classroom. All the above factors devaluate languages of former colonizers as sole languages of instructions in Africa and instead promote all languages spoken by the African child.

6.2 Benefits of Translanguaging Pedagogy

When Toboka said *"students [bebengaselafiye] were no longer afraid"*, she was registering a jovial mood and relaxed atmosphere created by the use of translanguaging in the classroom. According to Palmer et al. (2010), fear is an enemy to learning as it paralyses one's thinking capacity. This fear which negatively affects cognition is what Toboka avers was dispelled by translanguaging. Moreover, the jovial mood and relaxed atmosphere in a classroom not only encourages cognitive function but also helps the students to stay positive and focused on the learning process (Young 2014). These findings compare with the benefits of translanguaging found by Mgijima and Makalela (2016)'s quasi-experimental study in South Africa. The duo, too, registered improved learner participation. Thus, this study confirmed the findings that were registered by other studies in and out of Africa.

All the three data collection sets found improved content grasp as a benefit of using translanguaging. One lecturer even commented that *"concepts could be better explained in a familiar language"*. This means that lecturers also felt that the material they were teaching was explained better in a language the students understood better. Curriculum content grasp is the main aim of teaching and learning, and translanguaging seemingly attained that. Findings from many other studies that were conducted in and outside Africa are further corroborated by the findings of this study. These are studies that showed improved learning outcomes through the use of translanguaging. Some of these are by Brown (2019) who studied first-year students translanguaging with English to learn Arabic Language, Heugh et al. (2017) who studied a multilingual class of first-year students studying English as an Additional Language (EAL) at the University of Australia, Marie (2013) at the University of Rwanda, and Childs (2016) in South Africa. Adamson and Coulson (2014, in Adamson and Coulson 2015) also concluded from their study that translanguaging enhanced completion of tasks and improved output for a lot of university students who were lower in English proficiency where English is the language of instruction. Thus, the findings from this study further confirm the benefits of translanguaging in diverse multilingual educational settings which can profit Africa's sustainable education curricula.

The study also found translanguaging to be student-centred. The lessons that were conducted using translanguaging saw all students actively participating. As defined by Debele and Kelbisa (2017), being student-centred means involving all the students at the same time and toward the same outcome. This benefit is in line with the study conducted in Australia by Heugh et al. (2017). First year students from South Asia, South East Asia, Central Asia, Middle East and Africa were in one lecture room using translanguaging pedagogy in learning EAL. All these diverse linguistic backgrounds attained higher achievement in the EAL. Such student-centredness is embedded in the concept of singularization of plurality by Makoni et al. (2012). The findings from this study, therefore, further confirm the benefits that Africa can tap from shifting from exclusively using non-African languages for the African child.

Enhanced cross-referencing of the nature of translanguaging was registered as lecturers and students professed that translanguaging assisted students to cross refer to previously learnt content. In fact, a student, Zee, expressed that they did speech sound production practically linking what they learnt in Ndebele Ordinary and Advanced Level to the phonetics concepts being learnt in English. This benefit is in agreement with the findings made by Ngcobo et al. (2016) in their study which aimed at examining how translanguaging can assist in summary writing. The authors discovered that students reformulated and cross-referred what they had learnt on summary writing in another language and that language. When a lecturer in this case study commented that students could compare and contrast the same concepts from different languages, his sentiments were elaborating those by Garcia (2011: 2). She says, translanguaging "extends and deepens the thinking of students". This, as seen in the findings of the study is done through high order comparisons and contrasts. The African child from whatever linguistic background would, thus, access and understand the adapted curricula deeper and more extensively if translanguaging pedagogy is employed in learning and teaching instead of foreign languages.

Another mode of learning and teaching embedded in translanguaging was found to be the use of students as teachers. Commenting on the effect of students being teachers in their corroborative discussions, one lecturer said, *"other students can explain concepts better"* than the teacher. This observation registers peer interaction as being more beneficial than lecturer talk and promoted by translanguaging. Africa, can, thus benefit from this quality f the pedagogy.

Lecturers and students stated that translanguaging was also advantageous in that it was not restricted to the lessons observed but could be adopted in the teaching and learning of other aspects of the English syllabus and beyond. The findings specified poetry and literary studies. Such projection by both lecturers and learners treads on the researches made on the efficacy of translanguaging in different aspects of the curriculum. These include summary writing by Ngcobo et al. (2016), comprehension writing by Mgijima and Makalela (2016), and EAL by Heugh et al. (2017). Beyond English studies, translanguaging has been registered to be feasible in Mathematics and Science across educational levels in Africa (Charamba 2019; Gandara and Randall 2019, and Nhongo and Tshotsho 2019). This translatability shows the versatility of translanguaging as an advantage to learning many components and many subjects across all educational levels and multilingual settings. Its feasibility has been recorded also in diverse African countries which include Botswana (Bagwasi 2016), Democratic Republic of Congo (DRC) (Gandara and Randall 2019), Kenya (Mwaniki 2016), Malawi (Colin 2019), Namibia (Nghifingiwanga 2014), Rwanda (Marie 2013), South Africa (Makalela 2013), Uganda (Kyeyune 2010), and

Zimbabwe (Charamba 2019). The versatility of translanguaging evidences that the pedagogy can be adopted in the learning and teaching of diverse skills of sustainable education curricula in different African countries.

6.3 Limitations of Translanguaging Pedagogy

Lecturers and students found translanguaging to be inadequate in a number of ways. Firstly, translanguaging pedagogy was noted to take up a lot of time. After the lecturer had presented a stage of a concept, the students would group and discuss that stage, then report their understanding before the whole class moves on to the next stage of the same concept in the same manner. Mazak (2016, cited in Ngcobo et al. 2016) also details translanguaging as a winding method. Thus, it is evident that the approach uses up a lot of time. The translation and discussions detailed by the author are proof of the research finding on the time-intensive factor. On the other hand, however, time-intensity can be proof of effectual engagement and is better than having to re-teach entirely poorly grasped concepts taught in the foreign language.

Translanguaging is also attained through grouping people according to their common languages, and ethnicity. Lecturers and students saw this as threatening assessment. An emerging finding explained by a lecturer and a student is that the lecturer, being not well-versed in a language used by some students, might not have a way of assessing the congruence of the students' discussion to the primary instructions given or even assess students' grasp of the concept. This finding is contrary to what Mazak (2016, cited in Ngcobo et al. 2016) observed. According to Mazak, gauging and assessing concept comprehension by a lecturer who would not understand the students' language would be done at every feedback stage. Thus, Mazak dismisses chances of translanguaging bringing in difficulties in lecturers gauging students' understanding.

This research also found that the "non-standard English Language" used in translanguaging might negatively affect students' expression in writing. This contradicted an investigation by Champlin (2016). Instead of translanguaging damaging one's self-expression in the English language, Champlin's study revealed that translanguaging actually promotes the development of English language. Chaplin's findings further back up the observation by Garcia (2014, cited in de Oliveira and Clark-Gareca 2017). Garcia found that translanguaging leverages English as a Second Language (ESL) learning; hence it may promote and not disturb the language. Learners could then match the equivalent of the foreign language with their home language, thereby aiding writing.

Another limitation of translanguaging from lecturers and students who participated in the case study was that being determined by numbers, the approach may not be possible when there is only one speaker of a language in the class. This limitation means translanguaging pedagogy is not applicable in all situations save for those where a language that is used is shared. This finding echoes the observation by Robinson et al. (2018) that when there is only one speaker using a language in a setting, translanguaging is impossible. However, the concept of autonomous learning which emphasizes a student's individuality in creating his/her learning strategies (Little and Legenhausen 2017), contradicts such a limitation. This is because the individual makes his/her own strategies in learning.

In this study, lecturers and students agreed that translanguaging may threaten the students' grasp of the content. This confusion echoes the findings made by Robinson et al. (2018). These researchers found that translanguaging complicated students' understanding of the concept. CK, a student participant, explained the threat from another angle. He was quoted as saying that, *"some of the terms can't be substituted to mother language"*. This finding resonates with the limitation identified by Wolff (2018) that many African languages have limited vocabulary and terminology to handle some curriculum content. A more apt echo of the findings on translanguaging distorting content is recorded by Hu, Li and Lei (2014). The trio studied content translation by some professors in some universities in China. They found that content was watered down by translation. Therefore, the findings on disturbing content prove the claims tabled by other studies elsewhere in the world.

The last limitation found by the case study was that translanguaging may create social labels. The fear is that, as students are grouped along ethnical lines, labelling may be fostered. However, this limitation contradicts comments by Kabuto (2017). He sees translanguaging as a way of promoting an inclusive set up in class; one that values and respects every individual through his/her language. Most likely the labelling could be impelled by the levels of tolerance prevailing in different settings due to underlying socio-political circumstances in a particular country.

7 Conclusion

It seems highly probable that effective access of sustained development curricula may be impeded if Africa continues to exclusively use second or foreign languages in educational instruction. The African children can better understand the adapted curricula when using translanguaging than when exclusively using a language that is foreign or secondary. Joining other literatures from Africa and the world, this chapter implores African countries to officially consider co-opting translanguaging pedagogy from primary through to tertiary levels to enable African children access of the sustainable education curricula.

This chapter recommends that African languages be used strategically alongside the second or foreign languages in Africa's multilingual classrooms. Learning materials that are in English, French and/or Afrikaans should be supplemented by materials that are in local languages. However, due to the challenges noted to be inherent in translanguaging, the chapter recommends individual African states to further explore how the pedagogy can be implemented in their particular multilingual settings which may be different from those in the neighbouring country.

References

Adamson JL, Coulson D (2015) Translanguaging in English academic writing preparation. Int J Pedagogies Learn 10(1):24–37

Anwaruddin SA (2018) Translanguaging transformative pedagogy: towards a vision of democratic education. Rev Bras Linguist ApL 18(2):301–312

Bagwasi MM (2016) A critique of Botswana's language policy through translanguaging perspective. Current Issues Lang Plan 18(2):199–214

Bakamba EG (2015) Multilingualism as a sociolinguistic phenomenon: Evidence from Africa. In: Zsiga EC, Boyer OT, Kramer R (eds) Languages in Africa: Multilingualism, Language Policy, and Education. Georgetown University Press, Washington, pp 21–48

Baker C (2008) Becoming bilingual through bilingual education. In: Auer P, Wei L (eds) Handbook of multilingualism and multilingual communication. DeGruyter Inc., Berlin/Boston, pp 131–152

Banda F (2017) Translanguaging and English-African language mother tongues as linguistic dispensation in teaching and learning in a black township schooling Cape Town. Current Issues Lang Plan 19(2):198–217

Best JW, Khan JV (2010) Research in education. Basford, Boston

Briggs S (2014) Why self esteem hurts learning but self confidence does the opposite? https://www.opencolleges.edu.au/informed/features/self-efficacy-and-learning/

Brock-Utne B (2010) Policy in the language of instruction issue in Africa. Int J Educ Dev 30(6):636–645

Brown A (2019) Monolingual immersion versus translanguaging in the Arabic foreign language classroom. https://www.zu.ac.ae/main/en/research/lhebc/_blog/_blog-pages/monolingual-immersion-vs-translanguaging-in-the-arabic-fl-classroom.aspx

Champlin MJ (2016) Translanguaging and bilingual learners: a study of how translanguaging promotes literacy skills in bilingual students. Education Matters. Paper 323

Charamba E (2019) Translanguaging: developing scientific scholarship in a multilingual classroom. J Multilingual Multicultural Dev https://doi.org/https://doi.org/10.1080/01434632.2019.1625907

Childs M (2016) Reflecting translanguaging in multilingual classrooms: harnessing the power of poetry and photography. Educ Res Soc Change 5(1):22–40

Colin R (2019) Language in Malawian universities: An investigation into language use and language attitudes amongst students and staff. PHD Thesis. University of Glasgow. https://eleanor.lib.gla.ac.uk/record=b3346319

Creswell JW (2014) Research design: Qualitative, quantitative and mixed methods approaches. SAGE, London

Cskillik E, Golubenia I (2020) Translanguaging practices in early childhood classrooms from an intercultural perspective. In: Huertas-Abril C, Gomez-Parra ME (eds) Intercultural perspectives on modern developments in early childhood education. IGI Global, pp 15–39

Danielson C (2011) Charlotte Danielson's Framework for teaching. https://www.k12.wa.us/TPEP/Frameworks/Danielson/2011_Danielson_SMART_Card.pdf

Debele ET, Kelbisa EM (2017) The role of active learning methods for classroom participation: the case of first year students of Sociology in Samara University. IOSR-JHSS 22(7)13:11–18

DeLamater JD, Myers DJ, Collett JL (2018) Social psychology. Routledge, Boulder, New York

de Oliveira LC, Clark-Gareca B (2017) Collaboration between NESTs and NNESTs. In: Martinez AJDD (ed) Native and non-native teachers in english language classrooms: professional challenges and teacher education. De Gruyter inc., Boston

DeWalt KM, DeWalt BR (2010) Participant observation: a guide for fieldworkers. AltaMira Press, Blue Ridge

Filho WF, Castro P, Bacelar-Nicolau P, Azul AM, Azeiteiro UM (2016) Biodiversity and education for sustainable development (ESD) tendencies and perspectives. In: Castro P, Azeiteiro U, Bacelar-Nicolau P, Leal Filhow W, Azul AM (eds) Biodiversity and education for sustainable development, World Sustainability Series. Springer, Cham, pp 1–10

Gandara F, Randall J (2019) Assessing mathematics proficiency of multilingual students: the case for translanguaging in the Democratic Republic of Congo. Comp Educ Rev 63(1):58–78
Garcia O (2011) Educating New York's bilingual children: Constructing a future from the past. Int J Bilingual Educ Bilingualism 14(2):133–153
Garcia O (2017) Translanguaging. https://www.youtube.com/watch?v=511CcrRrck0
Garcia O, Vogel S (2017) Translanguaging. In: Noblit G, Moll L (eds) Oxford Research Encyclopedia of Education. Oxford University Press, Oxford. https://oxfordre.com/education/view/10.1093/acrefore/9780190264093.001.0001/acrefore-9780190264093-e-181?print=pdf
Gray E (2009) Doing research in the real world. SAGE Publications, London
Hassan N, Ahmed K (2015) Exploring translanguaging: a case study of a madrasah in Tower Hamlets. Res Teacher Educ 5(2):23–28
Heugh K, Li X, Ying S (2017) Multilingualism and translanguaging in the teaching of and through English: rethinking linguistic boundaries in an Australian University. In: Fenton-Smith B, Humphries P, Walkinshaw I (eds) English Medium Instruction in Higher Education in Asia-Pacific: issues and challenges from policy to pedagogy. Springer, Dordrecht, pp 259–279
Hillman S, Graham KM, Eslami ZR (2018) Teachers' translanguaging ideologies and practices at an international branch campus in Qatar. Eng Teach Learn 43(2019):41–63
Hu G, Li L, Lei J (2014) English-medium instruction at a Chinese university: rhetoric and reality. Lang Pol 2014(13):21–40
Kabuto B (2017) A socio-psycholinguistic perspective on biliteracy: the use of miscue analysis as a culturally relevant assessment tool. Read Horizons 56(1):25–44
Kivunja C, Kuyini AB (2017) Understanding and applying research paradigms in educational contexts. Int J Higher Educ 6(5):26–41
Knee S (2017) "I use half Nepali, half English, the hybrid language!" –Translanguaging for pedagogy in a Nepali literacy class. https://epapers.bham.ac.uk/2143/1/Sarah_Knee_Translanguaging_paper_for_conference_proceedings_12th_Jan_2016.pdf
Krashen S (1985) The input hypothesis: issues and implications. Longman, New York
Kyeyune R (2010) Challenges of using English as a medium of instruction in multilingual contexts: a view from Ugandan classrooms. Lang Cult Curriculum 16(2):173–184
Leung SKY (2019) Translanguaging through visual arts in early childhood: a case study in a Hong Kong kindergarten. Asia-Pacific J Res Early Childhood Educ 13(1):47–67
Little D, Legenhausen L (2017) Language learner autonomy: theory, practice and research. Multilingual Matters, Bristol
Makalela L (2013) Translanguaging in Kasi-Taal: Re-thinking old language boundaries for new language planning. Stellenbosch Pap Linguist plus 42:111–125
Makalela L (2015) Translanguaging as a vehicle for epistemic access: cases for reading comprehension and multilingual interactions. Per Linguam J Lang Learn 31(1):15–29
Makoni S, Makoni B, Abdelhay A, Mashiri P (2012) Colonial and postcolonial language policies in Africa: historical and emerging landscapes. In: Spolsky B (ed) Handbook of Language Policy. Cambridge University Press, Cambridge, pp 523–543
Mareva R (2016) Learners' code-switching in English as a Second Language (ESL) lessons: perspectives of four secondary school teachers in Zimbabwe. IRA Int J Educ Multidiscip Stud 4(1):71–90
Marie KA (2013) Coping with learning through a foreign language in higher education in Rwanda. https://www.diva-portal.org/smash/get/diva2:612253/FULLTEXT01.pdf
Mgijima VD, Makalela L (2016) The effects of translanguaging on the bi-literate inferencing strategies of fourth grade learners. Perspect Educ 34(3):86–93
Milligan O, Tikly L (2016) English as a medium of instruction in postcolonial contexts: moving the debate forward. Compar Educ 52(3):277–280
Mwaniki N (2016) Translanguaging as a class/lecture-room language management strategy in multilingual contexts: Insights from autoethnographic snapshots from Kenya and South Africa. Southern African Linguist Appl Lang Stud 34(3):197–209

Mwinda N, van der Walt C (2015) From “English-only” to translanguaging strategies: exploring possibilities. Per Linguam. 31(3):100–118
Ngcobo S, Ndaba NE, Nyangiwe BL, Mpungose NC, Jamal R (2016) Translanguaging as an approach to address language inequality in South African higher education: summary writing. Critical Stud Teach Learn 4(2):10–27
Nghifinhiwanga SL (2014) Integration of translanguaging lessons: An approach to teaching and learning in Namibian junior secondary schools. A qualitative case study in three regions in Namibia. https://brage.inn.no/inn-xmlui/handle/11250/225830
Nhongo R, Tshotsho BP (2019) Translanguaging as an instructional method in science and mathematics education in ESL classroom contexts. Independ J Teach Learn 14(3):57–71
Nyika A (2015) Mother tongue as the medium of instruction at developing country universities in a global context. South African J Sci 111(1/2):1–5
Oluoch EA (2017) Language instruction in Kenya: Focus on lower primary in schools in rural areas. Int J Educ Learn Dev 5(1):17–23
Palmer PJ, Zajonc A, Scribner M (2010) The heart of higher education: a call to renewal. Jasey-Bass, New Jersey
Qorro M (2006) Does language of instruction affect quality of education? https://hakielimu.org/files/publications/document101does_lang_instruc_fct_qual_edu_en.pdf
Robinson E, Tian Z, Martinez T, Qarqueen A (2018) Teaching for justice: introducing translanguaging in an undergraduate TESOL course. J Lang Educ 4(3):77–87
Shohamy E (2012) A critical perspective on the use of English as a medium of instruction at universities. In: Doiz A, Lasgabuster D, Sierra JM (eds) English-medium instruction at Universities. Multilingual Matters, Boston, pp 196–210
Solphie A, Leglise I (2018) Multilingualism and translanguaging as a resource for teaching and learning in French Guiana https://hal.archives-ouvertes.fr/hal-01692137
United Nations Educational Scientific and Cultural Organisation (UNESCO) (2018). Languages in education. https://en.unesco.org/themes/gced/languages
United Nations Educational Scientific and Cultural Organisation (UNESCO) (2019a) *Education for sustainable development* https://en.unesco.org/themes/education-sustainable-development
United Nations Educational Scientific and Cultural Organisation (UNESCO) (2019b) Education for sustainable development. https://en.unesco.org/themes/inclusion-in-education
Wang W, Curdt-Christiansen XL (2019) Translanguaging in a Chinese-English bilingual education programme: a university classroom ethnography. Int J Bilingual Educ Bilingualism 22(3):332–337
Wolff HE (2018) Africa failing to address imperialism. University World News, 23 February 2018
Young J (2014) Encouragement in the classroom: How do I help students stay positive and focused? ASCD, Alexandria

Sustainable Development Through the Lens of Partnerships for Inclusive Education in Africa: A Case Study from North West Province, South Africa

Ellen Kakhuta Materechera

Abstract Inclusive education is based on the principles of human rights and social justice, and emphasizes equality of opportunity and universal access to education for all learners. Sustainable Development Goal 17 (SDG 17), ties successful development agendas to inclusive partnerships—at global, regional, national and local levels—built upon a shared vision and goals centred around people and the planet. Equally important, SDG 4 aims to ensure inclusive and quality education for all while promoting lifelong learning. Quality education cannot be actualized outside of inclusivity which requires partnerships across all levels of governance in order to facilitate sustainable development. The aim of this chapter is to investigate teachers' experiences on how the implementation of inclusive education through partnerships could be transformed to be relevant for local school while contributing to sustainable development at the same time. Underpinned by complexity theory, this chapter reports on a qualitative research study in which an emergent case study design was utilized to inquire into how teachers at two selected schools in North West Province, South Africa experienced the implementation of inclusive education for sustainability. Data was collected from a sample of nine teachers, one principal, two learners and one visiting lecturer. The results clearly indicate predicaments of the teaching and learning process and serves as confirmation of the complexity of inclusive education and the need for partnerships when implementing inclusive education for sustainability. The chapter contributes to broadening the understanding of how inclusive partnerships are instrumental in creating quality education systems that will benefit the African continent.

Keyword Sustainable development · Inclusive education · Quality education · Partnerships in inclusive education · Africa

E. K. Materechera (✉)
Faculty of Humanities, North-West University, P/B X2046, Mmabatho 2735, South Africa
e-mail: ellen.materechera@nwu.ac.za

W. Leal Filho et al. (eds.), *Sustainable Development in Africa*, World Sustainability Series,
https://doi.org/10.1007/978-3-030-74693-3_14

1 Introduction

It can be argued that education is key to sustainable development (SD) in any sector of human experience. Although the concept of inclusive education is complex, the common denominator is that of embracing diversity. In harmony with the concept of SD, this chapter resonates with Slee (2011: 14) who explains that 'inclusive education invites us to think about the nature of the world we live in, a world that we prefer, and our role in shaping both of those worlds'.

It is expected that inclusive quality education (QE) be offered in order to achieve sustainable education, more so in an African context. One of the many ways to achieve sustainable education is through partnerships in inclusive education. Macfarlane et al. (2014) argue that the philosophical beliefs that drive inclusion and inclusive practices appear to be in tandem with, among others, the principle of partnerships. Despite the fact that education is key to SD, implementation of QE is marred with challenges (Lotz-Sisitka and Lupele (2017). This chapter argues that it is not so easy to achieve sustainable QE in the context of the African continent, where there are several factors contributing to the complexity of simply being able to successfully educate all learners.

The aim of this chapter is to investigate teachers' experiences on how the implementation of inclusive education through partnerships could be transformed to be relevant for local schools, while simultaneously contributing to sustainable development. After this introduction the chapter covers the theory underpinning this research and then explores the concept of SD, the Sustainable Development Goals (SDGs), SDG 4, partnerships and inclusive education. Finally, all is contextualized by a case study in the North West Province of South Africa and concluding remarks on the relevance for SD in Africa.

1.1 Theoretical Framework

This research is guided by complexity theory as used by Szekely and Mason (2019). A complexity approach acknowledges that on all levels of focus, whether individual, class, school, national or international, humans and human endeavours are complex. To deal with this complexity, a focus on only one specific level will not lead to a comprehensive understanding of the multi-dimensionality, non-linearity, interconnectedness and unpredictability that are encountered. In an education setting, Kuhn (2008) argues that learners, educators and administrators, classes, schools, universities, educational associations, and human endeavours such as educational research are all multi-dimensional, non-linear, interconnected, far from equilibrium and unpredictable. In the case of this research, both inclusive education and SDG 4 are regarded as non-linear and requiring interconnectedness and partnerships in order to achieve successful implementation that will contribute to the achievement of QE.

2 Sustainable Development (SD)

While sustainability has the connotation of being cautious in not compromising the ability of future generation to meet their needs while meeting the needs of the present generation (United Nations 1987), development entails progressive and visible transformation of the economy and the society. SD contains within it the two key concepts of 'needs' and 'limitations' (imposed, for instance, by technology and social organization, which limits the ability of the environment to meet present and future needs). SD is therefore concerned with the satisfaction of human needs and aspirations towards the objective of development, but with consideration of the environment (Amador and Martinho 2015). This applies to both developed and developing countries. In developing countries, the essential needs of vast numbers of people are focussed on food, clothing, shelter and jobs. In addition, beyond their basic needs these people have legitimate aspirations for an improved quality of life. Unfortunately, these are not being met (Dube 2019).

In the context of SD, development should contribute to the progressive transformation of the economy and society. But unless development policies pay attention to considerations in the way resources are distributed and used, benefits of physical sustainability cannot be secured. At the core of sustainability is also the concern for social equity between and within generations. Hence in sustainable development, when change is a requirement, it should be done with the future generation in mind. The issue of intergenerational consideration when dealing with SD, is also supported by Hanemann et al. (2017), although in their case application of the intergenerational approach is focused on family literacy and family learning.

The concept of sustainability resonates with the Chinese philosopher, Confucius, who purportedly once argued that, "If your plan is for one year, plant rice. If your plan is for ten years, plant trees. If your plan is for one hundred years, educate children" (Goodreads.com 2021). His argument is in line with those of sustainability advocates who recognise the need for long-term, complex solutions to the challenges facing humanity. In agreement with the Chinese philosopher's argument, is the fact that "learning begins at birth", although many of its benefits and sometimes even costs are experienced only later in life, when becoming a worker/professional, a parent or a senior citizen. More so, it is accepted that human beings do not stop to engage in the process of learning throughout their lifespan (UNESCO 2007). This argument confirms the lifelong learning aspect of SDG 4.

Therefore, due to the intergenerational dimensions presented by SD, Knif and Kairavuori (2020:75) argue that "sustainability calls for the culture of caring and critical thinking so that activities we do today may call for such commitment and communality that will provide an atmosphere of hope for future generations". Like SD, inclusive education is known to be a tool for hope and empathy in its embracing nature. It takes into consideration the avoidance of social exclusion and marginalization, among others, and instead, embraces diversity and participation of ALL (Kamenopoulou 2018).

3 Sustainable Development Goals (SDGs)

The year targeted by the United Nations (UN) for the fulfilment of its Millennium Development Goals (MDGs) was 2015, but it went by while most countries had not yet fulfilled these goals (Shizha and Makuvana 2017). The African continent was not exceptional to this shortfall. Hence, a new framework, the SDGs, which will run until 2030, has been accepted by the United Nations to replace the MDGs. The 17 SDGs were meant to strategically address a comprehensive set of conditions and human needs required for SD, including aspects such as education, poverty, climate change, equality, economic growth, etc. The agenda was launched in 2015 under the title "Transforming our world: The 2030 Agenda for Sustainable Development" (United Nations 2015a). It may be possible that the UN frameworks are globally and internationally embedded but the different contexts including political and cultural differences may be a challenge in implementation, especially for SDG 4 (Bengtsson et al. 2018).

To this end, Shizha and Makuvana (2017) argue that the political and economic realities of African Nations cannot be compared to those of the developed countries of the North. This argument is associated with the implication of the complexities of uniquely African challenges that call for uniquely African solutions. Bengtsson et al. (2018:39) state that "of importance to note is that the 17 SDGs have been nationalized and implemented in many countries". One may ask, are African countries managing or is there a need to do more in order to successfully achieve the SDGs? The reason for this question is because success in one context may be regarded as a failure in another.

It is at the 70th Session of the United Nations General Assembly in September 2015 where member states adopted the new global development agenda, Transforming our world: the 2030 Agenda for SD. At its heart is the 17 SDGs. This includes SDG 4 on education. The SDGs establish development priorities to 2030 and succeed both the Millennium Development Goals and the Education for All (EFA) goals, whose deadlines expired in 2015 (UNESCO 2016). Long-term success of some SDGs depends on the achievement of one or more of the SDGs (Vladimirova and Le Blanc 2016). Adams (n.d.) is also of the view that the SDGs cannot be achieved without collaboration. This leads to the next section of the chapter which specifically concerns SDG 4.

4 Quality Education – SDG 4

The focus of SDG 4 is QE, which is at the core of Education for Sustainable Development (ESD). QE has been defined as '*ensuring inclusive and equitable quality education and promoting lifelong learning opportunities for all*' (UNESCO 2016). In recognising its importance, Manu and Didham (2014) categorized QE as a top priority for achieving SD. Education in general has also been placed at the centre of the realisation of many of the other SDGs in Agenda 2030 of the United Nations. It is seen as an enabler of upward socioeconomic mobility key to escape poverty. Therefore, it is not surprising that Nelson Mandela is famously quoted, "Education is the most powerful weapon which you can use to change the world" (Nelson Mandela Foundation 2015). Others (Lutz 2018; Sifuna and Sawamura 2020) confirm this by stating that along with education comes cognitive empowerment, which is key to many of the other desirable developments that are addressed by the SDGs. It is further argued that education is essential for various reasons including:

- Enabling the achievement of most of the other SDGs.
- Helping to reduce inequalities and to reach gender equality.
- Empowering people everywhere to live more healthy and sustainable lives.
- Fostering tolerance between people and contributing to more peaceful societies. (United Nations 2015a).

In the global MY World 2030 survey run by the United Nations (United Nations 2014) during the consultation process for the SDGs, the emphasis on the significance of education is also reported. Respondents (women and men, of all age groups, from countries at all levels of national development, and at all levels of individual education) consistently ranked "a good education" as the single most important issue that would make a difference in their lives, in comparison to better healthcare and job opportunities.

Even though QE is crucial to the achievement of all the SDGs, it faces challenges in most African countries. Challenges include the aftermath of colonisation in Africa, economic factors and inequality, among others. For example, the colonial heritage of sub-Saharan Africa impacts on the governments that constitute it as well as programmes on social development, including the provision of education at all levels. To this end, Shizha and Makuvana (2017: vii) believe that "a framework-gap-filler, different from the MDGs or SDGs, is required, not from somewhere else, but from African governments and educational policy makers". This requires a re-imagination of QE for the African continent. Whilst admitting and submitting to the dynamics as well as demands of the global village and furthermore admitting that sub-Saharan Africa cannot do it alone, the best that can be done in the interim is to actively think and re-think about education in Africa.

Continuing with the narrative of challenges faced by Africa in providing QE, it is important to firstly note that paradoxically years of steady growth in enrolment rates in schools did not change the high rates of non-proficiency. For instance, while "more than 55% of the global total of children and adolescents of primary and lower secondary school age lacked minimum proficiency in reading and mathematics in 2015, more than 80% of children of primary and lower secondary school age were not proficient in reading in sub-Saharan Africa" (United Nations 2015a).

Secondly, on the one hand adequate infrastructure and teacher training play a critical role in the quality of education (Manrique et al. 2019), while on the other hand sub-Saharan Africa faces significant challenges in providing schools with basic resources. The situation is extreme at the primary and lower secondary levels, where less than one half of the schools in sub-Saharan Africa have access to drinking water, electricity, computers and the internet. Thirdly, there is continued discrimination against women and girls in accessing education. While about one-third of countries in developing regions have not achieved gender parity in primary education, the case in sub-Saharan Africa is that girls still face barriers to entering both primary and secondary school. These disadvantages in education translate into lack of access to skills and limited opportunities in the labour market for young women (United Nations 2015a). Consequently, this situation jeopardises the benefits of lifelong learning emphasised in SDG 4.

Fourthly, Lotz-Sisitka and Lupele (2017: 3) claim that African societies are characterised by the continuing effects of a long history of colonial intrusion, and the challenges associated with establishing new societies and governance structures following the post-1950s' independence period. The achievement of independence in Africa in the 1960s failed to alter the colonial state arrangements. That is, education continues along with the western models and paradigms that have limited relevance to African needs and aspirations. Fifthly, Bengtsson et al. (2018: 41) report that "education remains an under-financed, and under-prioritised sector, with many development experts dismissing it as less important than other sectors". There is also "a widespread legacy of considering education a mere substitution of income despite significant evidence to the contrary" (Bengtsson et al. 2018: 42). Since the list of these challenges is almost endless, this chapter will not attempt to list all of them. The question is, what is the way forward for the African continent? This leads to a re-think and re-imagination of the concept of QE in Africa.

Sifuna and Sawamura (2020: 74) state that "we humans have driven our planet and ourselves into eco-social crisis. It is our responsibility to find solutions to the problems we have caused". In this regard, sub-Saharan African countries and the African continent at large, need to put a higher premium in improving the quality of their education systems. Teachers need the means to educate socially engaged, responsible, future cosmopolitans who understand the complexity of the global society and know about and want to operate with others to create a better world. The aim of Education for Sustainable Development (ESD) is to educate people to live and act sustainably and understand environmental, social, and economic aspects of SD (UNESCO 2009). Bengtsson et al. (2018) argue that there is need to pay greater attention to the full span of the human lifecycle than has been done in the past. The following section of this chapter suggests possible actions to achieve SDG 4 through the implementation of partnerships and inclusive education.

5 Partnerships and Inclusive Education

5.1 Partnerships

Partnerships play a crucial role in fostering a sustainable society. Waitoller and Kozleski (2013) conceptualize partnerships as fertile ground for learning and identifying development opportunities, thereby enhancing the achievement of the SDGs. Because professionals work and interact across institutional boundaries, it is to be expected that they will face tensions and contradictions which are created by the overlap of different communities of practice, their policies and their mediating tools. As theorized by complexity theory, this combination of circumstances results in increased levels of complexity to be dealt with. However, the benefits towards achieving SDG 4 are more important than the complexities. State, civil society and private sector partnerships provide some examples (Alkaher and Gan 2020). Thus, Alkaher and Gan (2020) recommend to those who wish to promote ESD in their societies that they should establish school–state–civil society–business partnerships for their schools. They therefore view schools as examples of complex societal systems that can interact with other systems.

More examples of partnerships include family–school partnerships, HEI–community partnerships, and partnerships across the borders of families and schools. Crucial to all partnerships are elements such as cooperation, coordination, collaboration, multiplicity, interdependence, diversity and degree of heterogeneity. All these aim at enhancing opportunities and success for children and adolescents in the delivery of ESD (Evans and Radina 2014; Maurrasse 2002; Sheridan et al. 2012). Similarly, there is sharing of critical resources, knowledge and abilities, among others, in order to improve organisations' efficiencies in what Alves et al. (2016) refer to as "relations capabilities". In this case families, schools and government departments of education are organisations worthy of partnerships and the benefits thereof.

The closest to education, more so for learners with disability is family–school or school–family–community based partnerships. *Family–school partnerships* is defined as a child-focused approach wherein families and professionals cooperate, coordinate, and collaborate to enhance opportunities and success for children and adolescents across social, emotional, behavioural, and academic domains (Albright and Weissberg 2010; Downer and Myers 2010; Lines et al. 2010 cited in Sheridan et al. 2012). *Parent involvement* is defined as the participation of significant caregivers (including parents, grandparents, stepparents, foster parents, etc.) in activities promoting the educational process of their children in support of their academic and social well-being (Fishel and Ramirez 2005). Ultimately, learner centeredness is the focus of partnerships. However, Evans and Radina (2014) argue that family, school, and community partnerships are a critical part of student achievement, but the successful establishment of meaningful partnerships with low-income and minority populations remains elusive, implying that it is less inclusive.

Some partnerships involve HEIs and communities. These partnerships are meant to benefit both parties. Coincidentally, on the topic of mutual gain, higher education is gradually becoming more conscious of the cost of unhealthy relations with local communities. Where partnerships are successful for example, HEIs recognize how they benefit to the point where they attempt to centrally incorporate community partnerships into their missions and operations or stand committed to long-term change in communities (Maurrasse 2002).

5.2 Inclusive Education

Inclusive education is the result of a global movement that emerged in response to "systemic exclusion of students who are viewed as different (e.g., students with ethnically and linguistically diverse backgrounds, students from low socioeconomic backgrounds and students with disabilities) from meaningful access and participation in education" (Waitoller and Kozleski 2013: 36). With regard to SDG 4, it is obvious that exclusions of such students would deny them the benefits of lifelong learning and ESD.

Inclusive education is a continuous struggle towards (a) the redistribution of quality opportunities to learn and participate in educational programs, (b) the recognition and appreciation of the value of differences as reflected in content, pedagogy, and assessment tools, and (c) providing the opportunities for marginalized groups to represent themselves in decision-making processes that advance and define claims of exclusion and the respective solutions that affect their children's educational futures. In fact, equality of opportunities and working towards achievement of SD should not only be for so-called 'normal' people, but also for people with disabilities. Their needs and the needs of their children in the next generation ought not to be compromised. For QE to be realised in line with SDG 4, inclusive education embraces the accommodation of learners with barriers to learning by *inter alia*, considering curriculum modification, individualised teaching programmes and the design of appropriate teaching aids.

This notion of inclusive education as a continuous struggle reflects the view that people exist in dynamic contexts (Waitoller and Kozleski 2013: 36) and that inclusive education is a complex and challenging concept that raises many questions. Mitchell (2005) asks "Is inclusive education a Western idea which cannot work in developing countries? Is it even working in Western countries?" Issues including the social, political, economic and cultural contexts determine whether or not inclusive education is implemented in range of regions and countries around the world. Consequently, it is defined differently according to different contexts. Inclusive education is one of the most dominant and controversial issues confronting educational policymakers and professionals around the world today. For example, Phasha et al. (2017: 1) claim that "we all have our different conceptions of inclusion but not every school, family, community or institutional setting can claim to have understood it". Yet, if implemented successfully, this chapter argues that it can be used as one of the tools through which SD can be actualised. The argument comes along with bearing in mind that "inclusion—whether a program, a philosophy, or an administrative mandate—has triggered many emotions across school communities and classrooms, and among teachers, students, and parents" (Phillips et al. 1995: 262).

6 The Empirical Study to Contextualise the Argument

This chapter reports on a part of the findings of a larger study (PhD thesis) that investigated the implementation of inclusive education at two schools (one semi-urban and one rural) in North West Province, South Africa. The findings of the PhD are used in this chapter not only to contextualise the practical realities of the importance of partnerships in implementing inclusive education, but also to encapsulate the lived experiences of participants in the implementation of inclusive education.

Although two schools, five learners presenting varying barriers to learning, nine teachers, one school principal and one visiting lecturer originally participated in the larger study, a further purposeful selection of one school, one principal, one teacher, one learner experiencing barriers to learning and one visiting lecturer has been employed for this chapter. These were selected based on the relevance of their contributions to this chapter. All research ethics considerations were observed.

The study posed the following question: How can teachers' experiences in the implementation of inclusive education through partnerships be transformed to be relevant for local schools and contribute to sustainable development? The following sub-questions were formulated:

a) Do you perhaps have some children who do out of the ordinary things?
b) What support is needed at an inclusive school?
c) How does the teacher relate to the learners who experience barriers to learning in an inclusive classroom?

6.1 *Methodology and Research Design*

The study adopted a qualitative approach and employed an emergent nested case research design. The nested case research design was modified from Langer (2000). If research intentions are to be met, qualitative research demands that research plans must change as research progresses (Tom 1996). An emergent design means that data collection and analysis occurred as simultaneous and ongoing activities (Merz 2002). This design enabled the researcher to be more receptive to unanticipated yet relevant events that unfolded during data collection (Ary et al. 2002) so that research procedures could be adjusted according to what emerges during the study.

6.2 Participants and Data Collection

As stated earlier, the participants for the research findings reported in this chapter came from one school. Responses and experiences of these purposefully selected participants echo the majority of participants' responses in the larger study. Data collection methods included observations, interviews and informal conversations. Interviews were semi-structured and conducted face-to-face using an interview guide. Participants were interviewed twice including member checking (Creswell 2014). Thirty classroom observations were conducted in a total of nine classrooms. Informed by the emergent case design, more observations shifted to two particular classrooms where specific learners who experienced barriers to learning were placed. With some selected ethnographic principles employed, data were collected over a period of three 3 school terms spanning two academic years. The lengthy time resulted in more rich, in-depth data collected especially through observations and informal conversations. Most of the data from the learner were provided by the adults around him.

6.3 Data Analysis

Data were transcribed verbatim and the constant comparative method was used for analysis. Observations were recorded in a field notebook. This method, which is non-mathematical and thematic, is designed to identify themes and patterns in qualitative data (Creswell 2014). Simultaneous comparison of all units of meaning obtained, were combined with inductive category coding. In keeping with the emergent design, data analysis commenced simultaneously with the field work and continued throughout the study.

6.4 Results

After all the data were analysed, the results were developed, but of which only a selection of findings relevant to this chapter are presented here. These findings were selected based on their relevance for the implementation of inclusive education for SD. The selected themes that emerged during the analysis of the data are shown in Table 1.

Table 1 Emerging themes

Theme	Category
Frustrations	• How do I/we deal with a learner presenting barriers to learning?
	• Lack of knowledge
Partnerships	• Parents
	• Professionals
Commitment to care	• One-on-one instruction

7 Discussion and Analysis

The themes and their categories and the viewpoints and perceptions captured during the interviews and other interactions are discussed in this section. This discussion encapsulates an analysis based on all the questions asked. Pseudonyms have been denoted for participants as follows: principal Jane, teacher Jude and learner John.

7.1 *Frustrations*

Lack of Knowledge: How do I/we deal with a learner presenting barriers to learning?
The lack of knowledge to understand a learner with barriers to learning was frustrating for the school community. More especially for principal:

> *"Yes, there is, for example, this boy in Grade One. If he can just come here, he must touch everything here – the computer, the phone, the books, even yourself. He is not afraid of anyone"* (Interview with principal Jane)
>
> --------
>
> *"This boy can't sit down for a moment to learn. He moves from class to class, sometimes he is outside while his peers are in class engaged in the learning process. It looks like the child is in a world of his own. The teachers – even the learning support teacher – do not know what to do with him, even his mother doesn't know what to do with him".*
>
> (Interview with Principal Jane, continued)

From this narrative it is clear that learner John is not understood by the school community. Principal Jane does not understand learner John, and the principal confirms that the teachers of the school do not understand him either. In the excerpt above, the mention of "*his mother*" and the "*teachers*" alludes to family-school partnerships as reported by Evans and Radina (2014).

Having tried all that can be done, teacher Jude, one of the teachers teaching learner John, wrote a report to the school management. The following is what teacher Jude had to write:

> *"This learner (mentioning him by name) is disruptive in class. He always sings or takes other learners' pencils or writing materials while they are busy writing (doing activities).*

He lacks the writing and reading skills. He is always outside the classroom, visits other classes while the other learners are in their respective classes. He disrupts the class with a lot of noise in such a way that other learners cannot concentrate or do what is expected of them."

Analysing this report, teacher Jude was in need of partnership with the school management team. This emphasises the fact that partnerships are required for successful implementation of sustainable inclusive education.

Learner John's behavioural problems lead to a report which gives the impression of a teacher who is on the verge of giving up. Unlike teacher Jude who was committed to care despite lack of knowledge, some teachers literary utter sentiments that signify giving up including, "I do not feel I am properly trained to help them!" (Hansen-Thomas et al. 2016:308). Perhaps this could have been attributed to the fact that positive attitudes have greatly been influenced by the training of teachers in special education and inclusive education (Rodriguez et al. 2012). Teacher Jude and also the majority of teachers in the larger study, had not yet attended a workshop on inclusive education or received training about it. As such, their experiences might have influenced their attitudes negatively, which in turn could affect their relationships with learners, especially those with disabilities.

During data collection, classroom observations revealed that most of the time during lesson, learner John does not sit down like the other learners. At one stage he is sitting down, and at another he is standing up. His behavioural disability denies himself of constructive participation in the lesson. Principles of inclusive education such as embracing diversity and creating opportunities to teach (Engelbrecht et al. 2015), are unintentionally contradicted. Learner John was included and yet excluded from constructive participation in the lesson.

As argued by Cole (2015), learners with disabilities who are admitted into the mainstream school may be internally excluded for various reasons such as the case of learner Jude. The issue of education for sustainability is in direct contradiction with the statement because inclusive education should be able to offer QE that is sustainable for participation in the future economies. Based on the experiences of teacher Jude and my own observations, teacher Jude had this to say:

". . . and they say that they want these children in special schools to come into the normal school. Do you think it will work? Eh eh eh I don't think so [teacher answers her own question]. I think teachers will resign!" (Unstructured interview with teacher Jude)

Teacher Jude autonomously analysed and criticised the inclusive education discourse. According to her, inclusive education was seemingly imposed upon her by enrolling learner John who was included by default without formal assessment. Such thinking may imply that inclusive education cannot successfully be implemented especially in developing countries like South Africa or even Africa at large, where assessment and referrals are sometimes scarce.

Though teacher Jude loved her teaching job, learner John's behaviour made her wonder and she did not know how to go about teaching this boy. The situation initially hindered teacher Jude in her attempts to offer QE for SD. The class teacher does not have access to support staff to assist with the care of learner John. Such kind of support to the teacher was not possible at the time of the study at a rural school, which could be one of the characteristics of developing countries as those in Africa. The observations in this study confirm that inclusive education is a continuous struggle reflecting the view that we exist in dynamic contexts (Waitoller and Kozleski 2013). Achieving QE in this case may seem far-fetched.

"*They*" in the excerpt above refers to the Department of Education that published a policy document on inclusive education (Department of Education 2001). According to teacher Jude, training for inclusivity by the Department of Education had not yet reached teachers and yet they were expected to teach learners as John. When the teacher asked the question, "*do you think this will work?*", it implied doubt whether inclusive education is sustainable. The contribution of inclusive education to SDG 4 may therefore be challenged, as the response by teacher Jude to this question was "*I don't think so*". Promoting SD through partnerships is also at stake. The teacher felt that if she was trained for inclusive education, it would be an enabler to enhance her ability to offer QE for SD. Lack of training is contrary to Manrique et al. (2019) who emphasize teacher training as a contributor to enhance the successful implementation of inclusive education. However, the comment "*I think teachers will resign*" is contradictory to ESD and the achievement of lifelong learning for SD.

7.2 *Partnerships*

Parents

Parental involvement is crucial to the education of learners. More so, learners with disabilities/barriers to learning as illustrated by teacher Jude:

> *"The parents seem not to be co-operative because despite my efforts to get learner John's report (compiled by a psychologist), till now they have not yet submitted it even to my head of department (HOD)."* (Interview with teacher Jude).
>
> ----------------
>
> *"I understand that the family has the right to privacy but it is unfortunate on our side because based on the recommendations by the psychologist; the school could have been assisted on how to handle the little boy or even further assistance beyond the school could have been sought or in process. Management team of our school is aware of this learner."* (Informal conversation with teacher Jude).

Teacher Jude's reference to the parents and the psychologist adds weight to the partnership argument. This also confirms the argument that inclusive education or efforts to achieve QE cannot be actualised without partnerships. This is true even at the level of local schools, as for this case study. The findings of this study are in harmony with Shulla et al. (2020), who confirm that partnerships for SD are considered crucial to the implementation of the SDGs, as explicitly expressed in Goal 17. It is not surprising that teacher Jude mentioned parents because the most common partnerships at school level are family-school partnerships, where parental involvement is paramount. The importance of partnerships is emphasized in this study by teacher Jude when mentioning "*the school could have been assisted by the psychologist report*". Educational policies and practices can enable collaboration between schools and parents (El Nakali Bachman and Votruba-Drzal 2010). This can involve parents who share what they know about their child in the family context, while the teachers share what they know about the parents' child in the school context.

Professionals
Partnerships in inclusive education involve stakeholders at hand that schools might collaborate with. This excerpt from the report compiled by teacher Jude refers to the psychologist that was used to assess learner John.

> *"We sent him to a psychologist for observation, but the report is not yet here. We are still waiting."* (Informal conversation with teacher Jude)

Again, the need for partnership is emphasised. The informal conversation with teacher Jude confirms the need for partnerships with professionals as the psychologist in order to be able to assist with teaching and learning of learner John so that successful implementation of inclusive education for SD may be achieved.

7.3 Commitment to Care

One-on-One Instruction
In line with the custom amongst primary school teachers to practice a lot of caring Hargreaves (1994), teacher Jude was committed to care and offered one-on-one-teaching to learner John despite her lack of knowledge and initial frustration:

> *"When he is given activities to do, he just mixes everything up. He needs more attention and extra care. He must be monitored fully in every situation, i.e. around the school because anything can happen to him because of his curiosity".*

In classes with heterogeneous characteristics in terms of gender, age, prevalence of barriers to learning, as was the case in this study, some teachers may struggle to attend to individual needs. In this study teacher Jude tries to involve learner John after class or at times during breaks. In this way teacher Jude and learner John started to build a relationship. In so doing, teacher Jude tried to meet learner John's academic, social and behavioural needs. Researchers as Pierangelo and Giuliani (2011), Morrison-Valfre (2017) and Huberty (2012) have similar views. Teacher Jude tried to work towards QE in the context at hand. She demonstrated the need to offer QE despite the difficulties of large class and diversity of characteristics of learners in the class. Teacher Jude was determined to offer QE and therefore had to find time for one-on-one instruction with learner John. This implies understanding by the teacher what it entails to contribute towards realisation of SDG 4 through the implementation of inclusive education:

> *"At least now, this term [third term], John is better. It was not good at all when school started in the first and second terms. At least he can follow instructions now and will try to do some work, although not really finishing it. He needs more time to grasp a concept."* (Informal conversation with teacher Jude).

This is illustrated by the following instruction given to learner John by teacher Jude: "John, go and write "one" to "ten" on the chalkboard". Learner John follows instructions, walks to the chalkboard, takes a piece of chalk, and writes the numbers on the board. He does this quickly, but he writes the numbers 'six', 'seven', and 'nine' facing the other way around. The teacher comments, "*You see, he has the concept of numbers. He only needs more time*". The teacher in this study is therefore determined to offer QE, no matter what it takes. This shows commitment to care. Care is one of the elements of both SDG 4 and inclusive education. This finding corroborates the argument by Knif and Kairavuori (2020) that sustainability calls for a culture of care and critical thinking so that the activities engaged with speak to commitment and communality and thus create an atmosphere of hope for future generations. Similarly, inclusive education is known to be a tool for hope and empathy in its embracing nature. It takes into consideration the avoidance of social exclusion and marginalization, among others, and instead embraces diversity.

8 Concluding Remarks

8.1 Implications for Sustainable Development in Africa

This study reported on in this chapter investigated teachers' experiences on how the implementation of inclusive education through partnerships could be transformed to be relevant for local schools while simultaneously contributing to the attainment of SD. The findings showed that partnerships, teacher training and understanding of learners with barriers to learning and/or disabilities are crucial elements to the implementation of inclusive education that contributes to SD.

The implications of the findings for SD are presented here using complexity theory, where complexity is a science of learning systems, and learning understood in terms of the adaptive behaviours of phenomena that arise during the interactions of multiple agents (Davis and Simmt 2003). To this end SDG 4 focus on QE, while QE is an enabler of the achievement of most of the other SDGs. QE, however, involves complex systems that are multi-dimensional, non-linear and interconnected. Use of one element to sort out or solve problems experienced with QE will not reduce the inherent multi-dimensional nature of problems. Therefore, in the case of this study, both inclusive education and SDG 4 are regarded as non-linear processes requiring interconnectedness through partnerships. In the absence of this, it will not be possible to successfully implement inclusive education. This will compromise the achievement of not only SDG 4, but also SD at large at the national, regional and global levels, not forgetting the African continent. Thus, a lack of partnerships is a serious dent that may contribute to the downfall of the achievement of SD.

In addition, teachers are the people responsible for the implementation of inclusive education. If they are not trained properly and lack knowledge of learners with barriers to learning and/or disabilities whom they are supposed to teach, then the implementation of sustainable inclusive education will be unattainable. This will result in poor delivery of education to learners and consequently a poor contribution to SD. Also, if teachers are frustrated and desperate for knowledge, they will slowly develop negative attitudes towards teaching learners with barriers to learning and/or disabilities. This scenario adds another layer to the potentially poor contribution by inclusive education to SD. Teachers therefore need to be trained for inclusivity.

The implications for inclusive education suggest that there is a great need to strengthen and broaden partnership networks across all stakeholders, on the African continent, in order to realise the achievement of SD. Family–school partnerships and school–professional partnerships, as referred to in this chapter (literature and empirical investigation) go some way towards broadening the understanding of how inclusive partnerships are instrumental in creating quality education systems that will contribute to SD and benefit the African continent.

8.2 Conclusion

This chapter argued that SDG 4, with its focus on QE and lifelong learning, cannot be actualized outside of inclusivity, which requires partnerships across all levels of governance in order to facilitate SD. In the context of the African continent, where there are a number of impacting factors contributing to the associated complexity of being able to successfully educate all learners, it is a necessity to form partnerships that yield mutual benefits. Inclusive education is a complex and problematic concept that raises many questions. Quality Education is a similarly complex concept. Some teachers, found in different contexts on the African continent, are still grappling with the definition of QE, in as far as implementation is concerned, due to inequalities in the distribution of resources that exist.

Finally, the limitations of the work reported in this chapter have to be considered. It needs to be emphasized that the chapter reported findings from a larger study whereby it was possible to extract some spin-offs relevant to this chapter. The small sample size and the fact that the study was conducted in only one province of South Africa, contribute to the limited applicability of the results. However, Danforth and Morris (2006) still find localised and context specific studies, as reported here, useful and argue that the findings of a qualitative study provide "rich, local descriptions, detailed narratives, and critical insights", which will "invite readers into a complex ongoing dialogue for practical considerations" on the journey towards, in this case, achieving SDG 4 through sustainable inclusive education.

Therefore this chapter provokes thoughts to re-think and re-imagine QE for the African continent. Otherwise, it is left to speculation what the future holds at the onset of 2030 which is the deadline for the UN 2030 agenda. It is therefore suggested that further research involving larger samples employing quantitative approaches be undertaken nationally and continent wide in order to investigate successful partnerships that may contribute towards achieving not only SDG 4 but SD at large.

References

Adams CA (nd) The sustainable development goals, integrated thinking and the integrated report | Integrated Reporting. Integratedreporting.org: https://integratedreporting.org/resource/sdgs-integrated-thinking-and-the-integrated-report/. Accessed 19 Jan 2021

Albright MI, Weissberg RP (2010) School-family partnerships to promote social and emotional learning. In: Christenson SL, Reschly AL (eds) Handbook of school-family partnerships. Routledge, New York

Alkaher I, Gan D (2020) The role of school partnerships in promoting education for sustainability and social capital. J Environ Educ 51(6):416–433

Alves FS, De-Carli E, Segatto AP (2015) Theoretical framework about relational capability on inter-organizational. J Ind Integrat Manage 1(4):1–17

Amador F, Martinho AP (2015) Education for sustainable development in higher education: evaluating coherence between theory and praxis. Assessment Eval Higher Educ. https://doi.org/10.1080/02602938.2015.1054783

Ary D, Jacobs LC, Razavieh A (2002) Introduction to research in education. Wadsworth, Belmont

Bengtsson SEL, Barakat B, Muttarak R (2018) The role of education in enabling the sustainable development agenda. Routledge, New York

Cole T (2015) Mental health difficulties and children at risk of exclusion from schools in England: a review from educational perspective of policy, practice and research 1997–2015 Emmanouil Agianniotakis, Oxford

Creswell JW (2014) Research design: Qualitative, quantitative, and mixed methods approaches, 4th edn. Sage, Thousand Oaks

Danforth S, Morris P (2006) Orthodoxy, heresy and the inclusion of American students considered to have emotional/behavioural disorders. Int J Inclusive Educ 10(2–3):135–148

Davis B, Simmt E (2003) Understanding learning systems: Mathematic education and complexity science. J Res Math Educ 34(2):137–167

Department of Education (2001) Building an Inclusive Education and Training System. Education White Paper 6. Government Printers, Pretoria

Downer JT, Myers SS (2010) Theoretical and empirical bases of partnerships understanding systems theory. In: Christenson SL, Reschly AL (eds), Handbook of school-family partnerships. Routledge, New York, pp 3–9

Dube E (2019) Rarity or reality of some theories of poverty in explaining the poverty in Africa. J Rural Commun Dev 14(3):48–60

El Nokali NE, Bachman HJ, Votruba-Drzal E (2010) Parent involvement and children's academic and social development in elementary schools. J Theoret Soc Psychol 81(3):988–1005

Engelbrecht P, Nel M, Nel N, Tlale D (2015) Enacting understanding of inclusion in complex contexts: classroom practices of South African teachers. South African J Educ 35(3):1–10

Evans MP, Radina R (2014) Great expectations? Critical discourse analysis of title I school–family compacts. School Commun J 24(2)

Fishel M, Ramirez L (2005) Evidence-based parent involvement interventions with school-aged children. School Psychol Quart 20:371–402

Goodreads.com (2021) A Quote by Confucius. https://www.goodreads.com/quotes/79127-if-your-plan-is-for-one-year-plant-rice-if. Accessed 19 Jan 2021

Hanemann U, McCaffery J, Newell-Jones K, Scarpino C (2017) Learning Together Across Generations: Guidelines for family literacy and learning programmes. UNESCO Institute for Lifelong Learning, Hamburg

Hargreaves A (1994) Changing teachers, changing times. Teachers Colleges Press, New York

Huberty TJ (2012) Anxiety and depression in children and adolescents: assessment, intervention and prevention. Springer, New York

Kamenopoulou L (2018) Introduction: setting the scene. In: Kamenopoulou L (ed) Inclusive education and disability in the global south. Springer, London

Knif L, Kairavuori S (2020) Student teachers building a sustainable future through constructing equality in visual arts education. Disc Commun Sustain Educ 11(1):74–90

Kuhn L (2008) Complexity and educational research: a critical reflection. In: Mason M (ed) Complexity theory and the philosophy of education. Wiley-Blackwell, Oxford, pp 169–180

Langer JA (2000) Excellence in English in middle and high school: how teachers' professional lives support student achievement. Am Educ Res J 37(2):397–439

Lines C, Miller G, Arthur-Stanley A (2010) The power of family-school partnering (FSP): a practical guide for school mental health professionals and educators (school-based practice in action). Routledge, New York

Lotz-Sisitka H, Lupele J (2017) ESD, learning and quality education

Lutz W (2018) Universal education as the root cause of sustainable development. In: Bengtsson SEL, Barakat B, Muttarak R (eds) The role of education in enabling the sustainable development agenda. Routledge, New York

Macfarlane A, Macfarlane S, Gillon G (2014) Inclusion, disability and culture: the nexus of potential and opportunity for policy development. In: Will R, Morton M, McLean M, Stephenson M, Slee R (eds) Tales from school: learning disability and State education after administrative reform P. 255–270

Manrique A, Dirani E, Frere A, Moreira G, Arezes P (2019) Teachers' perceptions on inclusion in basic school. Int J Educ Manage 33(2):409–419

Manu PO, Didham RJ (2014) Quality education for sustainable education: a priority in achieving sustainability and well-being for all. In Policy Brief; Institute for Global Environment Strategies: Kanagawa, Japan

Maurrasse DJ (2002) Higher Education-community partnerships: Assessing progress in the field. Nonprofit Voluntary Sector Quart 31(1):131–139

Merz AH (2002) A Journey through an emergent design and its path for understanding. Reflect Pract 3(2):141–152

Mitchell D (ed) (2005) Contextualizing inclusive education: evaluating old and new international perspectives. Routledge, London

Morrison-Valfre M (2017) Foundations of mental health care, 5th edn. Elsavier, Canada

Nelson Mandela Foundation (2015) UN's My World Survey Reveals What The World Wants. https://www.nelsonmandela.org/news/entry/uns-my-world-survey-reveals-what-the-world-wants. Accessed 30 Aug 2020

OECD (2017) Education at a glance 2017: OECD indicators. OECD Publishing, Paris

O'Manique C, Fourie P (2016) Gender justice and the millennium development goals: Canada and South Africa considered. Politikon 43(1):97–116

Phillips L, Sapona R, Lubic B (1995) Developing partnerships in inclusive education. Intervent School Clinic 30(5):262–272

Phasha N, Mahlo D, Dei G (2017) Inclusive education in African Contexts: a critical reader. Sense Publishers, Rotterdam

Pierangelo RG, Giuliani, (2011) Teaching students with autism spectrum disorders: a step by step guide for educators, Digitized. Sage Publishers, Virginia

Rodriguez IR, Saldana D, Moreno JF (2012) Support, inclusion, and special education teachers' attitudes toward the education of students with autism Spectrum disorders. Autism Res Treat 1–8

Sheridan SM, Kim EM, Coutts MJ, Sjuts TM, Holmes SR, Ransom KA, Garbacz SA (2012) Clarifying parent involvement and family-school partnership intervention research: a preliminary synthesis (CYFS Working Paper No. 2012–4). Retrieved from the Nebraska Center for Research on Children, Youth, Families and Schools website: cyfs.unl.edu

Shizha E, Makuvana N (eds) (2017) Re-thinking postcolonial education in Sub-Saharan Africa in the 21st century: post-millennium development goals. Sense Publishers, Rotterdam

Shulla K, Leal Filho W, Lardjane S, Sommer JH, Borgemeister C (2020) Sustainable development education in the context of the 2030 Agenda for sustainable development. Int J Sustain Dev World Ecol 27(5):458–468

Sifuna DN, Sawamura N (2020) Challenges of quality education in Sub-Saharan African countries. Nova Science Publishers Inc., New York

Slee R (2011) The irregular school: exclusion, schooling and inclusive education. Routledge, London

Szekely E, Mason M (2019) Complexity theory, the capability approach, and the sustainability of development initiatives in education. J Educ Pol 34(5):669–685

Tom A (1996) Building collaborative research: living the commitment to emergent design. Int J Qual Stud Educ 9(3):347–359

UNESCO (2007) UNESCO and microsoft lead 'next gen' teacher seminar in Asia-Pacific. https://www.unescap.org/sites/default/files/UNESCO%20presentation%20(july%2020).pdf. Accessed 11 Dec 2020

UNESCO (2009) Review of contexts and structures for education for sustainable development 2009. UNESCO, Paris

UNESCO (2016) Global education monitoring report summary 2016: education for people and planet: creating sustainable futures for all. https://www.voced.edu.au/content/ngv%3A77813. Accessed 14 Nov 2020

United Nations (1987) Report of the world commission on environment and development: our common future. Oslo, Norway: United Nations General Assembly, Development and International Co-operation: Environment

United Nations. (2014) U.N. Global Survey Education top priority despite wealth. https://www.msnbc.com/msnbc/united-nations-my-world-survey-education-priority-msna339766. Accessed 22 Nov 2020

United Nations (2015a) Sustainable Development Goal 4 (SDG 4) | Education Within the 2030 Agenda for Sustainable Development. https://www.sdg4education2030.org/the-goal. Accessed 30 Aug 2020

United Nations 2015b Transforming Our World: The 2030 Agenda for sustainable development … Sustainable Development Knowledge Platform. <https://sustainabledevelopment.un.org/post2015/transformingourworld/. Accessed 29 Nov 2020

Vladimirova K, Le Blanc D (2016) Exploring links between education and sustainable development goals through the lens of UN flagship reports. Sustain Dev 24:254–271

Waitoller FR, Kozleski EB (2013) Working in boundary practices: identity development and learning in partnerships for inclusive education. Teach Teach EDuc 31:34–45

Hansen-Thomas H, Richins LG, Kakkar K, Okeyo C (2016) I do not feel I am properly trained to help them! Rural teachers' perceptions of challenges and needs with English-language learners. Prof Dev Educ 42(2):308–324

Dr Ellen Kakhuta Materechera graduated with a PhD in Education at the University of the Witwatersrand, Johannesburg, South Africa. She is currently working at the writing centre in the Faculty of Humanities, North-West University. Her research focus is in the field of inclusive education and teaching and learning in higher education.

Fostering Sustainability - The University of South Africa Experience

Johannes Machiel Dreyer

Abstract It is the purpose of this chapter to indicate how an African university, the University of South Africa (Unisa) goes about fostering sustainability in a uniquely African way. Reference is made to a module on Environmental and Sustainability Education (ESE) that thousands of student teachers have to complete as part of their qualification that should help them to foster sustainability in their classrooms by integrating it in their school subjects. There is an account of how a whole menu of qualifications in ESE was developed from a certificate course, initial teacher qualifications and Hons BEd to masters and doctorate level. Individual and collaborative ESE research activities are illuminated which include 11 Best of Both Worlds international conferences held worldwide with ESE themes. A description of the extremely active community outreach programmes that reach many communities in a range of contexts in many areas of South Africa and beyond follows. The chapter concludes by reflecting on how fostering sustainability is different in an African university, such as Unisa, where all it undertakes is rooted in Ubuntu values and reflects the essence of its African-ness.

1 Introduction

At higher education institutions, the instinctive first step to integrate any topic or theme is probably to include it in qualifications either as a theme, a module or a course in a qualification or multiple qualifications or even as a dedicated specialised qualification or qualifications.

Fostering sustainability at the University of South Africa (Unisa) required a comprehensive effort by a dedicated team of academics. This took place in the face of constant opposition from colleagues who wanted to include their preferred themes and topics rather that ESE in curricula and in the context of a country in which sustainability is only one of many pressing issues all competing to be included in curricula

J. M. Dreyer (Deceased) (✉)
University of South Africa, PO Box 43339, Theresapark, Pretoria 0155, South Africa

W. Leal Filho et al. (eds.), *Sustainable Development in Africa*, World Sustainability Series,
https://doi.org/10.1007/978-3-030-74693-3_15

of already oversubscribed qualifications. This is even more of an issue for educational qualifications where very rigid and prescriptive curricula are mandated by the Minimum Requirements for Teacher Education Qualifications (MRTEQ 2018) for inclusion. Therefore, inclusion of any additional topics requires a hard sell and a lot of lobbying. At Unisa we have managed to do so for ESE in an array of qualifications at different levels.

Integrating ESE into curricula is however just one aspect of what is needed to institutionalise it (Leal Filho et al. 2015). Lozano et al. (2015) indicated seven possible ways of implementing sustainability in higher education. I will indicate how we tried to nurture a culture of sustainability at the institution and foster sustainability in communities in many areas in South Africa while focusing on teaching and researching ESE. By concurrently advancing the case for sustainability on so many fronts had its rewards and broadened the impact that included many people in a variety of settings (Dreyer 2017).

The aim of this chapter is to relate how a group of dedicated lecturers at one higher education institution in South Africa went about establishing a multipronged approach to foster sustainability in and outside the institution.

When referring to sustainability we define it as a composite concept that is context-bound, firstly as part of the phrase "fostering sustainability" and secondly as associated with "Environmental and Sustainability Education" since this is how we, as an institution, is employing it. This aligns with the discussion by Vos (2007) regarding definitions of sustainability that should be linked to specific times and contexts. Vos also suggests that the definition be derived from practicing sustainability in a specific context.

Fostering sustainability, in our opinion, is a continuous process of trying to preserve enough critical resources for future generations to sustain themselves. This may sound very familiar, as it is the sentiment expressed in the Brundtland report (1987). Although hundreds of efforts have been made to define sustainability in a hundred different ways since then (Vos 2007) we think many of them are often self-serving to suit own interests or suited to different contexts than ours. We have therefore tried to convey our understanding of sustainability through everything we do; the example we set and the activities we use at our workshops and in our courses. This pragmatic approach is our way of trying to convey a very simple, easily understood message, namely 'live and act in such a way that your environmental handprint is larger than your environmental footprint' (or, actively work to benefit the environment more than you damage it).

Since we started our quest at Unisa for fostering sustainability through ESE in the early 1990s, we based our way of doing this on achievement of the UNEP (1977) principles: knowledge, awareness, attitudes, skills and participation. For us this is visible when we help our learners and community members to become aware of environmental issues; make knowledge available to understand the issues; inculcate a caring attitude towards the environment; foster a need to take action to prevent further environmental degradation and provide skills, tools and resources needed to do so.

We agree with Hungerford and Volk (1990) that to achieve sustainability we need to change the behaviour of people to become environmentally responsible. We are aware of all the debates and perceptions about the ways in which this should or should not be accomplished. We are also aware of theoretical approaches that distance themselves from behaviour change as a goal, such as the Action Competence Approach (ACA) as interpreted by Jensen and Schnack (2006). Although we do not agree with their opposed view to changing behaviour, we align ourselves to aspects of ACA, such as making people action competent and working as a group towards a goal rather than as individuals for the good of all. The latter aligns directly with Unisa's assimilation of the philosophy of *ubuntu* in all its activities in which caring for one another and recognising the humanness of people are primary values. This, in turn, is associated with Freire's 'humanizing pedagogy' (Freire 1970) with which underprivileged societies in South Africa can relate well as the objective is to assist people through education to become completely human in the face of prejudice, mistreatment, and oppression. In our opinion it is well-expressed by Mandikonza and Lotz-Sisitka (2016) who state that "reflexive ESE praxis provides a sensitising focus, initiating quality education with humanising properties necessary for the common good."

We believe in connecting people with their own environment and getting them to take responsibility for its sustained existence into the future, fully aware that in communities there are people who have different priorities (Breiting and Mogensen 1999). This chapter highlights that our modus operandi is to identify a community or to react to a need in a community where ESE can make a difference. We then go to the community and work with them in finding common ground for issues in their own environment and assisting them with knowledge, skills and resources to deal with such issues to improve their own lives. We make ourselves available for follow-up visits, workshops or support to try to ensure the sustainability of actions and projects.

Although some might feel that our approach and its underlying beliefs are a bit old fashioned, we are convinced of its soundness as we have continually received feedback and evidence of its effectiveness in practice as will be demonstrated in the rest of the chapter

2 The University of South Africa (Unisa) and Its Efforts to Foster Sustainability

Unisa was established in 1873, then known as the University of the Cape of Good Hope. In 1946 it became the world's first exclusively distance learning public university. Registrations for 2020 have not been audited yet but student number were close to 400 000 which is about a third of all South African tertiary education students (Unisa 2020a). Enrolments in the College of Education for 2019 were 115 234 (Unisa 2020b). The 2020 numbers are not yet finalized, but it would be more. In 2019 there were1 802 (31%) academics and 4 092 (69%) professional and support staff at the institution (Unisa 2019).

Currently (2020) 4 716 students are enrolled for the EE module in the two BEd qualifications where the module is included as a compulsory course (Unisa 2020c).

This means that over 4 700 future teachers are being exposed to sustainability concepts, issues and activities as it is required of each of these student teachers to learn about and apply these in the lesson plans of their specific school subjects. Although we do not know how many of these student teachers will eventually include sustainability in their teaching practice, the potential impact of fostering of sustainability on school learners could be huge. Every year another cohort with similar numbers is prepared for practice and the potential impact keeps on growing.

In the next section, I will relate how teaching ESE came about and how this was integrated in the educational qualifications at Unisa.

3 Integrating ESE in Qualifications and Developing Focused Qualifications to Foster Sustainability

It is understandable that Higher Education Institutions (HEIs) that pursue the fostering of sustainability see sustainability first and foremost as an educational agenda (Ravio 2011). Many HEIs worldwide have been involved in fostering sustainability at their institutions (Corcoran and Wals 2004; Brodie 2007; Barlett and Chase 2004; McNamara 2010). In Africa and in South Africa in particular this is also the case where some HEIs have done so to a greater or lesser degree. This chapter will focus on how sustainability is being fostered in the only dedicated distance education university of the country. I will highlight how our capacity to serve our students wherever they find themselves in our country through our distance education courses to provide universal access to people at all education levels and in all contexts and our ability to work with huge numbers of students has enabled us to reach an extremely large and varied audience.

The whole endeavour of fostering sustainability at our institution started when a now retired academic, Professor CP Loubser, completed his doctorate in Environmental Education (EE) and had the vision to introduce teaching about the environment to promote sustainability. His first step was to gather like-minded staff members with an interest in environmental matters around him and to start an EE interest group at the then Faculty of Education in 1992. The objective was to promote EE at the institution.

I was a member of that group of EE activists and our first visible act was to organise a tree planting ceremony on Arbor Day. In hindsight, we did not do well in generating interest in the event, but after doing so for a few years the institution adopted it as official university activity.

The first endeavour into the more formal educational realm was to establish a short learning programme (SLP) in EE. The need for this programme was established during community-based EE activities (40 h workshops) undertaken on campus and off campus for interested individuals and groups. These workshops consisted of a series of different EE activities spread over several weeks. At the end of every series of workshops the attendees were asked for feedback to reflect on, change and improve the workshops. One common theme in feedback from students was their expression of interest in doing 'something' more in this field; some asked what the possibility was of enrolling for some or other formal EE qualification.

Nothing of the kind existed at Unisa at that time, so we conceptualised, designed and developed the EE SLP in 1993 (Loubser 1995) which consisted of two 12-credit modules requiring 240 notional hours of learning. It had a very practical approach and students had to attend a five-day (40 h) workshop where they had to develop an EE programme for their contexts as partial requirement to graduate from the programme. The above mentioned SLP found wide acclaim and soon attracted so many students that the five-day practical workshop, which formed part of the programme, had to be presented in main centres across South Africa and even in neighbouring Namibia.

> The SLP attracted teachers, EE officers at botanical gardens, national parks, zoos, government departments (Environmental Affairs, Agriculture), EE centres and ordinary citizens who felt compelled to educate others about ways to ensure a sustainable future. Feedback from students was in general very positive and the students indicated the usefulness of the knowledge, skills and values acquired in many of their own practical contexts. Charatsari and Lioutas (2018) similarly found that their short course in EE inspired students to continue their learning in EE in whatever ways they found possible. A similar need for follow-up and/or more advanced EE studies became apparent on completion of the EE SLP, some graduates requested a follow-up course and a so-called ten module, 120 notional hour "Advanced Programme for EE" was developed (Dreyer 1996).

A watershed opportunity to extend the inclusion of EE into formal teacher education qualifications presented itself through two events. The first was the publication of a Government White Paper on Education and Training (Department of Education 1995, p. 18), which specified the following:

> Environmental education, involving an interdisciplinary, integrated and active approach to learning, must be a vital element of all levels and programmes of the education and training system, in order to create environmentally literate and active citizens and ensure that all South Africans, present and future, enjoy a decent quality of life through the sustainable use of resources.

The second was when the former Faculty of Education (now the College of Education) decided to introduce a BSecEd degree for teacher education (Unisa 1996). This presented us with an opportunity to include EE as a 12-credit module in the four-year qualification comprising 40 modules. We used the Government White Paper mentioned above as motivation for this inclusion to pave the way for us to also include EE in qualifications at other levels. An EE specialisation in the Honours BEd followed and soon after that a specialised EE MEd (10 modules plus a mini-dissertation). Eventually we moved from that to a full research master's degree and completed the hierarchy of EE qualifications with a doctor's degree in EE.

These qualifications presented students with a career ladder that started with a four-day workshop, progressed to a SLP programme certificate course and eventually culminated in a doctor's degree. Alumni that followed this career path became life-long friends and most loyal supporters that channelled others in their workplaces to our EE qualifications. We also ventured outside the College of Education when we were approached to develop a module on ESE for the BSc degree in Environmental Management, at second-year level. It is perhaps necessary to explain that, since sustainability as a concept and a goal became prominent, we integrated it in the EE courses and did not develop separate courses for it. In most cases we kept EE as course name although in some instances we started using the term "Environmental and Sustainability Education" (ESE) In this chapter I will refer to EE and ESE interchangeably because ESE became increasingly important in our courses, qualifications, research and community engagement.

Courses are updated regularly in an endeavour to keep them relevant and useful. One of the ways we try to stay relevant is to ask feedback from students. In 2014 we undertook a comprehensive impact study to determine how the course material and pedagogical strategies were perceived by former students (Brandão et al. 2015). Participants in the interviews expressed a high degree of satisfaction for the relevance of the content and effectiveness of teaching methods employed. Many indicated that they could now teach ESE with confidence and could influence people to adopt more sustainable work practices and personal lifestyles.

Many participants spoke about their personal involvement in environmental groups and projects and how they could experience the change they made in peoples' lives for the better and how the environment benefitted from changed attitudes, values and actions. The most satisfying aspect of the findings was that there was clear evidence of sustained efforts to foster sustainability in their own spheres of influence in different contexts. This correlates well with what Wamsler (2020) calls inner transformation that is needed to ensure sustainability and the contention of Galpin et al. (2015) that fostering sustainability requires of you to create a culture of sustainability that can be sustained.

At the University of Tokyo a similar evaluation of environmental sustainability courses done by Arce et al. (2017) found comparable satisfaction among students and very aptly referred to it as fostering a next generation of sustainability professionals. We at Unisa can strongly associate with that as can be demonstrated with a few examples of extremely successful graduates that made their mark in different spheres related to environmental sustainability.

Case 1: An EE master's graduate student who had been a teacher became seriously interested in EE when he did the SLP. Later he became the head of environmental education at a metropolitan municipality where he introduced an extremely active EE programme. Furthermore, he started and managed numerous sustained community projects across the city. Ordinary citizens were activated to participate in efforts to conserve their environments, live, and operate more sustainably.

Case 2: An EE doctorate graduate used to manage the EE centre for another very large metropolitan municipality before joining a university where he is instrumental in developing and presenting an EE programme that reaches large numbers of students, working on EE projects, collaborating with international partners and hosting an international seminar.

Case 3: An EE master's graduate coming from an impoverished background and taught in a poor township school studied further in EE and eventually progressed to become influential in a zoo conservation education programme that reaches thousands of school learners year after year.

Case 4: An EE master's graduate who had a very active EE programme at the school where she taught (even taking her learners to Malaysia to interact with others in the EE field) joined the Department of Basic Education where she is working tirelessly to get EE entrenched in all school subjects.

Case 5: A student who completed only an EE module as part of her teacher qualification was so inspired by it that she left teaching to join the EE section in a very large metropolitan municipality, where she is now the Head She manages what is arguably the most comprehensive and wide-ranging EE programme with a plethora of projects and initiatives across the metropole.

There are furthermore many thousands of our alumni who might not be in leadership positions but still influence people in their own contexts. Their actions help people to make responsible decisions to help ensure sustainability in their communities.

4 Unisa Operations to Foster Sustainability

The fostering of a culture of sustainability was given a boost at Unisa when the Principal and Vice-chancellor, Professor Mandlenkosi S. Makhanya, in his inaugural address on 16 February 2011 said (Makhanya 2011): "You can't have research that only concentrates on human beings and neglects the environment and ecological wellbeing" implying that both are equally important. A plethora of initiatives, articles and projects followed on this in an effort to "green" the university including green buildings, better waste management, indigenising gardens with South African plants and saving energy. Colleges and departments joined in with e-waste and recycling initiatives and several curriculum adjustments to promote sustainability. Loubser and Dreyer (2012) evaluated these strategies, using UNEP (2012) indicators for biophysical sustainability at universities as well as management indicators from the University Leaders for a Sustainable Future Sustainability Assessment Questionnaire for Colleges and Universities (ULSF 2015).

In the 2012 evaluation by Loubser and Dreyer to establish the effectivity of strategies undertaken by Unisa to promote sustainability it was found that there are many commendable initiatives from the part of the university estates section and from different schools and departments that may contribute to fostering sustainability. It was also found that there is no central unit or management structure available to tie together, manage, drive and monitor these activities, which means their continued existence may be at risk and the sustainability of such efforts may be compromised. This is unfortunately still the situation but nonetheless many of the initiatives continue.

In 2012, the College of Education undertook perhaps the first Ecological Footprint Analysis at a distance HEI to determine where the college stood in this regard in partnership with the University of Sao Paulo in Brazil (Brandão et al. 2015). Interesting findings from this analysis were that electricity was the largest component responsible for the Unisa College of Education Ecological Footprint at almost 93% followed by paper, transportation, built-up land and water at rather low percentages.

While the environmental impact of staff members was very similar to that of contact universities, student profiles differed radically. Unisa students do not drive to and from the university, which results in a much smaller impact in terms of transport. The university took measures to limit the huge impact of electricity use but there are limits to what can be achieved when one has such a huge staff component as mentioned earlier (Brandão et al. 2015).

5 Community Outreach and Engagement

5.1 National Botanical Gardens

One of the first things we did to establish ESE at Unisa was to do workshops at Botanical gardens, which started in March 1993. These were undertaken on Saturdays that stretched over a month. The EE staff of the botanical gardens attended the workshops but it was also open to the wider community around them. The format of the workshops was very practical by nature where short introductions were used before attendees had to complete various activities such as making their own paper from wastepaper, creating environmental games and develop a short, individualised ESE programme for a specific interest group or age group in their contexts. The purpose was to empower them with knowledge, skills, attitudes and the physical resources to foster sustainability among people in their work and living spaces.

We also visited all the botanical gardens in South Africa during 2015 and 2016 and scrutinised their EE programmes; interviewed their environmental educators and observed some of them working with visiting groups. The objective was to determine the effectivity of the EE programmes in practice.

The findings were reported to the head of EE and contained the following:

- Most gardens had a well-planned and active EE programme that was offered to visiting groups.
- A shortage of staff members hampered the smooth functioning of EE at some gardens.
- Some staff members had no official EE training.
- In many instances, the actual activities could at most be considered as Biology teaching with very limited, if any, ESE for the conservation of biodiversity.
- Some recommendations were made that could be considered if they revised the programme.

5.2 *National Zoo*

We were invited to present a workshop for the very established and well-trained EE staff at the national zoo in Pretoria in October 2011 during a time when they reflected on their existing EE activities. We assisted them in creating and developing conservation education strategies and activities that would better foster long-term sustainability. The focus was not just presenting a one-off very short-lived experience of animals, but also to try to inculcate the message of making responsible decisions and taking responsible actions to ensure a sustainable future. To determine if this was indeed happening through the zoo experience we helped them to create fun assessment activities.

One of our doctorate students worked as an environmental educator at the zoo at that time; and we assisted her in contacting the national zoo in Kenya and Zoo Negara (national zoo) in Malaysia so that she could go there and observe their EE programmes in action. This enabled her to do a comparative study of the three zoo EE programmes for her study that was finalised in April 2014 (Venter and Dreyer 2017).

5.3 *South African National Parks (SANPARKS)*

We had intermittent but continuous involvement in EE programmes of SANPARKS over at least 20 years.

At first, in 1999 and 2000 we presented workshops to neighbouring communities in and around Kruger National Park which is the crown jewel in the national park system. The objective was to focus on the role of national parks to conserve the biodiversity of the individual national parks, ensuring sustainable futures for animal life and habitats.

In 2002, we were asked to present EE workshops to train or update the training of environmental educators in Kruger National Park. We observed that most of their activities were focused on just keeping children busy by making cardboard animals and colouring in animal pictures rather than conveying the conservation message through their activities. We shared some of our tried and tested activities that fostered sustainability and provided them with some resources they could use in this regard.

Personally, I also became involved in the Senior Honorary Rangers (SHRs) of SANPARKS where there is a group of volunteers that serve and support the parks by making resources available; performing some administrative park duties and setting up information kiosks to answer questions from tourists. I was tasked to plan and implement a Junior Honorary Ranger (JHR) programme in all SHR regions in the country in February 2000 (Dreyer 2010). The objective was to involve young children between 12 and 18 in ESE and to equip them with the knowledge, skills, attitudes and resources to foster sustainability among their peers. They did some duties in parks under supervision of SHRs, but mostly worked on environmental projects such as clean-ups, eradication of invader species and recycling in their own communities.

After quite rigorous training where they completed several courses on ESE and conservation education JHR's were able to become involved in the training of their peers and community members to foster sustainability. The JHRs have been going strong for 20 years now and keep on involving the youth in their efforts to foster sustainability. Research (not published yet) has also shown that many JHRs later followed careers related to the environment to pursue a life-long involvement in the environment to help ensure a sustainable future.

One of our colleagues, Professor Gherda Ferreira, was involved with the Kids in Parks initiative. The "kids-in-parks" (KIP) programme was conceptualised in 2004 and implemented in 2005 and aims to promote access for teachers and learners to national parks in South Africa. The programme is managed by South African National Parks (SANParks) with the intention to develop a commitment among learners and teachers to contribute towards sustaining South Africa's national heritage and providing meaningful environmental learning. A further intention is to equip learners with knowledge, skills and values that are required to enable them to take action for the environment and to develop the competence and capacity of participating teachers. Finally, the plan was to 'take' the parks to schools and surrounding communities and give previously disadvantaged learners and teachers the opportunity to visit their nearest national park.

Two research projects were undertaken as part of the KIP programme. The first project focused on assessing the contribution of the programme to the development of the teaching competences of teachers who participate in the programme. Consequently the main objective of the research was to determine whether the programme contributes to the development of teacher competences; and whether specific needs that teachers may experience, could possibly be met in future programmes. The findings of this research suggested that teachers do benefit, particularly those in the Natural Sciences field. Though teachers who teach other subjects participate in the programme, the study focussed on offerings at one specific national park, which only invites Natural Sciences teachers. These teachers participated actively in the programme and indicated that they did not only acquire knowledge, but also gained ideas on how to approach the teaching of topics in the Natural Sciences curriculum.

The second project focussed on the efficiency of the KIP programme in developing and acquiring environmental education skills. The intention was to determine whether the existing KIP programme contributes to the development of environmentally related skills and competences of learners to help them tackle environmental issues. The findings indicated that the programme failed in its intention to develop learners' environmentally related skills as they focussed more on basic needs and on wild animals in the park. However, the learners were fascinated by what they saw; the programme was therefore successful in introducing the parks to the learners.

5.4 *National History Museum*

We were involved in a project in collaboration with museum staff and the HRs of the Pretoria region in 1998 and 1999 where groups of school learners had a sleepover at the National History Museum in Pretoria. During the evening, they worked on EE activities related to exhibits of animal life in the museum and reported back on these activities the following morning. Our involvement focused on making sure they understood the importance of conserving animal life and habitats so that their children would not have to visit a museum to see extinct species, thereby fostering sustainability.

5.5 *West Coast Fossil Park (WCFP)*

In 2016 we became involved in the EE programmes offered at the WCFP near Langebaan in the Western Cape through a student that was doing our EE SLE. The WCFP is situated on fossil beds laid down during the late Mioscene and the Pioscene (the last from about 5,2 million years ago). Interesting fossils include a three-toed horse, a carnivorous bear, several species of short-necked giraffes, two types of saber-toothed cats, a large type of wolverine, two type of giant pig, gomphotheres (from which elephants were descendants and several other giant animals that were the forebears of some mammals we know today. Most of the animal life from that time became extinct because of climate change gradually altering the landscape from tropical to the very dry climate of the Cape west coast. This provides fertile ground for conversations around habitat loss resulting from climate change, which caused the extinction of animal life.

We worked with their environmental educators and helped them reflect on their existing EE programme. We found that they had capable, dedicated and very active environmental educators that hosted numerous school groups on a regular basis. We were however a bit worried that their focus was on teaching Biology or Science related to the school curriculum almost exclusively. We suggested that they rather also focus more on fossil education where they could convey the message that habitat loss (caused by natural climate changes over time or now happening because of our unsustainable use of resources and pollution) will lead to the extinction of many species.

We made a recommendation on how they could add this focus to what they are doing to foster responsible behaviour and decision-making, thereby ensuring sustainability. We also supported them in developing activities suitable to pursue this as a priority.

5.6 Scouts Movement of South Africa

One of our former master's students is heavily involved in the Scouts movement in South Africa and asked us in 2018 to assist in taking a fresh look at ESE content in their booklets and manuals to determine where they stand in terms of fostering sustainability. We were also interested to hear from scout troops themselves and former Springbok scouts on what happens in practice and how the ESE knowledge, skills and attitudes gained through environmentally related activities impacted on them or influenced their career decisions.

At this stage (2020) we completed the research and preliminary indications are that there are a variety of commendable projects and activities fostering sustainability. Many former scouts have fond memories of their experiences in nature and several indicated that they are in environmentally related careers or that they are actively promoting sustainable decision-making and responsible behaviour at their workplaces.

A report to the Scouts movement will be made and followed up with a scientific article.

5.7 Schools Programme

Today there is probably consensus that fostering sustainability is something that can be achieved with school learners and this has been the approach in many countries to a higher or lesser degree. In 1991 already the Organisation for Economic Co-operation and Development (OECD) countries (which included most of Europe as well as Canada and the USA) endorsed what they expressed as fostering a 'deeply rooted environmental awareness' in school learners and published a book on *Environment, school and active learning* (OECD 1991) to support this initiative.

At Unisa we share the belief that school learners should be exposed to environmental issues with the objective to help shape informed, skilled and responsible citizens who will work towards a sustainable future. We therefore developed some booklets in 1998 and 2000 with numerous activities (different ones for different grades such as one on the responsible utilisation of resources for Grade 8 by Loubser et al. (1999).

The workbooks contained useful information but the main objective of the programme was to nurture a caring attitude towards the environment through active participation. A teacher was identified in each school that showed interest in teaching and learning about the environment that could act as champion and facilitate the learning. On request, we visited schools to do short workshops on the topics for the different grades.

For the first time but not the last, we came up against a phenomenon that gave us endless headaches with several of our projects where we were dependent on a champion to sustain the initiatives. Everything ran smoothly but when a champion retired or left, the projects could not be sustained for long. In later years, we tried to make sure that the champions collaborated with someone that could take over when they left, but at some stage, the chain is still often broken with devastating results.

We were nonetheless satisfied that the learners who participated were sufficiently equipped with knowledge, skills, attitudes and resources to influence their peers and others in their contexts to strive for a sustainable future. Clean-ups; adoption of derelict pieces of land to improve and conserve them; recycling projects and beatifying school grounds through regular planting of indigenous trees and eradication of alien vegetation were among the spinoffs for the wider school communities who participated.

Since 2016, one of our colleagues, Dr Soul Shava, has represented Unisa in the 'Fundisa for Change' initiative that encourages teachers to teach for change. He participated in the development of the programme for transformative environmental learning through teacher education. The Department of Basic Education approved the project and teachers in several parts of the country have been involved in this training.

5.8 *Working for Water*

We were approached by the managers of the Working for Water programme in the Pretoria region to present EE workshops to stimulate interest in ESE which they might pursue as a career once the Working for Water programme came to an end. There was quite a bit of interest as the programme always only temporarily employed local jobless people to complete a specific project near them, after which they were laid off. We do not know if people did pursue EE as a career afterwards.

6 Research

I have already referred to quite a number of ESE-related publications that stemmed from our ESE work in this chapter. Except for many scientific articles in academic journals and book chapters related to ESE, over 100 full scientific papers were published in the proceedings of the Best of Both Worlds international conference series. The conferences focused on ESE and the inaugural conference was held in South Africa in 1998. Attendees came from 12 countries. Over the years ten additional international conferences followed, another four in South Africa, three in Malaysia, two in Brazil and one in the USA. From the start sustainability was a strong focus with conference themes such as 'Milestones in the Decade of Education for Sustainable Development' (Dreyer et al. 2012) and 'Ensuring Effective Delivery for a Sustainable Future' (2015).

Professor Callie Loubser involved ESE lecturers from several universities in writing an academically oriented textbook on ESE in Southern Africa with the title:

Environmental Education and Education for Sustainability - Some South African Perspectives (Loubser 2014). The book has been widely used and prescribed in several universities, as it is the only one of its kind in Southern Africa.

7 International Partnerships

Part of the vision of Unisa is to form 'partnerships in Africa and worldwide' to enable us to assist the 'the people of Africa in achieving their dreams' (Unisa 2020a). We took this vision to heart and several partnerships were forged with ESE lecturers in other countries. These partnerships were very enriching and beneficial to our own teaching, community engagement and research efforts to foster sustainability. International perspectives and practices provided valuable insights that could be integrated into our practices. The collaborations with partners in other countries did not only create a community of EE practitioners that was hugely advantageous, but lifelong friendships were also formed with some of them.

7.1 England

A partnership was started with a regular attendee of the Best of Both Worlds international conferences, Professor Tony Shallcross, from Manchester Metropolitan University (MMU) in the United Kingdom. Reciprocal visits and collaboration on matters of interest took place, notably regarding the Eco-school movement. Eco-schools in the two countries were visited and information was exchanged. We also assisted a student from MMU in accessing eco-schools in South Africa as part of her research for a master's degree.

ESE presentations were made at each other's universities and collaborative ESE research between the partners resulted in scientific publications (Shallcross and Loubser 2010; Shallcross et al. 2006).

7.2 Brazil

A partnership was started with Dr Patricia Leme and Professor Wellington Delitti from Brazil who attended the Best of Both World international conferences held in South Africa and Malaysia and then offered to host the 6th conference in Brazil in 2014. A lecturer exchange programme was initiated that resulted in several reciprocal scholarly visits.

The partnership was extended when collaborative research was undertaken on the Ecological Footprint of the College of Education at Unisa. The lead researcher was Dr Leme from the University of São Paulo and a scientific publication resulted from this (Brandão et al. 2015). Dr Leme also spent six months at Unisa as a visiting lecturer, lecturing, speaking at conferences and presenting workshops.

7.3 Malaysia

An enduring partnership developed between Unisa and the Forest Research Institute of Malaysia in the person of Dr Noor Azlin Yahya, Director of their EE section. This comprehensive and multifaceted partnership included numerous reciprocal visits that included workshops, presentations and conference attendances.

The research included several projects and resulted in a few scientific publications (Loubser et al. 2012,2014; Dreyer et al. 2019).

A spin-off from this partnership was collaborative assistance between Unisa and Mr Gary Lim from EDUTREE Services (an initiative to organise nature camps for learners to foster sustainability in Malaysia). Mr Lim became our first international student of the EE SLP. He assisted us in presenting the EE SLP in Malaysia and he helped to host a school group that visited Malaysia for an EE experience there. A scientific output resulted from this partnership (Loubser and Lim 1997).

8 Involvement in National Bodies and Initiatives

Unisa lecturers have been involved in a number of bodies and initiatives. Professor Loubser was part of the *Environmental Education Policy Initiative* as well as the *Council of the Environmental Education Association* while he and I served on the *Standards Generating Body for Environmental Education* where all the unit standards and standardised qualifications for EE were developed on behalf of the *South African Qualifications Authority*.

9 Unisa ESE from an African Perspective

The events and activities regarding ESE taking place at Unisa, which are reported on in this chapter up to this point, are not unique as there are many other universities that undertake the same or similar actions to foster sustainability. What makes Unisa different is that as, an African university, we have a different ethos and approach that embodies our African-ness. Part of our being and part of our consciousness lies in our vision towards being the African university in service of humankind. In attaining our vision, we base our relationships with our students and with communities on the values of Ubuntu such as fairness, caring and humanness and infuse our offerings with them.

Ubuntu is an ancient African philosophy captured in the Nguni (isiZulu) aphorism: *Umuntu ngumuntu ngabantu* - "A person is a person through people" (Fraser-Moloketi 2009, 243). Letseka (2011) stresses the importance of applying Ubuntu in Education in South Africa (specifically at Unisa). One of the core values of Ubuntu we ascribe to is caring for others (Tutu 2007). This caring attitude however also extends to caring for our environment as expressed in the words of one of our iconic leaders, Archbishop Emeritus Desmond Tutu (Tutu 2007):

> We are totally dependant on the ecosystems which support us, so to see those ecosystems as 'us', as a system that we are embedded in, as an extension of our being is pragmatic, and accurate. As is the traditional practice of Ubuntu.
>
> If Ubuntu encourages us to cultivate and care for ourselves, for our families and our brethren, so too should we care for our larger, extended body – the veld, bush, soil, air, water, and the wetlands.

The fostering of sustainability is also undertaken in the spirit of Ubuntu and includes valuing African life views and treasuring indigenous knowledge, especially when it concerns the sustainability of the environment. Our approach is known as Eco-Ubuntu to emphasise that caring goes further than just people in that it also includes the environment we live in. Ancient societies lived in harmony with their environment and took care of it so that it could take care of them.

In a nutshell, we can capture this approach to fostering sustainability in an African context in the following words by Tutu (2007) "Let us practice planetary Ubuntu, widen and deepen the circle, act with love to all that we are, to our entire community, to our extended planetary being. I am because you are. We are because the planet is".

10 Conclusion

Part of the objective of this book is to look at the potential of African countries and institutions to foster sustainability and indicate how we are contributing to ways in which this can be achieved that is truly African and may be worthwhile to replicate elsewhere. We think that we managed to do what we did to achieve the goal of fostering sustainability because of four interrelated reasons. I discuss these reasons below.

Firstly, we created multiple pathways and engaged in a large variety of ways with numerous groups and individuals in a diversity of contexts to try to foster sustainability. Some would say we spread ourselves too thin, but we felt obliged to seize every opportunity and react to every request to carry the message of sustainability to as many regions as we could in our country and even beyond.

Secondly, we were blessed with having an inspired leader whose superpower is to gather like-minded people around him, motivate them and to lead them in the quest for fostering sustainability.

Thirdly, we decided from the very beginning to have a strong academic pathway for people to qualify themselves as ESE practitioners and to concentrate on being as practical as possible in our approach when working with community groups and individuals to ensure that they leave our workshops with the knowledge, skills, attitudes and resources to educate others and foster sustainability.

Fourthly, our approach to foster sustainability is rooted in our African-ness and underlain by truly African values of Ubuntu, caring for people and caring for the environment (Eco-Ubuntu).

I am the last of the old guard who established ESE in all its facets at Unisa, but we have a new bunch of dedicated lecturers who are now teaching the ESE modules and who have shown themselves as fully capable of continuing the ESE research and community outreach activities to foster sustainability well into the future. The ESE modules are entrenched in the new BEd qualifications, which should contribute thereto that all student teachers are capacitated to foster sustainability in their subjects once they enter teaching for some time to come. The role of Unisa in fostering ESE should go from strength to strength.

References

Barlett PF, Chase GW (2004) Sustainability on campus: stories and strategies for change. MIT Press

Brandão MS, Leme PCS, Loubser CP, Dreyer JM, Shava S, Delitti WBC (2015) The application of ecological footprint analysis to distance learning university contexts: a case study of the college of education, University of South Africa. In: Leal Filho W, Brandli L, Kuznetsova O, do Paço AMF (eds) Integrative approaches to sustainable development at university level, pp 155–167. Springer. https://doi.org/10.1007/978-3-319-10690-8_11

Breiting S, Mogensen F (1999) Action competence and environmental education. Camb J Educ 29(3):349–353. https://doi.org/10.1080/0305764990290305

Brodie A (2007) Environmental sustainability programs in higher education: policies, practices and curriculum strategies [unpublished PhD dissertation]. Bernard School of Education, University of the Pacific

Brundtland GH (1987) Report of the world commission on environment and development: our common future. Brundtland United Nations Commission. Oxford University Press. https://sustainabledevelopment.un.org/content/documents/5987our-common-future.pdf

Charatsari C, Lioutas ED (2018) Environmental education in university schools: a study in a logistics faculty. Appl Environ Educ Commun 18(2):124–135. https://doi.org/10.1080/1533015X.2017.1366884

Corcoran P, Wals A (2004) Higher education and the challenge of sustainability: problematics, promise, and practice. Kluwer Academic, Dordrecht. https://doi.org/10.1080/02607470600782427

Department of Education. (1995) White paper on education and training. https://www.gov.za/sites/default/files/gcis_document/201409/16312gen1960.pdf

Dreyer JM (1996) Only guide for Module 8: environmental education in practice. In: Advanced programme for environmental education. Centre for community Training and Development, Pretoria

Dreyer JM (2010) The contribution of junior environmental volunteers towards sustainability: a case study from Africa [Paper presentation]. In: Proceedings of the conference on best of both worlds. 4th best of both worlds international conference: effective environmental education for biodiversity conservation, Kuching, Malaysia (pp. 109–121). Forest Research Institute Malaysia, 8–10 November 2010

Dreyer JM (2017). Using student input to assess the impact of environmental education courses at an open distance education institution [Paper presentation]. In: Proceedings of the ninth best of both worlds international conference. 9th best of both worlds international conference: partnerships and practices: together we make a difference, Kruger National Park, South Africa. Unisa, 5–8 November 2017, pp 69–87

Dreyer JM, Le Roux CL, Loubser CP, Madakizels-Madiya N (eds). Proceedings of the 5th best of both worlds international conference: environmental education and education for sustainable development – milestones in the decade of education for sustainable development, Bela Bela, South Africa. Unisa, April 16–20 2012

Dreyer JM, Noor Azlin Y, Nik Azyyati AK (2019) Visitor's perceptions of the Forest research Institute of Malaysia (FRIM) as an urban open space for environmental learning: results of a qualitative study. Environ Dev Sustain 21(4):1933–1945. https://doi.org/10.1007/s10668-018-0112-4

Fraser-Moloketi G (2009) Towards a common understanding of corruption in Africa. Public Policy Adm 24(3):331–338

Freire P (1970) Pedagogy of the oppressed. Continuum

Galpin T, Whitttington JL, Bell G (2015) Is your sustainability strategy sustainable? Creating a culture of sustainability. Corporate Governance 15(1):1–17. https://doi.org/10.1108/CG-01-2013-0004

Hungerford HR, Volk TL (1990) Changing learner behavior through environmental education. J Environ Educ 21(3):8–21. https://doi.org/10.1080/00958964.1990.10753743

Jensen BB, Schnack K (1997) The action competence approach in environmental education. Environ Educ Res 3(2):163–178. https://doi.org/10.1080/1350462970030205

Jensen BB, Schnack K (2006) The action competence approach in environmental education. Environ Educ Res 12(3/4):471–86

Leal Filho W, Brandli L, Kuznetsova O, do Paco AMF (eds) (2015) Integrative approaches to sustainable development at university level. Springer, Heidelberg

Letseka M (2011) Educating for ubuntu Botho: lessons from Basotho indigenous education. Open J Philos 3(2):337–344. https://doi.org/10.4236/ojpp.2013.32051

Loubser CP (1995) Only guide for environmental education certificate course. Centre for Community Education and Training, Pretoria

Loubser CP (ed) (2014) Environmental education and education for sustainability: some South African perspectives, 2nd ed. Van Schaik Publishers

Loubser CP, Bornman M, Dreyer JM, Whitlock E (1999) Environmental education grade 8: Our natural resources. Unisa Department of Further Teacher Education, Pretoria

Loubser CP, Dreyer JM (2012) The process of greening of Universities – the University of South Africa as a case study. [Paper presentation] In: Proceedings of the 5th best of both worlds international conference. 5th Best of both worlds international conference: environmental education and education for sustainable development – milestones in the decade of education for sustainable development, Bela-Bela, South Africa, pp. 49–62. Unisa, 16–20 April 2012

Loubser CP, Dreyer JM, Noor Azlin Y, Azyyati AK (2014) The effectiveness of environmental education workshops for teachers, learners and schools in Malaysia. Environ Dev Sustain 16(6):1163–1176

Loubser CP, Dreyer JM, Noor Azlin Y, Azyyati AK, Naimah CL (2012) Addressing shortcomings in environmental education: assessing a Malaysian wetlands programme [Paper Presentation]. In: Proceedings of the 5th Best of both worlds international conference: environmental education and education for sustainable development – milestones in the decade of education for sustainable development, Bela Bela, South Africa, Unisa, 16–20 April 2012, pp. 201–210

Loubser CP, Lim G (1997) Planning and conducting an environmental education workshop in a rural Malaysian community: a case study of participation and research. Educare 28(1):49–60

Lozano R, Ceulemans K, Seatter CS (2015) Teaching organisational change management for sustainability: designing and delivering a course at the University of Leeds to better prepare future sustainability change agents. J Clean Prod 106(1):205–215

Makhanya MS (2011) Inauguration and investiture speech, Unisa. https://www.unisa.ac.za/static/corporate_web/Content/About/Executive%20management/Unisa%20Principal%20and%20Vice-Chancellor/documents/prof-makhanya_inaug-invest_16022011F.pdf

Mandikonza C, Lotz-Sisitka H (2016) Emergence of environment and sustainability education (ESE) in teacher education contexts in Southern Africa: a common good concern. Educ Res Soc Change 5(1):107–130. https://doi.org/10.17159/2221-4070/2016/v5i1a7

McNamara KH (2010) Research article: fostering sustainability in higher education: a mixed-methods study of transformative leadership and change strategies. Environ Pract 12(1):48–58. https://doi.org/10.1017/S1466046609990445

MRTEQ (2018) Revised policy on the minimum requirements for teacher education qualifications. National Qualifications Framework Act, 2008 (Act No. 67 of 2008). Department of Higher Education and Training, Pretoria

Organisation for Economic Co-operation and Development (1991) Environment school and active learning. OECD Publication Service, Paris

Ravio KO (2011) Sustainability as an educational agenda. J Clean Prod 19(16):1906–1907. https://doi.org/10.1016/j.jclepro.2011.07.009

San Carlos Arce RO, Yoshida Y, Kudo S (2017) Fostering the next generation of sustainability professionals – assessing field-based courses in a sustainability science graduate program. Challenges Sustain 5(1):52–61. https://doi.org/10.12924/cis2017.05010052

Shallcross T, Loubser CP (2010) Ideological and pedagogical tensions in an environmental sustainability education project: a review of adult education practices around Lake Fundudzi. Int J African Renaissance Stud Multi-, Inter- Transdiscip 5(2):290–302

Shallcross T, Loubser CP, Le Roux C, O'Donaghue R (2006) Promoting sustainable development through whole school approaches: an international, intercultural teacher education research and development project. J Educ Teach Int Res Pedag 32(3):283–301

Tutu D (2007) Eco – ubuntu. Enviropaedia. https://www.enviropaedia.com/topic/default.php?topic_id=336

UN Environment Programme. (2012) Greening Universities Toolkit. https://wedocs.unep.org/bitstream/handle/20.500.11822/11273/Greening_unis_toolkit_Single_Page.pdf?sequence=1&isAllowed=y

UN Environment Programme. (1977) Tbilisi Declaration. Intergovernmental Conference on Environmental Education Final Report. UNEP, Tbilisi. https://www.gdrc.org/uem/ee/tbilisi.html

Unisa. (1996) Calendar Part 3: undergraduate subjects and qualifications. Unisa Press, Pretoria

Unisa (2019) Unisa annual report 2019: enhancing our footprint in Africa. https://www.unisa.ac.za/static/corporate_web/Content/News%20&%20Media/Publications/Annual%20reports/docs/15850_Unisa%20IR_D14%20(3).pdf

Unisa (2020a) Who we are. https://www.unisa.ac.za/sites/corporate/default/About/Who-we-are#:~:text=Unisa%20is%20Africa's%20leading%20distance%20learning%20institution%20nurturing%20inspiring%20leaders%20of%20tomorrow.&text=Our%20vision%20%E2%80%9Ctowards%20the%20African,Africa's%20educational%20and%20developmental%20problems

Unisa (2020b) The leading ODL university. https://www.unisa.ac.za/sites/corporate/default/About/The-leading-ODL-university

Unisa (2020c). Student enrolments. https://www.unisa.ac.za/sites/corporate/default/About/Facts-&-figures/Student-enrolments

Unisa (2020d). Student system: study unit registration number. Online: Only accessible by staff members

University Leaders for a Sustainable Future (ULSF) (2015) Sustainability assessment questionnaire. https://ulsf.org/sustainability-assessment-questionnaire/

Venter E, Dreyer JM (2017) Perceptions of the characteristics of effective conservation education programmes in three countries [Paper presentation]. In: Proceedings of the ninth best of both worlds international conference. 9th Best of both worlds international conference: partnerships and practices: together we make a difference, Kruger National Park, South Africa. Unisa, pp. 69–87, 5–8 November 2017

Vos RO (2007) Defining sustainability: a conceptual orientation. J Chem Technol Biotechnol 82(1):334–339

Wamsler C (2020) Education for sustainability: fostering a more conscious society and transformation towards sustainability. Int J Sustain High Educ 21(1):112–130

Development and Education that Sustain: Expanding Value Chains in Agriculture, Mining and Higher Education in Southern Africa

Eureta Rosenberg, Overson Shumba, Justine Ngoma, and Leigh Cobban

Abstract Deep sustainability, as defined in this chapter, requires a radical conceptual shift, and aligned practical strategies for creating livelihoods and value in new ways. This chapter explores an expanded value chain approach through which industries, communities and higher education institutions in Africa can transition towards development that sustains ecosystems and society. An analysis is presented, based on examples from agriculture and mining, two of the most significant industries on the continent, as well as higher education. The analysis is done by applying key concepts to existing case studies from Zambia and South Africa, and then exploring implications for higher education institutions. The chapter concludes that possibilities for more sustainable and shared value creation are already evident in the selected industries, but achieving the shifts required is challenging. Higher education emerges as having a very significant role to play, through new curriculum and research directions, with greater local relevance, to enable graduates and industries to critically and creatively engage with the challenges on the road to greater sustainability. In this way, higher education institutions would become partners in expanded, shared value creation towards deep sustainability.

Keywords Expanded value chains · Deep sustainability · Agriculture · Mining · Higher education in Africa

---

E. Rosenberg (✉)
Environmental Learning Research Centre, Rhodes University, Grahamstown, South Africa
e-mail: E.Rosenberg@ru.ac.za

O. Shumba
Centre for Academic Development and School of Mathematics and Natural Sciences,
The Copperbelt University, Kitwe, Zambia

J. Ngoma
Department of Bio-Materials Science and Technology, School of Natural Resources,
The Copperbelt University, Kitwe, Zambia

L. Cobban
African Climate Development Institute, University of Cape Town, Cape Town, South Africa
e-mail: leigh.cobban@uct.ac.za

W. Leal Filho et al. (eds.), *Sustainable Development in Africa*, World Sustainability Series,
https://doi.org/10.1007/978-3-030-74693-3_16

1 Introduction

1.1 The Contribution the Chapter Seeks to Make

Development that sustains people and their environments in Africa, requires significant conceptual and practical shifts away from 'business as usual'. This chapter explores how we can practically foster sustainable development through the expansion of the value chains in the economies of the continent. Value chain expansion towards restorative, regenerative and circular economies involves both conceptual shifts, and some very practical changes to the ways in which we create livelihoods and wealth. The chapter will identify possibilities for such shifts towards sustainability, but will also conclude that these shifts require stronger partnerships between industry, government, communities and higher education. This contention is based on the interrogation of selected case studies from the field in Zambia and South Africa.

Universities have started to respond to the need for sustainable development, and a number of research reports are available to probe sustainable development and associated skills needs in various sectors (e.g. Cobban et al. 2017; Jenkin et al. 2016; Rosenberg et al. 2015; Shumba and Kampamba 2013; Van Staden 2019; Ward et al. 2016) along with studies from influential agencies such as the World Bank (2019a; b).

When analysing these attempts at sustainable development, as this chapter seeks to do, it is important to consider the history and nature of the concept. Development and environmental concerns have often been presented as opposites, but as far back as 1992, the United Nations Conference on Environment and Development, in Brazil, already linked environment, development and sustainability. O'Donoghue (1993) was among the scholars who historically 'connected the dots' by depicting the interrelated aspects of sustainable development (Fig. 1). His figure illustrates the importance of political responses or governance to enable sound social and economic practices, along with the protection of biophysical resources, the floor on which the walls of economic development and social well-being are built.

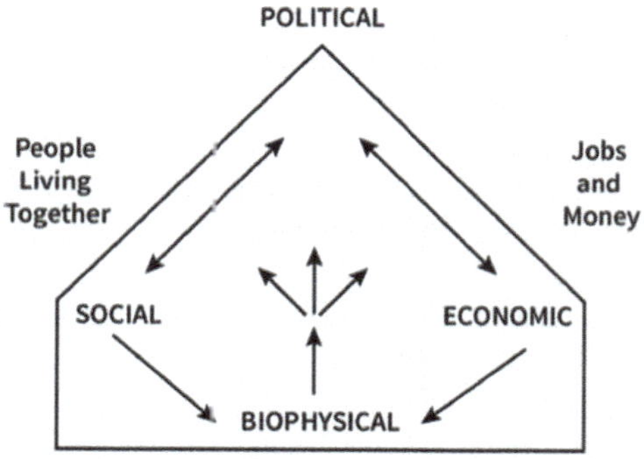

Fig. 1 The interconnected nature of sustainability. *Source:* O'Donoghue (1993)

Since 1992 a range of sustainable development policies have been developed around the world and in Southern Africa, but when the 2002 World Summit on Sustainable Development (https://sustainabledevelopment.un.org/milestones/wssd) reviewed progress, there was a general admission that the region has not succeeded well in the intentions to implement sustainable development.

There could be many reasons for this, from ambiguities in the notion of sustainable development and related concepts, as noted by Mohamed and Ramsarup (2020), to practical implementation challenges. This chapter aims to explore actual examples of attempts to identify and implement sustainable development with a view to understand better how it can be fostered, focusing in particular on the implications for higher education, the context in which all four authors work.

1.2 Context in Zambia and South Africa

Section 1.2 sketches some features of sustainability issues in the respective countries, as a backdrop to the case study analysis. The threat of climate change applies to all of Southern Africa, with increasing temperatures and disease burdens, floods, droughts and weakening food security evident in both Zambia and South Africa. Both countries also experience failures of conventional industrial pursuits to create livelihoods and shared wealth for the majority of citizens. This will be demonstrated next.

In Zambia a profile of the Kafue region's development path (ADB and WWF 2015) is instructional. Historically the resource abundance of the Kafue River Basin has been a source of economic opportunities in the form of mining, water supply, agriculture and hydropower. The resultant development has however been inadequate. As many as 75% of Zambians practice subsistence farming and many also depend on the *miombo* woodlands for forest products as livelihood supplements. The introduction of cash crops like tobacco and mining destroyed large areas of forest, and with them, the livelihoods of those who have not been accommodated in large scale agriculture or mining.

Since the opening of its first copper mine in 1928, Zambia has relied heavily on mining for economic growth. Mining in the Kafue region brought the construction of dams to supply hydro-electric power to the industry and the capital of Lusaka. The impact on downstream wetlands, biodiversity and farmers, has been mitigated with a system of releases, but rural communities benefit less from the power generated. As these communities clear the forests for charcoal production to meet their energy needs, soil erosion contributes to the siltation of the rivers which impact the ability of the dams to generate power, demonstrating the interconnected nature of the bio-physical environment, economic development and social issues like benefit sharing (as in Fig. 1). After Zambia declared independence, it invested mine earnings into social development. This positive trend was reversed when the commodity price fell in the 1970s. Environmental controls were relaxed, resulting in significant environmental damages and impacts on communities. The copper levels in the Kafue headwaters now exceed safe limits and farmers and their stock suffer related health issues.

Zambia's chosen development model has seen mining, agriculture, and the service sector in urban areas grow and expand, but so did indigent rural communities. Based on their analysis summarised here, the African Development Bank and WWF (2015) state that new approaches to development and resource management are urgently needed. The Zambian government have recognised the challenges. Noteworthy is the domestication of the UN Agenda 2030 and the Sustainable Development Goals in the Seventh National Development Plan (Republic of Zambia 2017).

In South Africa, mining and agriculture have long been regarded as the 'backbone' of the economy, but in the past decades both sectors have been shedding jobs, as part of a trend in reduced economic growth and rising unemployment, and both are affected by energy and water shortages (Cobban et al. 2017; SACMRMI 2013). Water demand in South Africa is currently estimated to be 17% above available supply (DWS 2018). Attempts to expand the inadequate electricity generation capacity through renewables have met with socio-political and economic challenges (Fakir 2017).

Following the 17th United Nations Framework Convention on Climate Change Conference of the Parties that took place in South Africa, the government, labour and industry agreed on a Green Accord towards a low-carbon, job intensive and inclusive growth path (EDD 2011). More recently, in the context of the novel coronavirus pandemic, the commitment to sustainable growth was reiterated by the President, when he launched a National Climate Change Adaptation Strategy and declared that while the economy required urgent recovery measures,

> *"we should not merely return to where we were before the pandemic struck. We are instead looking at actions that will build a new, inclusive economy that creates employment and fosters sustainable growth. An important aspect of this new economy is that it must be able to withstand the effects of climate change. A climate-resilient economy is necessary to protect jobs, ensure the sustainability of our industries, preserve our natural resources and ensure food security."* (Ramaphosa 2020)

This background, but also other sources (ADB and WWF 2015; Mohamed 2019; Mohamed and Ramsarup 2020; World Bank 2019a, b) indicate that the need for alternative development approaches, whether termed sustainable development, green growth or climate-resilient pathways, is recognised in South Africa and Zambia. It also suggests that shifting towards sustainability is difficult. This will be confirmed in the closer analysis of the case examples of efforts to move towards sustainability, including climate-smart agriculture, and forest-smart mining, undertaken in this chapter. These case examples deepen our understanding by demonstrating inter alia that:

- In both industries, opportunities for expanding value creation are already known in principle, and being further developed;
- The ability to achieve the practical shifts towards 'business unusual' in such expanded value chains is hamstrung by inadequate accountable partnerships between communities, industries and government;

- Higher education has a particular role to play in the expanded value chains of the selected industries, providing new knowledge for sustainability challenges, and building new capacity to tackle them; this in turn requires greater attention to local relevance, along with new content and research directions.

2 Methodology: Case Studies, Conceptual Framing and Analysis Details

The chapter is based on a review of existing case studies in agriculture and mining, in the two African countries where the authors are based. In addition, an analysis is done of a higher education example from Zambia, with references to higher education responses in South Africa.

The two selected industries are among the most important on the continent. In both Zambia and South Africa, mining and agriculture together contribute the bulk of Gross Domestic Product; employ significant numbers of people; and also use a very significant portion of natural resources like water, causing much concern over their sustainability. These points are elaborated by Rosenberg et al. (2015) and the World Bank (World Bank, 2019a, b) in relation to mining, and by Cobban et al. (2017) in relation to agriculture.

In both countries and both industries there are many more examples of efforts towards sustainable development that could have been reviewed. However, the case study methodology called for a reasonably in-depth treatment of each case, and chapter length did not allow exploration of more cases. Hence the authors confined themselves to one agricultural study from South Africa; two mining case studies, one each from South Africa and Zambia; and a higher education case from Zambia. The choice was influenced by cases of which the authors carried in-depth knowledge, as participating or lead researchers in the primary studies and/or in the cases being analysed.

In agriculture the case study analysed focused on climate-smart agriculture in the Western Cape Province of South Africa (Cobban et al. 2017; Rosenberg 2020a). In mining, one case study is on coal mining in South Africa (Rosenberg et al. 2015; Rosenberg 2020a) and the other, on copper mining in Zambia (Mondoloka 2017; World Bank 2019b). For an understanding of related developments in higher education, the African Centre of Excellence for Sustainable Mining in Zambia (Copperbelt University 2020) is the focus for the higher education case, while reference will also be made to educational responses to sustainability needs in agriculture, and in South Africa.

The chapter does not promise comprehensive coverage, nor is it a meta-review of all available cases in the field. Its aim is to conduct an illustrative exploration of sustainable development initiatives, opportunities and challenges through an in-depth qualitative analysis, using an innovative conceptual lens. For that reason it is important to expand on the key concepts applied in the analysis of these studies, which is done in the following paragraphs. For the methodology followed in the research that produced the case reports that were analysed, the reader is referred to those reports, as referenced.

The approach to sustainability applied in this chapter has been described as 'deep' or strong sustainability (Hattingh 2004) and treats environmental protection not as the opposite of development, but as the integral starting point of it. The embedded model of sustainable development produced by Hattingh (Fig. 2) is an alternative to the 'three pillars' model, a balancing act that describes social, economic and environmental development as the three pillars of sustainability (Hattingh 2004). By contrast, the embedded model of sustainable development is a radical idea. Much like Fig. 1, it implies that the economic sphere is held by the ecological sphere and shaped by what happens in the socio-political sphere—political decisions and social practices—and not the other way round. This means that it is possible—and necessary—to change how the economy is approached, by giving radical attention to ecological boundaries. For this reason, it could also be argued that the ecological sphere is the ceiling or the planetary boundary (Rockström et al 2009) that should determine how much and how far we develop in any one particular direction, and the social-political sphere our beliefs, policies and partnerships—are actually the foundation from which we build our economies, the walls that should protect all from hardship and deprivation.

The economy is often equated with 'markets' (from markets for raw material like copper to financial markets) but Raworth (2017) argues that economic activity involves four interlocking realms: markets, households (or communities, who provide both labour and consumers for the market), the Commons (including natural resources and waste sinks) and governance (by people and governments).

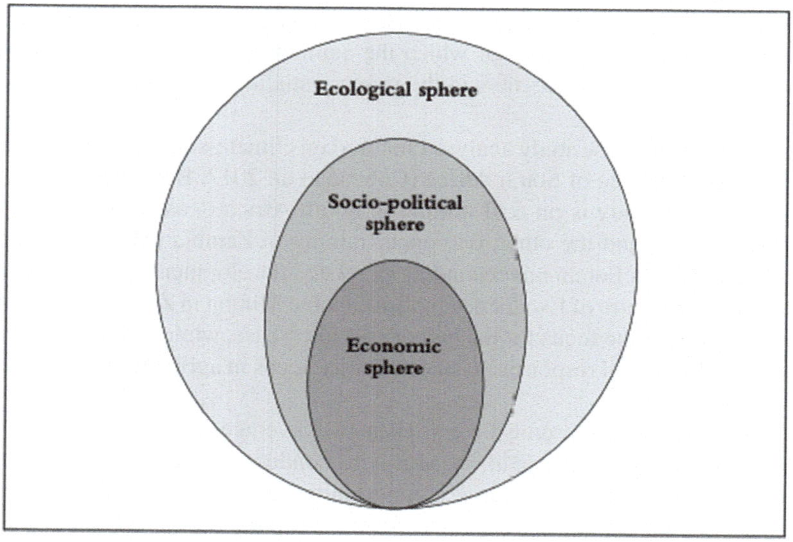

Fig. 2 Embedded model of sustainable development. *Source* Hattingh (2004)

These four realms all have to be attended to, and align well with each other, for a local area or larger region to thrive economically.

The analytic lens used here employs the notion of an industry value chain, but expands it according to Raworth's expanded version of the economy. A value chain traces a commodity (such as copper or poultry) or a service (such as research or education) from the start to the end, looking at the value required and produced along the way (Herr and Muzira 2009). For roast chicken on the table, the start of the value chain would include hens and broiler chicks, water, food and electricity, and the 'end' for the industry may be the point of sale, while for the consumer it may be bones in the bin. Along the way, various forms of waste are created and other inputs are made, e.g. transport, packaging. Value created besides the profit from sales include direct employment and side businesses. The value chain analysis in this chapter has been informed by earlier analyses of industry value chains (Cobban et al. 2017; Jenkin et al. 2016; Rosenberg 2020a; Ward et al. 2016) which have identified leverage points to foster greater sustainability and social justice, using a multi-factorial analysis that extends standard economic tools which look mostly at optimizing profits, avoiding legal snares and competition, and exploiting gaps (Aguilar 1967). This is a conceptual reframing of an industry standard, the value chain, adding a stronger environmental angle and a social justice angle, that is, looking at an industry value chain through a strong sustainability lens as outlined by Hattingh (2004) and others (Mohamed and Ramsarup 2020). An important concept in this expanded value chain model is the critical realist notion of 'absence', from Bhaskar (1993), who argued that research should look not only at what is obviously evident, but also at what should be there, but is not. This conceptual framing of the value chain is detailed in Rosenberg (2020c).

In this study, these concepts were applied in a close reading of the referenced documentation that comprise the case studies (Table 1, column 1), using the questions listed in Table 1 (column 2) to identify new insights on how to foster sustainable development, including the role of higher education.

3 Findings from Agriculture

3.1 *Case Study from South Africa*

For the agriculture case in South Africa, the analysis is based on the research of Cobban et al. (2017) in the Western Cape Province. The investigation by Cobban et al. (2017) focused on climate-smart agriculture, which aims to integrate the three dimensions of sustainable development—economic, social and environmental—by jointly addressing food security and climate challenges. In addition, beyond the climate-smart lens, the study by Cobban et al. (2017) looked at conservation agriculture, and also found examples of agro-ecology practices.

Cobban et al. (2017) focused on three commodities: winter grain; deciduous fruit; and poultry (broilers and eggs). Within each commodity's value chain, stages of

Table 1 Data sources (case documents) and analytic questions applying the key concepts

Case documents	Questions asked per case
Agricultural case study (for 3 sub-sectors as described in Sect. 3) Cobban et al. (2017) Rosenberg (2020b)	1. What are the industry value chain/chains? 2. What are the opportunities for change based on … • ecological drivers related to ecological ceilings and planetary boundaries; existing shortages, future risks • social and political drivers related to history, inequality, lock-ins, policy and legislation • economic drivers related to poverty, unemployment, growth, changing commodity prices and demand? 3. What are the absences, ambiguities and contradictions in the current value chains, e.g. who is contributing but not receiving much value, where is value wasted, where is value that is meant to be achieved, not achieved, what opportunities for value creation are overlooked or under-resourced? 4. What are the social, economic, political and ecological opportunities for change associated with … a. regenerative, restorative and circular economies b. expanded connections between markets, households and governance of the Commons? 5. What are the bottlenecks, lock-ins or challenges in achieving the opportunities for expanded value creation? 6. What are the skills and capacity needs to achieve these opportunities for expanded value creation? 7. How are higher education institutions responding?
Mining case studies Mondoloka (2017) World Bank (2019b) Rosenberg et al. (2015) Rosenberg (2020a)	The same questions as above
Higher education case (Copperbelt University 2020)	1. How is higher education responding to the expanded value chain opportunities identified in the other cases? In what way do current offerings differ from what was done before? 2. How can learning, capacity development and research address the challenges identified?

production that appeared to have the potential for impactful positive change were identified. A striking finding across the studied commodities was the differential access to resources and support in the agricultural sector; put differently, inequalities. For example, while the farmers bore the bulk of production risks, they were often not adequately compensated for their produce, and narrow profit margins meant that they were pressured to expand production, thus to use more water, fertilizer and land. Cobban et al. (2017) reported that farmers were however aware that nature is at risk and that many expressed an interest in conservation agriculture, which aims, for example, to protect biodiversity corridors, riverine vegetation and wetlands, and the ecosystem services these systems provide.

Another inequality was the differential access to resources like extension and information, including information about climate change adaptations, between large and small scale farmers, and between new entrants and established enterprises. The study suggested that this has negative implications for employment creation in new and smaller farming enterprises. Also noted was the increasingly precarious nature of the employment of seasonal farm workers as the farms on which they were employed, were rendered less profitable by unfavourable climatic and economic conditions.

The review of trends and drivers within the three commodity value chains revealed opportunities for sustainable development in the form of green enterprise development (i.e. new types of businesses), greening of existing practices and radically new ways of farming (strong sustainability). In the wine industry, for example, farmers have started acknowledging the value of biodiversity as natural pest control. By cutting out pesticides, tiny predators like the ladybird can thrive in vineyards, and keep aphids in check. Marketers have started using 'green' and 'organic' credentials for a competitive edge in increasingly health- and environment conscious markets.

Cobban et al. (2017) found that commercial grain farmers, too, were seeing the need for green technologies, services and products. The production of winter grain is affected by reduced rainfall and temperature increases. Farmers identified the need for specialist machinery and knowledge for conservation agriculture, including new practices like no-till, soil and water management, crop rotation, soil cover and new cover crops, new market opportunities and new income-generating activities (such as on-farm processing of wheat and barley).

In the production of pome fruit, there was also evidence of a demand for technologies and tools for on-farm water use efficiency; and post-harvest, for greater energy efficiency along the value chain, especially at pack houses and cold storage facilities. There is a need to reduce and reuse solid waste, and opportunities to create compost from on-farm solid waste.

Farmers are starting to invest in the diversification of income streams, for themselves and for workers. This includes selling produce and value-added items directly to the public, as well as accommodation and visitor facilities to connect to the tourism value chain. The income of seasonal fruit pickers is supplemented with out-of-season activities like pallet making, tool and machine maintenance, and childminding skills are being formally developed. A particularly sustainable initiative is a partnership between land owners and the government's public works programme *Working for*

Water, where invasive plant clearing teams are paid to restore water catchments on private land, with government support.

In addition to such new partnerships for sustainable development in agriculture, this case identified not only an emphasis on increasing production for the *market* despite dwindling resources (the focus of climate-smart agriculture) but an *expansion* of the value chain to more sustainably benefit the '*households*' of farmers and farm workers, through side-stream value chain expansion and diversification. It also showed up the need for greater equity in terms of access to resources.

The study by Cobban et al. (2017) also found examples of what is often called an agro-ecology approach, with a stronger focus on social justice and reparation. This was particularly evident in the small-scale production of poultry. Some commercial farmers moved away from the conventional large scale 'factory' style poultry farming to grass-fed poultry production on a smaller scale, producing higher value products with less environmental impact in what can be seen as a radical departure from business as usual, towards strong sustainability.

It was also noted that poultry farming is a popular entry point for new entrants into the agricultural sector, including the beneficiaries of land claim settlements. These new farmers struggle to sustain their businesses because they cannot compete with the big producers, nor absorb the high input costs. Cobban et al. (2017) also identified inefficiencies and waste in the poultry industry. Introducing *circularity* into the value chain could potentially make poultry farming more profitable even at a small scale. They found evidence of farmers taking this step towards sustainability, by using chicken feed produced from food waste. The simple circular economy is: waste is used to grow fly larvae, which is in turn fed to chickens. If the larvae cakes are locally produced, this can reduce the input cost and waste management cost. Transport cost to large abattoirs, and the need to produce bulk so as to access the slaughtering facility in the first place, can be reduced by introducing more local, smaller (community scale) abattoirs (with better waste management practices). This can also address the waste and cruelty in the mass-scale breeding, transportation and slaughtering of chickens. The innovation is in radically rethinking the economy where smaller (*household*/community) scale as well as *the Commons* are valued as important (not requiring the environment to simply absorb waste); and creating value-adding and cost-reducing circularities into the poultry chain. The introduction of the local or mobile abattoir also means that using male chicks and spent hens for food could potentially be made profitable by local production, again reducing waste and cruelty—as they are otherwise killed and discarded as waste, rather than sold on.

The study by Cobban et al. (2017) revealed cross-cutting opportunities and barriers to greening agricultural value chains. Importantly too, the opportunities, skills and specialism for unlocking sustainability were found not only at the farm level, but spread across agricultural value chains, and in linked sectors, such as water and manufacturing. This suggested a need for *partnerships* between entities operating at various points in the agricultural value chain, and in parallel value chains (such as marketing, water, energy, manufacturing and logistics).

3.2 *Implications for Education*

Cobban et al. (2017) identified that commercial grain farmers are looking for expert advice on environmental issues (including associated policy and regulations), on land management related to fire, invasive plants and riparian zone management, as well as high level technical expertise in more sustainable farming practices. Such knowledge was already to some extent available, but scarce, in high demand and expensive. In pome fruit production, skills required were those of water pump efficiency engineers and auditors; energy efficiency auditors and systems engineers. New skills for climate-smart fruit production included the ability to determine, implement and monitor energy management and energy efficiency plans; knowledge of water stewardship practices such as maintaining irrigation systems; and knowledge of suitable cultivars for changing climatic conditions.

In commercial agriculture, skills and specialisms for unlocking sustainability were found and needed not only at the farm level, but spread across agricultural value chains, and in linked sectors, as noted in 3.1. To make small-scale and environmentally-friendly poultry farming sustainable, a radical new approach was needed, and farmers seemed to need to work out the new practices for themselves; Cobban et al. (2017) reported that agricultural training institutions were not yet providing the required curricula.

4 Findings from Mining

This analysis is based on two case studies. The first is a study done in 2015 on green skills demand in coal mining in South Africa (Rosenberg 2020a; Rosenberg et al. 2015) and the second is a study relating to the introduction of forest-smart mining of copper in Zambia (Mondoloka 2017; World Bank 2019a).

Generally, the studies showed that mining value chains already feature multiple positive connections to other sectors (e.g. energy, chemicals, transport). However, mining can also adversely affect the ability of other sectors like agriculture, nature conservation and tourism to create value, and can affect households and existing livelihoods through air and water pollution, deforestation and land degradation, and encroachment into traditional lands. Although mining can score high on some economic indicators while commodity prices are favourable, it scores less on indicators concerned with environmental integrity and quality of life, in the short term and over the longer term.

4.1 The South African Coal Mining Case

In South Africa, mining has long been key in industrial development. In the past decade however, the industry has been in decline and jobs have been shed (Rosenberg et al. 2015; SACMRMI 2013). Civil society has expressed concerns about the large number of abandoned mines on the South African landscape, which are polluting water sources and preventing further development in areas that have been 'mined out'. Communities affected by mining are also concerned about loss of livelihoods. The study by Rosenberg et al. (2015; Rosenberg 2020a) found that the standard mining value chain for coal has many connections with parallel and downstream value chains, explaining why it has been sustained for so long, and how a 'locked-in' industry (Fakir 2017) may find itself resistant to change.

The study by Rosenberg et al. (2015) identified opportunities to expand the existing coal mining value chain, and to distribute benefits more equally, through enterprise and employment creation in remediation (of pollution like acid mine drainage) and restoration (of mined land). It also showed up the importance of governance to unlock the financial resources to employ retrenched workers and affected communities in restoration projects and businesses. This suggests that in an expanded value chain, governance (in this case applying the law that companies must have funds for concurrent rehabilitation and restoration after closure) actually takes on value, as it allows for the extension of the value chain, so that the end of value creation is not the point where the company stops mining. This observation is backed by the Framework Agreement from South Africa's 2018 Presidential Jobs Summit, which called for "challenges with the financial provisions for mining rehabilitation [to] be resolved" in an engagement between the Departments of Mineral Resources, Environmental Affairs and Water and Sanitation (Republic of South Africa 2018). This presents a clear step towards sustainable development: repairing environmental damage and in the process, creating jobs. It also demonstrates (as the agricultural case study did) a need for *coordination* among government entities, or accountable cooperative governance.

There are precedents for such sustainable development initiatives in the South African government's public works programmes, mentioned in 3.1, which provide short term employment in restoring water catchments, wetland rehabilitation, beach clean-ups, and the management of wildfires. The restoration of catchments and wetlands regenerates the integrity and the productive capacities of ecosystems, including their ability to buffer communities and their livelihoods from climate related stressors, as well as pollution.

Also significant is a more recent thrust of appointing citizens as environmental monitors to gather climate change related data, sample water quality including siltation levels in rivers, reporting pollution spills, controlling grazing loads on erodable land, and more. These monitors have been usefully connected to research in higher education institutions (e.g. https://www.ru.ac.za/tsitsaproject/). Expanding such restoration and monitoring work to mining and post-mining activities, that is, further expanding and connecting value chains, hold further sustainable development potential.

4.2 The Zambian Forest-Smart Mining Case

As outlined in 1.2, copper mining in Zambia has an impact on the *miombo* forest, and on the environmental services and forest products on which many communities depend. The unsustainability of mining which does not contribute to livelihoods is linked to an overreliance on foreign direct investment and limited local participation in the industry (World Bank 2019b). The actions of the poor further contribute to environmental destruction through, e.g., charcoal burning, encroachment into wetlands, and overfishing with dynamite. As noted by the African Development Bank and WWF (2015), more sustainable development strategies are needed.

One possible avenue, promoted by the World Bank (2019a, b), is forest-smart mining. Its aim is minimal damage to indigenous forests, through new technologies that extract with efficiency and precision and therefore require a smaller area of land to be disturbed. This reflects a trend towards mechanisation of copper mining elsewhere in the world as well (Araneda, 2017).

The study by the World Bank (2019a, b) noted that due to it being more mechanised, forest-smart mining need fewer workers, which has the advantage of new mines not causing huge influxes of people or the establishment of big temporary settlements in ecologically sensitive areas. On the down side, this means fewer job opportunities, and so the expansion of the mining value chain becomes important, to create work opportunities in restoration, rehabilitation, remediation, and mine supplies. Forest-smart mining puts an emphasis on creating conditions for new livelihoods, and enhancing existing livelihood activities, rather than detracting from them. Mining value chains could be expanded when companies support local livelihood activities by buying local products from the people, and paying for training. Examples include:

- plant propagation for rehabilitation purposes (such as vetiver grass)
- beekeeping, which ensures conservation of forests resources;
- mushroom growing and fish farming.

Forest-smart mining is an example of extending the value chain to include down- and side-stream industries that benefit from mining operations, help to mitigate the impact of mining operations, and restore the ability of mined areas to sustain people once mining is completed; that is, shifting from simply extractive, to a combination of extractive and regenerative value-adding processes.

A number of mining companies have attempted to adopt forest-smart mining around the world and in Zambia. One example is the Kalumbila mine which was opened in a previously remote forest area (Mondoloka 2017; World Bank 2019b). The mining company undertook to adopt a forest-smart approach to reduce impacts on the still intact forest biome. The mine concession is flanked by West Lunga National Park and three forest reserves; and three game management areas provide additional protection for wildlife. Ecological restoration features in the company's strategy in the region. This is important for biodiversity and for ecosystem services: three major river systems—the Zambezi, Kabompo, and Kafue—have their headwaters in the Kalumbila area. The company formed the Trident Foundation which established partnerships with the Forestry Department and the Department of National Parks and Wildlife, to manage an off-set area including corridors between protected forests, in partnership with government and local communities. The Foundation also developed forest-smart and conservation agriculture and other livelihoods projects to reduce impacts on the forest, working with local communities on e.g. bee keeping, commercial timber production and woodworking.

Forest-smart mining has however proved to be challenging in practice, due to several factors. Law enforcement (governing both the industry and local communities) has not been adequate, despite the country's relatively strong legislative frameworks that support forest-smart mining; according to the World Bank study (2019b) and Mondoloka (2017) the existing legal frameworks require greater capacity and enablement. Furthermore, authority over land is shared between several entities (including the Zambian Wildlife Authority, the Forestry Department, district councils, and traditional leaders), which can cause duplication of efforts and unclear roles, requiring coordination, which has not always happened. There were inadequate consultation and co-decision making between the Ministry of Mines, Energy and Water Development and other ministries, while the Ministry of National Development Planning has a key integrating and planning role (ibid p151). Signs of good governance are evident in a combination of land use planning and judicial enforcement of these land use plans, but are still inadequate. The absence of local (community) property rights also seems to jeopardize the sustainable management of natural resources (Mondoloka 2017). A history of centralization means that community, local, and district administrative bodies generally lack the capacity to comprehensively regulate land use and manage natural resources.

The analysis suggests that the mining company was struggling to shift to 'business unusual'. The environmental footprint of the Kalumbila mine and attendant developments was unnecessarily large (World Bank 2019b p154), causing large-scale fragmentation of the forest system. The site includes an abundance of roads, and there is no evidence of application of the mitigation hierarchy. An extensive social impact assessment and resettlement plan resulted in compensation of loss of livelihoods and land to local communities, but in the process large additional areas of forest were destroyed. Local communities experienced major social and demographic changes, including an influx of new job seekers. As a driver of economic development, the company has made a significant investment to the area. There has been cooperation with local and national authorities, but relationships have not been without incident, and at times lack of consultation and legal permissions clouded an otherwise massive investment in forest protection in Zambia.

The case of Kalumbila mine, and others around the world (World Bank 2019a, b), indicated that sustainable development innovations like forest-smart mining require strong partnerships between government, industry and communities. Mining has to work hand in hand with the forestry sector to ensure that practical forest management techniques are implemented and that ecologically important areas are protected. Zambia is a signatory to a number of international conventions and protocols on the protection of the environment and biodiversity conservation (Republic of Zambia 2017). The World Bank report (2019b) suggests that government needs to play a stronger role in service delivery so that communities do not rely solely on industry for their basic needs; the alignment between the economic realms of *market, household, Commons* and *governance* becomes evident here. Strategies are required to enable commitment from the key role players. For example, a national policy for forest-smart mining would ensure strong regulations that:

- Protect biodiversity and ensure no nett loss of biodiversity
- Encourage investment in long-term programs required for restoration
- Require mining companies to set aside part of their profits for rehabilitation purposes.

As in the South African coal mining case, a key challenge is to unlock the finances to pay for this work on a sustained (rather than ad hoc) basis. The World Bank (2019b) report indicates that forest-smart mining requires the internalization of environmental risks in the Bankable Feasibility Assessment of a mine, in order to take account of the costs of all avoidance, minimization, mitigation, and offsetting options.

4.3 Implications for Education

In the South African coal mining case study Rosenberg et al. (2015) identified that new skills and knowledge were required across the value chain, in mining and adjacent sectors (like water), at a range of levels, and spanning the technical, environmental, social and economic knowledge domains. This was borne out by the Zambian case study. Forest-smart mining of copper must be informed by technical, environmental and social knowledge, including a deep understanding of the relationship between mining, forests and people. While expertise in new mining techniques must be developed, equal efforts are needed to produce knowledge and capacity to address the challenges of implementing forest-smart mining, related to better governance, and cooperative governance. This includes building monitoring and resource management capacity among local communities and authorities.

5 Findings from the Higher Education Case

It is obvious that sustainable development requires new knowledge and skills. The case studies analysed provided insights into what kinds of knowledge areas and skills are important. These encompass technical, environmental, social, economic and governance related areas. (More detail on ‘green skills needs’ in these sectors can be found in Rosenberg et al. 2015 and Cobban et al. 2017). Research is needed in all these domains, to find comprehensive solutions for the multi-dimensional challenges associated with moving to 'business unusual', and new capacity has to be built through initial education, upskilling and work-based learning.

To respond to the knowledge and capacity needs, higher education institutions in Zambia are collaborating with the mining industry to promote and support forest-smart mining. One example is the Africa Centre of Excellence for Sustainable Mining (ACESM; Copperbelt University 2020). The ACESM's efforts to address sustainable mining as a developmental challenge is done through re-visioning courses and research foci, to address five thematic areas:

- Occupational health and community vulnerability in mining environments
- Restoration of mined and derelict environments and waste management
- Energy use, mining and process design
- Biodiversity and ecosystem impacts
- Livelihoods, entrepreneurship and corporate social responsibility.

Based at the Copperbelt University, the ACESM collaborates with various mines in conducting multidisciplinary research that facilitate measurement and mitigation of impacts, and drive improvement in the progressive restoration of degraded areas. The university also plays a role in building skills and capacity for sustainable mining. The thematic areas for courses and research show an expanded focus, addressing not only the traditional training areas like Occupational Health and Safety, but also community vulnerability. In addition to mining skills, workers and managers now learn about restoration, waste management, and biodiversity impacts, and retrenched workers are supported to gain entrepreneurial and life skills. Thus, the university becomes part of co-creating value in expanded and more sustainable value chains.

The ACESM, with its courses aimed at supporting environmental and social sensitivity in mining, is itself a sustainable development initiative. It is noteworthy that establishing the Centre gave greater value to the university's offerings; making them more locally relevant, and responding to injunctions that higher education must deliver value to society as a public service.

In South Africa, the Centre for Sustainability in Mining and Industry at the University of the Witwatersrand (www.wits.ac.za/csmi/) plays a similar role. Other higher education institutions, e.g., Rhodes University, have programmes expanding their capacity to support rural communities to conduct environmental monitoring and participate in cooperative governance (e.g. https://www.ru.ac.za/tsitsaproject/) These are just some examples of higher education institutions adding value and relevance to their offerings, by supporting and capacitating industries, communities and government to move towards sustainability. (For further examples, see PAGE 2016).

In agriculture, the study by Cobban et al. (2017) by contrast found that South African higher education institutions are not yet responding to the identified need for knowledge about conservation agriculture practices, agro-ecology and climate-smart agriculture. This has also been the observation in other studies, e.g. Van Staden (2020) noted that extension services are inadequate, and in high demand. Metelerkamp and Schiffer (2020) found that the new knowledge of organic farming practices are being generated by small teams and shared in informal knowledge networks. This means that marginalised rural communities with limited networking ability, may be the last to learn about more sustainable farming practices. If the challenge of large-scale sustainable development, and attendant social justice, is to be achieved, higher education institutions must start to add to the expanding value chains in agriculture—and across sectors—with greater urgency.

6 Synthesis and Conclusions

The case studies analysed showed that there are movements towards sustainability in the agricultural and mining sectors in the contexts in Zambia and South Africa. This is illustrated by the introduction of circularity in the poultry value chain, that reduces waste and input costs and increases viability for small businesses and new entrants. In mining, expanded value chains would include sustained restoration and remediation efforts, and value-add for communities whose livelihoods are disrupted by mining, as in the case of forest-smart mining, which also promises a smaller environmental impact. In both mining and agriculture, there are examples of doing more with less, saving natural resources and reducing waste. In all cases, there is also considerable room for expansion in alternative development paths. The will to do this is evident in statements by state leaders like Ramaphosa (2020) and influential institutions such as the World Bank (2019a, b).

Secondly, the analysis suggests that there are different conceptual approaches to sustainability. In mining, for example, while examples were shared of 'forest-smart' mining as an instance of sustainable development, it could be argued that no mining should be allowed in Zambia's intact forests, and that other forms of income generation should be found if strong sustainability is to be achieved. In the case of agriculture, the South African case study demonstrated that a climate-smart agriculture approach, which optimises income for markets through greater efficiencies on a large scale, may differ from the stronger social justice thrust of agro-ecology, which will take small scale and local markets and the redistribution of resources (e.g. local slaughterhouses) into account to address inequality.

Drawing on the notions of strong sustainability, circular and regenerative economies and expanded value chains, as well as the lens of absences (Bhaskar) the authors have noted a conceptual approach that extends beyond the standard market-based value chain to include recognition of the need for the protection of the Commons and the livelihoods of households, along with accountable, cooperative governance to protect the Commons. Mining that reduces its impact on existing livelihoods, and supports new livelihood opportunities, is possible in theory, as the forest-smart mining concept demonstrated. To make this happen, however, innovative and accountable partnerships and financing are needed.

Key chapter insights are that:

- Practical possibilities for expanding value chains and co-creating value as alternative development pathways are already known in industries in Africa, and being developed on an ongoing basis;
- Achieving the practical shifts required towards sustainability is however challenging, with the achievement of forest-smart agriculture, for example, still mostly an ideal rather than a reality in the case we examined;
- All the cases analysed showed the need for more and stronger accountable partnerships between communities, industries and government, and high levels of coordination among different entities;
- Higher education had a clear and important role to play, in providing new knowledge and building a range of capacities, across existing and new knowledge domains.

The chapter ends with an expansion on the latter point. The roles of higher education institutions include developing new sustainability skills and conducting research on both technical and social challenges, building research and monitoring capacity, and providing conceptual guidance on how best to approach sustainable development. Given the need for both new concepts and practical strategies, African universities and their partners should be sites for critically examining given conceptions of development, and for spearheading the conceptual and practical research to configure the most appropriate approaches to sustainable development on the continent. Involving students in such research will build their critical, creative and conceptual faculties.

Too often, Africa's higher education institutions promote an elitist education not fully localised to create value, and the agricultural case study showed that the integration of education for sustainable development is still far from adequate. This chapter hopes to contribute to discussions on how higher education, along with industry, communities and government, can support the much needed growth of sustainable development opportunities in Africa.

References

Araneda H (2017) From natural resources to innovation: an opportunity for TVET in Chile. In: Fundacion chile presentation at the UNESCO greening of TVET conference, Tangshan, China, 4– 6 July 2017

ADB (African Development Bank) and WWF (World Wide Fund for Nature) (2015) African Ecological Futures Report 2015. ADB and WWF Regional, Nairobi

Aguilar FJ (1967) Scanning the business environment. Macmillan, New York

Bhaskar R (1993) Dialectic: the pulse of freedom. Verso, London

Cobban L, Visser Z, Claassens K, Dlamini K, Hall A, Wicksteed E, Rosenberg E, Methner N, Ramsarup P, New M (2017) Green skills for ClimateSmart agriculture: a case study of poultry, winter grains and deciduous fruit value chains in the Western Cape. African Climate Development Centre, University of Cape Town, Cape Town

Copperbelt University (2020) African Centre of Excellence for Sustainable Mining. https://www.cbu.ac.zm/centre-of-excellence/. Accessed 6 Dec 2020

DWS (Department of Water and Sanitation) (2018) National Water and Sanitation Master Plan: Call to Action. Vol 1. Version 10.1. DWS, Pretoria

EDD (Economic Development Department) (2011) New Growth Path: Accord 4 Green Economy Accord. EDD, Pretoria

Fakir S (2017) Transition realism: The implications of rent-seeking to achieving South Africa's low-carbon technology ambitions. WWF-South Africa, Cape Town

Hattingh J (2004) Viewpoint: speaking of sustainable development and values … a response to Alistair Chadwick's viewpoint responding to destructive interpersonal interactions: a way forward for school-based environmental educators. Southern African J Environ Educ 21:157–165

Herr M, Muzira T (2009) Value chain analysis for decent work. ILO, Geneva

Jenkin N, Rosenberg E, Ramsarup P and Molebatsi P (2016) Green skills in the South African surface coatings sector: A focus on paint. Research report for the Chemical Industries Education and Training Authority. www.greenskills.co.za. Accessed 6 Dec 2020

Metelerkamp L, Schiffer E (2020) Net-Map: a frontline tool for the development and diffusion of sustainability skills. Southern African J Environ Educ 36(2):73–93. https://doi.org/10.4314/sajee.v36i1.11

Mohamed N (2019) (ed) Sustainability Transitions in South Africa. Earthscan/Routledge, London and New York

Mohamed N, Ramsarup P (2020) Green economy transitions and skills: global and South African perspectives. In: Rosenberg E, Ramsarup P, Lotz-Sisitka H (eds) Green skills research in South Africa: models, cases and methods. Earthscan/Routledge, London and New York, pp 15–33

Mondoloka A (2017) Approaches to supporting local and community development: The view from Zambia. 2017/41. WIDER Working Paper. https://www.wider.unu.edu/sites/default/files/wp2017-41.pdf. Accessed 6 Dec 2020

Donoghue R (1993) The environment, development and environmental education sources and perspectives for policy and curriculum initiatives in formal education. Share-Net, Howick

PAGE (Partnership for Action on Green Economy) (2016) Green Economy Learning Assessment South Africa: Critical competencies for driving a green transition. DEA, DHET, UNITAR, Rhodes University. https://www.un-page.org/files/public/green_economy_learning_assessment_south_africa.pdf

Ramaphosa C (2020) From the Desk of the President, 24 August 2020. The Presidency https://www.thepresidency.gov.za/from-the-desk-of-the-president/desk-president%2C-monday%2C-24-august-2020. Accessed 6 Dec 2020

Raworth K (2017) Doughnut Economics: Seven ways to think like a 21st century economist. Random House Business, London

Republic of South Africa (2018) Presidential Jobs Summit Framework Agreement. https://www.gov.za/sites/default/files/gcis_documents/Jobs_Summit_FrameWork_Agreement.pdf. Accessed 9 Jan 2021

Republic of Zambia (2017) Ministry of National Development Planning, Lusaka

Rockström J et al (2009) A safe operating space for humanity. Nature 461:472–475

Rosenberg E (2020a) Mining. A laminated, dialectic methodology for identifying not-yet-obvious green skills demand. In: Rosenberg E, Ramsarup P, Lotz-Sisitka H (eds) Green skills research in South Africa: models, cases and methods. Earthscan/Routledge, London and New York, pp 34–50

Rosenberg E (2020b) Green skills for agriculture. A method for focusing demand analysis and prioritisation. In: Rosenberg E, Ramsarup P, Lotz-Sisitka H (eds) Green skills research in South Africa: Models, cases and methods. Earthscan/Routledge, London and New York, pp 52–64

Rosenberg E (2020c) Synthesis and elaboration of critical realist methodology for green skills research. In: Rosenberg E, Ramsarup P, Lotz-Sisitka H (eds) Green skills research in South Africa: models, cases and methods. Earthscan/Routledge, London and New York, pp192–207

Rosenberg E, Lotz-Sisitka H, Togo, M, Ramsarup P, Mphinyane A (2015) Green Skills for Mining. Report on research for the Mining Qualifications Authority. https://www.greenskills.co.za/wp-content/uploads/2015/07/MQA-Report-Green-Skills-for-Mining.pdf. Accessed 5 Dec 2020

SACMRMI (South African Coal Mining Road Map Initiative) (2013) The South African Coal Mining Road Map. Fossil fuels South Africa and SANEDI, Johannesburg

Shumba O, Kampamba R (2013) Mainstreaming ESD into science teacher education courses: a case for ESD pedagogical content knowledge and learning as connection. Southern African J Environ Educ 29:151–166

Van Staden W (2020) Climate responsive innovation within the agricultural curriculum and learning system. Southern African J Environ Educ 36(1):73–87. https://doi.org/10.4314/sajee.v36i1.6

Ward M, Jenkin N, Rosenberg E, Ramsarup P (2016) Occupationally directed skills development for green public supply chain management. Rhodes University Environmental Learning Research Centre and Wits University Centre for Researching Education and Labour, Johannesburg

World Bank (2019a) Forest-Smart mining: identifying factors associated with the impact of artisanal and small-scale mining on forests. World Bank, Washington DC

World Bank (2019b) Forest-Smart Mining: Identifying factors associated with the impact of large-scale mining on forests. World Bank, Washington DC

Professor Eureta Rosenberg is the Chair of Environment and Sustainability Education at Rhodes University, South Africa, and the editor-in-chief of the Southern African Journal of Environmental Education. Her experience spans research, teaching and consulting in environmental education and skills for greener economies; strategic planning, transformative research methodology and evaluation. She jointly led a national Green Skills System Building project, which resulted in a volume on green skills research, cases and methodology, and conducted a range of studies and strategic projects on skills for sustainability.

Professor Overson Shumba is a professor of science education in the School of Mathematics and Natural Sciences and Acting Director for the Centre for Academic Development at the Copperbelt University, Zambia. He is conversant with global discourses on ESD and Education 2030 and pursues research interests in mainstreaming ESD, curriculum and instruction, and environmental education. He is the technical advisor on education and training matters for the Copperbelt University African Centre of Excellence for Sustainable Mining.

Dr Justine Ngoma is a lecturer and a researcher at Copperbelt University in Zambia. She focuses her research on understanding the interaction between vegetation and the atmosphere through tree growth ring analysis and application of Global Dynamic Vegetation Models. She also develops below and above ground biomass models and assesses forests' biomass and carbon sequestration ability. Justine holds a PhD in Environmental Science, MSc in Environmental Science, ACCA Professional level III, and BSc in Forestry.

Ms Leigh Cobban is a doctoral candidate in the African Climate Development Institute at the University of Cape Town in Cape Town, South Africa. She participated in the national Green Skills System Building project in that country by conducting research into climate-smart agriculture in the Western Cape, and the associated skills and educational needs. Her scholarly interests include curriculum innovation in higher education in response to sustainability concerns.

Fostering Sustainable Development in Africa Through Multidisciplinary Management Education: The Case of Sustainability Marketing Knowledge-Base

Neha Purushottam, Pantaleo Daniel Rwelamila, and Ozias Ncube

Abstract Sustainability challenges are dynamic, complex and interlinked. Education for sustainability (EfS) under target 4.7 of the Sustainable Development Goal 4 (SDG 4) is considered the key driver to promote sustainable development and lifestyles. This is a critical target for African countries where awareness and progress towards EfS are slower. The importance of EfS is undeniable for South Africa, where the educational system still carries the historical impacts of colonialism and apartheid. Moreover, the multifaceted nature of sustainability issues like sustainable consumption, poverty, the water-energy-food nexus, food security and blurred boundaries of possible sustainable solutions of each of these issues, necessitates that EfS initiatives be comprehensive and integrative. The same is true for management education which faces additional challenges due to its focus on profit. Orthodox marketing in management education is often considered a contributor to unsustainability, by propagating consumption and materialism. Sustainability marketing emerged as an alternative which has the potential to change the attitudes and behaviours of marketers, intermediaries, consumers and stakeholders by making them sustainability oriented. In this chapter, a team of three academics from the Graduate School of Business Leadership of the University of South Africa, through subjective personal introspection (SPI), examine their journey of developing and implementing a sustainability marketing module. The aim of this chapter is to share our journey to establish the importance of facilitators of multidisciplinary EfS in the higher education institutions (HEIs). These facilitators are the sustainability orientation of universities; an environment conducive for sustainability-champions, multidisciplinarity and team cohesiveness; a clear framework and boundaries for integration of each discipline; context-driven course contents which improves responsiveness towards SDGs attainment. The academics believe that highlighting these facilitators will help in promoting multidisciplinary EfS in management education, with higher education institutions in Africa intending to integrate EfS in their offerings thus benefitting from it.

N. Purushottam (✉) · P. D. Rwelamila · O. Ncube
Graduate School of Business Leadership, University of South Africa (UNISA), Midrand, South Africa
e-mail: purusn@unisa.ac.za

W. Leal Filho et al. (eds.), *Sustainable Development in Africa*, World Sustainability Series,
https://doi.org/10.1007/978-3-030-74693-3_17

Keyword Facilitators · Sustainable development · Multidisciplinary Management education · Sustainability marketing knowledge base · Africa

1 Introduction

Limited resources, the sink capacity of the earth, climate change and disruptions like COVID-19 are creating difficult scenarios for sustainable development (Krishnamurthy 2020; Slater and Dixon-Fowler 2010). Humans, business, ecosystem and society are part of one system and this factor contributes complexity and multi-dimensionality to sustainability challenges (Dentoni and Bitzer 2015). Therefore, addressing these issues needs interconnectedness and multiplicity of interventions (Starik 2004). It also requires active participation of higher education institutions (Blewitt 2010). These institutions are involved in generating and disseminating knowledge, and in the process, empower students; hence, are vital in leading radical change towards sustainability (Bloom et al. 2006). Business schools are an important part of higher education (Napier et al. 2008) and started making changes in their offerings under the pressures of sustainability-oriented consumers, public and private sectors, accreditation agencies, civil society and under the initiatives of sustainability champions in the business schools itself (Borin and Metcalf 2010; Stephens et al. 2008; Filho 2000). Although effective education for sustainability (EfS) can make a real difference, it might be a long and difficult journey for business schools as it requires changing their values and practices (Holbrook 2018).

This chapter is developed against this backdrop, with the conceptualisation, development and implementation of a sustainability marketing module at the Graduate School of Business Leadeship (GSBL) of the University of South Africa (Unisa) as case in point. The chapter starts with introducing education for sustainability (EfS) and establishes the need for higher education institutions (HEIs), including business schools, to participate actively. The second section presents country and institutional contexts to provide an understanding of the system. The third section covers methodology, where introspective methods are discussed and SPI is introduced. This is followed by Sect. 4, where narratives are discussed under three phases, namely (a) pre-module development phase to display the role of backgrounds, collective values and worldviews in motivating academics towards EfS; (b) module development phase which broadly covers development of the module and presents the framework of module and value addition by multidisciplinarity, including the way the module integrates issues and practices of sustainable development in Africa; (c) post module development phase which covers reflections from implementing the module and future directions. The section on narratives concludes with a summary and implications. Section 5 concludes the chapter by summarising our journey, suggesting that multidisciplinary EfS is vital in addressing sustainability challenges in Africa. Multidisciplinary sustainability marketing education can play a key role in integrating EfS in management education, which can further orient the public and private sectors to gear towards sustainability. While doing so, we also build a case for a strong integration of sustainability in management education.

2 Education for Sustainability (EfS)

Knowledge is a powerful resource in empowering one to think and act, directing wealth creation and ensuring human development (UNDP 2010; Bloom et al. 2006). In addressing issues of sustainable development and sustainability, education is identified as a key driver. It is important to mention that education integrating sustainability and the Sustainable Development Goals (SDGs) is often covered under the terms ESD (education for sustainable development) and EfS (education *for* sustainability), which have blurry boundaries. These terms are differentiated from the perspective of strong and weak sustainability, and also from the perspective of political ideologies affecting sustainability efforts (Springett 2005; Stough et al. 2018). These differences are well captured elsewhere (Springett 2005) and beyond the scope of this chapter. For the purposes of this chapter, ESD and EfS are not differentiated.

According to Rieckmann et al. (2017), it is clear that EfS is wider in scope, diverse in content, transformational in nature and it calls upon building on interlinkages. EfS seeks to develop an understanding of the context, complexity and interconnectedness of issues; to build a progressive mindset; to ensure optimal use of available resources and to innovate value systems, strategies and ways of living (Krishnamurthy 2020; Décamps et al. 2017; Rowe and Wehrmeyer 2010). Therefore, EfS requires multidisciplinary approaches, context-driven contents, action-oriented learning, and a practical focus on local and regional issues (Décamps et al. 2017; Rowe and Wehrmeyer 2010; Yen-chun Jim et al. 2010).

Businesses fuel consumption and are considered to be growth engines of economies (World Economic Forum 2013). Over the years, management education, in the effort to become customer centric, perpetuated such thinking (Holbrook 2018). However, under pressures from stakeholders, sustainability related skills were identified as critical for future business leaders, which motivated business schools to opt for integration of EfS in their curriculums (Rundle-Thiele and Wymer 2010). Although, the way EfS is introduced and the degree of its integration remains hugely diverse among business schools. Integration of sustainability in management education ranges from ad-hoc inclusion in any course to avoid risk of non-compliance (bare minimum) to integrating sustainability holistically across campuses (Edwards et al. 2017). Interestingly, integration efforts in management education are categorised as weak and as a result, sustainability is far from becoming mainstream (Viswanathan 2012; Wilhelm 2008; Coopey 2003). Researchers (Cornuel et al. 2015; Varey 2010) opined that the ontological assumption under which organsations focus on monetary definition of value can be one reason for weak integration. Other reasons include the focus of management education on competitiveness rather than on cooperation (Springett 2005), and on making students ready 'for work' rather than on their readiness 'for life' (Clarke et al. 2006).

From a sustainability perspective, the discipline of marketing within management education is often considered as a propagator of consumption and materialism and, hence, faces a lot of epistemological challenges as far as integration of sustainability is concerned (Varey 2010; Kemper et al. 2020). Moreover, orthodox marketing prioritises profit over social and environmental goals and, therefore, is not an apt vehicle to propagate sustainability goals. On the other end, sustainability marketing aims to achieve triple bottom line objectives, provides sustainable solutions, facilitates changes in consumer attitudes and directs behaviours and consumption practices towards sustainability (Borland and Lindgreen 2013; Belz and Peattie 2012). This makes a strong case for the sustainability marketing discipline to lead the way in integrating EfS in management education. For the purposes of this chapter, the focus will be on integrating multidisciplinarity in sustainability marketing in management education, its facilitators and their role in fostering the SDGs in Africa.

3 Context

Progress of EfS in HEIs is influenced by established operating dynamics (internal structure and institutional frameworks), stakeholders' awareness and involvement, resources, and the higher education environment of the country which need to be examined at individual, group, organisation, and system levels (Blanco-Portela et al. 2017). It was realised that the worldviews, backgrounds, attitudes and motivations of academics involved in these courses as well as the country and institutional contexts, affect their engagement with EfS and its progress (Kemper et al. 2020; De Witt et al. 2016; Shrivastava 2010).

South Africa is a young democracy where pre-1994, under the apartheid regime, education was discriminatory towards the masses (Badat 2009). The University of South Africa (Unisa) was established in 1873 and received its current name in 1916. Historically, Unisa was an exception, among previously white universities, by allowing access to black students which was unusual during the apartheid era (Arnold 2017). Even up to today, higher education in South Africa is still affected by the historical burden of apartheid era education policies and education inequalities, which in turn affected access to income, wealth, opportunities and lifestyle across racial profiles (Badat 2009). Post 1994, the idea of distance and later on online education attracted the masses and Unisa, being an open distance e-learning (ODeL) institution, became the university of choice (GSBL 2016). The Graduate School of Business Leadership (GSBL) was established in 1969. Currently, learning environments in GSBL are either fully online or hybrid, depending on the programmes (Doctoral of business leadership—DBL, Masters programmes—MBA and MBL, Post graduate diploma programmes, executive education short learning programmes, etc.). In 2010, Unisa became part of the UN's sustainability drive under the Millennium Developmental Goals (MDGs, now the SDGs). The movement was focused on university operations, community engagement and integration of sustainability into the curriculum, the latter first at the College of Agriculture and Environmental Sciences

(Mawonde and Togo 2019). However, this has created a conducive environment for EfS initiatives across other colleges as well (Unisa 2019).

4 Methodology

The methodology applied for this chapter is referred to as 'subjective personal introspection' (SPI), which is basically the 'researchers' introspection method' with elements of '*interactive introspection*' because each of us was the subject as well as the researcher (Holbrook 2018; Banbury et al. 2012). '*Researchers introspection methods'* belongs to a broader category of '*introspective methods'* and its variants are termed differently in various disciplines, like '*autoethnography*' in sociology, anthropology and philosophy; '*self-experimentation*' in psychology, behavioural and brain science; and as '*SPI*' in studies of consumer behaviour and marketing (Xue and Desmet 2019). We acknowledge the limitations the selection of this method brings, like falling short on conventional research criteria, dependence on our own recollection of memories and limited generalisability. However, we chose this methodology to (a) unearth epistemological challenges of integrating EfS in management education due to its strong ties with capitalism, and (b) to capture the richness of emotions and experiences of academics who advance EfS through their efforts. The emotional engagement and backgrounds of academics are identified as strong motivators in promoting EfS (Shrivastava 2010). SPI gave us an opportunity to create insightful narratives through introspection based upon our own experiences and, in so doing, we were better positioned to respond to the very way in which EfS is framed.

We present our narratives, in the next section, by using three phases. They are the pre-module development, module development and post module development phases. It is important to mention that in the pre-module development phase, the narratives are independent in nature as they provide the brief socio-cultural and academic journeys of three team members. In the development phase, these narratives are integrative in nature, viewed from the perspectives of the three discipline experts. In the post module development phase, the narratives are cohesive in nature and futuristic in orientation.

5 Narratives.

5.1 *Pre-development Phase*

5.1.1 Narrative 1

Born and raised in a joint family, in a small city of western Rajasthan (predominantly a desert area) in India, has interestingly, contributed to my love for the natural environment. Growing up with liberal Indian worldviews and being a science graduate, has somehow ensured that the wellbeing of the environment and society will remain a part of my life journey. During my post graduate works, my discomfort started to build with the heavily profit focused philosophies of conventional management education. This feeling of discomfort grew deeper when I joined the corporate world, where profit focus is accepted as mainstream. Later, during my doctoral studies, I got opportunities to get involved in work on real social and environmental problems, through my involvement with the sustainability centre at the university, which made me realise the potential of business schools in addressing sustainability challenges.

Post doctorate, South Africa became my work field and I joined GSBL. When I started at GSBL, my association with sustainability remained limited to teaching a few sustainability related topics within the syllabus of orthodox marketing, and research and supervision on the topics of sustainability marketing and community engagement projects. The opportunity to move from teaching orthodox marketing to sustainability marketing came when new programmes were designed in 2015–2016, under a major change in South Africa's educational landscape (Arnold 2017; UNISA 2017). During this time, through a decade long collaborative works with my colleagues in project management (PM) and supply chain areas, we identified that (a) sustainability marketing strategies face challenges when the environment is complex, dynamic and highly inter-linked (Dentoni and Bitzer 2015), (b) they display traits of projects and (c) African contexts suffer from contradictory and complex sustainability issues and, therefore, will benefit from utilising PM knowledge and tools (Purushottam and Rwelamila 2011). Similarly, my work on sustainable consumption has made me realise the importance of supply loops in stopping wastage and ensuring sufficiency across the value chain. As a result, I requested my colleagues to become part of the team and so our journey began.

5.1.2 Narrative 2

Growing up in a humble village in the northwest of Lake Victoria in Tanzania, an agricultural environment predominantly characterised by permanent crops and closer to the equator, naturally pushed the understanding of the centrality of the environment to every inhabitant of my village. Without reading any book or journal, my understanding of the environment which directly influenced the sustainability of agricultural activities, was and still is total. I moved away from the village, to seek higher education and greener pastures outside of my comfort zone, to the central region and later to Dar es Salaam, which was the capital city at the time. Thereafter, I could clearly experience a plethora of economic dynamics and its inferences to all the pillars of sustainability. As a foundation of my expertise in project management, appreciation of the sustainability should be considered as a foundation to what has characterised what I know today.

Travelling and teaching across continents consolidated my understanding of total sustainability management (TSM) and, more specifically, its importance in any project management domain as evident in my research work more than 15 years ago (e.g. Rwelamila et al. 2000; Dlungwana et al. 2002).

Being an academic, and a practicing project manager and procurement systems expert, I have had situations where I have witnessed sustainable development (SD) considerations being ignored in practice, by considering them as 'impositions' and 'distractions'. I have, thus, realised that if various professional programmes at universities could be revised to embrace SD fundamentals, graduates taking these programmes will be in a better position to embrace SD principles during their professional practices because they will see them as necessary and normal.

An invitation by my colleague, as the module lead person, to join a module development team automatically pushed me into a state of 'celebration'—'celebration' of contributing to the 'full death' or 'partial death' of orthodox thinking in marketing—*a chance to embrace TSM in marketing projects through preparing future marketers/marketing managers.*

5.1.3 Narrative 3

Growing up in the city of Bulawayo, Zimbabwe, I had a mixed experience and introduction to sustainability issues. Within each year of my childhood, I would have resided in the urban area (Bulawayo) for schooling purposes, followed by residence in my rural area, Insuza Village, during school holidays. This way of life exposed me, at an early age, to the different challenges of sustainability from both an urban and rural perspective. In the urban area, sustainability issues were waste management and pollution; while the rural folk were more concerned with land degradation, deforestation, livestock health and livelihood issues. It is these experiences, I believe, that have provided me with a modern-day sustainability orientation.

When I started tertiary education, I choose the fields of applied mathematics, in general, and operations research, as a major. I was fascinated by how best to utilise the power of mathematics to make optimal decisions that better the lives of humanity in highly constrained environments, specifically constraints like climate change, resource scarcity and their efficient utilisation, pollution and life cycle considerations. However, slowly, it became clear that focusing on optimal solutions alone is not sufficient. I had to consider the managerial practice of those entrusted with making such decisions. Consequently, I embraced the discipline of supply chain management, which provides a holistic understanding of the value creation process of products and services, and the impact that either (1) the environment has on this process, or (2) this process has on the environment.

Therefore, when my colleague invited me to participate in the development and delivery of this module, I embraced the opportunity as it provided a platform to re-orient managers around the concepts of value, convenience and social capital for entities, and how closed loop supply chains could be utilised as enablers of sustainable consumption through the '*cradle to the next cradle*' perspective.

5.2 *Module Development Phase*

5.2.1 Framework

The module is a second-year elective in a two year MBL (blended learning) programme. The module (Strategic Sustainable Marketing [SSM]) has five study units with the first four providing the core knowledge base on SSM, while the last one presents Sustainable Project Management (SPM) and Sustainable Supply Loop Management (SSLM) as tools for developing, implementing and managing SSM initiatives. When we started to shape the module, we knew that we had a difficult task at hand as (a) we were designing a multidisciplinary module and needed to ensure sufficiency for the core discipline knowledge base; (b) the design was for blended mode (delivery comprises of face-to-face study schools with an online mode used for other engagements and feedback on formative assessment and venue based final exams); (c) our intention to equip students with a coherent, contemporary and comprehensive understanding of sustainability marketing values, strategies and tools (GSBL 2020) where they can see business as a part of a living system, and to develop skills to deal with long term issues of triple bottom line (Bridges and Wilhelm 2008; Filho 2000) and (d) to develop their ability to use the knowledge base in contextual issues (an African context in this case) which is a known challenge in the area of sustainability marketing (Brower 2011; Amran et al. 2010).

To handle challenges of multidisciplinarity, we ensured that (a) integration does not remain limited to the course contents but is extended to delivery, module activities, all formative and summative assessments and to reflections; and (b) the balance and boundaries are maintained by ensuring that the core knowledge base of sustainability marketing was given due weighting (covered in four out of five study units), while project management and supply loop management gets integrated as tools to improve the efficiency and effectiveness of sustainability marketing and, thus, took a supporting role (covered in one out of the five units). We wanted students to understand the temporary, transitory and dynamic nature of projects, and its role and importance in implementing sustainability marketing strategies for projects, like collaborations to manage the 'cradle to grave' product life cycle, reverse logistics, closed loop supply chains, consumer engagement, collaborative consumption and so forth. This knowledge of sustainability marketing projects intends to orient students to handle networks of multiple organisations and stakeholders, especially in dynamic

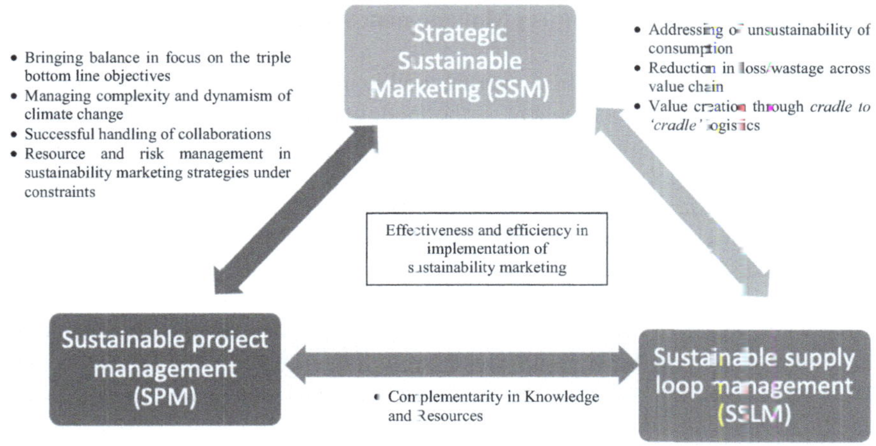

Fig. 1 Module building blocks in practice. *Source* the authors

and complex situations (Rwelamila and Purushottam 2016). In this regard, Fig. 1 summarises the interactions and enablement among the three disciplines.

The challenge of the blended mode was addressed by leveraging study schools, where all three lecturers actively engage face-to-face with students to support them understand the uniqueness of the module, integration of three disciplines, boundaries and their blurring as well.

To ensure that students see business as part of a living system and contextualise it appropriately, we build on our common understandings which are derived from our own sustainability journeys, to ensure that the module motivates students to reflect and question their own and organisational practices, to initiate a 'dialectical discourse'; to apply the gained knowledge in their own organisations (Springett 2005) and to make informed and responsible decisions (Rieckmann et al. 2017). This helps them in understanding contextual uniqueness and limitations in solving sustainability issues and makes them think innovatively to find solutions in constraints (as we faced and solved during our journeys).

We developed the module based on a philosophy that sustainability is our collective heritage, responsibility and future. We need to ensure that the module enables students to get ready to solve SD challenges in the African context. Therefore, integration with the African context is achieved by (a) ensuring all our outcomes have explicit assessment criteria covering the regional and local contexts; (b) ensuring that African sustainability marketing case studies are given in all formative and summative assessments; (c) by providing reading material in each study unit on the African issues, practices and scholarly work in the relevant areas and (d) developing module specific open source library guides with UNISA library staff to provide students with a dedicated resource to find information on sustainable development covering challenges and solutions in Africa.

5.2.2 Value Addition by Project Management Knowledge Base

As indicated above, the integration of sustainability in the module, through the project management curriculum component of the module, was considered a fundamental move. When designing the SPM component of the SSM module, it was necessary to aim at developing the sustainability mindset. Hence, one of the first objectives was to embrace a *sustainability focus* (Muff 2013). This was depicted by arguing for sustainability marketing initiatives within areas such as water, energy, food, soil and air contamination, waste, consumption, social aspects, population, climate change, and species extinction.

The whole idea of a structured approach in embracing sustainability in marketing education, by applying SPM tools, was strongly influenced by what Edwards et al. (2017) refer to as holistic sustainability-integrated management education (SiME) principles—moving away from the 'business as usual' paradigm. SPM is one of the three components (the other two components are SSM and SSLM, as indicated in Fig. 1), hence, when articulating its fundamentals, it is important to make sure that commonalities between the three components are maintained, in order to allow students to appreciate the relevance of each component. The following have been integrated in the SPM knowledge offering, while maintaining harmony between the other components:

- SSM provides frameworks, main module content and contexts;
- SPM provides appropriate tools for the management of sustainable marketing initiatives parameters—quality, schedule, utility, cost, stakeholders, health and safety, and an environment for product-based marketing projects and for the most service-based marketing projects;
- The need to make sure that students understand the importance of combined efforts of SSLM and SPM;
- Success of SSM initiatives are strengthened by a combination of applying SSLM and SPM tools, and also by how the marketing firm, or unit/directorate within an organisation conforms to PBO characteristics, described elsewhere in Purushottam and Rwelamila (2015), and Rwelamila and Purushottam (2016);
- Summative and interim assignments in SSM for students are primarily strengthened by a seamless combination of SSM, SSLM and SPM; and
- What Edwards et al. (2017) refer to as SiME is achieved through the framework presented in Fig. 1.

5.2.3 Value Addition by Supply Loop Management Knowledge Base

Unsustainable consumption in a continent like Africa is unacceptable because the majority of the population is not able to fulfill their basic needs. To address issues of growing criticality of resources and issues of unsustainable consumption, supply loop knowledge was integrated in the module. This knowledge builds the understanding of students about the way supply loops contribute to making consumers creators of value during the post consumption process. Based on this motivation, it was clear that integrating closed supply loops in the module are very important, as it will provide opportunities for heightened understanding of sourcing and distribution (movement and storage) strategies by utilising sustainable supply loops. Therefore, when designing the sustainable supply loop portion of the module, we adopted a practice approach by emphasising on identifying supply chain management practices that facilitate movement towards sustainability in Africa. Figure 1 illustrates, clearly, that while the three components of SSM, SPM and SSLM are distinct and unique, their complementarity is extremely important in pursuing sustainability oriented business goals.

In integrating SSLM into the development of the module, the emphasis was on creating competences in graduating students that enables them to shift their organisations towards SDGs. In doing so, they are expected to utilise project management methodology in the execution and/or delivery of their strategies, including those for managing sourcing, operations and distribution options. Currently, this is becoming critical as customers/stakeholders are rejecting products and shunning organisations that do not conform to expected sustainable practices, especially with regards to activities at sourcing, value creation and distribution stages. The graduating students are encouraged to consider the following in their practice:

- The design of the supply loop (*being very explicit in the nature and character of suppliers, distribution options in Africa*)
- Identification of supply (chain) loop drivers in Africa—*both logistical and cross-functional to be deliberate in terms of focus in maximising opportunities for sustainability*
- Adopting a risk-based view in assessing vulnerability in Africa to put contingency and/or mitigation measures in place
- Sustainable supply (chain) loop management practices—*using suitable metrics, practice and performance (*Das 2017*) can be measured.*

5.3 Post Development Phase Reflections

EfS, in our case, is integrated at module level (vertical level), not at the programme or business school level (horizontal level) and the sustainability marketing module was offered for the first time in 2018 as an elective. From a sustainability marketing discipline perspective, it was noted that, sustainability integration not being at programme level made students think that these issues are only for certain types of organisations which are away from the mainstream. It is observed that the majority of students come from organisations which emphasise orthodox marketing frameworks. This creates a challenge in making them open up to the sustainability marketing philosophies—they are dominated by '*the devil you know syndrome*'. Therefore, it is important to tease out the understanding of differences between orthodox and sustainability marketing during face-to-face lectures, online discussions and feedbacks. We also found that it is important to highlight that sustainability is not just for some people, but for everyone in Africa.

Multidisciplinary approaches make EfS effective (Wilhelm 2008) but the incorporation of such an approach is difficult (Filho et al. 2015). Multidisciplinarity makes the module heavy on content. Our students often find it difficult to understand the scope, boundaries and interlinkages among sustainability marketing, project management and supply loops. Therefore, we try to engage students, in study schools, to specify what aspects of SPM and SLLM are considered and how they make SSM effective and efficient; we also address this through extensive and intensive feedback after each formative assessment. We also ensure that all three module team members are present during face-to-face lectures to address any doubts in this regard. We bring the attention of our students to interlinkages of the SDGs and why multidisciplinarity will be effective in a community oriented Africa.

When the team started working, we ensured that SSM, SPM, SSLM knowledge is fused with a sense and objective of helping students regain what is lost in 'orthodox marketing'. Since most of the strong voices on SD, backed by financial muscle, seem to come from industrialised countries (in the north), our conversations with students about SD deliberately draw upon African indigenous knowledge and culture, based on Mugo's (1999) guiding questions: *Where is the person? Where is the human being?*—the emphasis being on owning the SD principles. In our conversations with students, about various SD practical challenges and opportunities, we make sure that we emphasise that SD fundamentals are not foreign but African, too.

We believe that SD fundamentals and principles should be aimed at developing the whole student—avoiding what Hoppers et al. (1999) refer to as production of one-legged, one-armed or one-eyed graduates, who are unable to relate to the other parts of themselves and to others, as well as to life itself. We try to ensure that our roles include shaping the direction towards a culturally relevant knowledge base. Therefore, we emphasise that students should see themselves as people who are capable of taking initiatives, especially to create a new SD perspective in the marketing knowledge base that will challenge the dominant marketing perspective (the orthodox) that has engulfed Africa and most other developing countries.

We believe that the value system adopted by graduating students influences their worldview, hence, their practice and behaviour towards SD as well. Our anticipation is that, apart from the business school environment, in a small way, we have contributed to the identification and capacitation of *sustainability champions* in industry and practice, potentially spanning all industry sectors within each of the countries that our students come from.

In the future, we would also like to incorporate African epistemologies and philosophies in the domain of sustainability marketing. Another area where we are exploring possibilities is how to use ODeL (blended learning mode) environments to facilitate the involvement by students in sustainability projects at research and at community engagement levels.

5.4 Summary and Implications

The narratives and reflections covered in the pre-module development, module development and post module development phases provide insights about the facilitators of EfS. From introspection at the pre-development phase, it emerged that (a) multicultural and social backgrounds of sustainability champions play an important role in building sustainability values and motivations; (b) sharing philosophies (Indian and African) ensured team cohesiveness; (c) involvement with sustainability beyond tuition activities in research, community engagement and institutional initiatives provides context oriented relevance; and (d) multidisciplinary exposure in your career and discomfort with status-quo management education nudges innovative collaborations. The discussions in context (Sect. 3) along with this phase highlight that the factors contributing to designing a multidisciplinary module came from the lecturers' personal motivations, their research domains, their previous work experiences in the area, the university's move towards the SDGs and changes in the higher education sector of the country.

The integrative introspections at the module development phase indicate that having clear expectations, clear discipline boundaries and a module framework (*like in our case, we were clear that this is SSM module which is calling upon the knowledge base of SPM and SLLM to enrich its efficiency and effectiveness*) helped in (a) seamless integration of different knowledge bases; (b) ability to build on each discipline's strengths; (c) opening the potential of multidisciplinarity in addressing complex SD issues; and (d) making the module context relevant.

The cohesive introspections at the post development phase highlight the realities—like sustainability and sustainability marketing which are far from being mainstream in organisational practice. Therefore, multidisciplinary EfS efforts need to be engaging, open and repetitive. It also emerged that there is a need to (a) integrate sustainability at programme level in HEIs to make EfS effective in achieving the targets of SDG 4; (b) engage students and bigger stakeholder communities in building an understanding that progress towards SDGs needs inclusivity; (c) use ODeL more effectively to make EfS accessible. Any future attempts towards multidisciplinarity in EfS needs to consider (a) issues of content heaviness in multidisciplinary modules; (b) clear framework of integration; (c) the consistent need to explain the roles of each discipline, their boundaries and blurriness; (d) ensure context relevance (which in this case is the African context) for appropriate and quality EfS.

6 Conclusion

EfS is an important step towards making sustainability mainstream and the SDGs attainable. This understanding is motivating African HEIs, including business schools, to integrate EfS in their portfolios. However, sustainability issues cannot be solved in isolation and, therefore, multidisciplinarity in EfS is essential. We shared our journeys to motivate a case for multidisciplinary sustainability marketing education and in the process, we also build a case for promoting conducive environments for such efforts, in HEIs in Africa, by highlighting the role of facilitators, based on our own journeys. Through introspection, we found that our personal and professional backgrounds motivated us to integrate sustainability in our modules through multidisciplinarity. We realised the importance of working in HEIs which strive towards the SDGs; opportunity to deliver the module in a blended environment of ODeL; having a clear framework to facilitate integration and boundaries; and team cohesiveness. We also feel that strong integration with the African context, through inclusion of African issues, voices, practices and scholarly works, also helped us in improving relevancy of our course. Therefore, we strongly recommend that to promote EfS in HEIs, it is important to build and promote a conducive environment for multidisciplinary EfS modules and programmes. We would also like to mention that, African HEIs will be able to expedite the progress towards SDGs in Africa by identifying sustainability champions and by creating conducive environments for multidisciplinary EfS and by ensuring that the EfS is grounded in an African context. We also realised that efforts to integrate EfS should not be limited to certain disciplines or at module level, but should be promoted at programme, college and university levels, where inter-college, even inter-university, collaborations can be promoted. We would like to conclude by saying that it is time for the HEIs to guide the journey towards accomplishing SDGs and to facilitate the realisation of a sustainable future for Africans.

References

Amran A, Khalid SA, Razak DA, Haron H (2010) Development of MBA with specialisation in sustainable development: the experience of Universiti Sains Malaysia. Int J Sustain High Educ 11(3):260–273

Arnold MW (2017) Higher education in management: the case of South Africa. In: Dameron S, Durand T (eds) The future of management education. Palgrave Macmillan, London, pp 255–276

Badat S (2009) Theorising institutional change: post-1994 South African higher education. Stud High Educ 34(4):455–467

Banbury C, Stinerock R, Subrahmanyan S (2012) Sustainable consumption: introspecting across multiple lived cultures. J Bus Res 65(4):497–503

Belz F, Peattie K (2012) Sustainability marketing, 2nd edn. Wiley, West Sussex

Blanco-Portela N, Benayas J, Pertierra LR, Lozano R (2017) Towards the integration of sustainability in higher education institutions: a review of drivers of and barriers to organisational change and their comparison against those found of companies. J Clean Prod 166:563–578

Blewitt J (2010) Higher education for a sustainable world. Educ Train 52(6/7):477–488

Bloom D, Canning D, Chan K (2006) Higher Education and Economic Development in Africa. https://www.edu-links.org/sites/default/files/media/file/BloomAndCanning.pdf. Accessed 13 Oct 2019

Borin N, Metcalf L (2010) Integrating sustainability into the marketing curriculum: learning activities that facilitate sustainable marketing practices. J Market Educ 32(2):140–154

Borland H, Lindgreen A (2013) Sustainability, epistemology, ecocentric business, and marketing strategy: Ideology, reality, and vision. J Bus Ethics 117(1):173–187

Brower HH (2011) Sustainable development through service learning: a pedagogical framework and case example in a third world context. Acad Manag Learn Educ 10(1):58–76
Bridges CM, Wilhelm WB (2008) Going beyond green: the "why and how" of integrating sustainability into the marketing curriculum. J Mark Educ 30(1):33–46
Clarke P, Gray D, Mearman A (2006) The marketing curriculum and educational aims: towards a professional education? Mark Intell Plan 24(3):189–201
Coopey J (2003) Sustainable development and environmental management: the performance of UK business schools. Manag Learn 34(1):5–26
Cornuel E, Hommel U, Painter-Morland M (2015) Philosophical assumptions undermining responsible management education. J Manag Dev 34(1):61–75
Décamps A, Barbat G, Carteron JC, Hands V, Parkes C (2017) Sulitest: a collaborative initiative to support and assess sustainability literacy in higher education. Int J Manag Educ 15(2):138–152
Dentoni D, Bitzer VC (2015) Managing wicked problems in global agriculture: the role of universities in multi-stakeholder engagements and value creation. J Clean Prod 106:68–78
De Witt A, de Boer J, Hedlund N, Osseweijer P (2016) A new tool to map the major worldviews in the Netherlands and USA, and explore how they relate to climate change. Environ Sci Policy 63:101–112
Das D (2017) Development and validation of a scale for measuring sustainable supply chain management practices and performance. J Clean Prod 164:1344–1362
Dlungwana WS, Nxumalo XH, Van Huyssteen S, Rwelamila PD, Noyana C (2002) Development and implementation of the South African construction excellence model (SACEM). In: Ahmad I, Ahmed SM, Azhar S (eds) Proceedings of construction in the 21st century: challenges and opportunities in management and technology, Florida International University, Miami, USA
Edwards M, Benn S, Starik M (2017) Business cases for sustainability-integrated management education. In: Arevalo JA (ed) Handbook of sustainability in management education. Edward Elgar Publishing, pp 45–66
Filho WL, Shiel C, Paço AD (2015) Integrative approaches to environmental sustainability at universities: an overview of challenges and priorities. J Integr Environ Sci 12(1):1–14
Filho WL (2000) Dealing with misconceptions on the concept of sustainability. Int J Sustain High Educ 1(1):9–19
GSBL (2020) MBL Programme modules. Strategic Sustainable Marketing. Module overview. https://www.unisa.ac.za/sites/sbl/default/Programmes/MBL-Master-of-Business-Leadership/MBL-programme-modules. Accessed 20 Mar 2020
GSBL (2016) History. https://www.unisa.ac.za/sites/sbl/default/About/History. Accessed 10 July 2020
Holbrook MB (2018) A subjective personal introspective essay on the evolution of business schools, the fate of marketing education, and aspirations toward a great society. Australas Mark J 26(2):70–78
Hoppers CAO, Moja T, Mda T (1999) Making this our last passive moment: the way forward. In: Makgoba MW (ed) Mafube Publishing Limited, Sandton, South Africa, pp 233–239
Kemper JA, Ballantine PW, Hall CM (2020) The role that marketing academics play in advancing sustainability education and research. J Clean Prod 248:119229
Krishnamurthy S (2020) The future of business education: a commentary in the shadow of the Covid-19 pandemic. J Bus Res 117:1–5
Mawonde A, Togo M (2019) Implementation of SDGs at the University of South Africa. Int J Sustain High Educ 20(5):932–950
Muff K (2013) Developing globally responsible leaders in business schools: a vision and transformational practice for the journey ahead. J Manag Dev 32(5):487–507
Mugo MG (1999) African culture in education for sustainable development. In: Makgoba MW (ed) African renaissance. Mafube Publishing Limited, Sandton, pp 210–232
Napier NK, Harvey M, Usui K (2008) Management education in emerging economies: the impossible dream? J Manag Educ 32(6):792–819

Purushottam N, Rwelamila PD (2015) Business schools: a platform for sustainable empowerment of informal sector through a project management framework. In: Rwelamila PD, Spring A (eds) Towards sustainable African business development: integrating formal and informal business, proceedings 16th annual international conference, International Academy of African Business and Development (IAABD), Kenya, Nairobi, pp 241–250
Purushottam N, Rwelamila PD (2011) Strategies in a Multi-Polar World: Opportunity for Application of Project Management Approaches Bus Stud J 3(1):109–117
Rieckmann M, Mindt L, Gardiner S (2017) Education for Sustainable Development Goals. Learning Objectives. UNESCO Publishing
Rowe A, Wehrmeyer W (2010) Education for sustainability developing MBA students' critical reflective and action learning in their work context. Rev Bus Res 10(2):145–149
Rwelamila PD, Purushottam N (2016) Strategic project management as an innovative approach for sustainable green campus buildings in Africa: the need for a paradigm shift. Smart Sustain Built Environ 5(3):261–271
Rwelamila PD, Talukhaba AA, Ngowi AB (2000) Project procurement systems in the attainment of sustainable construction. Sustain Dev 8(1):39–50
Rundle-Thiele SR, Wymer W (2010) A census of stand-alone ethics, social responsibility and sustainability course requirements for marketing graduates in Australia and New Zealand. J Mark Educ 32(1):5–12
Shrivastava P (2010) Pedagogy of passion for sustainability. Acad Manag Learn Educ 9(3):443–455
Slater DJ, Dixon-Fowler HR (2010) The future of the planet in the hands of MBAs: an examination of CEO MBA education and corporate environmental performance. Acad Manag Learn Educ 9(3):429–441
Springett D (2005) 'Education for sustainability' in the business studies curriculum: a call for a critical agenda. Bus Strategy Environ 14(3):146–159
Starik M (2004) Toward a multi-level, multi-systems approach to sustainability governance and outcomes: an essay. Innovation 6(2):167–169
Stephens JC, Hernandez ME, Román M, Graham AC, Scholz RW (2008) Higher education as a change agent for sustainability in different cultures and contexts. Int J Sustain High Educ 9(3):317–338
Stough T, Ceulemans K, Lambrechts W, Cappuyns V (2018) Assessing sustainability in higher education curricula: a critical reflection on validity issues. J Clean Prod 172:4456–4466
UNDP (2010) Human development report 2010: The Real Wealth of Nations: Pathways to Human Development. https://hdr.undp.org/en/reports/global/hdr2010/. Accessed 20 Mar 2019
Unisa (2019) United Nations global compact: University of South Africa communication on engagement 2018/19: towards sustainability and social transformation. https://www.unisa.ac.za/static/corporate_web/Content/UNGC/Documents/2018_2019%20UNISA%20-%20UNGC%20COE%20Report%20-%20Final%20-%20Approved%20(006)%20(002).pdf. Accessed 20 June 2019
Unisa (2017) MBL teach out plan. https://www.unisa.ac.za/sites/sbl/default/News-&-events/Announcements/MBL-teach-out-plan. Accessed 20 Oct 2020
Varey RJ (2010) Marketing means and ends for a sustainable society: a welfare agenda for transformative change. J Macromarketing 30(2):112–126
Viswanathan M (2012) curricular innovations on sustainability and subsistence marketplaces: philosophical, substantive and methodological orientations. J Manag Educ 36(3):389–427
Wilhelm W (2008) Marketing education for sustainability. J Adv Mark Educ 13:8–20
World Economic Forum (2013) Sustainable Consumption: Stakeholder Perspectives. https://www3.weforum.org/docs/WEF_ENV_SustainableConsumption_Book_2013.pdf. Accessed 20 Oct 2020
Xue H, Desmet PM (2019) Researcher introspection for experience-driven design research. Des Stud 63:37–64
Yen-Chun Jim W, Shihping H, Lopin K, Wen-Hsiung W (2010) Management education for sustainability a web-based content analysis. Acad Manag Learn Educ 9(3):520–531

Neha Purushottam is a full professor of Sustainability Marketing at the Graduate School of Business Leadership (GSBL), University of South Africa (UNISA). She has a bachelor's in Biology, a master's degree in Economics, another Master's in Business Administration (MBA) and a Ph.D. in Management Science. Her twenty plus years of work experience is comprised of teaching, research and corporate consulting in India and Africa. Her research focus areas are Sustainability Marketing, Sustainable Consumer Behaviour, Sustainability in consumer policy and Sustainability in Management Education. Through these research areas, her work mainly focuses on SDG 12 (Sustainable Consumption and Production) and SDG 4 (Quality Education).

P. M. D. Rwelamila is a full Professor of Project Management at the Graduate School of Business Leadership (GSBL), University of South Africa (UNISA). He graduated in Construction Economics and Project Management. He has a master's degree in Construction Management from Brunel University and a Ph.D. in Project Management and Procurement systems from the University of Cape Town. He has worked in a number of countries, including Tanzania, Kenya, Uganda, Botswana, Zambia, Australia, United Kingdom and Sweden. He has experience of more than 30 years in consulting in project management, construction economics, policy development. He has more than 190 published and peer-reviewed journal papers, chapters in books and conference proceedings.

Ozias Ncube received his master's degree in Mathematics from the University of Zimbabwe in 1998. He is currently pursuing his PhD in Industrial Systems specialising in supply chain management from the University of Pretoria, South Africa. He is currently based at the Graduate School of Business Leadership (GSBL) and is responsible for supply chain management and related modules. His research interests are on sustainable supply chains, supply chain vulnerability and resilience analysis, the application of decision support systems for supply chain competitiveness and supply chain ethical leadership. He has more than 50 MBL\ MBA research supervision to completion in the area of supply chain management.

Resources, Environment and Agriculture

Gender, Agricultural Productivity and Sustainability in Cameroon

Grace Mehe Tchinda, Cyrille Bergaly Kamdem, and Francis Andrianarison

Abstract This study analyzes the impact of gender on the agricultural productivity and sustainability in Cameroon using data from the third survey of households in Cameroon called ECAM3. This study uses two models. To show the effect of gender on the partial productivity of producers, the model of the productivity function is used. To show the effect of gender on the agricultural sustainability, the decomposition model is used to explain the differences that exist between men and women in their involvement to sustainable agriculture. The results show that being man has a significant and positive impact at 1% on overall agricultural productivity. Moreover, the women are less involved than men in the agricultural sustainability, despite their poor access to technology, to use of labor hand, in credit, to the land, which makes their productivity weak.

Keywords Gender · Sustainability · Agricultural productivity · Discrimination

1 Introduction

The link between gender and agriculture is old in the literature. This reality is still recognized today as it exists the relationship between farmers attitudes towards environment and their farming practices (Fairweather and Campbell 2003; Rezaei-Moghaddam et al. 2005; Karami and Mansoorabadi 2008). Over the last decades, problems of agricultural sustainability are become a permanent concern, because the

G. Mehe Tchinda (✉)
Faculty of Economics and Applied Management, University of Douala-Cameroon, PO. Box 7818, Douala – Bassa, Cameroun
e-mail: grace.tchinda@yahoo.com

C. B. Kamdem
Faculty of Economics and Management, University of Yaoundé II-Soa, Cameroon, PO. Box 1365, Yaounde, Cameroun
e-mail: bergaly@yahoo.fr

F. Andrianarison
United Nations Development Programme (UNDP), P. O. Box 836, Yaounde, Cameroon
e-mail: francis.andrianarison@undp.org

W. Leal Filho et al. (eds.), *Sustainable Development in Africa*, World Sustainability Series,
https://doi.org/10.1007/978-3-030-74693-3_18

destruction of environment can not only slow economic growth, but also the health and well-being of local people. Moreover, it is evident that the farmers' systems are the humans systems in which women in developing countries are important contributors to small-scale agriculture, rural labor and food security. Agriculture in sub-Saharan Africa (SSA) is essential for human well-being and economic growth: it contributes around 25% to gross domestic product (GDP) and employs a huge chunk of the population (FAO 2006). Indeed, Cameroon is one of the countries of Sub-Saharan Africa which concentrates 58.4% of men against 68.6% of women in agricultural production (ECAM 2007). However, despite the high proportion of women in agriculture, productivity is even lower for women farmers compared to their male counterparts. And Despite these important roles, women have more difficulty than men in accessing resources such as land, credit, agricultural inputs and services that increase the productivity of their plots and also their potential to improve their welfare (FAO 2011; World Bank 2008; Deere and Leal 2001). In addition, the literature has generally found that women are less productive than men in agriculture despite their high rate of involvement in agricultural production. This low productivity is therefore at the origin of the considerable decline in agricultural productivity. The figures show that if these women had the same access to productive resources than men, they could increase the yields of their farms by 20 to 30%, raising the total agricultural production of developing countries. FAO (2011) estimates that closing the gap in agricultural productivity between men and women around the world could increase agricultural production in low-income countries by 2.5% to 4%, and reduce undernourishment of women from 12 to 17% in the world (FAO 2011). However, although the women are enormous contributors, their importance in the development of agriculture is often ignored. Yet, they are closer to the nature in raison of their role in satisfaction the daily needs of households (Karami and Mansoorabadi 2008). To face these problems, groups have started to preconize a new agriculture. Thus, in this context of difference in agricultural productivity between men and women, this study aims to answer the following fundamental question:

What is the impact of gender on agricultural productivity and on sustainability in Cameroon?

The main objective of this study is to evaluate the effect of gender on farmers' productivity on the one hand and on the other hand on agricultural sustainability in Cameroon.

2 Methodology

To achieve the goal set in this study, two models were used. The first model is the expression of the productivity function as a function of variables related to gender as well as those related to environmental sustainability and economic sustainability, whose constraint on the explained variable allows the use of the Tobit model. The second model is the Oaxaca and Blinder decomposition model (1973) to explain the differences that exist between men and women in their involvement in sustainable agriculture.

The data used come from the third ECAM survey of households in Cameroon. They were collected as part of a study on the living conditions of populations and the poverty profile in Cameroon, carried out by the national statistics institute through a questionnaire drawn up by researchers from the said institute. The survey in question took place during the month of August 2007. Regarding the size of the sample, ECAM shows that a sample of 8,000 to 12,000 households at the national level is sufficient to have key indicators on poverty and household standard of living. In total, a sample of 12,609 households, typed according to some of their characteristics was taken on the national territory, with a satisfactory degree of precision at the stratum level. ECAM3 being, like the previous ones, a survey with multiple objectives, its questionnaire is structured around a number of modules corresponding to the areas of investigation according to the objectives selected. The different sections covered touch, among others, the following areas: health of household members, education, employment and income, domestic work and child labor, household status and equipment, patrimony, household environment, residential mobility and migration, accessibility to basic infrastructure, agriculture and rural activities, governance and corruption, domestic tourism, household spending and final consumption.

The data collected was analysed using the stata 12 software which allowed us to make estimates and obtain its results.

This section will be subdivided into two sub-sections: the first presents the model of the productivity function and the second presents the decomposition model.

2.1 The Productivity Function Model

In the literature, to analyze the production function, some authors (Fuglie 2015; Oladeebo and Fajuyigbe 2007; Udry 1996; Tiruneh et al. 2001) use the log function to estimate the production function in the form:

$$\log Yi = \beta X + \delta G + \varepsilon, \tag{1}$$

where Yi is productivity, X is the set of characteristics associated with the farmer and the farmer, G is the variable associated with the head of the farm's sex, and ε is the error term.

The econometric expression of the production function is as follows:

$$\text{"}ln\,Yi = \alpha 0 + \alpha 1\,ln\,landi + \alpha 2\,ln\,technoi + \beta\,ln\,labori + \gamma\,ln\,agro - ecologyi + \delta\,ln\,Genderi + \chi\,ln\,sustainablei + \varepsilon i\text{"} \tag{2}$$

where Yi is productivity, $land_i$ refers to the status of ownership; $techno_i$ is the set of technology associated with production such as seeds and equipment; $labor_i$ represents the labor force used in production measured by the labor cost per unit of land; $agro\text{-}ecology_i$ the indicator variable for agro-ecological zones; $Gender_i$ is a characteristic variable for the sex percentage of the head of household, $sustainable_i$ the variable associated to the agriculture sustainability such as economic sustainability and environmental sustainability and ε is the error term. Since the equation is linear, we can apply the OLS for the estimation.

Given the values of the productivity variable, the Tobit model seems appropriate for estimating the productivity function with left censoring. Tobit models were initiated by Tobin (1958) and generally refer to regression models in which the defining domain of the dependent variable is constrained in one form or another. Given these factors, the following Tobit model may be the most appropriate estimation procedure, given the left-censoring of the dependent variable at zero:

$$\ln Y_i^* = Gender_i + X_i\beta + \eta_i, \tag{3}$$

$$\ln Y_i = \begin{cases} \ln Y_i^* si \ln Y_i^* > 0, i = 1 \ldots N \text{ et } \eta_i \sim N(0, \sigma^2), \text{ iid} & (4a) \\ 0 \text{ si} \ln Y_i^* \leq 0 & (4b) \end{cases}$$

where Yi is an indicator of production (productivity), gender is the gender indicator, X is the set of control variables, N is the number of observations, and ηi is the error term assumed to be independent and identical to zero and fixed variance. Thus Yi follows a normal centered and reduced law. The productivity function which is the explained variable admits a conditional distribution with respect to the explanatory variables X whose conditional density or conditional probability is noted:

$$f(Y, \sigma) = \frac{1}{\sqrt{2\sigma}} exp\left\{-\frac{1}{2\sigma}(y - \sigma)^2\right\} \tag{5}$$

The set of Eqs. 4 and 4a indicates the expected value of productivity (given the observable characteristics) which is a function of the probability that Yi is non-zero (uncensored) and positive productivity. The X vector of control variables is divided into five groups: (1) land, (2) technology, (3) labor, (4) the agro-ecological zones indicator and (five) sustainable. The description of the explanatory variables is presented in Table 1.

2.2 Decomposition Model

Decomposition methods have been widely used in the literature on the gender wage gap and unions. In addition, they have been used to understand what factors account for changes in inequality and explain growth (Women 2015) This method is also useful for explaining differences in inequalities in agricultural productivity (Palacios-López and Lopez 2014; Aguilar et al. 2014).

This study draws on the method of Oaxaca and Blinder in analyzing the most relevant elements to explain any different between men and women in their involvement towards sustainable agricultural. We used five groups of variables. The first represents the set of socio-economic characteristics. These socio-economic characteristics include indicators of age and education level, areas fished, health of head of household and household, number of persons in the household, input costs per unit of land, marital status and access to credit. The second variable represents technology such as seeds and equipment. The third is capital representing labor and/or labor. The fourth variable is the indicator of agro ecological zones or even spatial variable, and the fifth is the indicator of sustainability that content the economic sustainability and environmental sustainability. The social sustainability has not been evaluated because it can't be perceptible in socioeconomic characteristics. Economic sustainability is equal to value of production minus the different cost. And environmental sustainability or ecology is equal to quantity in value the pesticide used.

$$Y = f(S, K, W, S) \tag{6}$$

The expression of the different productivity functions of men and women are as follows:

$$\log Y_h = \alpha_o^h + \alpha_1^h X_i^h + \varepsilon_i^h \tag{7}$$

$$\log Y_f = \alpha_o^f + \alpha_1^f X_i^f + \varepsilon_i^f \tag{8}$$

where α_o^h and α_o^f are points at the origin of the production functions respectively men and women; α_1^h and α_1^f are vector coefficients of men and women; X_i^h and X_i^f are vectors of individual characteristics respectively of men and women ε_i^h and ε_i^f the error terms respectively for men and women. With $\alpha_o^h \neq \alpha_o^f$ since there is a productivity gap between men and women.

The least squares regression is as follows:

$$E(Y_h/X^h) = \tilde{\alpha}_0^h + \tilde{\alpha}_1^h \ddot{X}^h \tag{9}$$

$$E(Y_f/X^f) = \tilde{\alpha}_0^f + \tilde{\alpha}_1^f \ddot{X}^f \tag{10}$$

Hence the following equation:

$$E(Y/X) = \tilde{\alpha}_0 + \delta_o D_i + \tilde{\alpha}_1 \ddot{X} + \delta_1 D_1 \ddot{X} \tag{11}$$

D_i is the variable dummy which takes the value 1 if the individual is a woman and 0 if it is a man. We can therefore establish the gap in productivities between men and women. Then we have the following decomposition:

$$E(Y_f) = E[E(Y_f/X^f)] = \tilde{\alpha}_1^f \ddot{X}^f \quad (12)$$

$$E(Y_h) = E[E(Y_h/X^h)] = \tilde{\alpha}_1^h \ddot{X}^h \quad (13)$$

Thus, the counterfactual is: $E(Y_c) = \tilde{\alpha}_1^h \ddot{X}^f$.

The specification that will be used to generate an alternative measure of the structural effect and the composition effect will be obtained by adding and subtracting: E(Yc)

Thus we obtain the following equation:

$$E(Y_f)\ E(Y_h) = [E(Y_f) - E(Y_c)] + [E(Y_c) - E(Y_h)]$$

$$E(Y_f)\ E(Y_h) = [\tilde{\alpha}_1^f \ddot{X}^f - \tilde{\alpha}_1^h \ddot{X}^f] + [\tilde{\alpha}_1^h \ddot{X}^f - \tilde{\alpha}_1^h \ddot{X}^h] \quad (14)$$

$$= \underbrace{[\tilde{\alpha}_1^f - \tilde{\alpha}_1^h]\ \ddot{X}^f}_{(1)} + \underbrace{\tilde{\alpha}_1^h\ [\ddot{X}^f - \ddot{X}^h]}_{(2)}$$

This equation gives two effects. (1) the estimated effect of differences in the returns of men and women or the part that evaluates the decomposition; (2) the estimated effect of differences in the distribution of men's and women's characteristics that explain their comportment towards sustainability, or the part not explained of decomposition effect.

This study proposes an Oaxaca decomposition method based on Tobit models to show which elements are most relevant to explain any gap between men and women in agricultural production.

In the context of Cameroon in particular and African in general, African customs just recognize the right of women to cultivate a piece of land without having the right to appropriate it. Although it is she who practices essentially the agricultural activity while ensuring the food security. Indeed, land and ethnicity are intimately linked. Depending on the ethnic group, the perception of the earth can take different forms. Access to land ownership is strongly linked to the strength of the customary regime. Land tenure insecurity is a significant threat to the sustainability of space and resource management. The traditional wisdom has a lot to offer in terms of life in harmony with nature or society, and this is one the fundamental principles of concept the sustainability. So integration of human knowledge deeply rooted when they are applied by qualified farmers, leads to natural resource management that altruistically and not just for economic purposes, which makes agriculture significant and sustainable. The stake of the land as invaluable wealth is considered a property of succession which is excluded to the woman except at the death of her husband. So we are going to attribute possession of the land only to widowed women.

2.3 Data

The data used in this study come from the third Cameroon household survey called Ecam3. They were collected as part of a study on the living conditions of populations and the poverty profile in Cameroon, carried out by a national statistical institute through a questionnaire. The ECAM3 survey was conducted during August 2007 on a sample of 12,609 households in the national territory, with a satisfactory degree of precision at the stratum level. Moreover, the expected effect of the explanatory variables used in this study is presented in Table 1:

Table 1 List of variables used in the analysis

Group of variables	Variables	Definition	Unit of measure	Expected effect
Result variable	Farmer productivity	Log of farmer productivity	In kg per hectare	+
Farmer characteristics	Farmer sex	Sex of the farmer or head of household	1 = Man 0 = woman	+
	Farmer age	Age of the farmer or head of household	In year	±
	Age square	Farmer Age square		
	Household size	Number of persons in the household		±
	Marital status	Marital status of farmer	1 = if farmer is married 0 = others	+
	Education	Level of education	1 = educated 0 = non educated	+
	Credit access	Credit access	1 = Credit access 0 = non credit accesss	+
	Total number of assets owned by members	Total number of assets owned by members of household	1 = house, car 0 = none	±
	Land certificate	Possession of land certificate	1 = Possession of land Certificate 0 = none	±

(continued)

Table 1 (continued)

Group of variables	Variables	Definition	Unit of measure	Expected effect
	Annuity product	Annuity product	1 = To be annuity product 0 = to be food product	+
	Health	Farmer assessment of his current state of health	1 = good 0 = bad	±
	Environmental sustainability	Involvement the exploiting in environmental sustainability	1 = Involved 0 = none	±
	Economic sustainability	Involvement the exploiting in economic sustainability	1 = Involved 0 = none	±
Farm characteristics	Farm size	Farm size in hectare	In hectares (ha	±
	Labor cost	Labor cost during the last 12 months per hectare	In francs cfa per ha	±
	Seedling cost	Seedling cost during the last 12 months per hectare	In francs cfa per ha	±
	Pesticides cost	Pesticide cost during the last 12 months per hectare	In francs cfa per ha	±
	Fertilizer cost	Fertilizer cost during the last 12 months per hectare	In francs cfa per ha	±
	total area exploited by household	total area (ha) exploited by household	In hectare	±
	Estimated value of production	Estimated production value during the last 12 months per hectare	In francs cfa	
	Innovation	Seedlings and equipment	1 = at least one innovation 0 = none	+

(continued)

Table 1 (continued)

Group of variables	Variables	Definition	Unit of measure	Expected effect
	Seedling	Using of improved Seedlings	1 = improved Seedlings 0 = Traditionnal improved Seedlings	+
	Equipment	Using of modern Equipments	1 = modern Equipments 0 = traditionnal équipements	+
	Unimodal	Agro-ecological zone of the mangrove grouping: the Littoral and the South-west et le Sud-ouest	1 = Belong to unimodal area 0 = none	
	Bimodal	Agro-ecological forest zone regrouping: Center, South and East	1 = Belong to bimodal area 0 = none	
	Highlands	Agro-ecological zone of the humid savannah regrouping: West and North-west	1 = Belong to Highlands 0 = none	
	High savannahs	Agro-ecological zone of the low savanna grouping: Adamawa	1 = Belong to High savannahs 0 = none	
	Sudano sahelian	Sahelian agro-ecological zone grouping: Extreme-North and North	1 = Belong to Sudano sahelian 0 = none	

3 Results and Discussion

3.1 Descriptive Statistics of Explanatory Variables

Descriptive statistics for explanatory variables are presented in Table 2.

Table 2 summarizes the average of all the explanatory variables of the model and the outcome variable. Our sample consists of 198,175 typical producers in rural areas of our various cities in Cameroon.

Table 2 Descriptive statistics of the variables

Variable	Obs	Mean	Std. Dev.	Min	Max
Farmer sex	198877	0.723	0.447	0	1
Farmer age	198813	45.876	15.163	15	98
Household size	198877	5.175	3.225	1	43
Education	19877	2.599	1.838	0	8
Farm size	15072	2.468	11.58	0	570
Total number of assets owned by members	198525	0.821	0.8	0	20
Labor cost	152479	10.589	111.032	0	7100
Seedling cost	152671	5.157	48.941	0	2000
Pesticides cost	152607	3.438	11.859	0	275
Fertilizer cost	152543	6.492	27.178	0	1050
Health	198845	1.806	0.999	1	9
Total area exploited by household	197951	1.903	3.89	0	95
Cultivated products	203165	16.498	9.234	1	32
Estimated value of production	43027	39.558	215.526	0	30125
Land certificate	173931	0.094	0.292	0	1
Innovation	203165	0.132	0.338	0	1
Unimodal	203165	0.255	0.436	0	1
Bimodal	203165	0.113	0.317	0	1
Highlands	203165	0.223	0.416	0	1
Sudano sahelian	203165	0.095	0.294	0	1
Credit access	194047	0.025	0.157	0	1
Marital status	198877	0.52	0.5	0	1
Farm productivity	34783	2.044	1.456	−3.912	8.006
Seedling	43107	0.055	0.229	0	1
Equipment	189247	0.13	0.337	0	1
Environmental sustainability	198175	6.685	25.395	0	610
Economic sustainability	42225	−10.595	384.822	−18198	30115

3.2 Estimation of the Agricultural Production Function in Cameroon

The results of the Tobit model (Table 3) allow to better separate overall productivity and specific productivities according to cash products and cash crops.

Table 3 Comparing Tobit's results on overall and specific productivities

VARIABLES	Overall productivity	Productivity specific to annuity products	Productivity specific to foods products
Farmer Sex	0.0785***	0.0999	0.0502***
	(0.0193)	(0.0781)	(0.0192)
Farmer Age	-0.00171***	0.000689	-0.00172***
	(0.000518)	(0.00184)	(0.000517)
Household size	0.0307***	0.00287	0.0306***
	(0.00235)	(0.00821)	(0.00236)
Education	0.000168	0.0692***	-0.00531
	(0.00469)	(0.0165)	(0.00469)
Farm size	0.00526***	-0.00808	0.00550***
	(0.00175)	(0.0150)	(0.00170)
Total number of assets owned by members	0.0108	0.0198	0.00314
	(0.00884)	(0.0296)	(0.00888)
Labor cost	0.000513***	0.000274	0.00123***
	(0.000198)	(0.000641)	(0.000201)
Seedling cost	0.000200	-0.000658	0.000899***
	(0.000285)	(0.000885)	(0.000290)
Pesticides cost	0.0150***	0.0242***	0.0136***
	(0.00144)	(0.00392)	(0.00148)
Fertilizer cost	0.00125***	0.00327	0.00113***
	(0.000371)	(0.00201)	(0.000366)
Health	-0.0609***	-0.0412	-0.0597***
	(0.00742)	(0.0255)	(0.00742)
Total area exploited by household	-0.176***	-0.116***	-0.184***
	(0.00371)	(0.0133)	(0.00380)
Cultivated products	-0.0544***	-0.129***	-0.0485***
	(0.000819)	(0.00772)	(0.000841)
Estimated value of production	0.00296***	0.00141***	0.00455***
	(9.97e-05)	(0.000318)	(0.000108)
Land Certificate	0.0451*	-0.0363	0.0369
	(0.0264)	(0.107)	(0.0261)
Innovation	-0.218**	-0.171	-0.0445
	(0.0948)	(0.178)	(0.109)
Unimodal	-0.175***	-0.186***	-0.181***
	(0.0180)	(0.0662)	(0.0180)
Bimodal	0.466***	0.285***	0.443***
	(0.0297)	(0.107)	(0.0299)
Highlands	0.353***	0.180**	0.324***
	(0.0212)	(0.0729)	(0.0214)
Sudano sahelian	0.237***	0.382***	0.177***
	(0.0293)	(0.124)	(0.0290)
Credit access	0.0515	0.281	0.0335
	(0.0535)	(0.187)	(0.0535)
Marital status	-0.0563***	-0.0673	-0.0379**
	(0.0161)	(0.0534)	(0.0162)
Seedling	0.715***	0.388**	0.538***
	(0.0883)	(0.168)	(0.101)
Equipment	0.0839	0.196	-0.0917
	(0.0924)	(0.161)	(0.107)
Environmental sustainability	-0.00435***	-0.00353**	-0.00437***
	(0.000651)	(0.00164)	(0.000681)
Economic sustainability	0.000231**	-0.000196	0.000602***
	(9.20e-05)	(0.000315)	(9.33e-05)
Constant	3.068***	3.519***	2.971***
	(0.0397)	(0.147)	(0.0398)
Observations	27,712	2,016	25,696

Standard errors in parentheses
*** p<0.01, ** p<0.05, * p<0

Table 3 presents the results of the Tobit model regressions on agricultural productivity in Cameroon. These results show a positive significance of the gender variable at 1% for overall productivity and specific productivity when considering cash crops. This results in the fact that being a man contributes positively to overall and specific productivity. So being a man significantly increases productivity.

For the land title variable, we also observe a positive significance at 1% on all global and specific productivities. This means that those with titles are very involved in agricultural production.

Regarding innovation, negative significance is observed at 1% for the overall productivity, specific to annuity products and for the specific productivity of foods products. This is reflected in the fact that innovation is not used in overall productions, specific to annuity products and specific productivities of foods products.

The cost of labor variable is significant and positive for all productivities. This implies that labor is important in all productions.

For Agro-ecological zones, unimodal zone has negative significance at 10% for overall productivity and specific productivity of foods products, and not significance for specific productivity of annuity product. This translates the fact that annuity products are grown in very low proportions that other products. Bimodal zone has positive significance at 1% for overall productivity and specific productivity of annuity product, and for specific productivity of foods product. Which indicate foods product and annuity product are more grown in this zone that other products. For Highlands zone, we observe positive significance for overall productivity, for specific productivity of foods products and for specific productivity of annuity product. Reflecting the fact that foods products and annuity products are more grown in this zone that other products. Sudano-Sahelian zone has positive significance in the different productivities. Which that indicate all products are more grown in that zone.

Concerning environmental sustainability, negative significance is observed at 1% for all productivities. This means that in the production of these products, environmental sustainability is valued in lesser proportions. For economic sustainable, we observe positive significance at 1% for overall productivity, specific productivity of foods products and for specific productivity of annuity product. Which indicate the economic sustainable is more intense in the overall productivity, specific foods products and annuity products.

These results enable to assess the effect of land, technology, labor, the spatial effect and the agricultural sustainability in the analysis of agricultural productivity. It is now necessary to explain the different involvement men and women in agricultural sustainability to identify the causal effect of the impact of gender on agricultural productivity in Cameroon.

3.3 Explanation of Involvement Differences According to Gender in Sustainability by Oaxaca-Blinder Decomposition

The estimate of Oaxaca and Blinder decomposition in 1973 consists in estimating separately for the groups of men and women. This estimation is a productivity equation that relates yield to the different observable productive characteristics of producers such as agricultural sustainability as well technological characteristics used in agro-ecological zones.

Table 4 shows the results of the regression including all Oaxaca variables.

The results in Table 4 reveal a number of findings.

Age has a negative significance at 1% both in the decomposition of men and women. In addition, age has a negative impact as well in the yield of women. This implies the more women aged the more they are not enough strength to produce. Educational attainment has an impact positive and non-significant in the yield of women and negative and significant at 1% in the yield of men. The cost of labor has a positive and significant influence at 1% for to women and for men. This implies that the men as the women use the workforce to increase their yield. Marital status has a negative significance to the women. This means that most of the women farmers are widowed or divorced.

Innovation or technology has a meaningful influence and negative at 5% to the men. This implies innovation is less used in production by men.

With regard to the unimodal zone, we observe a negative significance at 1% by women and men. What implies that women and men practice less agriculture in this zone. The bimodal zone has a significant and positive influence at 1% on men and women. Which implies that agriculture is less practiced this zone. In the highland zone, we observe positive significance at 1% by women and men. Which means that women and men are more implicated in agriculture. The Sudano-Sahelian has a significant and positive influence at 1% only for women.

Concerning environmental sustainability, we observe a meaningful influence and negative at 1% by women and a positive significance at 1% by men. This means that the women are less implicated in the environmental sustainability than men. For economic sustainability, we observe a positive significance at 1% only for men. This indicates that men are more implicated in economic sustainability in their agricultural practice.

The estimated results of the decomposition of Oaxaca and Blinder show that the real causes of productivity differences between men and women are largely explained by sustainability agricultural, innovation and spatial distribution.

Table 4 Detail decomposition

VARIABLES	Women	Men
Farmer Age	-0.00191***	-0.00110
	(0.000603)	(0.00106)
Household size	0.0266***	0.0425***
	(0.00266)	(0.00539)
Education	0.00219	-0.0434***
	(0.00511)	(0.0123)
Farm size	0.00382*	0.00346
	(0.00201)	(0.00380)
Total number of assets owned by members	0.0120	-0.0169
	(0.0109)	(0.0148)
Labor cost	0.00157***	0.0158***
	(0.000283)	(0.00137)
Seedling cost	-0.000157	0.0175***
	(0.000300)	(0.00154)
Pesticides cost	0.0169***	0.00116
	(0.00154)	(0.00440)
Fertilizer cost	0.00280***	0.0113***
	(0.000534)	(0.00139)
Health	-0.0640***	-0.0477***
	(0.00880)	(0.0134)
Total area exploited by household	-0.177***	-0.147***
	(0.00426)	(0.00784)
Cultivated products	-0.0564***	-0.0473***
	(0.000963)	(0.00151)
Estimated value of production	0.00302***	-0.00311***
	(0.000105)	(0.000692)
Land Certificate	0.0321	0.0903*
	(0.0319)	(0.0462)
Innovation	-0.154	-0.581**
	(0.103)	(0.249)
Unimodal	-0.141***	-0.189***
	(0.0220)	(0.0310)
Bimodal	0.483***	0.416***
	(0.0329)	(0.0745)
Highlands	0.404***	0.239***
	(0.0248)	(0.0415)
Sudano sahelian	0.289***	0.0366
	(0.0331)	(0.0663)
Credit access	0.106	-0.00655
	(0.0731)	(0.0763)
Marital status	-0.0548***	-0.0600
	(0.0175)	(0.0423)
Seedling	0.644***	0.870***
	(0.0957)	(0.234)
Equipment	0.0145	0.485**
	(0.0999)	(0.245)
Environmental sustainability	-0.00542***	0.0166***
	(0.000699)	(0.00232)
Economic sustainability	-3.52e-05	0.00813***
	(9.74e-05)	(0.000682)
Constant	3.168***	2.860***
	(0.0462)	(0.0815)
Observations	20,431	7,281

Standard errors in parentheses
*** p<0.01, ** p<0.05, * p<0.1

Implication of gender in agricultural sustainability is confirmed by fact that the proportion of agricultural productivity agricultural enhancing agricultural sustainability is attributed to men.

The use of innovations by men considerably increases their productivity. However, women do not even have much access of same that credit access, which makes their productivity low. This exclusion of women is partly explained by discrimination.

In most agro-ecological zones, agriculture is practiced mainly by men, with the exception of the highland zone, whose the women are more implicated, and whose the agricultural sustainability is more intense.

4 Conclusion

The objective of this study is to analyze the effect of gender on agricultural productivity and sustainability in Cameroon. For this two models were used.

The first model focused on the expression of the productivity function has shown that gender, land title, innovation, labor, agro-ecological zones and agricultural sustainability have a significant but negative and positive impact on all productivities.

The second model centered on Oaxaca and Blinder decomposition allowed showing the differences between men and women in their implication in agricultural sustainability, as well that in their productivity. The agricultura sustainability is majoritarily praticated by men despite their low access to the technology.

These results also show that, to being discriminated against, women experience spatial segregation linked to the distribution of agro-ecological zones.

In view of the main findings of this study, our recommendations are in line with the Sustainable Development Goals for eradicate the poverty and hunger, to protect the planet and make sure that all human beings live in peace and prosperity.

We recommend policies to facilitate women's access to productive resources such as: land, agricultural inputs, irrigation, agricultural credit and others, by sensitizing public authorities to promote access to land equitably between men and women through laws or decrees. Fight against unfavourable and discriminatory cultural practices to empower women and girls.

Facilitate access to improved seeds and equipment of good quality adapted to the type of soil and agro-ecological zones, while developing and multiplying the structures or supply stations on the national territory.

To train farmers and women farmers to adopt more sustainable approaches and to favour agricultural system, and foods more efficient and sustainable such as: selection of species and varieties well adapted to the site and of exploitation condition, crop diversification, and cultural practices for improve biological stability and farm economic, soil management to improve and protect the soil quality, efficient and humane use of inputs, good soil conservation and biodiversity.

Stimulate the participation of the female gender in agricultural production, and gender mainstreamed into agricultural programs and research in the context of ecological needs as well as regional cultural differences.

Guide efforts in term of sustainable agricultural development on women as well as programs focused on increasing agricultural productivity. Achieve a good distribution of agro ecological zones to allow a good representation of women in agriculture.

Make agricultural technology accessible to all for sustainable agriculture and agriculture for all. Promote education and scholarship of women and provide school accessible to all and for all.

These measures will contribute to ending hunger, ensuring food security and promoting sustainable agriculture.

References

Aguilar A, Carranza E, Goldstein M, Kilic T, Oseni G (2014) Decomposition of Gender Differentials in Agricultural Productivity in Ethiopia. The World Bank

Deere and Leal (2001) Empowering Women Land and Property Rights in Latin America.docx. University of Pittsburgh Press

ECAM (2007) Rapport de la troisième enquête auprès des ménages (ECAM3)

Fairweather John R, Campbell Hugh R (2003) Environmental beliefs and farm practices of New Zealand farmers contrasting pathways to sustainability. Agric Hum Values 20(3):287–300. https://doi.org/10.1023/A:1026148613240

FAO (2006) Securite alimentaire et developpement agricole en afrique subsaharienne: dossier pour l ... Place of publication not identified: Food & Agriculture Organi

FAO (2011) La situation mondiale de l'alimentation et de l'agriculture 2010-2011: le rôle des femmes dans l'agriculture : combler le fossé entre les hommes et les femmes pour soutenir le développement. FAO, Rome

Fuglie K (2015) Accounting for growth in global agriculture. Bio-Based Appl Econ 4(3):201–34. https://doi.org/10.13128/BAE-17151

Karami E, Mansoorabadi A (2008) Sustainable agricultural attitudes and behaviors: a gender analysis of Iranian farmers. Environ Dev Sustain 10(6):883–98. https://doi.org/10.1007/s10668-007-9090-7

Oladeebo JO, Fajuyigbe AA (2007) Technical efficiency of men and women upland rice farmers in Osun State, Nigeria. J Hum Ecol 22(2):93–100. https://doi.org/10.1080/09709274.2007.11906006

Palacios-López A, Lopez R (2014) Gender Differences in Agricultural Productivity: The Role of Market Imperfections (No. 164061). Department of Agricultural and Resource Economics, the University of Maryland

Rezaei-Moghaddam K et al (2005) Agricultural Economics.docx

Tiruneh A, Tesfaye T, Mwangi W, Verkuijl H (2001) Gender Differentials in Agricultural Production and Decision-Making among Smallholders in Ada, Lume, and Gimbichu Woredas of the Central Highlands of Ethiopia. Cimmyt

Tobin J (1958) Estimation of relationships for limited dependent variables. Econometrica J Econometric Soc 26:24–36

Udry C (1996) Gender, agricultural production, and the theory of the household. J Polit Econ 104(5):1010–46

UN Women (2015) The cost of the gender gap in agricultural productivity in Malawi, Tanzania, and Uganda
World Bank La (2008) L'agriculture au service du développement, 1st edn. De Boeck Supérieur. http://www.cairn.info/l-agriculture-au-service-du-developpement--9780821372999.htm

Gender Differentials in Livelihood Security among Aquaculture Fish Farmers in Southwest Nigeria

Lawrence Olusola Oparinde

Abstract Despite several studies on livelihood security, there is a dearth of studies on livelihood security in relation to gender among aquaculture fish farmers. Therefore, this study examined gender differentials in livelihood security among aquaculture fish farmers in southwest, Nigeria. This is important because of the relevance of aquaculture, livelihood security and gender issues to the Sustainable Development Goals (SDGs). The selection of 192 respondents was achieved using a multistage sampling procedure. Identified security pointers from the data were grouped under four security domains and standardised in line with the method used to measure life expectancy in reports of human development. Also, descriptive statistics and two-stage least squares regression were employed to analyse the data collected. The empirical findings revealed that male aquaculture fish farmers were more secure economically in terms of food, health and education than their female counterparts in the study area. Furthermore, the results of the analysis showed that, overall, the livelihood security status of female aquaculture fish farmers was poorer relative to their male counterparts. Moreover, the effect of economic security and educational security on overall livelihood security is greater among male aquaculture fish farmers than among their female counterparts. Therefore, there is a need to close the gap between men and women in terms of livelihood security through economic and educational empowerment programmes if sustainable development is to be achieved, especially for female aquaculture fish farmers in southwest Nigeria.

Keywords Gender · Aquaculture · Livelihood · Security · Two-stage least squares regression

1 Introduction

There is no doubt that world hunger is increasing no matter how well the prospects for economic growth (Food and Agriculture Organisation [FAO] 2008; FAO et al.

L. O. Oparinde (✉)
Department of Agricultural and Resource Economics, The Federal University of Technology, PMB 704, Akure, Nigeria
e-mail: looparinde@futa.edu.ng

W. Leal Filho et al. (eds.), *Sustainable Development in Africa*, World Sustainability Series,
https://doi.org/10.1007/978-3-030-74693-3_19

2018). The poor are becoming more susceptible with the recent sudden increase in the prices of food globally. The situation of people who are deficient in resources has been made worse with the current recession. The increasing menace of food shortage is one of the challenges to be addressed. Also, how to ensure access to nutrition, healthcare, safe water and sanitation and resources should be addressed. All these are important in making a decent living affordable for the poor (Rahman and Akter 2010). Food security was identified as the most off-track Millennium Development Goal (MDG) (European Union 2010). The importance of the problem of food insecurity and poverty is illustrated by its position as topmost goal of the seventeen Sustainable Development Goals (SDGs) formulated by the United Nations (UN) to end poverty and hunger by 2030.

About 86.9 million (about 50%) out of 180 million Nigerians now live in extreme poverty (Kazeem 2018). Nigeria was given 96th position out of 113 countries in the 2018 Global Food Security Index (The Economist Group 2018). This makes food insecurity an issue of great concern in Nigeria. With Nigeria having one of the fastest growing populations worldwide, it is also characterised by some of the worst health indicators in Africa (United States Agency for International Development [USAID] 2019). The Nigerian government has therefore been called upon by researchers and other stakeholders to consider inventive means to address healthcare issues (Asakitikpi 2019). About 10.5 million children within the age group 5 to14 years are out of school in Nigeria, despite the fact that the policy of free and compulsory primary education is in place (United Nations Children's Emergency Fund [UNICEF] 2019).

These realities motivated the need to thoroughly examine the level and nature of livelihood security in Nigeria. Livelihood security refers to the abilities, resources and actions essential for a source of revenue (Hertz et al. 2010). It can also refer to the need for a sufficient and maintainable reach to revenue and other factors to allow families to have their basic needs fulfilled (Frankenberger 1996). This comprises of sufficient access to food, drinkable water, health services, education, shelter, and opportunities for community involvement and social incorporation (Frankenberger et al. 2002) .

In Nigeria, the livelihood of the poor is supported by aquaculture fish farming, which provides opportunities for improved employment and income generation (Ahmed and Lorica 2002). Aquaculture industry is growing fast in Nigeria, while providing food security, economic growth, employment and serving as strategy for poverty reduction (Asiedu et al. 2017). This includes illiterate women, who can earn revenue for their households through participation in aquaculture fish farming (Rajee and Mun 2017). If the aquaculture fish farming potential of Nigeria is fully harnessed, the country can save over USD 1.1billion being spent annually on fish importation, while creating employment, improving livelihoods and increasing foreign exchange earnings (Adebayo 2019). This aligns with the decision by the UN to make sustainable management of fisheries one of the ten targets of SDG 14. Smallholder producers as well as their households benefit from the production and consumption of fish (Azzarri et al. 2014; Weeratunge et al. 2014). However, research that examines the importance of aquaculture fish farming for the empowerment of women, specifically in Nigeria, is scarce (Harper et al. 2013; Azzarri et al. 2014).

There is a disparity in the encouragement provided to men versus women for their labour and capital investment in time in aquaculture fish. This is an indication of the existence of gender differences in accessing and controlling factors of production as well as the benefits realised from aquaculture (Shanthi et al. 2010). Gender refers the separate roles, rights and responsibilities of men versus women as defined by socio-cultural norms and institutions (Gutierrez-Montes et al. 2012). Over the past ten years gender is increasingly recognised as one of the factors influencing the attainment of livelihood security (Okali 2011). It therefore makes sense to have SDG 5 with a focus on gender equality and women empowerment, as it is vital to link gender equality and sustainable development (UN 2014).

The existence of lines of authority between men and women can be used to explain gender and social relations (FAO 2017), which affect accessibility and control of resources as well as delivery of gains of livelihood activities (Stacey et al. 2019). Locke et al. (2007) explained that it is necessary to address entrenched gendered disparities, since men are usually involved in efforts to increase household income, whereas women are usually concerned about poverty reduction and household food sufficiency. The ability of men and women to be involved in governance and policy is therefore influenced by gendered norms and relations (Leisher et al. 2016). This has implications for the sustainability of natural resource as well as the livelihoods that depend on it (Kawarazuka et al. 2017).

More significance can be attached to the involvement of women in aquaculture than is usually presumed. Increased recognition is being given to women who participate in aquaculture fish farming in various ways, as they make significant contributions to the welfare of their households. However, little is realised as benefits by these women because of gender inequalities in the socio-cultural and economic spheres (Weeratunge-Starkloff and Pant 2011). As a result, there may be a reduction in productivity and inefficient labour allocation at the micro and macro levels because of gender disproportions in aquaculture fish farming (Weeratunge-Starkloff and Pant 2011).

Attention has been given in the literature on the role of women in aquaculture, with numerous studies reporting on the means through which women involve themselves in this industry. Yet, there is insufficient information on way that aquaculture fish farming influenced the status of women as well as gender relations within families and communities. Also, there is inadequate information on the influence of the stated relationships on the usefulness of aquaculture fish farming to enhance livelihoods and address poverty (Shanthi et al. 2010) and thereby supporting sustainable development.

The extant literature in the area of livelihoods is skewed towards livelihood diversification and/or activities and impacts on the welfare status of farmers (Aderinoye-Abdulwahab et al. 2015; Rajee and Mun 2017; Gani et al. 2019). There is a dearth of studies, however, on livelihood security in the context of multiple security domains in relation to gender among aquaculture fish farmers worldwide, but specifically in Africa and in Nigeria. Hence, this study analysed gender differentials in livelihood security among aquaculture fish farmers in southwest Nigeria. The study specifically examined the livelihood security status of male and female aquaculture fish farmers and identified the determinants of livelihood security. This study contributes to the body of knowledge on gender, livelihood security and sustainable development by providing a detailed analysis on the livelihood security status of aquaculture fish farmers using four security domains (economic, food security, health, and education) across gender. The study is relevant to sustainable development as it covers SDGs 1,2,3,4 and 5 of the UN.

2 Literature Review

2.1 Gender and Livelihood Security in Nigeria

A discussion on the issues of gender and livelihood security requires an understanding of what is meant with gender equity and gender equality. According to the International Labor Organization (ILO), gender equity means impartial treatment for women and men in accordance to their individual needs. However, gender equality refers to having equal rights, opportunities and treatment for men and women and for male and female children in every aspect of life (Consultative Group on International Agricultural Research [CGIAR] 2017).

Gender disparity between men and women in getting access to and controlling the resources of production is not fair to women and their children. It also creates a bad economy, leading to biased allocation of scarce resources, higher healthcare expenses, decreased efficiencies, and poor human development trends (Oniang'o and Mukudi 2002). As a consequence women have fewer opportunities and reduced revenue from aquaculture compared to their male counterparts—including being left in the position of poverty. It is established that increased gender equity in the fishery subsector will contribute significantly to the attainment of the SDGs on poverty reduction and food and nutrition security. Specifically, gender equity in aquaculture fish farming is capable of realising benefits such as increased fish productivity, increased household revenue and improvement of nutritional outcomes (CGIAR 2017).

Gender equality and women empowerment have been recognised to be part of the drive to combat poverty, hunger and diseases in order to enhance sustainable development. Variations according to gender in access to resources are one of the rudimentary reasons for inequality in food security and livelihood among households headed by men versus women (Babatunde et al. 2008). Gender equality is recognized as a crucial matter in the economic development of developing economies. Countries that give the least recognition to gender equality consistently fail to provide the same opportunities for women than for men. In this regard, Nigeria occupies the ninth position among the ten lowest countries ranked in terms of gender equality by the World Economic Forum (WEF) (Archibong et al. 2018).

Archibong et al. (2018) stated that in terms of economic and educational opportunities, health and survival, political empowerment and representation, and health outcomes, women were far worse off than men in countries that scored the worst in the gender gap. Women and girls in Nigeria clearly have worse life opportunities than men when compared to other countries (Department for International Development [DFID] 2012). Irrespective of their academic attainment, women still have less than 30 percent of public sector positions. Only 15 percent of 80.2 million Nigerian women have bank accounts. They are also less likely to receive bank loans than men and are also less likely to be land owners than men. This shows the negligence of the Nigerian government in solving the problems of gender inequality in spite of efforts from numerous quarters to lessen its persistence (Archibong et al. 2018) and is one of the factors contributing to the non-attainment of SDG 5.

Evidence increasingly points towards the existence of interactions between gender equality and the sustainability of the economy. For instance, when women partake in public administration, public resources are more likely to be assigned for investment in human development, including aspects as the health of children and women, nutrition and employment (Buckingham-Hartsfield 2013). Achievement of food security and sustainable livelihood greatly depends on access to and control over agricultural assets and production resources by women. This implies that the design of approaches that are more environmentally, socially and economically sustainable in the administration of local landscape, food production and water security requires the knowledge of and collective action by women (FAO 2011). Furthermore, economic growth is positively influenced by women education and by the full participation of women in decision making and employment (Agarwal 2010).

2.2 *Fisheries, Aquaculture and Gender*

According to the FAO and ECOWAS Commission (2018), women significantly contribute to the dietary requirements of households. The evidence obtainable shows that the involvement of women in aquaculture fish farming is speedily growing, and that this involvement is contributing to family food security in Nigeria. The gender functions of women in fish production are acknowledged in fishing, processing and marketing. Odulate et al. (2011) stated that women are strictly involved in artisanal fishing activities in coastal wetland communities. The activities involve moving fish out of canoes, processing of fish as well as the marketing thereof. The dominant work of women in fisheries around Lake Kainji area is the processing aspect, involving almost 60 percent of women (Nwabeze et al. 2013). Women have received little recognition despite the significant role they play. Going by all economic, political and social pointers, women have a low status in the fishery sector. Therefore, women in Nigerian fish production should be given incentives if their role in the production of fish is to be enhanced (Olufayo 2012). Also, poor access of women to funds for value chains in fish must be tackled in order for their capacity to be developed. In Nigeria, the most common technique of fish production involves rearing of fish in earthen ponds. The earthen pond system that is now popular was due to increased awareness about the system given by the Fisheries Department of the Federal Ministry of Agriculture and Fisheries Departments at various Nigerian universities. According to the farmers, the technique is easy to implement and leads to faster growth in fish. Gender functions in fisheries show that men dominate in production of fingerlings, artisanal fishing, fish processing, storage as well as marketing. The major activities for women include only processing and marketing (FAO and ECOWAS Commission 2018).

2.3 Gender Mainstreaming for Sustainable Development in Nigeria

Gender mainstreaming is an approach or procedure aimed at achieving gender equality (Abali et al. 2014). Creation of avenues for men and women to be equally relevant in the scheme of affairs in every country, is a positive step towards the attainment of sustainable development (Archibong et al. 2018). Gender mainstreaming implies the inclusion of women in all areas of decision making and development programmes as a policy matter. In a nut shell, it is an approach for taking gender matters into the mainstream of society in order to ensure gender equality (Kanjere and Rachid 2014). This approach was established and brought to attention in the Beijing Platform for Action at the 1995 United Nations World Conference on Women (Archibong et al. 2018). The United Nations Entity for Gender Equality and the Empowerment of Women (2014) stated that a solid and sustained commitment to gender mainstreaming is the only way of guaranteeing that women and men partake in and benefit from development efforts equally. They explained that the accomplishment of gender inequality is beyond just enacting laws and adopting policies.

Gender mainstreaming may involve variations in goals, approaches and actions so that women and men can have an effect on, partake in, and gain from development procedures in an equitable way. The aim of mainstreaming gender equality is to change uneven socio-institutional arrangements into even arrangements for men and women (International Labor Office [ILO] 2010). Gender mainstreaming is important because it is a way of making sure that the implications of gender variations are considered in every aspect of any project, involving aspects as design, implementation, monitoring and evaluation. This is to ensure that resources are directed at women for activities conducted by women and at men for activities conducted by men. Therefore, in order to reduce rural poverty, encourage food security and raise efficiency in, for example, the agricultural sector, a rigorous effort is required to have an exact understanding of the state and functions of women and men in agriculture in any given area (Abali et al. 2014).

2.4 *Gender Mainstreaming Situation in Nigeria*

Two basic issues are discussed in this section and some problems relating to gender mainstreaming with regards to each of these two issues. The issues refer to the position of women in the constitution and the allotment of thirty-five percent affirmative action.

2.4.1 Position of Women in the Constitution

Gender gaps in the constitution of Nigeria are evident via the language used in the constitution. The language of the constitution is in line with the patriarchal tradition of Nigerian society. The pronoun "he" surfaces in 235 places in the 1999 constitution (Constitution of the Federal Government of Nigeria 1999), while "women" as a word appears in just two places (Archibong et al. 2018). Concerning indigene-ship, there is a sustained ambiguity about the "origin" of a woman who gets married to a man from a different ethnic or geographical area to hers. The truth for most of these women is that they do not have any decisive right either to the community they left or to which they are married into. In some occasions, women have been deprived of their rights to political positions because they can neither claim their original places of origin nor that of their husbands (Morley 2012).

2.4.2 Thirty-five Percent Affirmative Action

Despite the fact that women constitute about half of the estimated population of Nigeria, this numerical strength has not been given an equivalent representation in Nigeria's political life. The 1995 Beijing Conference endorsed allotment of 35 percent of political positions, power as well as decision making to women. In Nigeria, this is by far not the case as women can hardly claim five percent of these positions. However, the negligible increase in gender involvement in the country is tied to appointed positions, since the country has failed to deliver in terms of women in elected positions (Conyok 2015). Electoral and power allotment activities fail to reflect the interests of women. As a result they are politically powerless despite the fact that about two-thirds of Nigerian voters are women. As far as the index for political participation of women is concerned, Nigeria is far behind among most African countries (Archibong et al. 2018).

3 Data and Description of Variables

The primary data used in this research were supplied by 192 aquaculture fish farmers using a structured questionnaire between January and February, 2019 in the south-western part of Nigeria. There is a high demand for catfish in the study area which attracts many people to aquaculture fish farming. A multistage sampling procedure was employed in the selection of respondents. The first stage involved random selection of two out of the six states in the region, with the Ondo and Osun States which were selected. The second stage involved purposive selection of six Local Government Areas (LGAs) in each of the chosen states, with consideration of the level of urbanisation and prevalence of aquaculture fish farming. Random sampling was employed in the third stage, during which four communities were selected in each of the chosen LGAs. In the fourth stage, random sampling was used to select four aquaculture fish farmers in each of the selected communities, which made up a total of 192 respondents. This was done using the list of registered fish farmers in each of the selected states. However, only 190 copies of questionnaire were used for the analysis. The remaining 2 copies of questionnaire were not used in the analysis due to unsatisfactory data provided.

Table 1 presents the definition, description and mean differences of the variables used in the analysis. The results of the analysis indicate statistically significant differences between the male and female aquaculture fish farmers with respect to some of the variables used. For example, the average economic security index for male fish farmers was 0.53, while female aquaculture fish farmers recorded 0.43 for the same index. Also, there is a significant difference between the food security index of male and female respondents, respectively 0.42 and 0.28. In the case of the health security index, male fish farmers had a higher value than their female counterparts, respectively 0.24 and 0.15. The overall livelihood security status of female aquaculture fish farmers was poorer relative to their male counterparts with 0.34 and 0.25 for male and female respondents, respectively.

3.1 Analytical Technique

In this study, 22 security pointers from the data set were identified and categorised under four security domains (economic security, food security, health security and educational security), following Rahman and Akter (2010). After this, Household Livelihood Security (HLS) indices were calculated in line with Hahn et al. (2009). It is assumed that each of the identified pointers has the same weight relative to the overall HLS index. The next step was the standardisation of the pointers in line with the method used in computing life expectancy in reports of human development. The expression for the identified pointers is given as

Table 1 Definition and mean differences of variables used in the analysis

Variable	Description	Male	Female	Difference
Economic security index	Economic security pointers	0.53	0.43	0.10**
Food security index	Food security pointers	0.42	0.28	0.14***
Health security index	Health security pointers	0.24	0.15	0.09**
Education security index	Education security pointers	0.32	0.24	0.08*
Overall livelihood security index	Index of the overall livelihood security	0.34	0.25	0.09*
Age	Age of the respondents	50.6	48.99	1.61***
Marital status	1 if the respondent is married; 0 otherwise	1.84	1.84	0
Level of education	Number of years spent in school	12.68	12.37	0.31**
Household Size	Number of persons living in the family and share meals	4.43	3.52	0.91***
Dependency	Dependent members of the family	0.28	0.21	0.07***
Non-farm revenue	Revenue from non-farm activities in naira	60,632.08	58,626.37	2,005.71***
Instruments				
Stdpond	Standardised monetary value of pond	0.18	0.47	−0.29***
Stdfudgrp	Standardised value of number of food groups	0.48	0.62	−0.14**
Stdsickday	Standardised value of number of days of other sickness	0.14	0.11	0.03**
Std7plus	Standardised value of number of 7 plus literate people	0.17	0.004	0.17***

Note: *, ** and *** represent significance at 10%, 5% and 1% levels respectively

$$Standardized\ pointer_j = \frac{Pointer_j - Minimum_j}{Maximum_j - Minimum_j} \quad (1)$$

where minimum and maximum values of the pointers have bearing on the same state where the households are situated. After standardising each of the pointers representing a certain livelihood security domain, the required HLS index is calculated by taking the mean value of the standardised pointers:

$$Household\ Livelihood\ Security(HLS)_i = \frac{\sum_{j=1}^{n} Standardized\ pointer_j}{J} \quad (2)$$

where J represents the number of pointers used to calculate the index. After defining each HLS index, calculation of the composite household's overall Livelihood Security (LS) index follows using Eq. (3):

$$Livelihood\ Security(LS)_i = \frac{\sum_{i=1}^{n} w_i HLS_i}{\sum_{i=1}^{n} w_i} \quad (3)$$

where w_i represents the weights determined by the number of pointers used to calculate each HLS index. Weights differ from one household to another because of household level differences in the number of pointers.

3.2 Econometric Estimation

A two-stage least squares (2SLS) regression model was employed in order to surmount the problem of endogeneity, since household livelihood security (HLS) is potentially endogenous in the model specified for the determinants of livelihood security. Another estimation procedure that could be used is ordinary least squares (OLS), but which is not capable of resolving the problem of endogeneity and OLS estimators would be biased and inconsistent. According to Angrist and Krueger (2001), the use of instrumental variables in 2SLS leaves room for the estimation of consistent coefficients that are void of asymptotic unfairness from lost variables and measurement errors. However, this approach can lead to unfair and erratic estimates if the instruments used are invalid or weak and the model is not stated correctly, leading to over or under-identification of issues. In order to take care of such challenges in this study, Durbin-Wu-Hausman test statistics and the joint significance test of instrument validity were conducted. The model is specified as follows:

$$Overall\ Livelihood\ Security_i = \propto_i + \rho_i \sum_{i=1}^{4} HLS_i + \sum_{k=1}^{n} \beta_{ik} X_{ik} + \varepsilon_i \quad (4)$$

where HLS represents household livelihood security indices, Xs are the explanatory variables representing socio-economic characteristics of households, $\propto, \rho\ and\ \beta$ are the parameter estimates of interest and ε is the error term. The 2SLS regression approach was used for the purpose of replacing the real HLS variables causing problems in the equation by counterpart variables that are purged of their stochastic component to ensure that the OLS procedure could be used. In order to achieve this, a reduced form equation was specified as a function of all the explanatory variables in Eq. 4 and a set of instrumental variables as follows:

$$HLS_i = \gamma_i + \sum_{k=1}^{n} \beta_{ik} X_{ik} + \delta_i Z_i + \varepsilon_i \quad (5)$$

where Z_i represents a set of instrumental variables which influence HLS but not the overall livelihood security indices directly and γ, β *and* δ are the parameter estimates. The predicted values from the OLS-estimated reduced form Eq. 5, defined as HLS_i, is introduced into the structural equation for the purpose of replacing the HLS variables causing the problems. Therefore, the equation can be reduced to Eq. 6 that can be estimated by using the OLS approach.

$$Overall\ Livelihood\ Security_i = \tau_1 + \tau_2 HLS_i + \sum_{k=1}^{n} \tau_{3ik} X_{ik} + \epsilon_i \qquad (6)$$

where τ_i represents parameter estimates and ϵ_i represents an error term. The vector of instruments Z_i used in this study include the standardised monetary value of ponds, the standardised value of number of food groups, the standardised value of number of days of other sicknesses and the standardised value of the number of seven years plus literate people in the household.

4 Empirical Results and Discussion

4.1 Economic Security

Eight economic security pointers were selected as shown in Table 2. The t-test was employed to check the significance of the difference between the two categories. The higher the values of the pointers the better off the households are economically and more secure economically as well. Considering all the economic security pointers, male aquaculture fish farmers were better off economically than their female counterparts. The difference in terms of all the pointers is statistically significant. Many factors (for instance, inequality in access to production resources) could be

Table 2 Economic security pointers

Pointers	Male	Female	Difference
Income per person in naira	60,546.29	57,966.64	2,579.65***
Pond value	293,502.50	273,755.5	19,747.00***
Equipment value	58,700.51	54,751.09	3,949.42***
Active population (15–59 yrs population/family size)	3.90	2.08	1.82***
Share of 15–59 years people in employment	0.82	0.79	0.03**
Income per woman in naira	30,273.15	28,983.32	1,289.83***
Current loan per person in naira	150,682.10	134,518.40	16,163.70***
Current savings per person in naira	12,109.26	11,593.33	515.93***

Note: ** and *** represent significance at 5% and 1% levels respectively

Table 3 Food security pointers

Pointers	Male	Female	Difference
Number of food groups consumed per day	5.93	4.47	1.46**
Number of meals and snacks consumed per day	2.87	3.00	−0.13**
Household food grain stock per person	2,421.85	2,318.67	103.18***
Number of food convenient months per year	8.87	8.79	0.08
Number of meals consumed per woman	2.90	3.00	−0.10

Note: ** and *** represent significance at 5% and 1% levels respectively

responsible for the differences recorded. This has caused male fish farmers to be more economically secure compared to their female counterparts. This confirms the assertion of Weeratunge-Starkloff and Pant (2011) that women in agriculture may possibly constitute a greater percentage of the poor within the sector. This serves as further confirmation for the existence of SDG 5, which is to attain gender equality and women empowerment. As stated by the United Nations (2014), connecting gender equality and sustainable development is of key importance.

4.2 *Food Security*

Table 3 presents food security pointers. Five pointers are considered. The t-test was used to check the significance of the difference between the two categories. The differences in three pointers between male and female fish farmers are statistically significant. These pointers are number of food groups consumed per day, number of meals and snacks consumed per day and household food grain stock per person. Male fish farmers consumed more food groups per day and had more food grain stock per person relative to female fish farmers. However, female fish farmers consumed a higher number of meals and snacks per day than male fish farmers. All these, except number of meals and snacks consumed per day, resulted in female aquaculture fish farmers being less food secure than their male counterparts. This is an indication that the SDGs about zero hunger and gender equality are yet to be achieved. To know how important women are in any economy, the FAO (2011) stated that accessing and controlling agricultural and production factors are key to the accomplishment of food security and a sustainable livelihood.

4.3 *Health Security*

Concerning health security, the t-test was again employed to arrive at the significance of the differences between the two categories in Table 4. The number of days household members suffer from diarrhoea, the number of days household members

Table 4 Health security pointers

Pointers	Male	Female	Difference
Number of days members suffered from diarrhea in the last 30 days	0.21	0.13	0.08**
Number of days members suffered from other sickness in the last 30 days	0.33	0.28	0.05**
Number of days lost to sickness	0.75	0.26	0.49***
Number of antenatal consultations	0	0.29	−0.29***

Note: ** and *** represent significance at 5% and 1% levels respectively

Table 5 Education security pointers

Pointers	Male	Female	Difference
Number of 7 years plus literates	0.17	0.04	0.13***
Number of 15 years plus adult male literates	1.12	1.03	0.09**
Number of 15 years plus adult female literates	2.31	2.48	−0.17**
Number of 11–15 years boys enrolled in the school	0.45	0.35	0.1**
Number of 11–15 years girls enrolled in the school	0.19	0.04	0.15***

Note: ** and *** represent significance at 5% and 1% levels respectively

suffered from other sicknesses and the number of days lost due to sickness in the last 30 days were significantly higher among male than among female fish farmers. On the contrary, the number of antenatal consultations among female fish farmers is higher than for their male counterparts. This is expected because men do not go for antenatal consultations. This result directly addresses SDG 3, which is about making sure that people are healthy to the extent that child mortality is reduced and life expetancy is increased.

4.4 *Education Security*

Five pointers were employed to measure education security, as shown in Table 5. The t-test was used to check the significance of the differences between the two categories. All the pointers are significantly higher among male than female fish farmers, except number of 15 years plus adult female literates. This is an indication that there is still plenty evidence of inequality between male and female school enrolment. This result could be linked to lower economic power of female-headed households, which might be preventing them from enrolling their children in school. Going by this outcome, SDG 4 which is about having access to comprehensive and unbiassed quality education, has not yet been achieved. The FAO and ECOWAS Commission (2018) confirms that a wide and noticable difference in level of education can be observed between men and women in Nigeria.

4.5 Determinants of Livelihood Security

The overall test of possible endogeneity of household livelihood security (HLS) among male respondents provided the following results: Durbin (score) χ^2 (4) of 16.087 ($p = 0.0029$) and Wu-Hausman F(4,90) value of 4.02563 ($p = 0.0048$) at a 1% level of significance. For female respondents, the test result gave a Durbin (score) χ^2 (4) of 60.1267 ($p = 0.0000$) and aWu-Hausman F(4,75) value of 36.5161 ($p = 0.0000$) at a 1% level of significance. Therefore, the null hypothesis which state that every variable used in the male and female categories is exogenous, is rejected. This implies that endogeneity should be controlled in the estimation procedure, since HLS is highly endogenous in the model for both the male and female categories. Also, the first-stage OLS regression estimates of F-statistics for the instruments' joint significance for male and female categories are as follows: F(4,92) value of 210.70 ($p = 0.0002$) and F(4,78) value of 160.30 ($p = 0.0001$), respectively. This leads to rejection of the null hypothesis, namely that the standardised values used by HLS for pond, number of food groups, number of days of other sicknesses and number of 7 years plus literate people in the household are weak.

OLS and 2SLS regression estimates are presented in Table 6. All security variables (that is, economic, food, health and education) are highly significant factors that influenced household livelihoods security (HLS). They all have expected positive relationships with the signs of the coefficients. The effect of economic security and education security on the overall livelihood security is greater among male aquaculture fish farmers than their female counterparts, indicating that SDGs 1 and 4 are yet to be attained. However, there is little or no difference in the effect of food security and health security on overall livelihood security between male and female fish farmers. Specifically, a unit increase in the economic security would bring about a 44% increase in the overall livelihood security in the male category, while it would bring only a 34% increase in the overall livelihood security in the female category. Also, a unit increase in education security, food security and health security would lead to respectively 21, 26 and 19% increases in the overall livelihood security for male fish farmers, while it would bring about respectively 6, 27 and 19% increases in overall livelihood security for female fish farmers.

Level of education and non-farm revenue had direct and significant correlations with overall livelihood security. This demonstrates the importance of education and livelihood diversification in achieving higher overall livelihood security and SDGs 1 to 4 for the households of male fish farmers. However, dependency burden and overall livelihood security showed an indirect but significant association in the female category. This is in agreement with Rahman and Akter (2010) who reported that dependency burden had a significantly negative effect on overall livelihood security.

Key: Ecosecindex = Economic security index, Foodsecindex = Food security index, Healthsecindex = Health security index, Educationsecindex = Education security index.

Table 6 Ordinary least square and two-stage least square regression estimates

Variable	Male		Female	
	OLS	2SLS	OLS	2SLS
Constant	−0.011 (1.02)	−0.016 (1.23)	−0.104 (3.91)	−0.269 (3.00)
Ecosecindex	0.427*** (14.70)	0.436*** (11.71)	0.330*** (12.12)	0.339*** (3.12)
Foodsecindex	0.238*** (29.78)	0.255*** (11.73)	0.251*** (14.78)	0.265*** (4.37)
Healthsecindex	0.180*** (26.31)	0.187*** (19.64)	0.193*** (11.52)	0.188* (1.82)
Educationsecindex	0.222*** (22.15)	0.207*** (19.67)	0.056*** (5.26)	0.058*** (2.61)
Age	−3.60E−05 (0.38)	−2.88E−05 (0.19)	3.982E−04 (1.45)	0.003*** (2.57)
Marital status	0.005* (1.78)	0.003 (1.21)	0.024*** (3.96)	0.021 (0.07)
Level of education	3.241*** (3.87)	3.248** (1.85)	0.001 (1.25)	0.002 (0.68)
Household size	−0.002 (1.39)	−0.003** (2.08)	−0.012*** (4.41)	−0.014 (1.30)
Dependency	−1.961E−04 (0.03)	−1.077E−04 (0.11)	−0.049*** (3.25)	−0.069** (1.95)
Non-farm revenue	3.76E−08*** (2.78)	4.08E−08* (1.74)	1.77E−07 (1.11)	1.52E−07 (1.35)
Fingerlings	3.246E−04 (0.66)	4.491E−04 (0.91)	0.002*** (2.08)	0.042 (0.23)
N	100		90	
R^2	63.83	66.73	72.72	54.54

Figures in parenthesis are t-values. *, ** and *** represent significance at 10, 5 and 1% levels respectively

5 Conclusion and Recommendations

This study examined gender differentials in livelihood security among aquaculture fish farmers in southwest Nigeria. From the results, it is concluded that male aquaculture fish farmers were more secure economically, in terms of food, health and education than their female counterparts in the study area. Also, the effect of economic security and education security on overall livelihood security is greater among male aquaculture fish farmers than their female counterparts.

Following the findings of this study, a number of recommendations are made, if sustainable development in aquaculture and Nigeria at large will be achieved. There is the need to close the gap between men and women in terms of livelihood security through economic and educational empowerment programmes among female aquaculture fish farmers by development agents. Government, at all levels, should develop and implement gender friendly provisions that respect rights of women and economic steadiness for women. To this end governments need to have their resources rearranged and priorities modified such that the SDGs concerned with gender equality and education can be achieved. The concept of education for sustainable development should be given topmost attention in Nigeria, since sustainability considers futuristic, economic, social and environmental issues, while incorporating the need for gender equality. Also, non-governmental organisations, governments and individuals need to be encouraged to embrace a sustainable livelihood approach that addresses poverty and food insecurity concerns in an all-inclusive manner. This approach basically searches for better strategies to assist people and communities in ways they believe address their daily lives and needs instead of imposition of intervention measures on them. Furthermore, since non-farm revenue has been realised to be one of the vital factors that affect overall livelihood security, female fish farmers should be advised to adopt income diversification strategies. This will assist them to gain from the value of additional non-farm revenue in achieving increased overall livelihood seccurity. Moreover, land/ponds and equipment should be made more accessible to female aquaculture fish farmers in order to achieve higher productivity, economic security and food security status. With regard to this study, a larger sample size would have given better representativeness of results. The sample size used was due to limited fund availability. Also, findings from this study cannot necessarily be used to draw conclusion for the country or Africa as a whole, because different cultural and operational practices for fishing are found between regions and from one geopolitical zone to the other. Further studies that extend the scope of the research to cover the whole of Nigeria and other countries in Africa should therefore be conducted as well.

References

Abali I, Emerhirhi E, Okoromah B (2014) Gender mainstreaming in agricultural production: a review. Nigeria J Rural Sociol. 15(1):135–141

Archibong EP, Bassey GE, Nwagbara EN (2018) Gender mainstreaming and attainment of sustainable development goals: does nigeria have the political will to successfully work the talk? Global J Soc. Sci 17:21–29

Adebayo OT (2019) Aquaculture: a catalyst for sustainable fish production. Inaugural Lecture Series 105. Federal University of Technology, Akure, p 73.

Aderinoye-Abdulwahab SA, Nwaachukwu SC, Salawu OL, Popoola PO (2015) Assessment of livelihood activities of rural farmers in kwara state, Nigeria. Ethiopian J Environ Stud Manag 8(2):120–129

Agarwal B (2010) Gender and Green Governance: The Political Economy of Women's Presence Within and Beyond Community Forestry. Oxford University Press, Oxford, p 528

Ahmed M, Lorica MH (2002) Improving developing country food security through aquaculture development—lessons from Asia. Food Policy 27(2):125–141

Angrist JD, Krueger AB (2001) Instrumental variables and the search for identification: from supply and demand to natural experiments. J Econ Perspect 15(4):69–85. https://doi.org/101257/jep 15469

Asakitikpi AE (2019) Healthcare Coverage and Affordability in Nigeria: An Alternative Model to Equitable Healthcare. https://wwwintechopencom/online-first/healthcare-coverage-and-affordability-in-nigeria-an-alternative-model-to-equitable-healthcare-delive. Accessed 30 Jul 2020

Asiedu B, Adetola J, Kissi IO (2017) Aquaculture in troubled climate: farmers' perception of climate change and their adaptation. Cogent Food Agriculture 3:1296400. https://dxdoiorg/101080/233 1193220171296400

Azzarri C, Cross E, Haile B, Zezza A (2014) Does livestock ownership affect animal source foods consumption and child nutritional status? Evidence from Rural Uganda. World Bank Policy Research Working Paper 7111, Washington, DC: 1–48.

Babatunde RO, Omotesho OA, Olorunsanya EO, Owotoki GM (2008) Gender differences in resources allocation among rural households in Nigeria: implication for food security and living standard. Euro J Soc Sci 5(4):161–173

Buckingham-Hartsfield, (2013) Gender Equality: a pre-republishing for sustainable development. Geography 87(3):227–233

Conyok R (2015) The critical role of women in Nigeria politics. declarations and reservations. United Nations Treaty Collection

Consultative Group on International Agricultural Research (CGIAR) (2017) Why gender equality matters in fisheries and aquaculture. https://fishcgiarorg/impact/stories-of-change/why-gender-equality-matters-fisheries-and-aquaculture. Accessed 30 Jul 2019

Department for International Development (DFID) (2012) Human Right Approach to School Protection. Oxford University Press, Oxford, p 36

European Union (2010) A twelve-point EU action plan in support of the millennium development goals. Communication from the Commission to the European Parliament, the Council, the European Economic and Social Committee, and the Committee of the Regions. Brussels, 21.4.2010 COM (2010) 159 final

Food and Agriculture Organisation (FAO) (2008) The State of Food Insecurity in the World

Food and Agriculture Organization of the United Nations, Rome, Italy, p 56. https://www.fao.org/3/a-i0291e.pdf. Accessed 30 Jul 2019

Food and Agriculture Organisation (FAO) (2011) The state of food security in the world: The multiple dimensions of food security. Italy, Rome

Food and Agriculture Organisation (FAO) (2017) Towards gender-equitable small-scale fisheries governance and development - a handbook In support of the implementation of the voluntary guidelines for securing sustainable small-scale fisheries in the context of food security and poverty eradication, by Nilanjana Biswas, Rome, Italy. https://www.fao.org/3/i7419en/I7419EN.pdf. Accessed 30 Jul 2019

FAO and ECOWAS Commission (2018) National gender profile of agriculture and rural livelihoods – Nigeria. Country Gender Assessment Series, Abuja. p 92

Food and Agriculture Organisation [FAO], International Fund for Agricultural Development [IFAD], United Nations Children's Emergency Fund [UNICEF], World Food Programme [WFP], World Health Organisation [WHO] (2018) The state of food security and nutrition in the world 2018. Building climate resilience for food security and nutrition. FAO, Rome. p 181

Frankenberger T (1996) Measuring household livelihood security: an approach for reducing absolute poverty. Food Forum, No. 34. Washington, DC, USA

Frankenberger TR, Luther K, Becht J, McCaston MK (2002) Household Livelihood Security Assessments: A Toolkit for Practitioners. Prepared for the PHLS Unit by TANGO International Inc., Tucson, Arizona. https://www.alnap.org/system/files/content/resource/files/main/hls-assessment-a-toolkit-for-practitioners.pdf. Accessed 30 Jul 2019

Gani BS, Olayemi JK, Inoni OE (2019) Livelihood diversification strategies and food insecurity status of rural farming households in North-Eastern Nigeria. Econ Agric 66(1):281–295

Gutierrez-Montes I, Emery M, Fernandez-Baca E (2012) Why gender matters to ecological management and poverty reduction. In: Ingram J, DeClerck F, Rumbaitis del Rio C (eds) Integrating Ecology and Poverty Reduction. Springer, New York, NY. https://doi.org/10.1007/978-1-4614-0186-5_4
Hahn MB, Riederer AM, Foster SO (2009) The livelihood vulnerability index: a pragmatic approach to assessing risks from climate variability and change—a case study in Mozambique. Global Environ Change 19(1):74–88
Harper S, Zeller D, Hauzer M, Pauly D, Sumaila UR (2013) Women and fisheries: contribution to food security and local economies. Marine Policy 39:56–63. https://doi.org/10.1016/jmarpol20 1210018
Hertz D, Bhatia S, Mudahar M, Sally M (2010) Guidance notes on recovery: livelihood. International Recovery Platform (IRP), United Nations Development Programme (UNDP)
International Labour Office (ILO) (2010) Gender mainstreaming in local economic development strategies: a guide, LED programme, ILO Bureau for Gender Equality, Geneva. pp 4–31
Kanjere M, Rachid M (2014) The integrative approach to gender mainstreaming as a vehicle to women development in Africa. Mediterr J Soc Sci 5(27):1431–1434
Kawarazuka N, Locke C, McDougall C, Kantor P, Morgan M (2017) Bringing analysis of gender and social–ecological resilience together in small-scale fisheries research: challenges and opportunities. Ambio 46:201–213. https://doi.org/10.1007/s13280-016-0814-5
Kazeem Y (2018) Nigeria has become the poverty capital of the world. Quartz Africa. https://qz.com/africa/1313380/nigerias-has-the-highest-rate-of-extreme-poverty-globally/. Accessed 29 Jul 2019
Leisher C, Temsah G, Booker F, Day M, Samberg L, Prosnitz D Agarwal B, Matthews E, Roe D, Russell D, Sunderland T, Wilkie D (2016) Does the gender composition of forest and fishery management groups affect resource governance and conservation outcomes? Syst Map Environ Evid 5(6):1 – 10
Locke C, Muljono P, McDougall C, Morgan M (2017) Innovation and gendered negotiations: Insights from six small-scale fishing communities. Fish Fish 18:943–957
Morley L (2012) International trends in women's leadership in Higher Education. British Council. https://studylib.net/doc/5883259/morley--l.--international-trends-in-women-s-leadership-in. Accessed 11 Nov 2019
Nwabeze GO, Ifejika PI, Tafida AA, Ayanda JO, Erie AP, Belonwu NE (2013) Gender and fisheries of Lake Kainji, Nigeria: A review. J Fish Aquat Sci 8:9–13
Odulate DO, George FAO, Idowu AA (2011) Role of women in fisheries in coastal wet land areas of Oqun State Southwest Nigeria. In Martins O, Meshida EA, Arowolo TA, Idowu OA, Oluwasanya GO (eds) Managing Coastal and Wetland Areas of Nigeria. Proceedings of the Environmental Management Conference, September 12–15, 2011, Federal University of Agriculture, Abeokuta, Nigeria, pp 123–131
Okali C (2011) Searching for new pathways towards achieving gender equity: Beyond Boserup and 'Women's roles in economic development'. ESA Working Paper No 11–2011. p 25
Olufayo MO (2012) The gender roles of women in aquaculture and food security in Nigeria. In Shriver AL (Ed) Visible Possibilities: The Economics of Sustainable Fisheries, Aquaculture and Seafood Trade. Proceedings of the Sixteenth Biennial Conference of the International Institute of Fisheries Economics and Trade, July 16–20, Dar es Salaam, Tanzania
Oniang'o R, Mukudi E (2002) Nutrition and gender. In nutrition: a foundation for development. Geneva: ACC/SCN. https://www.unscn.org/files/Publications/Briefs_on_Nutrition/Brief7_EN.pdf. Accessed 20 Jul 2019
Rahman S, Akter S (2010) Determinants of livelihood security in poor settlements in Bangladesh. NAF International Working Paper Series, No 10/01
Rajee O, Mun ATK (2017) Impact of aquaculture on the livelihoods and food security of rural communities. Int J Fish Aquat Stud 5(2):278–283
Shanthi B, Krishnan M, Chandrasekaran VS (2010) Socio-economic and gender analysis in aquaculture. CIBA Spec Publication No 48:59

Stacey N, Gibson E, Loneragan NR, Warren C, Wiryawan B, Adhuri D, Fitriana, R (2019) Enhancing coastal livelihoods in Indonesia: an evaluation of recent initiatives on gender, women and sustainable livelihoods in small-scale fisheries. Maritime Studies pp 1–13

The Economist Group (2018) Global Food Security Index 2018: Building Resilience in the Face of Rising Food-Security Risks. A report from The Economist Intelligence Unit. p 47

United Nations (2014) World survey on the role of women in development: gender equality and sustainable development. A Report of the Secretary-General, United Nations General Assembly(A/69/156). p 119

United Nation Entity for Gender Equality and the Empowerment of Women (2014) Gender Mainstreaming. https://www.un.org/womenwatch/osagi/gendermainstreaming.htm. Accessed 20 Dec 2019

United Nations Children's Emergency Fund (UNICEF) (2019) Education: The Challenge. https://www.unicef.org/nigeria/education. Accessed 20 Dec 2019

United States Agency for International Development (USAID) (2019) Global Health-Nigeria. https://www.usaid.gov/nigeria/global-health. Accessed 11 Nov 2019

Weeratunge N, Béné C, Siriwardane R, Charles A, Johnson D, Allison EH, Nayak PK, Badjeck MC (2014) Small-scale fisheries through the wellbeing lens. Fish Fish 15(2):255–279. https://doi.org/101111/faf12016

Weeratunge-Starkloff N, Pant J (2011) Gender and aquaculture: sharing the benefits equitably. The WorldFish Center, Penang, Malasia, Issues Brief 2011–32:12

Assessing the Household Preference Level for Sustainable Clean Cooking Energy in Lagos State, Nigeria: Case Study of Biofuel

Olaniran Anthony Thompson, Daniel Akinrinlola Olawamide, and Mosunmola Lydia Adeleke

Abstract There is need to assess household preference level for sustainable clean cooking energy such as biofuel gel to achieve Sustainable Development Goals (SDGs) 7 of affordable and sustainable clean energy. The study determines the factors affecting respondents' awareness, identified factors influencing the preference level and factors influencing respondents' willingness to pay for biofuel gel in Nigeria. Data were obtained from one hundred and eighty biofuel gel users and one hundred and eighty non-users leading to a total of three hundred and sixty respondents' through a multi-stage sampling technique. Descriptive statistics, 5-likert scale type, probit regression model, tobit regression model and double-bound logit were used to analyze the data collected. The result of the probit regression model revealed that age, years of consumption of biofuel gel, marital status, knowledge of renewable energy and familiarity with greenhouse gas were the variables affecting the respondents' awareness of biofuel gel. The tobit regression model result revealed that years of consumption of biofuel gel, household size and price of biofuel gel, income, familiarity with climate change and gender were the variables influencing the preference level of biofuel gel. The result of the double-bound logit revealed that bid amount, years of education, household size, income and awareness of biofuel gel were variables influencing the respondents' willingness to pay for biofuel gel in the study area. The study recommends that government at all levels should sensitize the households on the usage of biofuel gel in Nigeria to achieve goal seven of SDGs.

Keywords Clean cooking energy · Biofuel gel · Sustainable environment · Perception · Tobit regression model · Nigeria

O. A. Thompson (✉) · D. A. Olawamide
Department of Agricultural and Resource Economics, Federal University of Technology, P. M. B. 704, Akure, Nigeria
e-mail: oathompson@futa.edu.ng

M. L. Adeleke
Department of Fisheries and Aquaculture Technology, Federal University of Technology, P. M. B. 704, Akure, Nigeria
e-mail: mladeleke@futa.edu.ng

W. Leal Filho et al. (eds.), *Sustainable Development in Africa*, World Sustainability Series,
https://doi.org/10.1007/978-3-030-74693-3_20

1 Introduction

A large proportion of the world's population including most of the developing world generate the energy they need for cooking from solid fuels. Even though most of these fuels are known to pose a real threat to human health (Perera 2018). According to World Health Organization (WHO), "more than 40% of the world's population—which includes 1 billion children under 15—is exposed to high levels of household air pollution from mainly cooking with polluting technologies and fuels." In developing countries, women frequently rely on coal and biomass fuels for cooking and heating, putting them and their children at higher risk to the effects of home pollutants (United Nation (UN) Climate Change Summit 2019). Particularly in the developing countries, approximately three billion people cook using polluting open fires or simple stoves fueled by kerosene, solid fuels such as biomass, charcoal and animal waste for their daily cooking purposes (WHO 2018). These fuels are combusted in poorly designed and inefficient cookstoves resulting in emission of noxious gases and products of incomplete combustion (Makonese et al. 2018), which are the major human source of greenhouse gases and spurs climate change by releasing carbon dioxide into the atmosphere (Perera 2018). In turn, climate change damages agricultural production and subsequently threatens the nutritional health of human beings (Muller and Yan 2018).

The continued use of solid fuels has been linked to increased morbidity and mortality (Makonese et al. 2018). It has a negative effect on the atmosphere and consequently people's lives (Nlom and Karimov 2014). The World Health Organization (WHO) estimates that over 4 million people die prematurely from illnesses caused by household air pollution from inefficient cooking practices using polluting stoves paired with solid fuels and kerosene (Neira and Ramanathan 2020). In 2016, household air pollution was responsible for 3.8 million deaths and 7.7% of the global mortality (WHO 2018). Most of these deaths occur in low and middle-income countries (LMIC) and are attributable primarily to pneumonia in children and non-communicable cardiorespiratory diseases in adults. Access to clean, affordable and efficient energy has become a challenge for the majority of the low to medium income households in developing countries (Kimemia and Annegarn 2012; Makonese et al. 2018).

Household fuel transition from dirty fuels towards clean fuels in the developing countries has received growing research attention, due to the recent concerns about the negative externalities of the use of traditional fuels on the environment and health (Muller and Yan 2018). These challenges coupled with the degradation of global environment caused by the Greenhouse Gas (GHG) emissions and the prediction of the depletion of the fossil fuel resources by scientists have all encouraged the global community to search for alternative sustainable (renewable) and environmentally-friendly energy resources (Kurnia et al. 2016). Amidst other alternative energy sources, biofuel have emerged and earned more attention across the world because biofuels are considered the most sustainable (renewable) and environment friendly energy source (Demirbas 2010), and have been adopted by a large population of the developed world. One of these biofuels is the biofuel gel or ethanol gel, a cooking fuel which is been touted as a suitable replacement for firewood and kerosene (Dioha et al. 2012; Lloyd 2015).

However, depending on the price of fossil fuels, biofuels is relatively more expensive than the fossil fuels, due to the high costs of production and purification (Dioha et al. 2012; Ajala et al. 2015), and they will have to be marketed at higher prices (Sivashankar et al. 2016). Therefore, developing biofuels market has been strongly connected to intervention from the government. This intervention can be applied to both the supply and the demand side, providing incentives to producers or conversely, promoting the use of biofuels by the consumers (Goldemberg and Guardabassi 2009; Van der Horst and Vermeylen 2010). Direct market support could also be achieved if consumers value the benefits provided by biofuels. Thus, understanding whether consumers will be willing to pay these higher prices and why, is a key issue that should be considered when designing policies to increase biofuel gel use in Nigeria and other Sub-Sahara African Countries.

2 Biofuel (Renewable Energy) Energy and Conventional Energy Used for Cooking in Nigeria

Plate 1a shows the picture of the conventional cooking stove using kerosene used by an average Nigeria in a middle-class family. Plate 1b shows the picture of charcoal pot used by some family also to cook in Nigeria and other parts of Africa. While Plate 1c shows the common cooking method using firewood in rural Nigeria and other rural parts of Africa. Plate 2a shows the source of cooking energy for the firewood cooking method which is firewood from the forest. This is one of the courses of deforestation in Nigeria and Sub-Sahara Africa (SSA). Plate 2b shows the coal which is the source of energy for the charcoal pot. Likewise, the coal is gotten from burning wood. Plate 3a is the picture of biofuel stove and Plate 3b shows the biofuel gel which is the source of energy for biofuel stove.

(a) (b) (c)

Plate 1 **a** Kerosene Stove. **b** Charcoal Pot. **c** Firewood Cooking

(a) (b)

Plate 2 **a** Firewood for Cooking. **b** Charcoal Used in Cooking

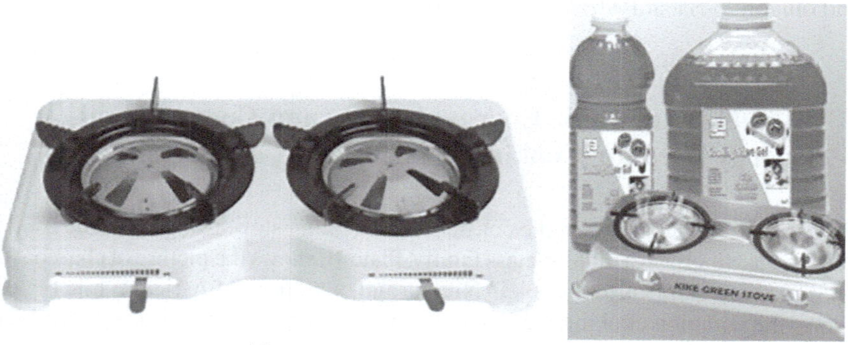

(a) (b)

Plate 3 **a** Biofuel Stove. **b** Biofuel Gel *Source* https://images.search.yahoo.com/

3 Materials and Methods

3.1 Study Area

The study was conducted in Lagos state Nigeria in 2019. Lagos state is in southwest Nigeria on the coastal plain of the Blight of Benin. There are 20 Local Government Areas (LGAs) in Lagos state. The state lies on the geographical coordinates of 6°35′N, 3°45′E, the state is bounded in the north and east by Ogun State. In the west, it shares boundaries with the Republic of Benin and stretches over 180 km along the coast of the Atlantic Ocean in the south. (National Population Commission (NPC) 2007). The state has a land area of 3,577 km^2. (NPC 2007). According to Lagos state government the population of Lagos state resident is approximately 16 million as at 2015 and according to projections will hit 30 million by 2025. As described by administrative division, Nigerian congress, Lagos State is arguably the most economically important state of Nigeria, it is the nation's largest urban area. For decades Lagos has been the epicenter of Nigeria's economic and social development and continues to be, it is a major financial center and would be the fifth-largest economy in Africa if it were a country (Ekundayo 2013).

3.2 *Sample and Sampling Techniques*

The population for this study were both biofuel gel users and non-biofuel gel users. A Multi-stage sampling procedure was used in the selection of respondents for this study. The first stage involved the purposive selection of three (3) out of the twenty Local Government Areas (LGAs) in Lagos state noted for biofuel gel production. The LGAs are Eti-osa, Lagos mainland and Ikeja. In the second stage, three (3) communities from each of the LGA were randomly selected. The third stage involved the purposive selection of 20 households that were using biofuel gel by snowball technique and random selection of 20 households that were not using from each of the community. This led to a total of 360 respondents divided into 180 biofuel gel users and 180 non-biofuel gel users.

3.3 *Nature and Source of Data*

Primary data were used for this study. Primary data were obtained through the administration of well-structured questionnaire. The questionnaire was used to obtain relevant information on numbers of variables such as socio-economic profile of the selected biofuel gel users and non-biofuel gel users such as the age of consumers, household size, education, gender and occupation, willingness to pay criteria, their knowledge on biofuel gel and estimated annual income.

3.4 *Data Analysis and Model Specification*

The data collected for this study were analyzed using the following methods: Descriptive statistics, 5-likert scale type, Probit Regression Model (PRM), Tobit Regression Model (TRM) and double-bound logit.

3.5 Socio-economic Characteristics

Descriptive statistics such as frequency distributions, percentages, and mean distributions were used to describe the socio-economic characteristics of the respondents in the study area.

3.6 Respondents Awareness of Biofuel Gel

Probit model was used to determine the factors affecting respondents' awareness of biofuel gel in the study area. Probit model is appropriate when the response takes one or two possible outcomes.

This is expressed as

$$P_i[Y = 1] = [Fx_i] \quad \text{(Gujarati 2003)}$$

Where

$$Y = \beta_0 + \beta_i X_i \tag{1}$$

Explicitly,

$$Y = \beta_0 + \beta_1 X_1 + \beta_2 X_2 + \beta_3 X_3 + \beta_4 X_4 + \ldots + \beta_{10} X_{10} + U \tag{2}$$

Where,

Y_i = Response with respect to awareness of biofuel gel. (Dichotomous variable 1 = yes; 0 = no)
β_i = vector of unknown coefficients
X_i = Independent variables

The independent (explanatory) variables that was used in examining the awareness of respondents for biofuel gel were as specified below:

X_1 = Age of consumer (years)
X_2 = Gender of the consumer (male = 1, female = 0)
X_3 = Consumption of biofuel (years)
X_4 = Marital status (married = 1, 0 otherwise)
X_5 = Years of formal education (years)
X_6 = Household size (number of persons)
X_7 = Estimated monthly income (₦)
X_8 = Main occupation of consumer (Government Staff = 0, Private Staff = 1, Self Employed = 3)
X_9 = Knowledge of renewable energy (1 = Yes, No idea = 0)
X_{10} = Familiarity with climate change (1 = Yes, No idea = 0)

X_{11} = Familiarity with Green House Gas (GHG) (1 = Yes, No idea = 0)
X_{12} = Membership of environmental association
U_i = Error term.

3.7 Respondents Preference Level of Biofuel Gel

Tobit regression model was used to examine the preference level of respondents for biofuel gel in the study area. The level of preference was measured using a scale of 1–5 (not preferred = 1, least preferred = 2, indifferent = 3, preferred = 4, most preferred = 5) based on if the consumers prefer the products or not. The following attributes were measured using the above scale: price, heat intensity, clean cooking, emission, accessibility, easy to light up, price stability. The Tobit regression model is specified thus;

$$Y_i^* = X_i \beta + U_i \quad (3)$$

$$U_i = N(0, \delta^2)$$

Where

Y_i^* = latent variable representing levels of preference for biofuel gel
X_i = vector of consumers' characteristics relevant in explaining the levels of preference for biofuel gel
β = vector of parameters to be estimated
U_i = normally distributed error term

Where

Y_i = the dependent variable that was used in examining the preference level of consumers for biofuel gel
$Y_i^* = y_i$ if $0 < y_i < 1$ (for those that prefer based on biofuel attributes)
$Y_i^* = y_i$ if $y_i = 0$ (for those that do not prefer)

The independent (explanatory) variables that were used in examining the preference level of consumers for biofuel gel were as specified below:

X_1 = Age of Consumer (years)
X_2 = Consumption of biofuel gel (years)
X_3 = Marital Status (married = 1, 0 otherwise)
X_4 = Gender of the consumer (male = 1, female = 0
X_5 = Year of Formal Education (years)
X_6 = Household size (number of persons)
X_7 = Estimated Annual income (₦)
X_8 = Main Occupation of consumer (Government Staff = 0, Private Staff = 1, Self Employed = 3)

X_9 = Price of products (₦/Litre)
X_{10} = Familiarity with GHG (1 = Yes, No idea = 0)
X_{11} = Familiarity with climate change (1 = Yes, No idea = 0)
X_{12} = Knowledge on renewable energy (1 = Yes, No idea = 0)
X_{13} = Membership of Environmental Association
U_i = Error term

3.8 Respondents Perception of Biofuel Gel

The likert-type scale was used to determine the perception of biofuel gel users' and non-users' in the study area. Respondents were asked to respond to statements relating to the economic attraction of biofuel gel usage, using Strongly Agreed (SA), Agreed (A), Indifferent (I), Disagreed (D), Strongly Disagreed (SD). The responses were scored as 5, 4, 3, 2 and 1 for SA, A, I, D, and SD respectively. The mean from each statement was obtained and used to classify the responses on each statement into SA (>4.5), A (3.5–4.4), I (2.5–3.4), D (1.5–2.4), SD (<1.5). The grand means for all the statement were calculated to be able to place the responses on a scale that enable a conclusion to be drawn on what the perception of the respondents is on biofuel gel.

3.9 Factors Influencing Respondents' Willingness to Pay for Biofuel Gel

Double-bounded logit model was used to analyze factors influencing respondents' willingness to pay for biofuel gel. Following Hanemann et al. (1991), the following response probabilities were obtained for the logit model.

$$P_i^{NN} = 1 - 1/\left(1 + e^{-\alpha+\beta \text{ LOW BID}}\right) \quad (4)$$

$$P_i^{YN} = 1/\left(1 + e^{-(\alpha+\beta \text{ HIGH BID})}\right) - 1/\left(1 + e^{-(\alpha+\beta \text{ FIRST BID})}\right) \quad (5)$$

$$P_i^{NY} = 1/\left(1 + e^{-(\alpha+\beta \text{ FIRST BID})}\right) - 1/\left(1 + e^{-\alpha+\beta \text{ LOW BID}}\right) \quad (6)$$

Where,

FIRST BID—Starting bid Value
LOW BID—Follow-up Lower bid value
HIGH BID—Follow-up Higher bid Value

The double-bound log-likelihood function is

$$L^{DB} = \sum I_i^{YY} \log P_i^{YY} + \sum I_i^{YN} \log P_i^{YN} + \sum I_i^{NY} \log P_i^{NY} + \sum I_i^{NN} \log P_i^{NN} \quad (7)$$

$$i = 1 \ldots n$$

Where Li, indicates the response category of each respondent i

With respect to four possible outcomes of dichotomous choice the double-bounded logit.

The model is explicitly stated as:

$$L_n \frac{WTP}{1 - WTP} = \beta_0 + \beta_1 X_1 + \beta_2 X_2 + \beta_3 X_3 + \beta_4 X_4 + \beta_5 X_5 + \beta_6 X_6 + \beta_7 X_7 + \beta_8 X_8 + \beta_9 X_9 + \beta_{10} X_{10} + \beta_{11} X_{11} + \beta_{12} X_{12} + \beta_{13} X_{13} + \varepsilon_i \quad (8)$$

X_1 = BID amount (₦)
X_2 = Gender of the consumer (male = 1, female = 0)
X_3 = Age of consumer (years
X_4 = Consumption of biofuel (years)
X_5 = Year of formal education (years)
X_6 = Marital status (married = 1, 0 otherwise)
X_7 = Household size (number of persons)
X_8 = Estimated annual income (₦)
X_9 = Awareness of the product (Aware = 1, Otherwise = 0
X_{10} = Main occupation of consumer (Government Staff = 0, Private Staff = 1, Self Employed = 3)
X_{11} = Knowledge on renewable energy
X_{12} = Familiarity with GHG
X_{13} = Familiarity with climate change
X_{14} = Membership of environmental association
μ = Error term

4 Results and Discussion

4.1 Socioeconomic Characteristics of the Respondents

The mean age of biofuel gel users was 43.5 years while that of the non-biofuel gel users was 46.6 years. This showed that biofuel gel users were younger than non-biofuel gel users. This result is in line with the study carried out by Gracia et al. (2011) on "Consumers' willingness to pay for biodiesel in Spain" where younger respondents preferred and were willing to pay for biofuel. This implies that younger people were more likely to prefer and pay for biofuel gel in the study area. As shown in Table 1, the gender distribution of the respondents shows that 42.8% of the biofuel gel users were male and 57.2% of them were female, while 55.6% of the non-biofuel gel users were male and 44.4% of them were female. This result is in line with Rahut et al. (2014) who opined that female-headed households prefer modern fuels to traditional fuels. This may be since females are more likely to be the decision-maker in the choice of cooking fuel to be used at home.

Also, from Table 1, the majority (76.7.0%) of the biofuel gel users and 73.4% of non-users of biofuel gel were married. This implies that most of the respondents were matured and responsible, had a family to maintain and have a clear knowledge of their wellbeing. Hence the marital status of respondents is believed to influence their decision-making ability. Again, from Table 1, most respondents (biofuel gel users, 85.0% and non-users, 64.4%) have ten years formal educational. The mean years of schooling were 16.5 years and 14.5 years for both biofuel gel users and non-biofuel gel users respectively. This implies that most of the respondents had a formal education. This may be attributed to the fact that the study was carried out in the urban populace where the majority of the populace were highly educated and these set of people were also willing to respond to the research upon administering questionnaires. Education has always played a huge role in increasing awareness of technology and consequently the welfare of the people (Obayelu et al. 2014). It is expected that education would provide people with opportunities to access information and understand the benefits of renewable fuel such as biofuel gel. This corroborates the fact that high level of education will raise the awareness of the benefit that renewable energy inherits (Yuan 2011; Hast et al. 2015). This implies that the year of formal education by the respondents might influence the preference and willingness to pay for biofuel gel.

The household size comprises of the head, wife or wives, children and other dependents that live under the same roof (Adekunle et al. 2016). The mean household size for users of biofuel gel was four persons per house while that of the non-user of biofuel gel was six persons per house suggesting an urban moderate family size (Obayelu et al. 2014). This simply means that biofuel gel users had lesser family size than non-biofuel gel users. This implies that respondents with small household size may prefer and be willing to pay for biofuel gel. Furthermore, the result in Table 1 reflected more private staffs (35.6%) among the biofuel gel users while government staffs (37.8%) were more among the biofuel gel non-users. Also, a lower proportion of 30.6% and 26.7% of self-employed individuals were found to be users and non-users of biofuel gel respectively. This shows that those who work in the private sector and who were self-employed appreciates biofuel gel in the study area.

The mean estimated income for users of biofuel gel was ₦253,811.76 ($705.03) while that of the non-users was ₦159,356.40 ($442.66). This implies that the biofuel gel users earned more than non-biofuel gel users, this could imply that biofuel gel users had the financial endowments to afford biofuel gel. This finding is in accordance with the findings of Sivashankar et al. (2016) who reported that higher-income earners are more likely to pay a higher price for biofuel. This might be due to the assumption that people with high income will have extra money to spend on renewable energy (Liu et al. 2013). Again, from Table 1, the mean year of consumption of biofuel gel by

Table 1 Socioeconomic characteristics of respondents

Variables	Biofuel users		Non-users	
Age (Years)	**Frequency**	**Percentage**	**Frequency**	**Percentage**
Less than 20	–	–	–	–
21–30	12	6.7	26	14.4
31–40	60	33.3	40	22.2
41–50	70	38.9	26	14.4
51–60	26	14.4	86	47.9
Above 60	12	6.7	2	1.1
Total	**180**	**100.0**	**180**	**100.0**
Sex				
Male	77	42.8	100	55.6
Female	103	57.2	80	44.4
Total	**180**	**100.0**	**180**	**100.0**
Marital status				
Single	21	11.6	34	18.9
Divorced	9	5.0	8	4.4
Widow	12	6.7	6	3.3
Married	138	76.7	132	73.4
Total	**180**	**100.0**	**180**	**100.0**
Years of formal education				
1–5	–	–	16	8.9
6–10	27	15.0	48	26.7
> 10	153	85.0	116	64.4
Total	**180**	**100.0**	**180**	**100.0**
Household size				
Less than 5	137	76.1	94	52.2
5–10	43	23.9	86	47.8
Total	**180**	**100.0**	**180**	**100.0**
Main occupation				
Public Sector	61	33.8	68	37.7
Private Sector	64	35.6	64	35.6
Self Employed	55	30.6	48	26.7
Total	**180**	**100.0**	**180**	**100.0**
Monthly income				
< 100,000	10	5.6	60	33.3
101,000–200,000	28	15.5	70	38.9
201,000–300,000	76	42.2	34	18.9
301,000–400,000	48	26.7	14	7.8
401,000–500,000	18	10.0	2	1.1

(continued)

Table 1 (continued)

Variables	Biofuel users		Non-users	
> 500,000	–	–	–	-
Total	**180**	**100.0**	**180**	**100.0**
Years of consumption				
<5	66	35.3		
6–10	114	64.7		
Total	**180**	**100.0**		

Source Computed from Field Survey Data, 2019

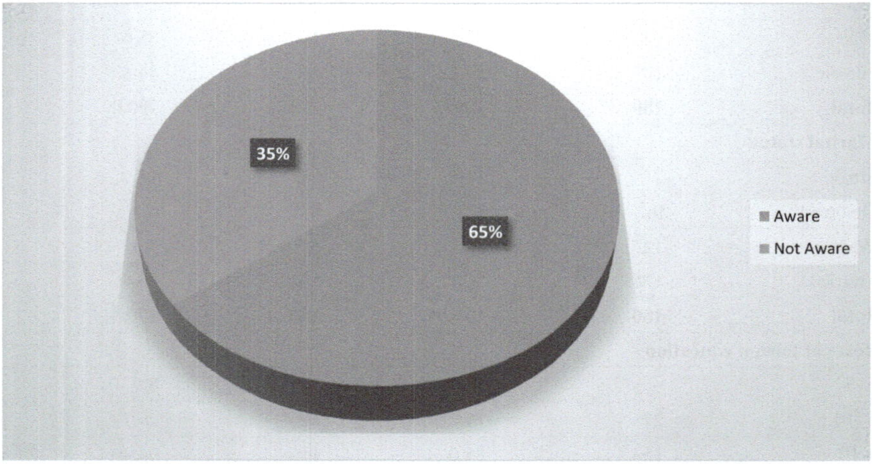

Fig. 1 Distribution of respondents according to the awareness of biofuel gel. *Source* Computed from Field Survey Data, 2019

the users was 9.89 years in the study area. So, biofuel gel users are not novice in the usage of the products, hence, they can reasonably compare it with the conventional cooking fuel such as kerosene, charcoal and firewood.

Figure 1 revealed that larger proportion (65.0%) of the respondents in the study area were aware of the product while 35.0% of the respondents were not aware of the product. The implication of this is that those who do not have information on the product are likely not to know the value of biofuel gel product and this may negatively influence their willingness to pay for the product.

Table 2 shows how information about biofuel gel was been disseminated to the respondents in the study area. From Table 2, media was responsible for 26.6% of the awareness of biofuel gel according to the respondents. Sale agents (21.2%) were ranked second as the most important channels of information by the respondents in the study area. Magazine was the least (1.8%) commonly used sources of information

Table 2 Distribution of respondents by source of awareness

Source of awareness	Percentage	Rank
Media	26.6	1st
Sales agents	21.2	2nd
Friends/Relatives	18.6	3rd
Others	11.5	4th
Markets/Stores	10.6	5th
Academic studies	9.7	6th
Magazine	1.8	7th
Total	**100.0**	

Source Computed from Field Survey Data, 2019

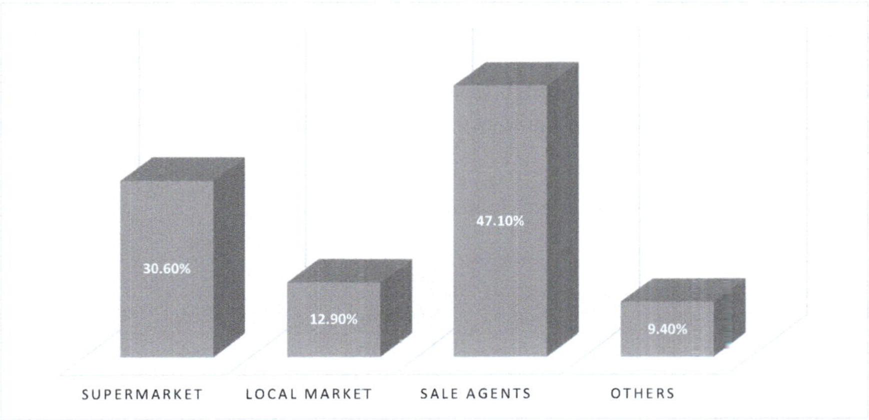

Fig. 2 Distribution of respondents' based on the source of purchase of biofuel gel *Source* Computed from Field Survey Data, 2019

on biofuel gel. Other sources of information as regards the awareness of the product include seminars and online store.

The source of purchase of the biofuel gel are shown in Fig. 2. The figure revealed that 47.1% of the users claimed they purchased the product from sales agents and 30.6% from the supermarket, 12.9% of the users purchased theirs from the local market. 9.4% of the respondents purchase theirs from other sources which include: online stores and retailer shops.

4.2 *Determinants of Respondents' Awareness Level of Biofuel Gel*

Table 3 shows the determinants of respondents' awareness level of biofuel gel in the study area. The loglikelihood ratio chi-square of 195.16 obtained in the study implies that parameters included in the probit model were significantly different from zero at the 1 per cent level of significance. The likelihood function was −16.18 indicating the model had strong explanatory variables that influenced the awareness of biofuel gel in the study area. A total of twelve explanatory variables were included in the model, out of which five of the variables influenced the awareness level of biofuel gel among the respondents in the study area. The age of the respondents negatively influenced the awareness level of biofuel gel in the study area and significant at 5%. The negative coefficient indicates that a year increase in the age of the respondents reduces the probability of awareness level of biofuel gel in the study area. This could because younger respondents may likely know about biofuel gel than older respondents, because younger people have more opportunities to receive information on environmental education through various communication channels such as social media. This is in line with the findings of Karytsas and Theodoropoulou (2014); Du

Table 3 Probit estimates of consumers' awareness of biofuel gel

Variables	Coefficients	Std. Error	P > /z/
Age of consumers	−0.1014436	0.0462875	0.028**
Gender of consumers	0.0610611	0.5937873	0.918
Consumption years	1.585475	0.5023544	0.002***
Marital status	0.5867187	0.3452763	0.089*
Year of education	−0.1003058	0.1682796	0.551
Household size	−0.4279081	0.2678604	0.110
Monthly income of consumers	1.20e-06	4.55e-06	0.792
Main occupation of consumers	0.1339501	0.3569151	0.707
Knowledge on renewable energy	1.958124	0.7858179	0.013**
Familiarity with climate change	−0.4041885	0.6664572	0.544
Familiarity with GHG	1.812632	0.7344129	0.014**
Membership of environmental association	−0.2690037	1.201424	0.823
Constant	3.377017	3.060351	0.270
Number of observations	175		
Log likelihood	−16.18105		
LR chi^2 (12)	195.16		
Pseudo R^2	0.8578		
Prob > chi^2	0.0000		

Source Computed from Field Survey Data, 2019
***Significant at 1%, **Significant at 5%, *Significant at 10%

et al. (2018) who stated that environmental awareness had a negative correlation with the respondents' age.

At 1% level of significance, the coefficient of years of consumption of biofuel gel by the respondents in the study area had a positive and significant influence on their awareness level of biofuel gel. Implying that a year increase in consumption of biofuel gel will probably increase the awareness level of biofuel gel in the study area. This implies that respondents that had been using biofuel gel for years tend to be more aware of the product. Marital status was statistically significant at 10% level of significance with a positive coefficient. The positive coefficient which was in favor of the married means respondents' who were married were more likely to be aware of biofuel gel than others who were either single, widow or divorced. It simply means most of the respondents were mature and responsible, had a family to maintain and have a clear knowledge of their wellbeing. This may be because the marital status of respondents is believed to affect their decision-making ability.

At 5% level of significance, the coefficient of knowledge of renewable energy had a positive and significant influence on the level of awareness. This implies that a unit increase in the knowledge of renewable energy will probably increase the awareness level of biofuel gel product in the study area. This result confirms the findings of Sivashankar et al. (2016) who established that awareness of renewable energy has a positive association with the awareness level of the respondents. Furthermore, at 5% level of significance, the coefficient of familiarity with GHG had a positive and significant influence on their level of awareness. This implies that a unit increase in the knowledge of GHG will probably contribute to the awareness level of biofuel gel product in the study area. This result also agrees with the findings of Sivashankar et al. (2016) who established that familiarity of GHG has a positive influence on respondents' awareness level of biofuel.

4.3 *Consumer Preference Level for Biofuel gel based on Selected Attributes*

Table 4 shows the preference level of biofuel gel based on selected attributes (price, heat intensity, clean cooking, emission, price stability, ease to light up and accessibility). It was revealed that the product is mostly preferred by the respondents based on its emission (environmental benefits) with preference index of 0.59, followed by its clean cooking, price stability and heat intensity with each having the same index of 0.52 respectively. This implies that the respondents in the study area mostly prefer biofuel gel base on the following attributes (emission, clean cooking, heat intensity, price stability, easy to light up) when compared to conventional fuel. Findings indicate that the biofuel gel attribute has a positive effect on consumers' utility. This is in line with the findings of Giraldo et al. (2010) who stated that consumer's preference and willingness to pay extra for biofuel is influenced by the positive attributes of biofuel. However, biofuel gel product is least preferred because of its high price

Table 4 Distribution of respondents preference level based on selected attributes of biofuel gel

Selected attributes	MP (5)		P (4)		I (3)		LP (2)		NP (1)		Preference index
	F	%	F	%	F	%	F	%	F	%	
Price	6	5.3	21	18.6	14	12.4	61	54.0	11	9.7	0.37
Heat intensity	8	7.1	84	74.3	3	2.7	11	9.7	7	6.2	0.52
Clean cooking	15	13.3	65	57.5	20	17.7	6	5.3	7	6.2	0.52
Emission	35	31.0	64	56.6	9	8.0	3	2.7	2	1.8	0.59
Price stability	12	10.6	69	61.1	15	13.3	13	11.5	4	3.5	0.52
Ease to light up	15	13.3	65	57.5	14	12.4	13	11.5	6	5.3	0.52
Accessibility	2	1.8	37	32.7	35	31.0	35	31.0	4	3.5	0.43

Source Computed from Field Survey Data, 2019
Key: MP—Most Preferred, P—Preferred, I—Indifferent, LP—Least Preferred, NP—Not Preferred, %—Percentage, F—Frequency

and for being not easily accessible with an index of 0.37 and 0.43. This result also relates to findings of Giraldo et al. (2010) that the main reason for respondents not buying biofuel was proximity. This means that respondents would consider changing to biofuel if it were available in a wider range (Chatterton et al. 2016).

4.4 Determinants of Respondents' Preference for Biofuel Gel

As shown in Table 5, the determinants of preference for biofuel gel was analyzed using tobit regression model. The years of consumption of biofuel gel, household size and price of the product were significant at 1%, income and familiarity with climate change were significant at 5% while gender was significant at 10%. An additional increase in the year of consumption of biofuel gel increases respondents' preference level for biofuel gel by 7.80%. This implies that an increase in the years of consumption of biofuel gel had a positive influence on the preference for biofuel gel by the respondent in the study area. These means respondents that have been consuming biofuel gel for a longer period tends to have a higher preference for the product. Household size also plays an important role in energy choices. Israel (2002); Abebaw (2007); Zhang and Kotani (2012) all find evidence of negative association of household size with per capita energy consumption. Household size of respondents had a negative coefficient of −0.2640922 and it is statistically significant at 1%. This means a unit increase in the household size will reduce their preference for biofuel gel which is regarded as a clean fuel by 26.41%. This is in line with the assertation of Reddy (1995); Ouedraogo (2006); Rao and Reddy (2007); Pandey and Chaubal (2011); Özcan et al. (2013); Danlami et al. (2017) indicate that larger households prefer dirty (conventional) fuels over clean (modern) fuels. One possible reason is

Table 5 Tobit estimates of the preference level for biofuel gel

Variables	Coefficient	Z-Value	Std. Err
X_1 age of respondents	−0.0068807	−0.20	0.034722
X_2 years of consumption of biofuel	0.0779668	3.66***	0.0212939
X_3 marital status	−0.033924	−1.22	0.0278207
X_4 gender	−0.106909	−1.87*	0.0573236
X_5 year of education	−0.1231847	−1.35	0.0913249
X_6 household size	−0.2640922	−3.24***	0.0815712
X_7 monthly income	0.0685959	2.04**	0.0336148
X_8 main occupation	0.0055313	0.16	0.0355972
X_9 price of the product	−0.1438837	−5.68***	0.0253138
X_{10} familiarity with GHG	−0.0375243	−0.58	0.0642327
X_{11} familiarity with climate change	0.13019	2.05**	0.0634401
X_{12} knowledge of renewable energy	−0.0051298	−0.04	0.1175727
X_{13} member of environmental association	−0.0462873	−0.45	0.1029485
Constant	1.000438	4.53***	0.2208262
Number of observations	113		
Log likelihood	−27.257326		
LR chi^2 (13)	104.74		
Pseudo R^2	0.6577		
Prob > chi^2	0.0000		

Source Computed from Field Survey Data, 2019
***Significant at 1%, **Significant at 5%, *Significant at 10%

that household size is often larger in poorer households that cannot afford modern fuels.

The price of the product was statistically significant from zero at 1% level of significance and had a negative influence on the preference level for biofuel gel in the study area. This indicates that a unit increase in the price of the product will lead to 14.39% decrease in the preference level of biofuel gel in the study area. This is in line with the findings of Gracia et al. (2011); Zhang and Kotani (2012); Lay et al. (2013); Nlom and Karimov, (2014) that an increase in the price of a cooking fuel significantly reduces the probability of choosing and using such cooking fuel respectively. At 5% level of significance, the coefficient of familiarity with climate change had a positive and significant influence on preference level. The positive coefficient of 0.13019 implies that a unit increase in the knowledge of climate change will contribute to the increase in preference level for biofuel gel by 13.02% in the study area. This result also agrees with the findings of Sivashankar et al. (2016) who established that familiarity with climate change has a positive influence on respondents' preference level for biofuel.

With respect to household income, this implies a unit increase in household income will lead to 6.86% preference level for biofuel gel in the study area. Revealing that income is a major determinant of biofuel gel consumption among the households in the study area. This implies that as households' disposable income increases, more money may be devoted to the purchase and consumption of clean fuel. Numerous studies point to income as the major driver behind the usage of clean (modern) fuels. This is in line with the study by Lay et al. (2013) carried out in Kenya that rising income of the households increased the patronage of electricity and solar energy compare to using wood and kerosene for cooking. Baiyegunhi and Hassan (2014) show in rural Nigeria that the transition from dirty fuel to clean fuel occurs along to rising income. All these studies seem to corroborate the energy ladder concept, which emphases income in explaining the transition from 'inferior' traditional fuels to 'normal' modern fuels (Muller and Yan 2018).

Gender was statistically significant at 10% probability level with a negative coefficient. The negative coefficient of −0.106909 which was in favor of the female gender means female-headed household are more likely to prefer biofuel gel as their source of fuel. This corresponds with findings of Farsi et al. (2007); Rao and Reddy (2007; Mensah and Adu (2013); Rahut et al. (2014); Danlami et al. (2017) that female-headed households prefer modern (clean) fuels to traditional (dirty) fuels. This may be attributed to the fact that women are often responsible for household cooking and thus they are directly affected by the air pollution emitted from the burning of the dirty fuels.

4.5 Respondents Perception of Biofuel Gel

The respondents' perception of the benefits, value and the environmental advantages of consuming biofuel gel was generally positive. Drawing an inference from the class boundaries of means as shown in Table 6, the respondents agreed that biofuel gel was the most appropriate source of energy when compared to conventional fuels with a mean score of 3.7. The respondents agreed that biofuel gel when used for cooking, do not dirty cooking gadgets when compared to conventional fuel with a mean score of 4.0. The respondents disagreed that biofuel gel was easily accessible compared to conventional fuel with mean score of 2.4. However, with mean score of 3.2, the respondents were indifferent in their perception about biofuel gel faster than conventional fuel when used for cooking. The respondents agreed that biofuel gel was an important source of energy with the mean score of 3.9 but were indifferent that its supply was reliable in the study area.

Equally, the respondents agreed that biofuel production could be done sustainably, and that biofuel gel use was good for the environment with mean scores of 4.1 and 4.4 respectively. Respondents show a positive perception toward the use of biofuel gel because they mostly agree that biofuel gel was an appropriate source of household cooking, do not dirty cooking gadgets and good for the environment. The implication of these was that consumers were very concerned about their health, their choice of

Table 6 Respondents perceptions of biofuel gel consumption

Perception statement	Strongly Agree (5)		Agree (4)		Indifferent (3)		Disagree (2)		Strongly Disagree (1)		Mean score	Inferences based on mean class
	Freq	%	Freq	%	Freq	%	Freq	%	Freq	%		
Biofuel gel is the most appropriate source of energy when compared to conventional fuel	17	15.0	66	58.4	15	13.3	14	12.4	1	0.9	3.7	Agreed
Biofuel gel when used during cooking do not dirty cooking gadgets compared to conventional fuel	17	15.0	82	72.6	12	10.6	2	1.8	–	0.0	4.0	Agreed
Biofuel gel is easily accessible compared to conventional fuel	–	0.0	21	18.6	13	11.5	63	55.8	16	14.2	2.4	Disagreed
Biofuel gel is faster than conventional fuel when used for cooking?	10	8.8	69	61.1	20	17.7	13	11.5	1	0.9	3.2	Indifferent

(continued)

Table 6 (continued)

Perception statement	Strongly Agree (5)		Agree (4)		Indifferent (3)		Disagree (2)		Strongly Disagree (1)		Mean score	Inferences based on mean class
	Freq	%	Freq	%	Freq	%	Freq	%	Freq	%		
Do you think biofuel is an important source of energy?	11	9.7	85	75.2	14	12.4	1	0.9	2	0.8	3.9	Agreed
Is the supply of the biofuel energy reliable?	4	3.5	60	53.1	18	16.0	25	22.1	6	5.3	3.3	Indifferent
Do you think biofuel production can be done sustainably?	18	15.9	84	74.3	10	8.8	1	0.9	–	0.0	4.1	Agreed
Do you think that biofuel gel use is good for the environment	50	44.2	59	52.2	4	3.5	–	0.0	–	0.0	4.4	Agreed

Source Computed from Field Survey Data, 2019

cooking fuel, place a high value on biofuel gel and were likely to be willing to pay for biofuel gel products. This in line with the findings of Giraldo et al. (2010) who reported that willingness to pay for biofuel gel was influenced by consumers perception on the benefits of consuming the biofuel.

4.6 Factors Influencing Respondents' Willingness to Pay for Biofuel Gel

The factors influencing respondents' WTP for biofuel gel are presented in Table 7. The result revealed that five variables significantly determine the consumer's willingness to pay for biofuel gel in the study area, which were bid amount, years of education, household size, income and awareness of biofuel gel. The Bid amount had

Table 7 Factors influencing respondents' WTP for biofuel gel

Variables	Coefficient	Std. error	P > /z/
BID amount	−0.1075058	0.0379283	0.005***
Gender	5.69472	7.569554	0.452
Age	−0.074847	0.4596627	0.871
Years of biofuel consumption	1.034713	3.879882	0.790
Year of education	0.5290556	1.977308	0.007***
Marital status	−3.28556	3.861409	0.395
Household size	−0.2170215	3.261299	0.000***
Income	0.04383	0.0000432	0.000***
Awareness	0.2305132	12.67429	0.069*
Occupation	−3.875401	4.722383	0.412
Knowledge of renewable energy	−7.526723	11.85301	0.525
Familiarity with GHG	−3.133414	8.845236	0.723
Familiarity with climate change	−7.992842	8.56780	0.351
Member of environmental group	15.91207	13.39205	0.235
Constant	326.3472	41.33674	0.000***
Number of observation	175		
Log likelihood	−99.125807		
Prob > chi^2	0.0000		
Wald $chi^2(14)$	264.41		

Source Computed from Field Survey Data, 2019

a significant negative relationship on the respondents' willingness to pay for biofuel gel. This implies that as bid amount increases by one unit, the respondents' willingness to pay also decreases by 10.75%. The bid amount was significant at 1% level of significance which implies that the bid amount highly affects the decision of the respondents in the study area. This is in line with the assertion of Kim et al. (2018) that the negative sign indicates that as the bid amount increases, the respondents would be less likely to pay for biofuel gel, providing credence for the plausibility of the willingness to pay responses. This implies that a higher bid amount induces a lower likelihood of saying yes to an offered bid.

Years of education was statistically significant at 1% probability level with a positive coefficient. A unit increase in the years of education of a respondent tends to increase the willingness to pay for biofuel gel by 52.91%. This implies that an increase in the years of education had a positive influence on the willingness to pay for biofuel gel by the respondents in the study area. Likewise, more education generally implies higher income. It may thus be that the estimated education effect is just an ill observed income effect, which is consistent with typical rankings of fuels according to necessities and luxuries (Muller and Yan 2018). Household size of respondents had a negative coefficient of 0.217021 and it is statistically significant at 1%. This means a unit increase in the household size will reduce their willingness to pay for biofuel gel by 21.70% in the study area. This is because more individuals would be competing for the few available resources, this is in line with the assertions by Abebaw (2007); Zhang and Kotani (2012) that there is a negative relationship between household size and per capita energy consumption.

The result further revealed that income was significant at 1% and had a positive relationship in influencing households' willingness to pay for biofuel gel. The coefficient of 0.04383 implies that a unit increase in household income would increase the likelihood of the respondents' willingness to pay for biofuel gel by 4.38%. Showing that income was a major determinant of biofuel gel consumption among households' in the study area. This implies that as households' disposable income increases, more money will be available for the purchase and consumption of biofuel gel. Numerous studies; Brew-Hammond (2010); Peng et al. (2010); Kowsari and Zerriffi (2011); Lay et al. (2013); Baiyegunhi and Hassan (2014); Sivashankar et al. (2016)) point to income as the major driver behind the willingness to pay for modern fuels. When incomes increase respondents are more likely willing to pay for biofuel gel.

Respondents' awareness of biofuel product increases their willingness to pay for the product. The result indicates that awareness was statistically significant at 10% and had a positive relationship with the respondents' willingness to pay for biofuel gel. With the coefficient of 0.2305132, which implies that a unit increase in the awareness of the environmental benefit of the product will have a positive influence on the households' willingness to pay for the product by 23.05%. This is in line with the findings by Gracia et al. (2011) who concluded that increasing consumer awareness about biofuel gel would increase the willingness to pay for the biofuel gel. The Log-likelihood Ratio (LR) statistics exhibited appropriate signs and was significant at 1% probability level, meaning the explanatory variables included in the model explained the probability of willingness to pay for biofuel gel in the study area.

4.7 Major Constraints Militating Against the use of Biofuel Gel

Table 8 shows the major constraints militating against the use of biofuel gel by respondents in the study area. Various constraints were militating against the use of biofuel in the study area. They include inadequate supply and unavailability of the product, high cost of the product, poor quality of the product, not having a fixed price and poor awareness of the product among others. Not having easy access to the products was ranked first (93.7%) among the problems of the respondents. Gracia et al. (2011) in his analysis of consumers' willingness to pay for biodiesel in Spain, identified poor accessibility as the most severe constraint of respondents willing to purchase biodiesel. This implies that the convenience aspects of purchase have similar value to consumers.

The study also revealed that the high price of the product is ranked second (92.0%) as another constraint. This is in accordance with the study of Moula et al. (2017) that a significant decrease in the price of biofuel gel would be appreciated by the consumers. It was also observed that 90.3% of the respondents identify poor awareness of the product as a hindrance towards the use of biofuel gel in the study area. They do not have in-depth knowledge about the benefit of the product. This is in line with the findings of Van de Velde et al. (2011) that public knowledge about green energy is limited and there is a perceived lack of information, particularly on biofuel gel. A proportion of 81.1% and 78.9% of the respondents identified an inadequate supply of the product and poor quality of the product as a major constraint encountered.

Table 8 Major constraints militating against the use of biofuel gel

Constraints militating against the use of biofuel gel	Yes		Rank
	Freq.	Rel. Freq	
Accessibility of the product	164	93.7	1st
High price/Cost of the product	161	92.0	2nd
Poor awareness of the product	158	90.3	3rd
Inadequate supply of the product	142	81.1	4th
Quality of the product	138	78.9	5th
Poor financing and income	129	73.7	6th
Not having a fixed price	120	68.6	7th

Source Computed from Field Survey Data. 2019
* Multiple responses

5 Conclusion and Policy Implications

The study assessed the household preference level for sustainable clean cooking energy in the study area using biofuel gel as the case study. The study revealed that the key variables influencing the level of awareness of biofuel gel in the study area were age of the respondents, the years of consumption of biofuel gel, marital status, knowledge of renewable energy and familiarity with greenhouse gas. The determinants of the preference level of biofuel gel were the years of consumption of biofuel gel, household size, price of the product, income, familiarity with climate change and gender. The variables that significantly determine the consumer's willingness to pay for biofuel gel in the study area were bid amount, years of education, household size, income and awareness of biofuel gel. The study revealed that the product was mostly preferred by the respondents based on its emission (environmental benefits), by its clean cooking, price stability and heat intensity. Hence, the study has contributed empirically in the area of knowing those variables that the policy maker should take cognizant of with respect to factors influencing awareness, preference level and willingness to pay for biofuel gel and other clean energy to achieve the goal seven of SDGs in Nigeria and other Sub-Sahara Africa.

Therefore, from the study, the following policy implications can be inferred; since age is a critical factor in determining the awareness of biofuel gel, then the product should be promoted more among the younger generations using the social media as a key promotional vehicle. Likewise, the users of the products should be encouraged to assist in promoting the product based on its attributes such as environmental benefits and clean cooking. So, such should be engaged as sale agents of the product. Government at all levels, national and international non-governmental organizations should sensitize the citizens on the benefits of renewable energy and the effects of greenhouse gas emission. Hence, citizens should be made aware of the benefits of biofuel gel while sensitizing them about the ill-effects of conventional fuel, fostering a sense of personal environment responsibility, greater motivation and commitment towards the protection of the environment through media, promotions product and technology demonstration, house-to-house canvassing, trade fairs, educational outreach, shows at educational institutes, and public places. This will enhance the awareness and preference level for biofuel gel in Nigeria and Sub-Sahara Africa generally. Biofuel gel should be promoted more among the women and price policies can be implemented in the form of price taxes to discourage the consumption of conventional fuels and price subsidies to foster the consumption of biofuel gel (clean fuels) in Nigeria and Africa.

References

Abebaw D (2007) Household determinants of fuelwood choice in urban Ethiopia: a case study of Jimma Town. J Developing Areas 41(1):117–126

Adekunle CP, Akerele D, Adekunle AK, Amodemaja TS (2016) Consumers' willingness to pay for organic leafy vegetables in Abeokuta southwest Nigeria: evidence from double bounded dichotomous choice approach. Niger J Agric Food Environ 12(1):17–23

Ajala OE, Aberuagba F, Odetoye TE, Ajala AM (2015) Biodiesel: sustainable energy replacement to petroleum-based diesel fuel – a review. ChemBioEng Rev 2:145–154. https://doi.org/10.1002/cben.201400024

Baiyegunhi LJS, Hassan MB (2014) Rural household fuel energy transition: evidence from Giwa LGA Kaduna State, Nigeria. Energy Sustain Dev 20:30–35

Brew-Hammond A (2010) Energy access in Africa: challenges ahead. Energy Policy 38(5):2291–2301

Chatterton TJ, Anable J, Barnes J, Yeboah G (2016) Mapping household direct energy consumption in the United Kingdom to provide a new perspective on energy justice. Energy Res Soc Sci 18:71–87

Danlami AH, Applanaidu SD, Islam R (2017) From biomass cooking fuel source to modern alternative for Bauchi State households: a preliminary analysis. J Biofuels 8(3):323–331. https://doi.org/10.1080/17597269.2016.1226724

Demirbas A (2010) Social, economic, environmental and policy aspects of biofuels. Energy Sci Technol 2:75–109

Dioha IJ, Ikeme CH, Tijjam N, Dioha EC (2012) Comparative studies of ethanol and kerosene fuels and cooking stove performance. J Nat Sci Res 6(1):43–54

Du Y, Wang X, Brombai D, Moriggi A, Sharpley A, Pang S (2018) Change in environmental awareness and its connection to local environmental management in water conservation zones: the case of Beijing, China. Sustainability 10:2087–2094. https://doi.org/10.3390/su10062087

Ekundayo MO (2013) Out of Africa: Fashola: Reinventing Servant Leadership to Engender Nigeria's Transformation. Author House, p 135. ISBN 9781481790406

Farsi M, Filippini M, Pachauri S (2007) Fuel choices in urban Indian households. Environ Dev Econ 12(6):757–774

Giraldo L, Gracia A, Do Amaral E (2010) Willingness to pay for biodiesel in Spain a pilot study for diesel consumers. Spanish J Agric Res 8(4):887–894. https://doi.org/10.5424/sjar/2010084-1382

Goldemberg J, Guardabassi P (2009) Are biofuels a feasible option?. Energy Policy 37(1):10–14. https://doi.org/10.1016/j.enpol.2008.08.031

Gracia A, Barreiro-Hurlé J, Pérez y Pérez L (2011) Consumers' willingness to pay for biodiesel in Spain. In: International congress of the european association of agricultural economists conference, Zurich, Switzerland, ETH Zurich, 30 August–2 Sept 2011. https://hdl.handle.net/10532/1719

Gujarati DN (2003) Basic econometrics, 4th edn. Tata McGraw-Hill Edition, New Delhi, p 1002

Hanemann WM, Loomis J, Kanninen BJ (1991) Statistical efficiency of double-bounded dichotomous choice contingent valuation. Am J Agric Econ 73(4):1255–1263

Hast A, Alimohammadisagv B, Syri S (2015) Consumer attitudes towards renewable energy in China – the case of Shanghai. Sustain Cities Soc 17:69–79

Israel D (2002) Fuel choice in developing countries: evidence from Bolivia. J Econ Dev Cult Change 50(4):865–890

Karytsas S, Theodoropoulou H (2014) Socioeconomic and demographic factors that influence publics' awareness on the different forms of renewable energy sources. Renew Energy 71:480–485

Kim G, Lee H, Yoo S (2018) Willingness to pay for substituting coal with natural gas-based combined heat and power in South Korea: a view from air pollutants emissions mitigation: a contingent valuation study. Sustainability 10:1554–1567. https://doi.org/10.3390/su10051554

Kimemia DK, Annegarn HJ (2012) Productive uses of basic energy and fuel transitions in urban South Africa. Energy Environ Res 2:103–112

Kowsari R, Zerriffi H (2011) Three-dimensional energy profile: a conceptual framework for assessing household energy use. Energy Policy 39(12):7505–7517

Kurnia JC, Jangam SV, Akhtar S, Sasmito AP, Mujumdar AS (2016) Advances in biofuel production from oil palm and palm oil processing wastes: a review. Biofuel Res J 3(1):332–346. https://doi.org/10.18331/BRJ2016.3.1.3
Lay J, Ondraczek J, Stoever J (2013) Renewables in the energy transition: evidence on solar home systems and lighting fuel choice in Kenya. Energy Econ 40:350–359
Lloyd P (2015) A pilot test of ethanol gel as a paraffin replacement for low income urban environment. J Energy South Africa 25(3):45–56
Liu W, Wang C, Mol A (2013) Rural public acceptance of renewable energy deployment: a case of Shandong in china. Appl Energies 102:1187–1196
Makonese T, Ifegbesan AP, Rampedi IT (2018) Household cooking fuel use patterns and determinants across southern Africa: evidence from the demographic and health survey data. J Energy Environ 29(1):52–65. https://doi.org/10.1177/0958305X17739475
Mensah T, Adu G (2013) An empirical analysis of household energy choice in Ghana, Uppsala Working Paper Series No 6
Moula ME, Nyari J, Bartel A (2017) Public acceptance of biofuels in the transport sector in Finland. Int J Sustain Built Environ 45(2):45–61. https://doi.org/10.1016/j.ijsbe.2017.07.008
Muller C, Yan H (2018) Household fuel use in developing countries: review of theory and evidence. Energy Econ 70(3):429–439. https://doi.org/10.1016/j.eneco.2018.01.024
Neira M, Ramanathan V (2020) Climate change, air pollution, and the environment: the health argument. In: Al-Delaimy W, Ramanathan V, Sánchez Sorondo M (eds) Health of people, health of planet and our responsibility. Springer, Cham, pp 93–103. https://doi.org/10.1007/978-3-030-31125-4_8
Nlom JH, Karimov AA (2014) Modeling fuel choice among households in northern Cameroon. WIDER Working Paper Series 038.45
NPC (2007) National Population Commission, Nigeria Annual Census, 2006. https://www.population.gov.ng/. Accessed 08 Oct 2017
Obayelu OA, Agboyinu OM, Awotide BA (2014) Consumers' perception and willingness to pay for organic leafy vegetables in urban Oyo State Nigeria. Eur J Nutr Food Saf 4(2):127–136
Ouedraogo B (2006) Household energy preferences for cooking in urban Ouagadougou Burkina Faso. Energy Policy 34(18):3787–3795
Özcan KM, Gülay E, Üçdoğruk Ş (2013) Economic and demographic determinants of household energy use in Turkey. Energy Policy 60:550–557
Pandey VL, Chaubal A (2011) Comprehending household cooking energy choice in rural India. Biomass Bioenerg 35(11):4724–4731
Peng W, Zerriffi H, Jiahua P (2010) Household level fuel switching in Rural Hubei. Energy Sustain Dev 14(3):238–244
Perera F (2018) Pollution from fossil-fuel combustion is the leading environmental threats to global pediatric health and equity: solutions exist. Int J Environ Res Public Health 15:16–23. https://doi.org/10.3390/ijerph15010016
Rahut DB, Das S, Groote HD, Behera B (2014) Determinants of household energy use in Bhutan. Energy 69:661–672
Rao MN, Reddy BS (2007) Variations in energy use by Indian households: an analysis of micro level data. Energy 32(2):143–153
Reddy BS (1995) A multilogit model for fuel shifts in the domestic sector. Energy 20(9):929–936
Sivashankar P, Weerahewa J, Pushpakumara G, Galagedara L (2016) An estimation of the willingness to pay for biodiesel: a pilot study of diesel consumers. Bio-based Appl Econ J. 5(1):32–41. Italian Association of Agricultural and Applied Economics (AIEAA). https://doi.org/10.13128/BAE-15283
Van der Horst D, Vermeylen S (2010) Spatial scale and social impacts of biofuel production. Biomass Bioenergy 35(6):2435–2443
Van de Velde L, Verbeke W, Popp M, Van Huylenbroeck G (2011) Trust and perception related to information about biofuels in Belgium. Public Underst Sci 20(5):595–608

United Nation (UN) Climate Change Summit (2019) Cities and pollution contribution to climate change. Climate reports. https://www.un.org/en/climatechange/cities-pollution.shtml
WHO (2018) World Health Organization: Household air pollution and health. Report on Indoor air pollution and household energy. https://www.who.int/news-room/fact-sheets/detail/household-air-pollution-and-health
Yuan X, Zuo J, Ma C (2011) Social acceptance of solar energy technologies in china – end users' perspective. Energy Policy 39:1031–1036
Zhang J, Kotani K (2012) The determinants of household energy demand in rural Beijing: can environmentally friendly technologies be effective? Energy Econ 34(2):381–388

‘But We Cannot Do It All’: Investors’ Sustainability Tensions and Strategic Selectivity in the Development of Geothermal Energy in Kenya

Chigozie Nweke-Eze and Eric M. Kioko

Abstract Several studies have questioned investors’ adequate consideration of the three pillars of sustainable development in the construction of development projects in the Global South. Other studies have proposed and developed frameworks for fostering their adoption and application in the planning and construction of such projects. However, relatively little attention has been directed to understanding the intricate processes and dynamics involved in investors’ adherence to these sustainability triad. This paper explores these sustainability adherence processes, and their associated challenges and imponderability in the context of large-scale geothermal projects development in Kenya. We argue that investors’ commitment to the sustainability framework in the development of such projects is characterized by *sustainability tensions,* reflected in conflicting interests, dilemmas, and power struggles that investors face as they attempt to simultaneously apply the three principles of sustainable development in delivering their projects. In order for investors to manage these tensions, the study shows that they engage in *strategic selectivity*, whereby the extent of adherence to certain components of the sustainability principles are based on winning interests, priorities and convenience. These processes are explored by drawing on perspectives from sustainable development, Triple-Bottom-Line and corporate sustainability discourses. Expert and informal interviews, document analyses, ethnographic fieldwork and field visits are used to track and illustrate these processes, using the case of large-scale geothermal project developments in Olkaria, Kenya.

Keywords Sustainability · Tensions · Strategic selectivity · Geothermal · Kenya

---

C. Nweke-Eze (✉)
Institute of Geography, University of Bonn, Meckenheimer Allee 166, 53115 Bonn, Germany
e-mail: cne@uni-bonn.de

E. M. Kioko
Department of Environmental Studies and Community Development, Kenyatta University,
P O Box. 43844-00100, Nairobi, Kenya
e-mail: kioko.eric@ku.ac.ke

W. Leal Filho et al. (eds.), *Sustainable Development in Africa*, World Sustainability Series,
https://doi.org/10.1007/978-3-030-74693-3_21

1 Introduction

In "The Future We Want" (UN 2012), world nations reaffirmed the need to achieve sustainable development, in its three dimensions of economic, social and environment, in the face of new and emerging challenges. At continental level, African states also reaffirmed their commitment to a new more inclusive and sustainable path for attaining industrialization and economic development in the Agenda 2063 "The Africa We Want" (AUC 2015). African countries, along with most other countries of the Global South, essentially grapple with challenges in all the three dimensions of sustainable development (UN 2013). The breadth and linkages of these challenges and how to simultaneously address them, present sustainability dilemmas, tensions and conflicts (Jayanti and Gowda 2014, Brix-Asala et al. 2018, Newig et al. 2007, Shove and Walker 2007).

There is much skepticism on the possibility to simultaneously address the three dimensions of sustainability given the diverse country-specific political, institutional, and economic capacities and priorities (Romijn et al. 2010). This is particularly true for Africa where resource-intensive investments and industrial expansion remain a key priority for the achievement of rapid economic growth despite their negative socio-environmental impacts. Pressing challenges such as high poverty levels, population growth, and inadequate employment opportunities continue to undermine efforts towards sustainable development (Ahenkan and Osei-Kojo 2014). With these dynamics, reconciling economic growth with social equity and environmental protection becomes an almost elusive endeavor given the many conflicting goals, priorities, and trade-offs (OECD 2013; Ramos-Mejia et al. 2018; Leach et al. 2010).

Such choices, trade-offs and decisions are especially reflected in Kenya's transition to green energy solutions through the massive investment in geothermal power. Geothermal energy development is expected to play a major role in the country's transition to a green economy (GoK 2007). In Kenya, investment in renewable energy has increased fourfold from USD 88 billion in 2005 to USD 349 billion in 2015, most of which has gone into geothermal energy development (Koissaba 2018). Olkaria geothermal plants, located in Kenya's Rift Valley, continues to witness massive investments for expansion and upgrading, while other sites in Menengai and Baringo-Silali are in plant construction and exploration phases. Investors in geothermal in Kenya include the national government (through the Kenyan Electricity Generation Company and Geothermal Development Company), international development financial institutions and private independent power producers (Klagge and Nweke-Eze 2020).

Despite these massive investments by diverse investors and the expected economic potential, geothermal development in Kenya raises concerns over the extent of investors' commitment to social and environmental dimensions of sustainable development (Mariita 2002; Schade 2017). There is therefore need to take a closer look at *what counts as sustainable? for whom?* and *by whom?* (Scoones et al. 2015). Answering these key sustainability questions requires understanding the processes, interplays and conditions in which green projects materialize. In this paper, we explore the processes in which both public and private investors seek to simultaneously adhere to the sustainability triad in delivering large-scale renewable energy projects, using Kenya' largest geothermal projects in Olkaria as case study. In so doing, we reveal the challenges, conflicts and tensions that characterize investors' application of the sustainability triad in the Global South context and how these sustainability tensions and dilemmas are managed through strategic selectivity—that is, through adhering to certain sustainability components while neglecting others, based on winning interests, priorities and convenience.

Going forward, we first set out a conceptual framework on the interface and tensions between different players and perspectives in adherence to sustainable principles, based on sustainable development, corporate sustainability and Triple-Bottom-Line discourses. Thereafter, we present the Olkaria geothermal case study, describing the project site and plants and the socio-economic and environmental contexts of the area. In our analysis, we examine investors' relations with other stakeholders (in this case, local project-host communities and conservationist institutions/groups) in their efforts to apply the sustainability triad in the development of Olkaria geothermal. In so doing, we reveal the conflicting interests, dilemmas and power struggles that investors face and how they manage these tensions through strategic selectivity and trade-offs in the development of such projects. In the discussion and conclusion, we draw the main lessons from the paper and outline future prospects necessary for fostering sustainable development in Africa.

2 Sustainable Development, Triple Bottom Line, and Corporate Sustainability

Despite the popularity and importance of sustainable development, the concept is difficult to define with precision and, therefore, difficult to measure (UN 2008). It is also characterized by tensions, disconnect, ambiguity, contradictions, and paradoxes, which define the incompatibility of the interplay of economic, social, and environmental factors of development, otherwise known as the sustainability triad (see McIntosh 2003). In most developing countries context, there is the tendency to prioritize the economic dimension of sustainable development over the social and environmental dimensions (Enns and Bersaglio 2020; Romijn et al. 2010). This is especially the case for most investors in the developing world where the desire for profits often outweigh sustainability concerns, epitomizing what Alexander (2015) and Lange and Washburn (2012) call Corporate Social Irresponsibility. Given this tendency, the corporate sustainability idea and the Triple Bottom Line (TBL) approach have been forefront in guiding investors in assessing the sustainability of their businesses and the scope of their corporate social responsibilities.

The corporate sustainability concept presents economic, environmental and social concerns as systematically interconnected and interdependent at different levels, and enjoins firms to address them simultaneously (Hahn et al 2015). Ideas of corporate sustainability are traceable to Carroll's work of the late 1990s, which largely draw from the report of the Brundtland commission. The justification for corporate responsibility, according to Carroll (1999), stems from an understanding that actions of corporations touch community members at many points and, as a result, corporates or businesses should be responsible for the consequences of their actions to communities and the environments in which they operate. The TBL concept (introduced by Elkington (2004)), on the other hand, is a traditional accounting framework adopted by corporations in examining the extent of their broader company value (Lee 2007). Under the TBL framework, corporate sustainability is anchored on environment, economics and equity dimensions, and with the position that corporation should ideally commit to social and environmental concerns in their operations as they do on profits (Gray and Milne 2004).

However, as Hahn et al. (2015) show, the expectation of companies and businesses to simultaneously address the three pillars of sustainability faces tensions at different levels and scales (also see Brix-Asala et al. 2018 and Ramos-Mejia et al. 2018). In adherence to sustainability in the development of renewable energy projects in the Global South, we identify that these tensions manifest in form of conflicting interests, projected expectations, leading to agitations, contestations and dilemmas, which are managed through negotiations and strategic selectivity (see Fig. 1). The sustainability outcomes are based on interest hierarchies and power pulls among public and private sector investors at international and national levels, and the local

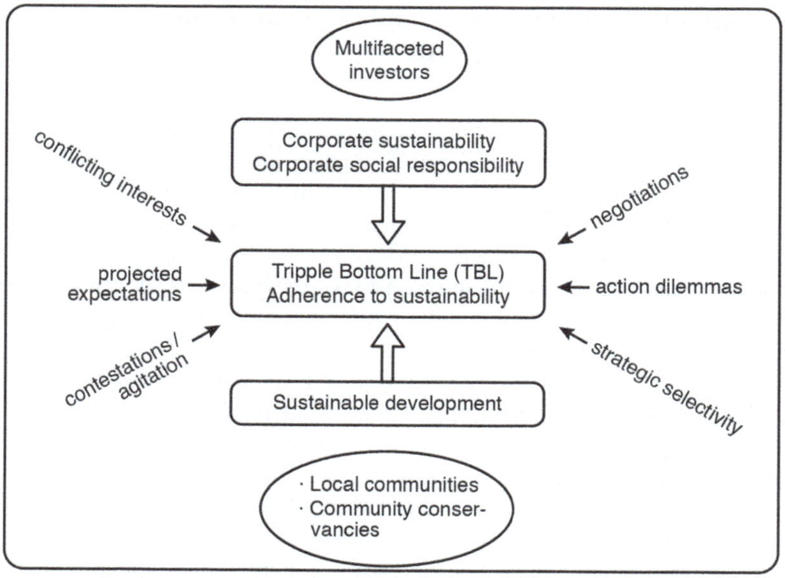

Fig. 1 Adherence to sustainability principles: players, perspectives, and external factors (Authors' own)

communities as well as wildlife and bio-diversity conservationist institutions and groups, at the TBL—sustainability adherence interface (see Fig. 1). The local host-communities and conservationists act as face-off actors, facing-off with the investors, in the development of renewable energy projects.

In the following sections, we will empirically explore stakeholder characteristics and their conflicting interests as well as the extent to which tensions in adherence to sustainable development influence decision-making and investors' choices in Olkaria geothermal development in Kenya. Moreover, we explore the extent to which these choices are based on the balanced needs of all stakeholders including investors/shareholders, and government, communities and environmental conservationists (Jamali 2008).

3 Study Area and Methodology

3.1 *Study Area*

The Olkaria geothermal field is located in the central part of the Kenya Rift Valley, south of Lake Naivasha and about 120 km northwest of Nairobi. The field is divided into seven blocks, which include Olkaria East field serving Olkaria I power plant, Olkaria Northeast field serving Olkaria II, Olkaria West field serving Olkaria III, and Olkaria Domes field serving IV & V (see Fig. 2). Except for Olkaria III,[1] All Olkaria power plants (Olkaria I, II, IV) are operated by a partly government-owned parastatal, Kenya Electricity Generating Company (KenGen). For this study, we will focus on these KenGen geothermal projects. KenGen's Olkaria I, Olkaria II, Olkaria IV currently generate a total of 185 MW,[2] 105 MW and 140 MW of electricity, respectively. Olkaria V is under construction within the Olkaria Dome field, with estimated generation of 165.4 MW; and Olkaria VI of 140 MW is planned for 2021, under a Public Private Partnership (PPP) project development model

Olkaria I and Olkaria II are located within the Hells Gate National Park—a fact that has elicited some environmental management concerns. Hell's Gate National Park lies on the south of Lake Naivasha and about 120 km north-west of Nairobi. The state gazette the park in 1984, three years after the commissioning of Olkaria I Power Station. The park features rare flora, fauna, and exquisite sceneries. Olkaria IV power plant is located on Kedong Ranch, a property acquired by KenGen, and is about 15 km from the Olkaria I power station. Furthermore, KenGen utilizes water from the nearby Lake Naivasha, which is a wetland of international importance according to the Ramsar Convention on Wetlands.

[1]Olkaria III is 48 MW power plant, owned and operated by the privately owned U.S. company OrPower 4, a subsidiary of Ormat Technologies Inc., under an IPP arrangement.

[2]With the plans to rehabilitate Olkaria I, units 1–3 and to build a new plant (Olkaria I unit 6, which is estimated to be 83 MW), total capacity of Olkaria I is estimated to reach 237.7 MW from its current 185 MW by 2021.

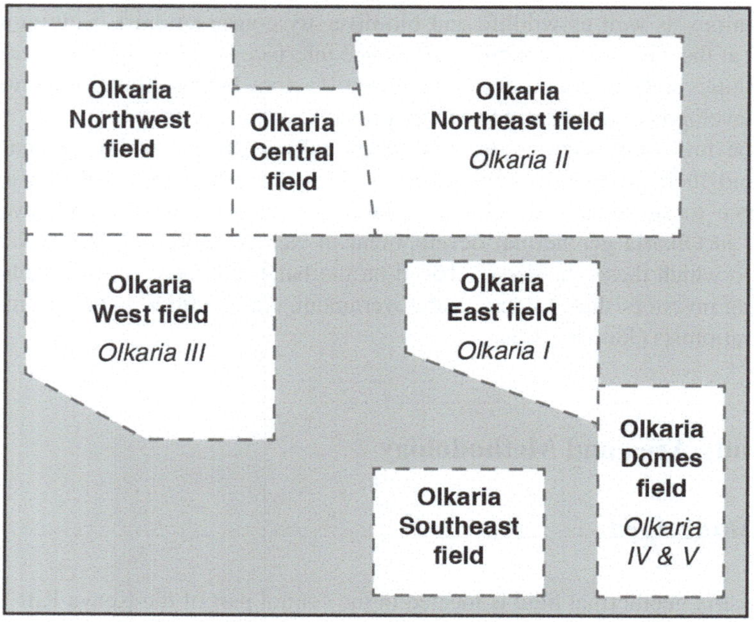

Fig. 2 Structural map of Olkaria geothermal fields and plants in Kenya (Field Data 2018 and 2019; Omenda et al. 2014)

The cumulative impacts of the existing Olkaria I & II, the proposed Olkaria IV and Olkaria I Units 4 & 5 power stations on air quality and noise pollution prompted KenGen to earmark the following nearby villages for resettlement: Cultural Centre, Olo Nongot, Olo Sinyat, Olo Mayiana. Interviews with KenGen officials revealed that a Resettlement Action Plan (RAP) to relocate the Project Affected Persons (PAPs) was developed as a mitigation measure against the impacts of prolonged exposure to air and noise pollution promising to provide PAP with social amenities like land, water, modern residential houses, a school, churches, and a health facility.

3.2 *Methodology*

The Olkaria geothermal projects were chosen as case study because of its provision of a high representative material, involving diverse stakeholders and capturing the stakeholder interplays and dynamics in investors' adherence to sustainability in the development of the largest geothermal projects in Africa. The analyses of the study are based on several expert and informal interviews (2018–2019), document analyses (reports, work papers and proceedings), ethnographic fieldwork and field visits (2018–2020). The expert interviewees work at government agencies and parastatals at different levels (Ministry of Energy (MoE), Ministry of Environment and Natural

Resources (MoEN), Kenya Wildlife Service (KWS), National Land Commission (NLC), the National Treasury, County Commission, County Government). They also consist of project developers (KenGen) and staff members in development finance institutions (European Investment Bank (EIB), German Development Bank (KfW), Agence Française de Développement (AFD)). For anonymity reasons, we refrain from providing the identities of our informants. Ethnographic fieldwork and field visits (2018–2020) featured observations, informal interviews and group discussions with community members at the fringes of the geothermal sites (namely, Cultural Centre, Olo Nongot, Olo Sinyat, Olo Mayiana), who had to be displaced to allow for further geothermal development in Olkaria. All the expert interviews were recorded using an electronic audio device, while observations, informal interviews and group discussions were recorded through note-taking. The interviews were transcribed and thematic analysis was carried out. This included coding of data prior to the identification, review and analysis of key themes. Each theme was explored in order to develop an understanding of the perception, interests and motivations of the participants. The interviews provided rich data on the investor-community-conservancy relations, which is vital for our analysis of investors' adherence to the sustainability triad. Ethnographic fieldwork provided evidence on behaviors, expectations, social structures and shared beliefs of the case study communities. Furthermore, in order to augment, validate and triangulate the findings of the study, several project reports, working papers and proceedings were reviewed and analyzed.

4 Results

This section discusses the results and the findings of the study in two main clusters. In the first cluster (Sect. 4.1), it discusses the constitution of stakeholders, their characteristics and interests in the development of the Olkaria geothermal project. The second cluster (Sect. 4.2) draws from the first to discuss the conflicting nature of stakeholder interests, sustainability tensions and investors management of these tensions through strategic selectivity.

4.1 Stakeholders, their Characteristics and Interests in the Development of Olkaria Projects

Stakeholders in the geothermal sector in Kenya can be more generally grouped into two: namely, the investors and the local communities and conservationists. Investors, here, covers both public and private sector project developers, managers, financiers and consultants. The stakeholders, their relationships and interests in development of the Olkaria geothermal projects are discussed along the lines of these groups, in the sub-sections below.

4.1.1 Investors

Investors in the development of the analyzed Olkaria geothermal projects come from public and private sectors, at international and national levels. They consist of the Government of Kenya (GoK), the project developer (KenGen) and development financial institutions (DFIs). The roles, characteristics and interests of the Olkaria geothermal investors in Kenya are discussed as follows.

Government of Kenya (GoK), Line Ministries and Parastatals

The Government of Kenya (GoK) is involved in the projects through its various ministries and parastatals, the foremost of which are the ministry of energy (MoE), and the National Treasury (NT). On behalf of the Kenyan government, the Ministry of Energy (MoE) is in charge of all aspects of the energy sector including the development of Olkaria geothermal projects.

Geothermal resource development ranks high in GoKs national development strategy, as contained in its Vision 2030 (GoK 2007). The transition from hydropower to geothermal and other energy sources strengthens Kenyas energy resilience to unpredictable precipitation shocks of which hydropower was often prone to, as well as imported oil-price fluctuation shocks[3] (GoK 2013). Despite the focus on expansion of renewable energy sources to achieve energy sufficiency and security, GoK is also keen on expanding the non-renewable energy sector through exploration of oil fields in northern Kenya and investment in nuclear energy power plants (Field Data 2019 and 2020).

The Kenya Electricity Generation Company (KenGen) and Their Consultants

KenGen is a 70% government and 30% private sector-owned company, mandated with the responsibility of developing, managing and operating power plants for electricity generation in the country. KenGen is the largest electricity generator in Kenya with 62% of national installed capacity amounting to approximately 1796 MW, with geothermal leading in the installed capacity mix (Maino 2019). Key to KenGen's plan for further expansion of its installed capacity is the development of geothermal energy (Omenda et al. 2014).

KenGen is the sole project developer of the studied Olkaria geothermal projects. It did so in cooperation with actors in the public sector, namely GoK and Development Financial Institutions (DFIs), as well as actors in the private sector—IPPs and consultants (such as GIBB Africa[4]). KenGen has a Geothermal Development Office responsible for all administrative aspects of all geothermal projects. It's regulatory Affairs office plays an interfacing and coordinating role in the RAP implementation process, constantly engaging with the MoE, and KenGen's executive committee

---

[3]The GoK had relied on Hydropower for energy generation in the past.

[4]GIBB Africa is a consultancy firm hired by KenGen to conduct the census to determine compensation and to draft the Resettlement Action Plan (RAP).

Table 1 DFIs and roles in the development of Olkaria projects

DFIs	Roles
EIB	Loans, technical assistance in due diligence
AFD	Loans, project procurement, technical assistance in the resettlement of project affected persons (PAPs) and in the implementation of environmental and social safeguards
KfW	Loans
JICA	Loans, technical assistance in field exploration
World Bank	Loans, technical assistance in the resettlement of PAPs

Sources: Field data (2019), Schade (2017)

(which is the highest decision-making organ) and the lenders (all of which are DFIs). The Environment and CDM office regularly supervised the implementation of the social and environmental sustainability frameworks and requirements of the project and regularly reported to the DFIs on progresses in RAP implementation. It achieves this in cooperation with the Social Safeguards Office[5] managed by the community liaison officer, who is responsible for day-to-day implementation of social safeguards and coordination of the Project Affected Persons (PAPs), the Resettlement Action Plan Implementation Committee (RAPIC), and the local administration. The Environment and CDM office also works with the Property and legal manager's office in issues relating to land transfers and settlements; and with the Project Execution Office for the technical and infrastructural aspects of the RAP implementation.[6]

Development Financial Institutions (DFIs)

The development financial institutions (DFIs) involved in the development of Olkaria fields include the European Development Bank (EIB), the French Agence Française de Développement (AFD), the German Kreditanstalt für Wiederaufbau (KfW), Japan International Cooperation Agency (JICA), and the World Bank. The DFIs play an important catalytic role in the development of Olkaria projects by providing financing, in their various forms (loans, grants and mezzanine) as well as technical assistances. Table 1 summarizes the DFIs and their specific roles in the development of one or more of the Olkaria projects.

In order to organize, coordinate and better focus their actions, some of the DFIs often form groups, which meets periodically (three to four times in a year) for discussions on a broadened and deepened cooperation and coordination. A peculiar example of such groups in the studied context is the Energy Sector Development Partners Group, which includes all DFIs working in the Kenyan energy sector, and whose

[5]The Social Safeguards Office compiled monthly progress reports to share with KenGen, IEP, and the county administration. It also assumed the role of secretary in the Resettlement Action Plan Implementation Committee (RAPIC). The RAPIC was a deliberation platform, which involved all important stakeholders involved or affected by the RAP, including the investors and the Project Affected Persons (PAPs).

[6]Including site layout and tendering, and supervision of contractors doing construction.

convener are rotated from time to time among the member DFIs (Field Data 2018 and 2019). In the energy group, DFIs share experiences of project development as well as indicate interests for cooperation in energy projects development. In addition, the Mutual Reliance Initiative (MRI) was initiated in 2009, among the European DFIs[7] with a view to establishing an efficient division of labor in order to fulfill the development obligations set out in the Paris Declaration on Aid Effectiveness and the Accra Agenda for Action, of which development of renewable energy is among the key priority areas. The operational guideline of the MRI dictates approaches and delegates functions for involvement among member DFIs in a project (OECD 2011). In the case of the development of Olkaria IV, for instance, the AFD was the assigned lead financier and project materials procurer for Olkaria IV as well as lead for the implementation of environmental and social safeguards. EIB was placed in charge of technical due diligence, while AFD together with the World Bank administered resettlements (Field Data 2019 and Schade 2017). Other than the European DFIs, the Japanese JICA and the World Bank has played signification financing and technical assistance roles in the development of Olkaria projects. The RAP was administered based on the World Bank's OP 4.12 on involuntary resettlement. As a result, the World Bank wielded a relative high influence in the resettlement processes. The DFIs got quarterly updates on RAP implementation from KenGen.

4.1.2 Local Communities and Conservationist Institution

In the development of large-scale renewable energy projects, investors have to deal with other stakeholders who are equally interested or affected by the development of projects. For Olkaria, these are the project-host and neighboring communities (including the Project Affected Persons (PAPs)) and state conservationist institution—the Kenyan Wildlife Service (KWS). The local communities are interested in ensuring that their human rights are adhered to and have the expectation that their socio-economic livelihood will be improved as a result of the presence of the projects. The KWS, on the other hand, act as mediums for the wildlife and biodiversity, ensuring that the habitats of the wildlife and forest covers are maintained during and after the development of the project. The characteristics and interests of these stakeholders are discussed in details below.

Local Communities and Project Affected Persons (PAPs)

During the development of Olkaria IV and Olkaria I unit 1 & 2, it was determined that the total impact of the overall existing and planned geothermal facilities in the area will adversely affect certain neighboring communities. As a result, four communities (Cultural Centre, OloNongot, OloSinyat, and OloMayiana) were resettled. The four local communities were initially convinced and viewed the resettlement as an opportunity to improve their socio-economic livelihoods (Field Data 2018).

The Cultural Centre is a permanent village and a business center, which until resettlement mainly subsisted on tourism. The other three communities are predominantly pastoralists who lived and still live on a livestock-based economy. Other than

[7]Including EIB, AFD and KfW.

these main sources of socio-economic livelihood, members of the four communities (mainly men) also gained employment opportunities offered by KenGen and their contractors, as well as by flower farms. The communities could sell pumice stones and women could engage in other small-scale trading. In addition, the communities also engaged in charcoal burning, which was illegal, as they had no license to do so (GIBB Africa 2012, pp. 5–6). Until the resettlement, the Cultural Centre, founded about 30 years ago as a place of a cultural and spiritual significance to the Maasai, consisted of one manyatta.[8] Most members of the Cultural Centre lived in the manyatta. The manyatta conjoined with each other in a circle, creating a space in the middle, where the village performed traditional dances and spiritual rites. There were also few members of the Cultural Centre who lived close to the Olkaria Primary School, away from the manyatta. The OloNongot households lived in manyattas in relatively close proximity to each other, while the OloSyniat households lived further from each other. In OloMayana Ndogo, the most of the manyattas were arranged in a linear form, with some also in clusters.

According to the 2009 GIBB Africa survey (GIBB Africa 2009a and b), the level of socio-economic wellbeing in the four communities was generally considered poor. In 61.8% of the PAP households, water was collected from a distance, mainly by women and girls. Only the Cultural Center had a public water pump, which was installed with the help of a French NGO. Furthermore, 45% of the PAPs were believed to have no access to sanitation services and none of the households had electricity (GIBB Africa 2012). Prior to resettlement, the standard of formal education among the PAPs was also generally poor. According to data compiled for the Environmental and Social Impact Assessment (ESIA) of Olkaria IV (GIBB Africa 2009a and b), 51% of household heads and wives had no education, 22% had some amount of primary education, 12% had some secondary education, 8% had some technical training, and just 3% had attended university (EIB-CM 2015, p. 29). Most of the older members of the community spoke the local Maa language, while some younger literate members spoke Kiswahili and English (Field Data 2018 and 2019). Some community activists, however, contend with this survey finding stating that it was skewed to paint a picture of bringing hope of better welfare among the local communities through the resettlements.

According to the updated edition of the 2009 GIBB Africa Census (GIBB Africa 2009a and b), 1,209 community members were qualified for compensation, of which 948, made up of landowners with properties and homes, were qualified for resettlement. A total of 284 persons from OloNongot, 139 persons from OloSinyat, 299 persons from the Cultural Center and 226 persons from OloMayana were deemed suitable for housing on the RAP site (GIBB Africa 2009a and b). The Maasai had some polygamous households, which implied that each spouse of a one-male household head and their respective children had their own house unit. The husband of these wives rotated among them and was liable for providing for them and their dependent children (Field Data 2018, 2019, Schade 2017). On this basis, the overall

[8]Traditional Maasai housing units or clusters

number of housing units to be constructed was estimated to be 164 (GIBB Africa 2012), as opposed to the planned 150 housing units based on the first RAP estimates.

The resettlement land of 1700 acres is located within the Olkaria geothermal block. Upon relocation, the PAPs were, for the first time, eligible to get their title deeds, making them formal landowners—a move, which served as a major incentive for the communities to accept the relocation (Field Data 2019). At the RAP-land, KenGen provided the communities with a primary and early childhood school for 320 pupils and a health facility[9] (Field Data 2018 and 2019). All public facilities, the schools, and the dispensary were connected to the newly installed electricity grid, with cost covered by the communities (Field Data 2019).

The Kenya Wildlife Service (KWS)
The Kenya Wildlife Service (KWS) is a GoK parastatal created in 1990, under the Ministry of Tourism and Wildlife. Its mandate, according to a senior KWS official, is "to take care of wildlife, for the prosperity and for the people of Kenya, the citizens of Kenya and for the next generation" (Field Data 2019). Before its creation as a parastatal, the wildlife department within the Ministry of Tourism and Wildlife of the GoK initially undertook the activities of the current KWS. It was created at a time when the GoK was committed to strengthen conservation through law and policies for the protected areas (Field Data 2019).

KWS manages most of the National Parks and Reserves in Kenya. It's activities are run from its own generated revenues from tourist fees, complemented with supplementary funds from the GoK, public and private donors and other partners. In order to better achieve its mandate, KWS often works together with the Ministry of Environment, Water and Natural Resources, as well as the Kenyan Police and the National Intelligence Service to combat environmental crimes. For ease of reach, effectiveness, and resource allocation, KWS functions are devolved into eight regions consisting of the northern region, the mountain region, the central rift region, the southern region, the Tsavo's, the coast region, the western region and the eastern region. The Hell's Gate National Park, where most of the studied Olkaria geothermal projects are located, is within the Rift Valley region.

4.2 Conflicting Interests, Sustainability Tensions and Strategic Selectivity

As shown in the previous Subsect. 4.1, the stakeholders in Olkaria's geothermal plants are multiple, diverse and with different interests. The interests of the investors are aligned, to some extent, because of their perceived goal of ensuring the development of geothermal energy as an engine to socio-economic growth and development in Kenya. DFIs, in line with their sustainability agenda, required that KenGen adhere to the their social and environmental sustainability principles, which are enshrined in

[9] The communities previously relied on local herbal medicines for curing ill health.

their Environmental and Social Impact Assessment (ESIA) frameworks as a condition for accessing financing for project development. However, as we will see in the sub-sections below, this mechanism alone is inadequate to fully enforce and investigate the extent to which KenGen and other investors adhere to the sustainability principles. ESIAs do not guarantee whether concerns for possible negative environmental impacts will be addressed or whether the need to invest sustainably will be adhered to at the post-finance unlocking phase (that is, during the project implementation phase).

KenGen also has to abide by the GoK laws and regulations during and after the projects' development. These laws and regulations are meant to protect the human rights of PAPs as well as help in the conservation of biodiversity and promotion of tourism (a sector that the government also earns revenues from). The extent to which KenGen adheres to these sustainability frameworks, laws and regulations is, however, questionable. There are allegations that ministry representatives collude with KenGen to push agendas that mainly favor cost-effective and face-paced development of the projects to the detriment of thorough consideration of social and environment concerns (Field Data 2019).

For the GoK, such projects are often political projects that are meant to inspire public confidence that the government "is doing something" (with reference to development). They inspire high expectations of change and are therefore popularized as success stories in a bid to limit any possible counter narratives. For DFIs, the financial support are interpreted as being "part of a big thing" (with reference to transition to a green and clean alternatives). These DFIs are the channel through which climate mitigation financing reach the projects at the local levels. Local communities (individuals and groups) have their immediate concern on expected socio-economic benefits, whether perceived or real. Their hopes are that the existence of such projects directly improves their socio-economic wellbeing. On its part, KWS prioritizes environmental conservation and preservation of biodiversity. However, KWS is state-owned and therefore does not have the capacity and legitimacy to change state decision with regard to the implementation of a project.

These conflicting interests and expectations between investors and their face-offs (local community and the KWS) generate tensions at the interface of sustainable project implementation as will be discussed in details below. These tensions and dilemmas created by conflicting sustainability interests among stakeholders are managed through sustainability trade-offs; that is the choice to implement a certain sustainability principle over others—a practice we refer to as strategic selectivity.

4.2.1 Investors and the Local Communities

As stated earlier, the construction of Olkaria projects necessitated the evacuation and relocation of a number of Project Affected Persons (PAPs). The Resettlement Action Plan Implementation Committee (RAPIC) was the key forum for the PAPs to engage in decision-making on the relocation process. It was the place where recommendations and negotiations on the execution of the RAP were taken in consultation with selected representative members of the PAP (Field Data 2018 and 2019).

The PAP had 24 participants included in the RAPIC, consisting of five gender-balanced representatives from each of the communities (three men, two women) (GIBB Africa 2012, p. 10). The youths, the vulnerable and disadvantaged members of the communities, the Council of Elders and the Administrative managers of the Cultural Center, each had one selected representative member in the RAPIC (GIBB Africa 2012, p. 10). Other RAPIC representatives included the Naivasha Deputy County Commissioner[10] who chaired the RAPIC, the county-level heads of line ministries, county-level heads of line ministries, one provincial-level administrative delegate,[11] and the KenGen implementation team. Since the World Bank's guideline on social safeguard was adopted, the framework and functions of RAPIC was therefore greatly informed by the guidance of the Bank's local social security experts. In addition to the RAPIC, there was also the institution of the Community Advisory Council (CAC). The CAC consisted of two elders from each village who had been chosen from among their peer and age groups. It was the first stage of the operational-level dispute mediation process, where complaints and grievances relating to resettlements and resettlement process were laid (Field Data 2018 and 2019). It was meant to provide recommendations and feedback to the developers and investors on how best to deal with problems such as land registry issues and how to navigate or compensate for activities that require meddling with certain culturally or spiritually sensitive sites (Field Data 2018 and 2019).

For the resettlement of the PAPs, KenGen built only 150 houses, although a later census suggested that it should be 180 houses. To explain this discrepancy, the Tacticus report explained that the "forgotten cases" claimed by some of the communities, in particular the Cultural Centre, did not change the number of PAP under the category of Land Owners with Assets, and thus did not affect the number of houses to be constructed (Tacitus 2012). Tenants who built their own houses were liable for monetary compensation at cost of construction but not to a house on the resettlement land (Tacitus 2012, p. 57). The same compensation mechanism applied to landowners who owned extra houses that they rented out (Tacitus 2012, p. 58). Business community members of the Cultural Centre that did not live in the village were considered not to be liable for compensation. This is because the Cultural Centre continued to exist in its original location as a business and commercial centre (Tacitus 2012, p. 33). The non-resident opportunists, who falsely claim abode among the PAPs, were also not considered (Field Data 2018 and 2019).

The PAPs claimed that the settlement was not culturally appropriate. The physical geography of the land, which had steep-sided valleys, restricted the accessibility and movement of polygamous men from one house-wife to another, a situation which was further exacerbated during the rainy seasons when the valleys and gullies are flooded. Since some of the households were at Narasha (—a village not available for resettlement), some polygamous families were divided. Furthermore, members of the Cultural Center, who were accustomed to living together in a clustered and circular environment in close proximity, considered it psychologically unacceptable

[10]The Naivasha District Commissioner had this position prior to the devolution of government.

[11]This position no longer existed after the devolution of government.

to live in isolated family groups at comparatively wide distances from each other (also see Schade 2017).

Concerning the allocation of houses, the PAPs claimed that there was insufficient consideration given to the needs of vulnerable persons (elderly, orphans, some female-headed households, and disabled). The Tacitus report stressed that GIBB Africa did not identify this category of PAPs and encouraged KenGen to quickly identify them and their needs in order to determine the type and level of support to be offered to them (Tacitus 2012, p. 59). Until the time of this study, such grievances persisted. In the same manner, several of the rights of indigenous people were not sufficiently addressed. The PAPs seemed not to be aware of these rights and therefore did not raise this, although conditions for addressing these needs were contained in the World Bank OP 4.10 on indigenous people (also see Schade 2017).

Furthermore, the Muslims among the PAPs complained about not been offered a mosque in the resettlement land, nor a travel cost compensation alternative for visiting the mosque located in Naivasha (Field Data 2018 and 2019). When confronted with this complaint, KenGen claimed that they were uninformed about this situation, stating that the census recorded the existence of only one Muslim family in the area and that they had limited funding to delve into religious requirements outside of the RAP compensation guidelines (Schade 2017).

Since relocation, the accessibility of the PAPs to each other decreased considerably due to expanded distances and limited access to existing transport facilities, with the Cultural Centre being the most affected (Field Data 2018 and 2019). On a regular basis, those PAPs who had previously stayed at the Business Center now had to cover a distance of 14 km in either direction (Schade 2017). The Tacitus study, had however, foresaw and listed this issue in their report (Tacitus 2012, p. 49), but it was not addressed (Field Data 2018–2020).

Regarding the question of the land's adequacy to maintain livestock and the absence of alternatives to address the issue, the Tacitus report makes the following statement:

> *"According to the GIBB report, the PAPs requested that even after the relocation, they should be allowed to continue grazing their livestock in the areas of their current settlements. KenGen, however, informed the PAPs that it could not commit itself to ensure the PAPs continue grazing in their current settlement areas since the land on which they are currently settled either belong to Kedong Ranch [a property associated with the current GoK president] or to KWS, and KenGen cannot make commitments on their behalf. KenGen hopes though, that because the resettlement site is in close proximity to the current PAPs settlements and grazing command areas, the status quo would be maintained. In this respect, if there was to be any interference by the legal landowners, it would not have been occasioned by the fact of the resettlement".*
>
> (Tacitus 2012 p. 16)

Another issue raised by the PAPs and the financiers was that it was necessary to make provision for a second piece of land for the Cultural Centre, in response to the land request made by the Centre, for the establishment of the Business Centre in a location elsewhere were it is possible to stay overnight (Schade 2017). KenGen agreed to the request and searched for suitable land north of the existing Cultural

Centre (Schade 2017). However, KenGen later rejected the identified parcel of land because of a pending court case and because certain development activities were already planned on the land (Field Data 2018 and 2019).

4.2.2 Investors and the Kenya Wildlife Service (KWS)

Before Hell's Gate was gazetted as a national park, KenGen had already acquired the land for geothermal development in the early 1960. Bearing this fact in mind, in addition to being a parastatal under the GoK, KWS was willing to negotiate and compromise where necessary to see to the development of the projects (Field Data 2019). KWS's key interests and concerns, however, bordered on water quality system for wildlife, air pollution, loss of food and forages for the animals as well as noise and vibrations (Field Data 2019).

In order to ensure that the interests of KWS were not completely compromised, KWS formed the Hells Gate Park Action Committee (PAC), which held monthly update-meetings. This provided a platform for the mediation of most sustainability concerns (Field Data 2019). The committee included all the concerned stakeholders, including KenGen, community representatives, GoK representatives, as well as Tourism hoteliers that are located around the Hell's Gate Park (Field Data 2019). KWS then held regular separate meetings with the DFIs and their consultants to report their sustainability concerns. This meeting between KWS and DFIs was of particular concern to KenGen as it got financing and technical support from DFIs on the condition that it address and adheres to environmental and social concerns stipulated by KWS (Field Data 2019). For KenGen, expression of dissatisfaction had the potential to slow the release of expected financial assistance from the DFIs.

Within these mediating frameworks, KenGen agreed to wear KWS recommended clothing colours, adhere to speed limits, move away from strategic animal habitats and other potential areas that many endanger the animals, while carrying out operations in the Hell's Gate park (Field Data 2019). When asked on the efficiency of the PAC, a KWS senior warden in charge of the negotiations noted:

> *"The committee really assisted. Because if you are dealing as one person against them, it is very difficult to achieve any results. There were local people [community representatives and Tourism Hoteliers] who asked very difficult questions that a government officer would not ask another government officer. They asked the hardest questions and that really helped to balance out some of the sustainability concerns".*
>
> (Field Data 2019).

When asked to rate the extent to which KenGen adhered to their stipulated sustainability concerns, the KWS senior warden replied:

> *"We did not achieve everything, but they were a very responsible organization...I would place them at 80%".*
>
> (Field Data 2019).

5 Discussion and Conclusion

There is no doubt that African countries appreciate the sustainable development approach as critical for the realization of complex development problems. This is visible in both regional and country-specific development blue prints and in the attempts towards transitioning to green economies. However, reconciling the three dimensions of sustainability—society, economy and equity—is rather difficult given the diverse continental and state-specific priorities. As OECD (2013:27) observes, developing countries have the greatest need and demand for economic growth and welfare improvement in the short term, which makes the balance between more long-term welfare gains from socio-environmental improvements, challenging. Our case study analysis reflects this observation. In the development of Olkaria geothermal projects, diverse investors face challenges in balancing the delivery of a cost-effective and an economically viable project with the socio-economic expectations of local communities and the environmental conservation concerns of conservationist institutions/groups. As such, the interests and priorities of the investors' often clash with those of the local host communities and conservationists. Investors are not able to meet all the expectations of their face-offs, thus the phrase "*But we cannot do it all*" from a senior staff in the project developing company, KenGen.

We also see that investors manage these sustainability tensions through strategic selectivity. They tend to focus on implementing certain components of the sustainability triad (mainly the economic components) while neglecting the others (mainly social and environmental components). In this manner, the investors interests and priorities, take the upper hand, and geothermal projects are pursued and advanced notwithstanding the existence of some unresolved sustainability issues, especially regarding land and compensation. When the sustainability principles are juxtaposed, we find that the economic dimensions (economic viability and cost effectiveness) of the sustainability triad are mildly compromised as against the largely pending and unaddressed socio-economic welfare, livelihood, bio-diversity and environmental issues and concerns, contained in the social and environment dimensions. These findings resonate more generally with current experiences of large-scale investments throughout Africa (Lind et al. 2020).

The study, however, shows that strategic selectivity is not acted out in a 'blunt', direct and purposeful manner. Investors do make efforts to balance-off these sustainability tensions. Such efforts include the establishment of rules, laws and frameworks (such as the ESIAs and the Community Land Acts), the implementation of Corporate Social Responsibility (CSR) projects and activities, and the formation of stakeholder groups (also see Klagge et al. 2020). These endeavors have, however, proven to be insufficient as the socio-economic concerns of local communities are not fully and sustainably addressed. In other to improve sustainability outcomes, there is need to prioritize in-person monitoring of the implementation of the projects to ensure that they go on according to stipulated standards as contained in the ESIAs, Laws and Acts (also see Muthuri and Gilbert 2011). Furthermore, mechanisms and spaces for interactions between the investors and the local communities and conservationists, should be increased, and made more inclusive, diverse and representative of affected communities and conservationists. This will foster proper deliberation and consideration of sustainability options among all stakeholders during the planning and implementation stages of the projects.

Acknowledgements We thank the interview partners for their time during fieldwork. We also thank the two anonymous reviewers for their helpful suggestions. This research was supported by the German Research Foundation (DFG) through funding for the project "Energy futures" as part of the collaborative research center "Future Rural Africa" (Project-ID 328966760—TRR 228).

Disclosure Statement: No potential conflict of interest was reported by the authors.

References

AUC – African Union Commission (2015) Agenda 2063: The Africa We Want. Framework Document. African Union Commission, Addis Ababa

Ahenkan A, Osei-Kojo A (2014) Achieving sustainable development in Africa: progress, challenges and prospects. Int J Dev Sustain 3(1):162–176

Alexander P (2015) Corporate social irresponsibility. Routledge, London

Brix-Asala C, Geisbüsch A-K, Sauer PC, Schöpflin P, Zehendner A (2018) Sustainability tensions in supply chains: a case study of paradoxes and theirmanagement. Sustainability 10(2):424

Carroll AB (1999) Corporate social responsibility: evolution of a definitional construct. Bus Soc 38(3):268–295

EIB-CM – European Investment Bank Complaints Mechanism (2015) Conclusions Report: Olkaria I and IV, Kenya, Complaint SG/E/2014/07 and SG/E/2014/08. EIB-CM, Luxembourg

Elkington J (2004) Enter the triple bottom line. In: Henriques A, Richardson J (eds) The triple bottom line, does it all add up? assessing the sustainability of business and CSR. Earthscan, London

Enns C, Bersaglio B (2020) On the coloniality of 'new' mega-infrastructure projects in East Africa. Antipode 52(1):101–123

GIBB Africa (2009a) Environmental and Social Impact Assessment (ESIA) Report (No. E2337V6). Nairobi, Kenya

GIBB Africa (2009b) KenGen Olkaria IV Project - Resettlement Action Plan. Nairobi, Kenya

GIBB Africa (2012) Olkaria IV (Domes) geothermal project in Naivasha District: resettlement action plan for Olkaria IV power station. GIBB Africa, Nairobi

GoK – Government of Kenya (2007) The popular version. Government of Kenya, Nairobi

GoK – Government of Kenya (2013) Kenya national climate change action plan (2013–2017). Government of Kenya, Nairobi

Gray R, Milne M (2004) Towards reporting on the triple bottom line: mirages, methods and myths. In: Henriques A, Richardson J (eds) The Triple Bottom Line does it all add up? Assessing the sustainability of business and CSR. Earthscan, London, pp 70–80

Hahn T, Pinkse J, Preuss L, Figge F (2015) Tensions in corporate sustainability: towards an integrative framework. J Bus Ethics 127(2):297–316

Jamali D (2008) A stakeholder approach to corporate social responsibility: a fresh perspective into theory and practice. J Bus Ethics 82(1):213–231

Jayanti RJ, Gowda MVR (2014) Sustainability dilemmas in emerging economies. IIMB Manage Rev 26:130–142

Jensen MW (2019) The impact of renewable energy projects on indigenous communities in Kenya: the cases of the Lake Turkana Wind power project and the Olkaria geothermal power plants. (IWGIA Report 28). The International Work Group for Indigenous Affairs (IWGIA), Copenhagen

Klagge B, Nweke-Eze C (2020) Financing large-scale renewable-energy projects in Kenya: investor types, international connections, and financialization. Geografiska Annaler: Series b Hum Geogr 102(1):61–83

Klagge B, Greiner C, Greven D, Nweke-Eze C (2020) Cross-scale linkages of centralized electricity generation: geothermal development and investor-community relations in Kenyas semi-arid north. Politics Gov 8(3):211–222

Koissaba B (2018) Green energy and climate change a threat to indigenous peoples in Kenya. Cross cultural consultants, Nairobi

Lange D, Washburn N (2012) Understanding attributions of corporate social irresponsibility. Acad Manage Rev 37(2):300–326

Leach M, Scoones I, Stirling A (2010) Dynamic sustainabilities: technology, environment, social justice. Pathways to Sustainability. Earthscan, Oxon/New York

Lee KM (2007) So what is 'Tripple-Bottom Line.' Int J Diversity Organ Commun Nations 6:1447–9583

Lind J, Okenwa D, Scoones I (eds) (2020) Land, investments & politics: reconfiguring Africa's pastoral drylands. James Currey, Woodbridge

Maino R (2019) Olkaria PPP investor conference 140MW PPP geothermal project. Request for qualifications stage. Kenya Electricity Generation Company (KenGen), Nairobi

Mariita NO (2002) The impact of large-scale renewable energy development on the poor: environmental and socio-economic impact of a geothermal power plant on a poor rural community in Kenya. Energy Policy 30(11/12):1119–1128

McIntosh M (2003) Raising a ladder to the moon: the complexities of corporate social and environmental responsibility. Palgrave Macmillan, New York

Muthuri J, Gilbert V (2011) An institutional analysis of corporate social responsibility in Kenya. J Bus Ethics 98(3):467–483

Newig J, Voβ J-P, Monstadt J (2007) Governance for sustainable development: coping with ambivalence, uncertainty and distributed power. J Environ Planning Policy Manage 9(3):185–192

OECD – Organisation for Economic Co-operation and Development (2011) Annex 8 - mutual reliance initiative (April). OECD, Paris

OECD – Organisation for Economic Co-operation and Development, (2013) Putting Green Growth at the Heart of Development, OECD Green Growth Studies. OECD Publishing, Paris

Omenda P, Simiyu S, Muchemi G (2014) Geothermal country update report for Kenya: 2014, Africa rift geothermal (ARGeo)-C5 proceedings, 29–31 October 2014. ARGeo, Arusha

Ramos-Mejia M, Franco-Garcia M-L, Jauregui-Becker JM (2018) Sustainability transitions in the developing world: challenges of sociotechnical transformations unfolding in contexts of poverty. Environ Sci Policy 84:217–223

Romijn H, Raven R, de Visser I (2010) Biomass energy experiments in rural India: insights from learning-based development approaches and lessons for Strategic Niche Management. Environ Sci Policy 13:326–338

Schade J (2017) Kenya 'Olkaria IV' case study report: human rights analysis of the resettlement process. Working paper 151. Centre on Migration, Citizenship and Development (COMCAD), Bielefeld

Scoones I, Newell P, Leach M (2015) The politics of green transformations. In: Scoones I, Leach M, Newell P (eds) The politics of green transformations. Routledge, London, pp 1–24

Shove E, Walker G (2007) CAUTION Transitions ahead: politics: practice and sustainable transitions management. Environ Plan A: Econ Space 39:763–770

Tacitus (2012) Short-term technical assistance to KenGen for the implementation of Olkaria IV Resettlement Action Plan (RAP): Olkaria IV RAP Implementation Process (No. RP883 V17). Tacitus Ltd., Nairobi

UN – United Nations (2008) Achieving sustainable development and promoting development cooperation. department of economic and social affairs, office for ECOSOC support and coordination. United Nations, New York
UN – United Nations (2012) The future we want: outcome document of the united nations conference on sustainable development. United Nations, Rio de Janeiro
UN – United Nations (2013) World economic and social survey 2013: sustainable development challenges. United Nations, New York

Systems Solutions for Enabling Off-Grid Rural Communities to Have Access to Water and Energy: Implications for Sustainability and Policy Reforms

Johan Meyer, Sune von Solms, and Taffy Makonese

Abstract Access to clean energy and water can make substantial contributions towards promoting rural development, particularly in poor off-grid communities in developing countries. To gain a better understanding of the relationship between clean energy access and sustainable development in rural communities, this chapter employs a systems thinking/dynamics approach toward the achievement of selected sustainable development goals (SDGs) through co-created solutions in an off-grid rural village in Northern Limpopo Province, South Africa. A systems model is constructed to evaluate the contribution that access to clean energy and water will make to rural livelihoods. Presented is a model for the development and implementation of renewable energy solutions in order to provide access to clean energy and water as enablers for the other SDGs. Photo Voltaic solutions are utilised to facilitate subsistence and small-scale commercial farming, provide energy to households for lighting and entertainment, and power a bakery to provide employment and income for the local community. Results highlight how poor, marginalised and off-grid communities could attain socio-economic, human and physical development by making sustainable use of locally available natural resources through solar energy and related technologies. The findings in this article have implications for the sustainability of and future energy policies and strategies for off-grid rural communities in South Africa and across Africa.

Keywords Dynamic systems modelling · Energy solutions · Sustainable development goals · Systems thinking · Water security

J. Meyer (✉) · S. von Solms
Department Electrical and Electronic Engineering Science, University of Johannesburg, Johannesburg, South Africa
e-mail: johanm@uj.ac.za

T. Makonese
Process Energy and Environment Technology Station, University of Johannesburg, Johannesburg, South Africa

W. Leal Filho et al. (eds.), *Sustainable Development in Africa*, World Sustainability Series,
https://doi.org/10.1007/978-3-030-74693-3_22

1 Introduction

Water and energy supply or lack thereof may be either drivers or constraints on human development. Food, water and energy security form the basis of a self-sufficient economy, but as a water-scarce country with little arable land, and 60% of rural households having no access to electricity, South Africa's economy is testing the limits of its resource constraints (Shezi 2015). These basic resources are intricately intertwined, and their interactions are inherent in the energy-water-food (EWF) nexus (Fig. 1). When formulating policy from an energy perspective, the fact that the production of water and food requires the use of energy, needs to be taken into consideration. Food and water supply require significant amounts of energy (Bazilian et al. 2011). Hence, access to energy in rural areas leads to improvements in the supply of clean water, improvements in food production (SDGs 2 & 6), as well as improved health and poverty reduction (SDGs 1 & 3) (Bieber et al. 2018). It is therefore a key driver for economic growth and good jobs (SDG 8). However, meeting the SDGs using coal-fired power generation instead of renewable energy sources will release more greenhouse gases (GHGs), which will impact climate change (SDG 13) thereby threatening longer term energy and water security, and consequently food supply (Bieber et al. 2018).

Globally, there are over 1.4 billion people without access to electricity (Keairns et al. 2016), 900 million without access to safe water, and 2 billion people without food security (Bazilian et al. 2011). In the sub-Saharan region of Africa specifically, about 650 million lack access to electricity and rely on solid fuels including biomass, plant residues, and animal waste to meet their basic energy needs for rudimentary and inefficient cooking devices (Mungwe et al. 2016). It is estimated that more than half

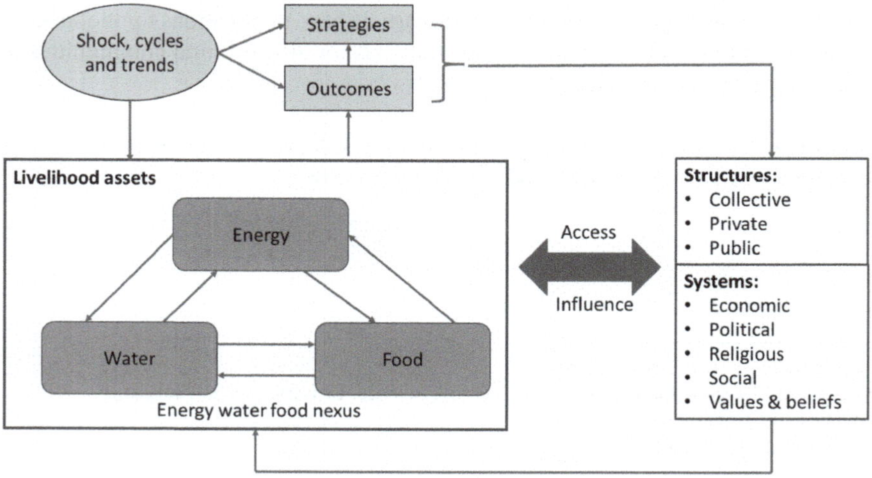

Fig. 1 Illustrating how basic resources are highly and intricately intertwined, and their interactions are inherent in the energy-water-food nexus within the sustainable livelihood framework

of the world's population will be residing in water stressed areas by 2030 (World Water Assessment Program 2012). According to the World Bank (2016), a boom in the global population is expected to impose a strain on food resources with a proportional increase in food demand by 2050. These forecasts point to the intrinsic interactions between various resources and the need to meet future demands without sacrificing any of the resources in the nexus (Bieber et al. 2018). Studying the EWF nexus is imperative for an understanding of how the resources interact in the system in order to provide balanced solutions to the challenges affecting communities (Chen and Chen 2016). Each of the three areas in this nexus affect the others in complex and substantive ways so that ignoring challenges in one could lead to significant impacts in another (Bazilian et al. 2011).

Rural electrification has been suggested as a policy instrument for improving energy access. However, studies carried out in South Africa have revealed that most rural households that are connected to the grid continue to use traditional fuels, with electricity being used for lighting, refrigeration, and entertainment (Madubansi and Shackelton 2006). The cost of electricity and related technologies have been identified as the major impediments to the uptake of electricity in marginalised communities and rural areas. In another study, it has been argued that the erratic nature of grid electricity supply leaves households with no choice but to revert to conventional energy sources (Masekoameng et al. 2005).

Notwithstanding the benefits of grid electricity to such communities, there is a great need to look at challenges the communities face from the point of view of sustainable livelihoods, which integrates some facets of systems thinking. Most of these households are energy poor as they tend to expend a higher proportion (>10%) of their disposable income on energy services (Lloyd 2014). The interrelation between energy and irrigation has become a key issue in South Africa given the massive year to year electricity hikes. Agriculture is likely to be worst affected by these tariff hikes due to its energy demand for irrigation (Bazilian et al. 2011). Decreasing irrigation and relying on rain-fed agriculture could have dire impacts on food production and food security owing to droughts being a function of climate change. This will be likely to push vulnerable communities deeper into poverty. An available solution, which is not exploited in most off-grid communities in South Africa, involves the use of distributed photovoltaic powered water pumps to supply enough water from boreholes for drinking and agricultural purposes (Bazilian et al. 2011).

Concerning the issue of improved livelihoods and poverty reduction, it should be noted that moving out of general poverty is not a straightforward process. It is a dynamic process that is influenced by several factors including natural, economic, and political ones (Apine et al. 2019). Water and food are "knock on sectors" when energy access is improved in off-grid and marginalised communities. As such, a systematic and coordinated approach becomes imperative (Bazilian et al. 2011) in the context of rural development. Thus, the promotion of access to energy should be done together with the determination and establishment of sustainable livelihood sources for marginalised and vulnerable communities. According to Apine et al. (2019), the sustainable livelihood approach is people-centred and has been employed to identify practical priorities in development programmes aimed at reducing poverty and empowering target communities. However, it should be noted that sustainable livelihood approaches are not a universal remedy and as such should not be used to replace other important tools in development programmes such as sector-wide approaches, participatory engagements, or integrated rural development (Serrat 2017).

Mungwe et al. (2016) believe that these challenges cannot be resolved through isolated technologies but a systems approach to a more comprehensive energy access strategy. A systems approach is required to address development and sustainability challenges in rural communities and evaluate optimisation strategies for infrastructure investment so that these communities can attain socio-economic, human, and physical development. Computational modelling, which is embedded in systems dynamics, is a robust tool that is used to analyse the interactions within the EWF nexus while providing solutions for sustainable decision-making (Bieber et al. 2018). Hence, the systems thinking approach is an interdisciplinary method that is applied to situations to understand complex inter- and intra-cyclical effects between variables in a system, unlike linear approaches that deal with cause-and-effect relationships (Makonese and Meyer 2018; Mavhura 2017). Causal-loop diagrams are used to illustrate feedback from interrelationships that exist between variables in a system (Banson et al. 2015). A systems thinking approach to rural energy access is needed considering that most of the factors affecting the uptake and use of energy carriers and related technologies are becoming increasingly covariant.

In modelling the EWF nexus, Johnson et al. (2015) opined that only 4% of the studies relating to this nexus addressed it in its entirety, with the majority focussing on only two of the three EWF sectors, for example, the optimization framework proposed by Dubreuil et al. (2013) on the interaction between water and energy. An emerging trend in EWF nexus modelling is to study the three systems together both for synergies and to identify improvements in resource utilization and resource efficiency (Bieber et al. 2018; Fang and Chen 2017; Kan et al. 2016). Furthermore, there is a paucity of information in the literature on energy access from solar-powered mini-grid systems and its associated links to water access and food security in the context of a rural village.

There is, therefore, a pressing need to analyse the levels at which factors influence energy access, water access, and food security in marginalised rural communities. It is possible, at the community level, to analyse interrelationships between shocks, stresses, vulnerability, resilience, and the well-being of households (Mock et al. 2015). However, households are a part of broader, complex, and interconnected sub-systems that interact with shocks and stresses. Information obtained from the household level helps us understand the nature and determinants of the absorptive, adaptive, and transformative capacities inherent in these social systems (Mavhura 2017).

Access to clean energy and water can make substantial contributions towards promoting rural development, particularly in poor off-grid communities in developing countries. The aim of this paper is to gain a better understanding of the relationship between clean energy access and sustainable development in rural communities, this chapter employs a systems thinking/dynamics approach toward the achievement of selected sustainable development goals (SDGs) through co-created solutions in an off-grid rural village in Northern Limpopo, South Africa. This paper presents the development and implementation of renewable energy solutions to provide access to clean energy and water as enablers for the other SDGs, using Photo Voltaic solutions for small-scale commercial farming, household energisation and economic development. The chapter follows with an introduction to the Gwakwani village leading to the development of systems dynamic models for the energy-water-food nexus in Sect. 3. Results from four case scenarios are presented and discussed in Sect. 4. The implications of the results obtained from the dynamic systems modelling are elaborated on in Sect. 5. The chapter ends with a conclusion contextualising the work in terms of sustainable development in Africa.

2 The Gwakwani Village Case

Gwakwani is a Venda word meaning "armpit", as the small village is in the armpit of the Mutala River in the Northern part of the Vhembe District of Limpopo Province in South Africa, which has international borders with Botswana, Zimbabwe, and Mozambique. One dirt road provides access to the village, which is located approximately 17 km from the nearest town and closest petrol filling station, and approximately 650 km from the industrial hub of Johannesburg, Gauteng Province. The geographic position of Gwakwani Village is shown on the map presented in (Fig. 2).

The village has a population of between 80 and 100 people, all of whom have Venda as first language. Very little English is spoken. As a result of the economic inactivity in the village most able household members are forced to find employment at a considerable distance, in industrial and urban areas. This leaves the village population made up mostly of the elderly, who look after the children. From a survey conducted by the University of Johannesburg in Gwakwani Village in 2016, the average age of the house owners was 51 years of age with an average household occupancy of 4 persons, as shown in Fig. 3.

The lack of a reliable and affordable electrical grid supply forces villagers to obtain illumination in the form of candles and open fires fuelled by wood collected in the surrounding bushes. Those who can afford to do so, use batteries to power radios or small television sets. The geographical location of the village is in a natural depression, resulting in severely restricted cellular signal coverage limited to fixed spots such as a specific tree or fence post where a sliver of the signal is available. Cellular phones require regular charging of batteries which can only be achieved by travelling to the nearby village to obtain an electricity point and paying between R5 and R10 for the charge.

In the Vhembe District, only 7.5% of people have piped water in their dwellings, 27.9% have access to piped water at a distance further than 200 m from their dwellings, 2.6% are reliant on borehole water, and sanitation for 60% of the inhabitants consists of a pit latrine (Limpopo State of the Environment Report 2004).

Fig. 2 Geographic location of Gwakwani Village in the Northern Limpopo province of South Africa

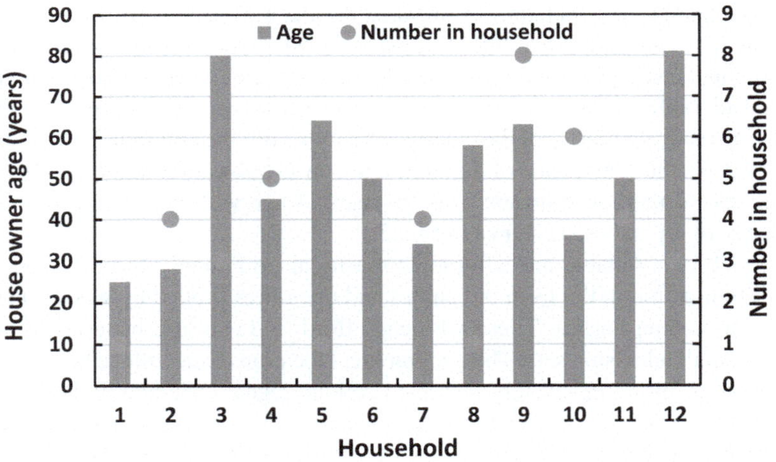

Fig. 3 Household occupancy and house owner age

The main water supply of Gwakwani Village is obtained from a diesel-driven borehole pump. Water from the borehole is pumped into a 10 000-L tank located approximately 500 m from the borehole and is situated on a geographically high point in the village. Water is distributed using gravity, and feeds into a pipeline running along the main road through the village. Communal tap points are located at three different locations along this water supply line. Diesel for the pump is obtained from the South African Department of Water Affairs depot located approximately 14 km away. Water is usually pumped into the tank once a week, but only as long as the diesel supply lasts. Once the monthly diesel supply runs out, the villagers are responsible for collecting money to procure and obtain diesel for the pump. This process results in a limited and unpredictable water supply.

Although the village is located next to the Mutale River, the water in the river is seasonal, with the rainy season being mostly in the late summer months. The water supply from the river is shared with the livestock which makes the water quality unsuitable for human consumption. Malaria poses a real threat to anyone collecting water from the river, especially during late evening or early morning, since the village is located in a moderate risk malaria region.

The community can be classified as part of the lower socioeconomic segment of the global population who make a living from subsistence farming. The village relies, for financial survival, on those in the household who are employed in the urban or industrial regions of the country and on the social grants scheme offered by the government. The deficit in income is supplemented by subsistence farming.

This study uses indicators developed from the Sustainable Development Goals (SDGs). The authors have used this as a guide to develop sustainable development for rural areas. As in Yuan et al. (2020), three broad perspectives are utilized, namely availability, access, and stability. Table 1 shows the attributes and indicators adapted from the SDGs for specifically rural areas. These attributes were considered during the development of the cocreated solutions in the case of the Gwakwani village.

3 Systems Dynamics Modelling

A dynamic systems model was developed which allowed for rudimentary system dynamics modelling of the rural village. Causal loop diagrams and stocks and flow diagrams were employed to ascertain the levels of association between the variables in the system. The model simulated the population numbers, the availability of useable water from natural resources, and village wealth as a result of farming activities. Stella professional desktop software was used to implement the dynamic modelling in the case study area of Gwakwani Village, Limpopo, South Africa. This software is developed by ISEE Systems. An advantage of this software is that it allows someone with minimal mathematical modelling skills to use it to utilise nested feedback loops and non-linearity models from differential equation models of a real-world system (Nguyen et al. 2017). The model for the village population is depicted in Fig. 4.

Table 1 Attributes and indicators for rural sustainability derived from the SDGs—based on and adapted from Yuan et al. (2020)

Sector	Attribute	Indicators and description
Energy	Access	*Access to electricity*—simple, stable access to electrical energy (electrification)
	Availability	*Energy consumption*—the amount of energy used to meet basic needs
	Stability	*Renewable source*—an energy source that cannot be depleted and can be replenished naturally
Water	Access	*Access to water*—stable access to clean, safe water on the premises, to enable an active and healthy lifestyle
	Availability	*Water consumption*—the amount of water consumed and not returned to the water source after being withdrawn
	Stability	*Water sources*—total water bodies including rivers, dams, boreholes, aquifers, springs, and wastewater collection etc
Food	Access	*Nutrition*—The minimum dietary requirements to maintain an active and healthy lifestyle
	Availability	*Food production*—the amount of food produced per unit area of arable land
	Stability	*Farmland*—Arable land including pastures, meadows, and permanent crops

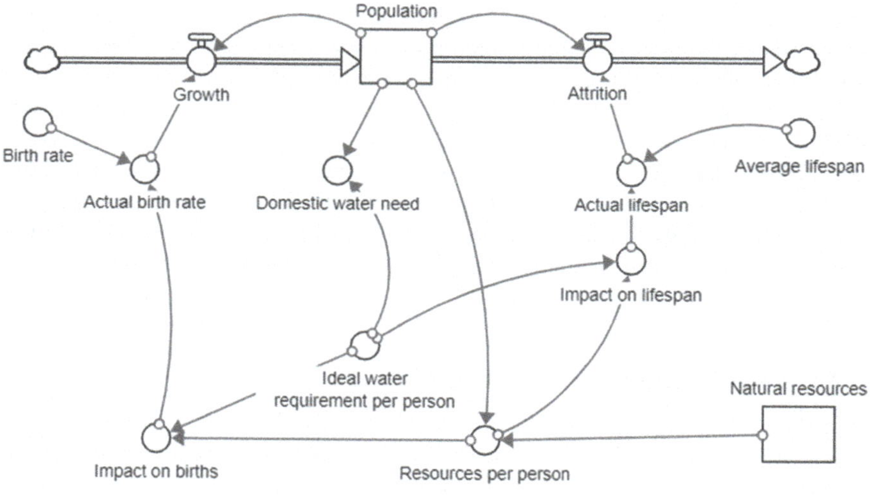

Fig. 4 Systems dynamic model for determining the population of the village

With reference to Fig. 4, the total population of the village is modelled as a stock, *Population,* with an inflow of *Growth,* which includes natural births according to a predefined *Birth rate*. The *Birth rate* is impacted by the availability of resources. The outflow from *Population* is through the *Attrition* rate which is impacted by the natural lifespan of a person and the availability of *Resources per person*. The idea is when more than the ideal amount of resources is available, the village will increase in *Population* by natural birth as well as by attracting people from neighbouring

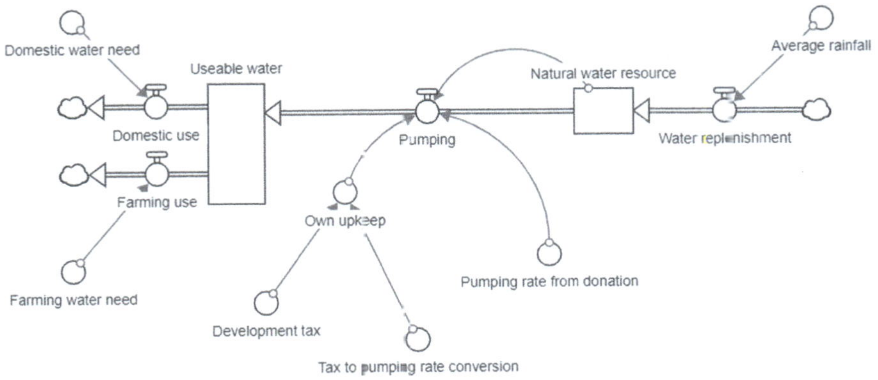

Fig. 5 Useable water availability model indicating the extraction of water from the available natural resources through pumping obtained from donation or re-investment

villages. The *Population* decreases as a result of the natural death rate, and by people leaving the village as a result of a drop in *Resources per person* to below the ideal number of resources per person.

The available water is modelled as shown in Fig. 5. The *Usable water* stock is obtained by pumping water from the *Natural water resource* stock, which is replenished by the *Average rainfall*. In the model, it is assumed that all the water obtained by replenishing of the *Natural water resource* remains available for sequestration into *Useable water*. For simplicity's sake, the model does not consider the loss of available *Natural water resource* through runoff, evaporation, and droughts. The outflow of the *Natural water resource* is effected by pumping of water. The pumping rate, *Pumping* is dependent on two main factors, firstly, the amount of water that can be pumped as a result of a once-off donation of solar-powered water pumping equipment and secondly, water pumped as a result of investing back into infrastructure development by means of some tax collection, *Development tax*. This mechanism allows for the modelling of an increase in ability to utilise more of the natural water resource as a result of the village's increases in wealth and population. The *Usable water* stock is increased through the inflow from *Pumping* water from the *Natural water resource* stock. The *Usable water* stock is depleted by two outflows namely *Domestic use* and *Farming use*. For the sake of simplicity, water losses through leaks and broken infra structure is not considered.

The *Donation ability* is the ability from the donation to fulfil its intended purpose. The *Donation ability* is increased through an inflow of *Equipment upkeep* based on the *Maintenance rate* provided by the equipment donor. The *Donation ability* is reduced by an outflow of *Equipment decay* as a result of natural equipment decay and failure over time. The model for the *Donation ability* is shown in Fig. 6.

The water needed for farming activities is modelled as shown in Fig. 7. The farming activities considered are the keeping of livestock and the planting of crops. The *Total number of livestock* is computed from the *Average number of livestock per person* and the *Population*, adjusted by the *Impact on births* index which is

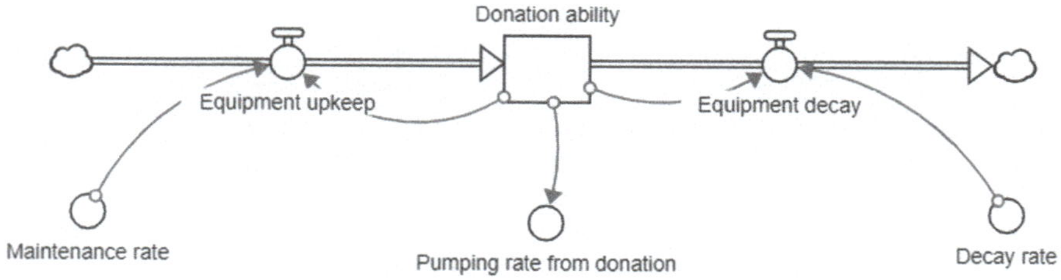

Fig. 6 Donation ability model

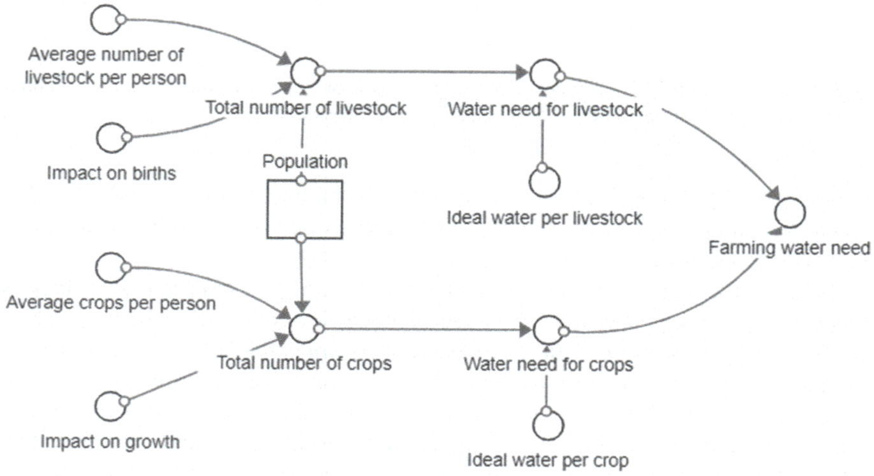

Fig. 7 Farming water need model, taking into account livestock and crop watering needs

dependent on the availability of resources. The *Total number of crops is* computed from the *Average number of crops per person* and the *Population,* adjusted by the *Impact on growth* index which is dependent on the availability of resources. *Ideal water per livestock* and *Ideal water per crop* values are based on the ideal amount of water required for crops and livestock of this nature.

The community wealth is modelled as shown in Fig. 8. The *Wealth* stock is increased by the inflow of *Wealth creation* which consists of two parts: namely, the *Livestock value* and the *Crop value.* The *Wealth* of the community is depleted by two outflows: namely, the *Cost of living* and a *Development tax* used for increasing the ability to utilise the natural resources. A variable *Tax rate* is used to control the percentage of tax allocated for reinvestment in development for utilising more of the natural resources. The *Cost of living* is assumed to be a fixed value per person.

The dynamic systems model was also used to gain a better understanding of the results of four cases. The aim of the modelling was not an attempt to predict absolute values of population numbers or wealth but rather to observe the trends resulting from these scenarios. The four cases that were modelled as scenarios are as follows:

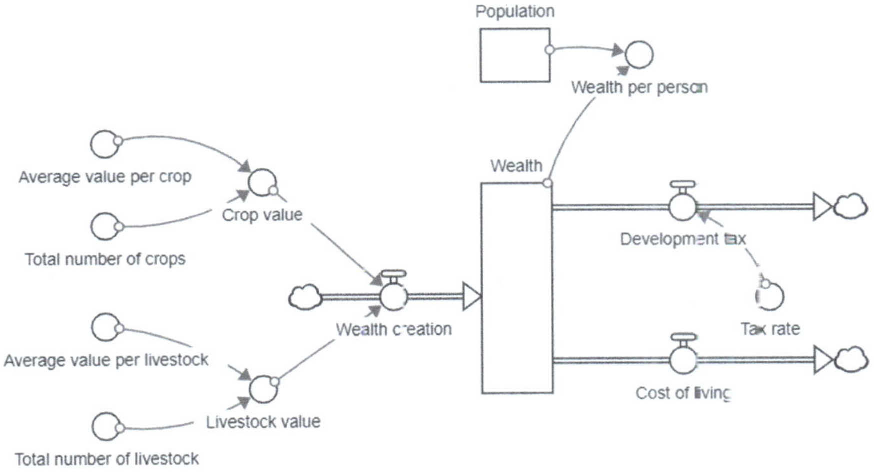

Fig. 8 Community wealth model

Case 1 aimed to confirm the results of what happens most often in these community engagement projects classified as "Give and Forget" projects. A once-off donation from a donor or benefactor organisation is employed to install a typical solar water pumping scheme in the village to harvest the natural water resource. The donation is a once-off donation meaning there is no follow up growth investment from the donor or any provision for maintenance of the equipment. The village residents are receiving the benefit for free without paying any service delivery fees or contributing to any maintenance costs incurred by the donated infrastructure. The installed infrastructure is subject to natural wear and tear over time, resulting in equipment degradation and eventual catastrophic failure.

Case 2 considered the scenario where the donor or benefactor organisation included a maintenance mechanism into the beneficiation agreements. The equipment is subsequently maintained and kept in the original capability state. There is, however, no provision made for any expansion or growth of the originally supplied donated infrastructure. This Case is typical of the Gwakwani village case where the University of Johannesburg and the project partners, as stakeholders in the project, are maintaining the donor equipment, since the university and project partners obtain some form of benefit from the project which offsets the costs incurred. The village residents are receiving the benefit for free without paying any service delivery fees or contributing to any maintenance costs incurred by the donated infrastructure.

Case 3 investigated the effect on village development when the village residents are required to invest back into the growth and maintenance of the installed infrastructure. This scenario investigates the ability of growth in the village as a result of a "tax" or service delivery fee rendered to allow for expanding the infrastructure's ability to utilise the available natural water resource. This would typically mean the installation of additional photovoltaic panels, additional boreholes, pumps, storage tanks, and distribution networks. The donor infrastructure is maintained by the benefactor to the original capability state.

Case 4 investigated the effect of overexploitation of the available natural water resource. As in Case 3 the village residents are required to invest back into the growth and maintenance of the installed infrastructure which allows for growth in the village as a result of a "tax" or service delivery fee rendered for expanding the infrastructure's ability to utilise the available natural water resource, but also models a scenario in which the replenish rate of the natural water resource is reduced to allow the rate of utilisation to reach the natural replenish rate. In this case, the village is using all the available natural water resource.

4 Results

A safe, clean, affordable and easily accessible water supply is needed for good health and sustainable food supply. In off-grid rural communities, access to energy is a critical input to the water supply. Before bringing solutions to the village, the households were dependant on the water supply from the nearby river and did not have access to running water. As the river is infested with crocodiles, and there have been reports of livestock being attacked by crocodiles at the watering holes, a direct consequence was that the villagers lost wealth, sank deeper into poverty, and became entangled further in poverty traps. Moreover, food supplies from animal products were diminished.

The perturbations to the systems dynamic modelling were largely limited around the supply of potable water as a fundamental requirement for sustainable development.

The effect of these four cases on the village population is shown in Fig. 9. It can be seen in Case 1, where the infrastructure is not maintained, that the village population declines rapidly. The infrastructure decays and can no longer supply the required amount of water per person, with the result that the village's attractiveness decreases, and, as a consequence, people leave the village to find a life somewhere else. Thus, the population decreases rapidly as a result of the inability of the deteriorating infrastructure to supply the watering demand.

As seen from the simulation, support for projects after the initial investment is crucial for the sustainability of the project. In many cases, a once-off donation of equipment or infrastructure is provided in a project where little to no support and maintenance are provided. Not supporting a project throughout the intended project life-cycle and also not providing a sustainability plan for the continuation of the project after the project time has come to a close, almost immediately negates all the growth obtained through the project as can be seen by the simulated model. It is therefore essential for the sustainability of the project itself that a service delivery approach must be followed where long-term services are provided, rather than a project stand-alone approach (Naidoo and Meyer 2015; Dickinson and Bostoen 2013).

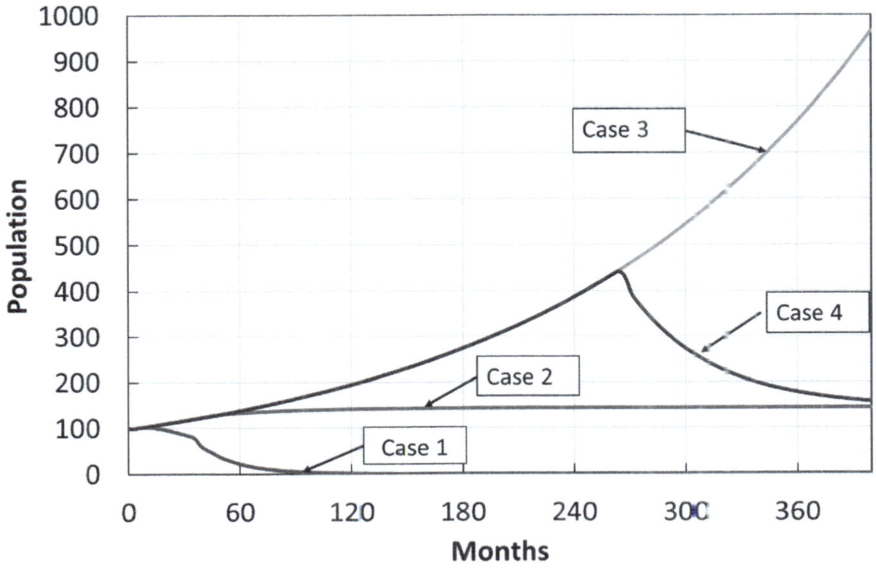

Fig. 9 Village population for the four cases evaluated

For Case 2, the infrastructure is maintained. This means that the capability to sequestrate the natural water resource into a useable supply of water is maintained at a constant level. There is an initial increase in population since the capability of supplying useable water from the natural resources exceeds the demand of the village which makes the village attractive, thus enhancing the growth in population. However, as soon as the water demand equals the ability of the infrastructure to supply, the village population stabilises without any further growth. The natural population growth as a result of births and immigration into the village is matched by people having to leave the village due to an insufficient basic water supply. The stabilised population level would be dependent on the ability of the donated infrastructure to provide the necessary water demand; however, the population will only grow to the point where the infrastructure can no longer meet the demand.

In Case 3, the ability of the infrastructure to supply the demand grows because of re-investment in the form of a service delivery fee, and the population numbers increase exponentially. The population would keep on increasing up to the point where the rate of water demand increase exceeds the natural water resource supply replenish rate as shown in Case 4. After the point where the supply exceeds the demand, the attractiveness of the village decreases with the result of people leaving the village. It is interesting to note that the rate of population decline is rapid after the point where the minimum amount of water needed per person can no longer be sustained. This would manifest in the reduction of crops and livestock per person, and eventually, the necessary exit of village people themselves from the village. It was noted from the model, that even with a very small service delivery fee, exponential

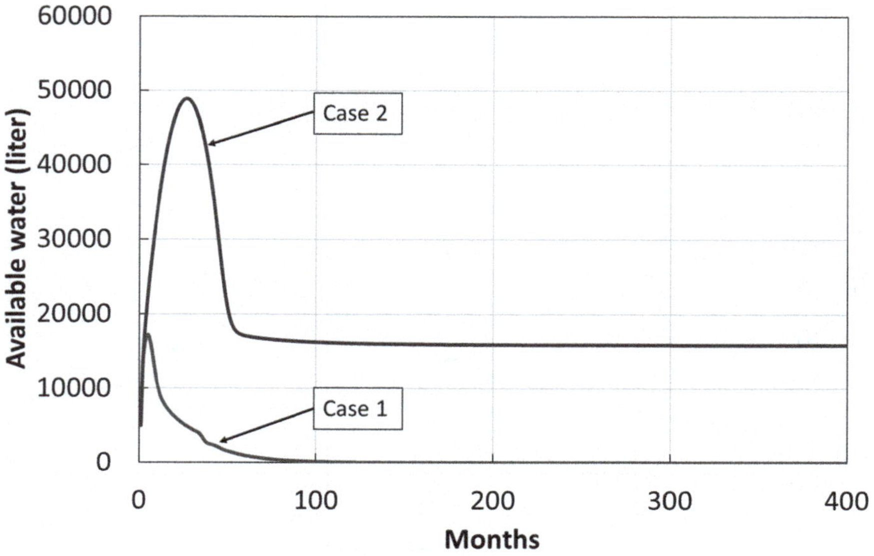

Fig. 10 The available water for Case 1 and Case 2

growth in population numbers is possible. It was also noted that an increase in the service fee does not produce a higher rate of population increase; it does, however, reduce the wealth per person. The dynamics associated with revenue collection and population was not the focus of this study and would require more intensive modelling which is beyond the scope of this research.

The available water for use as sequestrated from the natural water resource is shown in Fig. 10. In Case 1, the lack of upkeep and maintenance of the infrastructure results in the decline of the available water supply, which is expected. However, when the infrastructure is maintained as in Case 2, there is an initial surplus in available water which is taken up by the increased population and inherent farming water needs to the point where the water need and water supply is equal, and the population numbers stabilise. This again highlights the need for some form of donor agreement that includes the maintenance of the donated infrastructure. Without the ability to maintain the donor infrastructure, equipment degradation and eventual failure as shown in Case 1 (Fig. 10) results in available water supply decreases.

Under Case 4, where a service delivery fee is charged for the upkeep and growth of the infrastructure that is used to sequestrate the natural water resource into an available water supply, the population's water needs will increase with the associated growth in farming water needs up to the point where the ability to sequestrate water from the natural water resource exceeds the replenish rate of the natural water resource as shown in Fig. 11. The available water for use is consumed to the point where there is an insufficient supply for the minimum water per person, resulting in a rapid decline in population numbers which then relieves the pressure on the natural water resource, and allows the numbers to rise. The village wealth per person

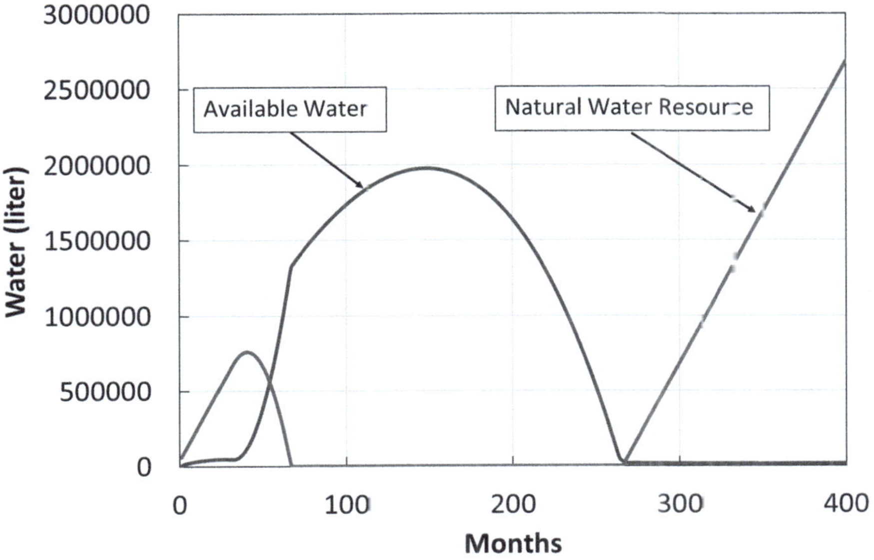

Fig. 11 Utilisation of natural water resource for Case 4

is investigated under the four cases considered. Under Case 1, where there is only a once-off donation with no maintenance, depending on the initial capability of the donation, the wealth per person does not exceed the assumed cost of living. When the infrastructure is maintained by the benefactor as in Case 2, there is an initial increase in wealth per person which exceeds the cost of living while the donation ability exceeds the water demand allowing for increased farming which then increases the community wealth. However, at the population stability point, there is an insufficient water supply to allow farming activities to exceed the cost of living. With a service delivery fee, as in Case 3, there is a significant increase in wealth per person, since the reinvestment creates more available water which allows for growth in farming which in turn, increases the wealth per person. There is, however, a maximum value of wealth per person attainable. The wealth per person is rapidly reduced when the water sequestration rate exceeds the natural water replenish rate, and the community essentially has grown itself into collapse. The rapid decline in population can be explained as follows: When the available water level exceeds the minimum value per person, the excess water is available for farming and wealth creation. When the available water is distributed amongst the population for their farming needs, eventually a maximum equal distribution is obtained which stabilises the wealth per person. However, when the available water, after farming needs are considered, is less than the minimum water per person amount, the farming activities are restricted and eventually, the population numbers decline with a significant reduction in wealth per person (Fig. 12).

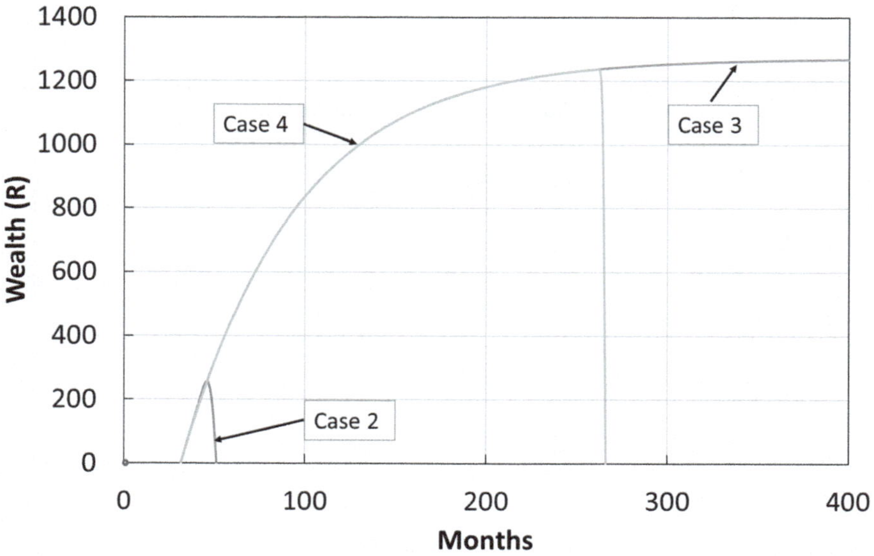

Fig. 12 Community wealth per person

5 Implications

Renewable energy supplies in remote and marginalised villages that are not connected to the grid have the potential to reduce "energy poverty" and take villagers out of poverty traps, a common feature in rural communities in developing countries. Rural energisation through solar-powered micro-grids allows off-grid communities to produce sufficient energy to meet their demand and needs and avoids the necessity of importing expensive and conventional fuels from urban areas. The development of off-grid renewable energy systems, especially for water pumping, food production, and microenterprise development is needed to address socio-economic, human, and physical development through food security and reduced poverty traps. This is because when a remote village has access to clean, safe, and dependable alternative solutions, this triggers economic and social development. If hybrid systems are introduced in the villages, there is an opportunity to attract energy-intensive sub-macro manufacturing and processing industries such as agro-food processing and smelting (Makonese and Meyer 2018). Integrated systems based on renewable energy have the potential to free rural communities from over-dependence on woody biomass and grid electricity (in parts where rural households are connected) and provide them with a more stable and cheaper power supply. Such a system could attract new businesses that range from accommodation to restaurants, leading to the development of self-contained, modern, and smart rural villages.

Qualitative analyses of the sustainable indicators (SI) (Table 1) showed that Gwakwani Village has very low sustainable development levels in the energy, water, and food dimensions. The village does not have supporting infrastructure for energy and water access as well as food production through agricultural mechanisation. However, Gwakwani Village has enough agricultural land to sustain high food production. (Yuan et al. 2020) contend that there are three perspectives needed for energy, water, and food sustainability, namely infrastructure development, technological innovation, and collaborative partnerships between various stakeholders. The low SI ranking is due to several factors including a lack of support from local government, a rural development policy that lacks incentives, and limited research funding investment to tackle rural development and sustainability challenges. There is not a pragmatic attitude towards scientific and technological innovations from enterprises, with some government efforts on technological development focusing mainly on the urban set up in a quest to build smart cities (Yuan et al. 2020).

Renewable energy and access to electricity are high on the villagers' priority list. (Sodiq et al. 2019) found that energy issues were most important, with water and food availability as fundamental support. Although the South African government is advocating for the penetration of renewable energy resources in marginalised rural areas, there has not been much support regarding infrastructure construction and subsidies. Eskom (the South African electricity utility) continues to hamper the uptake of cheaper alternative energy sources in these areas due to its business models of building, operating, and maintaining only coal-based power plants (except for one nuclear plant). According to Rennkamp et al. (2017) attempts to transform Eskom have failed dismally as the organization continues to maintain its status quo of being the only electricity producing and distributing utility in the country.

Results from this study have implications for development sustainability and policy reforms. In the light of the broad understanding of sustainability within the SDG framework of providing peace and prosperity for people while preserving the earth now and into the future, projects providing clean energy and water in communities requires solutions to make sustainable use of locally available natural resources. In support of the preservation of the environment for future generations, it is suggested that projects follow a service delivery approach where long-term services are provided as opposed to stand-alone projects; this strategy will provide and sustain a certain level of access to environmental resources (Dickinson and Bostoen 2013).

There is a great need for financial and technical support in the construction of infrastructure related to energy, water, and food. Technical innovations that require government support, and contributions from interested stakeholders, including local technology companies, should be advocated during the early stages of development (Yuan et al. 2020). The introduction of core technology and increasing the number of specialists with innovation abilities through training and education produces EWF nexus sustainability (Rehman et al. 2020).

Energy access strategies and rural development projects should be contextually appropriate—they should not be driven by national policies or agendas, but by local concerns or demands. Local networks, including cooperatives, community champions, tribal chiefs, and any other relevant and interested structures or institutions in that community should manage or oversee such projects to provide co-created solutions with the donors/funders (Dowling et al. 2018). Chaurey et al. (2004) argued that the provision of electricity through off-grid renewable energy might not significantly raise living standards unless it enables income generation. This could be achieved through developing a sustainable agro-business model that encompasses the EWF nexus.

The role of intermediate institutions such as universities and local governments in the dissemination of renewable energy technologies should not be undermined. Academia, including research institutions, as a stakeholder in a development project, brings immense benefits to the programme. Having access to topic experts and researchers ensures that the solutions implemented are always inclusive of the latest technologies and techniques. Academia also brings a more holistic viewpoint on the EWF nexus problem in rural and marginalised communities. Although academic institutions will take more time to implement plans, they will look at producing the most efficient solutions to a problem. By involving academic institutions in the rural development discourse, a plethora of opportunities are presented for communities and students to enhance their learning experience by participating in these projects to co-create solutions.

The results of this research-based systems thinking approach to the EWF nexus have implications for policy development, especially in the context of rural development. According to Shannak et al. (2018), integrating data from our case scenarios and practical experience into dynamic systems would result in more robust modelling tools that could influence policy in a positive and constructive direction and help to improve the institutions and approaches for managing these interlinked resources. The framework showcased in this chapter emphasizes the role that can be played by quick appraisals from participatory observers and from interaction with the communities. This gives information about the key issues affecting the communities without detailed modelling. The models and appraisal exercises complement each other in finding sustainable solutions to key challenges.

The role of stakeholders, especially the government, in the assessment process is emphasized. However, there are complexities inherent in addressing policies for energy, water, and food when they fall under the control of different government departments without similar mandates, unless these policies are governed by a single national strategy on the EWF nexus. The national strategy document should effectively deal with societal and the complex interactions of the nexus and replace silo-thinking with systems thinking approaches. Keairns et al. (2016) and Bazilian et al. (2011) believe that "*…forming constructive linkages across the boundaries that exist between the three areas will require strong political leadership, compelling visions, significant cooperation and humility.*"

6 Conclusions

This study presented a case study of a marginalised rural community in South Africa dealing in an integrated way with energy, water and food. Results showed that both access to clean and safe water and agriculture need energy when the target communities are off-grid. The case study was concerned with the promotion and adoption of solar mini-grid systems in off-grid rural communities to address EWF nexus challenges. It was demonstrated that an integrated systems thinking approach is needed to address the water, energy, and food nexus challenges due to the complex interrelationships of the sectors. Stakeholder involvement is imperative in clarifying issues and priorities/values, without which development programmes could fail. It was clear that community involvement is critical to the sustainability of community projects and programs. Community involvement is the key that ensures the understanding of the needs of the people and makes decisions that will meet those needs in the best possible way. Our findings illustrate how poor, marginalised, and off-grid communities could attain socio-economic, human, and physical development by making sustainable use of locally available natural resources through solar energy and related technologies.

The study also investigated the issues of donor funding and the sustainability of developmental projects in rural communities, in line with the energy, water and food nexus. Generally, it was concluded that rural development programmes should not provide standalone projects without the provision of basic support/upkeep. At the very least, support for the infrastructure that is provided in the project must be sustained through training of locals to maintain the infrastructure.

A limitation of this study is that the models employed might be seen as very basic and only considers fundamental relationships mostly between the supply of water for human consumption and a generalised consumption for agriculture and farming. The models use a lumped parameter approach rather than a detailed breakdown of all variables in the system. Although the models are basic, the results from them are used not in absolute values but rather in the trend prediction. Another limitation could be that the village is seen as a rural village in South Africa which was ranked in 2020 as second on the continent of Africa in terms of gross domestic product, however, the circumstances encountered in the Gwakwani village would be fairly representative of any rural village in Africa. The government social grant scheme offered by the South African government may also impact the considered wealth of the Gwakwani village differently to a village located in an African context where social grants may not be available, and funding from donors limited. This study does not consider the additional method for income generation utilised in Gwakwani, which stems from a pure Photo Voltaic powered bakery where villagers are employed as bakers in a bread bakery providing bread to the local community.

From an African perspective, the findings from this research are aligned with the work from Tumushabe (2018), which emphasizes the need to establish task teams on food security in each major country, and regional bodies to allow coordination on policy and promote adaptive strategies for sustainable development. They also resonate with the statement by Yang who established through a questionnaire survey to experts, that SDG 2 (Zero Hunger), SDG 1 (No Poverty) and SDG 6 (Clean Water) had the highest priority on a global and regional scale (Yang et al. 2020).

References

Apine E, Turner LM, Rodwell LD, Bhatta R (2019) The application of the sustainable livelihood approach to small scale-fisheries: The case of mud crab Scylla serrata in South west India. Ocean Coast Manag 170:17–28

Banson KE, Nguyen NC, Bosch OJH, Nguyen TV (2015) A systems thinking approach to address the complexity of agribusiness for sustainable development in Africa: a case study in Ghana. Syst Res Behav Sci 32:672–688

Bazilian M, Rogner H, Howells M, Hermann S, Arent D, Gielen D, Steduto P, Mueller A, Komor P, Tol R, Yumkella KK (2011) Considering the energy, water and food nexus: towards an integrated modelling approach. Energy Pol 39(12):7896–7906

Bieber N, Ker JH, Wang X, Triantafyllidis C, van Dam KH, Koppelaar RH, Shah N (2018) Sustainable planning of the energy-water-food nexus using decision making tools. Energy Pol 113:584–607

Chaurey A, Ranganathan M, Mohanty P (2004) Electricity access for geographically disadvantaged rural communities – technology and policy insights. Energy Pol 32:1693–1705

Chen S, Chen B (2016) Urban energy-water nexus: a network perspective. Appl Energy 184:905–914

Dickinson N, Bostoen K (2013) Using ICT for monitoring rural water services: from data to action. Triple-S Working paper 4. IRC International Water and Sanitation Centre: The Hague, Netherlands

Dowling R, McGuirk P. Maalsen, S (2018) Multiscalar governance of urban energy transitions in Australia: the cases of Sydney and Melbourne. Energy Res Soc Sci 44:260–267

Dubreuil A, Assoumou E, Bouckaert S, Selosse S, Maizi N (2013) Water modelling in an energy optimization framework in the water-scarce middle east context. Appl Energy 101:268–279

Fang D, Chen B (2017) Linkage analysis for the water-energy nexus of city. Appl Energy 189:770–779

Goepel KD (2018) Implementation of an online software tool for the analytic hierarchy process (AHP-OS). Int J Anal Hierarchy Process 10(3)

Johnson N, Burek P, Parkinson S, Havlik P, Nakicenovic N, Riahi K, Wiberg D (2015) Modelling the energy-water-land nexus: a review of opportunities and challenges for integrated assessment modeling. Int Inst Appl Syst Anal

Kan G, Zhang M, Liang K, Wang H, Jiang Y, Li J, Ding L, He X, Hong Y, Zuo D (2016) Improving water quantity simulation & forecasting to solve the energy-water-food nexus issue by using heterogeneous computing accelerated global optimization method. Appl Energy

Keairns DL, Darton RC, Irabien A (2016) The energy-water-food nexus. Ann Rev Chem Biomol Eng 7:239–262

Limpopo State of the Environment Report (Phase 1): Department of Finance and Economic Development (2004). www.soer.deat.gov.za/dm_documents/Limpopo_State_of_environment_reporting_phase_1___UrZKg.pdf

Lloyd P (2014) Challenges for household energisation and the poor. J Energy South Afr 25(2):2–8

Madubansi MC, Shackleton C (2006) Changing energy profiles and consumption patterns following electrification in five rural villages, South Africa. Energy Pol 34(18):4081–4092

Makonese T, Meyer J (2018) Household energisation in rural south africa: a systems approach towards energy access. In: 2018 international conference on the domestic use of energy (DUE), Cape Town, South Africa, 3–5 April 2018

Masekoameng KE, Simileng TE, Saidi T (2005) Household energy needs and utilisation patterns in the rural Giyani rural communities of Limpopo Province. Journal of Energy in Southern Africa 16(3):4–9

Mavhura E (2017) Applying a systems-thinking approach to community resilience analysis using rural livelihoods: the case of Muzarabani district, Zimbabwe. Int J Disaster Risk Reduction 25:248–258

Mock N, Béné C, Constas C, Frankenberger MT (2015) Systems Analysis in the Context of Resilience. Food Security and Nutrition Network (FSIN), Technical Series No. 6. Rome

Mungwe JN, Bandiera L, Accorona D, Colombo E (2016) Sustainable energization of rural areas of developing countries – a comprehensive planning approach. Energy Procedia 93:46–52
Naidoo R, Meyer J, von Solms S (2018) A systems engineering framework for small scale rural community engagement projects: a case study. In: Proceedings of the international conference on industrial engineering and operations management, South Arica, 29 October–1 November 2018
Naidoo R, Meyer J (2015) Cellular technology for the prevention of give-and-forget community engagement projects, SATNAC 2015. 6–9 September 2015. Hermanus, Western Cape, South Africa
Nguyen , Cook, S, Ireland V (2017) Application of system dynamics to evaluate the social and economic benefits of infrastructure projects. Systems 5(2):29
Rehman FU, Ding Y, Noman AA, Khan M (2020) The nexus between infrastructure and export: an empirical evidence from Pakistan. Glob J Emerg Mark Econ
Rennkamp B, Haunss S, Wongsa , Ortega A, Casamadrid E (2017) Competing coalitions: the politics of renewable energy and fossil fuels in Mexico, South Africa and Thailand. Energy Res. Soc. Sci. 34:214–223
Serrat O (2017) The sustainable livelihoods approach. In: Knowledge solutions. Springer, Singapore, pp 21–26
Shannak S, Mabrey D, Vittorio M (2018) Moving from theory to practice in the water–energy–food nexus: an evaluation of existing models and frameworks. Water Energy Nexus. https://doi.org/10.1016/j.wen.2018.04.001
Shezi L (2015) Nearly 60% of Africans have no access to reliable electricity. HyperText, May 2015
Sodiq A, Baloch AAB, Khan SA, Sezer N, Mahmoud S, Jama M, Abdelaa A (2019) Towards modern sustainable cities: review of sustainability principles and trends. J Clean Prod 227:972–1001
Tumushabe JT (2018) Climate change, food security and sustainable development in Africa. In: Oloruntoba S, Falola T (eds) The Palgrave handbook of African politics, governance and development. Palgrave Macmillan, New York. https://doi.org/10.1057/978-1-349-95232-8_53
World Bank, Food Security - Overview (2016)
World Water Assessment Program (WWAP): The United Nations World Water Development Report 4 (2012)
Yang S, Zhao W, Yanxu L, Cherubini F, Fu B, Pereira P (2020) Prioritizing sustainable development goals and linking them to ecosystem services: An expert's knowledge evaluation. Geogr Sustain. https://doi.org/10.1016/j.geosus.2020.09.004
Yuan MH, Chiueh PT, Lo SL (2020) Measuring urban food-energy-water nexus sustainability: finding solutions for cities. Sci Total Environ. https://doi.org/10.1016/j.scitotenv.2020.141954

Professor Johan Meyer is an associate professor at the Department of Electrical Engineering at the University of Johannesburg. He has a D.Ing in electrical engineering from the Randse Afrikaanse Universiteit. He worked in the aviation industry before returning to academia. His research interest is in renewable energy, rural development, and systems engineering. He is a senior member of IEEE and SAIEE and a registered professional engineer.

Professor Sune von Solms is an associate professor at the Faculty of Engineering and the Built Environment (FEBE) at the University of Johannesburg. She obtained her M.Eng and Ph.D. in Computer Engineering at the North-West University in the field of telecommunication and networks. She is registered as a Professional Engineer and a member of the IEEE and SASEE. She conducts research into the social and human aspects of engineering, engineering education, and the impact of technology in society.

Dr Taffy Makonese is an energy and air quality specialist with the Process Environment and Energy Technology station at the University of Johannesburg. He holds a PhD in Energy Studies from the University of Johannesburg. His current research interests are in air pollution and related health effects. He has also researched the development of culturally appropriate stove performance testing protocols, and their usefulness in wider stove dissemination programmes funded by the World Bank, the Global Alliance for Clean Cookstoves (GACC) and others.

Determinants of Access, Use and Sustainability of Improved Water Sources by Households in Cameroon

Lydie Bamou Tankoua

Abstract Despite the potential benefits of improving water sources, to ensure everyone's access to water and achieve Goal 6 of the Sustainable Development Goals (SDGs), millions of people still rely on unsafe water sources. This chapter examines the determining factors of access, use and sustainability of improved water sources by households in Cameroon. For this purpose, bivariate models are estimated in this chapter, where the dependent variables are the probabilities of using improved water sources for drinking or non-drinking purposes. The results suggest that the likelihood of using improved and sustainable water sources is positively associated with health, wealth, education, female gender, proportion of children in the household, waste and wastewater disposal systems, while spatial variables are negatively associated with access to water sources. Provision of subsidies for connection to individual taps, construction of more improved water stations in rural areas and installation of wastewater treatment for recycling are examples of actions suggested to the authorities to improve the situation. Such actions should increase the sustainability and the supply of quality water services, and thus, the likelihood of Cameroonians households to achieve Goal 6 of SDGs.

Keywords Bivariate models · Cameroonian households · Improved water sources · Sustainability

1 Introduction

Access, use and sustainability of improved water sources have many health and economic outcomes. Improved water sources are sources that, by nature of their design and construction, are capable of providing safe water. These include water from pipes, boreholes or tube wells, protected dug wells, protected springs and rainwater. On the other hand, unimproved sources include unprotected dug wells and unprotected sources as lakes, rivers, etc. (WHO 2017). Despite the potential benefits of improving water sources, millions of people are still relying on unsafe sources.

L. Bamou Tankoua (✉)
Faculty of Economics and Management, University of Yaoundé II Soa, Yaoundé, Cameroon

W. Leal Filho et al. (eds.), *Sustainable Development in Africa*, World Sustainability Series,
https://doi.org/10.1007/978-3-030-74693-3_23

According to a recent report on drinking water, sanitation and hygiene (WHO and UNICEF 2017), Worldwide 844 million people still lack basic drinking water services with 159 million of them depending on surface water to meet their safe water needs, notwithstanding the potential advantages of improving water sources. Thus, it is not surprising to see securing access to water and sanitation for all and ensuring sustainable management of water resources as the sixth goal of the 2030 Sustainable Development Agenda, adopted by UN member countries in 2015.

In Latin America, 80% of the urban population relies on groundwater as source of drinking water, as well as for irrigation and other basic needs. However, growing industrialization, intensive use of agrochemicals and inadequate wastewater treatment systems increase the risk of the contamination of water sources (WHO 2017). Africa is home to nearly 15% of the world's population but is endowed with only 9% of global renewable water sources, which are unevenly distributed across the continent. The gap between water availability and water demand is growing fast, especially in cities, where the urban population is expected to nearly quadruple by 2037 (World Bank 2012). The improvement of living standards and the change in consumption patterns are contributing to this growth in water demand. On the other hand, water availability is decreasing with the competing demands from agriculture, mining and industry, together with the deterioration of water quality. Large numbers of people depend on groundwater as their primary or alternative water source, but this source is threatened by pollution and over-extraction (World Bank 2012).

In Sub-Saharan Africa, 40% of the population still does not have access to improved water sources due to a high population growth rate, among other things. On the positive side, in Cameroon (a Central African country), the proportion of households using water from improved sources increased from 53% in 2004 to 71% in 2011 (NIS and ICF International 2012). However, it is increasingly observed that although available, people may not use improved water infrastructure if it does not meet their needs. Therefore, understanding the factors that could drive households to use improved water sources is of great importance.

Over the past decades, numerous studies attempted to identify the determinants of the choices of water sources made by households. Examples of recent studies in this domain include Rauf et al. (2015), Adams et al. (2016) and Rahut et al. (2016). However, most of these studies and even the ones before them focus on the primary water sources used by households. They ignore the fact that, in developing countries, households have access to more than one of the different types of water sources (Nauges and Whittington 2010). In addition, these sources must be treated and protected from chemicals and toxins from household and industrial waste to reduce the scarcity of drinking water sources and to improve their sustainability. At the micro-economic level, access to running water generates a significant time saving since it avoids long waiting hours at public water points, which can be regarded as one of its main advantages.

This chapter provides in-depth perspectives on the determinants of access, use and sustainability of water sources, with an analysis of specifically Cameroonian households. For that purpose, we use data from the 2006 multi-indicator survey on a sample of 9967 households. This data allowed consideration of the diversity of water sources used simultaneously by households. It was assumed that the selection of water sources depends largely on its intended use. This led to an analysis of the probability of using improved sources for consumption or other domestic uses (watering, personal hygiene, etc.). The bivariate probit model is used in an empirical analysis to consider the potential interdependence between water source choices.

The review of the literature in Sect. 2 of the chapter allows the identification of factors affecting the sustainability of water resources. Given that many rural people are poor, it is generally assumed that the water supplied in rural areas must be funded by government agencies (Briscoen et al. 1990). This is followed by a presentation of the data on which the chapter is based in Sect. 3, and the econometric model which has been used in Sect. 3. The chapter then turns to the presentation of the findings in Sect. 5, and some concluding remarks with policy implications in Sect. 6.

2 Literature Review

Contrary to developed countries where almost all households have a connection to a piped water network and use it as the primary source for all their water uses, households in developing countries often depend on a variety of water sources, including piped and non-piped, with different characteristics and levels of services (Nauges and Whittington 2010). Analysis of the economics literature on household water demand provides examples of several studies dealing with the determinants of choices of water sources. The approach adopted consists of the analyses of choices of water sources by households either as a main focus (Madanat and Humplick 1993; Persson 2002; Briand and Loyal 2013) or as an auxiliary objective in combination with a model of water use which depends on the choice of the source (Nauges and Strand 2017; Briand et al. 2010; Coulibaly et al. 2014). Moreover, these studies mainly used discrete choice models, such as binary logit/probit models, to examine the probability of using a given type of water source. This is the case with the study by Basani et al. (2008), who modeled the probability of using the piped water network in Cambodia and the study by Briand and Loyal (2013), who modeled the probability of using private water connections developed by small-scale private operators in the peripheral districts of Maputo (Mozambique). Other studies rather used models with dependent variables and more than two modalities (multinomial or bivariate probit models for instance) to study the probability of choosing water sources among available options (Mu et al. 1990; Irianti et al. 2016).

The studies referred to in the previous paragraph suggest that the choice of water sources by households might be affected by two broad categories of factors: water source characteristics and household characteristics. Regarding water source characteristics, they suggest that the accessibility to a water source, measured either by the distance from home to water source, the time of water collection or the connection fees to a piped network, might reduce/increase the probability of using that source or an alternative source. Moreover, the quality of the service at a given water source (water quality, pressure at the tap, frequency of water cuts, etc.) also affects the demand for that source. The higher the quality of the service at a given source, the higher the probability of choosing that source will be, all things being equal. Also, the higher the quality of the service at a given water source, the lower the probability of choosing other sources will be, all other matters being equal.

Factors having an impact on the choices of water sources that have been explored in the literature include various household characteristics. We review here the household characteristics which are most often highlighted. The literature mainly shed light on the positive impact of income/wealth and education on the choice of water sources. Indeed, authors as Nauges and Strand (2007), Basani et al. (2008) and Mahama et al. (2014) found that households with a higher income and better education are usually more willing to obtain improved water services. Briand and Loyal (2013) found that illiteracy of the heads of households decreases the probability of being connected to private water connections supplied by small-scale private operators in Maputo by 14.8%. Similarly, when the household wealth index increases by one unit, the probability to be connected increases by 6.6%. The positive effect of income/wealth is associated with the fact that in general, the safest and most reliable water sources are those involving the highest access costs, but which may be untenable for the poor. Likewise, concerning education, the link between education and access to improved water sources is because of the increased awareness of the importance of safe drinking water for healthy life on the part of educated people as well as the opportunity costs of not having safe drinking water (Rahut et al. 2016).

Spatial variables are also found to have significant effects on the choices of water sources (Rahut et al. 2015; Rauf et al. 2015). It is generally suggested that households in insecure or rural areas are inclined to collect water from sources associated with low quality services. Several other socioeconomic variables such as house ownership versus renting are also found to either positively or negatively affect access to piped networks and/or improved water sources (Briand et al. 2009 and Briand et al. 2010).

The economics literature also focuses on the determinants of rural infrastructure demand and willingness to pay for water (Briscoe et al. 1990). Access to improved water source is not only regarded as a contextual or a household variable, but also as an explanatory variable as one of the components of sustainable development indicators (Kobiané 1998). In recent decades, numerous studies attempted to identify the determinants of household access and sustainability of water sources (Adams et al. 2016; Rahut et al. 2016), while previous studies focused on the main sources of water used by households. At a macro level, access to running water is also an indicator of level of sustainable development achieved by a country (Hewett and Montgomery 2001). The construction of water reservoirs in urban areas to make water available for use is considered as an important factor concerning the sustainability of water sources (Zoungrana and Combelem 2016; Combelem et al. 2020). Household waste and wastewater discharged into water sources as rivers or lakes are considered as factors which compromise the sustainability of water sources (Corcoran et al. 2010).

As for demographic variables, the size of a household is considered as a significant factor that may positively or negatively affect the choice of water sources (Nauges and Van Den Berg 2009). Having many people in a poor household can be a demotivation to choose a connection to a private water sources, while in a wealthier household the same situation would serve as encouragement to indeed obtain connection to such water sources (Briand and Loyal 2013). However, most studies conclude that larger household sizes are usually associated with a higher probability of using private connections. The choice of water sources in terms of composition of households by sex and age are also documented in the literature. For instance, it is suggested that the fact that the head of a household is a woman, positively affects the choice of using improved or private water sources, while the opposite applies when it is a man (Totouom and Fondo 2012 and Irianti et al. 2016). However, some studies (i.e. Rahut et al. 2016), point towards the opposite.

A common characteristic of studies on the choice of water sources is that most of them do not account for the fact that households might use several sources. Examples of studies which indeed include this aspect in their analysis are Mahama et al. (2014), Briand et al. (2009), and Madanat and Humplick (1993). Briand et al (2009) consider the multiplicity of water sources used by households (private water connection at home and public standpipe). Mahama et al. (2014) and Madanat and Humplick (1993) conducted usage-specific studies that account for the fact that households may use different water supply systems for different needs of water. However, contrary to Mahama et al. (2014) who estimated separate probabilities of choosing water sources, Madanat and Humplick (1993) jointly analysed household choices of water sources and connection decisions to account for the interdependence between the two choices, to which the study as reported in this chapter aligns.

3 Data and Materials

Data used in this study are from the Multiple Indicator Cluster Survey (MICS) conducted in 2006 by the National Institute of Statistics (NIS). We did not use data from the recent Demographic Health Survey mixed with the Multiple Indicators Clusters Survey conducted in 2011 because of some missing information in the database. For the purpose of the MICS, the country was stratified into 12 regions (the 10 administrative regions of Cameroon and Douala and Yaoundé). These regions were identified as areas of study and stratified according to place of residence. The sample for this study was drawn using a two-stage design. In the first stage, enumeration areas were drawn in each study area with a probability proportional to the size of the population. In the second stage, a variable number of households were randomly drawn, after updating the maps of the selected areas (count of households in the area). Although the sample is stratified in each field of study, the sampling design does not allow obtaining a self-weighted sample. Therefore, the calculation of the coefficients of weighting was done in such a way as to allow robust estimates according to the national level, areas of study, and urban and rural settings. During the MICS, 9667 households were successfully surveyed. As part of this survey, there were two questions in the household section on the sources of water. The first question concerned the source of drinking water, while the second question concerned the source of water for other domestic uses. The MICS shed light on the diversity of water sources used in Cameroon. Table 1 provides the distribution of surveyed households according to their water sources, while Table 2 provides the distribution of surveyed households according to the quality of water sources (improved/unimproved).

The results show that public taps or standpipes are the main sources of drinking water among the surveyed households (19.49% of the sample). This is followed by tube wells or boreholes (14.03% of the sample) and unprotected springs (12.35% of the sample). Regarding the source of water used for other domestic uses, protected wells are the most used source (16.84% of the sample). This is followed by unprotected wells (16.53% of the sample) and surface water sources as rivers, dams, lakes, and streams (12.22% of the sample). The different sources of water were classified into two groups: improved and unimproved sources. Following the WHO (2017), this study regarded improved water sources to include piped water (into

Table 1 Distribution of households by water sources (NIS 2006)

Sources of water	Water source for drinking		Water source for other domestic purposes	
	Number	(%)	Number	(%)
Piped into dwelling	711	7.35	701	7.25
Piped to yard/plot	442	4.57	415	4.29
Public Tap/standpipe	1,884	0,019	1,129	0.012
Piped neighbour	750	7.76	407	4.21
Tube Well/borehole	1,356	0.014	896	9.27
Protected well	675	6.98	1,628	0.017
Unprotected well	1,013	0.010	1,598	0.017
Protected spring	669	6.92	430	4.45
Unprotected spring	1,194	0.012	1,097	0.011
Rainwater collection	17	0.18	112	1.16
Tanker truck	25	0.26	9	0.09
Cart with small tank	56	0.58	45	0.47
Surface water	830	8.59	1,181	0.012
Bottled water	32	0.33	0	0
Others	12	0.12	14	0.14
Missing	1	0.01	5	0.05
Total	9667	100	9667	100

Table 2 Distribution of households according to the quality of water sources (improved/unimproved) (NIS 2006)

Sources for drinking purposes	Sources for other domestic purposes		Total
	Improved sources	Unimproved sources	
Improved sources	5549 57.56%	946 9.81%	6495 67.37%
Unimproved sources	164 1.70%	2982 30.93%	3146 32.63%
Total	5713 59.26%	3928 40.74%	9641 100%

dwellings, compounds, yards or plots and to neighbours or public taps/standpipes), tube wells/boreholes, protected wells, protected springs and rainwater collection. The following water sources are regarded as unimproved: unprotected dug wells, unprotected springs, small carts with tanks, tanker trucks, surface water and bottled water. Bottled water is considered as an unimproved source not because of its quality but because of the limitation concerning the potential quantity of the supplied water.

The data in Tables 1 and 2 show that 6495 households (67.37% of the full sample) use an improved drinking water source. Of these households, 5549 (85.43%) use an improved source to satisfy their other domestic water needs, while 946 (14.57%) rely on unimproved sources. Of the 3146 households (32.63% of the full sample) that use an unimproved drinking water source, the majority (94.94%) still relies on unimproved sources to satisfy their other domestic water needs.

This research shows that population growth in cities is increasing the demand for water supply. In addition, it is observed that factors that can ensure the sustainability, well-being and access to water for future generations are still largely inadequate. Such factors in need of attention include the financing of water supply by government or non-governmental institutions, the proportion of water and waste treated to protect water sources, the construction of water reservoirs and the installation of saving materials in municipal sites. The data collected for this study did not specifically capture data on sustainability variables that could be incorporated into the model. This constitutes an important limitation of this study. An econometric approach would be required to assess the determinants of the sustainability of improved water sources.

4 Econometric Model

In this study the probabilities of using an improved source for each of the two identified water uses are jointly estimated with bivariate probit models to account for the potential interdependence of the two probabilities. The probability that household i chooses an improved water source corresponds to the probability that this alternative water source has a higher utility than the unimproved ones. The observed choice of the water sources of the ith household can be expressed as follows:

$$\boldsymbol{y}_{i1} = 1 \textbf{ if } \boldsymbol{y}^*_{i1} = \boldsymbol{x}'_{i1}\boldsymbol{\alpha} + \boldsymbol{\varepsilon}_{i1} > 0 \textbf{ and } \boldsymbol{y}_{i1} = 0 \textbf{ if } \boldsymbol{y}^*_{i1} \leq 0 \tag{1}$$

$$\boldsymbol{y}_{i2} = 1 \textbf{ if } \boldsymbol{y}^*_{i2} = \boldsymbol{x}'_{i2}\boldsymbol{\beta} + \boldsymbol{\varepsilon}_{i2} > 0 \textbf{ and } \boldsymbol{y}_{i2} = 0 \textbf{ if } \boldsymbol{y}^*_{i2} \leq 0 \tag{2}$$

where, $y_{i1} = 1$ and $y_{i2} = 1$ mean that an improved source is respectively used for drinking and for other domestic uses. y^*_{i1} and y^*_{i2} are the unobserved variables associated with y_{i1} and y_{i2} respectively. α and β are the estimators of the parameters. while $\boldsymbol{x}_{i1}$ and $\boldsymbol{x}_{i2}$ are the explanatory variables. The errors $\boldsymbol{\varepsilon}_{i1}$ and $\boldsymbol{\varepsilon}_{i2}$ are assumed to be normally and jointly distributed with:

$$\begin{pmatrix} \varepsilon_{i1} \\ \varepsilon_{i2} \end{pmatrix} \sim N\left[\begin{pmatrix} 0 \\ 0 \end{pmatrix}, \begin{pmatrix} 1 & \rho \\ \rho & 1 \end{pmatrix}\right] \quad (3)$$

If ρ, which measures the correlation between the residuals of the two equations, is statistically different from zero, then the two equations are indeed correlated and the choice of the bivariate probit is justified. If, on the other hand, it is not statistically significant, the use of standard probit is recommended. In general, ρ is not directly estimated, but rather Athro, defined by:

$$Athro = \frac{1}{2} ln\left(\frac{1+\rho}{1-\rho}\right) \quad (4)$$

The maximum likelihood method was used to estimate the specified models. The logarithm of the likelihood function estimated to have α and β is given by:

$$\ln L = \sum_{i=1}^{n} \ln \emptyset_2\left[(2y_{i1} - 1)\varepsilon_1, (2y_{i2} - 1)\varepsilon_2, (2y_{i1} - 1)(2y_{i2} - 1)\rho\right] \quad (5)$$

$$\ln L = \sum_{i=1}^{n} \ln \emptyset_2\left[q_{i1}\varepsilon_1, q_{i2}\varepsilon_2, q_{i1}q_{i2}\rho\right] \quad (6)$$

where $q_{i1} = (2y_{i1} - 1) = -1$ **if** $y_{i1} = 0$ and **1** if $y_{i1} = 1$. Also, $q_{i2} = (2y_{i2} - 1) = -1$ if $y_{i2} = 0$ and **1** if $y_{i2} = 1$. $\emptyset_2$ is the cumulative distribution function of the bivariate normal distribution.

4.1 The Explanatory Variables

The selection of the explanatory variables included in the models is based on the literature and the availability of related data in the MICS database. The first variable considered in this analysis measures the accessibility to water sources. **Distance from home to an improved drinking water source** was used in this study to capture accessibility to improved sources. However, there is a potential endogeneity issue to address with reference to inclusion of distance to water sources in the model developed for this study. Indeed, provision is explicitly made for the possibility that the distance between homes and improved sources is not random, but rather that characteristics such as unobserved preferences or unobserved group characteristics are associated with the proximity between place of residence and location of improved sources. Moreover, there is also a reverse causality that may exist between the two variables. Indeed, wealthier households for instance are likely to have both better water supply access and a greater preference for improved water sources. Finally, there is also the possibility of a measurement error which is common to survey data and may be explained by the fact that for instance, a respondent may be ashamed of appearing like a poor person when confessing that he/she collects water far from home.

To deal with the potential endogeneity issue associated with the variable 'distance from home to an improved drinking water source', the methodology would have to use an approach based on an instrumental variable (IV). However, the database used does not contain suitable instruments for the variable 'distance from home to an improved drinking water source' (i.e. variables which are correlated with the endogenous variable and which only affect the outcome variable through their effect on the endogenous variable). Following Briand et al. (2009) and Nauges and Strand (2017), the distance from home to an improved source is replaced for each household with the average distance calculated for the sample of households found in the same municipality. If the assumption is made that endogeneity arises from individual choices within communities, access to water can be considered as exogenous if community averages are used instead of household data (Nauges and Strand 2017). The large coverage of the database used for the empirical analysis (9667 households) guarantees a high heterogeneity in the characteristics of households and the variable 'distance from home to an improved drinking water source'. In the database used for this study, data on this variable are indeed available only for households using improved drinking sources. This means that the 32.63% missing data are related to households using unimproved source. This could affect the efficiency of the estimated coefficients and it is why the results of this study should be analysed with caution. Further research using primary data should address such limitations of the study.

Wealth is one of the household characteristics considered in this study as an explanatory variable. The wealth index used in the study was computed by the NIS of Cameroon. Its computation was based on the multiple component analysis that used housing characteristics and ownership of some durable goods. The index is further normalized from 0 to 1 based on the following formula:

$\mathrm{wealth}_N = \frac{wealth - wealth_{min}}{wealth_{max} - wealth_{min}}$, Where $wealth_G$ is the gross indicator, $wealth_{max}$ and $wealth_{min}$ are the maximum and the minimum of the wealth index. The normalized wealth score is further grouped into four wealth quartiles: bottom, 2nd, 3rd, and top wealth quartile. Another variable used in the model for this study is '**education of the heads of households**). It is classified as three dummies: no education (1 if the household head has no education, 0 otherwise), primary education (1 if the household head has gone at most through primary education, 0 otherwise) and secondary education or more (1 if the household head has gone at least through secondary education, 0 otherwise). The composition of households was also considered in terms of the variable '**proportion of children in households**'. This variable was obtained by dividing the number of children less than five in the households by the household size. The effect of the variable '**gender of the heads of households**' on the choices of water sources was also considered in this study: a binary variable equal to 1 was used if the head of household is a woman and 0 if otherwise. Other variables included in the regressions are spatial variables. Two spatial variables were considered: the **region of residence** of households is represented by 12 binary variables for each of the 12 regions considered during the MICS survey and the **place of residence** (1 for urban and 0 for rural). The estimations were performed using Stata 12 software and the results are discussed in the following section.

5 Results and discussion

This section presents the results from the different estimations conducted in this study and are provided in Table 3. Model 1 only includes distance to improved drinking water source and spatial variables. In Model 2, the proportion of children and the gender of the household head were added. Model 3 includes wealth quartiles and variables of education in addition.

The results firstly suggest the existence of strong associations between the different choices of water sources by households. Atrho is positive and statistically significant at 1%. The result of the Wald test of rho = 0 allows the rejection of the hypothesis that rho = 0. The bivariate probit model is therefore a better fit for the analysis than a single equation probit model. This means that households' choices of water sources for different uses are positively correlated. Households collecting drinking water from improved sources are also likely to collect water for other domestic uses from improved sources. In addition, households collecting drinking water from unimproved sources are also likely to collect water for other domestic uses from unimproved sources.

The results furthermore suggest significant negative associations between distance to improved drinking water sources and the probability of using these improved sources. This indicates that more consideration should be given to actions that facilitate increased accessibility of households to improved drinking water sources. Policies aimed at providing households with private connections at home, could be implemented. The construction of more improved water stations in rural areas could be encouraged as well. We suggest that the construction of such infrastructure in rural areas should be decentralized and entrusted to municipalities and local communities for better maintenance and sustainability over time.

The results provided in Table 3 suggest positive effects of the explanatory variables (effects of wealth, education, female gender, urban areas and proportion of children in the household) on the probability of using improved water sources. In contrast, living in regions other than Douala strongly reduces this probability.

The marginal effects evaluated in terms of the means of the explanatory variables are reported in Table 4. $P(y_1 = 1, y_1 = 1)$ and $P(y_1 = 0, y_1 = 0)$, referring to the probability of using respectively improved water sources and unimproved water sources for both drinking and other domestic uses. $P(y_1 = 1, y_1 = 0)$ and $(P(y_1 = 0, y_1 = 1))$ is the probability of using respectively improved sources (unimproved sources) for drinking and unimproved sources (improved sources) for other domestic uses.

Table 3 Determinants of the use of improved water sources

Dependent variables: probability of using an improved water sources (1=yes, 0=no)						
	Model 1		Model 2		Model 3	
Variables	Water source for drinking	Water source for other domestic uses	Water source for drinking	Water source for other domestic uses	Water source for drinking	Water source for other domestic uses
Distance to improved drinking water sources (in meter)	-0.00026***(0.000)	-0.00037***(0.000)	-0.00026***(0.000)	-0.00038***(0.000)	-0.00025***(0.0[illegible]0)	-0.00037***(0.000)
Wealth quartile (ref. Bottom wealth quartile)						
Second wealth quartile					0.489***(0.045)	0.373***(0.046)
Third wealth quartile					0.989***(0.055)	0.755***(0.053)
Top wealth quartile					1.509***(0.071)	1.341***(0.065)
The head of the household has the primary education					0.0858**(0.071)	0.0689*(0.040)
The head of the household has at least the secondary education					0.109**([illegible].047)	0.094**(0.046)
Proportion of children in the household			0.082(0.095)	-0.073(0.093)	0.203**([illegible].100)	0.036(0.097)
The head of the household is a woman			0.021(0.035)	0.064*(0.033)	0.107***(0.037)	0.141***(0.035)
The household lives in an urban area	0.941***(0.034)	0.714***(0.032)	0.942***(0.034)	0.713***(0.032)	0.436***(0.04[illegible])	0.257***(0.039)
Region (reference=Douala)						
Yaoundé	0.031 (0.120)	0.312***(0.082)	0.029(0.119)	0.312***(0.082)	-0.119(0.13[illegible])	0.196**(0.085)
Adamawa	-0.950***(0.100)	-0.937***(0.072)	-0.953***(0.100)	-0.933***(0.072)	-0.582***(0.10[illegible])	-0.617***(0.077)
Centre	-0.267**(0.104)	-0.245***(0.075)	-0.270***(0.104)	-0.247***(0.075)	-0.026(0.112)	0.005 (0.080)
East	-0.995***(0.101)	-0.914***(0.074)	-1.001***(0.100)	-0.910***(0.074)	-0.546***(0.10[illegible])	-0.494***(0.079)
Far North	-0.846***(0.103)	-0.520***(0.075)	-0.852***(0.103)	-0.518***(0.075)	-0.105(0.115)	0.149*(0.0885)
Littoral	-0.513***(0.104)	-0.406***(0.071)	-0.516***(0.104)	-0.411***(0.071)	-0.295***(0.10[illegible])	-0.165**(0.074)
North	-1.128***(0.102)	-0.811***(0.076)	-1.134***(0.102)	-0.799***(0.076)	-0.441***(0.1[illegible])	-0.183**(0.084)
North West	-0.679***(0.103)	-0.544***(0.076)	-0.682***(0.103)	-0.554***(0.076)	-0.306***(0.1[illegible])	-0.212***(0.079)
West	-0.666***(0.099)	-0.277***(0.072)	-0.671***(0.099)	-0.285***(0.072)	-0.438***(0.1[illegible])	-0.0254(0.076)
South	-0.830***(0.101)	-0.704***(0.075)	-0.833***(0.101)	-0.707***(0.075)	-0.672***(0.1[illegible])	-0.523***(0.078)
South West	-0.382***(0.103)	-0.014 (0.076)	-0.385***(0.103)	-0.017(0.076)	-0.351***(0.1[illegible])	0.066(0.079)
Constant	0.854***(0.093)	0.516***(0.062)	0.843***(0.094)	0.510***(0.063)	-0.036(0.109)	-0.253***(0.079)
Athro	1.689*** (0.035)		1.694*** (0.035)		1.542*** (0.036)	
Wald test of rho=0	chi2(1) = 2305.650 Prob > chi2 =0.000		chi2(1) = 2307.320 Prob > chi2 =0.000		chi2(1) = 21[illegible]4.83[illegible] Prob > chi2 =0.000	
Observations	9,402		9,402		9,360	

Robust standard errors in parentheses *** p<0.01, ** p<0.05, * p<0.1

The results shed light on the heterogeneity of the effect of accessibility to improved drinking water sources on the latter probabilities. The effect of distance to improved drinking water source is negative for $P(y_1 = 1, y_1 = 1)$ and $P(y_1 = 0, y_1 = 1)$ while it is positive for $P(y_1 = 0, y_1 = 0)$ and $P(y_1 = 1, y_1 = 0)$. Marginal effects are respectively −0.00013, −0.00001, 0.00009, and 0.00005.

Table 4 Marginal effects

Variables	$P(y_1 = 1, y_1 = 1)$	$P(y_1 = 0, y_1 = 0)$	$P(y_1 = 1, y_1 = 0)$	$P(y_1 = 0, y_1 = 1)$
Distance to improved drinking water sources (in meters)	-0.00013*** (0.000)	0.00009***(0.000)	0.00005*** (0.009)	-0.00001* (0 .000)
Wealth quartile (ref. Bottom wealth quartile)				
Second wealth quartile	0.145*** (0.016)	-0.13311***(0.011)	-0.00140 (0.009)	-0.01062*** (0.002)
Third wealth quartile	0.274*** (0.017)	-0.23975***(0.011)	-0.01626* (0.01)	-0.01842*** (0 .002)
Top wealth quartile	0.430*** (0.014)	-0.33921***(0.011)	-0.07170*** (0.009)	-0.01883*** (0.003)
The head of the household has the primary education	0.028* (0.015)	-0.02553**(0.012)	-0.00028 (0.008)	-0.00184 (0.002)
The head of the household has at least the secondary education	0.037** (0.017)	-0.02275*(0.014)	-0.002260 (0.021)	-0.00191 (0.003)
Proportion of children in the household	0.02660 (0.035)	-0.05212* (0.029)	0.03873* (0.0077)	-0.013121** (0.006)
The head of the household head is a woman	0.05089*** (0.012)	-0.03523**(0.010)	-0.01711** (0.0089)	0.00146 (0.002)
The household lives live in an urban area	0.11175*** (0.014)	-0.12281***(0.012)	0.2675*** (0.0402)	-0.01568*** (0.003)
Region (reference=Douala)				
Yaoundé	0.03637 (0.033)	0.00444(0.030)	-0.07587*** (0.023)	0.03506* (0.021)
Adamawa	-0.24127*** (0.028)	0.21023***(0.036)	0.03067 (0.033)	-0.00022 (0.008)
Centre	-0.00072 (0.029)	0.00605 (0.030)	-0.0078 (0.028)	0.00247 (0.009)
East	-0.20032*** (0.029)	0.18971*** (0 .035)	0.00343 (0.031)	0.00717 (0.010)
Far North	0.02864 (0.032)	0.00838 (0.029)	-0.06322*** (0.021)	0 .02620* (0.015)
Littoral	-0.07729*** (0.028)	0.08789***(0 .031)	-0.02471 (0.027)	0.01411 (0.012)
North	-0.10223*** (0.033)	0.12421***(0.033)	-0.05405** (0.023)	0.03207** (0.016)
North West	-0.9194*** (0.03k0)	0.09555***(0 .033)	-0.01394 (0.027)	0.01034 (0.011)
West	-0.06813** (0.030)	0.09643*** (0.027)	-0.08683*** (0.018)	0.05854*** (0.021)
South	-0.22237*** (0.029)	0.22909*** (0 .034)	-0.02353 (0.028)	0.01781 (0.013)
South West	-0.03302 (0.032)	0.06470** (0.026)	-0.08930*** (0.018)	0.05762*** (0.021)

Robust standard errors in parentheses *** p<0.01, ** p<0.05, * p<0.1

Table 4 wealth quartiles have positive coefficients only in the first column of the marginal effects. Wealth increases the probability of adopting improved water sources for all domestic uses and decreases the probability of adopting unimproved sources for any use. This result might be explained by the fact that contrary to unimproved sources such as wells and surface water being free of charge most of the time, many improved sources involve financial costs that potentially make them unaffordable for many households. Regarding the marginal effects of the different wealth dummies, it is observed that their magnitude in absolute value increase with the wealth quartile. For instance, the marginal effects of the second, third, and top wealth quartiles are respectively 0.145, 0.274, and 0.430 in the first column of marginal effects. This means that the effect of wealth on the choice of improved water sources is larger (in magnitude) for richer households.

Education is associated with positive effects on the choices of improved water sources for all domestic uses. In addition, it has negative effects on the use of unimproved water sources for any domestic use. This result is consistent with previous studies which suggest that better educated households have access to and use higher quality water sources (Adams et al. 2016; Rahut et al. 2015; Irianti et al. 2016).

Concerning gender, the results show that in general, when the head of the household is female, the probability of using improved water sources for drinking and other domestic uses increases, while the probability of using unimproved water sources for other domestic uses decreases. Although these studies do not all report exactly the results on the impact of the gender of the head of the household, most of them conclude that female-headed households are more likely to have access to and use improved sources or private taps. For instance, Totouom and Fondo (2012) found that compared to male-headed households, female-headed households are more likely to adopt private or collective taps as main water sources. Such results are not surprising since it is known that women generally take charge of caring for sick persons in households. Therefore, they feel more concerned about the detrimental health effects of consuming unsafe water than men. This is to the extent that in case of sickness, the time that they would usually have spent on productive and remunerative jobs, education of children or the preparation of meals, will rather be reallocated to sick persons (Totouom et al. 2018).

The impact of the composition of households on their choices of water sources was also analyzed. The results documented that the higher the proportion of children in a household, the higher (less) the probability will be of using improved sources (unimproved sources) for drinking, no matter the source used for other domestic uses of water.

Regarding place of residence, the results show that living in an urban area increases (decreases) the probability of using improved (unimproved) sources to satisfy all the domestic water needs by 0.11175 points (0.12281 points). Results also show that compared to the households of Douala, households in other regions of the country have a lower probability to use improved sources. Some robustness checks were carried out in addition to test the robustness of the results. Three alternative models were estimated:

The first model consisted of single probit equations. The results are provided in Table 5. They show that the sign and the significance of the different coefficients are almost the same than previously.

In the second model, the variable of accessibility to improved drinking sources was excluded from the demand equation for improved non-drinking sources. Indeed, it would have been reasonable to rather include the distance to improved non-drinking sources in this equation. Unfortunately, the lack of related data did not allow this. As a result, the strong assumption was made that in each municipality the average distance to improved drinking water sources is equal to the average distance to improved water sources for other domestic uses. In this model, the latter assumption was no longer considered. However, the existence of interdependence between the choices of water sources was still acknowledged by estimating a seemingly unrelated

Table 5 Results from the robustness check

Dependent variables: probability of using an improved water sources (1=yes, 0=no)						
	Probit		SUR Probit		Bivariate probit	
Variables	Water source for drinking	Water source for other domestic uses	Water source for drinking	Water source for other domestic uses	Water source for drinking	Water source for other domestic uses
Distance to improved drinking water sources (in meters)	-0.000252***(0.000)	-0.000339***(0.000)	-6.47e-05***(0.000)			
Time spent fetching water from improved drinking sources (in minutes)					-0.00192(0.001)	-0.00431***(0.001)
Wealth quartile (ref. Bottom wealth quartile)						
Second wealth quartile	0.477***(0.045)	0.351***(0.045)	0.489***(0.045)	0.372***(0.046)	0.485***(0.045)	0.363***(0.046)
Third wealth quartile	1.005***(0.056)	0.721***(0.053)	0.987***(0.055)	0.750***(0.053)	0.982***(0.055)	0.740***(0.053)
Top wealth quartile	1.585***(0.075)	1.322***(0.065)	1.514***(0.071)	1.341***(0.065)	1.511***(0.070)	1.330***(0.065)
The head of the household has the primary education	0.081*(0.041)	0.075*(0.040)	0.089**(0.041)	0.074*(0.040))	0.089**(0.0401)	0.0735*(0.040)
The head of the household has at least the secondary education	0.102**(0.049)	0.094**(0.046)	0.118**(0.047)	0.105**(0.045)	0.122***(0.047)	0.106**(0.045)
Proportion of children in the household	0.226**(0.103)	0.049(0.096)	0.206**(0.100)	0.043(0.096)	0.207**(0.099)	0.042(0.096)
The head of the household is a woman	0.122***(0.039)	0.135***(0.035)	0.111***(0.037)	0.142***(0.035)	0.112***(0.037)	0.145***(0.035)
The household lives in an urban area	0.464***(0.042)	0.257***(0.039)	0.426***(0.040)	0.252***(0.039)	0.415***(0.040)	0.240***(0.039)
Region (Reference=Douala)	-0.214(0.153)	0.251***(0.094)	-0.114(0.130)	0.209**(0.085)	-0.122(0.130)	0.187**(0.086)
Yaoundé	-0.816***(0.123)	-0.612***(0.080)	-0.602***(0.107)	-0.652***(0.076)	-0.616***(0.106)	-0.665***(0.076)
Adamawa	-0.222*(0.128)	-0.010(0.084)	-0.085(0.112)	-0.116(0.0774)	-0.098(0.112)	-0.080(0.078)
Centre	-0.724***(0.125)	-0.511***(0.083)	-0.592***(0.108)	-0.578***(0.078)	-0.590***(0.109)	-0.537***(0.079)
East	-0.329**(0.130)	0.140(0.087)	-0.105(0.115)	0.153*(0.085)	-0.096(0.116)	0.171**(0.085)
Far North	-0.532***(0.123)	-0.161**(0.078)	-0.300***(0.109)	-0.175**(0.074)	-0.309***(0.109)	-0.183**(0.074)
Littoral	-0.646***(0.128)	-0.182**(0.086)	-0.476***(0.112)	-0.247***(0.083)	-0.446***(0.115)	-0.155*(0.088)
North	-0.542***(0.126)	-0.208**(0.082)	-0.300***(0.110)	-0.195**(0.079)	-0.301***(0.109)	-0.203***(0.079)
North West	-0.659***(0.123)	-0.025(0.080)	-0.464***(0.106)	-0.074(0.075)	-0.464***(0.106)	-0.0528(0.076)
West	-0.864***(0.126)	-0.515***(0.083)	-0.730***(0.109)	-0.606***(0.077)	-0.752***(0.108)	-0.599***(0.078)
South	-0.553***(0.125)	0.070(0.082)	-0.361***(0.109)	0.051(0.079)	-0.036(0.112)	0.026(0.079)
South West	0.175(0.125)	-0.240***(0.082)	-0.068(0.109)	-0.321***(0.078)		-0.225***(0.083)
Athro			1.640***(0.036)		1.646***(0.036)	
Observations	9,371		9,360		9,360	

Robust standard errors in parentheses *** p<0.01, ** p<0.05, * p<0.1

bivariate probit. The findings show that the variable of accessibility to water sources remains negative and significant. In this way the interdependence of the probabilities of using improved sources is also confirmed.

In the third model, an alternative measure of accessibility to water sources was used. Instead of using distance, time spent fetching water from improved drinking water sources was used. The significant waiting time at the source can only be considered through this alternative measure. The results of the estimated bivariate probit model show that the coefficient associated with the time variable is also negative, although significant only for other domestic water uses.

In conclusion, the outcome of the robustness checks confirms the robustness of the results to different specifications and estimation methods.

6 Conclusion and Implications

In this study, the determinants of access, use and sustainability of improved water sources by households in Cameroon were analysed. This study contributes to the literature on the choices and sustainability of water sources. The focus is on the diversity of water sources used by households and the interdependence of the selection of these water sources on their intended uses. The findings indicate that accessibility to improved water sources constitutes a significant factor in terms of their use. Subsidies for private tap connections and construction of improved water stations in rural areas are examples of the kind of actions suggested. Such measures should mainly target the poor since the findings confirm that poverty is a key obstacle to the use of improved water sources. To increase the use of improved water sources, the poor should be the focus of policy makers.

The analysis also shows that living in urban areas increases the probability of using improved water sources. This result is partly explained by the fact that urban households have better access to improved and diversified water supply sources than rural households. For instance, among 26.5% of cameroonian households that take at least 30 min of round trips to collect water from drinking sources, 35% live in rural areas and 18% in urban areas (NIS and ICF International 2012). The negative sign of the regional variables obtained reflects the low diversification of water supply sources in the other regions of the country, and thus, the limited access of people to safe water in these regions compared to Douala. Although not the specific focus of this chapter, some studies have attempted to identify the determinants of the sustainability of water sources. This includes the construction of water reservoirs in urban areas (Combelem et al. 2020). Acess to running water is also an indicator of the level of sustainable development of a country (Hewett and Montgomery 2001), while direct household waste and wastewater discharge into rivers and lakes threaten the sustainability of water sources.

In this regard, the effective implementation of the Cameroonian government priority actions in the water sector which consist of providing 700 000 new connections to the public network in urban areas and 400 000 water supply stations in rural areas and the construction of processing stations of household waste and wastewater in urban areas is awaited and should not be limited to ad hoc actions. Such actions should increase the sustainability and the supply of quality water services, and thus, the likelihood of the Cameroonian households to use these services that will allow the achievement of SDG 6.

Appendix: Descriptive Statistics on the Explanatory Variables

Variables	Mean	Std. Dev.	Min	Max
Distance to improved drinking water sources (in meters)	279.964	418.623	21	7000
Time spent fetching water from improved drinking sources	22.138	13.126	1.667	75
(in minutes)	0.362	0.193	0	1
Normalised wealth score	0.261	0.439	0	1
The head of the household has no education	0.364	0.481	0	1
The head of the household has the primary education	0.374	0.483	0	1
The head of the household has at least the secondary	0.113	0.153	0	0.75
education	0.25	0.432	0	1
Proportion of children in the household	0.461	0.498	0	1
The head of the household is a woman	0.083	0.276	0	1
Urban	0.081	0.273	0	1
Douala	0.081	0.273	0	1
Yaoundé	0.085	0.279	0	1
Adamawa	0.081	0 0.273	0	1
centre	0.090	0.286	0	1
East	0.081	0.272	0	1
Far north	0.084	0.277	0	1
Littoral	0.084	0.278	0	1
North	0.084	0.278	0	1
North west	0.081	0.273	0	1
West	0.081	0.273	0	1
South				
South west				

References

Adams EA, Boateng GO, Amoyaw JA (2016) Socioeconomic and demographic predictors of potable water and sanitation access in Ghana. Soc Indic Res 126(2):673–687

Basani M, Isham J, Reilly B (2008) The determinants of water connection and water consumption: empirical evidence from a cambodian household survey. World Dev 36(5):953–968

Briand A, Loyal A (2013) La demande de raccordement des ménages auprès des petits opérateurs privés d'eau potable, le cas des quartiers périphériques de Maput. Revue Economique 64(4):685–719

Briand A, Nauges C, Travers M (2009) Les déterminants du choix d'approvisionnement en eau des ménages de Dakar. Revue D'economie Du Développement 17(3):83–108

Briand A, Nauges C, Strand J, Travers M (2010) The impact of tap connection on water consumption patterns: the case of household water use in Dakar Senegal. Environ Dev Econ 15(1):107–126

Briscoe J, Furtado de Castro P, Griffin C, North J, Olsen O (1990) Toward equitable and sustainable rural water supplies: a contingent valuation study in Brazil. World Bank Econ Rev 4(2):115–134

Combelem O, Zoungrana TD, Yonkeu S (2020) Analysis of the determinants of the technical efficiency of horticultural production in the face of the degradation of water reservoirs in urban centers in Burkina Faso. J Econ Sustain Dev 11(12)

Corcoran E, Nellemann C, Baker E, Bos R, Osborn D and Savelli, H (2010) Sick WATER? The Central Role of Wastewater Management in Sustainable Development. United Nations Environment Programme/United Nations Human Settlements Programme/GRID-Arendal (UNEP/UN-Habitat)

Coulibaly L, Jakus PM, Keith JE (2014) Modeling water demand when households have multiple sources of water. Water Resour Res 50:6002–6014

Hewett P, Montgomery M (2001) Poverty and public services in developing-country cities. Working Paper 154, New York, Population Council, Policy Research Division

Irianti S, Prasetyoputra P, Sasimartoyo TP (2016) Determinants of household drinking-water source in Indonesia: an analysis of the 2007 Indonesian family life survey. Cogent Med 3(1)

Kobiané JF 1998 Essai de construction d'un profil de pauvreté des ménages à Ouagadougou à partir des caractéristiques de l'habitat dans F Gendreau (ed.), Crises, Pauvreté et Changements démographiques dans les pays du Sud, Paris, pp 117–131

Madanat S, Humplick S (1993) A model of household choice of water demand supply systems in developing countries. Water Resour Res 29:1353–1358

Mahama AM, Anaman KA, Osei-Akoto I (2014) Householders' access to improved water in urban areas of Accra Ghana. J Water Health 12(2):318–331

Mu X, Whittington D, Briscoe J (1990) Modelling village water demand behavior: a discrete choice-choice approach. Water Resour Res 26:521–529

Nauges C, Den Berg V (2009) Demand for piped and non-piped water supply services: evidence from Southwest Sri Lanka. Environ Resour Econ 42(4):535–549

Nauges C, Strand J (2007) Estimation of non-tap water demand in central American cities. Resour Energy Econ 29:165–182

Nauges C, Strand, (2017) Water hauling and girls school attendance: some new evidence from Ghana. Environ Resour Econ 66(1):65–88

Nauges C, Whitttington D (2010) Estimation of water demand in developing countries: an overview. World Bank Res Observer 25(2):263–294

NIS, ICF International (2012) Enquête Démographique et de Santé et à Indicateurs Multiples du Cameroun 2011. Calverton, Maryland, USA

Persson HT (2002) Household choice of drinking-water source in the Philippines. Asian Econ J 16(4):303–316

Rahut DB, Behera B, Ali A (2015) Household access to water and choice of treatment methods: empirical evidence from Bhutan. Water Resour Rural Dev 5:1–16

Rahut DB, Ali A, Chhetri NB, Behera B, Jena PR (2016) Access to safe drinking water and human health: empirical evidence from rural Bhutan. Water Sci Technol. Water Supply 15(5):1349–1360

Rauf S, Bakhsh K, Hassan S, Nadeem AM, Kamran MA (2015) Determinants of a household's choice of drinking water source in Punjab, Pakistan. Pol J Environ Stud 24(6):2751–2754

Totouom FLA, Fondo S (2012) Determinants of the households: choice of drinking water source. J Sustain Dev Afr 14(3):86–97

Totouom A, Fouéka R, Ngouhouo J (2018) Determinants of household avoidance behavior to cope with unsafe drinking water: case study of Cameroon. Rev Agric Food Environ Stud 99(2):121–148

UN (United Nations). n.d. Wastewater Treatment. Sustainable Development Knowledge Platform, United Nations. www.un.org/esa/sustdev/natlinfo/indicators/methodology_sheets/freshwater/waste_water_treatment.pdf

WHO (2017) Safely managed drinking water - thematic report on drinking water 2017. Geneva, Switzerland. License: CC BY-NC-SA 3.0 IGO

WHO and UNICEF (2017) Progress on drinking water, sanitation and hygiene: 2017 update and SDG baselines. License: CC BY-NC-SA 3.0 IGO. https://www.who.int/mediacentre/news/releases/2017/launch-version-report-jmp-water-sanitation-hygiene.pdf. Accessed 7 Dec 2017

World Bank (2012) The Future of Water in African Cities: Why Wastewater? DC, World Bank, Washington

Zoungrana TD and Combelem O (2016) Determinants de la participation des populations riveraines à la protection des retenues d'eau en milieu urbain : cas des barrages no. 1,2 et 3 de la ville de ouagadougou au Burkina Faso, Edition Université Aube Nouvelle, JOASG 1(2)

Sustainable Cropping Intensification and Its Role on Profitability of Cassava-Based Farms in a Changing Climate: Evidence from Rivers State, Nigeria

Anthony Ojonimi Onoja, Helen Emuejevoke Obukonise, and Eunice Ngozi Ajie

Abstract The paper evaluated the level of cropping intensification (a sustainable development strategy in staple crops farming) and assessed its effect on farm profit using primary data from 140 cassava–based farming households in Rivers State of Nigeria. The data were obtained using a structured questionnaire and analyzed using descriptive statistics, Herfidahl index and OLS multiple regression models (with four functional forms). Results show that the crop production intensity index (Herfidahl index) ranged between 0.1 and 0.89 with a (mean = 0.52). Mean Gross Margin recorded by the farmers was $777.64 USD, indicating that the farms were profitable. The profitability levels were significantly determined by the magnitude of crop intensification, proxied by Herfindahl index (elasticity = 0.276, $p < 0.05$); household size (elasticity = 0.519, $p < 0.01$); farmer's educational status (elasticity = 0.085, $p < 0.01$); and gender of the farmer (1.523, $p < 0.01$). It was recommended that, cassava crop intensification, system should be adopted as an integral measure to promote cassava production for food security, employment generation and sustainable development strategy. To boost sustainable production farmers should be assisted by government and non-governmental institutions to diversify their production, boost higher productivity through supply of improved and affordable inputs, building of farmers' capacity in intensive crop management technologies, access to extension services in a gender inclusive manner.

Keywords Sustainability · Cropping intensity · Cassava · Africa · Rainwater · Multiple cropping

---

A. O. Onoja (✉) · H. E. Obukonise
Department of Agricultural Economics and Extension, University of Port Harcourt, Port Harcourt, Nigeria
e-mail: anthony.onoja@uniport.edu.ng

E. N. Ajie
Department of Agricultural Science, Ignatius Ajuru University of Education, Rumuolumen, Port Harcourt, Rivers State, Nigeria

W. Leal Filho et al. (eds.), *Sustainable Development in Africa*, World Sustainability Series,
https://doi.org/10.1007/978-3-030-74693-3_24

1 Introduction

The global food security challenge looks very daunting (World Food Programme, WFP 2020). A recent report by WFP (2020) indicates that eighteen countries stood at the risk of dipping further into crises that could pose serious threats to people's food security and livelihood. Nigeria is among these countries. Aboloma (2018) aptly noted that by 2050, the world will have to grapple with providing food for an estimated 9.7 billion people. This implies that the food demand will rise by 60% greater than its status in 2018. Unfortunately, Shimeles et al. (2018) aptly noted that in Sub-Saharan Africa (SSA), agricultural sector productivity trails behind other regions of the globe. The New Partnership for Africa's Development, NEPAD (2013) indicated that Africa had 33 million farms, 80% of which were operating on small scale land holdings of less than 2 ha. Owing to the fact that SSA farming systems production largely depend on family's capital and labour force, the region experiences poor farm productivity. A recent study by Onya et al. (2019) noted that land access for farming purpose is highly constrained in Nigeria owing to barriers from gender discrimination, poor access to credit, stringent land policies and communal land tenure systems. Africa Research Institute (2009) noted that these subsistence farmers ability to contribute meaningfully to food security at a national scale are highly restricted due to fragmentation of land they have access to which are too small. This now calls for innovative ways of utilizing the limited land they have in order to achieve higher productivity and contribute meaningfully to domestic, national and global food security needs especially under a climate change regime. NEPAD (2013) notes that agriculture utilizes and manages basic resources which include: land, water and energy. Thus it is impossible to divorce agricultural development from attainment of sustainable development of natural resources. They noted that in the past 27 years, the development of Africa's agricultural sector had mobilized more and more arable land. However, it is sad to note that till date the sustainability of some agrarian systems remain threatened by several factors including the effects of climate change, and population explosion, both putting pressure on land resources and have compromised the capacity of traditional farming systems to renew soil fertility. Hence, the report notes that in the near future, the challenge will be on how "to accelerate growth in production while controlling its impact on the environment and natural resources such as land, water and energy, which are the foundations for the development potential of future generations."

Cropping intensification has been recommended as a good adaptation strategy against risk of climate change (Onoja 2019; Degla et al. 2016). Climate change is a major threat to food security especially in the developing economies (Intergovernmental Panel on Climate Change, IPCC 2014; Lee et al. 2015; Kompas et al. 2018; and Asmus et al. 2019). According to Coulter (2019) it is becoming clearer that remarkable changes in climate change strategies and management are emerging yet there are no consensus over the best practices required to study uncertainties and impacts of both climate change and its policies including risk reduction, adaptation to climate changes and resilience building at different strata of cooperation.

So far, the global response to food insecurity can be seen in the United Nations' set goals of "ending hunger, achieving food security and promoting sustainable agriculture as the second of its 17 Sustainable Development Goals (SDGs) for the year 2030", added Aboloma (2018). Despite a strong dependence of Nigerians' population on farming and agriculture, malnutrition still pervades many regions of the country. Agriculture accounts for 70% of Nigeria's employment. It, has been a major contributor to Nigeria's economy for several decades but rising population (of over 200 million), urbanization and industrialization have increased the pressure on land resulting in reduced farm output that is not sufficient to meet national food demand (Fasoriyo and Taiwo 2012; Osu 2017). This impacts negatively on food security situation and exacerbates poverty in Nigeria as farms are being abandoned to the elderly or for poor paying off-farm jobs (Osu 2017).

Nigeria is the world's largest producer of cassava with cassava being a main staple food as well as her leading cash crop. According to FAO (2020), Nigeria produces about 50 million metric tons of cassava per annum from a land area of about 3.7 million ha. The country's cassava output accounts for approximately 20% of the global output with the country maintaining around 34% of Africa's output and contributing 46% to West Africa's output.

High productivity is crucial to both food security and poverty alleviation problems facing the country. Unfortunately the country records a dismal average yield of about 13.63 metric tons per ha, a far cry from its potential cassava yield of about 40 metric tons per ha (FAO 2020). This agrees with Adebayo et al. (2020) whose empirical study shows that cassava production in Nigeria has been facing gross inefficiency with resources being sub-optimally allocated on most farms despite the fact that the country remains the largest producer of cassava in the world About two-thirds (i.e. 66%) of the nation's total production comes from the southern part of the country where Rivers State lies, while in the about 30% is in the north-central, and 4% in other parts of the north. The crop is predominantly grown by smallholders on small plots for family consumption and local sale (FAO 2020). Farms operating on large commercial scale are few (Sahel 2016; FAO 2020). With climate change factors, conflicts over land and poor access to land owing to restrictive communal land tenure systems the productivity and supply of cassava in Nigeria had been badly constrained (Agunbiade 2005). Under such a scenario, one available option to maximize output from the limited land and produce sustainably is by adopting cropping intensification. This system of farming also helps farmers adapt to increasing impact of climate change. It is the realization that production of cassava in Nigeria had been inefficient and faces sustainable development issues that interventions such as Building an Economically Sustainable and Integrated Cassava Seed System in Nigeria (BASICS) was established by the Bill and Melinda Gates Foundation to provide a model innovative, commercially and economically viable cassava stem business model. This was a 4-year project that aimed at intensifying cassava production and sought to increase its yield by 40%. This was to be achieved by implementing the rapid propagation technology-autotrophic hydroponic system (SAH) which would enable smallholder farmers provide high quality cassava roots to industrial processors thereby enhancing their personal incomes (Sahel 2016). According to Uphoff (2012) and Uphoff (2011) there are emerging opportunities geared towards increasing agricultural production in ways that could directly abate the food insecurity facing several billion people which attain this in environmentally-friendly ways. Such systems can enable crops to adapt to or cope with biotic and abiotic stresses being currently exacerbated by climate change. One of such system is called sustainable intensification. Abraham et al. (2014) observed that while there exist diverse terminologies used by different organizations to represent sustainable intensification, there is a general consensus among the researchers and organizations that "agricultural sectors around the world need to pursue modified strategies for 'sustainable intensification' if global food security requirements are to be met throughout this century". FAO (undated) asserted that sustainable crop production intensification (SCPI) "aims to produce more from the same area of land while reducing negative environmental impacts, conserving natural resources and enhancing healthy ecosystem services." The system is eco-friendly and combines farmers' traditional knowledge with modern technologies adapted to small-scale farmers' needs.

Aim and Objective of the Study

Given the fact that land plays an important role in the livelihood of the majority of rural households, food security and poverty reduction cannot be achieved unless issue of access to land, security of tenure and the capacity to use land productively and sustainably are addressed (Osu 2017). Therefore, the objective of this paper is to: (i) evaluate the levels of cropping intensification, (ii) ascertain the profit and profitability levels of cassava based farms, and (iii) determine the role of crop intensification and other socio-economic factors on farm profit in cassava–based farm households of Rivers State, Nigeria.

Findings will be useful to provide evidence for policy making in the area of sustainable food production, climate change risk resilient building of small holder farmers, land intensification strategies and food security in Nigeria.

Empirical Literature Review

Adhikari et al. (2018) defined a system of crop intensification (SCI) as an agricultural production strategy which aims at increasing and optimizing the benefits accruing to farmers for making better use of available resources. Such resources, they noted, include: "soil, water, seeds, nutrients, solar radiation, and air." Garbach et al. (2017) stressed the need to always evaluate agricultural options in context, weighing the "factors and interactions of time and space so that field operations are conducted in a timely way, with land area optimally occupied by crops, and not just by a single crop." They also amplified the need to appraise the ecosystem services they provide. There is a global consensus that agricultural production needs to rise in order to sustain a rising world population and changing demand patterns of agricultural produce without violating the planetary boundaries (Springmann et al. 2018).The implication of this assertion is that farm productivity increases must be derived from the currently available land. This should be done without jeopardizing the drive to protect our living environment, conserving natural and agricultural biodiversity (The

Royal Society 2009). Pretty et al. (2018).asserted that sustainable agricultural intensification approaches is expected to come up with the strategies to achieve that but acknowledged that before that, a significant global redesign of agricultural systems would be required, Crop intensification has been practiced across the world as a response to land conservation needs over the ages. For instance, LEISA-India (2011) noted that in distressed lands under the impacts of drought and floods tribal farmers of Western Ghats in Maharashtra, India were adopting system of crop intensification (SCI), an agro-ecological method, to build enabling environment for the crops to grow to their potentials. It was further observed that the farmers' traditional knowledge and innovative capacities affirmed the effectiveness of the system "in building climate resilient cropping system of finger millet and chickpea."

A study carried out by Skarżyńska (2012) to analyze the economic outcome of crop production in Poland found that the scale of diversification is indicated by the level of the farm gross margin, income from activities of the farm management and production profitability index of the farms. It was observed that high-intensity in cropping is more associated with higher profit levels.

They further noted that using modern technological strategies could threaten sustainable crop development as they can "contribute to reducing the unfavourable impact of chemical agents on the natural environment, consistent with maintaining high economic efficiency of agricultural production." Some recent quantitative empirical analyses from various cropping systems have also indicated that systems of crop intensification can outperform chemically managed monocultures across a wide range of globally important ecosystem services while producing sufficient yields and reducing environmental externalities (Onoja 2019; Lundgren and Fausti 2015). As a consequence, cropland intensification and dietary changes have received much attention as potential solutions to reduce the decline of natural area (Alexander et al. 2015). The System of Rice Intensification (SRI) was empirically investigated for its economic performance by SRI-Rice (2014). The study found that in economic terms, *the costs of production per hectare* with intensification, went up *on average by about 50% per hectare initially, but*, "given the increases in yield, on average *the costs per unit of crop produced decline by about 40%* across the crops .." This, they noted resulted in more than doubling of doubling of farmers income per hectare.

Scientists are in agreement that biologically diversified cropping systems are often more able to meet global food security needs sustainably and efficiently (Cassman and Grassini 2020; DeLonge et al. 2016; and Gliessman 2014). In a review of some quantitative syntheses and meta-analyses Delonge et al. (2016) found that most of the researchers agreed that intensified cropping systems can outperform chemically managed monocultures farming systems over a broad "range of globally important ecosystem services while producing sufficient yields and reducing environmental externalities". These were also in line with the findings of Kremen and Miles (2012); Lundgren and Fausti (2015). They stressed that such agroecological farming systems (such as crop intensification), in some cases could produce equivalent or higher yields than the conventional and monoculture farm systems while at the same time enhancing ecosystem services and profitability. The findings agreed with Davis et al. (2012); Kremen and Miles (2012); and Prieto et al. 2015).

According to Hossain et al. (2017) multiple cropping can be described as a farming system "involving two or more crop species or genotypes growing together and coexisting for a certain period." It is gaining more relevance as farmers recognize the benefits in intensifying land use as a means to remain competitive and meet up with the increasing global demand for different crop originated products. Besides, it is popular among many "subsistence or low-input/resource-limited agricultural systems on the fringes of modern intensive agriculture." Hossain et al. (2017) noted that by providing real yield gains without much use of farm inputs, or maintaining greater stability of yield with decreased inputs, multiple cropping comes as a unique way of delivering sustainable intensification. Degla et al. (2016) noted that multiple cropping systems are sustainable development strategies that are largely influenced by frequency of contacts of farmers with extension services, farmers' education level, and farm size. They noted that effective adoption of the system of crop intensification was constrained by gender of the farmer, adult literacy level, perception of farmers about adaptation to climate change, years of experience under climate change impacts, and farmers' geographic location.

2 Research Methods

The study was carried out in Rivers State, Nigeria. The state's population as at 2016 stood at about 193,392,500 (City Population 2016). Rivers State, with coordinates of 4.8396 °N and 6.9112 °E, was largely involved in agriculture and fishing livelihoods before the discovery of oil in commercial quantity in 1951 in the state. Citing the Federal Ministry of Agriculture and Natural Resources, the Rivers State (2019) observed that approximately 40% of the state's rural inhabitants were committed to farming in 1983 and the state remained one of the leading states in the production of yam, cassava, cocoyam, maize, rice and beans. The report asserted that about 39% (760,000 ha) of the state's total land mass, especially in the upland area, was suitable for cultivation. Economic crops produced by the farmers in the area included cassava, oil palm products, rubber, coconut, raffia palm, jute and food crops including vegetables, melon, pineapples, mango, pepper, banana and plantain. The fishing industry is an important sector in Rivers State. Fishing still remained a favorite livelihood activity in the state with over 270 species of fish existing in many artisanal fishing grounds in the riverine areas.

A cross sectional survey in design, the study was based on primary data obtained from 140 farmers via the use of structured questionnaire. Data collected were analyzed using descriptive statistics, Farm Gross Margin, Herfindahl index and OLS multiple regression models. Four functional forms of the model, including linear, semi-log, double log and exponential log models were tried after which the best model was selected for the analysis based on theoretical and econometric criteria.

The first stage in the analytical approach was to compute the profit of cassava farming in various farms regardless of their combinations. This was done using Gross Margin analysis. Farm Gross Margin profit refers to the difference between the annual gross income of an enterprise and the variable costs directly associated with the enterprise (Rural Solutions SA 2017). To compute enterprise gross margins, fixed (overhead) costs are not considered, as it is assumed that they will be incurred irrespective of the level of the enterprise undertaken.

The Gross margin is given by:

$$GM = \text{Total Farm Revenue} - \text{Total Variable Costs}. \quad (1)$$

The gross margin was given in monetary values. Farm revenue was computed by multiplying the quantities of farm produce sold by the unit prices of sales of farm the produce. The variable costs include costs of Planting materials, family and hired labour, packaging, communication and transport costs.

After this, the Hefindarl index for various farms were computed following Chen et al. (2018), Maddison (2007), Nhemachena and Hassan (2007) and Lang and Stulz (1994). It was also used by Culas and Mahendrarajah (2005); Pope and Prescott (1980) for analysis of agricultural intensification. According to Pew Research Centre (2014) the Herfindahl-Hirschman Index can be described as a measure used in various fields to ascertain the degree of concentration of human or biological populations as well as organizations. The Pew Research Centre noted that "the Herfindahl-Hirschman Index is a widely accepted measure of concentration used by biologists, ecologists, linguists, economists, sociologists and demographers."

In order to ascertain the level of crop intensification, for each household, number of farming crop enterprises combinations in each category were counted and thereafter a normalized Herfindahl-Hirschman Index (HHI) was constructed as follows.

The normalized HHI is given as follows:

$$HHI = \frac{\Sigma (k - 1)^{Ks2k - 1/K}}{1 - 1/K} \quad (2)$$

Where k = 1, 2, 3 indexes a category. K = 3 is the total number of categories for our study, and sk is the ratio of the number of farming practices (out of the following enterprises: cassava monocrops = 1, cassava/maize = 2, cassava/maize/yam = 3, applied in category k) adopted to the total number of farming practices in all categories. Lower estimates of HHI implies higher diversity or cropping intensity while higher values of the index reflects lower intensification.

The profits of the various levels of cropping intensification cassava based systems were then estimated using Gross Margin analysis. The profitability index of each system was then estimated. Profitability ratios refers to financial metrics used by business analysts and investors to measure and evaluate the ability of a firm to generate income (profit) relative to its revenue, balance sheet assets, operating costs, and shareholders' equity during a specific period of time (Corporate Finance Institute, CFI 2020). Profitability ratios are used to evaluate the ability of a farm firm to utilize its assets in generating profits and monetary value to its owners. Higher ratios are often preferred by most farm firms, since this implies higher capacity of the business enterprise to generate revenues, profits, and cash flows. Profitability ratios are most useful when they are analyzed in comparison to similar enterprises or compared to previous periods. Apart from Net Present Value (NPV) and Benefit Cost Ratio (BCR), one of the most commonly used profitability ratios' formulae in farm business enterprises is the Production Profitability Index. In the present study, the profitability index of the farms were further evaluated using Production Profitability Index (PPI) following Skarżyńska (2012) as follows:

$$\text{PPI}(\%) = \text{Production Value (in USD)/Economics costs (TVC in Gross Margin)} \times 100/1 \quad (3)$$

Thereafter four functional regression analysis (Ordinary Least Squares) models, including linear, semi-log, double log and exponential log functions were estimated. The farm gross margin, GM, is maintained as the dependent variable to evaluate the possible role of crop intensification index variable (proxied by HDFLINDX) on the estimated profit level of the cassava farms.

The implicit form of the model is given as:

$$\text{GM} = \text{f(HDFLINDX, HHSIZE EDULV, SEX, FAMEXP, OCCUPTN, EXT, AGE)}. \quad (4)$$

Where, GM = estimated monetary value of gross margin as earlier specified; HDFLINDX = value of estimated Hefidarl index for each farm household (a ratio), EDULV = number of years spent in formal education; SEX = gender of the household head (dummy 1 = male, 2 = female), FAMEXP = years spent in cassava farming, OCCUPTN = type of full-time occupation practiced (Dummy, farming = 1, civil servant = 2, trading = 3, artisan = 4, fishing = 5, others = 6), EXT = frequency of agricultural extension visits in the past one year (count), AGE = age of household head in years, and TOTALFMSZ = farm size in hectares. The best performing regression model estimated was selected based on the alignment of most of the signs of the estimated coefficients' agreement with theoretical expectations, the size of the coefficient of determination, the F-ratio estimated and the significance of its estimates as well as the Akaike Information Criteria (AIC) consideration (Gujarati 2004). The smaller the AIC, the better the fitting of the models for explaining the relationship. The functional forms were estimated using E-Views econometric software.

Table 1 Rate of cropping intensity estimated for cassava-based crop farmers using Diversification Index (Herfindahl Index)

Range of Herfindafal Index	Frequency	Percentage frequency
0–0.10	17	12
>0.10–0.20	5	4
>0.20–0.30	7	5
>0.30–0.40	12	9
>0.40–0.50	12	9
>0.50–0.60	15	11
>0.60–0.70	26	19
>0.70–0.80	12	9
>0.80–0.90	17	12
>0.90–1	17	12
Total	140	100
Mean Diversification Index = 0.52		

3 Results and Discussion

3.1 Estimates and Frequencies of the Crop Production Intensity Index

The crop production intensity index recorded by the farming households ranged between 0.1 and 0.89 with a mean score of 0.52 (See Table 1).

The results in Table 1 indicate a relatively medium level of crop diversification level in the area. The farmers could be practicing these to spread risks of crop failure and possibly cope with possible crop failures in the advent of harsh or intolerable climate regimes. This corroborates the assertion of Chen et al. (2018) who noted that crop intensification and diversification are climate change adaptation strategies adopted by crop farmers in a changing climate.

3.2 Gross Margins and Profitability Estimates of Farms in Cassava Based Cropping Systems

Results of the Mean Gross Margin estimated in Table 2 indicate that the cassava farmers generally attained average profit levels of N122, 866.93 (US $777.64) per year from their farms indicating that the farms were profitable (See Table 2). The high profit generating capacity estimated in the various levels of crop intensification systems in the State indicate that various degrees of profit could be attained from the cassava intensification systems. Such profits can contribute to improved household incomes and reduction of poverty in most marginalized households adopting cassava

Table 2 Pooled mean of gross margins and profitability indices for various cropping intensification systems

	Pooled Mean of Costs and Returns		GM for Various cropping intensification systems		
	Costs and Returns in Naira for the average farm	Mean GM Estimates for all groups involved in crop diversification	Monocropping (No Diversification)	Cassava/maize =2,	Cassava/maize/yam
Item	Amount in Naira (₦)	Amount in US Dollars	Amount in US Dollars	Amount in US Dollars	Amount in US Dollars
Total Sales	154,533.18	978.06	767	987	1180
Cassava Revenue	58,722.61	371.66	290	375	450
Maize Revenue	33,997.30	215.17	168	217	260
Yam Revenue	61,813.27	391.22	305	395	473
Less Variable Costs:					
Planting materials	11,083.19	70.15	62	69	80
Total Labour cost	15,833.13	100.21	55	100	140
Cost of Packaging	1,583.31	10.02	9	9	13
Communication	1,583.31	10.02	7	11	13
Cost of transportation	3,166.63	20.04	18	20	23
TVC	31,666.25	210.44	151	209	269
Gross Margin per ha	122,866.93	767.61	616	778	911
Production Profitability Index (PPI)	**488**	**465**	**508**	**472**	**439**

farming as a way of livelihood. The fair level of profits corroborates the earlier assertion that cassava multiple cropping has potentials for contributing to household poverty reduction drive (Agunbiade 2005; FAO 2020). This is also in agreement with Royal Society (2009) who noted that sustainable intensification of crop production, like any agricultural sustainability practice must contribute to improving food crop production and be seen to "reduce rather than exacerbate global inequalities if they are to contribute to economic development."

With regards to the Gross Margin recorded, all the farms operating at various levels of intensification were operating profitably. The pooled mean Gross Margin for all levels was 767.61 USD per hectare. Meanwhile, for monocropping systems where there is zero intensification the Gross margin was lowest (616 USD per ha) compared to the cassava/maize intensification system which netted a Gross Margin of 778 USD per ha. The highest Gross Margin recorded in the Cassava/maize/yam cropping system (911 USD per ha) only confirmed the fact that, given a fixed size of land, the systems of crop intensification with higher diversity of crops stand to bring in more farm revenues per hectare of land. The PPI obtained from the study's estimate was 465%. This is a very high ratio indicating that most farmers in the study were very profitable. According to Manteufield (1984) and Skarżyńska (2012) if the

PPI falls below 100, the production is deemed unprofitable, but if it exceeds 100, the higher the profitability.

Regarding the level of economic costs the results obtained indicate that their development trend corresponds to the direct costs-related trends in consonance with the findings of Skarżyńska (2012). According to Skarżyńska (2012) farms with low cropping intensity recorded lower economic costs than farms with higher cropping intensity. This explains the lower levels of profitability (PPI) recorded for the cassava/maize/yam enterprises and cassava/yam enterprises whose PPIs were 439 and 472 respectively. The monocropping with zero diversification thus recorded the highest PPI of 508. However all the farms, as earlier noted were profitable as their PPIs were greater than 100. The lower profitability in the higher cropping intensive farms cannot only be attributed to direct costs, but rather to indirect costs and other production factor costs. According to Skarżyńska (2012) levels of direct costs, being mainly controlled by the farmer are the determinants of the level of economic costs.

3.3 *Effect of Cropping Intensification and other Socio-Economic Factors on Farm Profits*

The relevant parameter estimates for the regression equations aimed at determining the effect of cropping intensity alongside other factors on cassava based cropping systems are displayed in Table 3. The levels of profitability were significantly determined by the magnitude of crop intensification, proxied by Herfindahl index (elasticity = 0.276, $p < 0.05$); household size (elasticity = 0.519, $p < 0.01$); educational status of the farmer (elasticity = 0.085, $p < 0.01$); and gender of the farmer (1.523, $p < 0.01$) (See Table 3). The lead equation selected after econometric diagnosis and theoretical evaluation was the Cobb-Douglas model whose R2 was 0.87and had an estimated F-ratio of 53.49 ($p < 0.01$). The model also had the lowest Akaike Information Crieteria, AIC (See Table 3).

The fact that cropping intensity index significantly influenced the farm profits corroborates the assertions and findings of FAO (2020), Sahel (2016) and Uphoff (2012). The findings are similar to a recent findings from a study carried out in Poland. In the study, found that economic results of diversification scale is reflected by the level of gross margin and income from management activity, as well as by the production profitability index in the activity types analysed. When using low-intensity, as compared to high-intensity, cropping technologies, the economic results of the activities in question become more favourable. The profitability analysis of various production factors points to a prevalence of agricultural farms with low intensity levels of the activity conducted. Lower outlays of production means contributed to a more effective utilization of both the natural soil fertility, and labour combined with fixed assets. A lower use of chemical crop-enhancing agents forces the farmer to employ more environmentally-friendly methods to keep production at a profitable level. The results show that the use of modern technological achievements may

Table 3 Regression estimates of influence of crop intensification and socio-economic attributes on farm profitability in the study

Dep. Var: Gross Margin	Linear Model		Semi-Log Model		Double-Log Model		Exponential -Log Model	
Explanatory Variables	Coefficient	t-Statistic	Coeff.	t-Statistic	Coeff.	t-Static	Coeff.	t-Stat.
Intercept	-138283.9	-3.038***	8.722	24.107***	10.109	13.394***	-45898.71	-0.393
	-45520.02		-0.362		-0.755		-116894.4	
HDFLINDX	41667.64	1.696***	0.168	0.86	0.276	2.302**	5187.018	1.601
	-24574.43		-0.195		-0.183		-3239.937	
HHSIZE	24219.62	4.608***	0.143	3.427***	0.519	2.966***	91747.9	3.413***
	-5256.314		-0.042		-0.175		-26882.87	
EDULV	7142.147	2.769***	0.098	4.787****	0.085	6.307***	3405.751	1.642
	-2578.974		-0.02		-0.013		-2073.9	
SEX	93395.19	2.102**	0.559	1.581	1.523	3.302***	130536.9	1.839***
	-44435.4		-0.353		-0.461		-70985	
FAMEXP	-3246.65	-2.188**	-0.015	-1.246	-0.022	-0.188	3245.042	0.18
	-1483.913		-0.012		-0.117		-18015.17	
OCCUPTN	-2628.798	-0.536	0.106	2.709***	0.126	1.315	-14016.52	-0.942
	-4902.168		-0.039		-0.096		-14883.01	
EXT	152.844	0.018	0.023	0.35	0.088	0.633	2992.855	0.138
	-8401.532		-0.067		-0.139		-21643.36	
AGE	-527.572	-0.814	-0.004	-0.871	-0.115	-0.571	5933.064	0.19
	-648.428		-0.005		-0.201		-31197.07	
TOTALFMSZ	11521.22	1.261	0.032	0.444	0.098	0.605	23311.33	0.952
	-9136.226		-0.073		-0.163		-24483.89	
R-squared	0.83		0.86		0.87		0.77	
Adj. R. Sqr	0.81		0.85		0.85		0.74	
Akaike info criterion	25.02		1.54		1.46		25.36	
Schwarz criterion	25.32		1.84		1.76		25.66	
F-statistic (Prob(F-stat)	39.33 (p=0.00)		49.2 (p=0.00)		53.49 (p=0.00)		25.2 (p=0.00)	

NB: (***) = Figure(s) significant at < 1% a; (**) = < 5% ; (*) =<10% alpha level. Fig in parentheses are standard errors.

contribute to reducing the unfavourable impact of chemical agents on the natural environment, consistent with maintaining high economic efficiency of agricultural production.

4 Conclusion

The study adopted an econometric and farm management analysis approach to evaluate the performance of sustainable crop intensification system in cassava based farming systems in Rivers State, Nigeria, It analyzed the productivity from lands adopting sustainable cropping intensification and compared with those using non-sustainable practices such as monocropping. This was attained using gross margin and profitability ratios after which it determined the factors which influenced sustained profitability of the cassava based cropping systems the area. It was found that the rate of crop diversification, a sustainable cropping system was moderate while the farms operated profitably. One of the major lessons from the performance of the farm enterprises studied is that with respect to profitability levels, farms with lower diversity scores performed better. However, when we evaluate other long run benefits inherent in adoption of more diversified farming systems (higher crop intensive farms with more sustainable attributes), the advantage of profitability which is not too different from the other farming practices whittles out. For instance, it is common knowledge that more diversified farming systems encourages agro-biodiversity, help farmers spread risks of crop failure and mitigate green house gas emission. These added sustainability attributes which are of high advantages can help contribute to sustainable development goals besides actualization of economic gains. The study also found that farm profits in the study were majorly determined by the intensity of crop diversification, household size, educational status of the farmer and gender of the farm household head.

Based on the findings, it was recommended that, given the environmental benefits and income generating potentials of the cassava crop intensification, the sustainable cropping intensification farming system should be adopted as an integral measure to promote cassava production for food security, employment generation and climate change adaptation strategy. To boost sustainable production of cassava in Nigeria, farmers should be assisted and encouraged by government and non-governmental institutions to diversify their production base and encourage higher productivity through supply of improved and affordable inputs (e.g. improved seeds/seedlings and fertilizers). Given that education level of farmers contributed to sustenance of farm profit levels, sustainable development of the cassava farming systems must incorporate building of farmers' capacity in intensive crop management technologies. Other measures recommended are improvement of access to formal education and extension services in a gender inclusive manner by farmers. The fact that gender is a factor that influenced level of profits earned by the farmers in the study provides a lesson to development agencies and policy makers on the need to mainstream gender in the promotion of sustainable cassava production systems. Thus efforts to make cassava farming a poverty reducing strategy by policy makers must include gender mainstreaming in the ways farmers can access the necessary resources for production such as equal access to farm credit or finance, market access and access to land. A serious consideration of the above recommendations will elevate the steps towards attainment of sustainable development of staple crop farming in Africa's largest economy.

References

Aboloma OA (2018) Cultivating growth in Nigeria. https://www.iso.org/news/ref2300.html. Accessed 29 June 2020

Abraham B, Araya H, Berhe T, et al (2014) The system of crop intensification: reports from the field on improving agricultural production, food security, and resilience to climate change for multiple crops. Agric Food Secur 3(4):1–12. https://doi.org/10.1186/2048-7010-3-4. Accessed 22 May 2020

Adebayo O, Opeyemi O, Jacob A, Mayokun O, Mistura R, Omotayo AO, Oluseyi OK (2020) Resource-use efficiency and production constraints among cassava farmers in Nigeria: insight, linkage and pathway. J Dev Areas 54:1–12. https://doi.org/10.1353/jda.2020.0045. Accessed 22 June 2020

Adhikari P, et al (2018) System of crop intensification for more productive, resource-conserving, climate-resilient, and sustainable agriculture: experience with diverse crops in varying agroecologies. Int J Agric Sustain 16(1):1–28. https://doi.org/10.1080/14735903.2017.1402504. Accessed 17 June 2020

Africa Research Institute (November 2009) Waiting for a green revolution. Briefing Note 0902

Agunbiade LW (2005) Cassava production in Nigeria: constraints, solutions and prospects. Coll Rev 12:267–271

Alexander P et al (2015) Drivers for global agricultural land use change: the nexus of diet, population, yield and bioenergy. Glob Environ Chang 35:138–147

Asmus ML, Nicolodi J, Anello LS, Gianuca K (2019) The risk to lose ecosystem services due to climate change: a South American case. Ecol Eng 130:233–241

Cassman KG, Grassini PA (2020) Global perspective on sustainable intensification research. Nat. Sustain 3:262–268. https://doi.org/10.1038/s41893-020-0507-8. Accessed 22 May 2020

Chen M, Wichmann B, Luckert M, Winowiecki L, Förch W, Läderach P (2018) Diversification and intensification of agricultural adaptation from global to local scales. PLoS ONE 13(5):1–27. https://doi.org/10.1371/journal.pone.0196392. Accessed 12 May 2020

City Population (2016) Rivers state. Rivers state government: Port Harcourt. https://www.citypopulation.de/php/nigeria-admin.php?adm1id=NGA033

Corporate Finance Institute, CFI (2020) Profitability ratios: measures of a company's earning power. Training Module. CFI Education Inc. https://corporatefinanceinstitute.com/resources/knowledge/finance/profitability-ratios/. Accessed 2 July 2020

Coulter L, Serrao-Neumann S, Coiacetto E (2019) Climate change adaptation narratives: linking climate knowledge and future thinking. Futures. In press

Culas R, Mahendrarajah M (2005) Causes of diversification in agriculture over time: evidence from Norwegian farming sector. In: 11th international congress of the European association of agricultural economists, 2005, Copenhagen, Denmark

Davis AS, Hill JD, Chase CA, Johanns AM, Liebman M (2012) Increasing cropping system diversity balances productivity, profitability and environmental health. PLoS ONE 7(10):e47149. https://doi.org/10.1371/journal.pone.0047149

Degla P, Adekambi S, Adanhoussode P (2016) Drivers of multiple cropping-systems as adaptive strategy to climate change in Central-Benin (West Africa). J Agric Sci 8(8):48–58. https://www.ccsenet.org/journal/index.php/jas/article/view/59833. Accessed 30 June 2020

DeLonge MS, Miles A, Carlisle L (2016) Investing in the transition to sustainable agriculture. Environ Sci Policy 55:266–273. https://dx.doi.org/10.1016/j.envsci.2015.09.013. Accessed 28 June 2020

FAO (undated) Sustainable crop production intensification. https://www.fao.org/3/a-i6168e.pdf
Fasoyiro SB, Taiwo KA (2012) Strategies for increasing food production and food security in Nigeria. J Agric Food Inf 13(4):338–355. https://doi.org/10.1080/10496505.2012.715063. Accessed 22 June 2020
Food and Agricultural Organization, FAO (2020) FAO in Nigeria: Nigeria at a glance. https://www.fao.org/nigeria/fao-in-nigeria/nigeria-at-a-glance/en/. Accessed 28 June 2020
Garbach K, Milder JC, DeClerck FAJ, Montenegro de Wit M, Driscoll L, Gemmill-Herren B (2017) Examining multi-functionality for crop yield and ecosystem services in five systems of agroecological intensification. Int J Agric Sustain 15:11–28. https://dx.doi.org/10.1080/14735903.2016.1174810. Accessed 21 June 2020
Gliessman SR (2014) Agroecology: the ecology of sustainable food systems, 3rd edn. CRC/Taylor & Francis Group, London
Gujarati DN (2004) Basic econometrics, 4th edn. The McGraw-Hill Companies, New York
Hossain MS, Kader MA, Islam N (2017) Multiple cropping for sustainable and exaggerated agricultural production system. J Biosci Agric Res 14(02):1202–1209. https://doi.org/10.18801/jbar.140217.148. Accessed 27 June 2020
Intergovernmental Panel on Climate Change, IPCC (2014) Climate change 2014: impacts, adaptation, and vulnerability. Part A: global and sectoral aspects. Contribution of working group II to the fifth assessment report of the intergovernmental panel on climate change. https://www.ipcc.ch/report/ar5/syr/docs/ar5_syr_headlines_en.pdf
Kompas T, Pham VH, Tuong NC (2018) The effects of climate change on GDP by country and the global economic gains from complying with the Paris climate accord. Earth's Future 6(8):1153–1173
Kremen C, Miles A (2012) Ecosystem services in biologically diversified versus conventional farming systems: benefits, externalities, and trade-offs. Ecol Sci 17(4):40
Lang LHP, Stulz RM (1994) Tobin's q, corporate diversification, and firm performance. J Polit Econ 102(6):1248–1280
Lee C, Schlemme C, Murray J, Unsworth R (2015) The cost of climate change: ecosystem services and wild land fires. Ecol Econ 116:261–269
LEISA-India (2011) System of crop intensification – promoting a profitable crop sequence of finger millet and chickpea. AME Foundation (www.amefound.org) in collaboration with ILEIA (1999–2011) and MISEREOR. https://leisaindia.org/system-crop-intensification-promoting-profitable-crop-sequence-finger-millet-chickpea/. Accessed 25 May 2020
Lundgren G, Fausti SW (2015) Trading biodiversity for pest problems. Sci Adv 1(6):e1500558
Maddison DJ (2007) The perception of and adaptation to climate change in Africa. World Bank Policy Research Working Paper no. 4308
Manteuffel R (1984) Ekonomika i organizacja gospodarstwa rolniczego [Economics and organisation of agricultural farms]. PWRiL, Warszawa
Nhemachena C, Hassan, R (2007) Micro-level analysis of farmers' adaptation to climate change in Southern Africa. IFPRI Discussion Paper No. 00714. International Food Policy Research Institute, Washington, DC
Onoja, AO (2019) Economic benefits from cassava in peri-urban multiple-cropping systems in Nigeria. In: Simelton E, Ostwald M (eds) Multifunctional land uses in Africa (open access): sustainable food security solutions, 1st edn. Routledge (Taylor and Francis), London, 176 p, 47 B/W Illus. https://www.routledge.com/Multifunctional-Land-Uses-in-Africa-Open-Access-Sustainable-Food-Security/Simelton-Ostwald/p/book/9780367246440. Accessed 2 March 2020
Onya SC, Ugochukwu GC, Ejiba IV (2019) Farm-level determinants of access to land by arable crop farmers in Ikwuano local government area of Abia State, Nigeria. Agro-Sci J Trop Agric Food Environ Ext 18(1):50–55
Osu AK (2017) Population dynamics and food security in Nigeria. Int J Adv Stud Econ Public Sect Manag 5(2):129–189. https://www.internationalpolicybrief.org/images/2017/APRIL/EPSM/ARTICLE10.pdf. Accessed 26 June 2020

New Partnership for Africa's Development, NEPAD (2013) Agriculture and Africa – transformation and outlook. NEPAD, Johannesburg

Pew Research Centre (2014) Global religious diversity: methodology. Pew Research Centre, Washington DC. https://www.pewforum.org/2014/04/04/methodology-2/. Accessed 2 July 2020

Pope RD, Prescott R (1980) Diversification in relation to farm size and other socioeconomic characteristics. Am J Agric Econ 62(3):554–559

Pretty J, Benton TG, Bharucha ZP, Dicks LV, Flora CB, Godfray HCJ, Goulson D, Hartley S, Lampkin N, Morris C, Pierzynski G, Prasad PVV, Reganold J, Rockstrom J, Smith P, Thorne T, Wratten S (2018) Global assessment of agricultural system redesign for sustainable intensification. Nat Sustain 1(8):441–6. https://doi.org/10.1038/s41893-018-0114-0. Accessed 2 July 2020

Prieto I, Violle C, Barre P, Durand JL, Ghesquiere M, Litrico I (2015) Complementary effects of species and genetic diversity on productivity and stability of sown grasslands. Nat Plants 1(4):1–5

Royal Society (2009) Reaping the benefits: Science and the sustainable intensification of global agriculture. The Royal Society Science Policy, London, 172 p. https://royalsociety.org/-/media/Royal_Society_Content/policy/publications/2009/4294967719.pdf. Accessed 28 June 2020

Rural Solutions SA (2017) Farm gross margin and enterprise planning guide 2017: a gross margin template for crop and livestock enterprises. Stored Grains Project, Rural Solutions, South Australia, 88 p. https://www.grainandgraze3.com.au/resources/2017-Farm-Gross-Margin-and-Enterprise-Planning-Guide.pdf. Accessed 29 June 2020

Sahel (2016) Cassava: a staple crop in Nigeria. Sahel Capital Partners & Advisory Limited. Newsletter 13: December, 2016, 5 p. https://sahelcp.com/wp-content/uploads/2016/12/Sahel_Newsletter_Volume_13.pdf

Shimeles A, Verdier-Chouchane A, Boly A (2018) Understanding the challenges of the agricultural sector in sub-saharan Africa. In: Shimeles A, et al (eds.) Building a Resilient and Sustainable Agriculture in Sub-Saharan Africa. https://doi.org/10.1007/978-3-319-76222-7_1

Skarżyńska A (2012) Cropping intensity vs. profitability of selected plant production activities in Poland. Stud Agric Econ 114:31–38. https://doi.org/10.22004/ag.econ.122446. Accessed 29 June 2020.

Springmann M, Clark M, Mason-D'Croz D, Wiebe K, Bodirsky BL, Lassaletta L, de Vries W, Vermeulen SJ, Herrero M, Carlson KM, Jonell M, Troell M, DeClerck F, Gordon LJ, Zurayk R, Scarborough P, Rayner M, Loken B, Fanzo J, Godfray HCJ, Tilman D, Rockstrom J, Willett W (2018) Options for keeping the food system within environmental limits. Nature 562(7728):519–525. https://doi.org/10.1038/s41586-018-0594-0. Accessed 28 June 2020

SRI-Rice (2014) The system of crop intensification: agroecological innovations for improving agricultural production, food security, and resilience to climate change. SRI International Network and Resources Center (SRI-Rice), Cornell University, Ithaca, New York, and the Technical Centre for Agricultural and Rural Cooperation (CTA), Wageningen, Netherlands. Accessed 8 May 2020

Rivers State (2019) Rivers state: agriculture. Rivers state government. https://www.riversstate.gov.ng/state

Uphoff N (2011) Agroecological opportunities to help 'climate-proof' agriculture while raising productivity in the 21st century. Sustainable Soil Productivity in Response to Global Climate Change. In: Sauer TS, Norman JM, Sivakumar MVK (eds) pp 87–102. JW Wiley, Chichester

Uphoff N (2012) Raising smallholder food crop yields with climate-smart agroecological practices. Booklet supplement for World Bank Institute seminar, Washington. https://sri.ciifad.cornell.edu/aboutsri/othercrops/Other_Crops_Brochure_Uphoff101012.pdf. Accessed 4 April 2020

World Food Programme, WFP (2020) WFP Global hotspots potential flashpoints in 2020. WFP, 11 p. https://docs.wfp.org/api/documents/WFP-0000111565/download/?_ga=2.155538235.1096467937.1578552831-951656840.1578552831

Assessment of a New Weeding Technology as a Solution for the Sustainable Welfare of Rice Producers in Cameroon

Bamou Tankoua Lydie, Kaldjob Mbeh Christian Bernard, and Nso Ngang André

Abstract This article assesses the impact of the adoption of an herbicide as a new weeding technology on the sustainable welfare of rice producers in Cameroon. Weeding is one of the main technical constraints linked to the rice production in Cameroon. Using the stratified probability sampling technic, a sample of 553 rice producers was selected, based on the data from the survey carried out within the framework of the C2D-PAR-RIZ project, in three subdivisions in the West region of Cameroon, during the period going to May to July 2018. The main criteria that have guided the choice of these areas are essentially, the Socio-economic importance and the agricultural constraints of rice production and the organization of rice producers. The Probit model and propensity score matching (PSM) have been used to assess this impact. The results reveal that with a statistical difference of US $54.207, the adoption of the new weeding technology sustainably improve income of producer in Cameroon. Furthermore, rice productivity, the size of the rice plots, agricultural training, number of weedings per campaign and rice income are the main determinants of this adoption. To ensure the sustainable well-being of rice producers in Cameroon, it is therefore important that agricultural technology vulgarization policies such as weeding techniques implemented by the authorities be further amplified, taking into account the local conditions of producers.

Keywords Adoption · Herbicide · Welfare · Rice producers · Propensity score matching · Cameroon

1 Introduction

Rice has a significant place in the dietary traditions of people in sub-Saharan Africa. With 728.443 tons of imported rice worth 183.7 billion FCFA in 2017 Minader (2017), demand in Cameroon remains mainly covered by importations. However,

B. T. Lydie · K. M. Christian Bernard
University of Yaounde II Soa, PO: 14023, Yaounde, Cameroon

K. M. Christian Bernard (✉) · N. N. André
Institute of Agricultural Research for Development (IRAD), PO: 2123, Yaounde, Cameroon

W. Leal Filho et al. (eds.), *Sustainable Development in Africa*, World Sustainability Series,
https://doi.org/10.1007/978-3-030-74693-3_25

despite the agricultural policies developed by the state of Cameroon to boost production, which stands at around 100.000 tons of paddy produced on nearly 44.000 ha, it remains low (Minader 2017). Indeed, with the effects of climate change, rice disease infestations, an unfavorable economic context for producers and above all the invasion of weeds in the rice plots, sustainable rice productivity is currently one of the priorities to address food insecurity of population (Hossain et al. 2020; Nandan et al. 2020). Indeed, for Chhun et al. (2020); Tshewang et al. (2016) Popp et al. (2013), weeds are actually, one of the main production constraints for rice producers, as they affect yield, income, and in fine the welfare of rice-farming households. From an agronomic point of view, several researches show that they technically cause poor plant growth, which is responsible of the decrease in crop production (Dilipkumar et al. 2020; Rao et al. 2020; Tshewang et al. 2020). In Africa, the research of Johnson et al. (2004) for example, reveals that they are responsible for a loss of production of almost 74% of transplanted rice, 89% of lowland rice and up to 100% for the of the rainy variety.

Managing weed invasions is thus a huge challenge for sustainable rice productivity in Africa. All else equal, poor weeding can lead to lower yields or, in more serious cases, to the abandon of the plot. To overcome this obstacle, different weeding techniques, from the traditional (manual) to the mechanic and chemical are used (Pannacci et al. 2020; Juraimi et al. 2013). The traditional weeding technology in rice production is still more fastidious and requiring in terms of physical effort for hired labor because of the large rice areas, while mechanical and specifically chemical control with herbicides is the most practical, efficient and economically profitable method for a sustainable welfare of farmers (Antle 2015; Anwar et al. 2012; Roder and Phengchanh 1997). Indeed, researchers and rural farmers believe that herbicides can be considered as a viable alternative to traditional weeding (Chauhan and Johnson 2011; Mahajan et al. 2009). They control weeds in the early growth stages of the crop, especially if rice farmers have large paddy fields and little time to spend on weeding. This has motivated the development of an integrated approach to weeding based on the identification of the critical period of competition with rice plants, which has been widely adopted by producers to overcome this constraint in a sustainable way. (Juraimi et al. 2013).

The concept of sustainability in rice production therefore refers in to the adoption of innovative techniques that contribute to increase agricultural productivity and farmers' income, to strengthen human capital and also preserve natural resources. In other words, it refers to a rice production system that is economically profitable, ecologically and socially acceptable. (Roy et al. 2016; Gómez-Limón and Sanchez-Fernandez 2010). However, although this form of production has not achieved a high growth rate, it nevertheless raises significant health and environmental concerns. Because, the spread in the surrounding water tables, the resistance of parasites and the recurrent cases of intoxication of rice farmers are increasingly the subject of several debates and criticisms (Boraud et al. 2010; Konan et al. 2014; Lopes Ovejero et al. 2013). Despite this, the low productivity of the rice plots specifically in Cameroon can be mainly explained by the weakness of production factors and particularly by the inaccessibility of new weeding technologies. Indeed, most poor rural producers still use unsuitable and low-quality of traditional equipment (machete/chopper, slicers, hoe, daba), which considerably affects their performance and income level, and consequently gives a competitive advantage to imported rice, which remains more available to households in the markets.

This technological deficit in weed control mainly affects Cameroonian women producers, who are the most involved in this activity. Indeed, they consider that traditional weeding has a certain advantage over the development of rice plants, but leads to major physical problems in the short term, especially after the weeding activity. For them, weed control with herbicides is more appropriate for large areas and therefore better suited to the rice production process. According to Cooper et Dobson (2007) et Rao et al. (2007), the benefit of herbicides remains the productive added value for farmers who adopt these substances. For his part, Akobundu (2006) believes that herbicides have better weed control rates, which also adds decomposed weeds to the plot that can serve as nutrient inputs to the plants. For Rodenburg et Johnson (2009), herbicides control weeds in the early stages of crop development, despite the significant financial means they require, they guarantee better incomes for rice farmers in particular. The adoption of herbicides in the Cameroonian rice industry therefore remains a necessary problem to ensure that producers maximize their income, who is the main way to improve their welfare (Olina Bassala et al. 2015).

Unfortunately, rice production systems in Cameroon still have a significant level of technological innovation delay that cannot guarantee for the small family production units, a sufficient level of income to improve their welfare. However, several research studies have been carried out to illustrate the problem of weeds management in sustainable rice production. Indeed Sylla et al. (2017), evaluated weed management practices in irrigated rice cultivation in Ivory Coast. They show that weed control methods have a significant effect on the abundance-dominance index of rice weeds. Furthermore, Boraud et al. (2015) reveal that farmers' weed management practices on rice production in Ivory Coast had a significant influence on rice productivity. By comparing average yield and income levels, Noltze et al. (2013) analyze the impacts of the adoption of rice intensification systems (RIS) on rice yield and household income in Timor Leste. On the contrary, they believe that there is no significant difference between adopters and non-adopters of RIS and conclude that this system is more adopted by rice farmers with below-average yields.

Unfortunately, research on the problem of agricultural weeding in general and rice in particular in Cameroon remains embryonic. It is more interested in topics such as the multiplication and dissemination of improved varieties, the management of irrigation systems on rice farms, the maintenance of soil fertility and the marketing of agricultural products. We therefore consider that it is necessary to assess the potential of the new chemical weeding technology to determine its contribution to the income difficulties of Cameroonian rice farmers. This article thus assesses the impact of the adoption of a new weed control technology on the welfare of rice farmers in the West Cameroon region. In particular, it aims to measure the effect that the adoption of a weeding technology would have on the income level of rice producers in this region.

Fig 1 Study area *Source* Adapted from Guillermou (2007) ▲ Surveyed areasthe title and the source here are below the figure

2 Materials and Methods

2.1 Study Area

This study was carried out in the West region of Cameroon precisely in the localities of Tonga, Bandounga, and Santchou. The main criteria that guided the choice of these three zones were essentially the Socio-economic importance of rice production the soil and climate characteristics favourable to rice production, and the concentration and organization of rice producers (Fig. 1).

2.2 Data Collection and Sampling Technic

This study was conducted in the localities of Tonga, Bandounga, and Santchou in West Region of Cameroon. The data was collected from a survey carried out within the framework of the C2D-PAR-RIZ project. These data are essentially primary. Using the stratified probability sampling technique, semi-structured interviews and participatory observations, a sample of 553 rice farmers was selected, with 240 producers using obsolete technology and 313 producers adopting herbicides as a new weeding technology (Table 1). These data concerned the socio-economic characteristics of producers (age, gender, membership of a rural Organization, income, area, total production, number of weedings, yield…), and the adoption status of the rice farmers.

Table 1 Distribution of producers by locality

Localities	Obsolete weeding technology	New weeding technology	percentage	Total
Tonga	80	100	32.54	180
Bandounga	100	123	40.32	223
Santchou	60	90	27.12	150
Total	240	313	100	553

Source author's construction

It appears from Table 1 that the locality of Bandounga has a more significant number of producers than Tonga and Santchou. This is justified by the fact that, although Tonga and Bandounga are in the same subdivision with Tonga as the district capital, the majority of rice producers carry out their agricultural activities in the locality of Bandounga where land is more available, whereas in Tonga, the population has practically occupied most of the space for its dwellings.

Estimation Technique

The welfare indicator on which the impact of the adoption of a new weeding technology is observed is the income of the rice farmer. Thus, to assess this impact, we compare rice farmers who adopt herbicides as a new weeding technology with farmers who do not adopt herbicides in the the study was conducted in West region of Cameroon, not in Central region. To do this, we will use the propensity score matching method proposed by (Rubin 2006). This technique consists in the construction of a statistical comparison group that is based on the probability of adopting the new weed control technology such as:

$$P(X) = T = \frac{1}{X}$$

The effect of treatment is captured by cancelling out differences in individual characteristics and treatment in the two groups. Descriptive statistics will allow us to characterize our sample as well as the treatment (adopting weed control technology) and control (not adopting weed control technology) groups.

Estimation Method

The average effect of the treatment on treaties is such that:

$$\begin{aligned}\Delta_{ATT} &= \mathrm{E}(\mathrm{Y}_T - \mathrm{Y}_{NT}|1)\\ &= \mathrm{E}[\mathrm{Y} - \mathrm{E}(\mathrm{Y}|X, T = 0)|T = 1]\\ &= [\mathrm{E}(\mathrm{Y}_T|T = 1, X = \mathrm{X}) - \mathrm{E}(\mathrm{Y}_{NT}|X, T = 0, X = X]\end{aligned}$$

The estimator Δ_{ATT} is the average of the deviations of the treated and the counterfactual.

Therefore, it is necessary to estimate $\mathrm{E}(\mathrm{Y}_{NT} X = X_i, T = 0) = f(\mathrm{X}_i)$, for each rice producer treated for characteristics X_i. In order to do this, it is essential to match on the basis of the propensity score and to make sure that the statistical difference of the means of the outcome variable of the two groups. Our estimation sequence is therefore as follows:

- We estimate the propensity score (PS) or probability of adopting weed control technology from a Probit or Logit model and calculate the marginal effects;
- We conduct a statistical difference test of the means of income of the two groups;
- We determine the common support by Kernel parametric regression suggested by Heckman et al. (1998) and the nearest five neighbour's method;
- Finally, we calculate the standard deviation of the estimator using the Bootstrap method.

3 Results

3.1 Descriptives Statistics by Herbicides Adoption

This results are based on the statistical differences between rice producers depending on the adoption status of herbicides. Table 2 presents descriptive statistics for the quantitative variables. The results of this table show that the majority of rice farmers in the sample are young, with the average age of adopters being 41 years, while the average age of non-adopters is 39 years. These producers also have relatively low income levels. In fact, the average income of adoptive rice farmers is 440 US $, while the average income of non-adopters is 212 US $, with the lowest being 80 US $. On the other hand, the results indicate significant differences in income. The average rice cultivated area is 0.84 ha for the non-adopters and 0.86 ha for the adopters. This shows that rice farmers who use herbicides use larger areas, which reinforces the difficulty of weeding activity. We also note that rice production and productivity reflect the differences between these two groups. On average, farmers who do not use herbicides produce 530 kg of rice in a season, while those who adopt them produce just over 1100 kg, i.e. a productivity of 665 kg/ha and 1336 kg/ha respectively on Table 2, the modality of the variable "yield" is ha/kg.

On the other hand, the results in Fig. 2 show that, in practice, rice producers generally carry out between 1 and 3 weeding passes per season. In fact, 55.27% of adopters carry out no more than 2 treatments per season compared to 44.17% of non-adopters. In addition, they have on average completed a secondary level of education. We also observe that the non-adopters are less members of farmers' organizations, with 54% saying they do not belong to a group compared to 61% of the adopters who belong to at least one rural group. However, the majority have followed at least one training course in rice production, i.e. 61.25% of the non-adopters versus 91.69% of the adopters. In addition, the adopters stand out in the application of fertilizers, with 67.5% and 89.14% of the non-adopters and adopters respectively using fertilizers. However, as far as improved varieties are concerned, 49.58% and 67.40% of non-adopters and adopters use at least one improved variety in their plot, although not all of them have access to credit with 11.82% and 44.58% not having access to financing.

Table 2 Description of model variables

Categories	variables	Definition	Modalities
Socioeconomics variables	Sex	Gender of rice Producer	1= male 0= female
	Age	Age of Producer	in Number of year
	Area	Rice area	Per hectar (ha)
	produc	production	Per Kg
	yield	yield	Per Kg
	Group	Member of rural Organization	1= Yes 0= No
	acceseed	Acces to seed	1= Yes 0= no
	Marstat	Marital statut	1= single 2= Married 3= Divorced 4= Widow
Agricultural variables	Edulevel	Level of education	0= not scolarized 1= primary 2=secondary 3= university
	Agritrai	Agricultural training on Rice production	1= Yes 0= no
	Weeding	Number of weeding per campaign	1= 1 time 2= 2 time 3= 3 time
	Herbicide	Adoption of herbicide	1= Yes 0= no
	varieties	rice variety	1=improved 0= local
	Fertilizer	Fertilizer utilization	1= Yes 0= no
	Acccred	Access to agricultural Credit	1= Yes 0= no
	Income	Rice income in US $	

Source Authors' construction

At the end, there is a statistical difference between the two groups in terms of average income level Table 3 (Table 4).

3.2 Determinants of the Probability of Adopting the New Weed Control Technology

The results of the Probit estimation of the use of the new weeding technology show that: the size of the plots, rice productivity, agricultural training, number of weedings and income determine its adoption. The number of weeding negatively influences the probability of adoption. The marginal effect of this variable (Table 5) indicates that it sufficiently (39%) reduces the probability of a rice farmer adopting the new technology. In terms of area, it indicates that the larger the area, the more likely

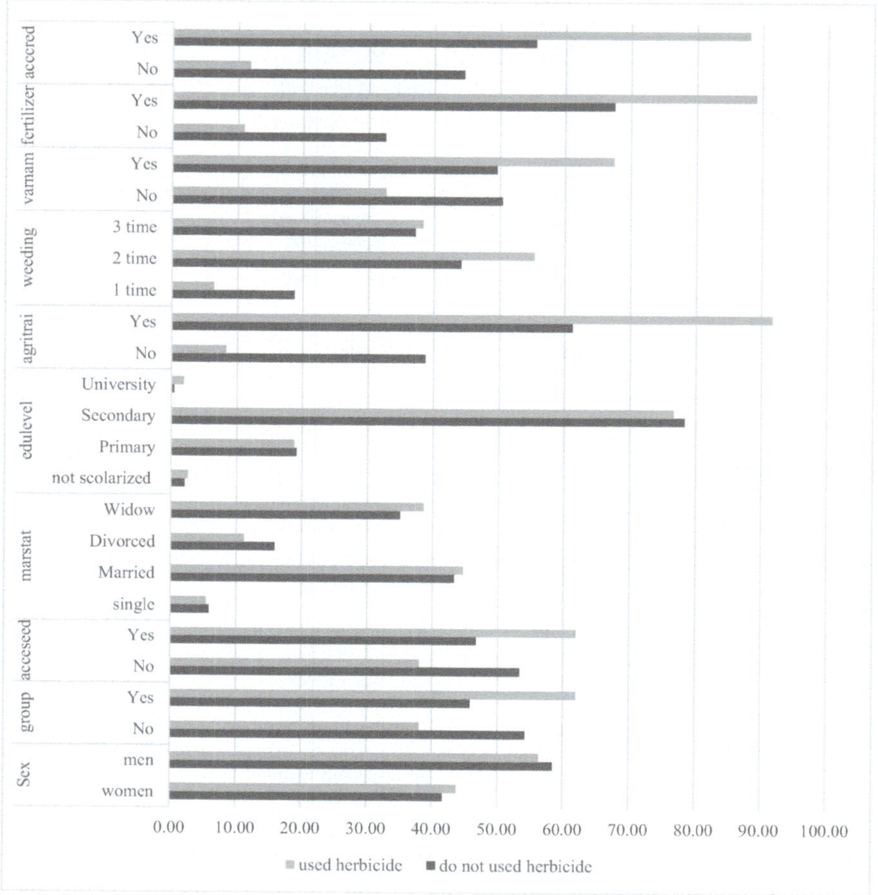

Fig 2 Distribution of qualitative variables *Source* authors from survey data

the farmer is to adopt herbicides. The marginal effect here indicates that plot size increases the probability of adopting the new technology by 39%. As for the training of the farmer in rice production techniques, the Probit estimate indicates that it has a positive effect on the probability of herbicide adoption. The marginal effect of this variable indicates that it increases the probability of adoption by 50% with the number of weedings. Ultimately, the income level positively determines the adoption of the new technology. Furthermore, the distribution of "Propensity scores" between the two groups of rice farmers is shown in Fig. 3, which clearly shows the difference in distribution between the treatment and control groups. All the statistics results of 0.000 need to be identified with (***).

Table 3 Descriptive statistical analysis

Var	Herbicide	Obs	Min	Max	Mean	Std-Err
Age	0	240	22.00	66.00	39.77	12.29
	1	313	22.00	72.00	41.02	12.27
Area	0	240	0.40	1.50	0.84	0.29
	1	313	0.40	1.50	0.86	0.28
Product	0	240	200.00	2000.00	530.97	239.15
	1	313	270.00	3000.00	1109.94	416.66
Yield	0	240	200.00	1500.00	665.18	264.14
	1	313	450.00	2666.67	1336.59	400.97
Income	0	240	80.00	800.00	106,194.17	47,829.31
	1	313	108.00	1200.00	443.97	166.66

Source authors from survey data

Table 4 Statistical difference test of the income averages of the two groups

diff = mean(0) - mean(1)
t = -19.1753
degrees of freedom = 55
Ho: diff = 0

Group	Obs	Mean	Std.Err
non-adopters	240	212.624	3098.05
Adopters	313	443.977	4710.151
combined	553	171904	3857.508
diff	-23.134	328.300	-127526

Source authors from survey data

3.3 *Average Effect of New Weeding Technology on Rice Income*

Based on Kernel and nearest five neighbor matches, the average treatment effect is US $54,207. Rice farmers who use the herbicides generate an additional income of US $54,207 compared to those who do not (Table 7). This effectively attests to the causal relationship between the use of herbicides as a weeding technology and rice income. This result is obtained with 213 observations. Indeed, 100 observations from the treatment group could not be matched with observations from the control group (Table 7). In the unmatched sample, the productivity already showed a strong difference between the two groups (Table 6). A naïve method would have been to directly compare the average incomes of the two groups of rice producers before matching.

Table 5 Probability to use the new weed control technology in rice production this title should be above the table

Probit regression

Number of obs = 553
Wald chi2(14) = 164.48
Prob > chi2 = 0.0000
Pseudo R2 = 0.6510
Log likelihood = -131.77858

Variables	Coef.	Std. Err	P>|z|
herbicide			
sex	-.2916461	.1914649	0.128
age	.0096296	.0081595	0.238
area	1.144849	.8851176	0.000
yield	.0037444	.0009081	0.000
group	-.4027759	.5954249	0.499
acceseed	.1291445	.5835231	0.825
marstat	-.1007991	.1034495	0.330
edulevel	.2489573	.1800777	0.167
agritrai	1.372936	.391141	0.000
weeding	-1.1642	.1978227	0.000
varnam	-.1586257	.2296875	0.490
fertilizer	.0471412	.2793486	0.866
acccred	.2807414	.3489673	0.421
income	6.03e-06	5.22e-06	0.000
_cons	-4.202164	1.073444	0.000

*** significant at 1%; ** significant at 5%; * significant at 10%

Source authors from survey data

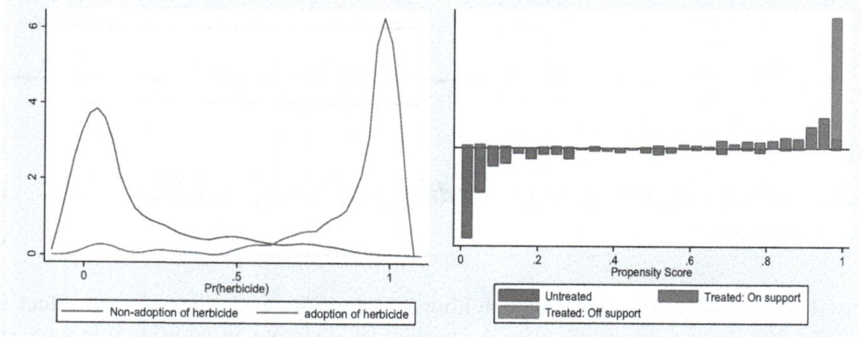

Fig 3 Distribution of propensity score between the two groups of rice producer this title and her source should be under the figure *Source* authors from survey data

Table 6 Marginal effects

Y = Pr (Part) (predict) 0.71221114	
Variable	dF/dx
sex*	−0.0980803
age	0.0032844
area	0.3904775
yield	0.0012771
group*	−0.1352351
acceseed*	0.044193
marstat	−0.0343799
edulevel	0.0849127
agritrai*	0.5014128
weeding	−0.3970777
varnam	−0.0535998
fertilizer*	0.0152025
acccred*	0.0939509
income	2.06e−06

Source authors from survey data

Thus we should have obtained a difference of US$ 23 (Table 3) for the use of herbicides. But the problem would have been to compare farmers with different individual characteristics. So, this remedied the matching by comparing producers with identical and/or relatively similar characteristics. The robustness of our result (US$ 54.20) is due to the fact that it is identical for kernel and the five closest neighbors (Table 7). The interest of this result lies in the possibility for small rice producers to improve their standard of living through income derived from improved plot productivity. This income depends on the price margin, i.e. the differential between the price of rice and the price of inputs (herbicides, fertilizers) and the salaried labor mobilized (Table 8).

4 Discussing the Conditions for Micro-economic Adoption: A Necessity of Technological Change

Increasing weed pressure and soils with low organic resource in rice producing areas are the serious threats to the sustainability of rice production. These difficulties further impact the productivity and welfare of producers who are already suffering from rural poverty and speculation on agricultural commodity prices. Given these challenges, the introduction of agricultural technologies could be an agronomic response to support producers against those constraints. Just as Danso-Abbeam (2018), the results of this research show that the introduction of herbicides into the agricultural

Table 7 Matching test of samples

variables	Unmatched samples		P-value	Matched by Kernel		P-value	Matched by the 5 closest neighbours method		P-value
	Mean			Mean			Mean		
	Treated	control		Treated	control		Treated	control	
herbicide	1	.	**0.000**	1	.	**0.000**	**0.000**	**0.000**	.
sex	.5623	.58159	38.5	.54491	.49735	0.87	.544	.633	0.100
age	41.019	39.745	.228	41.623	33.662	0.000	41.623	39.564	0.144
area	.862	.842	.394	.897	1.093	0.000	.897	.963	0.015
yield	1336.6	666.33	0.000	1046.5	1147.9	0.000	1046.5	1087.7	0.101
group	.619	.456	0.000	.443	.377	0.001	.443	.358	0.108
marstat	2.830	2.799	0.715	2.874	2.517	-8.0	2.874	2.971	0.366
edulevel	1.779	1.769	0.821	1.754	1.928	0.000	1.754	1.941	0.000
agritrai	.916	.610	0.000	.880	.546	0.000	.880	.725	0.000
weeding	2.319	2.179	0.013	2.245	2.684	0.000	2.245	2.583	0.000
varnam	.674	.493	0.000	.526	.268	0.000	.526	.374	0.005
fertilizer	.891	.673	0.000	.844	.582	0.000	.844	.827	0.680
acccred	.881	.552	0.000	.832	.523	0.000	.832	.705	0.006

Source authors from survey data

Table 8 Result of the impact of herbicide adoption on rice farmers' income

Statistics	Kernel	5 nearest neighbours
ATT	54.207***	54.207***
Standard deviation	783.644	783.644
Number of observations from the treated group	313(100)	313(100)
Number of observations from control group	553 (100)	553 (100)

***significant à 1%, **significant à 5%, *significant à 10%
Source authors from survey data

production process improves the welfare of producers. This has been previously confirmed by (Pimentel et al. 1992; Pimentel and Grieiner 2009) who believed that strategies based on agricultural pesticides in general have contributed to unprecedented growth in agricultural financial capital. Yet the reality remains that the use of these chemicals presents a potential health risk not only to the producer who handles them, but also to the consumer (pesticide residues) (Gouda et al. 2018). However, according to Cooper and Dobson (2007), There is a necessity to introduce technologies such as pesticides on farms to improve the quality of production, which would be beneficial to producers and consumers.

Specifically, our results of rice area as a determinant of herbicide adoption may seem surprising. It would indicate that small farmers would be the most likely to adopt herbicides. The explanation here would be linked to the fact that, since surface area is an indicator of a farmer's wealth, the dynamics of rice cultivation are sometimes driven by urban elites, who plant small plots using new equipment for the sake of modernity. However, while wealth generally has a positive effect on adoption, the variety of wealth indicators does not lead to the same result (Roussy et al. 2015). On the other hand, rice producers in the treated group have, on average, higher productivity than the control group. This may suggest that young producers with a longer planning horizon (and therefore more likely to adopt herbicides) have more land. This may also be related to land access conditions that have not been analyzed here.

Agricultural training before setting up a rice plantation predisposes to the knowledge and mastery of the information needed to run a rice plot and improve its performance. Producers trained in advance would therefore be more open to innovation. Finally, the fact that a little more than 55% of adopters have weeded at least twice their plot (compared to 45% of non-adopters) raises questions about the quality of herbicides, which are sometimes non-certified, inefficient and can therefore improve the weed resistance feared by producers. This can be explain by the fact that the adaptability of the active ingredients of this herbicides to the local ecosystem conditions of rice plants would not be guaranteed. These results thus confirm the fact that, as a consequence of liberalization policies and the withdrawal of the State through the various subsidies previously granted, ill-prepared producers have been abandoned in the hands of more sophisticated service providers. Globally, producers with little or no institutional support are less willing to adopt this type of innovation. However, other structural variables were not tested in this research on like institutional or organizational determinants of weed control technology adoption and their impact on welfare. These different parameters are as many research fronts for the future.

5 Conclusion

The present necessity to improve rice production, linked to the policy to figth again food insecurity, stimulated by the reduction of imports from the international market, is leading many countries, including Cameroon, to promote the use of herbicides to increase productivity and thus the sustainable welfare of local rice farmers. In this article we assessed the impact of a new weed control technology on the welfare of 553 rice farmers in the West Cameroon region. Using Probit methods and propensity scores, the results show that the use of herbicides leads to an increase in income of US $54.20.

From a sustainable development perspective, the result of this research is not sufficient to recommend the use of herbicides in sustainable rice production in Cameroon. Indeed, it is a question of verifying how the use of herbicides can have a long-term impact on increasing the income of producers. Nevertheless, other national rice production organizations believe that the level of production should indeed be increased to meet the national demand for local rice. Given these challenges, both institutional and organizational changes would be needed to enable a sustainable supply of agricultural inputs to promote growth in local production. This transfer of weeding technologies therefore implies a better co-construction between the different actors (producers-researchers-State) of the adoption of herbicides on the socio-economic living conditions and the sustainable welfare of rice farmers. In addition, the promotion of herbicides presents a health hazard related to the increased use of chemicals. It is therefore important to include in the extension process the analysis of the risks associated with the more or less excessive use of herbicides. Also, vulgarization policies need to consider adapting these weed control technologies to the local conditions of vulnerable producers.

Acknowledgements The authors are grateful to the C2D-PAR RIZ project of IRAD institute and especially to all the courageous rice producers of west region of Cameroon for providing all the informations in order to facilitate the achievement of this research.

Conflict of Interest The authors declare that there is no conflict of interest regarding this research.

References

Akobundu O (2006) Weed control in direct-seeded lowland rice under poor water conditions. Weed Res 21:273–278

Antle JM (2015) Pesticide policy, production risk, and producer welfare: an econometric approach to applied welfare economics. Routledge, London

Anwar MP, Juraimi AS, Puteh A, Man A, Rahman MM (2012) Efficacy, phytotoxicity and economics of different herbicides in aerobic rice. Acta Agric Scand Sect B - Soil Plant Sci 62:604–615. https://doi.org/10.1080/09064710.2012.681060

Boraud M, Aké-Assi E, Kassi N, Aké S, Gasquez J (2010) Impact agro écologique de simulation de culture transgénique de maïs résistant au glyphosate et effet répétitif d'un traitement herbicide sur la flore adventice en Côte d'ivoire. Sci Nat 7

Boraud N, Kouame K, Kla D (2015) Impact des pratiques de gestion des adventices sur le rendement du riz au centre de la Côte d'Ivoire. Int J Biol Chem Sci 9:1220

Chauhan BS, Johnson DE (2011) Growth response of direct-seeded rice to oxadiazon and bispyribac-sodium in aerobic and saturated soils. Weed Sci 59:119–122

Chhun S, Kumar V, Martin RJ, Srean P, Hadi BAR (2020) Weed management practices of smallholder rice farmers in Northwest Cambodia. Crop Prot 135:104793

Cooper J, Dobson H (2007) The benefits of pesticides to mankind and the environment. Crop Prot 12:1337–1348

Danso-Abbeam G, Baiyegunhi LJS (2018) Welfare impact of pesticides management practices among smallholder cocoa farmers in Ghana. Technol Soc 54:10–19

Dilipkumar M, Chuah TS, Goh SS, Sahid I (2020) Weed management issues, challenges, and opportunities in Malaysia. Crop Prot 134:104347

Gómez-Limón JA, Sanchez-Fernandez G (2010) Empirical evaluation of agricultural sustainability using composite indicators. Ecol Econ 69:1062–1075

Gouda A-I, Imorou Toko I, Salami S-D, Richert M, Scippo M-L, Kestemont P, Schiffers B (2018) Pratiques phytosanitaires et niveau d'exposition aux pesticides des producteurs de coton du nord du Bénin. Cah Agric 27:65002. https://doi.org/10.1051/cagri/2018038

Guillermou Y (2007) Organisations de producteurs et dynamiques paysannes dans l'Ouest-Cameroun. Afr Contemp 2007:251–271

Heckman J, Ichimura H, Smith J, Todd P (1998) Characterizing selection bias using experimental data. Econometrica 1017–98

Hossain K, Timsina J, Johnson DE, Gathala MK, Krupnik TJ (2020) Multi-year weed community dynamics and rice yields as influenced by tillage, crop establishment, and weed control: Implications for rice-maize rotations in the eastern Gangetic plains. Crop Prot 138:105334. https://doi.org/10.1016/j.cropro.2020.105334

Johnson D, Wopereis MC, Mbodj D, Diallo S, Powers S, Haefele S (2004) Timing of weed management and yield losses due to weeds in irrigated rice in the Sahel. Field Crops Res 85:31–42. https://doi.org/10.1016/S0378-4290(03)00124-2

Juraimi AS, Uddin K, Anwar P, Muda MT, Ismail MR, Man A (2013) Sustainable weed management in direct seeded rice culture: a review. Aust J Crop Sci 7:989–1002

Konan Y-R, Akanvou L, N'Cho S, Arouna A, Eddy B (2014) Analyse de l'efficacité technique des riziculteurs face à l'infestation des cultures par les espèces parasites striga en côte d'ivoire 12

Lopes Ovejero RF, Soares DJ, Oliveira WS, Fonseca LB, Berger GU, Soteres JK, Christoffoleti PJ (2013) Residual herbicides in weed management for glyphosate-resistant soybean in Brazil. Planta Daninha 31:947–959. https://doi.org/10.1590/S0100-83582013000400022

Mahajan G, Chauhan BS, Johnson DE (2009) Weed Management in aerobic rice in Northwestern Indo-Gangetic Plains. J Crop Improv 23:366–382

Minader (2017) Agriculture. In: Annuaire statistique du Cameroun. Institut National de la Statistique, Cameroun, p 24

Nandan R, Singh V, Kumar V, Singh SS, Hazra KK, Nath CP, Malik RK, Poonia SP (2020) Viable weed seed density and diversity in soil and crop productivity under conservation agriculture practices in rice-based cropping systems. Crop Prot 136:105210. https://doi.org/10.1016/j.cropro.2020.105210

Noltze M, Schwarze S, Qaim M (2013) Impacts of natural resource management technologies on agricultural yield and household income: the system of rice intensification in Timor Leste. Ecol Econ 85:59–68. https://doi.org/10.1016/j.ecolecon.2012.10.009

Olina Bassala JP, Dugue P, Granie AM, Vunyungah M (2015) Pratiques et perception des herbicides Olina et al 2015.pdf. IJASRA, pp 94–107

Pannacci E, Farneselli M, Guiducci M, Tei F (2020) Mechanical weed control in onion seed production. Crop Prot 135:105221. https://doi.org/10.1016/j.cropro.2020.105221

Pimentel D, Acquay H, Biltonen M, Rice P, Silva M, Nelson J, Lipner V, Giordano S, Horowitz A, D'Amore M (1992) Environmental and economic costs of pesticide use. Bioscience 42:750–760. https://doi.org/10.2307/1311994

Pimentel D, Grieiner AK (eds) (2009) Environmental and economic costs of the application of pesticides primarily in the United States. In: Integrated pest management: innovation-development process. Springer, Dordrecht

Popp J, Pető K, Nagy J (2013) Pesticide productivity and food security. A review. Agron Sustain Dev 33:243–255. https://doi.org/10.1007/s13593-012-0105-x

Rao AN, Johnson DE, Sivaprasad B, Ladha JK, Mortimer AM (2007) Weed management in direct-seeded rice. In: Advances in agronomy, pp 153–255. Elsevier

Rao AN, Singh RG, Mahajan G, Wani SP (2020) Weed research issues, challenges, and opportunities in India. Crop Prot 134:104451. https://doi.org/10.1016/j.cropro.2018.02.003

Rodenburg J, Johnson DE (2009) Weed management in rice-based cropping systems in Africa. In: Advances in agronomy, pp 149–218. Elsevier

Roder W, Phengchanh S (1997) Weeds in slasfa-and-bum rice fields in northern Laos. Weed Res 37:111–119

Roussy C, Ridier A, Chaib K (2015) Adoption d'innovations par les agriculteurs: role des perceptions et des preferences

Roy R, Chan NW, Xenarios S (2016) Sustainability of rice production systems: an empirical evaluation to improve policy. Environ Dev Sustain 18:257–278. https://doi.org/10.1007/s10668-015-9638-x

Rubin D (2006) Matched sampling for causal effects. Matched Sampl Causal Ef. https://doi.org/10.1017/CBO9780511810725

Sylla M, Traore K, Soro D (2017) Évaluation des pratiques de gestion des adventices en riziculture irriguee dans la localite de daloa, centre-ouest de la cote d'ivoire. Agron Afr 16

Tshewang S, Dendup C, Tshering P, Kristiansen P (2020) Weed research issues, challenges, and opportunities in Bhutan. Crop Prot 134:104273. https://doi.org/10.1016/j.cropro.2017.06.005

Tshewang S, Sindel BM, Ghimiray M, Chauhan BS (2016) Weed management challenges in rice (Oryza sativa L.) for food security in Bhutan: a review. Crop Prot 90:117–124. https://doi.org/10.1016/j.cropro.2016.08.031

Snail Farming as an Alternative Profitable Livestock System for Sustainable Development

Elise Stephanie Mvodo Meyo, Zita Akanyi Nkemasong, Godlove Shu, Jean Pierre Nguede Ngono, and Christopher Ngosong

Abstract The current population dynamics is projected to increase yearly, and expected to challenge the Sustainable Development Goals, especially in relation to food and nutrition security. While the demand for agricultural food products is increasing, the availability of cultivable land is either stable or decreasing due to various environmental factors such as drought, floods and climate change effects. Considering that sustainability can only be achieved when a system is resource conserving, socio-culturally supportive, commercially competitive, and environmentally friendly, urban and peri-urban snail (*Achatina fulica or Archachatina marginata*) farming systems could represent viable sources of essential dietary nutrients and income for many families in Africa. Snail farming is profitable, with minimal capital and land demands, and limited risks. Snails are handpicked in the forest and easily farmed in gardens, backyards, basins, or cages. They are omnivores and environmentally friendly without odorous wastes (unlike pigs and poultry). Snails have high protein content, low in fat and cholesterol, and they are cheaper than other meat products. In addition, their by-products are inputs for other industries such as cosmetics and medicines. However, the current rate of snail farming in Cameroon barely satisfies 25% of the national demand. Hence, this study aims to encourage snail farming by evaluating the profitability of snail business as a valuable urban and peri-urban livestock alternative for sustainable development in Cameroon. Data

---

E. S. Mvodo Meyo (✉) · Z. A. Nkemasong · G. Shu
Department of Agricultural Economics and Agribusiness, Faculty of Agriculture and Veterinary Medicine, University of Buea, P.O. Box 63, Buea, South West Region, Cameroon

G. Shu
Institute of Agricultural Research for Development (IRAD), P.O. Box 13, Njombe, Littoral Region, Cameroon

J. P. N. Ngono
School of Advanced Studies in Social Sciences, Institute of African Worlds, P.O. Box 93322, Paris, Aubervilliers Cedex, France

C. Ngosong
Department of Agronomic and Applied Molecular Sciences, Faculty of Agriculture and Veterinary Medicine, University of Buea, P.O. Box 63, Buea, South West Region, Cameroon
e-mail: ngosong.christopher@ubuea.cm

W. Leal Filho et al. (eds.), *Sustainable Development in Africa*, World Sustainability Series,
https://doi.org/10.1007/978-3-030-74693-3_26

were collected between March–May 2018 with the use of questionnaires administered to 60 urban and peri-urban snail farming households in the Buea Municipality, South West Region of Cameroon. Farm budgeting analysis enabled investigations of performance from farmers while semi-log functions enabled the evaluation of factors affecting the level of snail production in the study area. The descriptive statistics and gross margin results from farm budgeting and profit revealed a total production costs of only US$0.4 per kg, which produced fivefold return on investment and a profit of US$2 per kg of snail sold. Considering the environmental friendliness, profitability of snail business, and health value of snails in relation to food and nutrition security, urban and peri-urban snail farming is highly recommended within the scope of sustainable development in Cameroon.

Keywords Urban and Peri-urban · Farming systems · Snails · Sustainability · Cameroon

1 Introduction

Population growth, increasing demand for food, rural-urban migration, and threats from climate change have induced the need to intensify agricultural production systems to satisfy the increasing food demands (Carzedda et al. 2020). The global population is projected to reach 8.5 billion by 2030, exceed 9 billion by 2050, with a twofold increase by 2100 (Denning and Fanzo 2016). Feeding the projected 9.8 billion global population by 2050 is a major challenge that requires sustainable increase in agricultural productivity. Most of this population increase is envisaged in developing countries where the population might reach 7.9 billion in 2050, especially involving the population group of 15–59 years. Most Sub-Saharan African (SSA) countries are experiencing increased population with rates of rural exodus that come along with many opportunities and challenges for households, city planners, policymakers, local and national actors (Shackleton et al. 2020). Moreover, the population increase is showing strong positive correlation with food demands and human consumption trends as explained by three components; the population number, the affluence and usage of consumption, and the impact of unit of resource used. Coincidentally these countries are home to the highest number of poor people. The current population dynamics and food consumption patterns therefore pose critical sustainability issues for city dwellers. Sustainability is a current critical standpoint whereby the management of cities requires an all-inclusive approach that considers economic, environmental, social, and cultural dimensions (Modak 2020). In recent years, damages to ecosystems have undermined some selected Sustainable Development Goals (SDG) to end poverty, eliminate hunger and health issues, improve access to water, enhance the quality of cities and climate, and foster sustainable development. A viable solution for these problems involves obligatory changes by humanity, including sustainable food systems such as high value grain legume crops that deliver important nutritional and health benefits and alternative livestock production systems, reduction of meat consumption, and efficient waste management. Policymakers need to adopt strategies to enable the fulfilment of these challenges by reducing the misuse of resources, secure public investment pathways, position innovations in urban landscape, and harmonise institutional undertakings to boost both actual and future possibilities in order to satisfy human needs without jeopardising sustainability (Carzedda et al. 2020; Louzon et al. 2020). Meanwhile, the earth could attain sustained quality of human life if the urban carrying capacity continues to support ecosystem services, such as adequate food and energy supply, as well as urban infrastructural development. Overall, potential alternative farming systems such as urban and peri-urban snail production should satisfy the triple challenge of nutrition, low cost of investment, and conservation of the environment.

Snails are consumed by almost half of African countries and constitute an important export in some countries with strong profitability indicators, but snail farming has not been fully exploited as a profitable alternative for sustainable urban development. Munonye et al. (2019) conducted a profitability analysis of snail farming in Nigeria and found that snail farming requires low capital, it is highly profitable with a gross margin of US$1588.26, net profit of US$1581.77 and a return on investment of US$0.0079. They concluded that snail farming is profitable that will boost farmer's income and improve livelihoods, but lack of funds and religious bias were identified as major constraints. Also, Agbogidi and Okonta (2011) conducted a poverty reduction study through snail farming in Nigeria and highlighted the dietary value and high financial turn over, coupled with low capital requirement as a means to self-employment and food self-sufficiency. A similar study was conducted in Ghana by Mohammed et al. (2014) to evaluate the implication of snail production, and concluded that snail production is an important means of increasing income and protein supply in many communities, and also highlighted the low capital requirement. Grilla et al. (2016) studied the feasibility of snail farming as a model for small urban farms to expand into niche markets for increased profitability in United State of America, and presented the profitability challenges after the initial investment, with the potential snail farming to generate U$100 per kg. Hosseinifarhangi et al. (2019) conducted a technology-driven transition study in urban food production practices in Shanghai, which highlighted changes in the food production and supply pathways, involving the production of more food using fewer resources even in urban built up spaces and making agricultural production an integrated part of the daily life of urban residents. Moreover, Chang et al. (2017) posited that sustainability has become a critical perspective in managing firms via a holistic approach by considering economic, environmental and social dimensions, as direct implications for evolving theories. Current trends in innovative approaches and new theories emerged such as co-evolution theory and multi-level perspective, reflecting three directions of theory development.

There is increasing desire to assess the extent of global fluxes in livelihoods, and food security accomplishment (Fraval et al. 2018; Louzon et al. 2020). Sustainability induces reliable and proactive decision-making and intervention that reduces negative impact and stabilises ecological resilience, with economic growth, political stability, peace and security, as well as cultural enthusiasm, to ensure a sustainable environment for future generations. Urban sustainability entails considering alternative farming systems, which will feed the growing populations without deleterious effects on the environment. Urban development models should apply circles of sustainability management approaches to satisfy human needs at global and individual scales. Agricultural development through innovative research is a central element of sustainability, which ensures alternative farming practices that increase productivity per unit area of land. Non-conventional livestock farming systems (snails) in urban and peri-urban areas represent viable options for the provision of animal proteins and other essential nutrients as compared to traditional livestock production systems involving poultry, cattle and pigs (Adeyeye et al. 2020). This study seeks to employ a systems perspective to enhance the exploration of sustainable options that highlight the profitability of snail farming as a valuable alternative livestock farming system for sustainable development in urban and peri-urban areas in Cameroon.

2 Challenges to Sustainable Development Goals (SDGs)

Global population increase drives changes in demand for food commodities with the most drastic shifts recorded in developing economies (Herrero and Thornton 2013; Umberger 2015). Africa is a promising continent that registers 16% of the global population increase and it is the next most populated region behind Asia. The African population is projected to reach 42% by 2030 and double in 2050. Nigeria, the seventh-largest population in the world, and Cameroon's eastern neighbour, is projected to be the third-most populous country in the world from 2050. With this population increase, urbanisation would be a major force that affects the global demand for agricultural products, including staple crops and livestock products, which is expected to increase for the next 35 years (Umberger 2015; Olurankinse 2019). Urban livestock systems in Africa face numerous setbacks that affect livelihoods and environmental sustainability.

Urbanisation, prosperity, and food markets are affecting the quantity, quality, and preferences of food demands. Moreover, consumers are getting more concerned about food quality and safety, as well as its ethical and social impacts (Herrero and Thornton 2013; Denning and Fanzo 2016). It is possible to feed populations with livestock and limited budget, achieve healthy balanced diets while preserving the environment (Compassion in World Farming 2012; FAO 2018). These can be achieved through alternative urban and peri-urban livestock farming systems that enable the handling of numerous constraints such as food and nutrition security, economic growth, health, animal welfare, and conservation of natural resources.

In most low-income African countries, the demand for food, especially protein-rich foods often exceed local supply, which requires importation of large portions of their essential food products. However, this comes with implications on the foreign exchange and balance of payments, which are often negative, thus reinforcing the vicious cycles of poverty, malnutrition, hunger, and poor health. Therefore, it is necessary to consider cultural/traditional possibilities and integrate them into farming systems, and some non-conventional alternative meat products (Ebenebe 2000; Olurankinse 2019). In 2018, Cameroon's total SDG average score was 52.8, ranking 121 out of 156, and the second SDG, representing the eradication of hunger remains constant at 7.9% of population malnourishment, while the 11th SDG, standing for sustainable cities and communities dropped below 50 compared to the 2017 level. This has necessitated sustainable exit strategy for poverty that ensures resilient and affordable food systems, while guaranteeing the quality of cities. The need for diversification of food production systems and protein-rich food sources has ushered alternative small-scale livestock production systems for urban and peri-urban areas, with snail farming being one of the most favoured and fastest-growing in Cameroon. Moglia et al. (2020) called for modernisation and commercialisation of farming, especially for sectors such as livestock production that is predominantly under traditional extensive production systems. These production systems range from outdoor grazing of large and small ruminants to hunting and gathering of non-conventional livestock species like snail, rat, guinea pig, quail, duck, fish, bee, and rabbit (Alabi et al. 2019). Habitats from the wild are predominantly carried out in rural communities that do not satisfy the urban demand, these production systems are less resilient to the growing city challenge. Therefore, there is need for a shift to other production systems (e.g. urban and peri-urban livestock farming systems) that are economically and environmentally sustainable. Sustainability economics interprets ecological economics, whereby ecological variables and environmental issues are used in the multi-dimensional perspective that incorporates socio-cultural, health, and investment aspects. Alleviating poverty is an important SDG that requires actions that cause major ecological degradation with significant effect on sustainable development. Overall, it is impossible to separate economic development issues from environmental issues, and these principles underpin the concept of a bioregional economy on a broader scale.

3 Need for Sustainable Urban and Peri-Urban Livestock Alternatives

The concept of sustainability and sustainable development cuts across science and society (Harrington 2016), and the decline of arable land with increasing human population raises concerns about global food and nutrition security (Hosseinifarhangi et al. 2019). The most significant challenge for small-scale urban commercial farmers is to make profit after the initial cost of investment, which limits profitability and impedes the future of urban agriculture (Grilla et al. 2016). The lack of consistent stream of supportive funding may be one of the biggest obstacles for the advancement of agriculture in the urban landscape. The evaluation and management of sustainability cuts across many scales of economic sectors, landscapes, and at personal level.

Enhancing ecosystem resilience requires the use of all types of circular resources, just like the natural world does (www.investopedia.com). The natural environment has experienced unprecedented and accelerating rate of deterioration over the last 50 years. The main drivers of this deterioration are land use changes and exploitation by humans that are caused by societal behaviours from production to consumption coupled with the challenges of governance. Nonetheless, promoting heliculture (snail farming) as an alternative profitable urban and peri-urban livestock farming practice can enable the improvement of SDG on poverty, hunger and malnutrition, food and nutrition security, sustainable agriculture, and sustainable cities. To ensure human well-being, urban centres need to be made inclusive, secured, resilient and sustainable. Sustainable production and consumption patterns also have to be ensured, and implementation must be reinforced to revitalize global engagement for sustainable development (Angeli et al. 2020).

Increase in population dynamics and the high demand for animal food products is challenging urban landscapes and SDGs. Urban settings in Cameroon for instance are presently damaged by livestock systems as animals are either raised, transported, or slaughtered in urban areas, sometimes without due consideration for the environment. Moreover, the share of the family budgets allocated to food is insufficient for more than 50% of the population, as population increases without appropriate levels of income. In 2019, the country GDP per capita stands at US$1522.70 while the purchasing power parity is US$3652.70. Hence, the local nutritional demands for animal food products cannot be adequately satisfied, which highlights the need for non-conventional, non-soil, alternative integrated crop and livestock systems. This further aggravates the already precarious food and nutrition insecurity, associated health risks between animals and humans, and the deleterious effects on the environment. Previous evaluations studied particular aspects of urban livestock systems, without specific focus on interlinks in an all-inclusive manner, which has led to discrepancy of information regarding the management of urban farming systems, including the profitability of snail production.

4 Snail Farming as a Sustainable Profitable Alternative Livestock Farming

Snail farming in African urban areas represents a profitable business opportunity in urban and peri-urban areas with low-risk, requires low start-up capital, less space and labour, low demand for feedstock and professional knowledge (Blay et al. 2004; Ngenwi et al. 2010; Agbogidi and Okonta 2011; Mohammed et al. 2014). Snails can be picked from the forest and farmed in urban gardens, backyards, big basins, car tires, and large cages. They are omnivores and environmentally friendly with less offensive droppings as compared to pigs, cows, and poultry (Mohammed et al. 2014). Despite the enormous benefits associated with urban heliculture, the current snail production level cannot meet the market demand as a result of various biological and technical factors (Etchu et al. 2008). Besides seafood sources, Cameroon depends essentially on small and large ruminants (e.g. poultry, cattle, goat, sheep, and pig) to meet its animal protein needs. However, the demand for non-conventional protein sources is gaining importance due to various constraints in production and high cost of these food products. Hence, there is increasing demand for alternative animal food sources such as snails that are easily produced at low capital investment. Snail meat has low fat and cholesterol contents, it is cheaper than other meat products, and it is widely consumed in diverse local traditional dishes like *eru, sauce* and as a delicacy (snacks) in fast food. Nonetheless, the country can barely satisfy 25% of the national demand for snail products. Snails are important sources of animal protein in many parts of West and Central Africa with high contents of protein (69% dry weight), potassium, phosphorus, essential amino acids, and vitamins C and B complex (Baba and Adeleke 2006; Okpeze et al. 2007; Mohammed et al. 2014; Olurankinse 2019). Snail meat is very nutritious with medicinal, cultural/social values, coupled with economic benefits (Agbogidi and Okonta 2011; Mohammed et al. 2014). Consumers prefer snail meat to beef, pork, and other regular sources of protein as they are highly energetic, and important for nursing mothers and malnourished children (www.voanews.com).

Furthermore, the by-products of snail farming are widely used across allied industries such as health and cosmetics where the snail slime is sort after for its rich medicinal properties. In addition, snail shell is a rich source of calcium that makes it an important component for the production of concentrates to supplement animal nutrition. Snail species that are commonly domesticated in Cameroon include *Archatina archatina and Archachatina marginata* (Ngenwi et al. 2010). The production and consumption of snails can be up-scaled and apply in different contexts and heliculture is considered a viable means of producing nutrient rich food and generating income that fosters sustainable development. Snails have high fecundity and low mortality rates, and there are readily available domestic and international markets (Baba and Adeleke 2006; Titilola et al. 2015) Therefore, urban and peri-urban snail farming could represent an innovative-driven transition on urban planning and development in many low-income economies, which is a low-cost means of applying circular economy, biocapacity, and reduction of food wastes (Mohammed et al 2014). Heliculture allows the integration of agricultural practices inside urban centres as part of the daily life of the residents (Hosseinifarhangi 2019), which increases food and nutrition security, and reduces poverty and unemployment (Agbogidi and Okonta 2011; Mohammed et al. 2014; Titilola et al. 2015; Munonye and Moses 2019; Olurankinse 2019). Women in the suburbs of West and Central Africa including Nigeria, Ghana and Cameroon mostly practice heliculture, and the profitability could contribute in mitigating gender inequality (Alabi et al. 2019). The non-existence of a viable systems perspective for heliculture has also curtailed the explorations of sustainable options to increase the profitability. This research aims to evaluate the profitability of snail farming as a valuable urban livestock alternative for sustainable development in Cameroon.

5 Theoretical Overview

This study is supported by production systems, market demands, consumption trends, and the theories of sustainable development. The sustainable development theories essentially question the needs versus environmental problems, wealth distribution and the normative dimension of the ecological development. Ecological development reflects the adaptation of human activities in relation to their environmental and natural resources without jeopardising sustainability. Meanwhile, economic modernisation of production structures generally increases social inequalities, which are responsible for environmental deterioration (Urteaga 2011). Urbanisation is linked to market theory, and production factors such as land, labour, and capital are responsive to market conditions as the self-interest of producers often works for the benefit of consumers (www.britannica.com). The market economy that mostly applies to Cameroon is based on the law of demand and supply (www.encyclopedia.com). When scarcity occurs, entrepreneurs attempt new techniques, while unprofitable firms and initiatives go out of business. The permanent income theory of consumption also applies, as sources of income need to be permanent and demonstrate a growth pattern. According to Friedman, long-term expected income largely determines consumption levels not the current income (www.yourarticlelibrary.com). The Doomsday Scenarios withstands that there is a long-standing stiffness between prosperity and sociocultural stability, whereby markets will support and sustain economic growth but undermine the social aspect. Many projections have predicted a quickening of innovation change in the current century. If innovation has possibilities of raising the standard of living intensely, it also comes with important cultural break (www.yourarticlelibrary.com). Theories of sustainability endeavour to integrate and prioritise social responses to environmental and sociocultural problems. An economic model considers natural and financial capital, but an ecological model prioritises biological diversity and ecological integrity, while a political model focuses on social systems that ensure human self-esteem. Religion has also joined the economic growth and sustainability debate with critical and interesting resources for sociocultural change. Presently, various innovative methods and new theories have emerged such as the Co-evolution and Multi-level Perspective Theories. These new theories reflect the three main directions of the development theory: (1) shifting from "what" to "how", (2) increasing the use of multidisciplinary methodology, and (3) aiming for all-inclusive systems (Harrington 2016; Chang et al. 2017). Overall, this study integrates various pillars of the SDGs (1, 2, 3, 8, 9, 11, 12, and 15) to drive urban development and sustainability of developing countries in Africa.

6 Materials and Methods

Data were collected in the Buea Municipality, South West Region of Cameroon between March and May 2018. Questionnaires were used for interview to collect data from 60 randomly selected snail farmers operating in the urban and peri-urban areas. Data were collected on the farm sizes, production cost, productivity per farmer, and the income per farmer. Descriptive statistics were used to present the socio-economic characteristics, and farm budgeting analysis was used to derive and report farm performance standards. The Linear, Cobb Douglas and Semi-log Functions were used to assess the factors that are affecting the snail production level. These investigations are consistent with other studies (Baruwa et al. 2012; Ogunjimi et al. 2012; Chah and Inegbedion 2013; Titilola et al 2015; Munonye and Moses 2019).

Farm budgeting analyses were calculated using Eq. (1)

$$\mathrm{GM} = \mathrm{GR} - \mathrm{TC} \quad (1)$$

Where: GM = Gross Margin, GR = Gross Revenues, and TC = Total Production Costs.

A linear, Cobb Douglas and Semi-log Functions were used to investigate factors affecting snail production level in the study area; this is presented as Eq. (2)

$$\mathrm{Y} = \mathrm{a0} + \mathrm{a1X1} + \mathrm{a2X2} + \mathrm{a3X3} + \mathrm{a4X4} + \mathrm{a5X5} + \mathrm{a6X6} + \mathrm{Ut} \quad (2)$$

Where: Y = Production in Kg, X1 = size of farm (number of snails), X2 = Labour in hours, X3 = Capital in US$, X4 = Education level (years), X5 = Quality of feed used, X6 = Experience.

7 Results

The descriptive statistics indicated that majority (95%) of the urban snail farmers are women. More than half of the farmers (55%) are between the ages of 20–45 years, which indicates a high adoption rate by youths. Majority of the respondents are married (73%), which assures additional labour, while seven widows and two widowers affirm rearing snail in the urban area as a dietary complement, and as a source of income. About 75% of respondents have between 1 and 5 children, and snail meat represents a good alternative nutritional intake for large families that in return provide additional labour for the production activity. Only 7% of respondents had no formal education, while the educated population displayed a high adoption rate for innovation. Meanwhile, 30% of the farmers have been in the snail business for only 1–2 years, 18% for 3–4 years, and 12% for 5–6 years. None of the respondent had more than 7 years longevity in the snail business, which indicate that snail farming is a new activity in the study area, representing an innovative additional source of meat food, and income. Most respondents (97%) mainly feed their snails using household wastes. Majority of the farmers (83%) obtain their snail for rearing from the wild by simply handpicking without causing any environmental issues. Hired labour represents only 5% of the workforce as most of the required labour is done by family members. Most of the farmers (83%) obtain their starting capital from savings, relatives or friends, and 13% from farmers' association, only 2% obtained

Table 1 The farm size reflecting total number of snail reared and the output level of farmers that reflect the quantity of snails harvested

Farm size		Farm output—snails harvested (kg)	
Proportion of farmers (%)	Number of snails reared	Proportion of farmers (%)	Output (Kg)
31.7	500–900	30	100–500
51.7	1000–4000	58.3	600–1000
16.7	5000 and above	11.7	1000 and above

Table 2 Total production cost, total revenue, and gross margin calculation of snail farming

Expenditure			Revenue				Gross Margin	
Number of Farmers	Production cost (US$)	Total production cost (US$)	Number of farmers	Quantity of snails (kg)	Unit Price (US$)	Revenue (US$)		
35	15	525	19	35	2.4	1596	**Elements**	**Amounts (US$)**
20	25.5	510	31	125	2.4	9300	**Total production cost**	**1,185**
5	30	150	10	250	2.4	6000	**Total revenue**	**16,896**
TOTAL	**70.5**	**1185**	**TOTAL**	**410**	**2.4**	**16,896**	**Gross Margin**	**15,711**

it from financial institutions. A high proportion (56%) of the respondents produce snails for sale and consumption, 35% for sale only, and 9% for home consumption only. The highest proportion (51.7%) of farm size ranges between 1000–4000 snails, while 58.3% of farmers reported a monthly output level ranging between 600-1000 kg (Table 1); it is only 16.7% of farmers who rear above 5000 snails in their farms. Most of the harvests are often bought on site and are usually processed for sale in local and international markets, which demonstrates the availability of a ready market for urban snail farming.

Gross Margin Calculation and Profit of Snail Farming

Majority (58.3%) of the farmers have production costs of US$15 for 50 kg, while only 8.3% farmers reported topping US$30 for the production of 50 kg. On average, production cost per kg is US$0.4, while the revenue stream shows that one kg is sold at US$2.4, giving a profit of US$2 per kg (Table 2), a return of investment 5, and total revenue of US$16,896. From Table 2, Gross Margin is US$15,711, giving a gross profit margin of 93%. This is feasible as the activity does not require any huge investment as starting capital, since all the required items are basically tools and resources found in households.

Farm budgeting analysis revealed that total production cost mainly consist of feeding, transportation and labour. Regression results indicate that coefficients of feed type (0.054), education (0.042) and experience (0.723) have positive relationships

Table 3 Econometric results

Variables	Coefficients	T-calculated	T-critical
Education	0.042	0.619	1.684
Feed type used	0.054	1.864	1.684
Experience	0.723	0.492	1.684

with snail production level (Table 3). Increase in these factors will have a positive impact on the quantity of snails harvested.

8 Discussions

The handpicking of snails in the wild for setting up urban livestock production systems has infrequently been acknowledged or recorded, and land-based activities are not solely for rural settings, but also practiced by many urban households. Moreover, urbanisation reshapes some gender dimensions of roles with great innovative modifications (Shackleton et al. 2020; Adeyeye et al. 2020; Louzon et al. 2020). The majority of urban snail farmers are women, young, educated, married with a good number of children. Results indicate that snail urban farming is a significant incentive that boosts food/nutrition security, income generation, and eliminate household wastes. The results are consistent with Moglia et al. (2020) who reported the importance of gender-oriented approaches to understand inherent complexities within agrarian change. Sociocultural foods of some communities are theoretically sustainable, nutrient-rich, and can augment dietary diversity (Ghosh-Jerath et al. 2020). Urban inhabitants have confirmed that heliculture doesn't cause any environmental disturbance. The average price of US$2.4 per kg is far below the price per kg of other animal food products such as cattle, chicken or pork which cost US$6, US$4 and US$5, respectively. These calculations support the fact that snail farming is a better economic alternative for urban livestock farming. Overall, the results indicate that snail farming business requires low capital investment and is highly profitable (Titilola et al. 2015; Baruwa et al. 2012; Munonye and Moses 2019). According to Ogunjimi et al. (2012), the rate of adoption of heliculture depends on some factors including personality factor and economic status of individuals. The evolution towards sustainable urban farming practices requires synchronised preparation of supportive and compatible spatial development, adequate policy implementation, and monitoring and evaluation tools to augment the competitive opportunity (Hosseinifarhangi et al. 2019). Policymaking has to promote urban non-conventional livestock farming initiatives to boost urban consumption, income generation and environment protection. These initiatives will reduce production costs, ensure good nutritional habits and build sustainable cities. To enhance the socio-economic status of urban farmers, they should facilitate access to credit, capacity building, and demonstration/incubation centres for heliculture farmers should be built. In general, snail population in the wild is affected by indiscriminate hunting, bush burning, agro-chemicals, deforestation, and climate change, which highlight the need to enhance the domestication of snails. Interventions at the level of research, policy and regulation are therefore needed to leverage the potentials and opportunities presented by heliculture, especially under urban non-conventional livestock production systems.

9 Conclusion and Recommendations

Urban snail farming is a highly profitable business activity that guarantees sustainable development without jeopardising environmental productivity and sustainability. The observed increase in snail productivity and revenue of farmers with increasing production cost highlights the need to invest more capital in the snail farming business in order to generate more profit. Overall, this study demonstrates a significant profit of U$2 per kg of snail for every U$0.4 investment as the total production cost. Considering the high demand for snails in relation to the relatively low capital input and high revenues generated per capital, snail farming can be considered as a very lucrative business for urban dwellers. Moreover, snail farming demonstrated an invaluable contribution to household nutrition, which represents a significant step to ensuring food and nutrition security. Based on the environmental sustainability and economic viability of the urban snail farming system, it is recommended to increase urban snail farming to boost food and nutrition security, and generate additional income for urban dwellers. Hence, governments and private sector organisations, including the food industry should consider investing in the snail farming activity, by providing training on modern production and processing techniques in order to enhance productivity, leading to food and nutrition security that foster sustainable development. Further studies on the sustainability of snail farming should integrate more parameters in a holistic systems approach that includes socio-cultural factors, so as to effectively analyse the sustainability of urban snail farming systems. More researches are needed to develop innovative techniques to rapidly multiply reproductive snails, as inputs are highly required by farmers. Further emphasis should be on nutrition and consumption in the future, especially in the specific benefits for particular age groups or health categories.

References

Adeyeye SAO, Bolaji OT, Abegunde TA, Adesina TO (2020) Processing and utilization of snail meat in alleviating protein malnutrition in Africa: a review. Nutr Food Sci. https://doi.org/10.1108/NFS-08-2019-0261

Agbogidi OM, Okonta BC (2011) Reducing poverty through snail farming in Nigeria. Agric Biol J N. Am 2:169–172. https://doi.org/10.5251/abjna.2011.2.1.169.172 ISSN: 2151-7525

Angeli V, Silva PM, Massuela DC, Khan MW, Hamar A, Khajehei F, Graeff-Hönninger S, Piatti C (2020) Review: Quinoa (*Chenopodium quinoa* Willd.): an overview of the potentials of the "golden grain" and socio-economic and environmental aspects of its cultivation and marketization. Foods 9:216

Alabi OO, Shoyombo AJ, Ajala AO, Ogunjimi SI (2019) Animal agriculture: a viable tool for rural women empowerment and redemption from poverty. Scopus, Int J Civ Eng Technol (IJCIET) 10(2):2365–2373

Baba KM, Adeleke MT (2006) Profitability of snail production in Osun State Nigeria. J Agric Food Sci 4(2):147–155

Baruwa OI, Abogan PO, Kassali R (2012) Economics of raising African Giant Land Snail (*Archachatina marginata*) in Osun State, Nigeria. Int J Agric Econ Rural Dev 5(1) 46–55

Blay ET, Ofori BD, Heloo J, Ofori JB, Nartey E (2004) Managing agrodiversity the traditional way: lessons from West Africa in sustainable use of biodiversity and related natural resources. United Nations University Press, Tokyo, pp 203–214

Carzedda M, Nassivera F, Marangon F, Troiano S, Iseppi L, Bassi I (2020) Urban food security and strategic planning: involving Millennials in urban agriculture. In: Smart innovation, systems and technologies book series (SIST, vol 178) Proceedings from international symposium: new metropolitan perspectives NMP 2020: New Metropolitan Perspectives, pp 91–100

Chah JM, Inegbedion G (2013) Characteristics of snail farming in Edo South agricultural zone of Edo State, Nigeria. Trop Anim Health Prod 45:625–631. https://doi.org/10.1007/s11250-012-0269-z

Chang RD, Zuo J, Zhao ZY, Zillante G, Gan XL, Soebarto V (2017) Evolving theories of sustainability and firms: history, future directions and implications for renewable energy research. Renew Sustain Energy Rev 72:48–56

Compassion in World Farming (2012) Food security and farm animal welfare, 20 p. https://www.ciwf.org.uk/media/3758836/Food-security-and-farm-animal-welfare-report.pdf

Denning G, Fanzo F (2016) Ten forces shaping the global food system. In: Eggersdorfer M, et al (eds) Good nutrition: perspectives for the 21st century, Karger

Ebenebe CI (2000) Mini livestock production in Nigeria, the present and future. In: Proceedings of 5th annual conference of animal science association of Nigeria Port Harcourt, Nigeria, September 9th–12th. Academic Press, New York

Etchu K, Mafeni M, Ngenwi A (2008) Comparative performance of three edible snail species using intensive cage housing system in Cameroon. Bull Anim Health Prod Africa 56:345–352

Food and Agriculture Organisation (FAO) (2018) Shaping the future of livestock sustainably, responsibly, efficiently. The 10th Global Forum for Food and Agriculture (GFFA) Berlin, 20 p. https://www.fao.org/3/i8384en/I8384EN.pdf. Accessed 18–20 Jan

Fraval SH, James LM, Oosting SJ, Sayula G, Teufel N, Silvestri S, Poole EJ, Herrero M, van Wijk MT (2018) Livelihoods and food security in an urban linked, high potential region of Tanzania: changes over a three year period. Agric Syst 160(C):87–95

Ghosh-Jerath S, Kapoor R, Singh A, Downs S, Barman S, Fanzo J (2020) Leveraging traditional ecological knowledge and access to nutrient-rich indigenous foods to help achieve SDG 2: an analysis of the indigenous foods of Sauria Paharias, a vulnerable tribal community in Jharkhand, India. Front Nutr 7(61):1–26

Grilla AM, LaJeunesse CL, Morgan DM, McMaster DM (2016). Feasibility of snail farming as a model for small urban farms to expand into niche markets for increased profitability. https://digitalcommons.wpi.edu/iqp-all/3213

Harrington LMB (2016) Sustainability theory and conceptual considerations: a review of key ideas for sustainability, and the rural context. Pap Appl Geogr 2(4):365–382. https://doi.org/10.1080/23754931.2016.1239222

Heifer International Increases Support for Snail Farming in Cameroon. https://www.voanews.com/a/a-13-2007-01-29-voa37-66764512/563916.html. Accessed 11 Apr 2019

Herrero M, Thornton PK (2013) Livestock and global change: emerging issues for sustainable food systems. PNAS 110(52):5

Hosseinifarhangi M, Turvani ME, van der Valk A, Carsjens GJ (2019) Technology-driven transition in urban food production practices: a case study of Shanghai. Sustainability 11:6070. https://doi.org/10.3390/su11216070

Louzon AM, Pelfrêne B, Pauget F, Gimbert N, Morin-Crini F, Douay A, de Vaufleury A (2020) Bioaccessibility of metal (loid)s in soils to humans and their bioavailability to snails: a way to associate human health and ecotoxicological risk assessment? J Hazard Mater 384:121432

Market theory. https://www.encyclopedia.com/science/encyclopedias-almanacs-transcripts-and-maps/market-theory. Accessed 14 Aug 2020

Modak S (2020) Nutrition sensitive agriculture in Tripura, an overview. Int J Curr Microbiol Appl Sci 9(7):3580–3586

Moglia M, Alexander KS, Larson S, Dray AG, Greenhalgh G, Thammavong P, Thephavanh M, Case P (2020) Gendered roles in agrarian transition: a study of lowland rice farming in Lao PDR. Sustainability 12:1–20

Mohammed S, Ahmed AA, Adjei D (2014) Opportunities for increasing peasant farmers income through snail production in Ghana. Sch J Agric Vet Sci 1(4):195–200

Munonye JO, Moses P (2019) Profitability analysis of snail farming in Owerri agricultural zone Imo state. In: Contributed paper prepared for presentation at the 93rd annual conference of the agricultural economics society, April 15–17. University of Warwick, England

Ngenwi AA, Mafeni JM, Etchu KA, Oben FT (2010) Characteristics of snail farmers and constraints to increased production in West and Central Africa. Afr J Environ Sci Technol 4(5):274–278

Ogunjimi SI, Yusuf OJ, Ajala AO (2012) Factors influencing involvement of peri-urban farmers in mini-livestock farming in south-western Nigeria. Sci J Rev 1(3):58–69

Okpeze CN, Omole AJ, Ajayi FT, Adebowale EA (2007) Effects of feeding adult snails Stylosanthes guianensisor Lablab purpureus as substitute for pawpaw leaf. Afr J Biotech 6:1959–1962

Olurankinse CA (2019) Sustainable family well-being amid socio economic changes and challenges through food safety and nutrition. J Biol Sci Bioconservation 11(1):1–25

Shackleton CM, Drescher A, Schlesinger J (2020) Urbanisation reshapes gendered engagement in land-based livelihood activities in mid-sized African towns. World Dev 130(C):104946. https://doi.org/10.1016/j.worlddev.2020.104946

The Concept of Sustainability. https://www.investopedia.com/terms/s/sustainability.asp. Accessed 15 Aug 2020

Theory of Production. https://www.britannica.com/topic/theory-of-production. Accessed 14 Aug 2020

Three Important Theories of Consumption (With Diagram). https://www.yourarticlelibrary.com/economics/consumption-function/3-important-theories-of-consumption-with-diagram/37756. Accessed 11 Apr 2019

Titilola OL, Olaniyi OZ, Adebusola AA (2015) Mini-livestock farming as a strategy for food security in Oyo State of Nigeria. J Agric Sustain 7(2):171–186 ISSN 2201-4357

Umberger WJ (2015). Demographic trends: implications for future food demand. Paper prepared for agricultural symposium federal reserve bank of Kansas City, July 14–15

Urteaga E (2011) The alternative theories of sustainable development. Boletín De La Asociación De Geógrafos Espanoles 52(55):399–405

Experimental Study of Fire Behavior in Annually Burned Humid Savanna of West Africa in the Context of Bush Encroachment

Aya Brigitte N'Dri, Aka Jean-Noel Kpré, Koffi Prosper Kpangba, Tionhonkélé Drissa Soro, Koffi Vincent Kouassi, Kouamé Fulgence Koffi, Yao Anicet Gervais Kouamé, Ahou Blandine Koffi, and Louis N'Guessan Konan

Abstract Prescribed yearly mid-dry season fire is generally used in the Guinean savanna (a humid savanna of West Africa) as a vegetation management tool. Despite this fire, bush encroachment can be observed due to climate change and high sapling recovery rates with increasing CO_2 emissions. There is therefore a need to analyze other fire seasons. We analyzed fire behavior during a 5-year field fire in the Guinean savanna of Lamto reserve in Ivory Coast. Early (EDS), mid (MDS), and late (LDS) dry season fires were tested. Nine 0.5 ha plots were burnt annually to determine the rate of spread, fire intensity and the residence time above 60 °C. Fuel characteristics and weather conditions were measured to assess their impact on fire behavior. Recruitment from resprout to adult tree stage were assessed for an estimation of these fires' severity. Understory grass height, fuel load, and moisture content had greater values in EDS than in MDS and LDS. The rate of spread and intensity of both MDS and of LDS were greater than those of EDS fires. The best predictors for fire behavior were fuel moisture content and air humidity. In dry conditions, fires were faster and more intense. With climate change predicting increasingly longer and drier periods, we expect more and more intense fires. A longer time of exposure to lethal temperatures (>60 °C) and the phenological state of trees in the late dry season explains the higher severity of this fire (there was no recruitment to adult stage under LDS fire). In some years, LDS helped to reduce bush encroachment in the subject region. This data provides important insights into fire behavior and its severity in the Guinean savanna, informing fire management policies and procedures.

Keywords Burning season · Climate change · Fire behavior · Fire severity · Guinean savanna

---

A. B. N'Dri (✉) · A. J.-N. Kpré · K. P. Kpangba · T. D. Soro · K. V. Kouassi · K. F. Koffi · Y. A. G. Kouamé · A. B. Koffi · L. N. Konan
UFR des Sciences de la Nature, Unité de Recherche Ecologie et Biodiversité, Laboratoire d'Ecologie et de Développement Durable, Université Nangui Abrogoua, 02 BP 801 Abidjan 02, Abidjan, Côte d'Ivoire

W. Leal Filho et al. (eds.), *Sustainable Development in Africa*, World Sustainability Series,
https://doi.org/10.1007/978-3-030-74693-3_27

1 Introduction

Fire is an important modifier of ecosystems, influencing vegetation structure and composition, as well as the net primary productivity of aboveground plants (Bond 2008). Fire regimes are increasingly controlled, if not driven, by humans (Gill and Allan 2008), yielding dramatic changes in many ecosystems worldwide. This is occurring to the point that managers of protected areas sometimes have to revert to past fire regimes in order to maintain targeted ecosystem services or biodiversity (Swetnam et al. 1999). To understand different fire regimes, fire behavior is studied. Fire behavior is generally described by many parameters: the rate of fire spread, intensity, and residence time (Williams et al. 1998; Savadogo et al. 2007; Fontaine et al. 2012; Prat-Guitart et al. 2016; N'Dri et al. 2018). Fire behavior directly effects savanna plants (e.g. scorching, topkill, death and recruitment) (Werner and Franklin 2010; N'Dri et al. 2014; Smit et al. 2016). However, despite fire, woody encroachment has been observed across many savannas worldwide: in Australia (Sharp and Whittaker 2003; Lehmann et al. 2009); in America (Durigan and Ratter 2016; Barger et al. 2011), and in Africa (Belay et al. 2013; Stevens et al. 2016, Case and Starver 2017). Heubes et al. (2011) projected a tree density increase of 1–10% by 2050 in many West African countries. Such a bush encroachment can have negative consequences for soil carbon storage (Berthrong et al. 2012), grazing potential (Angassa and Baars 2000), biodiversity (Ratajczak et al. 2012), ground water recharge, and tourism (Gray and Bond 2013). Woody encroachment is related to climate change (Criado et al. 2020). Indeed, increasing levels of atmospheric CO_2 and climate change have emerged as important drivers of bush encroachment (O'Connor et al. 2014).

In wooded ecosystems where plants are long-lived, fuel loads are generally higher with longer intervals of time between fires. Thus, fire frequency becomes the primary driver of fuel accumulation. In most savanna and grassland ecosystems, where fire frequency is very high (Van Wilgen et al. 2003), the season of the fire plays a significant role on fuel quantity (Govender et al. 2006). While the effect of fire frequency on fire intensity is well documented (Govender et al. 2006; Robertson and Hmielowski 2014), few studies examine very high fire frequency and the impact of fire season on fire intensity (Rissi et al. 2017; Van Wilgen et al. 2007) and severity (Louppe et al. 1995; Laris et al. 2017). Moreover, quite recently, Prichard et al. (2017) reviewed fire regimes across the world's major bioregions and pointed out the relative scarcity of literature on these subjects when it comes to African savannas.

The most marked effects of fire on vegetation were found in the extremes: annual burning, or long periods of fire exclusion. These effects were greater in areas of higher rainfall (Van Wilgen et al. 2007). Among the African savannas, the most extreme fire conditions (annual burning, high rainfall) occur in the Guinean forest-savanna mosaic ecoregion, extending over 670,000 km^2 in West Africa (Kier et al. 2005). In this area, the fire return interval is generally between 1–3 years, with 21% of the area burned annually (Dwomoh and Wimberly 2017). Annual rainfall is in the range of 1000–1600 mm yr^{-1} (Mahé et al. 2012), yielding very high grass production (Gignoux et al. 2006; Koffi et al. 2018, 2019). On the continental scale, the Guinean ecoregion is a transition zone from forest to savanna: the transition between the two ecosystems is very sharp—often occurring in a little as four meters—while both the physiognomy and the respective floras are different (Soro et al. 2018). Fires almost never enter the forests, but annual fires are responsible for maintaining the adjacent open vegetation (Gignoux et al. 2016; Soro et al. 2018). Nevertheless, significant woody plant encroachment into this area has been reported despite the very frequent and intense fire regime (Gautier 1990; Dauget and Menaut 1992; Hochberg et al. 1994).

Moreover, few studies undertaken in the Guinean savanna of West Africa (Louppe et al. 1995) explored the impacts of fire treatment on vegetation. Overall, we do not know whether the severity of fires on trees is due to a higher fire intensity or not. It is therefore of primary importance, in the first instance, to understand how fire intensity and severity vary with the fire season.

To this end, we conducted a 5-year field experiment with three different fire treatments in a Guinean savanna to study fire behavior and severity in response to the fire season. This chapter aims to test the effect of fire seasons on fire intensity and severity in an extreme situation where fire frequency is at its maximum, with one fire each year. We hypothesize that the fire season and climate interact to influence fire behavior and severity and therefore affect savanna physiognomy (Fig. 1).

2 Materials and Methods

2.1 *Study Site, Experimental Design and Fire Treatments*

The study site is the Lamto Reserve (in central Côte d'Ivoire: 6°9 and 6°13′N, −5°15 and −4°57′W), a 2500-ha area of typical Guinean forest-savanna mosaic devoted to scientific research since 1962 (Abbadie et al. 2006). The average annual rainfall here is about 1,200 mm (Yapo et al. 2020; Tiemoko et al. 2020). The climate is sub-equatorial with four seasons based on monthly rain: a long rainy season from mid-March to July; a short dry season in August; a short rainy season from September to mid-November; and a long dry season from mid-November to mid-March. Fires in Lamto are set by managers in the middle of the long dry season (mid-dry season fire), following the practices of local populations. The main savanna type is a shrubby savanna with tree cover between 7 and 62% (Gautier 1990).

This study was conducted on three 230 × 120 m blocks of shrubby savanna, each separated from the surrounding similar savanna by a 10-m wide firebreak, and distant from each other by a few kilometers. Each block was divided into three 100 × 50 m plots separated from each other by 30 m-wide firebreaks. Each plot in a block received one of three fire treatments in a full factorial design: (1) early dry season fire (EDS) at the start of the long dry season (~18th November); (2) mid-dry season

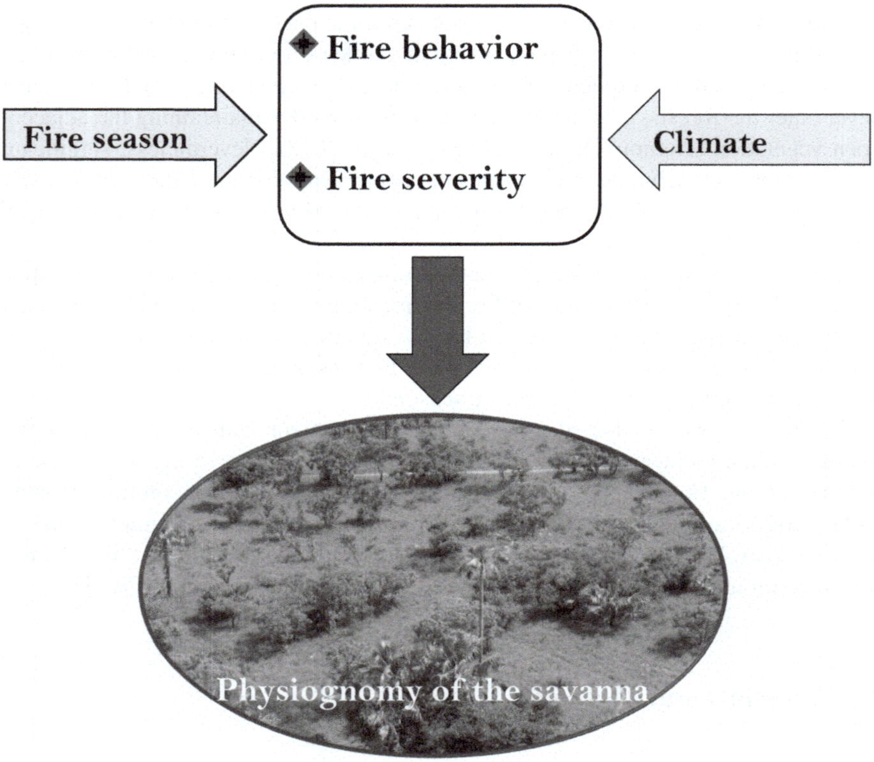

Fig. 1 Diagram of processes contributing to fire behavior, severity and therefore savanna physiognomy change after fires

fire (MDS) in the middle of the long dry season (~18th January); and (3) late dry season fire (LDS) at the very end of the long dry season (~15th March) (Fig. 2). In each plot, the designated fire treatment was applied five times from 2013 to 2018. All blocks were located on flat ground to eliminate the influence of slope. All plots had experienced annual mid-dry season fire for at least 50 years before the experiment.

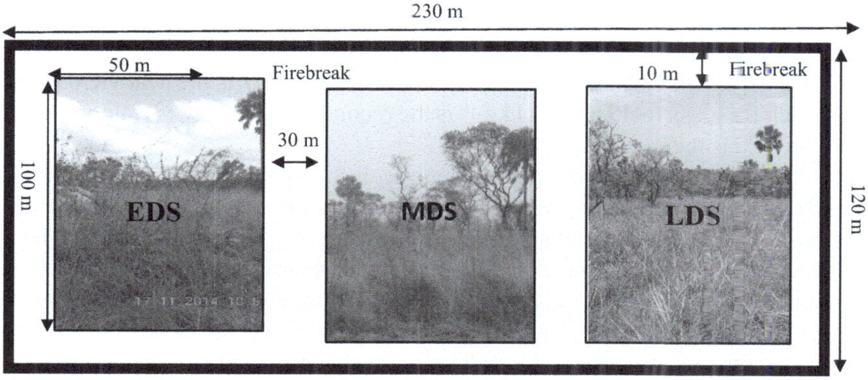

Fig. 2 Experimental set-up and fuel bed the day before each fire. EDS: early-dry season fire (November 2013), MDS: mid-dry season fire (January 2014); LDS: late-dry season fire (March 2014)

2.2 Fuel Characteristics

Fuel consisted primarily of grass and tree leaf litter. Fuel characteristics were described by the following variables before fires were applied: grass height (cm), fuel load (kg m^{-2}) and moisture content (%). Grass cover height (cm) was determined as the average vertical distance from the ground to the tip of the shoot of 50 randomly selected points. Fuel was collected from ten 1-m^2 quadrats per plot shortly before burning.

Fuel was separated into litter (fallen dead grass and tree leaves) and standing fuel (standing grass and stems), and immediately weighed for fresh mass. Samples were oven-dried at 80 °C to a constant weight and weighed for their dry mass. The moisture content was then calculated. Grass cover height (cm) was determined as the average vertical distance from the ground to the tip of the shoot of 50 randomly selected points.

2.3 Fire Behavior

Fire was ignited with a dry palm leaf along the shorter side of each plot in the direction of the wind, to rapidly establish a fire-line and ensure linear ignition. All burns were conducted in the morning (between 10:00–11:00 am) on the same day, in order to ensure a similarity of fuel moisture levels and weather conditions. We characterized fire behavior using three variables: rate of spread (m s^{-1}), fire intensity (kW m^{-1}), and residence time above 60 °C (s).

The rate of spread was determined by recording the time it took for a flame line to reach poles positioned every 10 m on both sides of each plot using stopwatches.

Fire intensity was estimated using Byram's (1959) equation:

$$I\left(\mathrm{kW\ m^{-1}}\right) = W\ H\ R$$

where: W (kg m^{-2}) is the mass of fresh fuel consumed; H is the heat yield of the fuel, which for grass fuel 16 890 kJ kg^{-1} is the recommended value (Trollope 1983); and R (m s^{-1}) is the rate of spread.
The fire residence time was defined as the time during which the air temperature at ground level, measured with a thermocouple, was above 60 °C—the widely accepted lethal temperature for most plant cells (Sidzabda et al. 2010). We used self-built synchronized fire temperature recorders (SFTRs) to measure this time. Each SFTR was comprised of a thermocouple (type K 250 mm steel probe, 0–1100 °C range) connected to a Raspberry Pi™ nano-computer (https://www.raspberrypi.org/), plus batteries. Prior to setting fires, the SFTRs were synchronized by sending a 'start' signal before burying them in the plots, with only the top 10 cm of the thermocouple probe protruding from the ground. They were placed at the ends of equilateral triangles with sides 20 m in length (in each plot). After the fires, the time series of temperatures were downloaded from the STFRs and residence times were computed from that data.

2.4 Weather Conditions

During each fire, relative humidity (%), wind speed (m s^{-1}) and air temperature (°C), were recorded every five minutes by using an in situ automatic weather station (reference number SM55PRO; Inovalley, Pontoise, France) placed 2 m aboveground in the fire break between two adjacent plots, a process similar to the methods of Kidnie and Wotton (2015). The recorded values were averaged over the period of fire propagation for each plot.

2.5 Recruitment from Resprout to adult stage

All the resprout of woody species (main stem dead and new sprouts grow from root crown) ≥0.5 m height in the experimental plots were permanently marked (in September 2013 at the end of the growing season previous to the application of the first fires) and monitored annually for survival over 5 years in each fire treatment plot.

Individually marked resprouts were relocated after the fires (to be revisited the next September in all plots), annually over 5 years. The responses (topkilled or resisted) of sub-adults to fire were assessed. After the initial fires (for each fire season), some resistant sub-adults could potentially resist again—or not—after the following fires. In fact, some sub-adults could resist continuously through two, three, four, or five fires (across years) and therefore recruited to adult stage.

We therefore calculated the proportion of sub-adults that were recruited (PR) to adult stages (continuously resisted all five fires).

PR = number of individuals resistant from fire 1 to 5/number of individuals continuously resistant of fires 1 to 4.

2.6 Data Analyses

All statistical analyses were performed using version 3.6.2 of the R software. Mixed-effect models (lmer R function) were developed to analyze the responses of fire behavior parameters; the relative importance of pre-burn fuel characteristics; weather parameters during fires; and the recruitment of trees from resprout to adult stage, relative to three explanatory variables: season (3 seasons), year (5 years) and plot (9 plots). Season was the fixed effect, while year and plot were treated as crossed random effects. A Tukey multiple comparison test (HSD) was used to compare the different modalities of significant fixed effects.

To examine the effect of fuel and weather variables on fire behavior, we used mixed-effect multiple regression models. Final models were selected using a stepwise procedure based on a minimum Akaike Information Criterion estimation. Year and plot were again treated as crossed random effects.

2.7 Seasonal Changes in Fuel and Fire Characteristics

Figure 3 shows that total fuel load and its moisture content were significantly greater in the early dry season than in the mid and late dry seasons ($P < 0.001$). There was no significant difference in fuel loads and moisture content between the mid and late dry seasons. The height of the grassy understory decreased significantly from the early to late dry season ($P < 0.001$).

There are few studies of the impact of the season on fuel and fire characteristics in humid savannas anywhere in the world; comparisons can yield important insights for savanna ecosystem management. The results of our study are consistent with the findings of Williams et al. (1998) in Northern Australia and of Higgins et al. (2000), and Van Wilgen et al. (2003) in South Africa that showed a decrease in moisture content as the dry season progressed. We also confirm results from Rissi et al. (2017) in Brazil that showed that the dry fuel load does not change during the dry season.

Fuel loads in the present study at Lamto are higher than those previously reported for other African savannas: about five times greater than those of a Sudanian savanna in Burkina Faso (Savadogo et al. 2007), and about ten times greater than in Chad (Guiguindibaye et al. 2013). These high fuel loads most likely explain the higher fire intensity we recorded compared to these savannas (Savadogo et al. 2007; Guiguindibaye et al. 2013). Frost et al. (1986) concluded that fire is more intense in humid savannas than in dry savannas due to the fact that differences in water availability leads to differences in grass biomass, i.e. fuel loads.

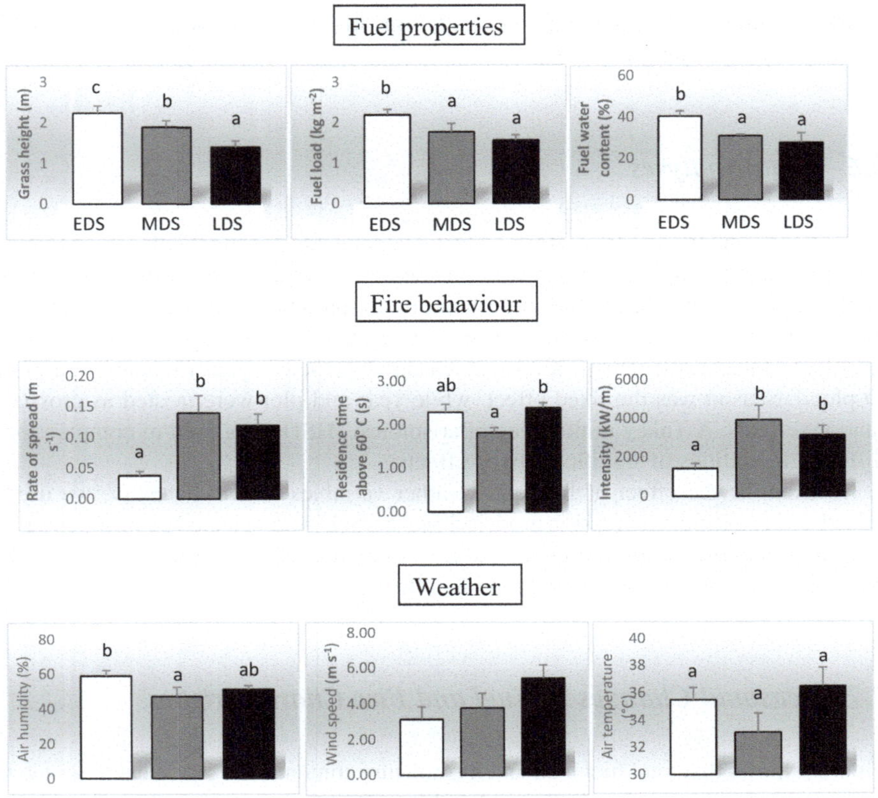

Fig. 3 Fire rate of spread **A**, intensity **B** and residence time **C** for the 2013–2018 period under the different fire seasons. EDS: Early Dry Season Fire, MDS: Mid-Dry season Fire; LDS: Late Dry Season Fire. Significant differences between fire treatments are indicated by different letters (Tukey HSD post-hoc tests)

All fire behavior variables—except residence time above 60 °C—showed the same pattern of response to the fire season: their values were lower during the early dry season (EDS) than during mid-dry season (MDS) and late dry season (LDS) fires ($P < 0.001$, $P = 0.003$, respectively for the rate of fire spread and fire intensity). There were no differences in the variables between MF and LF. The fire residence time was highest in the late dry season and lowest in the mid-dry season ($P = 0.021$). The early dry season fire presented the middle value (Fig. 3).

Based on fire intensity classes as defined by Morrison (2002), in Lamto, early dry season fire can be considered low intensity fire, while mid and late dry season fires can be classed as high intensity fires. In our study, both the rate of spread and intensity of fires increased as the dry season progressed.

2.8 Prediction of Fire Behavior with Fuel and Weather Characteristics

Table 1 shows that fire intensity (log I) and the rate of spread (log r) decreased with fuel moisture content and air humidity, explaining the 56% and 64% of the variability in these three aspects of fire behavior, respectively. Fire residence time above 60 °C (Rt) increased with fuel moisture content (FM), air humidity (AH) and air temperature (AT), and decreased with fuel load (FL) (51% of the variation explained). As for explanatory variables, fuel moisture content was a very important explanatory variable ($P < 0.001$) for 2 of the 3 variables that describe fire behavior, and somewhat important ($P < 0.10$) for a fourth variable (Table 1). Air humidity was also an important explanatory variable ($P < 0.01$ and $P < 0.05$) for 2 of the 3 variables describing fire behavior, and somewhat important ($P < 0.10$) for a fourth. These two explanatory variables were the sole significant predictors for fire intensity and rate of fire spread. They also appeared in models for fire residence time >60 °C (plus fuel load and air temperature). Fuel load was a significant predictor ($p < 0.01$) only in the model for fire residence time >60 °C.

The fuel characteristics and fire behaviors seen in our study are comparable to those reported by Afelu and Kokou (2015) in the transition zone between the guineo-congolese and the sudanian zones of Togo. The fuel loads in our humid savanna of Lamto were higher than those measured in the humid savannas of Northern Australia (Williams et al. 1998) and in the somewhat drier savannas of South Africa (Van Wilgen et al. 2007). However, in Lamto, the rate of fire spread was slower and the intensity of fire was lower than reported for the early and late dry season fires in those savannas. The difference in fire behavior could be due to the lower fuel moisture in Northern Australia and South Africa due to the single, much longer dry season. In Northern Australia, the very low moisture content of the fuel and the very dry air create favorable conditions for very rapid fire spread and higher fire intensity (Anderson and Anderson 2010; Fontaine et al. 2012). The comparison suggests that fuel moisture content is one of the main factors influencing fire behavior,

Table 1 Prediction of fire behavior variables from weather and fuel variables. Results show the best models found by stepwise backward multiple regressions with plot and year as crossed random effects (using the R *step* function of the *lmerTest* package). Response (predicted) variables: fire intensity (I), rate of fire spread (r), and fire residence time above 60 °C (Rt). Explanatory (predictors) variables: fuel load (FL), fuel moisture content (FM), grass height (GH), wind speed (W), air humidity (AH), air temperature (AT). *s.e.e.* = standard error of estimation. R^2 = coefficient of determination. F = F value

Response and explanatory variables	s.e.e.	R^2	F	*P-value*
$\text{Log}(I) = 10.61 - 0.0589\,FM^{***} - 0.0187\,AH$	0.5701	0.55	14.89	<0.0001
$\text{Log}(r) = 0.8036 - 0.0717\,FM^{***} - 0.0191\,AH^{*}$	0.5697	0.63	20.54	<0.0001
$Rt = 0.8694 - 0.7908\,FL^{**} + 0.0206\,FM. + 0.0221\,AH^{**} + 0.02496\,AT$	0.7276	0.51	5.74	0.002

$^{***}P < 0.001$; $^{**}P < 0.01$; $^{*}P < 0.05$; $.P < 0.1$; $P > 0.1$

further evidenced by the fact that fuel moisture is the most common and significant explanatory variable predicting various descriptors of fire behavior in our study. Our results are also consistent with those of Govender et al. (2006), who observed a doubling of fire intensity when fuel moisture decreased three times. In our study we observed the doubling of fire intensity when fuel moisture decreased by 12% (for 10% of fuel humidity, the fire intensity was 8800 kW/m; for 22% it was 4430 kW/m; and for 50% it was 540 kW/m).

Our results are also consistent with the conclusions of Govender et al. (2006) that the effects of seasonal moisture override that of fuel load. During the early dry season of our study, there were large quantities of live perennial fuel with high moisture content, but the fires had the lowest rate of spread and intensity (Fig. 3).

According to Hoffmann et al. (2012), the microclimate during fire determines flammability because of its effect on fuel moisture. In our study, air humidity was the only weather parameter that varied between seasons (with a higher average value during the early fire); i.e., this variable did impact fire behavior. In dry conditions, fires were faster and more intense while residence time was shorter. With climate change predicting increasingly long dry periods, we will expect more and more intense fires.

Contrary to reports from studies in other savannas that wind speed (Savadogo et al. 2007; Vega et al. 2012) and air temperature (Platt et al. 2015) strongly affect fire behavior, this was not the case in our study in the Guinean savanna of Lamto. These two parameters did not change over the seasons in our study, although fire behavior certainly did. Wind speed's lack of effect may be due to generally low wind speed in Lamto (Le Roux 1995), as reported in the Brazilian cerrado (Rissi et al. 2017); further, there are only minor variations in air temperature throughout the year (Le Roux 1995).

Our analyses indicating that seasonal fuel characteristics and weather conditions jointly contribute to determine fire behavior in these West African humid savannas is consistent with studies of Higgins et al. (2000) in southern African savannas, and Platt et al. (2015) in humid savannas of SE North America. In particular, in Lamto, fuel moisture and air humidity are highly predictive of fire behavior, and these variables are relatively easily measured in the field. The models presented in this study are novel and could be useful tools for assessing and monitoring various fire regimes in the humid savannas of West Africa. To date, no other work is available that can be used to predict fire behaviour in this region.

2.9 *Severity of Fire*

Figure 4 shows that sub-adults trees' recruitment to fire was significantly influenced by fire season ($P < 0.001$). The proportion of recruited to adult stage after five years was highest for EDS fire and there was no recruited after five successive LDS fires.

In our study, although there was no difference in fire behavior between fires set in mid-dry seasons and those set in late dry seasons, there were differences in fire

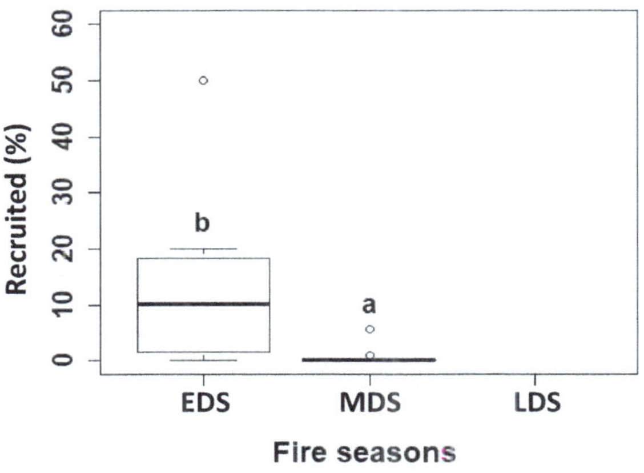

Fig. 4 Proportion of recruited sub-adults after five successive fires. EDS: Early dry season fire, MDS: Mid-dry season fire; LDS: Late dry season fire. Significant differences between fire treatments are indicated by different letters (Tukey HSD post-hoc tests)

severity (e.g., the effect on trees). Indeed, the late dry season fire prevented recruitment. This result is consistent with the previous studies in West African savannas, that showed the negative impact of late dry season fires on trees (Louppe et al. 1995; Laris et al. 2017). Similarly, the high impact of late dry season fires on trees is common in other savannas; most studies attribute this phenomenon to the higher fire intensity of late dry season fires (Williams et al. 1998; Savadogo et al. 2007; Hoffmann et al. 2019). We suggest that the differences in fire severity between mid and late-dry season fires (despite similar fire behavior) are most likely explained by the phenological state of trees. This was suggested as an explanation for humid savannas in Australia (Werner and Prior 2013) and SE North America (Robertson and Hmielowski 2014). In the Lamto savanna, all trees are leafless during the dry season so mid-dry season fire cannot cause significant damage to living tree tissues (i.e., the trunks and main branches are fire-resistant) (Gignoux et al. 1997; N'Dri et al. 2014). However, at the end of the dry season/beginning of the rainy season trees begin to set new leaves, so fire during this period could be very detrimental (Gignoux et al. 2009). In addition, the longer lethal temperature residence time during late dry season fires compared to mid-dry season fires could further explain the high impact of late dry season fire on trees.

Our results support the hypothesis that the generally observed high severity of late fires for savanna trees is not due to a higher intensity of fire.

3 Conclusion

In the humid savannas of Lamto that are representative of the Guinean savanna-forest mosaic ecoregion, up to 35–65% of the total variation of several metrics used to describe fire behavior can mainly be predicted by fuel and weather condition (fuel moisture and air humidity). In dry conditions, fires were faster and more intense.

With climate change predicting increasingly longer and drier periods, we expect more and more intense fires in the Guinean savanna. The specific qualities of the site (*e.g.* very low wind speed on average) may limit generalisation in the absence of a larger spatial and temporal data coverage, so further work using other sites is highly desirable. The ultimate goal is to improve fire management in West Africa.

Although there was no difference in fire behavior between fires set in the mid-dry season and those set in the late dry season, late dry season fires are more severe and will help to stop bush encroachment in the Guinean savanna. Further studies may be useful to analyze how the alternating application of mid-dry season fire (applied in Lamto since 1962) and late dry season fire (which prevents recruitment from resprouts to adult stage) can influence the recruitment of resprouts to an adult stage, in order to better manage vegetation in the Guinean savanna.

4 Recommendations

In the humid savannas of Lamto that are representative of the Guinean savanna-forest mosaic ecoregion, fire behavior can be predicted by fuel moisture and air humidity. The models presented in this study are novel and, to date, no other work is available that can be used to predict fire behavior in this region. These two parameters are relatively easily measured in the field and are related to weather. Therefore, climate change directly influences fire behavior. Late dry season fires in some years could help to reduce bush encroachment in that region.

References

Abbadie L, Gignoux J, Lepage M, Le Roux X (2006) Environmental constraints on living organisms. In: Abbadie L, Gignoux J, Le Roux X, Lepage M (eds) Lamto: structure, functioning and dynamics of a savanna ecosystem. Springer, New-York, pp 45–57

Afelu B, Kokou K (2015) Paramètres physiques d'évaluation du comportement des feux de végétation au Togo. Int J Biol Chem Sci 9(4):2091–2105

Anderson SAJ, Anderson WR (2010) Ignition and fire spread thresholds in gorse (Ulex europaeus). Int J Wildland Fire 19:589–598

Angassa A, Baars RM (2000) Ecological condition of encroached and non-encroached rangelands in Borana, Ethiopia. Afr J Ecol 38:321–328

Barger NN, Archer SR, Campbell JL, Huang CY, Morton JA, Knapp AK (2011) Woody plant proliferation in North American drylands: a synthesis of impacts on ecosystem carbon balance. J Geophys Res 116:G00K07. https://doi.org/10.1029/2010JG001506

Belay TA, Totland Ø, Moe SR (2013) Ecosystem responses to woody plant encroachment in a semiarid savanna rangeland. Plant Ecol 214:1211–1222

Berthrong ST, PIneiro G, Jobbágy EG, Jackson RB (2012) Soil C and N changes with afforestation of grasslands across gradients of precipitation and plantation age. Ecol Appl 22:76–86

Bond WJ (2008) What limits trees in C4 grasslands and savannas? Annu Rev Ecol Evol Syst 39:641–659

Byram GM (1959) Combustion of forest fuels. In: Davis KP (ed) Forest fire: control and use. Mc Graw-Hill, New York, pp 61–89

Case MF, Staver CA (2017) Fire prevents woody encroachment only at higher-than-historical frequencies in a South African savanna. J Appl Ecol 54:955–962

Criado MG, Myers-Smith IH, Bjorkman AD, Lehmann CER, Stevens N (2020) Woody plant encroachment intensifies under climate change across tundra and savanna biomes. Glob Ecol Biogeogr https://doi.org/10.1111/geb.13072

Dauget JM, Menaut JC (1992) Evolution sur 20 ans d'une parcelle de savane boisée non protégée du feu dans la réserve de Lamto (Côte d'Ivoire). Candollea 47:621–630

Durigan G, Ratter JA (2016) The need for a consistent fire policy for Cerrado conservation. J Appl Ecol 53:11–15

Dwomoh FK, Wimberly MC (2017) Fire regimes and their drivers in the upper Guinean region of West Africa. Remote Sens 9(11):1117. https://doi.org/10.3390/rs9111117

Fontaine JB, Vanessa CW, Enright NJ, Lade JC, Miller BP (2012) Fire behaviour in south-western Australian shrublands: evaluating the influence of fuel age and fire weather. Int J Wildland Fire 21:385–395

Frost PGH, Medina E, Menaut JC, Solbrig O, Swift M, Walker BH (1986) Responses of savannas to stress and disturbance. Biol Int Spec Issue 10:1–82

Gautier L (1990) Contact forêt-savane en Côte d'Ivoire centrale : Evolution du recouvrement ligneux des savanes de la réserve de Lamto (sud du V baoulé). Candollea 45:627–641

Gignoux J, Clobert J, Menaut JC (1997) Alternative fire resistance strategies in savanna trees. Oecologia 110:576–583

Gignoux J, Konaté S, Lahoreau G, Le Roux X, Simioni G (2016) Allocation strategies of savanna and forest tree seedlings in response to fire and shading: outcomes of a field experiment. Sci Rep 6(1). https://doi.org/10.1038/srep38838

Gignoux J, Lahoreau G, Julliard R, Barot S (2009) Establishment and early persistence of tree seedlings in an annually burned savanna. J Ecol 97:484–495

Gignoux J, Mordelet P, Menaut JC (2006) Biomass cycle and primary production. In: Abbadie L, Gignoux J, Le Roux X, Lepage M (eds) Lamto: structure, functioning and dynamics of a savanna ecosystem. Springer, New York, pp 115–135

Gill AM, Allan G (2008) Large fires, fire effects and the fire-regime concept. Int J Wildland Fire 17:688–695

Govender N, Trollope WSW, Van Wilgen BW (2006) The effect of fire season, fire frequency, rainfall and management on fire intensity in savanna vegetation in South Africa. J Appl Ecol 43:748–758

Gray EF, Bond WJ (2013) Will woody plant encroachment impact the visitor experience and economy of conservation areas? Koedoe 55(1):9. Art. #1106. https://doi.org/10.4102/koedoe.v55i1.1106

Guiguindibaye M, Belem MO, Boussim JI (2013) Caractéristiques des feux dans un incendie en savane soudanienne au Tchad. Int J Biol Chem Sci 7(3):1147–1156

Heubes J, Kühn I, König K, Wittig R, Zizka G, Hahn K (2011) Modelling biome shifts and tree cover change for 2050 in West Africa. J Biogeogr 38:2248–2258

Higgins S, Bond W, Trollope WSW (2000) Fire, resprouting and variability: a recipe for grass-tree coexistence in savanna. J Ecol 88:213–229

Hochberg ME, Menaut JC, Gignoux J (1994) The influences of tree biology and fire in the spatial structure of the West African savanna. J Ecol 82(2):217–226

Hoffmann WA, Jaconis S, Mckinley KL, Geiger EL, Gotsch SG, Franco AC (2012) Fuels or microclimate? Understanding the drivers of fire feedbacks at savanna–forest boundaries. Austral Ecol 37:634–643

Hoffmann WA, Sanders RW, Just MG, Wall WA, Hohmann MG (2019) Better lucky than good: how savanna trees escape the fire trap in a variable world. Ecology 101(1):e02895 https://doi.org/10.1002/ecy.2895

Kidnie S, Wotton BM (2015) Characterisation of the fuel and fire environment in southern Ontario's tallgrass prairie. Int J Wildland Fire 24:1118–1128
Kier G, Mutke J, Dinerstein E, Ricketts TH, Kuper W, Kreft H, Barthlott W (2005) Global patterns of plant diversity and floristic knowledge. J Biogeogr 32:1107–1116
Koffi KF, N'Dri AB, Lata JC, Konaté S, Srikanthasamy T, Konan M, Barot S (2018) Effect of fire regime on the grass community of the humid savanna of Lamto, Ivory Coast. J Trop Ecol 35(1):1–7
Koffi KF, N'Dri AB, Lata JC, Konaté S, Srikanthasamy T, Konan M, Barot S (2019) Effect of fire regimes on the demographic parameters of the perennial tussock grasses of a humid savanna. J Veg Sci 30:950–962
Laris P, Koné M, Dadashi S, Dembele F (2017) The early/late fire dichotomy: time for a reassessment of Aubréville's savanna fire experiments. Prog Phys Geogr 41:68–94
Lehmann CER, Prior LD, Bowman DMJS (2009) Decadal dynamics of tree cover in an Australian tropical savanna. Austral Ecol 34:601–612
Le Roux X (1995) Survey and modelling of water and energy exchanges between soil, vegetation and atmosphere in a Guinea savanna. Ph.D. Dissertation, Université de Paris 6
Louppe D, Ouattara N, Coulibaly A (1995) The effects of bush fires on vegetation: the Aubreville fire plots after 60 years. Commonw for Rev 74(4):288–292
Mahé G, Rouché N, Dieulin C, Boyer J-F, Ibrahim B, Crès A, Servat E, Valton C, Paturel J-E (2012) Carte des pluies annuelles en Afrique = annual rainfall map of Africa. Montpellier (FRA); Bondy: IRD; IRD, non paginé
Morrison DA (2002) Effects of fire intensity on plant species composition of sandstone communities in the Sydney region. Austral Ecol 27:433–441
N'Dri AB, Gignoux J, Barot S, Konaté S, Dembélé A, Werner P (2014) The dynamics of hollowing in annually burnt savanna trees and its effect on adult tree mortality. Plant Ecol 215:27–37
N'Dri AB, Soro TD, Gignoux J, Dosso K, Koné M, Koné NA, N'Dri JK, Barot S (2018) Season affects fire behavior in annually burned humid savanna of West Africa. Fire Ecology 14(5):1–11
O'Connor TG, Puttick JR, Hoffman MT (2014) Bush encroachment in southern Africa: changes and causes. Afr J Range Forage Sci 31(2):67–88
Platt WJ, Orzell SL, Slocum MG (2015) Seasonality of fire weather strongly influences fire regimes in South Florida savanna-grassland landscapes. PLoS ONE 10(1):e0116952
Prat-Guitart N, Rein G, Hadden RM, Claire M, Belcher D, Yearsley JM (2016) Propagation probability and spread rates of self-sustained smouldering fires under controlled moisture content and bulk density conditions. Int J Wildland Fire 25:456–465
Prichard SJ, Stevens-Rumann CS, Hessburg PF (2017) Tamm review: shifting global fire regimes: lessons from reburns and research needs. For Ecol Manage 396:217–233
Ratajczak Z, Nippert JB, Collins SL (2012) Woody encroachment decreases diversity across North American grasslands and savannas. Ecology 93:697–703
Rissi MN, Baeza MJ, Gorgone-Barbosa E, Zupo T, Fidelis A (2017) Does season affect fire behaviour in the Cerrado? Int J Wildland Fire 26:427–433
Robertson KM, Hmielowski TL (2014) Effects of fire frequency and season on resprouting of woody plants in southeastern US pine-grassland communities. Oecologia 174:765–776
Savadogo P, Zida D, Sawadogo L, Tiveau D, Tigabu M, Odén PC (2007) Fuel and fire characteristics in savanna–woodland of West Africa in relation to grazing and dominant grass type. Int J Wildland Fire 16:531–539
Sharp BR, Whittaker RJ (2003) The irreversible cattle driven transformation of a seasonally flooded Australia savanna. J Biogeogr 30:783–802
Sidzabda DD, Savadogo P, Zida D, Sawadogo L, Tiveau D, Oden PC (2010) Fire temperature and residence time during dry season burning in a Sudanian savanna-woodland of West Africa with implication for seed germination. J for Res 21(4):445–450
Smit IPJ, Asner GP, Govender N, Vaughn R, Van Wilgen BW (2016) An examination of the potential efficacy of high-intensity fires for reversing woody encroachment in savannas. J Appl Ecol 53(5):1623–1633

Soro Y, N'Dri AB, Bakayoko A, Gignoux J (2018) Analyse de la végétation dans un écotone forêt-savane d'Afrique de l'Ouest dans un contexte de boisement des savanes. Revue De L'environnement Et De La Biodiversité-PASRES 3:54–72

Stevens N, Erasmus BFN, Archibald S, Bond WJ (2016) Woody encroachment over 70 years in South African savannahs: overgrazing, global change or extinction aftershock? Philos Trans Roy Soc B 371:20150437. https://doi.org/10.1098/rstb.2015.0437

Swetnam TW, Allen CD, Betancourt JL (1999) Applied historical ecology: using the past to manage for the future. Ecol Appl 9(4):1189–1206

Tiemoko DT, Yoroba F, Diawara A, Kouadio K, Kouassi BK, Yapo ALM (2020) Understanding the local carbon fluxes variations and their relationship to climate conditions in a sub-humid savannah-ecosystem during 2008–2015: case of Lamto in Cote d'Ivoire. Atmos Clim Sci 10(02):86–205

Trollope WSW (1983) Control of bush encroachment with fire in the arid savannas of south-eastern Africa. Ph.D. Dissertation, University of Natal, Pietermaritzburg

Van Wilgen BW, Govender N, Biggs HC (2007) The contribution of fire research to fire management: a critical review of a long-term experiment in the Kruger National Park, South Africa. Int J Wildland Fire 16:519–530

Van Wilgen BW, Trollope WSW, Biggs HC, Potgieter ALF, Brockett BH (2003) Fire as a driver of ecosystem variability. In: du Toit JT, Rogers KH, Biggs HC (eds) The Kruger experience. Island Press, London, pp 149–170

Vega JA, Jiménez E, Dupuy JL, Linn RR (2012) Effects of flame interaction on the rate of spread of heading and suppression fires in shrubland experimental fires. Int J Wildland Fire 21:950–960

Werner PA, Franklin DC (2010) Resprouting and mortality of juvenile eucalypts in an Australian savanna: impacts of fire season and annual sorghum. Aust J Bot 58:619–628

Werner PA, Prior LD (2013) Demography and growth of subadult savanna trees: interactions of life history, size, fire season, and grassy understory. Ecol Appl 83(1):67–93

Williams RJ, Gill AM, Moore PHR (1998) Seasonal changes in fire behaviour in a tropical savanna in Northern Australia. Int J Wildland Fire 8(4):227–239

Yapo ALM, Diawara A, Kouassi BK, Yoroba F, Sylla MB, Kouadio K, Tiémoko DT, Koné DI, Akobé EY, Yao KPAT (2020) Projected changes in extreme precipitation intensity and dry spell length in Côte d'Ivoire under future climates. Theor Appl Climatol 140:871–889

COVID-19 Pandemic: A Threat Towards Achieving the Sustainable Development Goal on Combating Climate Change and Its Impacts in Zimbabwe's Rural Communities

Nyahunda Louis and Tirivangasi Happy Mathew

Abstract This chapter explores the threat of the coronavirus COVID-19 pandemic on achieving the Sustainable Development Goal (SDG, 13) on combating climate change and its impacts in the Makonde communal area in Zimbabwe. The outbreak of the corona virus disease-2019 (COVID-19) pandemic ignited the pronouncement of lockdown measures in Zimbabwe to curb the transmission of the virus while preparing for the management of the possible surgeNo. This posed some threats towards the implementation of SDGs dedicated to combat climate change and its impacts (SDG,13). This means a backlash on sustainable development progress made towards climate change adaptation and mitigation in Zimbabwe. The objective of the study was to explore the effects COVID-19 on climate action in Zimbabwe both at national and local levelsNo. The study was qualitative in nature where ten participants were selected through the purposive and convenience sampling techniques. Data collection was done through individual interviews with community members. The study also benefited from the review of secondary data, journal articles and national reports. The Discourse Content Analysis was followed to analyse the data from which the findings were derived. This chapter established that climate action in Zimbabwe's rural communities has been impeded by the outbreak of the COVID-19 pandemic. Thus, the COVID-19 pandemic dismantled efforts designed to combat climate change and its impacts. This poses a ripple effect on other SDGs, (1, ending poverty) and (2, ending hunger) that thrive when climate change and its impacts are well managed. The study recommends that, efforts towards combating climate change should also continue even in the wake COVID-19 climate change crisis poses danger to humanity in various ways if it remains unattended.

Keywords COVID-19 · SDGs · Climate change · Adaptation · Rural Zimbabwe

N. Louis (✉)
Unit of Institutional Change and Social Justice, University of the Free State, P O Box 339, Bloemfontein 9300, South Africa

T. H. Mathew
Department of Research Administration and Development, University of Limpopo, Private Bag X1106, Sovenga, Mankweng 0727, South Africa

W. Leal Filho et al. (eds.), *Sustainable Development in Africa*, World Sustainability Series,
https://doi.org/10.1007/978-3-030-74693-3_28

1 Introduction and Background

The outbreak of the COVID-19 pandemic when the world is grappling meeting targets of the United Nations Sustainable Development Goals (SDGs) also known as Agenda 2030 for sustainable development connotes that the SDGs process is being hampered. Since the outbreak of the COVID-19 pandemicNo, the SDGs process including the drive to combat climate change and its impacts (SDG 13) had been shelved, as countries prioritise contingent responses to the disruptions and threats posed by COVID-19 over other aspirations (Norouzi et al. 2020). The SDGs are a product of the collective global developmental initiatives by member states of the United Nations. These goals are premised on meeting the needs of the present and the incoming generation by laying a foundation where all transformation trajectories should be built from (Nilsson et al. 2016). As such, sustainable development is anchored on social inclusion, economic growth and environmental protection (Rosen 2017; Esquivel and Sweetman 2016; Leal Filho et al. 2020; Manzanedo et al. 2020). From the viewpoint of Peter and Jones (2020), SDGs represent how decent life for all people can be safeguarded. What should be noted is that, the SDGs are still to be evaluated by 2030 and a cursory assessment is going to be made based on what they are targeted to achieve, what they would have achieved and where they would have failed to yield expected outcomes (Oldekop et al. 2020). In this regard, it is important to note that the outbreak of the COVID-19 pandemic and its daunting impacts has clouded the public regard of climate crisis as one of the key priorities of the SDG 13.

Bashir et al. (2020) see climate change as a serious threat to human life just like COVID-19. As such, In this regard, taking urgent action to combat climate change and its impacts is should remain a key priority even in the wake of health emergencies such as COVID-19 (Oldelop et al. 2020). That being said, this chapter delved on the impacts of COVID-19 on climate change aspirations (SDG 13). The Intergovernmental Panel on Climate Change (IPCC 2016) revealed that climate change has a dramatic effect as it can lead to poverty, adverse health impacts, food insecurity and inequality (IPCC 2016). These variables constitute different goals for sustainable development (SDG 1, ending poverty, SDG2, food security, SDG 3, good health and SDG 10, ending inequality). What should be noted is that, almost every goal of the Agenda 2030 thrives when climate change and its impacts are well managed (Chiweshe and Mutondoro 2017).

What should also be highlighted is that the precarious impacts of COVID-19, climate change induced risks remain a threat to human life and livelihoods. As such, the manifestations of climate change through tornados, hurricanes, cyclones, veld fires, floods and drought has affected more than 50 million people worldwide between 2013 and 2019 causing deaths, injuries, destruction of property and human displacements (Handmer and Nalau 2019; Eckstein et al. 2019; Gori et al. 2020). The COVID-19 collided with these climate change emergencies and this has rendered climate action insufficient. The focus of this chapter on COVID-19 and climate change is informed by the acknowledgement that effective climate action begets proper management of health emergencies in the quest for sustainable development (Norouzi et al. 2020). In the foregoing, despite acute deaths and disruption of livelihoods, COVID-19 has also caused rapid socio-economic repercussions mostly in developing countries which poses a precarious threat towards their sustainable development aspirations particularly on combating climate change (Gori et al. 2020). Also, the economic crisis induced by COVID-19 is derailing the efforts of combating climate change towards achieving sustainable development especially in countries like Zimbabwe that are trapped under economic quagmires (Peters et al. 2020).

Before the COVID-19 pandemic, Zimbabwe was already grappling with the climate change impacts in every sector namely agriculture, health, water, energy, tourism and forestry resources (Chanza and Gundu-Jakarasi 2020). On the same note, Zimbabwe was saddled with economic woes where service delivery was in shambles owing to abysmal corruption, underinvestment in key infrastructures and institutions including health care facilities (Chanza and Gundu-Jakarasi 2020). It is of paramount importance to note that the economic challenges bedevilling the country serve as a backlash towards the realisation of the SDGs particularly SDG 13 on climate action. The outbreak of the COVID-19 pandemic exacerbated the worst-case scenario in a country without modest disaster risk preparedness modalities for health and climate related emergencies (Dzobo et al. 2020). Consequently, climate action aspirations were shelved as funding from the international community was redirected in revamping the health care facilities to cater for COVID-19 emergencies (Dzobo et al. 2020). Just like any disaster where vulnerable populations are at risk, COVID-19 is taking a toll on rural people in Makonde communal lands in the Makonde Rural District in the Mashonaland West Province of Zimbabwe. The people in this district have been grappling with climate change impacts for some time and their condition is aggravated by abject poverty and low adaptive capacities (Nyahunda et al. 2019). Their circumstances are like many other rural communities in Zimbabwe and this is aggravated by the government's failure to effectively implement the targets of SDGs towards climate change action despite the visibility of these complementary catastrophes (Nyahunda and Tirivangasi 2020). As such, the outbreak of the COVID-19 pandemic in the wake of climate change impacts is a compound disaster for the rural people who can't afford to recover from climate change aberrations alone. What this also means is that apart from COVID-19 threatening national efforts of combating climate change and its impacts, COVID-19 also presents numerous challenges to climate change adaptation efforts designed by rural people. This scenario has a devastating effect on the realisation of sustainable development. Against this backdrop, this chapter demonstrated how the COVID-19 pandemic is serving as a threat towards achieving sustainable development in Zimbabwe and its rural communities. The objective of the study was to explore the effects of COVID-19 on climate action in Zimbabwe at national level and in rural communities.

2 Literature Review

2.1 Overview of the COVID-19 Pandemic

The Corona Virus disease 2019 commonly known as COVID-19, caused by the severe acute respiratory syndrome (SARS-CoV-2) virus, originated in the Hubei province of Wuhan in China in December 2019. The symptoms of the COVID-19 include pneumonia, respiratory infection, respiratory failure, dizziness, sore throat, fatigue, fever and pulmonary failure and can cause death (Lu et al. 2020). This deadly virus is mostly transmitted in droplets released from an infected person's nose or mouth (Chen et al. 2020). Despite China being the cradle of the virus, in no time, COVID-19 had pervaded most parts of the world so much that, by 11 March 2020, more than 170 countries had recorded COVID-19 infected cases including deaths. At the time of writing this chapter, the global COVID-19 statistics were at 12.5 million confirmed cases with 6.89 recoveries and 560 000 deaths (WHO 2020). The lethality of the virus and its unpredictability prompted the World Health Organisation (WHO) to declare COVID-19 a global health pandemic. This was complemented by the pronouncements of health precautionary measures implemented across the world to curtail the transmission of the virus (WHO 2020; Chen et al. 2020).

In response to the health guidelines promulgated by the World Health Organization, Zimbabwe like many other countries pronounced its nationwide lockdown in March 2020. The aim was to curb the transmission of the disease among other contingent plans designed to manage the possible surge of the virus. As such, the nationwide lockdown in Zimbabwe resulted in restrictions on people's movements except for essential purposes, closure of various sectors and borders for human traffic (Dzobo et al. 2020). As at 10 July 2020, Zimbabwe had recorded 1362 cumulative cases with 23 deaths and 425 recoveries where most cases were identified on returnees from neighbouring countries especially South Africa which was the epicentre of the virus in the African continent (WHO 2020; Makurumidze 2020). Like many countries, Zimbabwe eased some of its lockdown restrictions in a bid to strike a balance between saving lives and the economy, but the blow of the pandemic had already taken a devastating posture in many sectors that are essential for the implementation of the SDGs.

What should be highlighted is, the outbreak of the COVID-19 pandemic and its unprecedented impacts in a country that was already marred with governance and developmental crisis means issues of sustainable development are far from being realised (Nyahunda and Tirivangasi 2020). In light of this, Mackworth-Young et al. (2020) opine that like in many developing countries, the impact of the COVID-19 pandemic on human lives and the economy in Zimbabwe is severe. This is also a concern for sustainable development particularly in rural communities where poverty levels are increasing and climate crisis such as food and water shortages are worsening. Leal Filho et al. (2020) denote that the scope of the COVID-19 pandemic and its economic repercussions pose a serious threat towards the realisation of the

SDGs. As mentioned earlier, the focus of this chapter then, is on the sustainable development facets related to climate change and its impacts.

3 Climate Change and Sustainable Development Goals

Zimbabwe like many other countries participated in the creation of the Agenda 2030 for sustainable development. As such, Zimbabwe recognizes the critical contribution of climate change in stalling its developmental efforts. In response, the country has mainstreamed the SDGs in its national developmental aspirations though the implementation process is still stymied with challenges (Mutangabende and Shava 2016; Chikweshe and Mutondoro 2017; Zhakata et al. 2017). According to Kokotovic et al. (2019), the SDGs are a set of 17 aspirational goals complemented by 169 targets adopted by the United Nations member states in 2015. Nilsson et al. (2016:4) denote that sustainable development refers to the developmental aspirations that meet the needs of the present without compromising the ability of future generations to meet their own needs. The focus of this chapter is on goal 13of the SGDs which is dedicated to taking urgent action to combat climate change and its impacts (Mugambiwa and Tirivangasi 2017; Kokotovic et al. 2019). Chanza (2014) affirms that climate change and sustainable development are indissoluble. As such, in crafting the SDGs, member parties firmly acknowledged the need to increase the fight against climate change as encapsulated in the primary international forums such as the United Nations Framework Convention on Climate Change (UNFCCC) and the Kyoto Protocol (Roberts and Huq 2015; Mutangabende and Shava 2016).

In addition, through the UNFCCC and the Kyoto Protocol which Zimbabwe ratified implore member states to employ strategic mechanisms to mitigate and adapt to climate change within available means. These conventions further implore member states to consider their social, economic, and environmental circumstances to intensify investments and actions pertinent for sustainable low carbon-based economies (Khan and Roberts 2013; UNFCCC 2014). Subsequently, the Paris Climate Agreement came into being in December 2015 to reinforce the pressing need to fight against climate change and its impacts. Accordingly, the Paris Climate Agreement is aimed at solidifying collective global efforts in addressing the inevitable climate change impacts within the ambit of the SDGs (Rogelj et al. 2016). In the foregoing discussion, Srivastava et al. (2020) highlights that summarily, the focus of SDGs is on five pillars namely; people, prosperity, peace, planet and partnerships as distributed across the 17 goals. Based on this, different governments allocate budget for different industries to achieve these goals. In this chapter the authors argue that the outbreak of the COVID-19 pandemic made governments to neglect climate issues by focusing on people only whilst other pillars are neglected particularly the one dedicated to caring for the planet where responsible consumption (goal 12), climate action (goal 13), life below water (goal 14) and life on land (goal 15) are indicated.

In support of the above exposition, Chanza (2014) opined that climate change and its impacts is adding to other entrenched drivers of underdevelopment in most parts of

Zimbabwe. What this means is that, the outbreak of the COVID-19 pandemic exacerbated the situation in a country grappling with economic woes, weak institutions and lack of meticulous disaster recovery mechanisms among other backlashes towards sustainable development (Nyahunda and Tirivangasi 2020). In this view, Jones and Comfort (2020) infer that the COVID-19 scourge dovetails a myriad range of environmental, social and economic catastrophes which have an impact on sustainable development. Similarly, Srivastava et al. (2020) posit that the COVID-19 pandemic has reinforced the connectivity between environment, health, and the economy. The threats posed by the COVID-19 pandemic towards sustainable development premised on taking climate action are delineated in the forthcoming section.

4 Materials and Methods

This study was a qualitative enquiry guided by the case study design. Purposive and convenience sampling techniques were adopted to select Makonde communal lands in Zimbabwe as the focal point of the study. The researchers purposively selected Godzi, Kamhonde and Hobwe communities in Makonde area as the focal points of the study. The same sampling techniques were followed to select ten participants from these communities to give their insights on the subject matter through individual interviews. As such, these participants provided insights on climate change adaptation activities that were affected by the outbreak of COVID-19. The health precautionary measures such as physical distancing, wearing of face masks and hand sanitizing were adhered to during the interaction with the participants. This chapter also benefited from the review of secondary data where the search of literature, national reports, journal articles and policy documents were guided by; COVID-19, climate change and SDGs as key words on google scholars search. This was meant to find out climate processes that were halted by the outbreak of the COVID-19 pandemic. The gathered data were analysed following the Discourse Content Analysis.

5 Findings and Discussion

5.1 Effects of COVID-19 Towards Climate Action

Climate action refers to mechanisms employed to minimise the occurrence of the climate change phenomenon or to manage its impacts through mitigation and adaptation as contemplated by SDG 13 (Chanza 2014). Notably, in the climate change discourse, the terms mitigation and adaptation are dominant as they harness actions or measures taken to combat climate change and its impacts (IPCC 2016). The UNFCCC (2014) defines climate change mitigation as strategies employed by human systems to reduce greenhouse gasses and emissions that cause climate change. Chanza and

Gundu-Jakarasi (2020) defines climate change adaptation as response strategies employed by human or natural systems to reduce exposure to climate change risks and to minimise harm. It should be highlighted that collective actions to combat climate change and its impacts as contemplated in Agenda 2030 for sustainable development are hinged on mitigation and adaptation at global, national and local levels. Arguably, the measures taken in responding to the COVID-19 pandemic were necessary, but climate action should have not been neglected. In the forthcoming subsections, the authors shall discuss the effects of COVID-19 on climate action.

Suspension of the Conference of the Parties (COP26)
Since the advent of the UNFCCC and the Kyoto Protocol in the 1990s, member parties have been meeting yearly on periodic conferences under the auspices of (COP 1–25) to assess the progress in dealing with climate change (Vanhala and Hestbaek 2016; Hermwille et al. 2017; Cozier 2019). It is important to note that, after the COP25 held in 2019, the COP26 of the UNFCCC for 2020 slated for November 2020 was suspended owing to the pandemonium caused by COVID-19. This development placed the agenda for fighting climate change under threat (Srivastava et al. 2020). This was done to protect the delegates and observers from contracting COVID-19 (WHO 2020). The suspension of the COP26 means then that, some countries are going to reprioritise their climate change commitments (Srivastava et al. 2020). Initially, the UNFCCC, the Kyoto Protocol which were later reinforced by the Paris Climate Agreement emphasised the importance of strategic measures to curtail climate change impacts through crafting, implementing and the regular publishing and updating of regional and national measures fostered to mitigate climate change effects complemented by facilitation of sustainable climate change adaptation (Zhakata et al. 2017; Nyahunda and Tirivangasi 2020).

Furthermore, the above-mentioned international conventions set out guidelines for the development of National Adaptation Programmes (NAPs), National Adaptation Programme of Action (NAPAs), Nationally Determined Contributions (NDCs) and Nationally Appropriate Mitigation Actions (NAMAs) by all member states in addressing the impact of climate change. On the same note, member parties should periodically update and submit their adaptation strides, describing their needs, priorities and actions (UNFCCC 2014; Rogelj et al. 2016). Reasoning from this, the suspension of this crucial forum connotes that some targets made during the advent of the agreements of keeping or reducing average temperatures below 2 degrees Celsius through the Reduced Emissions from Deforestation and Degradation (REDD) and REDD plus mechanisms; Land Use, Land Use Change and Forestry (LULUCF) among other measures by 2020 will remain hanging in a balance owing to the disturbances caused by the COVID-19 pandemic (Rajamani 2016; Nhamo et al. 2020).

In continuation, Srivastava et al. (2020) denotes that the delays or postponements might allow some countries to reprioritise their environmental commitments. This is because this conference would have compelled the member parties to develop COVID-19 economic recovery strategies in tandem with climate change goals. On the side-lines of the COP26, Landon (2020) posit that the World Conversation Congress which was slated for June 2020 was postponed to 2021 owing to the COVID-19 conundrums. This conference would have established mechanisms and global regulations dedicated to protecting wildlife and plants from climate change. For the above-mentioned reasons, the authors argue that COVID-19 stands to jeopardize the process of implementing the SDG 13. This is because world leaders won't be able to affirm their commitments to implement the SDGs as they are grappling with COVID-19 related exigencies.

Misdirected Priorities from Climate Change Policies
Lessons drawn from the emergence of the COVID-19 scourge is that long term aspirations of fighting against climate change and its impacts as contemplated in the SDGs have been halted in pursuit of short-term contingent measures targeted at combating the spread of COVID-19. Earlier observations made by Grubb (2014) reveal that negating the implementation of climate change policies would have adverse environmental consequences and sustainable development. According to Agrawala et al. (2020) the economic ramifications of COVID-19 in developing countries may lead to delays or underinvestment in climate change policies dedicated to low carbon economies and renewable energy sources. Notably, in Zimbabwe, the commotion created by COVID-19 is dampening the prospects of updating the country's NDCs and further implement the commitments to reduce or condense greenhouse gas emissions. This is because the processes of updating the NDCs involve broad-based consultation processes between different stakeholders which is not feasible during the COVID-19 induced lockdowns.

Furthermore, Zimbabwe like many other countries has a gamut of climate change policies promulgated in tandem within the dictates of the UNFCCC, Kyotol Protocol, Paris Climate Agreement and the SDGs which were devoid of implementation prior to the outbreak of the COVID-19 pandemic (Nyahunda et al. 2019; Nyahunda and Tirivangasi 2020). The net effect of the outbreak of the COVID-19 pandemic in the Zimbabwean context is that climate policy implementation prospects might be shelved or delayed. The socio-economic pressures dovetailed by COVID-19 conundrums in countries like Zimbabwe with archaic preparedness for health pandemics and natural disasters means funds dedicated for climate change policy interventions would be diverted to withstand the COVID-19 pressure. This means that climate change mitigation, adaptation and resilience which are the hallmark of climate action under the SDGs remain hanging in a balance. In the same vein, climate awareness programmes which are contact based might remain suspended in Zimbabwe owing to the lockdown restrictions of COVID-19.

In the ongoing discussion, the visibility of Non-Governmental Organisations and Civil Society Organisations charged with climate change related interventions (World Vision Zimbabwe, PLAN International and CARE International, to mention a few) is now scant. This is because these organisations have reprioritised their services in response to the COVID-19 scourge in the communities where they operate. The virtual consultative processes are not feasible due to technological backlashes particularly in rural areas where climate change actions need to be intensified. It is also important to note that most business firms and companies might miss their targets of declaring their carbon emissions reduction efforts. This could be attributed to the reprioritisation of commitments as they grapple with the economic backlashes posed by the pandemic. Lastly, a series of the Zimbabwean government's post cabinet media briefings aired by the Zimbabwe Broadcasting Corporation (ZBC) on every Wednesdays since the outbreak of the COVID-19 scourge pointed to decisions dedicated to reviewing of lockdown measures and other political grandstanding postures whilst disregarding other important issues such as climate action.

Disruption of Scientific Research
The commitments made by member parties to the UNFCCC, the Kyotol Protocol, the Paris Climate Agreement and other conventions to periodically update and submit their climate change adaptation and mitigation strides, needs and priorities thrive on scientific research (Khan and Roberts 2013; Duruji et al. 2018). However, the imposed lockdowns and stringent restrictions such as human travel bans, and gatherings means climate change scientists face challenges in collecting and collating data that informs climate change interventions and information towards the commitments to be made at national and global levels. The Zimbabwean government created Local Area Adaptations Plans (LAPs), District Adaptation Plans (DAPs) and Provincial Adaptation Plans (PAPs) in tandem with the dictates of the UNFCCC and the Paris Climate Agreement (Zhakata et al. 2017). These provide a database of weather conditions and events in every area, past and present climatic trends and their risks, the area vulnerability, projections of future climatic scenarios and resources in the areas that can be used for effective planning that builds resilient communities to climate change impacts (Dodman and Mitlin 2015).

The above-mentioned measures benefit from research collaborations between different stakeholders such as; government, civil society organizations, private sector, women, youths and farmers just to mention a few (Zhakata et al. 2017; Mugambiwa 2018). The impediments brought about by COVID-19 which necessitated lockdown restrictions annihilated research activities in Zimbabwe and beyond. This poses a threat by creating some gaps in data and important observations cardinal to inform the Paris Climate Agreement on progress made in cutting carbon gas emissions. The effect of this development is that, climate change impacts will remain ravaging communities unabated which in turn has a boomerang effect on sustainable development commitments. This is because low carbon-based commitments might remain unmet or delayed. Furthermore, rural communities that have been placed at the peripheries of developmental initiatives whilst impacted by climate change were the hardest hit by such calamitous developments evolving from COVID-19.

6 Impact of COVID-19 on Rural People's Climate Action Efforts

This chapter documented empirical details from Makonde communal lands on how the outbreak of COVID-19 threatened the attainment of sustainable development premised on combating climate change and its impacts. In Makonde communal lands, like in most rural communities, the impacts of climate change are visible and felt by smallholder farmers who depend on rain fed agriculture for livelihood (Sango and Nhamo 2015; Nyahunda et al. 2019). As such, in Makonde communal lands, climate change impacts are manifesting through decreased crop production leading to drought, heat waves, deaths of livestock, distribution of pathogens that cause diseases, erratic rainfall patterns, and decimation of water sources for household use and for livestock and depletion of the natural resources (Mugambiwa and Rukema 2019). The researchers documented evidence of climate change impacts in Godzi, Kamhonde and Hobwe communities where they established that these communities are vulnerable to the vagary impacts of climate change aggravated by limited economic resources, lack of technological advancements and institutional support, low adaptive capacity exacerbated by high poverty levels. The outbreak of COVID-19 and the lockdown measures meant that these people were surviving on limited resources and they could not diversify their income streams. As mentioned earlier, climate action contemplated by SDG 13 connotes combating climate change and its impacts through mitigation measures that reduce carbon gas emissions or managing its impacts through adaptation (UNFCCC 2014; Chanza and Gundu-Jakarasi 2020). It is of paramount importance to note that rural people in Makonde communal lands have not remained passive victims to climate change and its vagary impacts, but they have responded to it through several mitigation and adaptation measures either individually or collectively. Some of the common climate change response strategies in Makonde communal lands include; crop diversification, change of planting dates, migration, livelihood diversification, migration, use of indigenous knowledge systems for weather predictions and planting seasons and destocking of livestock through selling or slaughtering for consumption. This chapter established that, the outbreak of the COVID-19 pandemic dismantled efforts dedicated to combating climate change and its impacts in rural communities. In Makonde communal lands, the COVID-19 pandemic emerged at a time when people were dealing with climate change repercussions such as drought, water shortages and multifarious incidence of diseases (malaria) among other daunting calamities. What this means is that, efforts invested for climate action would be divided to withstand the impacts of COVID-19. At this point, the authors captured a few historical strategies that were dedicated for climate action which were impeded by the outbreak of the COVID-19 pandemic.

Livelihood Diversification

In a bid to manage the impacts of climate change and before the outbreak of COVID-19, rural people in Makonde communal lands were employing a cocktail of strategies to cushion their livelihoods from climate change impacts. In this regard, they diversified their livelihoods through; petty trading of farm produce, cross boarder trading, harnessing of social capital and conservation farming among other climate change response strategies. From this, what should be highlighted is that the usual livelihoods fostered to withstand the impacts of climate change were disrupted by the outbreak of the COVID-pandemic. Gukurume (2013) observes that the common climate change adaptation strategy in Zimbabwe's rural communities is livelihood diversification (Ndebele-Murisa and Mubaya 2015; Nyahunda et al. 2019; Nyahunda and Tirivangasi 2019). The IPCC (2016) denote that climate change is a threat to human lives and livelihoods. In light of this, Serrat (2017) define livelihoods as activities, assets and capabilities used as means of living. The COVID-19 induced lockdown restricted transportation and movement of people hence informal livelihood activities that involve hawking, vending and long-distance trading where most rural people are predominant were annihilated (Mhlanga and Ndlovu 2020). The net effect of such development is that, climate action dedicated to managing climate change impacts through diversification of activities was dismantled for rural people in Makonde communal lands.

It is also of paramount importance to note that managing climate change impacts is acknowledging that the climate system has changed and there is need to adapt to its risks. Rural people in Makonde communal lands submitted that they were taking various adaptation strategies because of failed harvests caused by climate change. The participants lamented that the outbreak of COVID-19 affected their livelihood diversification strategies and the government was passive to assist them to adapt to both COVID-19 and climate change impacts. According to Dzobo et al. (2020) the concomitant response to the outbreak of COVID-19 by the Government of Zimbabwe through the lockdown decrees which were not accompanied by safety nets and relief packages for the rural poor means they are suffering a double blow from COVID-19 and climate change. The disruption of livelihood activities in Makonde communal lands has lured economic insecurities, oscillating poverty levels and burgeoning food insecurity for rural people. This means that apart from SDG 13, SDG 1 for poverty and SDG 2 for hunger are also under threat from the COVID-19 scourge. Furthermore, economic security complemented by technoscience technologies are some of the important determinants for climate action in as much as they are missing in Zimbabwe's rural communities such as Makonde communal lands (Nyahunda et al. 2019). The outbreak of the COVID-19 pandemic exacerbated the situation and it has the potential to reverse the efforts made by rural people towards climate action.

In the ongoing discussion, the lethality and unpredictability of COVID-19 has instilled fear and anxiety especially in rural communities where there are high information gaps (Van Bavel et al. 2020). As a result, rural people in Makonde communal lands are reported to have suspended major activities that require intensive labour (weeding and forestry conservation activities) because they believe that by so doing, they were compromising their immunity. It is also important to alluce that health is a strong determinant for people to act against climate change be it through diversification of activities meant to mitigate or adapt to climate change. Having said that, the rapid spread and infection rate of the COVID-19 pandemic and its related fears has made people to disregard activities such as trading in highly populated markets that may place them at risk of infection. This implied that climate action is no longer treated as an exigency matter and this poses a serious threat towards the attainment of SDG 13.

Harnessing of Social Capital
Social capital is one of the resources available to individuals within Makonde communal lands. The effectiveness of social capital in fostering cooperation and associational activities towards finding common solutions to problems from climate change is widely acknowledged by most community members (Nyahunda and Tirivangasi 2021). Traditionally, rural people are on record of embracing social capital in taking action towards mitigation and adaptation responses to climate change and its impacts (Gukurume 2013; Chanza 2014; Nyahunda and Tirivangasi 2019). Gukurume (2013) defines social capital as a social web of networks created through interaction which strengthens trust and social norms. The social networks can be created through friends, kinship and neighbourhood. The utilization of social capital in rural communities as a determinant for climate change mitigation and adaptation in Makonde communal lands commonly manifest through people converging in groups to perform tasks such as planting of trees, ploughing and harvesting (Petzold and Ratter 2015). On the same note, rural people can harness social capital through sharing of information and scarce resources such as water, food, skills and problem-solving knowledge. Furthermore, participants submitted that they could harness social capital for sharing disaster management solutions and best practices aimed at natural resource conservation and creation of collective norms to build resilience against climate change (Thamga-Chitja and Tamako 2017). Contextually, COVID-19 is highly contagious, and it spreads through human to human transmission, hence, it requires physical distancing among other precautionary measures such as hand sanitisation and wearing of face masks (WHO 2020).

This means rural people can longer work together to find common solutions towards climate change threats. The instituted lockdown measures shattered the social cooperation and cohesion of rural people which is even rendering them docile to act against climate change and its impacts. Observations made by Musavengane and Simatele (2016) illuminate that social capital has a long-embedded record of playing a role in individual, household and community risk smoothing, sharing of best practices and post disaster recoveries which are vital components for adaptive capacity in the climate change discourse. Furthermore, social capital can be a vehicle through which the accumulation of different forms of capital can be achieved and contribute to environmental management (Nyahunda and Tirivangasi 2021). As indicated above, the outbreak of COVID-19 maimed most activities dedicated for climate action either through mitigation or adaptation. This means the realisation of sustainable development particularly SDG 13 is going to be a protracted process in rural communities.

7 Conclusion

This study demonstrated that the aspirations of fighting against climate change and its impacts have been halted since the outbreak of the COVID-19 pandemic in Zimbabwe and beyond. The unprecedented ramifications of COVID-19 have pushed countries to shelve meeting their climate change commitments in pursuit of saving lives and insulating economies in the wake of the precarious impacts of COVID-19. Health and climate change are captured in SDG 3 and SDG 13 as vital towards achieving sustainable development. Most importantly, other SDGs such as ending poverty, hunger and inequalities also thrive when climate change impacts are combated. It is worth mentioning that climate change is a health emergence that requires meticulous attention even in the wake of COVID-19. Also, it is undisputable that most plagues that have threatened humanity such as diseases, droughts, cyclones, hurricanes, veld fires and floods emanate from climate change. What this means is that efforts to avoid future health emergencies or pandemics become a renouncing success when climate action is pursued now. This is because the outbreak of the COVID-19 pandemic worsened the daunting impacts of climate change on humanity. Hence, the consequences of climate change require equal effort and attention when responding to the COVID-19 pandemic.

Governments across the world have to pay special attention to scientific warnings about the potential hazards of climate change and scale up their disaster management modalities extracting lessons from the abrupt outbreak of COVID-19. Thus, there is need to invest in infrastructural development to withstand climate change impacts the same way efforts are dedicated to health infrastructural development in response to the COVID-19 scourge. The trillions of dollars channelled towards the fights against COVID-19 require a similar effort for climate action to attain sustainable development. Lastly, the delay in implementing measures to fight against climate change due to COVID-19 stands to impede the realisation of sustainable development in developing countries. This would be the highest disregard of the precarious threats posed by climate change. Consequently, rural people where climate change impacts are taking a toll remain heavily encumbered and vulnerable to its risks.

References

Agrawala S, Dussaux D, Monti N (2020) What policies for greening the crisis response and economic recovery: lessons learned from past green stimulus measures and implications for the COVID-19 crisis. http://www.oecd-ilibrary.org. Accessed 4 June 2020

Bashir MF, Ma B, Komal B, Bashir MA, Tan D, Bashir M (2020) Correlation between climate indicators and COVID-19 pandemic in New York, USA. Sci Total Environ 728:138835

Chanza N (2014) Indigenous knowledge and climate change: insights from Muzarabani. Zimbabwe by Nelson Chanza Doctor of Philosophy in Faculty of Science. Submitted at Nelson Mandela University. http://www.nmmu.repository.ac.za

Chanza N, Gundu-Jakarasi V (2020) Deciphering the climate change conundrum in Zimbabwe: an exposition. In: Global warming and climate change. http://www.intechopen.org. Accessed 10 July 2020

Chen N, Zhou M, Dong X, Qu J, Gong F, Han Y, Qiu Y, Wang J, Liu Y, Wei Y, Yu T (2020) Epidemiological and clinical characteristics of 99 cases of 2019 novel coronavirus pneumonia in Wuhan, China: a descriptive study. The Lancet 395(10223):507–513

Chiweshe MK, Mutondoro FS (2017) Political corruption and the post 2018 agenda in Zimbabwe. J Public Adm Dev Altern (JPADA) 2(2):34–46

Cozier M (2019) The 24th United Nations conference of parties (COP 24) climate change summit. Greenhouse Gases Sci Technol 9(1):6–9
Dodman D, Mitlin D (2015) The national and local politics of climate change adaptation in Zimbabwe. Clim Dev 7(3):223–234
Duruji MM, Olanrewaju FO, Duruji-Moses FU (2018) From Kyoto to Paris: an analysis of the politics of multilateralism on climate change. In: Promoting global environmental sustainability and cooperation. IGI Global, pp 31–56
Dzobo M, Chitungo I, Dzinamarira T (2020) COVID-19: a perspective for lifting lockdown in Zimbabwe. Pan Afr Med J 35(Suppl 2):1–4
Eckstein D, Künzel V, Schäfer L, Winges M (2019) Global climate risk index 2020. German watch. https://germanwatch.org/sites/germanwatch.org/files/20-2-01e%20Global,20
Esquivel V, Sweetman C (2016) Gender and the sustainable development goals. Gend Dev 24(1):1–8
Gori T, Lelieveld J, Münzel T (2020) Perspective: cardiovascular disease and the Covid-19 pandemic. Basic Res Cardiol 115(3):1–9
Grubb M (2014) Planetary economics: energy, climate change and the three domains of sustainable development. Routledge, London
Gukurume S (2013) Climate change, variability and sustainable agriculture in Zimbabwe's rural communities. Russ J Agric Soc-Econ Sci 14(2):1–8
Handmer J, Nalau J (2019) Understanding loss and damage in Pacific Small Island developing states. In: Loss and damage from climate change. Springer, Cham, pp 365–381
Hermwille L, Obergassel W, Ott HE, Beuermann C (2017) UNFCCC before and after Paris–what's necessary for an effective climate regime. Clim Policy 17(2):150–170
Intergovernmental Panel on Climate Change (IPCC) (2016) Climate change 2014: impacts, adaptation and vulnerability. Summary for policymakers. Contribution of working group II to the fifth assessment report of the IPCC. Cambridge University Press, Cambridge. http://ipccwg2.gov/AR5/images/uploads/IPCC_WG2AR5_SPM_Approved.pdf. Accessed 15 July 2020
Jones P, Comfort D (2020) The COVID-19 crisis, tourism and sustainable development. Athens J Tour 7(2):75–86
Khan MR, Roberts JT (2013) Adaptation and international climate policy. Wiley Interdisc Rev Clim Change 4(3):171–189
Kokotović F, Kurečić P, Mjeda T (2019) Accomplishing the sustainable development goal 13–climate action and the role of the European Union. Interdisc Descr Complex Syst INDECS 17(1-B):132–145
Landon B (2020) Who governs the sea? An analysis of the regime complex in international shipping. http://www.academicworks.cuny.edu Accessed 12 July 2020
Leal Filho W, Brandli LL, Lange Salvia A, Rayman-Bacchus L, Platje J (2020) COVID-19 and the UN sustainable development goals: threat to solidarity or an opportunity. Sustainability 12(13):5343
Lu R, Zhao X, Li J, Niu P, Yang B, Wu H, Wang W, Song H, Huang B, Zhu N, Bi Y (2020) Genomic characterisation and epidemiology of 2019 novel coronavirus: implications for virus origins and receptor binding. The Lancet 395(10224):565–574
Mackworth-Young CRS, Chingono R, Mavodza C, McHugh G, Tembo M (2020) Here, we cannot practice what is preached: early qualitative learning from community perspectives on Zimbabwe's response to COVID-19. Bull World Health Organ. http://www.who.int.org Accessed 10 June 2020
Makurumidze R (2020) Empathy and use of evidence in handling travellers coming from COVID-19 high-risk countries. Pan Afr Med J 35(60):8–15
Manzanedo RD, Manning P (2020) COVID-19: lessons for the climate change emergency. Sci Total Environ 742:140563
Mhlanga D, Ndhlovu E (2020) Socio-economic implications of the COVID-19 pandemic on smallholder livelihoods in Zimbabwe. Preprints 2020, 2020040219. https://doi.org/10.20944/preprints202004.0219.v1. Accessed 24 June 2020

Mugambiwa SS (2018) Adaptation measures to sustain indigenous practices and the use of indigenous knowledge systems to adapt to climate change in Mutoko rural district of Zimbabwe. Jàmbá J Disaster Risk Stud 10(1):1–9
Mugambiwa SS, Rukema JR (2019) Rethinking indigenous climate governance through climate change and variability discourse by a Zimbabwean rural community. Int J Clim Change Strat Manag 11(5):730–743
Mugambiwa SS, Tirivangasi HM (2017) Climate change: a threat towards achieving 'sustainable development goal number two' (end hunger, achieve food security and improved nutrition and promote sustainable agriculture) in South Africa. Jàmbá J Disaster Risk Stud 9(1):1–6
Musavengane R, Simatele D (2016) Social capital and the pursuit of ecotourism as a land-use option in land reformed communities: a study of Kwa-Zulu Natal's tribal areas, South Africa. Afr J Hosp Tour Leisure 5(4):1–21
Mutangabende S, Shava E (2016) Assessing progress and pitfalls of the millennium development goals in Zimbabwe: a critical analysis. Afr Public Serv Deliv Perform Rev 4(4):573–605
Ndebele-Murisa M, Mubaya C (2015) Climate change: impact on agriculture, livelihood options and adaptation strategies for smallholder farmers in Zimbabwe. In: Beyond the crises: Zimbabwe's prospects for transformation, pp 155–198
Nhamo G, Dube K, Chikodzi D (2020) Restaurants and COVID-19: a focus on sustainability and recovery pathways. In: Counting the cost of COVID-19 on the global tourism industry. Springer, Cham, pp 205–224
Nilsson M, Griggs D, Visbeck M (2016) Policy: map the interactions between Sustainable Development Goals. Nature 534(7607):320–322
Norouzi N, de Rubens GZ, Choubanpishehzafar S, Enevoldsen P (2020) When pandemics impact economies and climate change: exploring the impacts of COVID-19 on oil and electricity demand in China. Energy Res Soc Sci 68:101654
Nyahunda L, Tirivangasi HM (2019) Challenges faced by rural people in mitigating the effects of climate change in the Mazungunye communal lands, Zimbabwe. Jàmbá J Disaster Risk Stud 11(1):1–9
Nyahunda L, Tirivangasi HM (2020) Barriers to effective climate change management in Zimbabwe's rural communities. In: African handbook of climate change adaptation, pp 1–27
Nyahunda L, Tirivangasi HM (2021) Harnessing of social capital as a determinant for climate change adaptation in Mazungunye communal lands in Bikita, Zimbabwe. Scientifica. https://doi.org/10.1155/2021/8416410
Nyahunda L, Matlakala FK, Makhubele JC (2019) Role of social workers in mitigating the effects of climate change in Makonde communal lands, Zimbabwe. e-Bangi 16(9):2–7
Oldekop JA, Horner R, Hulme D, Adhikari R, Agarwal B, Alford M, Bakewell O, Banks N, Barrientos S, Bastia T, Bebbington AJ (2020) COVID-19 and the case for global development. World Dev 134:105044
Peters GP, Andrew RM, Canadell JG, Friedlingstein P, Jackson RB, Korsbakken JI, Le Quéré C, Peregon A (2020) Carbon dioxide emissions continue to grow amidst slowly emerging climate policies. Nat Clim Chang 10(1):3–6
Petzold J, Ratter BM (2015) Climate change adaptation under a social capital approach–an analytical framework for small Islands. Ocean Coast Mana 112:36–43
Rajamani L (2016) Ambition and differentiation in the 2015 Paris agreement: interpretative possibilities and underlying politics. Int Comp Law Q 65(2):493–514
Roberts E, Huq S (2015) Coming full circle: the history of loss and damage under the UNFCCC. Int J Glob Warm 8(2):141–157
Rogelj J, Den Elzen M, Höhne N, Fransen T, Fekete H, Winkler H, Schaeffer R, Sha F, Riahi K, Meinshausen M (2016) Paris agreement climate proposals need a boost to keep warming well below 2 C. Nature 534(7609):631–639
Rosen MA (2017) Sustainable development: a vital quest. Eur J Sustain Dev Res 1(1):2
Sango I, Godwell N (2015) Climate change trends and environmental impacts in the Makonde Communal Lands, Zimbabwe. S Afr J Sci 111(7–8):1–6

Serrat O (2017) The sustainable livelihoods approach. In: Knowledge solutions. Springer, Singapore, pp 21–26

Srivastava A, Sharma RK, Suresh A (2020) Impact of Covid-19 on Sustainable Development Goals. Int J Adv Sci Technol 29(9 Special Issue):253–258

Thamaga-Chitja JM, Tamako N (2017) Does social capital play a role in climate change adaptation among smallholder farmers for improving food security and livelihoods. J Consum Sci 2:2–17

UNFCCC (2014) Report of the conference of the parties on its nineteenth session, held in Warsaw from 11 to 23 November 2013, decision 2/CP19, Warsaw, Poland. http://www.unfccc.int/resources/docs/. Accessed June 2020

Van Bavel JJ, Baicker K, Boggio PS, Capraro V, Cichocka A, Cikara M, Crockett MJ, Crum AJ, Douglas KM, Druckman JN, Drury J (2020) Using social and behavioural science to support COVID-19 pandemic response. Nat Hum Behav 4(5):460–471

Vanhala L, Hestbaek C (2016) Framing climate change loss and damage in UNFCCC negotiations. Glob Environ Polit 16(4):111–129

World Health Organization WHO (2020) Population-based age-stratified sero epidemiological investigation protocol for coronavirus 2019 (COVID-19) infection, 26 May 2020 (no. WHO/2019-nCoV/Sero epidemiology/2020.2). World Health Organization. http://www.who.int. Accessed 15 July 2020.

Zhakata W, Jakarasi VN, Moyo EN (2017) Zimbabwe's Actions towards climate resilience and low carbon development. Int J Green Growth Dev 3(1):101–106

Sustainable Development: Innovation, Industry and Cross-Cutting Themes

Advancing Resilience for Sustainable Development: A Capacity Development Approach

Per Becker

Abstract While Africa is one of the most promising continents in the world, it is facing mounting challenges to sustainable development in terms of disaster risk, climate change, environmental degradation and demographic and socioeconomic processes. Societies are thus increasingly conceived as having to be resilient to safeguard what they value over time. However, for resilience to have any practical meaning in relation to sustainable development requires further operationalization to facilitate both conceptual understanding and practical action. This chapter suggests such an operationalization, approaching resilience as an emergent property of a society's ability to perform different essential functions—anticipation, recognition, adaptation, and learning—and contends that advancing resilience is about developing the capacity to perform these functions. Sustainable development can therefore be pursued through deliberate capacity development efforts explicitly targeting these functions and their interconnections.

Keywords Sustainable development · Resilience · Capacity development · Emergence · Anticipation · Recognition · Adaptation · Learning

1 Introduction

The last several decades have seen the world changing at an astonishing and seemingly ever-increasing pace (see Heylighen 2008; Roth and Becker 2011; Rudel and Hooper, 2005), at the same time as risk has become a growing preoccupation of society (Beck 1992; Zinn 2008). The risk landscape is continuously changing, whilst the notion of sustainable development has become increasingly central (Becker 2014). Although much has happened since the seminal launch of the concept of

P. Becker (✉)
Division of Risk Management and Societal Safety, Lund University, Box 118, 22100 Lund, Sweden
e-mail: per.becker@risk.lth.se

Unit for Environmental Sciences and Management, North-West University, Private Bag X6001, Potchefstroom 2520, South Africa

W. Leal Filho et al. (eds.), *Sustainable Development in Africa*, World Sustainability Series,
https://doi.org/10.1007/978-3-030-74693-3_29

sustainable development by the World Commission on Environment and Development (WCED 1987)—and there has been many misconceptions about it (Leal Filho 2000)—sustainable development is still commonly conceptualized in relation to meeting the needs of both present and future generations (United Nations 2015). It thus spans as diverse issues as food (Godfray et al. 2010), energy (Chu and Majumdar 2012), water (Bhaduri et al. 2016), soil (Keesstra et al. 2016), education (Leal Filho et al. 2015; Sachs et al. 2019), health (Sachs et al. 2019), gender (Heise et al. 2019), and many more (United Nations 2015). However, many of these issues are interconnected and demand comprehensive action across them (Nilsson et al. 2016; Sachs et al. 2019). Moreover, sustainable development is inherently forward looking and requires conscious decisions and actions today for the future we want tomorrow. This introduces tremendous complexity and uncertainty concerning what could happen that would steer us away from our preferred and aspired development (Becker 2014); both in the sense of potentially depleted natural resources and the occurrence of various changes, disruptions, and disasters. Sustainable development is thus about development that can be upheld over time—in relation to available resources—and defended regardless of changes, trends, or shocks (Hallegatte et al. 2016; IPCC 2012). It is about the long-term governing of risk in relation to that development.

Although Africa is one of the world's most promising continents (UNDP 2019), it is facing mounting challenges to sustainable development in terms of increasing disaster risk and anticipated impacts of climate change (IPCC 2014), as well as of environmental degradation and demographic and socioeconomic processes (Becker 2014). It is in this precarious situation that the notion of resilience has emerged to grasp and address issues of sustainable development (Berkes 2007; Manyena 2006; van Niekerk 2015), largely based on arguments that sustainable development requires societies to be resilient to safeguard what they value over time (cf. Becker and Tehler 2013). However, there are numerous conceptualizations of resilience suggested by scholars from various disciplines; some of which more suitable in relation to the sustainable development of society than others. This chapter endeavours to provide an overview of the main categories of conceptualizations and argues that even the most suitable conceptualization must be further operationalized before the concept can have any real utility for actual efforts for sustainable development in Africa. Particularly if it is to have any role to play in relation to the capacity development stipulated as an essential instrument to reach the goals of the Agenda 2030 (United Nations 2015).

This chapter suggests such operationalization, approaching resilience as an emergent property of a society's ability to perform different essential functions, and contends that advancing resilience is about developing the capacity to perform these functions. It draws on Hollnagel's (2009) seminal work on his four cornerstones of resilience that have later been adapted to communities (Becker et al. 2011) and societies more generally (Becker et al. 2016); anticipation, recognition, adaptation, and learning. These cornerstones provide a foundation for operationalizing resilience, which is fundamental for both grasping and improving it (Becker 2014). Resilience is in this view an emergent property (Pariès 2006) based on society's ability to perform the four resilience cornerstones, or required generalized functions as I call them in this chapter. Advancing resilience for sustainable development is thus about developing the capacity of society to do so, and the chapter focuses on how to facilitate such development. It draws also on Rasmussen's (1985) functional hierarchy of systems to bridge between functional and structural perspectives needed to guide capacity development for advancing resilience in practice. This functional hierarchy spans different levels of abstraction; from purpose, through increasingly concrete levels of function, to the observable forms contributing in the real world to fulfil the functions and meet its purpose (Rasmussen 1985). It is expanded from its original technology-oriented focus to include all aspects of sociotechnical- (Patriarca and Bergström 2017) or socioecological systems (Becker 2014) and can be applied to organizations, communities, or societies in Africa (see Becker et al. 2015). The functional hierarchy thus assists us to connect the four required generalized functions not only upwards to the overall purpose of resilience, but also downwards to the numerous interconnected elements that together constitute the performance of the functions in practice (Becker 2014). It is on various assemblages of such elements and their interrelations that capacity development efforts must focus to advance resilience in the important efforts for sustainable development in Africa.

2 Resilience, Functions and Emergence

The concept of resilience has been used for different purposes in various scientific disciplines, resulting in a plethora of definitions. Many of these definitions describe resilience as (1) ability to "bounce back" to a single equilibrium (Cohen et al. 2011), as (2) a measure of robustness or buffering capacity before a disturbance forces a system from one stable equilibrium to another (Holling 1973) or as (3) ability to adapt in reaction to a disturbance (Pendall et al. 2010, p. 76). Although all three common approaches are useful for their intended purposes (Fig. 1), organizations, communities or societies are systems that adapt to changes over time (Bramlat and Woods 2010; Harvey 1968), making single- or multi-equilibrium approaches to resilience

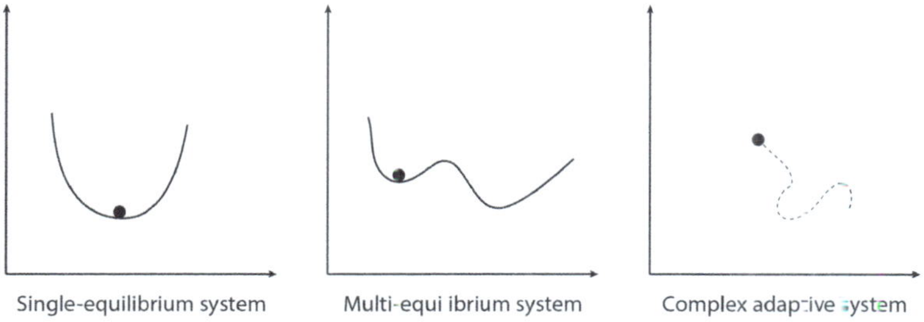

Fig. 1 Three common approaches to resilience represented as system states in state space, which is the set of all possible states of a system regardless of variables (Becker 2014)

less suitable in this context (Woods 2015). At least in relation to sustainable development that requires long-term perspectives by definition (Becker 2014). Moreover, organizations, communities or societies entail human beings with the ability not only to react to disturbances, but also to anticipate and learn from them (Becker 2014; Hollnagel 2009). The resilience of organizations, communities and societies is, in other words, not only including reactive qualities, but also proactive qualities (Becker 2014; Hollnagel 2018). This calls for modifications of conventional reactive notions of resilience in complex adaptive systems, when applying them in relation to sustainable development.

Sustainable development requires the ability of an organization, community, or society to achieve its objectives in as high proportion of cases as possible under varying conditions (cf. Hollnagel 2018). Sustainability can therefore not be conceived independently of its core activities and resilience is not only about how it addresses various adverse outcomes, but also about how it performs every day. Resilience can thus be defined as the "intrinsic ability of a system to adjust its functioning prior to, during, or following changes and disturbances, so that it can sustain required operations under both expected and unexpected conditions" (Hollnagel 2011, p. xxxvi). However, it is important to consider when applying this in relation to sustainable development that both the system and the environment it exists in are changing repeatedly, if not continuously, over time; altering what adjustments are feasible, functioning is intended, operations are required, conditions are unexpected, etc. The definition of resilience still holds, but, to recognize its temporality, resilience can be conceived as the ability to develop along a preferred development trajectory through the state space of the system (Fig. 2). At any point in time, the system has infinite possible futures out of which one (or a small range) is preferred over the others, often referred to as risk scenarios.

To grasp resilience as an emergent property, it is crucial to understand what emergence means. This is traditionally done by contrasting between resultants and emergents (Lewes 1875, p. 413). Whereas a resultant property of a whole is clearly traceable in the properties of the various components that constitute it, an emergent property is not, and emerge only in the interactions of the components. Resilience

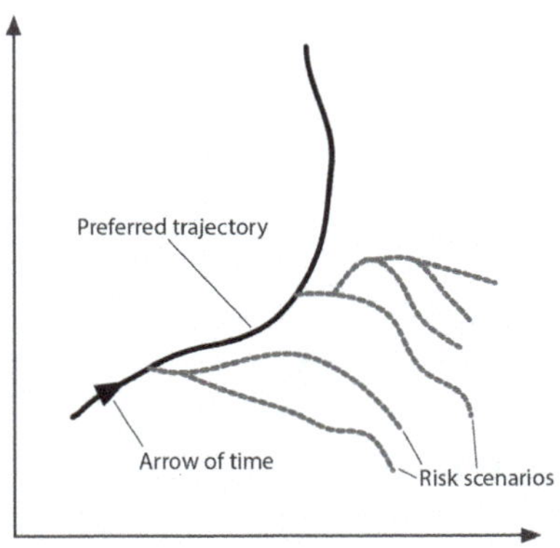

Fig. 2 Resilience as the ability to develop along a preferred development trajectory through the state space of a system

is thus simply not an additive matter, but emerge in the complex interactions of the components of the system. It is here that the work of Rasmussen, Hollnagel and Woods become invaluable to attempt to overcome the otherwise overwhelming complexity of both grasping and improving resilience (Bergström and Dekker 2019).

Bergström and Dekker (2019) argue convincingly for tracing the roots of Hollnagel's resilience cornerstones back to the collaborative work of Rasmussen, Woods and Hollnagel in the 1980s. In the festschrift to honour Rasmussen scientific contributions on his 60th birthday, Woods (1988) suggests ability to adapt, anticipate and detect as crucial to cope with complexity in a dynamic world. The similarities with Hollnagel's influential framework are obvious (Bergström and Dekker 2019), but it was Hollnagel who expanded the ideas and popularized them among both scholars and practitioners. Also Rasmussen's functional hierarchy of systems has its origin in their early efforts to grasp and address the problem of coping with complexity (Rasmussen and Lind 1981), and is also currently enjoying a revival (Becker et al. 2016; Patriarca and Bergström 2017; Patriarca et al. 2017).

Hollnagel (2009, 2017) suggests that resilience emerge in the performance of interdependent and dynamically coupled functions, which can be categorized under the four overall required generalized functions: anticipating, recognizing, adapting, and learning (Becker 2014; Becker et al. 2016). The performance of each generalized function is thus inherently dependent on the performance of the others. However, it is equally important to note that the performance of each also is dependent on the performance of numerous constituent functions further down the levels of abstraction of Rasmussen's (1985) functional hierarchy. It is, in other words, not only resilience that is an emergent property based on the performance of the four required generalized functions, but the performance of each is also emergent in itself (Fig. 3).

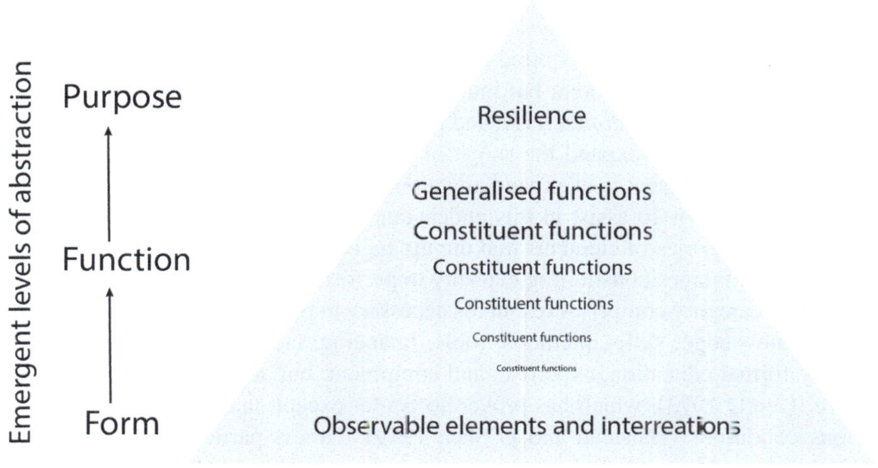

Fig. 3 An abstraction hierarchy of resilience (adapted from Rasmussen 1985)

The performance of *anticipation* emerges in the performance of different forward-looking constituent functions, such as risk analysis, forecasting, etc., and the performance of *recognition* emerges in the complex interaction of various functions for monitoring and impact assessment (Becker and Hagelsteen 2016). Similarly, the performance of *adaptation* emerges in the performance of functions preventing, mitigating, preparing for, responding to and recovering from changes, trends, or shocks that undermine sustainable development, and the performance of *learning* emerges in the performance of various functions for acquiring knowledge from experience (Becker and Hagelsteen 2016). The performance of the four required generalized functions are thus not *resultants*, but *emergents* of the performance of constituent functions, irrespective of how they are conceptualized or categorized. They are not simply layered, regardless how far down Rasmussen's (1985) functional hierarchy you go. This is of fundamental importance for advancing resilience for sustainable development, since that requires purposeful activities to improve the performance of the four required generalized functions but focusing on the level of abstraction that can be perceived and addressed directly. It is, therefore, critical to consider the emergent nature of the performance of any function and focus such activities not only on particular elements, but also on the interrelations between them. It is, in other words, a particular assemblage of them that constitutes the function. Identifying which elements and interrelations that can be strengthened in such assemblages is key to advance resilience for sustainable development in Africa.

3 Capacity and Resilience

To grasp the resilience of an organization, community or society, it is crucial to consider the complex assemblages of elements and interrelations that constitute its capacity to perform the four required generalized functions. Such capacity has been suggested to reside on different but interconnected levels, often categorized as the *individual level*, *organizational level*, and *the enabling environment* (CADRI 2011). It is thus as crucial to expand the scope of any attempt to understand or advance resilience in Africa, as it is to expand any attempt to understand a complex disaster (cf. Vaughan 1996). To assist in this endeavour, it has been suggested to consider four nested categories of elements making up, together with their interrelations, the complex assemblages constituting capacity to perform (Fig. 4).

The first category comprises resources necessary to perform a particular function, such as knowledge, skills, methods, tools, financing, etc. Such resources involve not only formal education, expertise, and equipment, but are also inherent parts of culture (Geertz 1973), which has proven to confer exceptional flexibility to handle various conditions (Halstead and O'Shea 1989). This is particularly pertinent in Africa, with established links between the rich indigenous knowledge of the continent and sustainable development (Boon and Hens 2007). The second category is the organization of these resources, which not only includes formal organizations, but also other forms of social organization according to interest, profession, friendship,

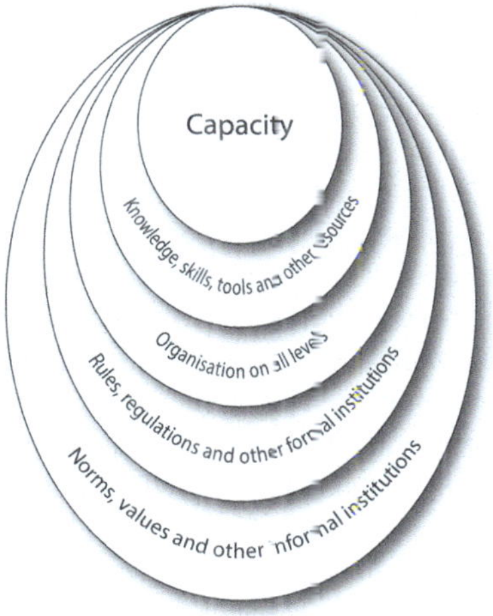

Fig. 4 Four nested categories of elements constituting capacity

ethnicity, etc. (Becker 2014). Organization is in itself a manifestation of sets of underlying formal and informal institutions (Handmer and Dovers 2007). The third category thus comprises existing rules, regulations and legislation, which can be codified or customary. Such formal institutions are rooted in informal institutions, which comprise the fourth and final category containing norms, values and other informal social structures. Although informal institutions have the same purpose as formal institutions—to provide lasting and predictable guidelines for interaction and behaviour, facilitating coexistence and collective activities by reducing the need for constant negotiation (Handmer and Dovers 2007)—they are without the same formal sanctions and rely instead on various social sanctions to discourage deviant behaviour (Helmke and Levitsky 2004); such as criticism, ridicule, shame and contempt. They are inherent parts of culture (Geertz 1973) and are both setting the framework for how actors can behave and the result of how actors behave over time (Giddens 1984). They are behind formal institutions, set boundaries and form of organization, and have influence over what knowledge, skills, tools and other resources are deemed practical, preferable, or even possible (Becker 2014). Although informal institutions are fundamental for the capacity to perform, they are notoriously difficult and slow to change and are therefore often ignored in capacity development (Becker 2014). However, advancing resilience for sustainable development necessitates developing capacity to perform emergent functions along the abstraction hierarchy, which in turn requires identifying and addressing the assemblages of all elements and interrelations constituting the functions together (Becker 2009).

4 Capacity Development for Resilience

There are many ways to advance resilience. This chapter focuses on capacity development as one approach that could benefit resilience for sustainable development in Africa. Particularly since it is identified as a central instrument for fulfilling the Sustainable Development Goals (SDGs) of the Agenda 2030 (Janowski 2016; United Nations 2015). Capacity development is, as such, considered a central process to facilitate sustainable development, but the link between the two is seldom explicitly elucidated. The main argument in this chapter is that resilience provides such link in our complex, uncertain, and dynamically changing world. Sustainable development can thus be achieved by deliberately developing capacities to anticipate, recognize, adapt to, and learn from the changes, trends, or shocks that could steer us away from the future we want.

Most definitions of capacity development express similar things, but with slightly different words (Becker and Hagelsteen 2016). First and foremost, more or less all definitions express capacity development as a process. Secondly, almost all of them indicate different levels where capacity development can take place. Thirdly, they describe capacity development in relation to the capacity to perform to achieve objectives. Fourthly, although the choices of words differ, more or less all definitions state that capacity development is about creating, utilizing and retaining capacity over time. Finally, the concept of capacity development is generally used in a context when an actor is supported by one or more other actors in the development of its capacity (OECD 2009). It involves, by definition, a partnership between internal partners (actors aspiring to develop capacity to perform) and external partners (actors supporting the process), which depends on the definition of the system boundary and could include a country being supported by a foreign country, a public authority being supported by a consultant, a community being supported by the Red Cross, etc. A common situation in Africa. A capacity development approach is thus about the process through which an organization, community or society can develop its capacity to perform the four required generalized functions, with the support of external actors.

It is important to note that a capacity development approach is not a prescription stipulating that an internal actor necessarily requires external actors to advance resilience. Not at all. It is simply an approach intended to facilitate achieving that, when there are external actors involved. Although this is not applicable to all circumstances of advancing resilience, it is surprising how often it is. That said, involving both internal and external actors may have positive effects in itself; not only by pooling required competences, but also by facilitating to overcome the insider/outsider dilemma (see Cedergren 2013; Dekker 2011). While Vaughan (1996) and Woods (2006), for example, caution against knowledgeable but "enculturated" insiders incapable of viewing their situation from alternative perspectives, Dekker (2011) suggests that outsiders can bring new perspectives but may lack requisite credibility to influence it. Finding a balance between insider and outsider perspectives is thus crucial (Cedergren 2013), and a core focus of capacity development (Hagelsteen and Becker 2019).

Developing capacity for resilience entails a number of practical aspects, many of which are interdependent, sequential, or nested. To develop capacity to perform a particular function, the involved actors must first identify the assemblage of elements and interrelations that constitutes it; the knowledge, skills, tools and other resources, the organization of them on all levels, the rules, regulations and other formal institutions, and the norms, values and other informal institutions. This demands acute attentiveness to the particulars of the organization, community, or society and its specific context, both in relation to the current status of relevant elements and interrelations as such, as well as to general opportunities and challenges for change. There are many available methodologies for participatory problem analysis, but this context poses additional demands. Here, all constituent parts must be taken into consideration, rather than only addressing one part just because it happens to suit a specific interest or expertise. Nor can a connected challenge be ignored just because it is difficult to address. For example, educating staff is not enough if an organization is unable to coordinate and utilize the newly acquired knowledge (Eade 1997), and technical equipment is useless if nobody knows how to use or maintain it (Becker 2009). This demands not only a participatory methodology, but a systemic one (Becker et al. 2016; Checkland 1981). As a practical recommendation to put these ideas into action, Checkland's (2000) Soft Systems Methodology (SSM) provides time-tested and concrete step-by-step guidelines for such analysis, which has been suggested for resilience (Wreathall 2009) and proven useful in relevant contexts (e.g. Gregory and Midgley 2000).

With the current performance of the particular function analysed, it is time for the actors to consider the objectives of the capacity development initiative, as well as criteria for evaluating its impacts. Here, it is vital for the actors to resist the urge to focus on getting quick, visible results to serve political purposes, which is a common drive behind decoupled and short-term activities (McEntire 2007). While such activities may be important for creating willingness to change and promote further investment, they need to hold together and be complemented by long-term objectives and activities. Although it is true that one specific capacity development initiative cannot do everything, it is crucial that it includes a mix of activities that addresses all the elements and interrelations needed to achieve the objectives of the initiative (Hagelsteen and Becker 2013). Otherwise, it would just be a waste of resources. Soft Systems Methodology (SSM) has the practical potential for assisting not only in identifying a particular assemblage of elements and interrelations together constituting the function requiring strengthening, but also in designing a mix of activities that address each identified shortcoming in it. Again, in a participatory and systemic manner.

In addition to considering the complexity and emergent nature of each analysed and addressed function, actors involved in capacity development must also consider the dynamic nature of the world they exist in. Many capacity development initiatives are unwittingly based on the fundamental assumption that everything else is kept constant during implementation (*ceteris paribus*). Even when actors recognize that this is not the case, capacity development initiatives are still often planned out in detail with few opportunities to adapt to changing circumstances over time (Hagelsteen and Becker 2019). As a practical guideline, it is important that the involved actors monitor the situation in and around the organization, community, or society developing its capacity for resilience continuously, and allow for flexibility in both the design and implementation of the initiative. Otherwise, there is a significant risk of either wasting resources on parts no longer needed or overlooking new crucial issues emerging. Effective capacity development for resilience is thus largely based on the ability of the involved actors to balance between parts and whole, as well as between control and flexibility (Hagelsteen and Becker 2019).

Capacity development requires the internal actors to have ownership of the processes of both design and implementation, and the external actors to support the process and not to take over (Hagelsteen and Becker 2013), as lack of local ownership has proven detrimental for performance over the years (Morgan 2002). Even if external actors can contribute to the performance of a particular function directly, the increased capacity cannot be sustained for longer than the lifespan of the initiative and will dwindle as soon as they leave. With local ownership, on the other hand, the internal actors may initially not have the capacity to adequately perform the function, but can develop that capacity to a certain level—with the support of external actors—and continue to perform the function on that level also after the initiative is over. This requires a well-functioning partnership between internal and external actors, which has proven challenging for a number of reasons; e.g. differences in organizational cultures, lack of ability to cooperate, tight time frames, ambiguous terminology, and power imbalances (Becker and Hagelsteen 2016). However, in order for such partnerships to work and local ownership to be facilitated, the involved actors need to explicitly and jointly define clear and mutually agreed roles and responsibilities of all partners. In more practical terms, simply sit down and talk this through at the very start of the partnership, and revisit the conversation regularly throughout the capacity development process. While the internal partner must be in the driving seat, external partners can have different roles, spanning from providing technical support to facilitating capacity development processes (Becker 2014). The role of an external partner ultimately depends on the needs of the internal partner and what support the external partner can provide (Stone Motes and McCartt Hess 2007). Finally, it is important that the role of an external partner in some way creates awareness, motivates and engages people, which help the internal partner to take responsibility and ownership of the processes developing the capacity to perform the required functions (Whitmore 2009).

It is important to note that developing capacity to perform a function involves not only technical abilities, focusing on the knowledge, skills, methods, infrastructure, regulations, and tools to perform that specific function. It also involves functional abilities to assess the situation and define objectives, formulate guidelines and strategies, budget, and implementing and evaluating activities (CADRI 2011). In addition to technical and functional abilities, sometimes referred to as hard abilities (cf. Becker and Hagelsteen 2016), capacity development also requires soft abilities to communicate with and engage other actors (Brinkerhoff and Morgan 2010) and to identify and navigate power relations between actors and to negotiate and resolve conflicts (Becker 2014). Soft capacities also include analytical and problem-solving abilities, as well as the ability to critically reflect upon, and learn from, experiences (Britton 2010). It is very much about self-image, self-esteem and self-confidence, where low self-esteem and poor self-confidence undermines the will to change (Angelöw 1991). Soft capacities include leadership, which is not just about recognizing needs and distributing resources (Stone Motes and McCartt Hess 2007), but also about having the courage to take risks, invest, and strengthen others (Lopes and Theisohn 2003). However, leadership is not a panacea for all challenges (Mitchell 2006), but only part of the whole (Becker and Hagelsteen 2016). Finally, soft capacities can also include the ability to deal with the enormous complexity of elements and interrelations constituting a function (Becker et al. 2016), as well as of the capacity development initiative in itself (Davies 2005). It is here more difficult to provide specific practical recommendations for putting these ideas into action. What is important, though, is to grasp the necessity of not only developing hard and often technical abilities, but also softer communicative, collaborative, analytical, and reflexive abilities. Such individual skills are transferrable from other situations, including quality academic education, but can be developed collectively in the group of partners by focusing explicitly on it when working to advance resilience for sustainable development together.

In short, capacity development for resilience is a long-term process through which an organization, community or society can develop its capacity to perform the four required generalized functions, with the support of external actors. It entails in practice:

- a partnership with clear and mutually agreed roles and responsibilities in which the internal partners drive the change and the external partners support the change;
- thorough analysis of the current capacity to perform the functions, acknowledging that it is emergent of an assemblage of elements and their interrelations, and each function interdependent with other functions;
- explicit consideration of all four nested categories of elements, comprising (1) knowledge, skills, tools and other resources, (2) the organization of them on all levels, (3) the rules, regulations and other formal institutions, and (4) the norms, values and other informal institutions.
- designing and implementing activities addressing all identified challenges, in the current capacity to perform the functions, necessary to meet defined performance objectives;
- examining predefined criteria to assess changes in the assemblages of elements and their interrelations constituting the functions;
- continuous monitoring of the organization, community or society and its environment, and flexibility to adjust the activities of the capacity development initiative to instigate the change needed to meet the performance objectives;

- explicit activities to promote local ownership of the process of change;
- developing not only hard abilities directly required to perform specific functions, but also soft communicative, collaborative, analytical, and reflexive abilities.

5 Conclusion

This chapter sets out to introduce the notion of capacity development as one approach to advancing resilience for sustainable development in Africa. It argues that resilience can provide a link between actual activities of capacity development and desired sustainable development in our complex, uncertain, and dynamically changing world, through its four required generalized functions of anticipation, recognition, adaptation, and learning. It is by developing capacity to perform and connect these functions that an organization, community, or society can become more resilient and achieve sustainable development regardless of the changes, trends, and shocks threatening to undermine it. The chapter is, as such, a novel contribution that introduces conceptual and methodological guidelines for deliberate development of resilience in situations with external actors involved, which is common in many important efforts for sustainable development in Africa. The chapter provides the theoretical underpinnings for operationalizing resilience in a way that facilitates linking the performance of the required generalized functions to the actual assemblages of perceivable elements and their interrelations that together constitute the capacity to perform them in practice. Such operationalization is necessary for intentional activities to advance resilience, since both the resilience and the performance of the functions constituting it, are emergent wholes arising from the complex interactions of their constituent parts. By recognizing the elements and interrelations that together constitute a particular function, regardless of level of abstraction, it becomes possible to design deliberate activities to increase the capacity to perform it. To assist in the identification of such assemblages of elements and interrelations, it is helpful to consider four nested categories of elements. Capacity development is then about designing and implementing such activities in an organization, community or society—when external actors are involved—and has been pointed out as a crucial process for meeting the Sustainable Development Goals of the Agenda 2030. Although there are obviously cases without any external engagement, it is surprising how often there is. This approach is therefore not a solution for all circumstances, but can facilitate practical initiatives to advance resilience for sustainable development in a broad range of real situations in Africa, by providing a framework for making the transition from analyses to concrete solutions and from plans to actual implementation.

References

Angelöw B (1991) Det goda förändringsarbetet: Om individ och organisation i förändring. Studentlitteratur, Lund

Beck U (1992) Risk Society: Towards a New Modernity. Sage Publications, London

Becker P (2009) Grasping the hydra: the need for a holistic and systematic approach to disaster risk reduction. Jàmbá J Disaster Risk Stud 2:12–24

Becker P (2014) Sustainability science: managing risk and resilience for sustainable development. Elsevier, Amsterdam

Becker P, Abrahamsson M, Tehler H (2011) An emergent means to assurgent ends: community resilience for safety and sustainability. In: Hollnagel E, Rigaud E, Besnard D (eds) Proceedings of the fourth resilience engineering symposium, 8–10 June 2011, pp 29–35 MINES ParisTech, Sophia Antipolis

Becker P, Abrahamsson M, Tehler H (2016). An emergent means to assurgent ends: societal resilience for safety and sustainability. In Resilience engineering in practice, vol 2, pp 29–40. As

Becker P, Hagelsteen M (2016) Kapacitetsutveckling för katastrofriskreducering. In: Baez Ullberg S, Becker P (eds) Katastrofriskreducering: Perspektiv, praktik, potential, pp 265–291. Studentlitteratur, Lund

Becker P, Tehler H (2013) Constructing a common holistic description of what is valuable and important to protect: a possible requisite for disaster risk management. Int J Disaster Risk Reduction 6:18–27. https://doi.org/10.1016/j.ijdrr.2013.03.005

Bergström J, Dekker SWA (2019) The 2010s and onward: resilience engineering. In: Dekker SWA (ed) Foundations of safety science, vol 1–11, pp 391–429. CRC Press, Boca Raton

Berkes F (2007) Understanding uncertainty and reducing vulnerability: lessons from resilience thinking. Nat Hazards 41:283–295

Bhaduri A, Bogardi J, Siddiqi A, Voigt H, Vörösmarty C, Pahl-Wostl C, Bunn SE, Osuna, VR (2016) Achieving sustainable development goals from a water perspective. Front Environ Sci 4. https://doi.org/10.3389/fenvs.2016.00064

Boon EK, Hens L (eds.) (2007) Indigenous knowledge systems and sustainable development: relevance for Africa. Kamla Ray Enterprises, Delhi

Branlat M, Woods DD (2010) How do systems manage their adaptive capacity to successfully handle disruptions? A resilience engineering perspective. In: Complex adaptive systems—resilience, robustness, and evolvability, proceedings from the association for the advancement of artificial intelligence conference, Arlington, 11–13 November 2010

Brinkerhoff DW, Morgan PJ (2010) Capacity and capacity development: coping with complexity. Public Admin Develop 30:2–10. https://doi.org/10.1002/pad.559

Britton B (2010) Self-reflection – monitoring and evaluation for personal learning. In Ubels J, Acquaye-Baddoo N-A, Fowler A (eds.) Capacity development in practice, pp 264–276. Earthscan, London

CADRI (2011) Basics of capacity development for disaster risk reduction. Capacity for Disaster Reduction Initiative, Geneva

Cedergren A (2013) Implementing recommendations from accident investigations: a case study of inter-organisational challenges. Accid Anal Prev 53:133–141. https://doi.org/10.1016/j.aap.2013.01.010

Checkland P (1981) Systems Thinking, Systems Practice. Wiley, Chichester

Checkland P (2000) Soft systems methodology: a thirty year retrospective. Syst Res Behav Sci 17(S1):S11–S58. https://doi.org/10.1002/1099-1743(200011)17:1+%3c :AID-SRES374%3e3.0.CO;2-O

Chu S, Majumdar A (2012) Opportunities and challenges for a sustainable energy future. Nature 488(7411):294–303. https://doi.org/10.1038/nature11475

Cohen L, Pooley JA, Ferguson C, Harms C (2011) Psychologists Understandings of Resilience: implications for the discipline of psychology and psychology practice. Austr Commun Psychol 23:7–22

Davies R (2005) Scale, Complexity and the representation of theories of change: Part II. Evaluation 11(2):133–149
Dekker SWA (2011) Drift into failure: from hunting broken components to understanding complex systems. CRC Press, Boca Raton
Eade D (1997) Capacity-building: an approach to people-centred development. Oxfam, Oxford
Geertz C (1973) The interpretation of cultures: selected essays. Basic Books, New York
Giddens A (1984) The constitution of society: outline of the theory of structuration. University of California Press, Berkley
Godfray HCJ, Beddington JR, Crute IR, Haddad L, Lawrence D, Muir JF, Toulmin C (2010) Food security: the challenge of feeding 9 billion people. Science 327(5967):812–818. https://doi.org/10.1126/science.1185383
Gregory WJ, Midgley GR (2000) Planning for disaster: developing a multi-agency counselling service. J Oper Res Society 51:278–290. https://doi.org/10.2307/254086
Hagelsteen M, Becker P (2013) Challenging disparities in capacity development for disaster risk reduction. Int J Disaster Risk Reduc 3:4–13. https://doi.org/10.1016/j.ijdrr.2012.11.001
Hagelsteen M, Becker P (2019) Systemic problems of capacity development for disaster risk reduction in a complex, uncertain, dynamic, and ambiguous world. Int J Disaster Risk Reduc 36:101102. https://doi.org/10.1016/j.ijdrr.2019.101102
Hallegatte S, Bangalore M, Bonzanigo L, Fay M, Kane T, Narloch U, Vogt-Schilb A (2016) Shock waves: managing the impacts of climate change on poverty. The World Bank, Washington, D.C.
Halstead P, O'Shea J (1989) Introduction: cultural responses to risk and uncertainty. In: Halstead P, O'Shea J (eds.) Bad year economics. Cambridge University Press, Cambridge
Handmer JW, Dovers S (2007) The handbook of disaster and emergency policies and institutions. Earthscan, London
Harvey E (1968) Technology and the structure of organizations. Am Sociol Rev 33:247–259
Heise L, Greene ME, Opper N, Stavropoulou M, Harper C, Nascimento M, Rao Gupta G (2019) Gender inequality and restrictive gender norms: Framing the challenges to health. The Lancet 393(10189):2440–2454. https://doi.org/10.1016/S0140-6736(19)30652-X
Helmke G, Levitsky S (2004) Informal institutions and comparative politics: a research agenda. Perspect Politics 2:725–740. https://doi.org/10.1017/S1537592704040472
Heylighen F (2008) Accelerating socio-technological evolution: from ephemeralization and stigmergy to the Global Brain. In: Modelski G, Devezas T, Thompson WR (eds) Globalization as evolutionary process: modeling global change. Routledge, Abingdon, pp 284–309
Holling CS (1973) Resilience and stability of ecological systems. Annu Rev Ecol Syst 4:1–23
Hollnagel E (2009) The Four cornerstones of resilience engineering. In: Nemeth CP, Hollnagel E (eds) Resilience engineering perspectives, volume 2: preparation and restoration. Ashgate, Farnham and Burlington, pp 117–133
Hollnagel E (2011) Prologue: the scope of resilience engineering. In Hollnagel E, Pariès J, Woods DD, Wreathall J (eds) Resilience engineering in practice: a guidebook. CRC Press, Boca Raton
Hollnagel E (2017) Safety-II in practice: developing the resilience potentials. Routledge, London
Hollnagel E (2018) Safety-I and safety-II: the past and future of safety management. CRC Press, Boca Raton
IPCC (2012) Managing the risks of extreme events and disasters to advance climate change adaptation (SREX). Cambridge University Press, Cambridge
IPCC (2014) Climate change 2014: impacts, adaptation and vulnerability. IPCC, Geneva
Janowski T (2016) Implementing sustainable development goals with digital government – aspiration-capacity gap. Govern Inf Q 33(4):603–613. https://doi.org/10.1016/j.giq.2016.12.001
Keesstra SD, Bouma J, Wallinga J, Tittonell P, Smith P, Cerdà A, Fresco LO (2016) The significance of soils and soil science towards realization of the United Nations Sustainable Development Goals. Soil 2(2):111–128. https://doi.org/10.5194/soil-2-111-2016
Leal Filho W (2000) Dealing with misconceptions on the concept of sustainability. Int J Sustain High Educ 1(1):9–19. https://doi.org/10.1108/1467630010307066

Leal Filho W, Manolas E, Pace P (2015) The future we want: Key issues on sustainable development in higher education after Rio and the UN decade of education for sustainable development. Int J Sustain High Educ 16(1):112–129. https://doi.org/10.1108/IJSHE-03-2014-0036
Lewes GH (1875) Problems of life and mind: first series – the foundation of a creed, vol 2. Trübner & Co, London
Lopes C, Theisohn T (2003) Ownership, leadership, and transformation: can we do better for capacity development? Earthscan, London
Manyena SB (2006) The concept of resilience revisited. Disasters 30:433–450
McEntire DA (2007) Disaster response and recovery: strategies and tactics for resilience. Wiley, Hoboken
Mitchell JK (2006) The primacy of partnership: scoping a new national disaster recovery policy. Ann Am Acad Polit Soc Sci 604:228–255. https://doi.org/10.2307/25097790
Morgan PJ (2002) Technical assistance: correcting the precedents. UNPD Develop Policy J 2:1–22
Nilsson M, Griggs D, Visbeck M (2016) Policy: map the interactions between sustainable development goals. Nature 534(7607):320–322. https://doi.org/10.1038/534320a
OECD (2009) The Paris declaration on aid effectiveness (2005) and the Accra agenda for action (2008). OECD, Paris
Pariès J (2006) Complexity, emergence, resilience…. In: Hollnagel E, Woods DD, Leveson NG (eds.) Resilience engineering: concepts and precepts. Aldershot and Burlington, Ashgate
Patriarca R, Bergström J (2017) Modelling complexity in everyday operations: functional resonance in maritime mooring at quay. Cogn Technol Work 19(4):711–729. https://doi.org/10.1007/s10111-017-0426-2
Patriarca R, Bergström J, Di Gravio G (2017) Defining the functional resonance analysis space: combining Abstraction Hierarchy and FRAM. Reliab Eng Syst Safety 165:34–46. https://doi.org/10.1016/j.ress.2017.03.032
Pendall R, Foster KA, Cowell M (2010) Resilience and regions: building understanding of the metaphor. Cambridge J Reg Econ Society 3:71–84
Rasmussen J (1985) The role of hierarchical knowledge representation in decisionmaking and system management. IEEE Trans Syst Man Cybernet 15:234–243
Rasmussen J, Lind M (1981) Coping with complexity. Risø National Laboratory, Roskilde
Roth A-S, Becker P (2011). Challenges to disaster risk reduction: a study of stakeholders' perspectives in Imizamo Yethu, South Africa. Jàmbá J Disaster Risk Stud 3:443–452
Rudel TK, Hooper L (2005) Is the pace of social change accelerating? Latecomers, common languages, and rapid historical declines in fertility. Int J Comp Sociol 46:275–296. https://doi.org/10.1177/0020715205059204
Sachs JD, Schmidt-Traub G, Mazzucato M, Messner D, Nakicenovic N, Rockström J (2019) Six transformations to achieve the sustainable development goals. Nature Sustain 2(9):805–814. https://doi.org/10.1038/s41893-019-0352-9
Stone Motes P, McCartt Hess P (2007) Collaborating with community-based organizations through consultation and technical assistance. Columbia University Press, New York
UNDP (2019) Human development data (1990–2018). Human Development Reports website https://hdr.undp.org/en/data. Accessed 16 June 2020
United Nations (2015) Transforming our world: the 2030 agenda for sustainable development (A/RES/70/1). United Nations, New York
van Niekerk D (2015) Disaster risk governance in Africa. Disaster Prevent Manage 24(3):397–416. https://doi.org/10.1108/DPM-08-2014-0168
Vaughan D (1996) The challenger launch decision. University of Chicago Press, Chicago
WCED (1987) Our common future. Oxford University Press, Oxford
Whitmore J (2009) Coaching for performance. Nicholas Brealey Pub, Boston
Woods DD (1988) Coping with complexity: the psychology of human behaviour in complex systems. In: Goodstein LP, Andersen HB, Olsen SE (eds) Tasks, errors, and mental models. Taylor & Francis, London, pp 128–148

Woods DD (2006) How to design a safety organization: test case for resilience engineering. In: Hollnagel E, Woods DD, Leveson NG (eds.) Resilience engineering concepts and precepts. Ashgate, Aldershot

Woods DD (2015) Four concepts for resilience and the implications for the future of resilience engineering. Reliab Eng Syst Safety 141:5–9. https://doi.org/10.1016/j.ress.2015.03.018

Wreathall J (2009) Measuring resilience. In: Nemeth CP, Hollnagel E (eds.) Resilience engineering perspectives, volume 2: preparation and restoration. Ashgate, Farnham and Burlington, pp 95–114

Zinn JO (2008) Introduction. In: Zinn JO (ed.) Social theories of risk and uncertainty: an introduction. Blackwell Publishing, Malden, vol 1, pp 1–17

Sustainability Management—A Conceptual Trade-Off

Muhammad Adil Rauf and Olaf Weber

Abstract The field of sustainability is facing challenges of conceptual openness, functional complexity, and operational ambiguity. A lack of consensus on these fronts is dividing intellect and resource bases, and it is failing to counter the sustainability challenges we face today and tomorrow. Sustainability management can be approached through normative, functional, and operational perspectives. Sustainability, being an integrated subject, will always bring in multidisciplinary aspects to address development challenges. On the other hand, the management perspective demands more operational discussion that may not be understood without its theoretical and conceptual base in a context. Similarly, in its normative terms, sustainability advocates ethical concerns based on the principle of fairness and justice towards nature, society, and future generations. However, the heterogeneity of the concept makes the normative objective of sustainability difficult to operationalize. Further to the normative heterogeneity, amalgamated remedial measures need to engage diverse stakeholders, perspectives, technology advancement, governance, and scale that render sustainable development even more complex and harder to operationalize. African countries' prevailing socioeconomic structure is important to study for its sustainability management's functional and normative approaches. Taking on a philosophical approach, we have discussed the prevailing theoretical and practical concepts in sustainability from a normative, functional, and contextual approach.

Keywords Sustainability management · Sustainable development · Contextual approach · Normative approach · Functional approach

M. A. Rauf (✉) · O. Weber
School of Environment, Enterprise, and Development (SEED), Faculty of Environment, EV 3, University of Waterloo, 200 University Ave W, Waterloo, ON N2L 3G1, Canada
e-mail: adil.rauf@uwaterloo.ca

O. Weber
e-mail: oweber@uwaterloo.ca

W. Leal Filho et al. (eds.), *Sustainable Development in Africa*, World Sustainability Series,
https://doi.org/10.1007/978-3-030-74693-3_30

1 Introduction

In the developing world, where poverty and scarcity are of a high scale, it would be difficult to justify and manage the equitable trade-off between environmental protection and the meeting of basic needs. Financial dependency, budget deficits, economic growth, and measurement of inflation are all subject to negotiation and compromise. However, the ecological component of sustainability being non-negotiable creates more confusion while dealing with socio-economic objectives (Karoly 2011). Therefore, the transition towards sustainability follows different trajectories across a variety of empirical and geographical contexts (Hansen et al. 2018). The complex linkages between objectives, drivers, and responses are subject to contextual definition, conceptual framework, relevant thresholds, and data availability. In such a case, the sustainability definition, objective, and measurement are more localized and differ from country to country (Verma and Raghubanshi 2018).

Developing countries' resource and capacity limitations are forcing them to prioritize and trade off between environment, economic, and social objectives. The scale and scope of the economic accumulation and respective environmental dissipation are forcing a complex trade-off and threatening an optimum and sustainable outcome (Verma and Raghubanshi 2018). Swain and Yang-Wallentin (2020) found a significant policy focus variation for sustainable development between developed and developing countries. The evidence suggests that socio-environment focus seems more beneficial for sustainable development in developed countries compared to socio-economic focus that benefits developing countries more (Swain and Yang-Wallentin 2020). In a similar context, Babu and Data (2015) found a bidirectional association between growth and development that prevented immediate translation of economic growth into social and environmental benefits.

In the African context, their colonial history and afterward perforated economic and political structure, under foreign influence, placed the Africans at a perpetual disadvantage. An unequal trade and exchange rate, financial dependence, capital and resource extraction, and inability to create production value have blocked Africa's broad-based development. This is quite evident from the African city prosperity index 2016, which was significantly lower than other major cities around the world (Nagendra et al. 2018). There is no one reason for this, and of course a blanket approach cannot solve the prevailing development challenges in Africa.

Most African countries lack the functional efficiency necessary for sustainable development. Functional efficiencies are driven by quality education, basic infrastructure, financial capital, and strong governance structure. Functional deficiencies lead to incapacity, inefficiency, unproductivity, and lack of integration across discipline and space (van Niekerk 2020), which means that strengths and weaknesses are not similar across geographies and functional components. For instance, environmental conditions, resource dependencies, varying liabilities, and regional non-convergence behavior in most African countries poses varying scale of sustainability challenges (Ulucak et al. 2020). Many African economies severely lack economic inclusivity which result in spatial inequality, uneven income distribution, and social disparities (van Niekerk 2020). Furthermore, weak governance and an underdeveloped financial structure make it difficult to materialize economic growth through financial development (van Niekerk 2020). In terms of contextual relevance, urban innovations in the context of developed countries may not fit well in the African context (Parnel and Robinson 2012). This is because scholarly influence from the global north has ignored capacities, geographic dynamics, and historic inequalities that predominantly influence perspectives in the global south (Nagendra et al. 2018).

Functional efficiency depends on the relationship between resources, financial and operational capacity, and standards. Resource and capacity limitations will define standards ethically justifiable in a contextual dimension. Standards such as defining prosperity, living standards, and social and environmental justice are normative. Similarly, a global approach to fair trade and burden-sharing of climate change may not be justifiable in a local context. In such a case, normative definitions are more local, influenced by socio-environmental conditions.

In such circumstances, sustainability challenges, particularly socio-environment sustainability objectives, cannot be achieved through common policies. Context-specific policies, for the relevant group of socio-environment components, are in order to facilitate convergence across sustainability objectives (Ulucak et al. 2020). In this review article, we have reviewed historical and current literature to identify emerging themes in sustainable development. We have provided a review on normative and functional approaches in sustainable development to enhance its theoretical and conceptual base in a context.

2 Sustainable Development—Emergence

Sustainability is about keeping the natural ecosystem intact, and human activities should be within the limits of the system's carrying capacity. It is about social justice and fair distribution of resources by maintaining intra and intergenerational equity (Costanza 1989; Odum et al. 1971). The economic system, fundamentally relying on increasing consumption and profit maximization, is not only extinguishing the natural resource capital of the planet, but also impacting the stability of its financial and political system (Jackson 2017). Furthermore, unconstrained economic growth and the prevailing financial system is disrupting the natural and social systems (Bansal and Song 2017). As described by Karl Polanyi (1944), instead of social values, markets are transforming societies by replacing the norms with rationality- and utility-raised ethical concerns.

The concept of sustainability first arose from the desire to determine the maximum sustainable yield in the forestry and fishing industry. However, an abundance of supply and slow economic growth kept the natural resources relatively inexpensive that ignored environmental and social consequences in the past (Smil 2000). In the mid-nineteenth century, Thomas Malthus diverted the attention towards the limitation of the earth's carrying capacity in a systemic approach, relating food production and population growth. Since industrialization, the increase in demand for natural resources has made fundamental changes to the global socio-economic and socio-ecological system, hence raising concerns around future uncertainties (Devezas et al. 2008).

Social and environmental concerns have emerged respectively from two distinct approaches of market social responsibility and sustainability of natural systems (Bansal and Song 2017). Market corporate responsibility surfaced as a counter-movement to unrestrained capitalism in an industrial era. Karl Polanyi (1944) was among those who highlighted the social consequences of the market. Meanwhile, environmental concerns were triggered after witnessing catastrophic changes in the earth's ecosystem. However, there was a fundamental difference between these two approaches. Market responsibility towards society was more normative, which further evolved into the concept of the triple bottom line and stakeholder management remained focused on corporate ethics for inclusive decision-making (Matten and Moon 2005). In contrast, environmental sustainability was more empirical, thus encompassing the broader system's perspective in which market, society, and natural systems interact (Bansal and Song 2017). These movements were somewhat against the contemporary management practices that were keeping humanity and truth away from nature and morality, respectively (Gladwin et al. 1995). However, both corporate responsibility and environmental protection had a narrow and disintegrated approach with the missing link to social equity and justice. The social and environmental disruptions caused by economic development gave birth to the notion of sustainable development that incorporated economic, environmental, and social perspectives. Brundtland (1987) came up with a conclusive idea and introduced the term of sustainable development by referring to the "development that meets the needs of the present without compromising the ability of future generations to meet their own needs" (p.6).

3 Sustainability Management—Concept

Sustainability management is expressed in terms of vision, changing values, moral development, social integration, and transitional process to achieve a better and desired future. Since the Brundtland Commission, various definitions and approaches have surfaced to define sustainability from economic, social, and environmental perspectives. The difference in these perspectives have led to varying objectives with distinct paths to achieving sustainability. In fact, the idea has remained fuzzy due to the complex form of interdependencies among its components, assertion of facts, normative valuations, and the thin line between description and prescription (Gladwin et al. 1995).

Ecology, economics, and sociology, each discipline has a different take on sustainability. Ecology and economics differ fundamentally on distinct focal points of dealing with the uncertainties and risks at the macro and micro levels (Ayres and Gowdy 2001). In economics, development equates income with human wellbeing and incorporates sustainability as a balance of income between present and future generations (Dietz and Neumayer 2007). Whereas, ecology looks at the functional stability of the earth's ecosystem, a precondition to human wellbeing and intergeneration equity (Griggs et al. 2013). The focus of sociology is on preserving community social and cultural values and their relationships (Ayres and Gowdy 2001). In addition to conceptual variation, Hammer and Pivo (2017) categorized the operational approaches to sustainability: Those linking economic growth and human wellbeing, those highlighting the importance of natural capital in sustaining economies, and those prioritizing the process of economic development. In short, economic objectives should not undermine natural resources. The challenge of sustainability management is to integrate these components into strategic and business goals and create a win–win solution.

3.1 Weak and Strong Sustainability

The debate over weak and strong sustainability revolves around the natural capital stock level – it either exists in abundance or needs protection—and its relation to well-being (Ayres and Gowdy 2001; Dietz and Neumayer 2007). Hartwick-Solow's model of substitutability assumed that both natural and manufactured capital stand equal in terms of the factor of production and generating well-being (Solow 1974). Furthermore, technological advancement can easily substitute natural capital faster than it depletes. Therefore, the aggregate stock of capital, that matters, should be maintained or reinvested to increase consumption later (Solow 1974; Hartwick 1977). However, this can stand true only if the natural capital is available in abundance, reproducible, or substitutable—perfectly elastic with produced capital—unlikely to sustain optimal growth in terms of perpetual provision of wellbeing (Dietz and Neumayer 2007; Daly 1990).

Ecological principles oppose substitution due to the irreversibility of the natural system and the qualitative difference that exists between natural and manmade capital (Ekins et al. 2003). Furthermore, the natural systems operate in cohesion to supply material, absorb waste, provide amenity service, and basic life support function for both humans and the system itself (Barbier et al. 1994). The life support function is of a direct determinant of human welfare, and it holds the primary value that embraces everything together (Turner and Pearce 1994; Dietz and Neumayer 2007).

Degrowth is another approach for sustainable transformation. Degrowth advocates decoupling growth from the economy, focusing more on socio-economic rather than a socio-technical approach. It works with the bottom-up approach, relying on social experiments, motivated to make large-scale changes, including rejection of capitalism, and reverting to localization of the economy (Khmara and Kronenberg 2020). In contrast, the technology dependence, delinking economy from nature and neglecting the scientifically proven limitations of resources, compromise natural and human equity. The ecological approach, on the other hand, is unable to integrate human and ecological integrity, resting purely on theoretical grounds that ignore social facts. Gladwin et al. (1995) were convinced that both technological dependence and strict ecological perspectives are unable to provide a strong base for sustainable development unless integrated to create a cohesion between ecology and human wellbeing.

3.2 Limits to Growth

The basic laws of nature have never been replaced; however, growing demand from an increasing population has changed the appearance and quantitative relations of the natural system (Odum et al. 1971). The continuously shrinking resources and increasing consumption restrict not only unlimited economic growth (Jackson 2017) but also its equitable distribution (Costanza 1989). To deal with scarcity and depletion of non-renewable resources, limits need to be applied to consumption (Costanza 1989; Daly 1990).

The laws of physics and the environment, such as conservation of energy and transformation, entropy, and the tolerance of the system, place physical limits on the growth. All these laws define constraints imposed by the ecosystem's function of energy creation, systems order, and maintaining production and consumption (Odum et al. 1971; Daly 1990). Furthermore, the human population structure depends on local habitat, weather conditions, reproduction process, and social attraction that result in aggregation. Similarly, Alee's principle and Shelford's law of tolerance emphasized the appropriate population size and relevant conditions necessary for sustainable development (Odum et al. 1971). These laws clearly describe the system requirements and functional constraints necessary for the ecosystem to maintain and grow.

The natural ecosystem response to varying external pressures is nonlinear, often abrupt, and sensitive around certain thresholds of key variables. Crossing these thresholds shifts the state of subsystems, such as changes in the monsoon cycle, land surface, rate of biodiversity loss, and concentration of greenhouse gases (Rockstrom et al. 2009). However, defining the thresholds for all the subsystems is indeed a difficult task (e.g., water and land degradation) (Steffen et al. 2015). Several critical processes, such as climatic changes, loss of biodiversity, and biogeochemical cycle, have already passed their threshold (Rockstrom et al. 2009), whereas land system was also later included in the list (Steffen et al. 2015). These are the clear indications of misaligned development trajectories followed in the past.

4 Sustainability Management—Prescription

There is a strong relationship between natural resource management, economic prosperity, and human wellbeing driven by production, consumption, and waste management phases of products and services (Tukker and Jansen 2006). These socioeconomic and environmental processes carry both normative and functional approaches and adopt distinct methodologies for sustainability. Some of the sustainability themes are categorized from the two approaches listed in Table 1.

4.1 Normative Approach

Researchers and policymakers are increasingly acknowledging the crucial role of normative dimensions and political contexts in addressing complex sustainability challenges (van der Hel 2018; Miller et al. 2014). Sustainability drives its normative aspiration from contexts of conflicting interests, the social base of the knowledge, researchers' bias, and the tussle between politics and science. However, these contexts play a part in problem identification and analysis (van der Hel 2018). However, the extent of the role of values and political aspects in shaping sustainability knowledge and transformation is still debatable (van der Hel 2018; Meinherz et al. 2020). This debate is then further extended to the role of values in the sustainability of present and future generations.

Any decisions made today will have their future environmental consequences. There could be more than one way to address this: act based on a future value judgment, prioritize present value, or adopt best practices. In welfare economics, the fundamental problem in judgment is a difference in valuation at any point in time

Table 1 Sustainability thematic approaches

Normative approach	Functional approach
Inter-generation equity	Regional management, resilience, and capacity
Environmental autonomy	Managing environment
Wealth, prosperity, and decoupling	Managing economy
Corporate responsibility	Managing business
Responsible investment, shared value	Finance and investment
Valuation—internalising	Assessment and measurement
Managing consumption	Material and production management
Tragedy of commons	Managing commons
Regulation	Governance

and space. Therefore, a conventional approach to future discounting seems controversial due to judgemental uncertainties, unknown behavior of future generations, and disparities among the present nations (Goulder and Stavins 2002). Temporal uncertainty and spatial disparities are thus preventing the discounting method to adopt a single formula, making it difficult to operationalize.

Bromley (1998) argued that intergeneration equity is not a matter of quantitative valuation of unknown future needs as the conventional sustainability approach follows; instead, a legacy of good practice of preserving the environment will provide equal opportunities to flourish. Ayres and Gowdy (2001) further added that the legacy of rights and opportunities for future generations can end the debate on substitution and marginal trade-offs. However, we disagree with Ayres and Gowdy; substitution and marginal trade-offs are functional approaches that cannot be abolished against a normative approach of legacy, just as a qualitative difference exists between material recycling and consuming less.

In terms of the present generation, sustainability also carries a normative foundation. Sustainability is to achieve a desirable state of the world, embodying social values and their relationship with the environment. However, human actions connect causality with logical necessity, adding descriptive and empirical connections (Kates 2001). Generally, scientific validity relies on empirical grounds, not a logical necessity. This was contested by Hume. Hume's analysis concluded that decisions are somehow driven by norms of necessities (Wang 2011), supporting the argument of efficacious human actions necessary for the normative foundation of sustainability (Daly 1990).

According to Hume (2003), the current state cannot be determined solely by facts, it carries social values and preferences. This means results and the process by themselves will be unable to prove outside the context, in which values stand relevant. Despite the difficulty of separating facts from values (Meinherz et al. 2020), it is, however, recommended to first acknowledge and then draw clear lines between facts and values (Potthast 2015).

Generally, public value definition and application has a limited empirical verification (Hartley et al. 2017). Public values are not universal, they represent those who define and value them. They are geographically dependent, influenced by local and regional policies, and institutional structures that influence normative context (Uyarra et al. 2019). From a critical social theory perspective, the normative dimension should be assessed in terms of its ideal and validity. Normative validity will determine the contextual applicability of the normative practices. For example, Cooke (2006) argued that the validity of a social or cultural norm is not just limited to a cultural attribute or subject to a specific authority. Normative validity should extend to the context-transcending capability to be accepted by a different cultural context. For example, equality as a social norm is embedded, enforced, or practiced while setting examples for others (Cooke 2006). Similarly, environmental norms may opt to different approaches to environmental protection, such as green growth and degrowth may or may not transcend the context (Sandberg et al. 2019). For instance, degrowth is further prioritized based on having a stronger normative justification (Sandberg et al. 2019; Cooke 2006), reflecting the difference invalidation.

In corporate sustainability, the normative dimension was also constructed first from social responsibility then extended to environmental values. Generally, corporate social responsibility (CSR) is adopted as reputation management. However, market normative and sociocultural systems may affect corporations' sustainable practices and disclosures (Tran and Beddewela 2020). In another approach, values also influence corporate decision-making. In conventional economic theory, a normative foundation assumes the capability of a decision-maker to make rational choices. For instance, values play a key role in supplier scrutiny and other procurement decisions (da Silva et al. 2020). Furthermore, according to da Silva et al. (2020), decision-making is considered as both empirical and behavioral, representing normativity and rationality.

4.2 Functional Approach

Complex and functionally distinct origins of sustainability make it highly contextualized and ontologically open. Therefore, the operationalization of sustainability depends on the way it is recognized and described (Purvis et al. 2019). Furthermore, sustainable development demands inclusive growth and a modified approach to design for environment, life-cycle assessment, full-cost estimation, system analysis, and purposiveness (Gladwin et al. 1995). Context-specific solutions may be useful in this quest to address variation in socioeconomic, environment, and geographical conditions (Gladwin et al. 1995). Similarly, to measure and operationalize sustainability, greater and practical decision-support tools are needed for the systematic inclusion of sustainability in management practices (Kates et al. 2005).

A function is generally defined as the intended use of something. This could be a device or a system. In both cases, the system will be formed from various elements in which relationships between the elements will define their values, significance, and even existence. Operation describes the method by which functions are performed and influence the produced value. In social, economic, and environmental interactions, functions will be described from their discrete elements and intended output. Whereas the operational framework will determine how the functional values will be realized, requiring careful consideration of both spatial and functional boundaries (Purvis et al. 2019). For instance, an electric vehicle's (EV) operational context will be changed by including or excluding the energy network system in the EV lifecycle analysis (Doufene et al. 2014).

In another example, political and normative disagreements often lead to a contested definition of justice, generally ill-defined (Dirth et al. 2020). According to Valentini (2011), the functional framework of justice is drawn from its normative base, empirical assumptions, and formal recognition by the social system it intends to serve. In this sense, the function of justice is inherently tied to the social structure that defines its contextual and operational constraints.

A process application depends on the delivery channels that hold public values. For instance, not all departments/institutions hold or are obligated to provide public values (Uyarra et al. 2019). The inclusion of definitions and decision-making is vital throughout the process of value judgment, institutional structure, embedded systems, and stakeholder engagement (Smith 2004). Uyarra et al. (2019) argued that conventional approaches to innovation policies are unable to solve societal challenges of poverty, climate change, and regional economic disparities. This has shifted the focus from sole reliance on market interventions to a more responsible and transformative approach, incorporating normative values in policymaking. For example, responsible research and innovation rely on purpose, rationales, social and political structures, and development trajectories and paths (Uyarra et al. 2019). The same is true in the African context. As discussed earlier, a level of regional integration, local capacities, resource limitations, production functions, income distribution, social disparities, and prevailing challenges will determine the development trajectories most compatible to the local governance, population, and environmental structures (Ulucak et al. 2020; van Niekerk 2020; Nagendra et al. 2018). Furthermore, Masocha (2019) suggested that normative values, derived from social realities, have significantly influenced small businesses' behavior towards sustainable development practices in South Africa.

In another example, urban transformation requires careful consideration of both sustainability and resilience. Urban sustainability and urban resilience are often poorly defined, sometimes used interchangeably. Elmqvist et al. (2019) argued that this confusion is affecting urban transformation efficiency because sustainability and resilience objectives sometimes contradict or even challenge each other. Urban sustainability's objective is to optimize and enhance urban systems, whereas resilience is more about the system's restoration capability in the aftermath of a disaster. In contrast to the normative approach of sustainability, resilience is more functional in that it relies on the system's functions and operational structure. In certain cases, increasing sustainability may often lead to reduced resilience (Folke 2016). For instance, energy efficiency through densification may compromise the urban ecosystem which is vital for urban resilience in climate change (Frantzeskaki and Tilie 2014). Similarly, transportation and communication efficiency may result in redundancy, a key feature of urban resilience (Elmqvist et al. 2019). Elmqvist et al. (2019) further argued that the systemic attribute of resilience depends on the functioning scale, and the outcome may or may not have a normative validation. They further added that any undesired resilience may also contradict sustainability objectives.

Furthermore, the process of transforming science into practice is also context dependent (Buyana 2020). Parnel and Robinson (2012) argued that a recalibration of scholarly work from the global north is inevitable to accommodate the African cities' contextual realities. For example, according to Buyana (2020), the waste management system in African cities mostly rely on an informal system of waste transformation. The local system of waste vendors, in combination with the tools and techniques sourced from local resources, creates a nexus of innovation, local material, and employment. Such systems may not be very adjustable to advance technologies used in developed countries (Buyana 2020). A local intellectual capacity building is necessary for alternative ideas to address the challenges of the rapidly transforming African cities (Parnel and Robinson 2012).

4.3 Trade-offs and Synergies

Mori and Christodoulou (2012) argued that environment, society, and economy are nested hierarchical functions and cannot be treated as parallel (Mori and Christodoulou 2012). Consequently, the benefit of one element could be a disadvantage to others. In sustainability objectives, particularly in the context of sustainable development goals, the synergies and trade-offs are driven by functional and operational dependencies between the objectives. These trade-offs and synergies are generally described in terms of opposing and favoring, respectively. In terms of Sustainable Development Goals (SGDs), the relationship between the indicators are highly contextualized and varies from country to country. These contextual relationships are highly dependent on trajectories, direction, governance structure, and technological framework. For instance, pursuing social goals will generally have negative environmental consequences. However, higher-income groups may have a greater advantage over others in investing in green technologies and reducing ecological footprints (Pradhan et al. 2017).

Efficiency improvement is generally regarded by sustainability. However, in a dynamic growth and with constant increase in emissions, improving efficiency is just not helping sustainability. In terms of the conceptual definition of sustainability, economic efficiency should be measured in terms of ecological impacts. Similarly, for social sustainability, income equality, prosperity, and wellbeing are the major indicators of sustainability. These indicators are driven by ecological sustainability. However, they are not directly influencing ecological sustainability (Karoly 2011). For instance, job loss or income inequality will have no direct implications on the environment unless the losses are connected to ecological damage due to a lack of infrastructure. In such cases, both economic and social sustainability are negotiable; however, environmental sustainability maybe nonnegotiable (Karoly 2011).

In terms of the human–environment relationship, human wellbeing is considered a multidimensional concept, yet we do not have a definitive answer as to what exactly is required to achieve it (O'Neill and Uebel 2015). It is sometimes defined as pleasure in the absence of suffering. In terms of the socioenvironmental relationship, maximizing return (wellbeing) and minimizing suffering (environmental impacts) will satisfy all concerns (Meinherz et al. 2020).

From the perspective of integrated decision-making, Raman et al. (2015) identified the in-capacity of a single large-scale policy option—such as the policies dealing with the production of biofuel in the United Kingdom—to deal with the nexus of technological motivations, social and cultural values, prevailing economic system, and conflicting perspectives of localization. In another example, a blanket approach to largescale afforestation campaigns can do more harm than good in terms of loss of biodiversity, reduced carbon sequestering, and decrease in the density of soil organic carbon (Heilmayr et al. 2020; Hong et al. 2020). Similarly, the goal of sustainable land-use practice is not finding a single best solution for the socio-ecological problems associated to the land system. It is to understand the process of decision-making, associated trade-offs and implications, underlying values and norms, and to assess operational and normative outcomes to determine winners and losers (Nielsen et al. 2019).

Concerning Sustainable Development Goals, Pradhan et al. (2017) found both significant negative and positive relationships across 169 SDG targets in 227 countries around the globe. Although, SDG indicators have shown relatively greater synergies across the world, however, still a majority of countries are dealing with trade-offs between 40–50% of their individual and collective targets (Nilsson et al. 2016; Pradhan et al. 2017). For example, access to electricity through non-renewable means may hamper renewable energy share. Similarly, addressing material footprint may impact economic production of a country. These interdependencies may cause conflicting and diverging results that could possibly lead to a relationship where one goal may restrict progress in other goals (Pradhan et al. 2017). The characterization of the interaction between SDG indicators, from both normative and functional approaches, is still a subject underexplored. Nilsson et al. (2016) have characterized SDG interactions in indivisibility, consistency, and canceling behaviors of indicators, proposing a seven-point scale framework. The same framework was adopted by ICSU (2017) to explicitly test causal and functional relations within the progress of goals and targets (ICSU 2017). The trade-off is not just limited to the input values of the function. It is equally important for the process it takes and the valuation of corresponding outputs. Consistency in approach throughout the functional process will help to eliminate normative and functional conflicts.

5 Discussion

The human ingenuity gap remains a key barrier to transformation to fully understand the complex dynamics of problems, intervention, and adaptive compensation (Tol 2016). Similarly, sustainability demands logic of appropriateness rather than a logic of consequence (Bernstein and Cashore 2007), clearly lacking in general policy trajectories. Furthermore, the lack of an integrated contextual approach is being held back from finding an appropriate development solution. In contrast, sustainable development is an aggregate outcome that could be achieved only through multiple sustainability management initiatives. Nisson et al. (Nilsson et al. 2016) suggested cross-sectoral and cross-goal collaborative policy development vital for the operationalization of sustainable development objectives. In addition to cross-sectional approaches, a directional perspective will help to determine the ideal decision necessary to be maintained throughout the process of implementation. Some of the sustainability themes are organized from two key directional perspectives—normative to functional and functional to normative- presented in Table 2.

Table 2 Sustainability thematic and directional approaches

Themes	Normative to functional	Functional to normative
Inter-generation equity	Future valuation, equal opportunities, consumption behavior, conservation, renewable/non-renewable	Present valuation circumstantial opportunities, investment behavior, best practices, alternatives, innovation, advancement
Managing economy	Development, degrowth, equal, distribution, cost vs value	Growth, inclusive growth, decoupling, green growth
Wealth and prosperity	Social wellbeing, social participation, immaterial and cultural engagement, physical and psychological growth, redefining prosperity	Income, employment, redefining necessity
Environmental management	Environment autonomy, nature as a precondition, embedded society and market	Human interests come first; Resource for economic growth; Wealth to protect environment; increased wealth will decrease degradation
Business and shared value	Regulation, CSR, externalities, social cost, motivations	Bottom of the pyramid; investing in people to enhance productivity; shared value, social innovation, actions (empirical), efficiency
Finance and investment	Divestment, responsible investment, devaluation, behavior	Innovation, opportunity cost, risk adjustment, rationality
Common resources	Community-scale, common interests, nonmarket solutions	National scale, pricing commons, market driven
Impact assessment	Internalize cost, non-market valuation, damage function, ecological footprints	Externalize cost, market valuation, fragility function, nonlinear and idiosyncratic, monetary value
Regulations and policies	Taxation, prevention, maximum yield, optimal state, same objective, unified goals, socio-environment specific, systemic	Revenue to combat, carbon incentives, adaptation, resilience, inclusivity, socioeconomic, systematic, difference in objective-specific goals
Material management	Reduce consumption, material costing, reduce supply, conservation	Substitution, material efficiency, recycling, substitution

Sustainability, in its normative terms, advocates ethical concerns based on the principle of fairness and justice towards nature, society, and future generations. However, the heterogeneity of the concept, associated with the fundamental challenges of valuation and generalization, makes the normative objective of sustainability difficult to operationalize. The normative objective of inter- and intra-generation equity, predominantly based on the assumptions drawn from the dominating behaviors, ignores the social facts and human volatility that depend on contemporary circumstances. Therefore, a generalization of social values is quite difficult for the present generation and makes it irrational for future generations. Similarly, in a dynamic socio-economic system, the valuation of natural capital is also quite challenging in achieving common ground. However, equity in terms of justice towards nature is a global concern. As Hawken (1993) said, "leave the world better than you found it" (p.139) helps in operationalizing environmental justice.

The depletion of natural resources, although empirical, has some uncertainties from the perspective of ethics, valuation, and distribution remain debatable. There are some commonalities between the depletion of natural resources and climate change, such as both are driven by socio-economic activities. However, the interaction of socioeconomic activities with the natural environment is complex and context dependent. Similarly, environmental norms may be generalized for the present generation, but may not be more than setting a precedent for future generations. However, climate change impacts are very much current and global, and that which carries a cost if neglected. Two approaches can be adopted in this regard: management or prevention. Management is responsive, incremental, and relies on capacity and technological advancement. On the other hand, prevention is predictive and enhances the system's capacity to be more resilient, adaptive, and transformative.

In socio-ecological systems, no one size fits all in developing strategies for sustainable development. It depends on the context, the current state, and the basin of attraction (Walker et al. 2004). For example, the definition of a region depends on its geographical and functional context altered by human behavior and community norms. The contextual approach will define the functional approach to management, such as dealing with common resources, material assessment, and finding socio-environmentally appropriate economic opportunities. According to Ivan (1997), rather than counterproductive, development should be context-specific and must be able to serve majority needs. Likewise, Sen's notion of capability and freedom of development advocates development valued by communities (Sen 1999). It reflects that sustainable development is not what is perceived, rather it is the one valued. Droz (2019) added further by referring to the human choice of lives as self-determination, but with general ethical restrictions to be maintained for environmental autonomy.

6 Conclusion

From the discussion above, it is quite evident the conceptual definition of sustainability will remain fuzzy unless dissected from normative and functional approaches in their respective context. The socioeconomic disparities and the distribution of natural resources make it quite challenging to achieve globally accepted principles for sustainability. It would be rather a moral relativism that could help to address the challenges of their respective normative and functional constraints. Environment, society, and economy are nested functions, and cannot be separated. The assertion of facts will determine priorities and consequent trade-offs. A sustainable trade-off requires not only functional input values, but the process, the path it takes, and the valuation of consequent outputs are equally important. A complete system's approach will help to eliminate the conflict between normative and functional approaches to sustainability. Furthermore, to achieve sustainable and relevant development, the recalibration of global knowledge to accommodate local facts is inevitable. Indeed, this will require developing local knowledge and operational capacity.

African countries' prevailing socioeconomic structure is an important case to study sustainability management from functional and normative perspectives. As suggested by Buyana (2020), the integration of science, policy, social norms, and prevailing practices is necessary to develop lasting solutions for the developmental challenges specific to the African context. This study provides a foundation to further develop sustainability management literature to address contextual challenges across geographies.

Economic growth can go hand in hand with sustainable development if the prevailing economic structure changes its parameters to be driven by socioecological values rather than driving them. This approach has more significance in terms of dealing with the type of challenges faced by Africa and other developing geographies across the globe. Overarching objectives, such as ethics and justice, can be adopted as a common approach, whereas a functional approach should incorporate ecological, geographical, and prevailing socio-economic conditions. In this context, sustainable development can be defined as development that is valued with freedom of choice and resilience (Local), without compromising the ability of the natural ecosystem to flourish (Global).

References

Ayres RU, Gowdy JM (2001) Strong versus weak sustainability: economics, natural sciences, and consilience. Environ Ethics 23:155–168

Babu S, Data S (2015) Revisiting the link between socio-economic development and environmental status indicators - focus on panel data. Environ Dev Sustain 17:567–586

Bansal P, Song HC (2017) Similar but not the same: differentiating corporate responsibility from sustainability. Acad Manag Ann 11(1):105–149

Barbier E, Burgess J, Folke C (1994) Paradise Lost? - the ecological economics of biodiversity. Earthscan, London

Bernstein S, Cashore B (2007) Can non-state global governance be legitimate? An Analytical Framework. . Regul Gov 1(4):347–371

Bromley DW (1998) Searching for sustainability: the poverty of spontaneous order. Ecol Econ 24:231–240

Brundtland G (1987) Our common future. Oxford University Press, Oxford

Buyana K (2020) Keeping the doors open: experimenting science-policy-practice interfaces in Africa for sustinable development. J Housing Built Environ 35:539–554
Cooke M (2006) Re-presenting the Good Society. MIT, Cambridge
Costanza R (1989) What is ecological economics? Ecol Econ 1(1):1–7
da Silva EM, Ramos MO, Alexander A, Jabbour CJ (2020) A systematic review of empirical and normative decision analysis of sustainability-related supplier risk management. J Clean Prod 244. https://doi.org/https://doi.org/10.1016/j.jclepro.2019.118808
Daly HE (1990) Toward some operational principles of sustainable development. Ecol Econ 2(1):1–6
Devezas T, LePoire D, Matias JC, Silva AM (2008) Energy scenarios: toward a new energy paradigm. Futures 40(1):1–16
Dietz S, Neumayer E (2007) Weak and strong sustainability in the SEEA: concepts and measurement. Ecol Econ 61(4):617–626
Dirth E, Biermann F, Kalfagianni A (2020) What do researchers mean when talking about justice? An empirical review of justice narratives in global change research. Earth Syst Gov. https://doi.org/10.1016/j.esg.2020.100042.
Doufene A, Dauron A, Chale HG, Krob D (2014) Model-Based operational analysis for complex systems - a case study for electric vehicles. INCOSE Int Symp 24(1):122–138
Droz L (2019) Redefining sustainability: from self-determination to environmental autonomy. Philosophies 4:42. https://doi.org/10.3390/philosophies4030042
Ekins P, Simon S, Deutsch L, Folke C, De-Groot R (2003) A framework for the practical application of the concepts of critical natural capital and strong sustainability. Ecol Econ 44:165–185
Elmqvist T, Andersson E, Frantzeskaki N, Phearson T, Olsson P, Gaffney O, Folke C (2019) Sustainability and resilience for transformation in the urban century. Nature Sustain 2:267–273
Folke C (2016) Resilience (Republished). Ecol Soc 21(4):44. https://doi.org/10.5751/ES-09088-210444
Frantzeskaki N, Tilie N (2014) The dynamics of urban ecosystem goverance in Rotterdam, The Netherlands. Ambio 43:542–555
Gladwin TN, Kennelly JJ, Krause T-S (1995) Shifting paradigms for sustainable development: implications for management theory and research. Acad Manage Rev 20(4):874–907
Goulder LH, Stavins RN (2002) Discounting: an eye on the future. Nature 419(6908):673–674
Griggs D, Stafford-Smith M, Gaffney O, Rockström J, Öhman MC, Shyamsundar P, Noble I (2013) Policy: sustainable development goals for people and planet. Nature 495(7441):305
Hammer J, Pivo G (2017) The triple bottom line and sustainable economic development theory and practice. Econ Dev Q 31(1):25–36
Hansen U, Nygaard I, Romijn H, Wieczorek A, Kamp L, Klerkx L (2018) Sustainability Transition in developing countries: stocktaking, new contributions and a research agenda. Environ Sci Policy 84:198–203
Hartley J, Alford J, Knies E, Douglas S (2017) Towards an empirical research agenda for public value theory. Public Manage Rev 19(5):670–685
Hartwick J (1977) Intergenerational equity and the investing of rents of exhaustible resources. Am Econ Rev 67(5):972–974
Hawken P (1993) The ecology of commerce: a declaration of sustainability. Harper Business, New York
Heilmayr R, Echeverria C, Lambin E (2020) Impacts of Chilean forest subsisdies on forest cover, carbon and biodiversity. Nature Sustain. https://doi.org/10.1038/s41893-020-0547-0
Hong S, Yin G, Piao S, Dybzinski R, Cong N, Li X, Chen A (2020) Divergent responses of soil organic carbon to afforestation. Nature Sustain. https://doi.org/10.1038/s41893-020-0557-y
Hume D (2003) A treatise of human nature. Courier Corporation, North Chelmsforth
ICSU (2017) A Guide to "SDG" Interactions: From Science to Implementation. International: International Council for Science (ICSU), Paris
Ivan I (1997) Development as planned poverty. In: Rahnema M (eds) The post development reader. Zed Books Ltd., London, pp 94–102

Jackson T (2017) Prosperity without growth: foundations for the economy of tomorrow, 2nd edn. Routledge, New York

Karnani A (2007) The mirage of marketing to the bottom of the pyramid: how the private sector can help alleviate poverty. California Manage Rev 49(4):90–111

Karoly K (2011) Rise and fall of the concept sustainability. J Environ Sustain 1(1). https://doi.org/10.14448/jes.01.0001

Kates R (2001) Queries on the human use of earth. Annu Rev Energy Env 26:1–26

Kates RW, Parris TM, Leiserowitz AA (2005) What is sustainable development? Environ Sci Policy Sustain Dev 47(3):8–21

Khmara Y, Kronenberg J (2020) Degrowth in the context of sustainability transitions: in search of a common ground. J Clean Prod 267. https://doi.org/https://doi.org/10.1016/j.jclepro.2020.122072

Masocha R (2019) Normative environmental configuration of SMEs within the sustainable development discourse in South Africa: an empirical study. Sustainability 11(23):6537 https://doi.org/10.3390/su11236537

Matten D, Moon J (2005) Corporate social responsibility education in Europe. J Bus Ethics 54(4):323–337

Meinherz F, Fritz L, Schneider F (2020) How values play into sustainability assessments: challenges and a possible way forward. In: Binder C, Wyss R, Massaro E (eds) Sustainability assessment of urban systems. Cambridge University Press, Cambridge, pp 65–86

Miller T, Wiek A, Sarewitz D, Robinson J, Olsson L, Kriebel D, Loorbach D (2014) The future of sustainability sceience: a solutions-oriented research agenda. Sustain Sci 9:239–246

Mori K, Christodoulou A (2012) Review of sustainability indices and indicators towards a new City Sustainability Index (CSI). Environ Impact Assess Rev 32(1):94–106

Nagendra H, Bai X, Brondizio E, Lwasa S (2018) The urban south and the predicament of global sustainability. Nature Sustain 1:341–349

Nielsen J, de Bremond A, Chowdhury R, Friis C, Metternicht G, Meyfroidt P, Thomson A (2019) Toward a normative land systems science. Environ Sustain 38:1–6

Nilsson M, Griggs D, Visbeck M (2016) Policy: Map the interactions between sustainable development goals. Nature 534:320–322

Odum EP, Odum HT, Andrews J (1971) Fundamentals of ecology, 3rd edn. Saunders Philadelphia, Philadelphia

O'Neill J, Uebel T (2015) Analytical philosophy and ecological economics. In: Martinez J (ed) Handbook of ecological economics. Edward Elgar Publishing, Cheltenham, pp 48–78

Parnel S, Robinson J (2012) (Re) theorizing cities from the globa south: looking beyond neoliberablism. Urban Geogr 33:593–617

Polanyi K (1944) The great transformation. Farrar & Rinehart, United States

Potthast T (2015) Ethics in the sciences beyond Hume, Moore and Weber: Taking epistemic-moral hybrids seriously. In: Meisch S, Lundershausen J, Bossert L, Rockoff M (eds) Ethics of science in the research for sustainable development. Nomos Verlagsgesellschaft, Baden-Baden, pp 129–152

Pradhan P, Costa L, Rybski D, Lucht W, Kropp J (2017) A systematic study of sustainable development goal (SDG) interactions. Earths Future 5:1169–1179

Purvis B, Mao Y, Robinson D (2019) Three pillars of sustainability: in search of conceptual origins. Sustain Sci 14:681–695

Raman S, Mohr A, Helliwell R, Ribeiro B, Shortall O, Smith R, Millar K (2015) Integrating social and value dimensions into sustainability assessment of lignocellulosic biofuels. Biomass Bioenerg 82:49–62

Rockstrom J, Steffen W, Noone K, Persson A, Chapin FS, Lambin EF, Foley JA (2009) A safe operating space for humanity. Nature 461(7263):472–475

Sandberg M, Klockars K, Wilen K (2019) Green growth or degrwoth? Assessing the normative justifications for environmental sustainability and economic growth through critical social theory. J Clean Prod 206:133–141

Sen AK (1999) Development as freedom. Oxford University Press, Oxford

Smil V (2000) Energy in the twentieth century: resources, conversions, costs, uses, and consequences. Annu Rev Energy Env 25(1):21–51
Smith R (2004) Focusing on public value: something new and something old. Austr J Public Admin 63(4):68–79
Solow R (1974) Intergenerational equity and exhaustible resources. Rev Econ Stud 41:29–45
Steffen W, Richardson K, Rockström, J, Cornell S, Fetzer I, Bennett E, Sörlin S (2015) Planetary boundaries: guiding human development on a changing planet. Science 347
Swain R, Yang-Wallentin F (2020) Achieving sustainable development goals: predicaments and strategies. Int J Sustain Dev World Ecol 27(2):96–106
Tol R (2016) The impacts of climate change according to the IPCC. Climate Change Econ 07(01):1640004
Tran M, Beddewela E (2020) Does context matter for sustainability discloure? Institutional factors in Southeast Asia. Bus Ethics Eur Rev 29:282–302
Tukker A, Jansen B (2006) Environmental impacts of products: a detailed review of studies. J Ind Ecol 10(3):159–182
Turner RK, Pearce W (1994) Environmetnal Economics: an elementary introduction. Harvester, Hemel Hempstead
Ulucak R, Kassouri Y, Ilkay S, Altintas H, Garang A (2020) Does convergence contribute to reshaping sustainable development policies? Insights from Sub-Saharan Africa. Ecol Indic 112. https://doi.org/10.1016/j.ecolind.2020.106140
Uyarra E, Ribeiro B, Dale-Clough L (2019) Exploring the normative turn in regional innovation policy: responsibility and the quest for public value. Eur Plan Stud 27(12):2359–2375
Valentini L (2011) Justice in a globalized wrold: a normative framework. Oxford University Press, New York
van der Hel S (2018) Science for change: a survey on the normative and political dimensions of global sustainability research. Global Environ Change 52:248–258
van Niekerk A (2020) Towards inclusive growth in Africa. Dev Southern Africa 37(3):519–533
Verma P, Raghubanshi A (2018) Urban sustainability indicators: challenges and opportunities. Ecol Ind 93:282–291
Walker B, Holling CS, Carpenter SR, Kinzig A (2004) Resilience, adaptability and transformability in social--ecological systems. Ecol Soc 9(2):5
Wang L-S (2011) Causal efficacy and the normative notion of sustinability science. Sustain Sci Pract Policy 7(2):30–40

South African Study Exploring Sustainable 3D Printing of Biofidelic Tailoring Mannequins

Reena Pandarum, Ilunga Jeanmark Tshimanga, and Wei Hua Ho

Abstract This chapter explores a methods-flow process for 3D printing Biofidelic (lifelike) virtual clones of South African women with heterogeneous body morphotypes using a 3D full body scanner. The current production of tailoring mannequins cannot reproduce the 3D complex geometries of a clothing retailers' live-fitting model' body types. Details, such as the adipose tissue on the body, are difficult to replicate when using matrials such as clay, foam, fiberglass or plastic that requires various modifications to the body oultine of the mannequin from concept to design. The output is an aesthetically pleasing but anatomically incorrect mannequin that does not represent the Biofidelic morphotype of a retailer's live-fitting model. Therefore, the authors are proposing an alternative method to manufacturing tailoring mannequins.in order to improve the sizing and fit of ready-to-wear garments. The lead times and production processes are compared to establish a method that is sustainable, and one that provides a protocol for 3D printing Biofidelic tailoring mannequins. The output is for use as a teaching aid in schools, academia and in the formal and informal clothing sectors. The study is novel and adds to the body of knowledge on sustainable manufacturing of Biofidelic tailoring mannequins on the African continent.

R. Pandarum (✉)
Department of Life and Consumer Sciences, University of South Africa, c/o Christiaan de Wet and Pioneer Avenue, Private Bag X6, Florida 1710, South Africa
e-mail: pandak@unisa.ac.za

I. J. Tshimanga
Department of Mechanical and Industrial Engineering, c/o Christiaan de Wet and Pioneer Avenue, Private Bag X6, Florida 1710, South Africa
e-mail: tshimil@unisa.ac.za

W. H. Ho
School of Mechanical Aerospace and Industrial Engineering, University of Witwatersrand, Johannesburg, 1 Jan Smuts Ave, Johannesburg 2000, South Africa

Department of Mechanical and Industrial Engineering, University of South Africa, c/o Christiaan de Wet and Pioneer Avenue, Private Bag X6, Florida 1710, South Africa

W. H. Ho
e-mail: weihua.ho@wits.ac.za

W. Leal Filho et al. (eds.), *Sustainable Development in Africa*, World Sustainability Series,
https://doi.org/10.1007/978-3-030-74693-3_31

1 Introduction

The 20th century saw the emergence of Black African female models, such as Alek Wek, Iman Abdulmajid, Liya Kebede, and Oluchi Onweagba. However, their body morphotypes and anthropometric measurements are dictated by the Western fashion industry norms (Pandarum and Yu 2015).The lack of a unique South African or African garment size chart, and the global village of free trade, have contributed to this. The garment sizing system currently in use in the South African clothing sector has been adopted from the British, mainland European, and American systems, and has been adapted over the years (Pandarum and Yu 2015). In 2015, the United Nations adopted the 2030 Agenda for Sustainable Development that is focused on "sustainable development for all" (2020:2). The tailoring mannequin manufacturing process, discussed, is from 3D scan to 3D printing and alludes to four of the United Nations Sustainable Development goals namely, Quality Education; Industry Innovation and Infrastructure; Responsible Consumption and Production on the African continent. One such initiative is to build capacity for the clothing sector by introducing students to new and emerging technologies that also address the industry, innovation, and infrastructure sustainable goals, and hence, this chapter.

Garment sizing and fit, especially for women, has been a challenge for the clothing manufacturing and retail sectors since the 1940s, and is largely attributed to the production of clothing for an hourglass morphotype. However, Gill (2014) has indicated that there are as many as nine to twelve different body types that can emerge in a country's population. Furthermore, in countries such as South Africa, the population is heterogeneous with different ethnic groups and cultures reporting incongruities with clothing sizes and fit. Connell et al. (2006:81) study attributes this to the clothing industry making incremental changes to sizing specifications and not on the variations in body morphotypes. The extracted anthropometric landmarks and thus, the target market representative tailoring mannequin, currently used in the sector, are generated from 1-D linear body measurements to manufacture the tailoring mannequin's body dimensions (Yu 2004). The measurement data is extracted from the primary body dimensions such as the shoulder, bust, waist and hip girths, inclusive of body height to manufacture a full-scale human 'figure' size mannequin. A tailoring mannequin is used by retailers to fit base-sized garments for a specific brand or target market. This then becomes the standard for the sample garment fitting trials prior to commencing full-scale production. This garment is, thereafter, test-fitted on a tailoring mannequin and/or a 'live-fit' model until the manufacturer is satisfied with the drape and fit. The sample garment style is thereafter mass customised for a retailer outlet in a range of garment label sizes. A tailoring mannequin is defined as a dummy, a lay figure or a dress form used by tailors and dressmakers to fit and size clothing in the manufacturing and retail sectors (Stively 2020:1).

However, one of the many challenges encountered by the South African clothing sector and in academia today, is to establish a common understanding on various issues related to the sizing and fit of garments (Pandarum and Yu 2015). More specifically, there is the question of how to measure and hence, fit garments on the human body using new and emerging 3D technologies as opposed to the traditional 1-D measurement systems, which is fraught with problems. The latter's shortcoming is that linear measurement does not take the curves and the adiposity levels of the subject's body into consideration. Therefore, this detail is difficult to reproduce using the current clay, foam, fiberglass and plastic moulding methods. Gribbin 2014 states that this results in incongruities in the garment size, drape and fit on women's bodies. Therefore, the aim of this chapter is to explore a methods-flow process for 3D printing Biofidelic tailoring mannequins, more specifically, for an hourglass morphotype generated from a 3D full body scanner of a South African woman subject (Pandarum et al. 2020). Veitch and Davis 2010 define Biofidelic as lifelike. The 3D printing process flow provides an integrated solution focused on Quality Education, Industry Innovation and Infrastructure using emerging and sustainable technology on the African continent.

2 Systematic Review of Reported Sizing and Fit Studies in South Africa and Africa

Since the 1990s, countries such as South Africa have attempted to develop body size charts reflecting body measurements taken from manual anthropometric studies and, more recently, using a 3D body scanner e-tape measurement. Winks' (1990) study defined the morphotypes extracted on control dimensions of bust girth and body height for the Black (14%), White (69%) and Coloured (17%), cultural groupings. This study defined three morphotypes such as the A morphotype with a mean drop value (the difference between the hip and bust girth measurements) of 12 cm. The M morphotype with a mean drop value of 5 cm and the H morphotype with a mean drop value of 0 cm. In 1998, Defty published size charts for pattern drafting for women, that were based on body heights, five girth and four length control dimensions.

However, Winks' (1990) study excluded the Indian population from his analysis due to the limited sample size, whilst Defty's (1998) book focused on size charts for the average, tall and short women's body dimensions, with and without shoes. Both these sizing and fit charts and tables were based on South African research that is now 23 and 31 years old respectively. Winks' (1990) data is still widely used and referenced both nationally and internationally in apparel sizing and fit studies as a technical report ISO/TR10652 (1990), standards sizing for clothes. This in turn influenced apparel sizing, sizing charts or tables and hence, fit. However, clothing manufacturers and retailers are said to adapt size charts for their target markets.

Strydom's (2006) study noted variability in the control dimensions and the body landmark identifications used by different South African retailers. The qualitative data from Strydom's study indicate that sizing systems currently used by South African retailers and manufacturers are based on averages rather than body morphotypes. However, the primary focus of Strydom's study was a general overview of the clothing-related population measures used by South African manufacturing companies and retailers, and not on mannequin development. A further sizing and fit collaborative initiative by MacDuff and Smith (n.d.) reported that Ergonomics Technologies is the custodian of the South African National Defence Force (SANDF) manual anthropometric database in South Africa. This means that the largest manual anthropometric database in the country is not in the public domain and thus inaccessible.

Recent South African body sizing and fit studies include Ola-Afolayan et al. (2013) using a dressmaker's tape measure on (n = 50) African women and Makhanya et al. (2014) using a [TC^2]—NX12-3D-full body scanner. The latter reported that for a student population, the most prevalent body types of (n = 234) females (125 Caucasian and 109 African) aged 18–25 years, were triangular, hourglass and rectangular. The study conducted by Ola-Afolayan et al. (2013) identified the women as pear-shaped wearing ready-to-wear dress sizes from sizes 16 to 24. Makhanya et al. (2014) focused on Caucasian and African students on campus, which resulted in a fairly limited age spectrum. It was also unclear whether the sample included the Coloured and Indian women on campus also being classed as African.

A further analysis of the 3D scanned e-tape data collected by Makhanya et al. (2014) was undertaken by Muthambi et al. (2015) using data for the African triangular figure shaped women wearing apparel sizes 30 to 38 to develop an experimental size specification for a basic sheath dress. However, the fit trials were, conducted on a non-representative sample of women who were not 3D scanned. The study concluded that the methodology used required improvement, as the experimental size specification did not offer the anticipated fit in all areas of the test apparel for a sheath dress.

In 2014, Kasambala et al. conducted a perception study on the emotional aspects of apparel sizing for women when rated against their value systems for fit, using a laddering interview technique. This study included a psychographic question (adapted from Liddelow 2011) illustrating six body types for (n = 59) women who classified themselves as being triangle, oval, rectangle, inverted triangle, hourglass and diamond-shaped. A further study conducted in 2016, by Sokhetye using 3D scanner e-tape measurements for the development of a jeans sizing system for young South African women (n = 60) black, pear- shaped women. This study found that there was on average a drop value of 39 cm between the waist and hip circumference when compared to the drop value of 24 cm for a standardised size used in the South African clothing industry. Phasha et al (2020) compared the fit of women's clothing on a traditionally sculptured and moulded tailoring mannequin. Her study reported incongruities between the bodies of the 3D scanned women when compared to the commercially used size 10/34 tailoring mannequin for petite South African women. None of the studies above reported or explored an alternative method of producing mannequins that are Biofidelic.

3 History of Tailoring Mannequins

The current mode of tailoring mannequin development and production processes is lengthy that begins with taking photographs, building moulds and skimming the body surface to obtain a somewhat natural looking tailoring mannequin based on the garment size it is intended to fit. However, a major criticism of tailoring mannequins manufactured today is that the adiposity levels and the location of the 'fatty layers', are difficult to reproduce using the current manual and labour intensive methods. According to Milliam 2020, pre-production mannequin development can span several months. Figure 1 is an example of a tailoring mannequin currently used in the South African clothing and retail sector to fit a ready-to-wear garment label size 10/34, classified as an hourglass morphotype.

As noted, several attempts were made by the mannequin manufacturer to improve the body outline from a traditional to a more 'natural' looking shape, but this still does not have the realistic level of detail or the adiposity that occurs in the human form. Tailoring and fitting mannequins are milled after moulding to obtain balance and symmetry, thus producing what is perceived to be an aesthetically pleasing body outline. However, the human form is not mathematically proportional therefore, these tailoring dress forms/mannequins are challenging to use for educators, the clothing sector and in related industries where human morphotypes, size and proportions are important to the end product.

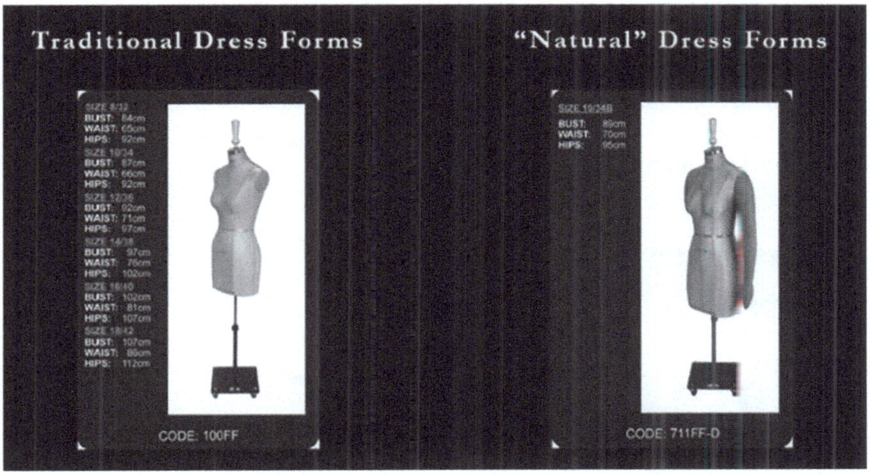

Fig. 1 An example of FigureForms tailoring mannequins (dress-form) currently used by the clothing sector in South Africa

4 Current and New Initiatives in Tailoring Mannequin Production

There is limited available information on tailoring and fitting mannequin manufacturing sector in South Africa and on the African continent. Manufacturers such as Dressmaker Dolls (2018) advertise and sell adjustable fitting and size-specific mannequins that resemble the fitting form illustrated in Fig. 1 above. Another manufacturer, FigureForms report that their fitting forms are developed from industry specifications and customised to aid in drafting pattern blocks extended to 'sample size checks' and for fit appraisal. This manufacturer currently also offers consultancy services in developing 3D virtual mannequins for use in fit assessment, virtual design and for visualization program software (FigureForms 2020:1).

However, none of these manufacturers report on 3D printing of tailoring mannequins nor on any sustainable production initiatives such as cleaner production manufacturing processes, recycling of waste materials or the use of sustainable materials and the impact of the fibreglass, foam and plastic on the environment as specified by the United Nations Sustainable Development Goals.

With the advent of 3D full body scanners interfaced with 3D printers, it is now possible to print realistic Biofidelic tailoring mannequins as a virtual clone of the human body Veitch and Davis (2010:1). The method is considered to be sustainable as the output is reproduced in a short space of time and is Biofidelic without any distortion to the true body outline as illustrated in Fig. 2. Furthermore, different morphotype categories and body frame sizes of virtual clones can be 3D printed quickly and cost-effectively, thus reducing the lengthy and costly development and manufacturing to market lead times. ISO 18825, 2016:3 defines a virtual clone as a virtual human body that is created by forming three-dimensional surface data from a 3D body scanned point cloud, using surface modelling processes, including noise elimination, hole filling and mesh generation.

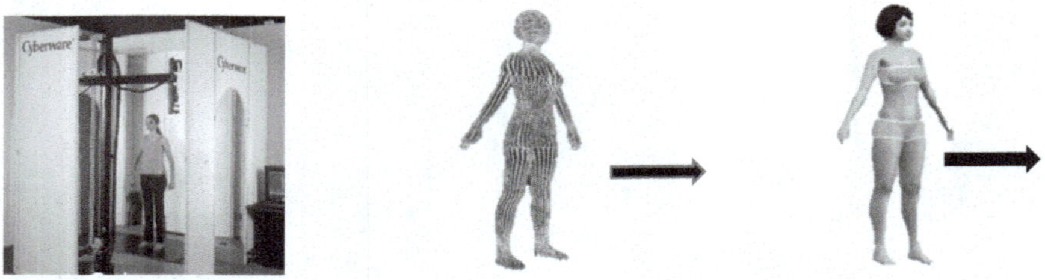

(a)3D body scanning **(b)** 3D scanned point cloud **(c)** Virtual clone

Fig. 2 Adapted from ISO 18,825 (2016)

Fig. 3 Adapted from www.tc2.com: Pattern Grating concept generated by the [TC^2]NX16 3D full body scanner

5 3D Full Body Scanner

The [TC^2] NX16-3D full body scanner is a custom apparel and design system used to generate the 3D point cloud topographical form of a subject morphotype. The system captures critical body landmarks and anthropometric data that guide the sizing of garments. Phase image profilometry and white light-based projection systems are used to generate the virtual clones or digital twin. Simmons and Istook (2003) explain that the profilometry method involves shifting the grating pre-set distances in the direction of the varying phase, thus capturing the subject's image in each position using structured white light (see Fig. 3). The triangulation data collected is a process of joining point data from the body scan, thereby creating a curved surface enabling the scan data to be visualized as a solid object.

6 Experimental-3D Anthropometric Data Acquisition

The 3D scan was collected from a sample of women classified as an hourglass according to the algorithms developed by a South African female study by Pandarum et al. (2020). The women were aged between 18 to 65 years, residing in Gauteng, South Africa. The sample size resulted from the women volunteering for the study and is not reflective of the heterogeneous grouping of the South African female population. The subjects completed a demographic questionnaire and a consent form prior to being 3D scanned captured under ethical clearance numbers [2011/CAES/044 and H15-SCI-TEX-001]. The female subjects were scanned in their undergarments as this is both hygienic and what one would be wearing whilst shopping for ready-to-wear garments in clothing retail stores. The scans were numbered consecutively and saved, with a printed copy handed to the women to take away as an incentive for taking part in the study.

The standing position in the [TC^2] NX16-3D scanner is erect, with the head in the Frankfurt-position and the feet parallel to each other, 350 mm apart, as indicated by the marking on the floor inside the scanning booth (ISO 20,685, 2010). The arms are outstretched, holding onto fixed handrails 1100 mm apart, with the right hand thumb hovering over the right handle on the button to activate the 3D scanner. This position optimizes the automatic body measurement of both the vertical and circumference areas of the upper and lower body.

7 Reverse Engineering

3D scanning and additive manufacturing (3D printing) is becoming prominent and rapidly advancing as a tool for modelling. This chapter investigated one such application, namely 3D printing an hourglass body morphotype Biofidelic tailoring mannequin for clothing design. In order to reproduce realistic mannequins, geometries must replicate the true figure shape for the hourglass morphotype and anthropometric measurements to a scale of South African female subjects. The scanning process, from which the tailoring mannequin was created, requires that the subject remains still in the scanning position for 54 s, as per the manufacturer specification.

Visual inspection of the point cloud scan extracted using structured white light often reveals missing and misplaced surfaces. These defects are generated during the capturing and generation of the 3D point cloud images using structured white light. Hidden areas such as under the armpits and the crotch are not captured, especially on the larger women, which result in missing points in the point cloud scan. If this missing data is present in the body scan when creating a solid object using direct 3D printing, the same defects will be present in the printed output. Hence, further design modelling of the 3D point cloud scan is often required to prepare the scanned file for 3D printing.

Reverse engineering is often applied in order to address these defects, it generally requires converting the 3D point cloud scan into surfaces or solids, which will serve as roots for further restructuring on the 3D point cloud scan. The conversion to a solid is generally executed by commercial computer aided design (CAD) packages. SpaceClaim design tool from ANSYS (ANSYS Inc.) offers a powerful and easy to use solution for reverse engineering applications and hence, the use in this chapter. The SpaceClaim design tool was used to enhance the 3D scan and correct for missing features (holes) in the scan by automatic surface fitting, using sketch curves along the facets, and blending of period loops in recreating complex geometries. (SpaceClaim 2011). The scan differences before and after reverse engineering are indicated in Fig. 4.

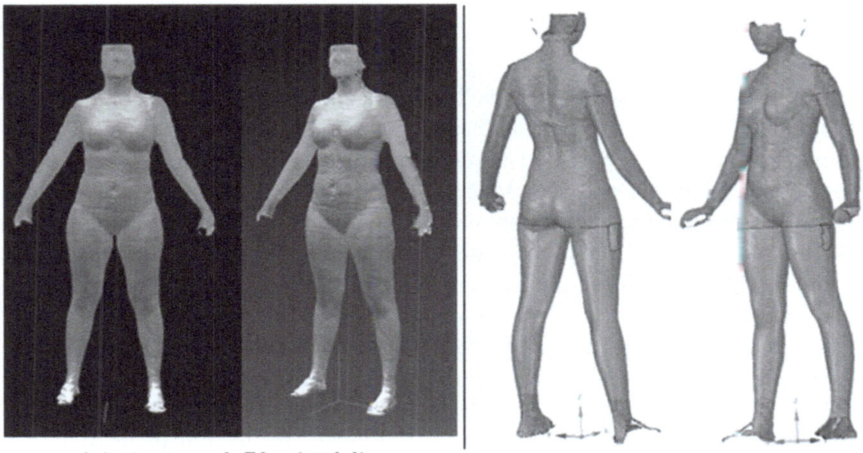

Fig. 4 **a** Scanned file; **b** CAD file

Data obtained in real body dimensions (RBD) from the [TC^2] NX16-3D was imported into an OBJ file format that is supported by the SpaceClaim design tool. The OBJ file was loaded in the SpaceClaim workbench and converted into a solid, and faces were merged during the conversion process. CAD packages employ boundary representation models or solid models for complex mesh generation. SpaceClaim is essentially centred on manipulating faces (SpaceClaim 2011). The Merge Faces tool helps to simplify the geometry by combining neighbouring faces into a single face; this results in a less complex and smoother mesh on the solid. After the solid file is generated, the prepare tab in the SpaceClaim workbench, with the ribbons and tools, is used to clean the imported file.

The Navigate Ribbon group provides tools that help to automatically detect and navigate problems in the geometry such as surface gaps in the imported file. The Missing Faces tool automatically identifies the gaps in the object, and upon command execution, the tool automatically fills in the missing faces on the imported object. Some missing faces that are not automatically detected are individually selected through their edges. After combining the fill and patch tools, additional mesh is generated on these selected regions. Subsequent sketching, which involves extrusion and combination, is added to improve areas that cannot automatically be filled correctly. Additional design work was conducted on the scanned object to imprint details that can identify models according to their body types by adding text to the scanned object. After filling in the gaps, small faces are removed to give a smoother finish before exporting the repaired object, namely, the Biofidelic mannequin, in STL format, which can subsequently be used by the 3D printer.

8 3D Printing and Method

3D printing is now considered as a well-established manufacturing technique. Two 3D printing processes are most prominent in the market, namely, the material extrusion process and the vat polymerisation process. Material extrusion includes technologies such as Fused Deposition Modelling (FMD), also known as Fused Filament Fabrication (FFF), while the vat polymerisation process includes technologies such as Stereolithography (SLA). In the material extrusion process, a solid thermoplastic material, usually in the form of a filament, is pushed through a heated nozzle, which melts the material, and deposits it on a build plate in an orderly fashion to create the desired object layer by layer. The vat polymerisation process make use of a light source such as a laser and selectively cures a photopolymer resin in a vat to produce the object layer after layer. FFF is a very popular and very cheap 3D printing technology, and the feeding material is equally affordable. This feeding material used in FFF can be made from recycled thermoplastic products, and this is the basis on which the FFF was used in this chapter. Details of the printer and the printing parameters are provided in Table 1. Furthermore, sustainable manufacturing technologies can be measured in the benefits it brings to the human, the environment, the social and to the economy. 3D printing offers countless ground-breaking opportunities to the human and one such opportunity has been explained on in this chapter. The alternative to 3D printing a lifelike mannequin will involve skilful handicrafts workers and wasteful subtractive manufacturing techniques with long production and manufacturing lead times.

The impact of 3D printing on the environment seems to have both comprehensive advantages and shortcomings such as there is a release of hazardous compounds in the environment and through the printing process. The material used in 3D printing very often depends on the status quo as is the case in any other manufacturing technique. The much-acclaimed sustainability aspect of 3D printing is in its availability, is easy

Table 1 MakerBot Replicator 2 printer setting parameters

Printer	MakerBot Replicator 2 From MakerBot Industries, LLC	
Maximum build volume	28.5 L × 15.3 W × 15.5 H cm	
Layer resolution	0.20 mm (standard)	
Materials/filament	MakerBot PLA Filament	
Filament diameter	1.75 mm	
Temperature of print	230 °C	
Scale of print	Scaled to fit	
Estimated infill	100% solid	
Estimated building time	5 h, 37 min	
Estimated material used	94.9 g	
Estimated cost of material in South Africa as of 2019 (PLA Filament)	Mass in kg	Price in USD
	0.9	75.4

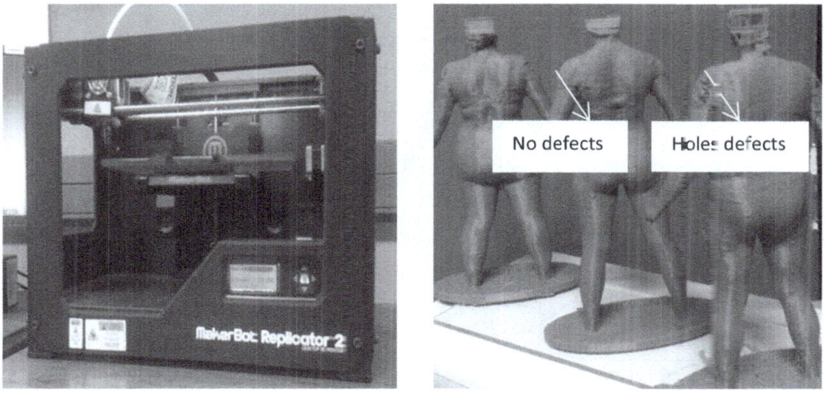

Fig. 5 **a** MakerBot Replicator 2 printer and **b** 3D printed tailoring mannequins

to use and in better waste management options. On the social and economic aspect, 3D printing is cost effective in special applications such as that discussed in this chapter as compared to other established manufacturing techniques.

The 3D object is constructed layer by layer fused in all XYZ directions Goyanes et al. (2014).

The 3D print is scaled to fit the maximum height of the MakerBot Replicator 2 printer without any distortion to the Biofidelic body morphotype. Figure 5 shows the 3D printed tailoring mannequin and the MakerBot Replicator 2 printer.

Physical measurements of the 3D printed tailoring mannequin was determined using a dressmaker's measuring tape, and the results were in agreement with the scanned data RBD file output. Figure 6 provides a comparison between a 3D printed mannequin and a point cloud scanned file.

As noted, the 3D printed scan image is more realistic Biofidelic tailoring mannequin for use in garment fitting when compared to what is currently used in the clothing sector.

Table 2 shows the output in centimetres (cm) for the body dimensions of the 3D printed tailoring mannequin and that of the 3D scanned body measurements of the virtual clone based on the actual body height of the 3D scanned subject of 160 cm. As noted, the output is accurate to scale of the Biofidelic virtual clone of the 3D point cloud scan and within acceptable limits for the end application.

9 Discussion

The 3D printed hourglass body morphotype reported in this chapter is referenced from a 3D scanner study conducted in Gauteng, South Africa by Pandarum et al. (2020). This study developed the algorithms for the different body morphotype categories emerging in the data-set of 509 women. The hourglass represented 24% of the women

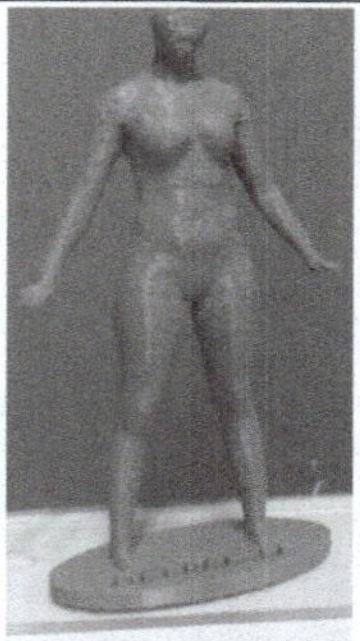

Half-scaled printed fitting mannequin of a South African female subject.

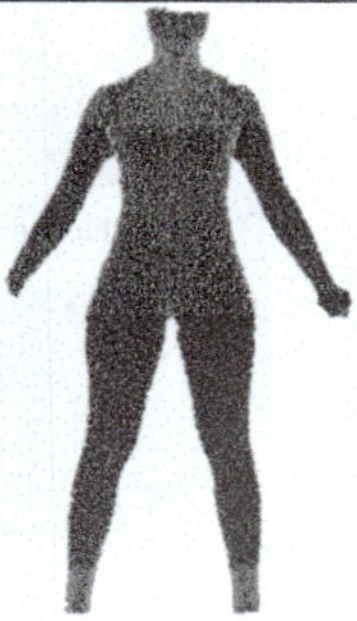

Scanned file – 3D point cloud of the same South African subject.

Half-scale hourglass tailoring mannequin currently used in academia in South Africa.

Fig. 6 MakerBot Replicator 2 printed scan indicating the adiposity and 3D point cloud of the scanned subject and the tailoring mannequin currently used by clothing retailers

Table 2 Primary body landmark dimensions based on body height of a 3D scanned female subject of 160 cm

	Bust (cm)	Waist (cm)	Hips (cm)
3D point cloud anthropometric measurements	88.92	68.94	100.98
3D virtual clone anthropometric measurements	89.00	69.00	101.00
Accuracy of the 3D printed point cloud and the virtual clone, key primary landmark-dimensions	99.91%	99.91%	99.98%

who were 3D scanned. The ratios of the hourglass morphotype are indicated in Figure 7.

The bust and waist to hip girth ratios for the hourglass morphotype are as follows: bust to waist girth ratio of **b > 1.24** and a hip to waist girth ratio of **h > 1.20**, where (b) is the bust girth and (h) the hip girth standardised to the South African National Standard [SANS 8559-1 (2019)] protocols.

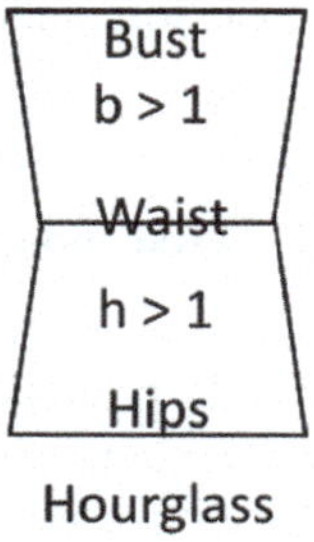

Fig. 7 The B/W (b) ratio = bust and H/W (h) = hip ratio (Pandarum et al. 2020)

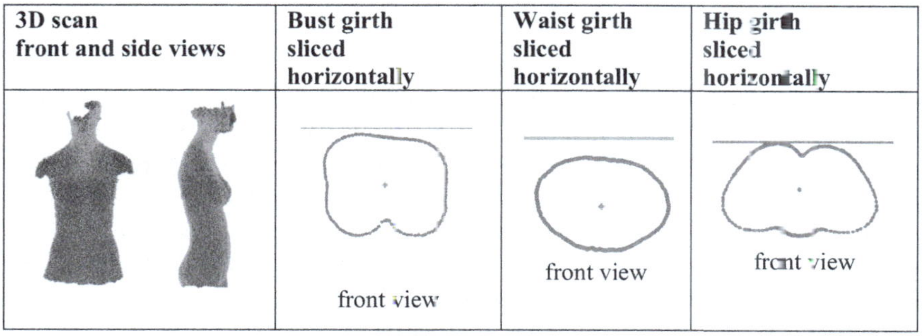

Fig. 8 Front views of the bust, waist and hip girth's perimeter slices for the hourglass morphotype 3D scan extracted at the 50th percentile (Pandarum et al. 2020)

To illustrate this further, the hourglass morphotype was sliced horizontally in the 3D scanner software interface. The contours of the bust, waist and hip girths were extracted at the 50th percentile to illustrate where on the body the adipose tissue is located. The irregularities at the perimeter of the slices can be explained in terms of the adiposity levels at that point and are present at different locations on the torso, thus providing an indication of the body morphotype for that particular female subject (see Fig 8).

The method flow process (MFP) for the 3D print and 3D body scans in point cloud from the scanner software interface was converted into a format that is user friendly and easy to transfer as a teaching aid. The authors also established the method flow process to reverse engineer the 3D scans to fill in any holes in the 3D scan, prior to 3D printing. This study found that the use of reverse engineering software enhanced the geometries of scanned 3D point cloud images. This process would not be required for a faultless 3D scan output. The dimensions of the MakerBot Replicator 2 printer used in the study are limited to the printing of a half-scaled fitting mannequin, which produces a Biofidelic virtual clone that is accurate to scale of the female subject 3D scanned hourglass morphotype. The output is for use in drafting pattern blocks and for fit appraisals in the clothing sector and also as a teaching aid in schools and for university level clothing practicals. Where students, both male and female, make and fit garments onto half-scaled Biofidelic tailoring mannequins to develop skills in reliably manipulating and altering pattern blocks and sample garments for fit-appraisals. The approach discussed in this chapter on sustainable 3D printing Biofidelic mannequin uses cost-effective resources that can be recycled. It also requires very little manufacturing floor space and has faster turnaround times that are sustainable. In contrast, the current mannequin manufacturing production processes and resources reported in the literature are non-renewable and unsustainable. Students entering the job market will be able to transfer the skills to their work environments, thus ensuring the widespread use of the method that is sustainable and easy to adopt and implement.

10 Conclusion

This chapter evaluated the feasibility of using 3D printing techniques to address contemporary issues relating to tailoring mannequin production and clothing incongruities resulting from the unique South African female body morphotypes and the non-renewable materials used in the current production process.

The authors have demonstrated that is it possible to 3D print a Biofidelic hourglass body morphotype from a South African female subject 3D scan. The manufacturing method of the Biofidelic tailoring mannequin was found to accurately represent the morphology of the 3D scanned hourglass point cloud scan after 3D printing. Furthermore, the chapter also established that it was easy to regulate and print the 3D scan, from point cloud scans, into a Biofidelic tailoring mannequin using cost effective resources requiring little manufacturing floor space. The method was found to be sustainable in that the printing lead times were approximately 15 h for printing a half-scale Biofidelic mannequin, as opposed to the current industry method that can take between one to six months.

The approach adopted in this chapter addressed four of the Sustainable Development Goals. The authors also established that the polymer used in the printing can be recycled and reused and hence, minimise waste and the impact it has on the environment as landfill. The output of the 3D printed tailoring mannequin is a solid object manufactured from premium quality material that is cost-effective, easy to obtain and is hardwearing with an increased lifespan that can be rapidly reprinted as and when the need arises. Clothing retailers can now produce a realistically shaped, sized, and proportioned 3D printed Biofidelic tailoring mannequin in their design studios or stores. The authors anticipate that using Biofidelic mannequins, ready-to-wear garments will require little or no alteration by customers and can be easily customised for a retailer's target market. Through the widespread use and adoption of this method, there will be a saving of resources, a decrease in fabric wastage and the clothing and related sectors' carbon footprint. Further work is recommended on 3D printing of other body morphotypes that emerge in populations on the African continent.

11 Future Trend

The presence of Covid-19 and the customers' aversion to fitting clothing in retail stores, has now necessitated the need for 3D body scanners and 3D Biofidelic virtual clones for use in online shopping. Future developments are envisaged in easy-to-use design integrated software developed from Biofidelic body morphotypes. This will optimise and enable the customisation of tailoring mannequins for both static and 4-D digital twins. The software is intended for use in industry sectors where body size, proportions and posture are important, such as: maternity wear; streetwear;

plus-size women and men's wear; bridal wear; kids wear; and apparel for those who are handicapped but not wheelchair bound and, in modular product development.

References

Connell L, Ulrich P, Brannon M, Presley A (2006) Body shape assessment scale: instrument development for analysing female figures. Cloth Textile Res J 24(2):80–92

Defty A (1988) The fascinating art of creating patterns, 2nd edn. Penwrite (PTY) LTD, Durban

Dressmaker Dolls. https://www.dressmakerdolls.co.za/. Accessed 17 July 2020

FigureForms. https://www.figureforms.co.za/. Accessed 17 July 2020

Gill S (2014) A review of research and innovation in garment sizing, prototyping and fitting. Textile Progress 47(1):1–123

Goyanes A, Asma B, Buanz M, Basit AW, Gaisford S (2014) Fused-fi lament 3D printing (3DP) for fabrication of tablets 476(1):88–92

Gribbin EA (2014) Body shape and its influence on apparel size and consumer choices. In: Faust ME, Carrier S (eds.) Designing of apparel for consumers: the impact of body shape and size. Woodhead Publishing, Philadelphia, pp 7–10

ISO/TR10652 (1990 [E]) Standard sizing systems for clothes. The international organization for standardization

ISO 20685 (2010 [E]) Method for reducing error in 3D scanning. The international organization for standardization

ISO 18825-1 (2016) Vocabulary and terminology used for the virtual human body. The international organization for standardization

Kasambala J, Kempen E, Pandarum R (2014) Female consumers' perception of apparel fit, personal values and emotions considering their body shape. Retail Market Rev 10(1):95–108

Liddelow J (2011) What' your body shape? https://www.style-makeover-hq.com/body-shape.html. Accessed 20 July 2020

MacDuff L, Smith JR (n.d) African body dimensions – a South African anthropometric initiative from uns.ac.id. Accessed 7 June 2020

MakerBot Replicator 2 from https://www.makerbot.com/3dprinters. Accessed 7 June 2020

Makhanya B, de Klerk H, Adamaki K, Mason A (2014) Ethnicity body shape differences and female consumers' apparel fit problems. Int J Consum Stud 38:1–9

Milliam R (2020) South African women body dimensions/size charts and body shapes, FigureForms, in conversation

de Muthambi A, Klerk H, Mason A (2015) Sizing for ethnicity in multi-cultural societies: validation of the size specifications for young South African women. J Family Ecol Consum Stud 44:1–10

Ola-Afolayan O, Mason A (2013) A customised size chart for the African pear shaped plus-sized South African women. In: DEFSA Conference Proceedings, Kenya

Pandarum R, Yu W (2015) Apparel sizing and fit. In: Nayak R, Padhye R (eds.) Apparel manufacturing technology, Woodhead Publishing, UK, pp 187–204

Pandarum R, Harlock SC, Hunter L (2020) Normative method for the classification and assessment of women's 3D scanned. Int J Cloth Sci Technol. https://doi.org/10.1108/IJCST-06-2020-0089

Phasha M, Harlock,SC, Pandarum,R (2020) Evaluating garment sizing and fit for petite women from 3D body scan e-tape measurements. J Consum Stud (48):93–111

SANS 8559-1 (2019) Size designation of clothes – Part 1: Anthropometric definitions for body measurements. South African Bureau of Standards

Simmons K, Istook C (2003) Comparison of three-dimensional body scanners. Textiles Mag 30(3):12–16

Sokhetye P (2016) Development of a jeans sizing system for young black pear-shaped South African women. Unpublished Master thesis. Durban University of Technology, Durban

SpaceClaim Tutorial: using 3D modelling software for 3D printing (2011). https://www.sculpteo.com. Accessed 5 July 2020
Stively M (2020) Dress Forms and Mannequins from https://www.seamwork.com/issues/2020/03/dress-forms-and-mannequins. Accessed 11 July 2020
Strydom M (2006) An evaluation of South African Apparel related population measures and sizing systems. Master's dissertation, University of Pretoria, Pretoria
Sustainable Development Goals: Young People and climate fighting for a stolen future from. https://www.za.undp.org/content/south_africa/en/home/sustainable-development-goals.html. Accessed 18 Aug 2020
[TC]^{2}-3 D body scanner. https://www.tc2.com/. Accessed 15 Mar 2020
Yu W (2004) 3-D body scanning. In: Fan J, Yu W, Hunter L (eds) Clothing appearance and fit: science and technology. Woodhead Publishing, Cambridge, pp 135–167
Veitch D, Davis B (2010) Use of 3D body scan data for apparel product development. The Technology Cooperation Programme, Sydney, May
Winks JM (1990) The development of standard sizing systems for clothes. Master thesis. University of Port Elizabeth, Port Elizabeth

Integration of Sustainable Development Goals into Leather Tanning Industries in Sub-Saharan Africa

R. O. Oruko, H. J. O. Ogola, J. N. Edokpayi, T. E. Volenzo, and J. O. Odiyo

Abstract Leather tanning industries are known to generate substantial amounts of toxic wastes that pollute ecosystems and pose environmental and health risks to the human populations living in the surrounding areas. The need to optimise and manage the leather tanning processes and wastes generated is therefore in line with the United Nations (UN), encompassing 17 sustainable development goals (SDGs), and 169 targets. This study sought to find out the contribution of United Nations Industrial Development Organisation (UNIDO) and other agencies mandated to assist the rapidly expanding leather tanning industries in sub-Saharan Africa (SSA). Specifically, the study focused on how these institutions helped create quality leather awareness for decent jobs, integrated green value chain, with eco-friendly tanning processes; information technology; as well as set up technical training, research institutes, and eco-industrial parks. It investigated how these entities promoted agroforestry, affordable energy, conserved indigenous knowledge and mainstreamed gender. It also looked at standards certification and labour laws that enable the tanning industries to attain sustainable development goals towards the 2030 agenda of the UN. The recent introduction of innovative bioprocessing and bioremediation techniques into this industry is also opening more opportunities for green leather development in line with the SDGs.

Keywords Green leather development · Integration · Sub-Saharan Africa · Sustainable development goals · Tanning industry

---

R. O. Oruko (✉)
Department of Ecology and Resource Management, University of Venda, Private Bag X5050, Thohoyandou 0950, South Africa

J. N. Edokpayi · T. E. Volenzo · J. O. Odiyo
Department of Hydrology and Water Resources, University of Venda, Private Bag X5050, Thohoyandou 0950, South Africa

H. J. O. Ogola
Department of Environmental Sciences, University of South Africa, UNISA 003, The Science Campus, P.O. Box 392, Pretoria, South Africa

W. Leal Filho et al. (eds.), *Sustainable Development in Africa*, World Sustainability Series,
https://doi.org/10.1007/978-3-030-74693-3_32

1 Introduction

The search for clothing material from animal hides and skins for use by early humankind faced challenges of putrefaction under damp conditions, stiff and cracking in very hot weather. These varying conditions reduced their suitability for clothing and other portable uses (Gromer et al. 2017; Oruko et al. 2014). The search for sustainable solutions to these challenges accidentally led to the discovery of tanning technology. The first tanning of leather is attributed to *Australopithecus habilis,* who roamed East Africa two million years ago (Infogalactic 2016; Thomson 2011). Thereafter, the technology expanded to the entire human race. However, in Africa, the tanning technique later became associated with specific communities. In the pre-colonial era, hides and skin trade as well as artisanal tanning, were practiced without policies and institutional structures that could guide quality and standards for posterity. This gap was partially addressed when the Europeans colonised the African continent. However, following the independence of most countries since the 60 s, the infant leather industries and their successors are still faced with a myriad of challenges such as poor quality, lack of skilled manpower and outdated machinery (Amde 2017).

The establishment of the United Nations Industrial Development Organisation (UNIDO) in 1966, as an agency of the United Nations, was to oversee industrialisation of member countries. This agency has helped realise the implementation of leather development programmes, such as quality improvement, upgrade of tanneries, the establishment of leather institutes, training of skilled manpower, transfer of technology, integration of environmental management, promotion of international standards (ISO), and mainstreaming of gender issues in the sub-Saharan Africa leather sector (UNIDO 2017). Currently, UNIDO is implementing the UN 2030 agenda encompassing, 17 SDGs and 169 targets. The 2030 Agenda of the UN recognises the importance of inclusive and sustainable industrial development (ISID) for achieving many of the interlinked SDGs. The nineth SDG focuses on building resilient infrastructure, promoting inclusive sustainable industrialisation and innovation that includes tanning industries (UNIDO 2017). The net result of the integration of these programmes for sub-Saharan Africa countries, in collaboration with UNIDO, amongst other agencies, has potentially sustained the expansion of the leather subsector in the region (Oruko et al. 2020). This chapter analyses in detail, the nature of leather, conservation of indigenous knowledge, job creation foot print, quality awareness, green value chain, technical training, research institutes, agroforestry, alternative energy, eco-friendly tanning processes, gender, labour laws, certifications, bioremediation and eco- industrial leather parks alignment, with SDGs as well as their targets in the region.

The Strategic Nature of Leather Industry and Conservation of Indigenous Tanning Knowledge

Although rubber and cotton from plants as well as synthetic material from carbon by-products, provide protection and flexibility, like leather, they are limited in their inability to prevent entry or allow the escape of water, heat, and sweat. Ability to

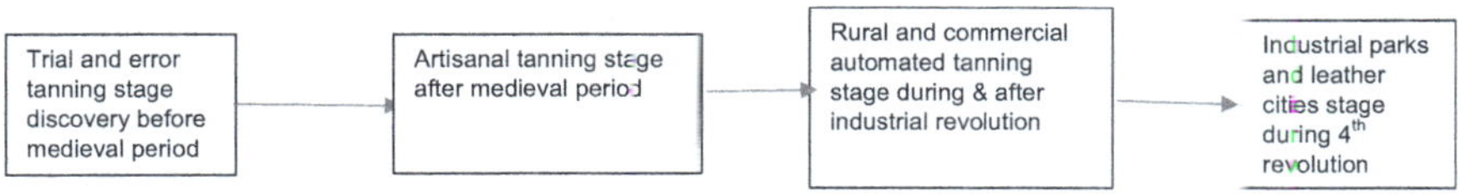

Fig. 1 Evolution stages of leather tanning technology in humankind civilisation

manage perspiration while preventing the water from entering is a unique property found only in the natural skin and processed leather (Virginija et al. 2012). This unique characteristic has sustained the leather industry from early humankind to date and is likely to continue into the unknown future in the civilisation of humankind. Sustainable evolution of the leather industry is depicted in Fig. 1.

The evolution of the tanning industry is strategic as it has enabled meat industry by-products to be utilised. This is a unique advantage that integrates with the industry innovation and infrastructure, to promote inclusive and sustainable industrialisation as well as foster innovation of leather tanning techniques in the region (UNIDO 2017; Okereke et al. 2019). The finished leather can be fashioned into different products, such as shoes, clothes, bags, and other leather articles across seasons to sustain humankind (Gromer et al. 2017), which is characteristic of resilience. As a result, the region has constructed over 175 tanneries, 650 leather goods companies, and 850 footwear companies (ALLPI 2020). This is, in addition to other uses of indigenous leather tanning knowledge among some communities. For instance, the Yirie method, that uses calcium powder from incinerated cattle bones, potash ash, quelea bird droppings, and *Acacia nilotica* pods, to tan skin into eco-friendly leather have been practised for a long time in Nigeria. Indigenous technology contributes towards sustainable value chains for local communities, as their products are deeply rooted in tradition, culture, geography, support rural development and promote the creation of job opportunities, which aligns to quality education, decent work, economic growth, and poverty reduction targets of the SDGs (Siyanbola et al. 2012).

Campaign for Quality Raw Skins, Leather Goods, Economic Growth, Information, and Communication Technology for Sub-Saharan Africa Countries

The quality of hides and skins in the pre and post-colonial era has been poor due to tick bites, knife cuts, poor shape, putrefaction and diseases among other problems (Kahsay et al. 2015). Such defects reduce the competitiveness of skins and hides in the local and international markets. However, skins from Africa have a comparative advantage. This is due to the tough fibres required for industrial belting leathers (Amde 2017). To address these constraints, co-ordinated awareness campaigns targeting the livestock farmers, slaughter points, and curing premises before the skins reach the tannery have been a permanent feature of improvement interventions between governments and UNIDO/FAO to date. The campaign has positively impacted poverty reduction, provided decent work, and contributed to economic growth in the region. As a result, producers of good quality skins are now guaranteed reliable income, while middle agents and tanners have assured sustainable supply of quality raw products (Frankema and Van Waijenburg 2019).

Currently, Africa produces 1,343,972,914 sq ft of bovine leather; 673,019,417 sq ft sheep leather and 602,510,873 sq ft goat leather. The current value from these semi-processed and raw hides and skins is estimated at $ 2billion (ALLPI 2020). Foreign exchange earnings from leather products have contributed to resilience-building and managing climate change risks for agriculture-based livelihoods (Adem 2019). Some of these tanning industries are even incorporating intelligent leather manufacturing processes, using information technology along the value chain, to improve energy efficiency, reduce production costs, promote a healthier working environment, security, and marketing. The industries attain responsible consumption, production and energy conservation (Mkhabela 2020). For Example, Namibia and Sudan have created platforms for information exchange, to improve their marketing and link the environment to an eco-leather label, as a target linking local and global markets (Ndhulukula and Plessis 2009). On its part, UNIDO has developed an animated e-learning course on footwear pattern engineering and conducted training of trainers for Tanzanians, to provide quality education for leatherworkers (UNIDO 2017).

The Job Creation Footprint in the Leather Industry

The value-chain analysis shows that a single animal skin can create sustainable employment, act as a source of income, reduce poverty and sustain livelihoods in a sub-Saharan Africa context, as illustrated in Fig. 2. It begins with the upstream (live animal, sale, slaughter and skin preservation) and ends with downstream (leather processing, utilisation and recycling of leather wastes, with potential negative environmental footprints).

Integration of Green Value Chain into Raw and Processed Leather in sub-Saharan Africa

Leather industries in sub-Saharan Africa continue to integrate environmental support functions, environmental rules, regulations, and market players in greening their value chain. They do this by introducing cyclical economic systems (DCED 2012). Green value chains (GVCs) reduce negative environmental footprints, with the sole aim being to make it easier for industries to export to European and American markets after meeting the international environmental agreements (UNIDO 2015). However, for a number of producers from the region who are still locked into low value-added activities, there are low economic returns and thus weak export incentives. Furthermore, there is a sustainable impact of the training on green value chains (GVCs) through UNIDO projects, including pollution control and improved quality for finished leather as well as shoe export into international markets (UNIDO 2015; Mwinyihija 2018).

GVC integration helps reduce hunger; creates responsible consumption, production, and reduces inequalities, as Africa now produces 1,702,677,082 sq. ft of processed leather into the global market (ALLPI 2020). In addition, industrialisation, innovation and infrastructure have also been promoted in the private sector. These companies have become engaged in industrial production, diversification of

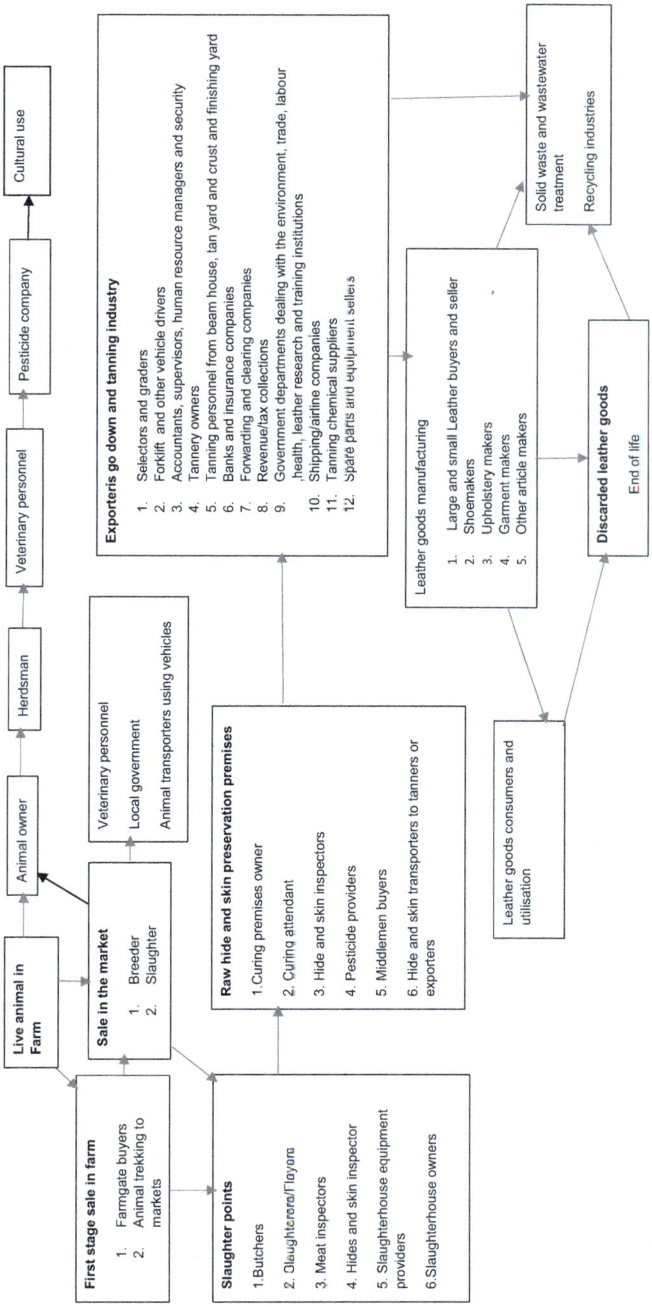

Fig. 2 Schematic illustration of how skin and processed leather influence job creation

the economy, local and global value chains, to guarantee decent work and economic growth. The establishment of leather institutes and a revolving fund have also promoted the training of skilled workers as well as the purchase of equipment in a revolving manner, for the private firms. These activities align with targets of quality education and transfer of knowledge through partnership into the region's tanneries (World Bank Group, Economic and Transformation Group 2015).

Technical Training, Partnership, and Establishment of Leather Institutions in sub-Saharan Africa

Lack of adequate skills among leather workers was established as one of the main problems facing the region's leather sector in most stages of production, design and marketing chain (Mwinyihija 2018). Thus, there is a need to establish sustainable training, research, and mobilise resource institutions, to enhance capacity-building skills. In order to achieve the right skills for their workers, Common Market for Eastern and Southern African (COMESA) countries have partnered with UNIDO and other agencies, to set up leather institutions at regional level. This gave birth to the chartered Africa Leather and Leather Products Institute (ALLPI), based in Addis Ababa and Eastern and Southern Africa Leather Industries Association (ESALIA), located in Lusaka (COMESA LLPI 2014). Leather stakeholder associations in the region have also promoted cooperation amongst themselves to overcome the economy of scale and attain coherence in policies and actions from local to the global arena. The other related leather institutions are government departments, research institutes, technical colleges, universities, leather councils, export promotion agencies, and various leather associations. Their establishment is aligned with quality education and technology exchange (ALLPI 2020). Some of these institutes have offered training to members of the associations on better grading systems at supply sources and price differentiation based on grades to promote quality hides and skins. This has resulted in higher income along the entire chain, with the overall result being reduced inequalities and economic growth in the region (Wanyoike et al. 2018).

Promotion of Agroforestry and Solar Energy Projects in Sub-Saharan Africa Tanneries

A few tanneries and research institutes have set up tree nurseries that supply tree seedlings to local communities. This is part of their social responsibility, climate change mitigation, and conservation of endangered species. The planted trees with and without their compounds are part and parcel of urban greening, carbon sequestration as well as sustainable cities initiatives (Missanjo et al. 2014). Although the ecological importance of *Acacia mearnsii,* which produces tannins for leather tanning, has not been positively connected to targets of climate action, affordable and clean energy supply in sub-Saharan Africa has those potential benefits (Wu and Yu 2019). Besides the tannins, *Acacia mearnsii* provides fodder for animals, wood for construction, fuel, a source of cash income, sustainability of soil nutrients, is a part of the water cycle and a buffer in climatic extremes. This is achieved through carbon sinks, to reduce greenhouse effects. Currently, the planted acreage in South Africa stands at 110,000 ha, while Zimbabwe, Kenya, Tanzania, Rwanda, Burundi account for 30,000 ha. In addition to the officially recorded planted area, extensive but largely unrecorded plantings in agroforestry systems in many of these countries exist (Chan et al. 2015). The plant species are important for taking up 2.3% of Sub-Saharan Africa's carbon emission, besides soil erosion control, soil fertility improvement, shade shelter, and aesthetics (Mbow 2020). To also reduce the high expenditures incurred in electricity in the tanning process, some tanneries are installing solar power plants of varying capacities, to tap the abundant solar resources, as alternative sources of green electricity. This lowers the carbon footprint while mitigating the effects of climate change and provides affordable clean energy (Gupta et al. 2018).

Integration of Modified Eco-Friendly Tanning Processes and Innovative Waste Bioremediation Techniques of Leather Industries in the Region
Leather tanning has had a negative image in society for a long while due to pollution, even though the industry creates employment and export earnings. In this region, a number of tanneries still use a conventional tanning process that does not meet the set standards (Oruko et al. 2020). Thus, there is a need to incorporate sustainable practices that are eco-friendly in all stages of processing liquid/solid/gaseous wastes, cleaner technologies and process modifications as well as introduction of innovative bioremediation techniques in regional leather industries (Gupta et al. 2018).

Currently, the management for wastewaters involves anaerobic treatment, based on lagoons, contact filter, upflow anaerobic sludge blanket (UASB) reactor, and sometimes high-rate biomethanation. These techniques convert organic waste into methane gas, lowers energy use and takes a small area, as the anaerobic treatment reduces the biological oxygen demand (BOD) and chemical oxygen demand (COD) loads by 50–60%. However, aeration methodologies are more advantageous because they use aerators, with and without the aid of aerobic microorganisms and achieve 30 ppm and 250 ppm BOD and COD respectively. These provide affordable clean energy, clean water, and sanitation in tanneries (Dixit et al. 2015; Thanikaivelan et al. 2005).

Wet air oxidation at ambient pressure and temperature, chemoautotrophic activated carbon oxidation, activated carbon filters, reed bed, and root zone techniques as well as reverse osmosis, have been developed to improve the tertiary treatment of tannery wastewaters. Furthermore, salt-bearing tannery wastewater is being treated with an improved high-rate transpiration system, while accelerated evaporation and crystallisation of saline streams using flat-plate collectors and sprinkler systems are also in use (Gupta et al. 2018). These help produce clean water and sanitation for the tanning industry, as their improvement saves life below water in wastewater receiving bodies.

Technological options for handling solid waste from the tanning process include the manufacture of glue, conversion of collagen-based biomaterials for medical and other applications from trimmings of raw hides/skins, pelts, and fleshings. The salt recovered from mechanical desalting and solar evaporation is reused for curing or pickling the hides and skins. Fleshings and solid sludge from primary and secondary treatments are used for biomethanation. The shavings, trimmings, and buffings are used in the manufacture of leather boards, as chrome shavings are further used as a reductant in the manufacture of basic chromate sulphate (BCS). Lastly, parchment from chrome shavings is used in the manufacture of home furnishing products, while low-quality hides and skins are made into gelatine and consumed as protein (Masilamani et al. 2016; Procter 2019). In some cases, organic compost is made in pits using raw flesh cuttings, shaving dust, and trimmings of the tanned skins, together with animal dung and plant materials. The resultant organic manure is used on farms to grow crops (Gupta et al. 2018; Mwondu et al. 2019). Gaseous wastes like hydrogen sulphide and ammonia from the tanning process, are ecologically removed through adsorption, using activated charcoal (Panda et al. 2014). Thus, recycling solid leather and gaseous wastes produces valuable products that help address poverty reduction, lowers hunger, creates decent work; leads to economic growth, good health and well-being as well as responsible consumption, industrial innovation and infrastructure; improved life on land, and fosters technology exchange for posterity.

The adoption of cleaner technologies involves several minimisation techniques along the tanning stages:- In pre-tanning, saltless or less-salt curing with solar drying, freeze-drying, microwave/dielectric drying, KCl instead of NaCl, borax–phenol, zinc chloride, silica gel, and low quantity of salt, metal oxinates, gamma, electron beam and irradiation techniques, while the transportation of green skins in refrigerated trucks reduces salt pollution. Although saltless curing processes are available, wet salting is still common practice in the region due to its low cost and high efficiency (Thanikaivelan et al. 2005; Oruko et al. 2014; Dixit et al. 2015). However, the tanners' long-term benefits are quality improvement for skins that meet market demand. This reduces poverty, promotes responsible consumption, production, and improves life on land and below water as less salt pollution is achieved.

The segregation of waste soak stream, by exposing them to solar evaporation to eliminate salt, involves co-current flow of skins and hides on the one hand, and soak liquor, on the other hand. The method was very effective, as it led to a reduction in the amount of water used by nearly 50–60% of tanneries in soaking operations. This offers industries an opportunity to reduce the hydraulic load by 20% in the total volume of wastewater and leads to the release of clean water that meets sanitation guidelines (Guche 2018). There are reports of dehairing using Sirolime (impregnation with hydrosulphide, oxidation of hydrosulphide and lime addition), hair recovery by filtration, and destruction of residual hair, using sulphide, lime, surfactant and recycled lime liquor (Ali et al. 2018). Continuous-flow unhairing technique is also in use to transform hairs into pulp and strip them with plastic blades, as pelts were fleshed and split. There are further reports of sulphide replacements with restricted application due to cost, efficiency, and toxicity, as well as non-enzymatic painting techniques, although the nickel compounds incorporated in them poses serious health problems (Jian et al. 2011). De Sousa (2014) promotes the use of proteases enzymes for dehairing because of merited hair-saving approach, as small amounts of sulphide and lime are applied as paint on the flesh side. Notable merits were higher area recovery (2–3%), smoother grain, better in-plant ecology, and higher efficiency of wastewater treatment. Other benefits are reduced BOD, COD, total dissolved substances (TDS), sulphide levels, and improved efficiencies of wastewater treatment plants. The overall implication of the above is a bearable environment for all living organisms on land and below water, as clean water and sanitation promote good health and well-being of populations (Kanagaraj et al. 2015).

Lime-free fibre opening methods using strong alkalis, *Lactobacillus*-based enzymes, lyotropic agents (salts/acids that cause lyotropic swelling), and bating enzymes are available but sparingly applied, due to the fibre bundles opening challenges. Furthermore, lime splitting for heavier hides lowers the cost of chemicals used and the amount of water consumed, along with processing time, in addition to area yield increases by about 5%. The washings of limed pelts are reused for liming, to reduce pollution load, the volume of wastewater discharged and ultimately clean water, as well as good sanitation, and a safe life below the water (Saravanan et al. 2014; Thanikaivelan et al. 2005). The use of ammonium chloride and sulphate in deliming contributes 75–80% of the ammonia in the effluent, which increases the COD of the wastewater. It also generates the dissolved ammonia– nitrogen-sulphate that produces a bad smell (Gajewski and Olszewski 2015). Thus, there is a need for alternatives, based on organic and inorganic acids, esters of carboxylic acids, and non-swelling aromatic acids. However, their cost is prohibitively high. Carbon dioxide deliming has been experimented with (Maina et al. 2019). This is because it has environmental benefits, which link well with the targets of clean water, sanitation, responsible consumption, production, and life below water.

During the tanning stage, conventional chrome tanning uptakes only 40–70% of the material used during tanning. This results in the wastage of material and ecological effects. The value of the loss in some countries has been estimated to be US$10,000,000 per annum. Furthermore, the physicochemical treatment methods result in chrome-bearing sludge, which is difficult to dispose of. Direct recycling of spent chrome liquor after preacidification to a pH of 1.0 produces leathers with good surface deposition of chromium. Others report membrane separation for the selective removal of chromium from neutral salts, as it uses a chrome-rich stream for tanning and a neutral-salt rich stream for pickling. The systems have economic viability for the tanner and integrate with responsible consumption of resources (Zhang et al. 2017; Oruko et al. 2020).

A solvent-extraction technique that recovers chromium from spent chrome liquor, using Cyanex 301–toluene, along with mono (2-ethylhexyl) phosphoric acid–n-hexane systems, have been experimented with. More research on the removal of chromium from tannery effluent using plants such as seaweed and *Sargassum species*, and reusing the chromium-bearing seaweed in the manufacture of basic chromate sulphate as a reductant, was reported as responsible consumption of resources from ecosystems (Nguyen et al. 2020; Arumugam et al. 2018). The use of zirconium oxychloride, titanium, and oxazolidine as single tanning agents have also been tested. Lastly, richly dyed garment suedes, based on vegetables and some organic tanning supplements, have also been made. These systems can be cost-effective, as they emerge as chrome savers, which meet the targets of industry innovation and realise good health and life on land and below water (Oruko et al. 2020; Gajewski and Olszewski 2015).

In post-tanning operations, retanning agents, fatliquors, dyes, and finishing chemicals, are used. The uptake levels of these chemicals vary, quantitatively depending on the level of applied technology. Treatability of post-tanning chemicals and the neutral salt content are of environmental concerns, as they significantly impact COD. However, the application of high-performing dyes can rationalise the entire wet finishing operations, to optimise the quantities of the chemicals used. Pollution load-reduction in finishing-yard effluent could also involve an array of measures, such as proper control of spraying machines (Minimise wastage and emission), use of roller coaters or low-pressure spray pistols (new application), elimination of lead and other toxic-metal-based pigment formulations (avoid toxicity) and the use of water-based formulations, in finishing (minimises volatile organic compounds). The use of newer and safer cross-links in protein finish, in place of formaldehyde, are current innovative approaches. The reduction in the discharge of chemicals in waste streams and gaseous form at this stage links with indicators of good health, wellbeing, clean water, sanitation, responsible consumption, production, life below water and on land (Hansen et al. 2020; Wang et al. 2019).

The end-of-pipe treatment technologies available for achieving wastewater discharge values in the region and some parts of the globe, still show some gaps that have led to the closure of some tanneries. Therefore, more advanced sustainable solutions that approach zero discharge of pollutants from leather industries are being developed as new industrial innovation and infrastructure to efficiently treat residual tanning pollutants. The strategies involve combining all possible known clean method approaches, to help meet the stringent regulations and standards required by the regional governments and international leather markets (Oruko et al. 2020).

In process modification, the zero-waste concept is designed in such a way that pre-tanning and tanning processes exclude soaking, because of diverse curing processes for green hides and skins (Thanikaivelan et al. 2005). Here, the working pH is set at a range of 4.0–8.5. To achieve the target, experimental research explored the liming and reliming chemical inputs and role of concentration, bioprocessing for sulphide and lime sludge replacement, lime-free enzyme assisted dehairing, target enzyme fibre opening for proteoglycans (protein conjugated with carbohydrates) in skin/hide. Others were sodium hydroxide-based fibre opening, to lower lime sludge as well as pickle and basification-free chrome tanning and integration of best practices from dehairing, fibre opening and pickle basification-free chrome tanning processes, to achieve zero discharge of pollutants, as innovative industrialisation (Oruko et al. 2020; Ali 2019).

The latest reported experimental outcomes that the regional tanneries have or may need to integrate for sustainable tanning in the industry are: an eco-friendly liming–reliming technique in the beam-house chemicals. The technique uses only one-tenth of the chemicals conventionally used, while maintaining the normal concentration gradient. The total dry sludge from the liming–reliming processes was reduced to 6 kg from 152 kg for 1 ton of raw skins (Giannetti et al. 2015; Thanikaivelan et al. 2005). Next was the use of lyotropic agents in fibre opening and pickleless chrome tanning system in a narrow pH range (4.0–8.0). This reduced COD and TDS loads by 67% and 64%, respectively. Another option that replaced water with ethanol, ethyl acetate, and ethyl lactate for chrome tanning was also found. The potential application of the technique was sustainable leather manufacturing, with solvent recycling and higher exhaustion up to 95% for a pickle-less process. There was also a reduction of COD; BOD, and TDS loads in the composite liquor of 14–26%; 21–28%, and 42–46%, respectively, thus meeting clean water and sanitation guidelines (Zhang et al. 2016; Silambarasan et al. 2015).

The potential of bioprocessing, to reduce environmental pollutant footprints, was significant in comparison to leather treated with lime. Waste sludge in the liming–reliming processes decreased from 152 to 12 kg, after processing 1 ton of raw skins with *galactosidases* enzymes from *Aspergillus terreus,* instead of lime. However, the amylase-treated hides, had to be soaked in a pickle bath consisting of a large quantity of salt and sulphuric acid, and then preserved with chrome tanning salt. The sludge had a significant increase of dissolved solids and chromium in the wastewater. Therefore, the model integration process using a three-step tanning technique was found necessary. In the technique, the enzyme-driven three-step tanning stage worked for thin, heavy, as well as thick skins, as their physical and tactile properties compared well with those of traditionally processed leather (Durga et al. 2016). These achievements align with sustainable climate action, clean water, sanitation and life below water and on land as targets of the SDGs.

Lastly, waste management in tanneries in the region is still a big challenge for tanners and local authorities, as they require innovative and eco-friendly bioremediation strategies. Thus, the exploration and exploitation of the techniques in our current work (yet to be published and under patent consideration) reveal that we were able to utilise a certain type of animal dung (name withheld due to patent regulations) for the first time to bioremediate the tannery chromium effluent. Applying the batch adsorption technique, the dung adsorption capacity was found to be 93% for Cr (VI). When dung use was upscaled to an anaerobic bioreactor technique, it was able to reduce high concentrations of total chromium in real tannery chromium wastewater by 99%, compared to 50% in the control. These achievements implied, that dung was a better adsorbent. It also has microbes with the potential to bioremediate chromium in anaerobic conditions. The microbes may have originated from the rumen of the animal. These findings are unique and align with the SDGs such as poverty reduction, lower hunger, good health, wellbeing, clean water, sanitation, industry innovation, infrastructure, affordable clean energy, life on land, and below water. This was because wastes in the bioreactor could generate biogas, while reducing chromium concentrations. Finally, the waste could be reused as organic manure on farms, to grow crops after the desorption of chromium in the waste.

Eco-industrial Leather Parks and Cities, Gender Mainstreaming, International Labour Conventions and Leather Certification in the Region

The Eco-industrial leather parks and cities concept has been embraced in the region. Currently, over five are operational, while the same number are proposed (ALLPI 2020). The concept integrates sustainable industrialisation and the transfer of technologies with environmental gains and social responsibility (UNIDO 2017). They are being mainstreamed in tanning industries, to promote a circular economy, water resource conservation, attract investment, generate employment, increase foreign exchange earnings, enhance technology transfer, create backward linkages, recycling and sound management of waste, as well as the utilisation of industrial synergies. This is done to increase resource consumption and reduce waste generation currently being witnessed in expanding leather industries in the region, to achieve sustainable cities and communities inhabited by all humankind (UNIDO 2017).

The UN-Women, FAO, and UNIDO have also made efforts to mainstream gender and young entrepreneurs' issues into the leather industry. Most of their projects have focused on youth and women entrepreneurship and met set targets of gender equality. They have also supported start-ups and existing enterprises, to help boost value addition for over 160 women in Tanzania; 42 women in Uganda; 450 women in Kenya; 50 in Zimbabwe; 60 in South Sudan, and part of 450 operators in Sudan (UNIDO 2017; Shrewsbury et al. 2007; Temsch and Senior 2002).

Many countries in the region have also ratified the International Labour Organisation (ILO) conventions, to respect, promote and realise the principles of elimination of all forms of forced and compulsory labour, effective recognition of the right to collective bargaining, effective abolition of child labour, and elimination of discrimination in respect of employment and occupation. Using the labour-centred development (LCD) approach, they have helped reduce inequality and gender bias, promoted peace, justice, and strong institutions for human rights in leather industries. The above is achieved as trade unions address decent work deficits in the informal economy, enforce contracts; champion adequate earnings, productive work, representation, secure employment, and social protection for workers (Schminke 2019).

The Leather Working Group also promotes quality certification of raw and processed leather products from Africa into the international market. This has prepared some companies for ISO 9001 certification, to meet CEN/TC 309 on global standardisation, directive 94/11/EC on labelling, and commission decision 724/99/EC on traceability of materials. These regulations protect the environment; support trade, and reduce unfair competition, as they integrate targets of industrial innovation and infrastructure goals (Memedovic and Mattila 2008; World Bank Group, Economic and Transformation Group 2015).

2 Conclusion

This study concludes that, the strategic processing of leather has led to a job creation footprint and quality awareness, integration of green value chain and information technology, promotion of agroforestry and affordable energy, conservation of indigenous tanning knowledge and modification of eco-friendly tanning processes in Sub-Saharan Africa. Furthermore, it has led to the development of eco-industrial parks and bioremediation of wastes, mainstreaming of gender, establishment of labour laws and certification of leather. These have moved this industry towards achieving SDGs and targets in sub-Saharan Africa. However, there is still a need for further exploration and improvement, to effectively and efficiently meet all SDGs for the industry by 2030. Therefore, we recommend further integration of information and communication technology for the dynamic and expanding future of leather markets. We also recommend an assessment and evaluation of environment and social safeguards, governance gaps and opportunities, and commitment of more resources by the tanning industry in agroforestry. This will help to conserve endangered vegetable tanning plants which are threatened with extinction, while exploring new ones to mitigate the effects of climate change and diversify the recycling of leather wastes. The latter can lead to the creation of more innovative industries as they carry further research on bioprocessing and bioremediation techniques. Eventually, these leather goods, techniques and knowledge need to be conserved, preserved and displayed in specially set up leather museums in sub-Saharan Africa, for intergenerational equity before extinction. Thus, there is hope that long-term integration of these approaches by the tanning industries in the region will contribute to the attainment of SDGs by 2030 and beyond, for socio-economic development in the region.

References

Adem M (2019) Production of hide and skin in Ethiopia; marketing opportunities and constraints: a review paper. Cogent Food Agric 5:1565078

Ali MA, Ali AAH, Abuuznien HMM (2018) Improvement of unhairing operations in tanning process in Sudanese Tanneries. J Environ Res 2:2–5

Ali AAEH (2019) Impact of unhairing liquor and chromium spent solution recycling as an alternative treatment option in leather industry. Am J Biomed Sci Res 2:000572

Africa Leather and Leather Products Institute (2020) Africa leather and leather products, Africa livestock, hides and skins and leather & leather products production data. Addis Ababa, Ethiopia

Amde B (2017) Major factors affecting hide and skin production, quality and the tanning industry in Ethiopia. Adv Biol Res 11(3):116–125

Arumugam N, Chelliapan S, Kamyab H, Thirugnana S, Othman N, Nasri NS (2018) Treatment of wastewater using seaweed: a review. Int J Environ Res Public Health 15:285

Chan JM, Day P, Feely J, Thompson R, Little KM, Norris CH (2015) *Acacia mearnsii* industry overview: current status, key research and development issues. South Forests J Forest Sci 77(1):19–30

Common Market for Eastern and Southern African–Leather and Leather Products Institute (2014) The charter establishing the COMESA –leather and leather products institute (LLPI) (incorporating 1993 amendments). https://www.euclid.int/documents/COMESA-LLPI-Charter.pdf

Donor Committee on Enterprise Development (2012) Green value chains to promote Green Growth June 2012. publish under the auspices of ILO. https://www.enterprise-development.org/wpcontent/uploads/Green_Value_Chains_to_Promote_Green_Growth.pdf or www.Enterprise-Development.org/page/ggwg

De Sousa MF (2014) Advances in understanding of enzymatic unhairing of bovine hides. A technical paper presented at the 110th annual meeting of the American leather chemists association, Gideon Putnam Resort, Saratoga Springs, NY. 18–20 June 2014

Dixit S, Yadav A, Dwivedi PD, Das M (2015) Toxic hazards of leather industry and technologies to combat threat: a review. J Clean Prod 87:39–49

Durga J, Ranjithkumar A, Ramesh R, Girivasan KTPV, Rose C, Muralidharan C (2016) Replacement of lime with *Carbohydrasesea* successful cleaner process for leather making. J Clean Prod 112:1122–1127

Temsch R, Senior MM (2002) Joint in-depth evaluation mission; UNIDO programs funded by Austria to strengthen the leather sector in Uganda. UNIDO Projects US/UGA/92/200, US/UGA/96/300, 26 October, pp 2

Frankema E, Van Waijenburg M (2019) The great convergence. Skill accumulation and mass education in Africa and Asia, 1870-2010. CEPR Discussion Paper No. DP14150

Gajewski R, Olszewski PK (2015) Promotion of the best available techniques and positive ecological solutions for leather industry in the light of efforts carried on within life+ programme. 2-Maintenance problems

Giannetti BF, Agostinho F, Luciano CM, Cecília MVB, Ulgiati SA (2015) Multicriteria cost–benefit assessment of tannery production: the need for breakthrough process alternatives beyond conventional technology optimisation. Environ Impact Assess Rev 54:22–38

Gromer K, Gabriela P, Saliari K (2017) Products of animal skin from Antiquity to the Medieval Period. Annalen Des Naturhistorischen Museums in Wien, Serie A 119:69–93

Guche (2018) MC Recycling of dusted salt from raw hides/skins for pickling operation using supernatant (waste liquor) from chrome recovery unit: a case of Dire tannery. A thesis. Addis Ababa university. Ethiopia

Gupta SK, Gupta S, Gayathiri S (2018) Pollution prevention is the key to drive sustainability Preliminary findings from a tannery unit in India. www.emeraldinsight.com/1477-7835

Hansen E, de Aquim PM, Hansen AW, Cardoso JK, Ziulkoski AL, Gutterres M (2020) Impact of post-tanning chemicals on the pollution load of tannery wastewater. J Environ Manage 269:110787

Infogalactic news: the planetary knowledge core (2016) Conservation and restoration of Leather. https://infogalactic.com/w/index.php

Jian S, Wenyi T, Wuyong C (2011) Kinetics of enzymatic unhairing by *protease* in leather industry. J Clean Prod 19:325–331

Kahsay T, Negash G, Hagos Y, Hadush B (2015) Pre-slaughter, slaughter and post-slaughter defects of skins and hides at the Sheba Tannery and Leather Industry, Tigray region, northern Ethiopia'. Onderstepoort J Veter Res 82(1):7. Art. 931

Kanagaraj J, Senthilvelana T, Panda RC, Kavitha S (2015) Eco-friendly waste management strategies for greener environment towards sustainable development in leather industry: a comprehensive review. J Clean Prod 89:1–17

Maina P, Ollengo MA, Nthiga EW (2019) Trends in leather processing: a review. Int J Sci Res Publ 9

Masilamania D, Madhana B, Shanmugama G, Palanivela S, Narayan B (2016) Extraction of collagen from raw trimming wastes of tannery: a waste to wealth approach. J Clean Prod 113:338–344

Mbow C (2020) Use it sustainably or lose it! the land stakes in SDGS for sub-Saharan Africa. Land 9(3):63

Memedovic O, Mattila H (2008) The global leather value chain: the industries, the main actors and prospects for upgrading in LDCs. Int J Technol Learn Innov Dev 1:4

Missanjo E, Chioza A, Kulapani C (2014) Effects of different pre-treatments to the seed on seedling emergence and growth of *Acacia polyacantha.* Int J Forestry Res 583069:6

Mkhabela M (2020) How 4IR manufacturing technologies can help turn South African industry around. Business report. Director at Antswisa Transaction Advisory Services

Mwinyihija M (2018) Leather and footwear value chain an east african perspective. Presented during the 2nd East African manufacturing business summit, held on 25th May 2017, Serena Hotel, Kigali, Rwanda

Mwondu J, Ombui JN, Onyuka A (2019) A survey of leather waste generation and disposal mode in selected counties in Kenya. J Substant Environ Peace 3(1):8–14

Ndhulukula K, Du Plessis P (2009) Green labelling, eco-certification and fair trade: threats and opportunities for Namibia (S. Montgomery, Ed.). Namibia, Windhoek

Nguyen VNH, Nguyen TH, Lee MS (2020) Review on the comparison of the chemical reactivity of Cyanex 272, Cyanex 301 and Cyanex 302 for their application to metal separation from acid media. Metals 10:1105

Okereke C, Coke A, Geebreyesus M, Ginbo T, Wakeford JJ, Mulugetta Y (2019) Governing green industrialisation in Africa: assessing key parameters for a sustainable socio-technical transition in the context of Ethiopia. World Dev 115:279–290

Oruko RO, Selvarajan R, Ogola HJO, Edokpayi JO, Odiyo JO (2020) Contemporary and future direction of chromium tanning and management in sub Saharan Africa tanneries. Process Saf Environ Prot 133:369–386

Oruko RO, Moturi WN, Mironga JM (2014). Assesment of tannery based solid wastes management in Asili, Nairobi Kenya. Int J Qual Res 3:227–238

Panda RC, Sivakumar V, Sreepradha C, Narayani T, Mandal AB (2014) Closed-loop control for odour-reduction during wastewater treatment in tannery. Future Eng Technol 10:2

Procter HR (2019) The Principles of Leather Manufacture. Good Press, pp 5589.

Saravanan P, Renitha TS, Gowthaman MK, Kamini NR (2014) Understanding the chemical free enzyme based cleaner unhairing process in leather manufacturing. J Clean Prod 79:258–264

Schminke TG (2019). Labour-centred development and decent work: a structuralist perspective on informal employment and trade union organizing in Uganda. Patrick Power Library, Saint Mary's University, Canada

Shrewsbury LF, Vestal TM, Jenswold JM, Scott JM, Claypool PL, Emslie AG (2007) Foreign aid and development. The leather industry in Tanzania. Master graduate College of the Oklahoma State University

Siyanbola WO, Egbetokun AA, Olueyi I, Olamade OO, Aderemi HO, Sanni M (2012). Indigenous technologies and innovation in Nigeria: opportunities for SMEs. Am J Ind Bus Manage 2:18846, 1–12

Silambarasan S, Aravindhan R, Rao JR, Thanikaivelan P (2015) Waterless tanning: chrome tanning in ethanol and its derivatives. RSC Adv 5:66815–66823

Thanikaivelan P, Rao JR, Nair BU, Ramasami T (2005) Recent trends in leather making: processes, problems, and pathways. Crit Rev Environ Sci Technol 35(1):37–79

Thomson R (2011) The role of leather science and technology in heritage conservation. Ph.D. thesis submitted to University of Northampton UK

United Nations Industrial Development Organisation (2015) Country programmes for inclusive and sustainable industrial development

United Nations Industrial Development Organisation (2017) Annual Report for 2016

United Nations Industrial Development Organisation (2015) e-learning training program on footwear pattern engineering in real professional education at the DIT Mwanza, Tanzania

Virginija J, Jiyembetova I, Gulbiniene A, Sirvaityte J, Beleska K, Urbelis V (2012) Comparable evaluation of leather waterproofing behaviour upon hide quality. Influence of retanning and fatliqouring agents on leather structure and properties. Materials science (Medžiagotyra) 18:2

Wang X, Yan Z, Liu X, Qiang T, Chen L, Guo P, Yue O (2019) An environmental polyurethane retanning agent with the function of reducing free formaldehyde in leather. J Clean Prod 207:679–688

Wanyoike F, Mugunieri LG, Mtimet N, Kiptoo E, Gulaid I (2018) An analysis of the hides and skins value chain in Somaliland. International Livestock Research Institute (ILRI) Research Report 50. Nairobi, Kenya

World Bank Group, Economic and Transformation Group (2015) Kenya leather industry Diagnosis, strategy and action Plan

Wu J, Yu S (2019) Effect of root exudates of *Eucalyptus urophylla* and *Acacia mearnsii* on soil microbes under simulated warming climate conditions. BMC Microbiol 19:224

Zhang C, Xia F, Long J, Peng B (2017). An integrated technology to minimize the pollution of chromium in wet-end process of leather manufacture. J Clean Prod 154:276–283

Zhang C, Xia F, Peng B, Shi Q, Cheung D, Ye Y (2016) Minimisation of chromium discharge in leather processing by using Methane sulfonic Acid: A cleaner pickling-masking-chrome tanning system. J Am Leather Chem Assoc

The Feasibility of an Environmental Awareness and Responsibility Programme in Rural South Africa: A Case Study from the South

Andrea Lombard, Melanie Nicolau, Sabelo Tshabalala, Anja du Plessis, Richelle Pienaar, Ryan Anderson, and Butho Khanye

Abstract Rural communities worldwide rely extensively on the natural resource base for their existence. In striving to fulfil basic human needs, the importance these communities attach to sustaining the natural environment is often overlooked or misunderstood. This chapter follows a case study approach to a non-formal environmental awareness and responsibility training programme emanating from a community engagement endeavour undertaken by a group of academic geographers in three rural villages of the North West Province of South Africa. In an attempt to accelerate the UN Sustainable Development Goals numbers six, seven, 12 and 15, the programme focused on the following topics: 1) identification and conservation of natural resources, 2) renewable energy, 3) recycling and 4) water management in rural communities. Training topics were identified keeping the specific local contexts in mind to reaffirm the existing sense of place of the respective communities. After several workshops and post-training evaluations, results of the training indicated that these programmes have the best chance to succeed when they are applied to and incorporated at a local level with community involvement. Results also indicate that the success of such a programme should not be measured solely from the perspectives of the trainers, but through the feedback received from the trainees in their own local, rural environment. The establishment of a partnership between those providing the training and the rural community members is of cardinal importance.

Keywords Environmental awareness · Environmental responsibility · Training programme · Rural communities · South Africa

A. Lombard (✉) · M. Nicolau · S. Tshabalala · A. du Plessis · R. Pienaar · R. Anderson · B. Khanye
Department of Geography, University of South Africa (UNISA), Private Bag X6, Florida, Roodepoort 1710, South Africa
e-mail: lombae1@unisa.ac.za

W. Leal Filho et al. (eds.), *Sustainable Development in Africa*, World Sustainability Series,
https://doi.org/10.1007/978-3-030-74693-3_33

1 Introduction

The concept of the "environment" is occasionally perceived as an extraneous entity—particularly in the rural areas of the Global South. Rural areas are often characterised by weaker economic performance owing to the "persistent out-migration of younger and better-educated people, lower educational attainment, lower average labour productivity and lower levels of public service provision" (Ward and Brown 2009, 1238). Furthermore, environmental debates within these areas are often neglected or overlooked. This is ironical as rural economies rely extensively on their natural resource base for sustainable economic activities (Li et al. 2019); thus, the need to raise environmental awareness and foster responsibility towards the environment, especially in these rural areas, is critical (Omisore 2018).

Although a debated document, Agenda 21, drawn up at the Rio Earth Summit in 1992, added value to this discussion through the assertion that:

> Education, including formal education, public awareness and training, should be recognised as a process by which human beings and societies can reach their fullest potential. Education is critical for promoting sustainable development and improving the capacity of the people to address environmental and development issues. ... Both formal [education] and non-formal education are indispensable to changing people's attitudes so that they have the capacity to assess and address their sustainable development concerns. (UNESCO 1992, 320)

Researchers and educators alike are constantly trying to find new ways to foster an environmentally responsible attitude in rural citizens. Reilly et al. (2008) and Wamsler (2020), argue that environmental education should be supported as it has the capacity to encourage residents to make informed decisions regarding their environmental behaviour based on awareness, skills, understanding and attitudes instilled in them through accomplished education.

The Millennium Development Goals (MDGs) provided the foundation for the most comprehensive anti-poverty movement in history (UN 2015a). The MDGs considered the interaction of humans within their physical environment to be cardinal in achieving sustainable development. By 2016, the Sustainable Development Goals (SDGs) replaced the MDGs through a resolution adopted by the General Assembly of the United Nations on 25 September 2015: *Transforming our World: the 2030 Agenda for Sustainable Development* (UN 2015b). The agenda of the SDGs presents a plan of action for people to improve and transform our world for the better, to foster prosperity, peace and partnerships. In all, there are 17 SDGs, with the sustainable use of the physical environment playing a dominant role in achieving most of the goals (Omisore 2018).

It could be argued that the key components in achieving the SDGs by 2030 are related, through environmental education, to the promotion of greater environmental awareness, making for more responsible citizens. In an attempt to contribute to the research in this field within the specific context of driving the agenda for more sustainable rural societies, through community empowerment, the case study presented by the authors of this chapter, is directed at an informal environmental awareness and responsibility programme, which was conducted over the course of two years in a rural area of South Africa.

2 Background: Identifying the Need for a Rural Environmental Awareness Training Programme

The national government of South Africa has tasked universities to be involved in responsible social change. The Department of Geography at the University of South Africa (UNISA) expresses this mandate in the form of community outreach projects and community engagement. Since 2012, the department has been actively involved in a variety of community projects in the rural area of Koffiekraal, Brakkuil and Pella, traditional rural communities historically linked to the Chiefdom of the *Bahurutshe* (Mpotokwane 1974) and lead by the *Bahurutshe-Boo-Mokgatlha* Traditional Authority. The villages are located in the Moses Kotane Municipality within the Bojanala Platinum District in the North West Province of South Africa (Fig. 1). The villages cover a combined area of 18.10 km^2, with an estimated population of 15 834 (StatsSA 2018). Households are predominantly female-headed and less than 14.2% of the population have completed formal schooling, while 29.3% of the population

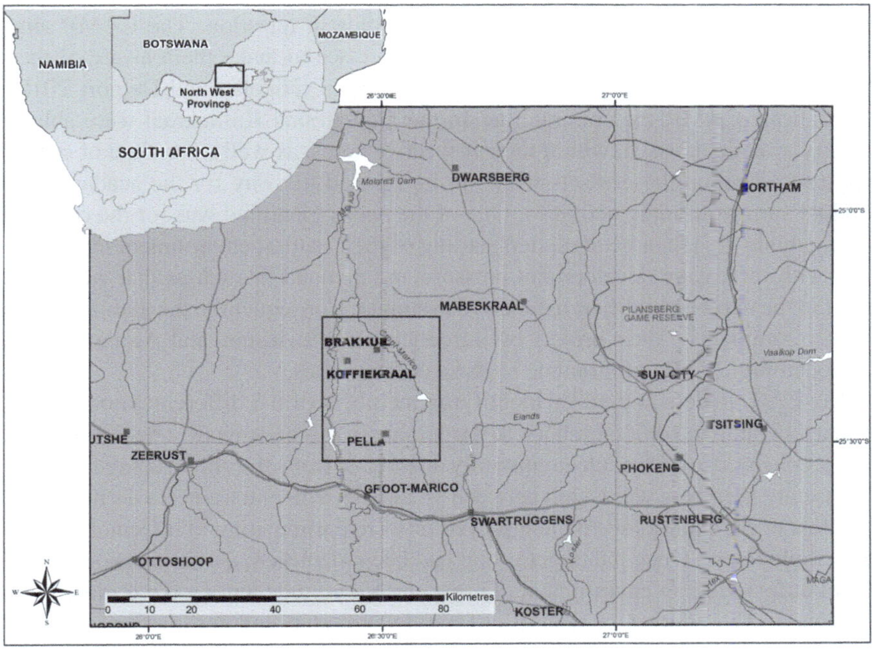

Fig. 1 Location of the rural villages under study

have not completed any form of formal schooling (Pretorius and Nicolau 2020). Only 2% of the households have piped water and flush toilets, and although most of the households have access to electricity, wood is still used for cooking (StatsSA 2018). The main road network linking the villages to the nearest towns is tarred, however the rest of the road network consists of predominantly dirt roads.

The villages are located in the Magaliesburg Biosphere, proclaimed by UNESCO in 2015 (UNESCO 2020), where there are two prominent African biomes, namely Grassland and sub-Saharan Savannah, with diverse fauna and flora. Geographically the villages are located near areas that have a wealth of minerals including platinum, nickel and chrome, and are constantly challenged by mining prospecting right applications, granting of such requests, investments and other related activities. The climate is suited to a variety of farming activities including, cattle, maize, citrus and tobacco. Most of the population rely on subsistence farming on communal land for their livelihood. Access to water remains a challenge in the three communities and their major source of water for household and agricultural use is from the Groot Marico River catchment area.

Activities reported on in this chapter are a result of a collaboration between the Department of Geography (UNISA) and a non-profit organisation (NPO), who first met with the three communities in 2012 to conduct a Community Asset Mapping Process (CAMP) workshop. Asset-based community development is a specific Participatory Rural Appraisal (PRA) technique focusing on community assets, rather than on the conventional needs approach, and on lifelong learning that is significant to and in the context of the specific communities in question. The CAMP aims at taking communities through a variety of exercises to make them aware of their existing social, human, cultural and physical capital (Nicolau and Delport 2015). Through the CAMP, the communities in the area around Koffiekraal were able to identify a range of assets within their respective areas, as well as a range of opportunities to develop these assets which in turn paved the way for the academics to conduct various training activities. One of the needs identified was for the community members to gain a better understanding of their natural environment in order to cope with increasing development pressure and demand in such as that would not result in the depletion of their natural environment. Consequently, the authors of this chapter commenced their project by introducing an Environmental Awareness and Responsibility (EAR) programme to these communities.

The EAR process started by identifying members of the different communities, primarily younger, adult members, to participate in this endeavour. The initial target group consisted of 15 adult community members from the three villages. These community members were aged between 18 and 35 with one trainee over the age of 50. Most of the trainees were unemployed or were participating in informal, temporary employment. Once the target group had been identified, a training programme, aimed at fostering an environmentally responsible attitude among the trainees, was developed. The chapter presents the training provided and the responses of the trainees to the programme.

3 Constructing an Environmental Awareness and Responsibility Programme

The training programme commenced with observational learning by targeting three of the local schools (one primary school and two secondary schools) and initially undertaking one day awareness workshops with the learners while the adult trainees from the villages acted as observers. The observational workshops addressed four key topics (identifying natural resources, renewable energy, recycling and water conservation and management) and included both a theoretical and a practical component. The practical part of each individual session offered learners the opportunity to apply the theory they learned by either making an item (like a poster of natural resources, usable/decorative items out of recyclables or water filters) or interactive activities (like testing water samples). The intention was that the trainees should be able to replicate the activities with other community members in the future. Figure 2 indicates the process followed over the course of one year of intensive training and subsequently a monitoring exercise undertaken one year later.

Following on the observational learning phase, the retention process included 5 workshops (two classroom training sessions which were informed by a range of activities and based on information contained in two guidebooks: a village walk which included cleaning up a section of Koffiekraal; installation of the litre lights

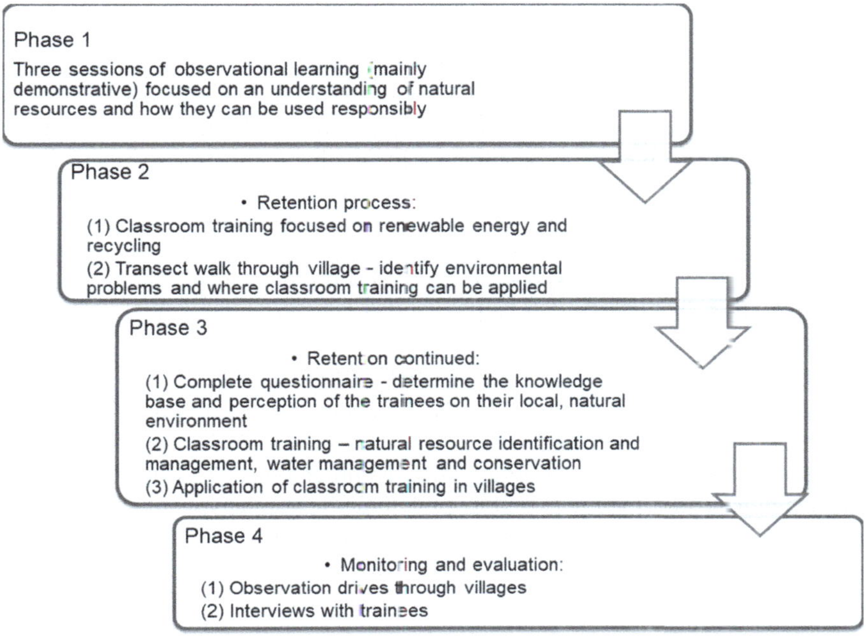

Fig. 2 Programme structure and time frame

Table 1 Summary of guidebook contents

Main topics	Information breakdown
Guidebook Volume 1	
1) Using renewable energy resources	- differences between renewable and non-renewable energy - electricity saving tips - instructions for building a solar oven - how to make a battery using a lemon - how to build a solar bottle bulb
2) Making items from recycled material	- defining recycling - how to recycle plastic (bottles) containers to make plant holders, bird feeders and waste scoops - using plastic bags to weave baskets
Guidebook Volume 2	
1) Making your own pesticides	- making natural, non-toxic pesticides for the garden using vegetables, soap and water - where to plant certain plants, which serve as pest repellants
2) Making your own compost	- making compost using green and brown material and sufficient moisture - what to add and what not to add to compost piles
3) Recycling and saving water	- using plastic bottles to water seed beds and vegetable gardens

at sites identified by the trainees; and a water supply audit). The guidebooks were compiled by the training facilitators from existing (mainly internet-based) sources and included only material applicable to the rural context of the trainees. The specific content of the two guidebooks are summarised in Table 1.

After the first classroom training session, the trainees had to complete a short questionnaire. The questionnaire aimed to determine the awareness of and perceptions of the trainees on their local, natural environment which informed the remainder of the programme. Although their study focused entirely on the perceptions of children and adolescents, Korhonen and Lappalainen (2004) reaffirmed the role that a sense of place plays in environmental education programmes in generating a concern for and an awareness of the local environment. This programme focused on various topics carefully included with the local environment in mind. In this chapter each one of the programme topics will be discussed individually before conclusions will be drawn on the successes and limitations of the programme as a whole and finally brief recommendations are provided to guide other researchers who might wish to replicate the programme in other Global South communities.

3.1 Natural Resources: Identification and Management

Goal 15 of the SDGs focuses on the protection and restoration of natural resources and the promotion of the sustainable use and management of these natural resources (UN 2015b:14) to promote sustainable development. The topic of community-based natural resource management has received growing attention over the last two decades (Leach et al. 1999; Blaikie 2006; Colvin et al. 2016; Musavengane et al. 2019). This reaffirms the importance of considering natural resources in promoting environmental awareness.

The five workshops, which were conducted for this study, introduced participants (learners of three local schools and 15 adult trainees) to key concepts relevant to natural resources management. These included, amongst others, defining the concepts of natural resources management and sustainability. The theme for the workshops on natural resource identification was—"Look after the environment so that the environment can look after you".

A walk around the area in which the workshops were held allowed for a visual exercise allowing the participants the opportunity to familiarise themselves with examples of natural resources in their local environment. This activity stimulated place-based learning as identified by Stern et al. (2013) as one of the best learning practices in an environmental education programme. Discussions conducted during this walk focused on how natural resources have been used in the construction of buildings and in developing infrastructure to highlight but a few. This approach is aligned to those of Vaughan et al. (2010) who noted that short courses play an important role in environmental education, and support the notion that learning needs to take place outside the classroom, where learners of all ages can make observations in the environment of what has been taught in the classroom.

After the introduction and the outside-the-classroom observational exercise, the participants were tasked with making a poster using the knowledge they have gained thus far. The poster project generated significant enthusiasm among the participants. Stationery, newspapers, and magazines were provided for participants to use in creating their posters (Fig. 3).

During the second classroom training session, trainees were introduced to sustainable food gardening as part of the natural resources' management topic. The rural communities under study are in remote areas and do not have adequate access to areas from which fresh fruit and vegetables are distributed on a regular basis. It is therefore crucial for them to be able to produce their own fresh produce.

In a practical application component of the classroom training session, the trainees had the opportunity to start their own vegetable garden at the training centre. We also arranged to meet with the staff members at a local crèche where the trainees showed them how to produce their own sustainable compost. This was one of the first steps to encourage the trainees to start interacting and disseminating their knowledge to other members of their greater communities.

Just before the second classroom training session a questionnaire was distributed to the trainees. When questioned about the current condition of their environment,

Fig. 3 Example of a poster created during the classroom training session

it became clear that the main issues are pollution and inadequate water supplies, coupled with the poor quality of the water. Most of the trainees noted the deteriorated state of the natural environment within which they live. Identifying the challenges experienced by the community is a key factor in developing the trainees' awareness of the environment and represents the first step in dealing with these challenges.

3.2 Renewable Energy

The villages in the study area do have access to the conventional, mainstream electricity supply of South Africa, but the main grid power does not reach all households in some more informal areas. Although SDG 7 aims at ensuring affordable, reliable and sustainable energy for all (UN 2015b), the resources to provide modern and sustainable energy resources to the rural communities are limited. The EAR programme, therefore, included a component on how simple and affordable renewable energy options can be used as an alternative source of electricity and also how to save on electricity costs by harvesting free natural energy. The use of alternative electricity sources has become increasingly important with South Africa's mainstream electricity supply and distribution network being under constant pressure with regular black-out periods.

The renewable energy training started with a discussion on the differences between renewable and non-renewable energy and why it is environmentally important to make use of renewable energy. General electricity-saving tips applicable to the specific rural areas were shared with the trainees. The practical component first focused on the construction of their own solar ovens using basic materials such as cardboard boxes and foil (Fig. 4).

The trainees showed significant interest in this activity and how the sun's energy can be used to cook food. The residents of the villages rely mainly on open fires to cook their food as many of them cannot afford to use electricity to cook every day. Apart from the activity where the trainees used basic materials such as cardboard

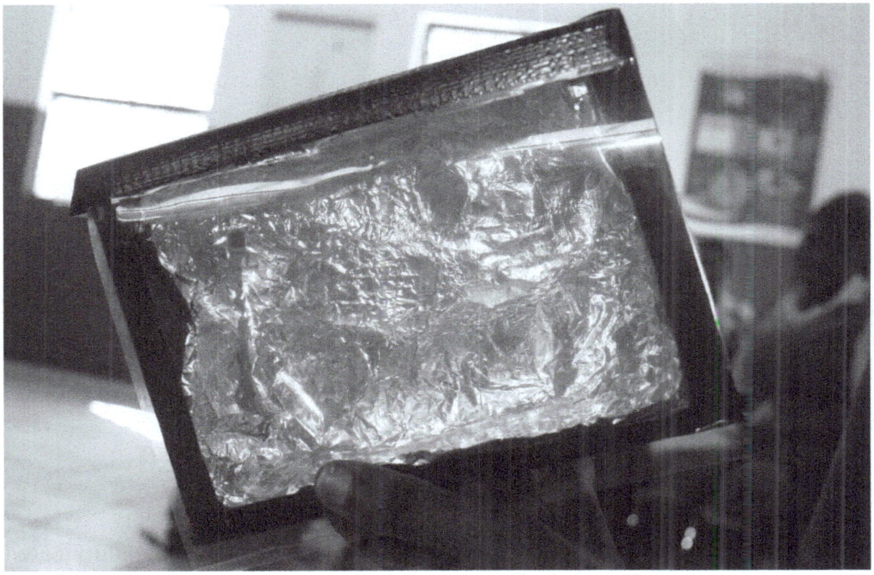

Fig. 4 Solar oven example made during the training

boxes and foil to construct their own examples of solar ovens, they were also shown pictures and videos of the real patents and which materials would be required to build a fully functional solar oven.

During the second practical component of the renewable energy training session, the trainees were shown how to create a battery using lemons, wire, copper coins, galvanised nails and a small LED bulb. Through this activity, an understanding could be cultivated that there are alternative ways to generate lighting, which can be used when conventional methods are not available.

The last part of the practical training session on this topic focused on the use of litre lights, devised in Brazil, and now considered to be a global practice. The *Liter of Light* project consists of a group of volunteers who teach marginalised communities how to use recycled plastic bottles to provide light in their homes (Literoflight 2016). The authors of this chapter tried to achieve a similar goal, albeit on a smaller scale, and focused specifically on the home villages of the trainees. The litre lights were not installed in their homes as most of the villagers had electricity in their houses. However, in these communities, the households make use of outbuildings, primarily for informal storage structures and restroom requirements. Most of these outbuildings have no windows and thus a very limited amount of natural light. This was identified as an area of concern. Five litre lights were therefore installed by the trainees in outbuildings at a crèche, community centre and one home to ensure illumination in the otherwise dark structures (Fig. 5). The litre lights consisted of a mixture of water and bleach which ensures the bottle stays clean for proper illumination over a longer period.

Fig. 5 Trainees installing litre lights in outbuildings

Rogers et al. (2008) and Ouedraogo (2019) concluded that community-based renewable energy projects are likely to gain more public acceptance than the top-down development of large-scale projects, but that they are unlikely to become widespread without greater institutional support.

A section of the questionnaire distributed to the trainees prior to the second classroom training session focused on renewable energy. After the two-month period following on the first classroom training session, 75% of the trainees could accurately define renewable energy. The main response to how to save electricity focused on unplugging electrical equipment not in use. In general, the renewable energy topic and associated activities were received well by the trainees and it was evident from their responses to the questionnaire that they understand the contribution renewable energy and other energy-saving methods can make in their local environment.

3.3 Recycling

Waste reduction is an essential ideology of sustainability, and recycling is an integral element of any solid waste management plan (UNEP 2013). Improving waste management is crucial for societies, especially for rural communities, where waste disposal service delivery is lacking. Recycling is a multifaceted scheme of environmental protection, the main objective of which is the reduction in raw material consumption and the quantity of waste (Haupt and Hellweg 2019). Poor solid waste management can lead to severe health and environmental problems (Limon et al. 2020).

The United Nations Environment Programme (UNEP) (2013) concluded that recycling is driven by the four R's (**r**ecycle, **r**e-use, **r**educe and **r**epair). Reducing waste, reusing materials and products, and recycling are some of the most influential ways that individuals, households, institutions and businesses can protect their communities and the environment. This is affirmed by SDG 12 aimed at reducing waste through, amongst other operations, recycling (UN 2015b). To be able to understand the impact of waste on the environment, education is required. Education for Sustainable Development (ESD) is a vigorous concept that incorporates a new knowledge system of education with the vision to empower people of all age groups about sustainable development. Education for sustainable development is intended to sensitize people into taking responsibility of the environment through creating and shaping a sustainable future for all (Khataybeh et al. 2010). The intention is to instil environmental protection values in people and a culture of environmental sustainability by protecting it, using resources sparingly and nurturing the environment. These measures can only be achieved through education, training and practical application. Khataybeh et al. (2010) argues that sustainable development through education can only be accomplished when educators and learners critically reflect on their communities, identify unsustainable elements in their lives and assess the challenges caused by conflicting values and goals. The studied communities have a responsibility to take care of their environment, especially in terms of properly disposing, handling and recycling waste material.

With the knowledge of ESD, these communities can positively influence the development of their environment, ecology, economic and socio-cultural aspects. Mihaela-Dima and Meghistan-Toma (2018) reiterate the importance of education for sustainable development because it changes individual behaviour which is a foundation for building strong local institutions and collaboration between different stakeholders. ESD is a key strategy that can be effectively implemented by rural communities in their recycling and solid waste management initiatives. The challenge that comes with ESD is that it necessitates some formalized structure to be successful. Continuous training is required whereby participants are taught and provided with learning material that is current and relevant. The communities are dynamic and also contribute to knowledge creation, which then creates a symbiotic relationship between participants and facilitators as everyone contributes to learning and knowledge creation.

During the classroom training activities, adult trainees were presented with key concepts such as the definition, importance, purpose and benefits of recycling while touching on the four R's linked to sustainability. The presentation of these concepts was interactive, and practical examples were provided on how recycling can be performed in rural areas using waste products from the local environment.

The communities of Koffiekraal and Brakkuil have access to intermittent waste removal services. Approximately 20% of the community have access to weekly refuse removal in the Ramotshere Moiloa Local Municipality (StatsSA 2011). A review of the 2011 census indicates that around 70% of the community make use of their own dumping areas. Community members expect weekly refuse removal, but according to the trainees, the service unreliable. Virtually no public waste bins are found on the streets of these communities which in its turn leads to littering—a matter of grave concern. One of the trainees commented that the community members are not informed about recycling which contributes to the littering. Solid waste in the

Fig. 6 Discarding of solid waste in holes and through burning

communities is commonly disposed of by discarding used items into a container or hole in their backyards and then burning the waste (Fig. 6). Due to the poisonous gasses released by plastics when burned with low heat fires (Obebe and Adamu 2020), such information could be beneficial in promoting the use of recycling as a method to promote health and safety in the communities. The manner of waste disposal by the communities practically demonstrates the truth in the sentiments "the degree to which a particular material is recycled depends on the income levels, the existence of local and national markets" (Wilson et al. 2006). In the case of the communities under study, there was no local market for solid waste and very little encouragement to embark on collecting solid waste for recycling purposes.

The EAR programme aimed to demonstrate and transfer some recycling skills through a variety of practical activities. The classroom training activities introduced the trainees to the concepts of littering as a form of pollution and its negative impact on the environment. The main challenge in rural areas is the absence of recycling infrastructure that can enable separation of waste streams to material recovery and buy back facilities (DEA 2009). The nearest town to Koffiekraal and Brakkuil does not have any recyclers which is an indication of the challenges faced by rural areas. Another general challenge facing recyclers, even in urban areas, is the limited waste treatment areas that are available to manage waste as most are outdated and lack infrastructure (DEA 2009). Training interventions play a vital role and help prepare communities while development of infrastructure progress as per governmental plans.

To further stimulate the trainees, the group was tasked to take a tour through the Koffiekraal village to identify and collect different types of waste. After the tour, a group discussion was conducted to highlight the apparent problems and there was consensus that littering, and pollution adversely impact upon the environment of the local village. The practical component of the recycling topic also focused on refashioning various items from the waste materials collected in the villages. These ranged from items which could possibly be sold to generate an income (e.g. using seeds and paper pulp to create animal shapes, and plastic bags to weave baskets) to other practical, everyday usable items (e.g. making dust scoops from plastic bottles).

Recycling is practised to some extent in rural communities, but in most cases, those who practise it do not realize the contribution they are making towards environmental sustainability (Limon et al. 2020; Wilson et al. 2006). For everyone to play a meaningful role in recycling, community members need to be equipped with the necessary skills. Trainees could easily identify pollution as the main problem in their communities. By being able to process the information shared by the trainers, trainees could visualise their own communities, were able to share their practical experience of the problem of pollution and took charge of ideas on bringing a solution to the problem constituting participatory action research (Kidd and Kral 2005).

After receiving guidance on recycling in the first classroom training session, the trainees showed that they could accurately define what recycling means. The questionnaire that they completed prior to the second classroom training session revealed that they could also determine which waste materials in their local villages could be recycled and/or refashioned. Through discussions we noted that there are several ways through which solid waste could be recycled even in communities where there were no formal structures for solid waste collection. Therefore, one could argue that the engagement with local communities provided them with possible solutions to the problem of solid waste and that the trainees now have new knowledge on how to deal with the associated problems (Ozanne and Saatcioglu 2008).

3.4 *Water Management and Conservation*

South Africa is characterised by limited natural water resources and is a water-stressed country and as such, SDG 5, calls for the sustainable management of water resources to ensure safe and affordable drinking water for all by 2030 (UN 2015b, 18). This is particularly vital in the rural areas of the country which are subjected to water supply and quality issues as well as poor service delivery.

The identified rural villages are located in an arid to semi-arid region, dominated by limited water resources and multiple water challenges such as inadequate water supplies and restricted access to drinking water. For the water management and conservation part of the programme, the facilitators consulted with community members and leaders to establish priority areas related to primary water threats. Most of the individuals within these communities are reliant on communal water sources for their survival. The water resources within these villages can therefore be regarded as scarce and unevenly distributed, as was confirmed by the trainees in their response to the questionnaire. Community members spend most of their time physically carrying water from a communal water source such as a tap or water tank to their respective properties (Fig. 7).

Since rural areas are generally dominated by an insufficient supply of water, which is also of a poor quality, effective water management is a necessity within the villages investigated in this case study (Mothetha et al. 2013). A lack of maintenance in the

Fig. 7 Communal water sources in the villages

existing water supply infrastructure in these areas and the current prolonged drought has exacerbated the need to expand knowledge and awareness on water management. This would promote better water consumption practices through conservation and the introduction of safety measures to ensure drinkability.

The identified rural villages are located in the Groot Marico River catchment which is a listed key strategic water resource for the entire North West Province and bordering countries. It has consequently been declared an international and national freshwater ecosystem priority area. The catchment is of prime importance to ensure both environmental and human health (Tempelhoff et al. 2012). The region has, however, been plagued by various prospecting mining rights applications since 2010 albeit known nationally and internationally as a key strategic water resource and national freshwater ecosystem priority area.

Since 2010, there were six applications for prospecting mining rights in this sensitive area. The applications ranged from mining for nickel and diamonds to gravel for road construction purposes. As a result of these applications, the "Mmutlwa wa Noko" (MWN) voluntary association was established driven by dedicated volunteers from various cultural backgrounds with the common goal of protecting the Groot Marico River catchment (MWN 2020). Due to the opposition from this community group and other interested and affected parties, five of the six applications have been unsuccessful to date. One application was withdrawn, one set aside by court ruling and the other three are still pending resubmission of the applications after addressing the concerns raised by interested and affected parties with regard to the sensitivity of the area. The one application that was successful unfortunately did not follow the required environmental and social impact assessment procedures prior to the commencement of the project. MWN opposed the validity of the application and as a result rehabilitation of the affected area commenced in 2016 (MWN 2020).

The already water stressed region, therefore, is consistently confronted and threatened by possible mining and other related activities. The, community representative of Koffiekraal who also happens to be the deputy chairman of the MWN association, realizes the importance of disseminating knowledge, creating water awareness

Table 2 Water-related issues that formed part of the training

Issues	Description
Availability	The amount of water available on a global, national and local scale with the emphasis on water as a limited natural resource
Degradation	The contribution of various sources of pollution. Pollution problems such as eutrophication, sedimentation, salinization, acidification and microbial pollution emphasizing the importance of safe clean drinking water
Conservation	Rainwater harvesting, with the emphasis being on water wastage, sustainable use and irrigation practices

within the communities as well as to get more community members involved in the voluntary association. His primary water concerns were considered in the development of the water management and conservation training sessions with the goal of creating water stewards within each community and forming water focus groups or water forums which could get involved in objecting to future applications for mining or other related activities.

In the training sessions, the focus of the water management and conservation component was primarily on creating awareness within these communities through fostering knowledge around the dominant theme of "Water is Life". The training focused on the dissemination of knowledge as reflected in the main water-related issues in Table 2. The issues were introduced through a range of presentation materials and real-life examples.

The classroom training sessions culminated in various practical activities where trainees received the opportunity to apply their knowledge related to water issues. With the assistance of the training facilitators, trainees completed a water supply audit within their respective villages. To establish the current state of water supply, the audit focused on the identification and capturing of water supply points and prevalent problems (Fig. 8).

Subsequent to the supply audit, a water quality audit was also completed by the trainees. A monitoring point was identified within the Koffiekraal community where observational monitoring would be completed by the trainees on a monthly basis. The results indicated that the trainees have the necessary understanding, knowledge and awareness to allow them to identify supply and quality water-related problems enabling them to form a successful water forum within their respective villages. The Koffiekraal trainees formed a successful water management focus group and became involved in the MWN association, representing the Koffiekraal village as interested and affected parties.

4 Successes and Challenges of the Programme

The training programme presented was not without its challenges. What is important to note is that success in terms of such a programme should not be measured purely by the standards of those providing the training, but instead be assessed in terms of

	Type of Water Source	Purpose of Use	Location	Notes/ Problems
13	Com tap	Drinking	078) 25° 15 12,4 S 26 28 47,6 E	No proplem
14	Com tap	Drinking	079) 25° 15 04,9' S 26° 28 58,4 E	Leaking
15	Com tap	Drinking	080) 25° 15 07,3 S 26' 29 09,7 E	No peoblem
16	Com tap	Drinking	081) 25° 15 14,4 S 26° 28 57,7 E	No problem
17		Borehole drinking TAP	082) 25° 15 19,2 S 26° 28 55,5 E	LEAking from jojo and water tap
18	Com tap	drinking	083) 25° 15 21,6 S 26 28 56,0 E	No problem

Fig. 8 Example of the Water Supply Audit completed by trainees

the feedback received from the trainees. It is still important to reflect on the successes and limitations of this programme from a facilitator's perspective as well to determine its feasibility for future endeavours of a similar nature.

The main challenges experienced were the following:

a) *Inconsistent attendance.* Throughout the training period, the number of trainees who attended the workshops differed significantly, but this cannot be attributed solely to a loss of interest. Because the target group incorporated a younger sector of the community, some of the members obtained employment in urban and suburban areas which meant that they had to leave their communities temporarily or, in some cases, permanently.
b) *Social struggles in the communities.* Rural community members should never be regarded as homogeneous entities. It is true that the communal structures in rural areas, and especially in those areas under traditional leadership, are sometimes stronger than those prevailing in urban areas; however, the individuals in these communities still tend to fight for their individual rights to a livelihood. During the training, it became clear that an external issue was preventing some of the trainees from working together which resulted in some of them terminating their attendance. Without involving themselves too much in the community's social struggles, the trainers should, therefore, not turn a blind eye to social problems such as these since neglecting to do so could inhibit the learning experience for those who are truly interested.

c) *Time management*. It is important not to get stuck on one component of such a programme. After the three participatory observation sessions with the school learners in the respective areas, it became clear that it was time for more intensive training sessions incorporating the adult trainees as active participants. Sufficient time should also be devoted to practical applications in the local environments of the trainees as such practical application truly empowers the trainees.
d) *Entry to the communities and compliance with traditional regulations*. Entry to the communities, which are under traditional leadership, was not a real concern for the facilitators of the discussed programme as the NPO had already ensured that we would be allowed to work in these villages. It should be noted, however, that if a similar programme were to be undertaken in a rural village ruled by a traditional chiefdom, the necessary permission to do the training should be obtained from the rulers before the entry of external facilitators to the community. Training facilitators also need to respect the local culture, traditions and rules of the rural communities in which they are operating.
e) *Sustainability of the programme*. The facilitators of this programme realised that to ensure a sustainable impact on the communities, the training should not be limited to a once-off workshop or activity. Regular training activities should be scheduled in the rural villages to ensure a lasting effect and the ideal would be to conduct several post-training visits for monitoring and evaluation purposes.

Overall, the challenges experienced did not inhibit the success of this informal training programme. From the trainee feedback received it became evident that the programme was more successful than originally anticipated. Table 3 indicates some feedback received from the trainees in their responses to three questions of the questionnaire aimed at assessing the aftermath of the training programme.

The key themes surfacing in these responses are water and waste management and how these management activities could play a significant role in creating jobs for the community members. The unemployment rate of the North West Province (in which these villages are located) is 33,2% almost 6% higher than the national average unemployment rate (StatsSA 2020). Employment, therefore, informs many of the decisions of the community members in these villages. If the trainees can present activities focusing on environmental awareness and responsibility as a form of generating a self-sustaining livelihood, the community members will become more interested.

The key successes of the informal training programme include the following:

a) *Printed guidebooks*. The trainees received the printed guidebooks positively. They showed their appreciation for the way in which the guidebooks had been compiled for them to be able to continue with the activities of the training in their larger communities. The trainees also regarded the practical application component of the training in the respective villages as an empowering exercise. They especially appreciated the fact that they could create items on their own and that the information was not merely demonstrated to them. This allowed for active participation which is seen as another characteristic of a best practice

Table 3 Trainees' responses to three specific questions in the questionnaire

Question	Key responses
What have you learned from the previous training session?	- "I have learned that I can reuse something for so many things and also save money and also make money for myself and create jobs for my community." - "How to save water and electricity. Things that I can recycle." - "That water is life and we need to save water and make sure that we must know how to use water safely." - "How important is my environment for present time and for the future."
What would you like to teach your community about the environment?	- "How to keep our environment clean and how to keep our resources saved especially our precious resources which is water." - "By saving water and saving electricity by switching lights off at night." - "Recycling and how to benefit from it." - "How to clean it, how to create jobs and making some project e.g. gardens."
How are you going to do this?	- "By showing them the importance of saving water and electricity." - "Call them in the same place and tell them about things they don't know." - "By having community meeting by my place and talking to young children to encourage them to create jobs for themselves." - "By sharing information and doing practicals." - "I will start by going house to house and teach them how to save water and have a clean one and then after I call the community to show to other so all can know about the safe and clean water."

programme (Stern et al. 2013). The trainees indicated that especially with regard to the recycling activities, some of them were aware of what recycling is, but that they had never been able to physically refashion waste into usable items themselves.

b) *Activities extended to the broader communities.* During the interviews conducted with the trainees one year after the commencement of the intensive training sessions, it became clear that the programme had achieved what it had initially set out to do. Although not necessarily at a macro scale, the training achieved certain ripple effects in the respective communities. In Koffiekraal, for instance, the trainees visited the local schools and talked to the learners about saving water and electricity. The Koffiekraal trainees also formed a successful water management focus group and got involved with the MWN association.

Similarly, the Brakkuil trainees visited the local municipality to report the case of leaking communal taps in their village. Some of these taps have since been repaired. Overall, the trainees indicated that they did not want anything to change in the programme. The only request was that they would appreciate a training session to show them how to repair their own taps as they sometimes have to wait extensive periods of time for the municipality to repair the taps.

c) *Regular reporting and reflection.* It is highly recommended that a technical report should round off each session of training. A feedback session, whether formal or informal, should also be conducted with the trainees after each session and the feedback should be recorded in the technical report. Reflection on each component of the training is required to adequately formulate the remaining aspects to be dealt with. This represents another best practice characteristic of an environmental education programme (Stern et al. 2013).

In general, from a facilitator's perspective, the results of the training programme might seem limited as the direct benefits were delivered to only a small group of community members, but from the trainee perspective, the programme exceeded its expectations and is already generating broader impacts on the respective rural villages.

5 Conclusion and Recommendations

The role of adult trainees in promoting environmental awareness within traditional leadership communities should not be overlooked (Doyle and Kransy 2003). They could forge connections between various social and organisational structures in the community (such as community leaders, the youth and adolescents). Within communities where resources are limited, lifelong learning opportunities that focus on ESD could be regarded as a critical aspect in achieving the SDGs in rural communities. Environmental education is essential for ensuring sustainable livelihoods and should be conveyed to all, especially to rural communities who have limited resources and access to information. It should never be overlooked that people in rural areas also have their own ways of taking care of their environment which have been inculcated in them through their indigenous knowledge systems. Therefore, in applying any environmental education and awareness training in an area where indigenous knowledge is valued by the respective community, facilitators should value and infuse this knowledge into the training programme (Mapira and Mazambara 2013; Sandoval-Rivera 2020).

When developing a similar programme, facilitators should note that involving school children and the youth as an interest group is valuable as this component of society are influential in adding to the intergenerational flow of knowledge. This will ensure environmental awareness in the majority of a community, stimulating an environmentally sustainable culture that can be further developed to the benefit of the community and the environment within which they reside.

Facilitators should also be aware that an important component of the eventual success of the programme will depend on the entry and introduction into the rural community. Facilitators should be constantly aware of the living conditions and economic lifestyles of the rural villagers, specifically those residing in areas under traditional leadership.

A roots-driven process such as CAMP is best suited to assist participants and facilitators in establishing the asset-based needs of the communities as well as the communities' inherent environmental awareness and their need for environmental education that will enable them to utilise their assets sustainably. Further advantages of a roots-driven process is that it allows for trainees to measure the success of their training within their own local environment, as opposed to a more top down approach of trainers determining the success of the programme. Finally, an advantage of CAMP is that it advocates for the establishment of a partnership between those providing the training and the rural community members resulting in a more sustainable result.

The informal training programme as presented through this case study was successful because of the balance between classroom training activities and practical applications in the trainees' local environments. Training programmes, such as the one described in this case study, can give community members the confidence and knowledge to identify more formal rural development opportunities—like *A Re Itireleng* (Let's Do It Ourselves) and the *Qhubeka Bicycle Project* (community oriented environmental and social programmes). The most important future prospect for the trainees of the programme is that they would be recognised by community leadership as green ambassadors in their respective communities. In this way they are able share their environmental awareness and responsibility knowledge with present and future generations in their community that is not only aligned to preserving the natural environment within a UNESCO proclaimed Biosphere Reserve, but also to ensure that sustainable living is possible within the sometimes challenging environment of their own communities.

Acknowledgements The authors of this chapter would like to acknowledge the now non-operational Greater Rustenburg Community Foundation for facilitating our entry into the three communities and to thank Ms Hellene Steenkamp for her invaluable administrative assistance during the duration of the project. We also thank the traditional leaders for allowing us to conduct this programme in their respective communities.

References

Blaikie P (2006) Is small really beautiful? Community-based natural resource management in Malawi and Botswana. World Dev 34(11):1942–1957

Colvin RM, Witt GB, Lacey J (2016) Approaches to identifying stakeholders in environmental management: insights from practitioners to go beyond the 'usual suspects.' Land Use Policy 52:266–276

Department of Environmental Affairs (DEA) (2009) National waste quantification and waste information system. Department of Environmental Affairs

Doyle R, Kransy M (2003) Participatory rural appraisal as an approach to environmental education in urban community gardens. Environ Educ Res 9(1):91–115

Haupt M, Hellweg S (2019) Measuring the environmental sustainability of a circular economy. Environ Sustain Ind 1. https://doi.org/10.1016/j.indic.2019.100005

Khataybeh AM, Subbarini M, Shurman S (2010) Education for sustainable development, an international perspective. Procedia Soc Behav Sci 5:559–603

Kidd SA, Kral MJ (2005) Practicing participatory action research. J Counsel Psychol 52:187–195

Korhonen K, Lappalainen A (2004) Examining the environmental awareness of children and adolescents in the Ranomafana region, Madagascar. . Environ Educ Res 10(2):195–216

Leach M, Mearns R, Scoones I (1999) Environmental entitlements: dynamics and institutions in community-based natural resource management. World Dev 27(2):225–247

Li Y, Westlund H, Liu Y (2019) Why some rural areas decline while some others not: an overview of rural evolution in the world. J Rural Stud 68:135–143

Limon M, Vallente J, Corales N (2020) Solid waste management beliefs and practices in rural households towards sustainable development and pro-environmental citizenship. Global J Environ Sci Manage 6(4):441–456

Literoflight. (2016) About us. Liter of Light. https://literoflight.org/about-us/

Mapira J, Mazambara P (2013) Indigenous knowledge systems and their implications for sustainable development in Zimbabwe. J Sustain Dev Africa 15(5):90–106

Mihaela-Dima A, Meghistan-Toma G (2018) Research on implementing education for sustainable development. In: Proceedings of the 12th international conference on business excellence. https://EconPapers.repec.org/RePEc:vrs:poicbe:v:12:y:2018:i:1:p:300-310:n:27.

Mmutlwa wa noko (MWN) (2020) Why protect the Groot Marico River Catchment? https://sunbirdmedia.wixsite.com/mwnoko/2010-2016.

Mothetha M, Nkuna Z, Mema V (2013). The challenges of rural water supply: a case study of rural areas in Limpopo Province. Council for Scientific and Industrial Research. https://researchspace.csir.co.za/dspace/handle/10204/7593.

Mpotokwane, J. (1974). A short history of the Bahurutshe of King Motebele, senior son of King Mohurutshe. Botswana Notes Rec 37–45

Musavengane R, Tantoh HB, Simatele D (2019) A comparative analysis of collaborative environmental management of natural resources in Sub-Saharan Africa: a study of Cameroon and South Africa. J Asian Afr Stud 54(4):512–532

Nicolau M, Delport C (2015) A community asset mapping programme for roots-driven sustainable socio-economic change in rural South Africa. Int J Soc Sustain Econ Soc Cult Context 10(1):1–11

Obebe SB, Adamu AA (2020) Plastic pollution: causes, effects and preventions. Int J Eng Appl Sci Technol 4(12):85–95

Omisore AG (2018) Attaining sustainable development goals in sub-Saharan Africa; the need to address environmental challenges. Environ Dev 25:138–145

Ouedraogo NS (2019) Opportunities, barriers and issues with renewable energy development in Africa: a comprehensible review. Current Sustain Renew Energy Rep 6:52–60

Ozanne JL, Saatcioglu B (2008) Participatory action research. J Consumer Res 35(3):423–439

Pretorius RW, Nicolau MD (2020) Empowering communities to drive sustainable development: reflections on experiences from rural South Africa. In: Leal Filho W, et al. (eds) Universities and sustainable communities: meeting the goals of the Agenda 2030, World Sustainability Series. Springer

Reilly S, Petrillo H, Demchik M (2008). Environmental education's role in sustainable development: three case studies from India, South Africa and the United States. NR523 – International resource management. https://www.uwsp.edu/forestry/StuJournals/Documents/IRM/Reilly.pdf

Rogers JC, Simmons EA, Convery I, Weatherall A (2008) Public perceptions of opportunities for community-based renewable energy projects. Energy Policy 36:4217–4226

Sandoval-Rivera JCA (2020) Environmental education and indigenous knowledge: Towards the connection of local wisdom with international agendas in the framework of the Sustainable Development Goals (SDGs). Diaspora Indigenous Minority Educ 14(1):14–24

StatsSA (2011) Statistics by place. https://www.statssa.gov.za/?page_id=993&id=ramotshere-moiloa-municipality
StatsSA (2018) Statistics by place. https://www.statssa.gov.za/?page_id=4286&id=10762
StatsSA (2020) Quarterly labour force survey, quarter 1: 2020. Statistics South Africa. https://www.statssa.gov.za/publications/P0211/P02111stQuarter2020.pdf
Stern MJ, Powell RB, Hill D (2013) Environmental education program evaluation in the new millennium: what do we measure and what have we learned? Enviro Educ Res 20(5):581–611
Tempelhoff J, Gouws C, Motloung S (2012) The river as an artefact: interpreting the Groot Marico and its people in the 21st century. South Afr Museums Assoc Bull 35:24–30
UNESCO (1992) United Nations conference on the environment and development: agenda 21. https://sustainabledevelopment.un.org/content/documents/Agenda21.pdf
UNESCO (2020) Journey to where it all began. https://magaliesbergbiosphere.org.za/.
United Nations Environment Programme (UNEP) (2013) Repair, recycle, reuse, reduce: The 4Rs – a way of life. TUNZA: the UNEP magazine for youth. https://www.ourplanet.com/pdfs/Tunza_11.1-Eng.pdf
United Nations (UN) (2015a) The Millennium development goals report 2015. https://www.un.org/millenniumgoals/2015_MDG_Report/pdf/MDG%202015%20rev%20(July%201).pdf
United Nations (UN).(2015b) Resolution adopted by the general assembly on 25 September 2015. https://www.un.org/en/ga/search/view_doc.asp?symbol=A/RES/70/1&Lang=E
Vaughan C, Gack J, Solorazano H, Ray R (2010) The effect of environmental education on school children, their parents, and community members: a study of intergenerational and intercommunity learning. J Environ Edu 34(3):12–21
Wamsler C (2020) Education for sustainability: fostering a more conscious society and transformation towards sustainability. Int J Sustain High Educ 21:112–130
Ward N, Brown DL (2009) Placing the rural in regional development. Reg Stud 43(10):1237–1244
Wilson DC, Velis C, Cheeseman C (2006) Role of informal sector recycling in waste management in developing countries. Habitat Int 30:797–808

From Millennium Development Goals to Sustainable Development Goals Through Informal Cross-Border Trading: An Escape for the Poor in a Suffocating National Economy in Gweru, Zimbabwe

Beauty Dzawanda and Mark Makomborero Matsa

Abstract Informal cross-border trading can help to achieve the sustainable development goals through positive macroeconomic and social ramifications, such as income creation and food security, especially for poor households. This research examines how the informal cross-border trading is achieving sustainable development goals in a challenging economic situation in Zimbabwe. Qualitative design was used as a method, as it allowed the collection of detailed information from the participants through semi-structured interviews, which was the major data collection tool. Twenty-five informal cross-border traders were selected through snowball sampling. Secondary data was used for comparative analysis in order to reveal the achievements of the millennium development goals and the sustainable development goals by the informal cross-border trading before and during the cash crisis, since this research was undertaken during Zimbabwe's cash crisis economy period. Data analysis was done from an interpretive approach using thematic analysis. The results revealed that the informal cross-border trade plays a vital role in achieving, particularly the millennium development goals 1, 2 and 3 and the sustainable development goals 1, 2, 4 and 5. The informal cross-border trading improves the lives of those who are engaged in it through poverty alleviation, income creation, employment creation and by improving food security. The income created from the informal cross-border trade assists traders in accessing different goods and services needed for survival. The research recommends a strong support from the Zimbabwean government to this sector as it is a potential medium for achieving the sustainable development goals.

Keywords Informal cross-border trade · Sustainable development goals · Millennium development goals · Challenging economic situation · Poor households

B. Dzawanda (✉) · M. M. Matsa
Department of Geography and Environmental Studies, Midlands State University, P Bag 9055, Senga Road, Gweru, Zimbabwe
e-mail: dzawandab@staff.msu.ac.zw

M. M. Matsa
e-mail: matsam@staff.msu.ac.zw

W. Leal Filho et al. (eds.), *Sustainable Development in Africa*, World Sustainability Series,
https://doi.org/10.1007/978-3-030-74693-3_34

1 Introduction

Informal cross-border trade is pervasive in Africa. Although estimates differ, they all indicate that it constitutes a significant part of total intra-African trade. Informal cross-border trade refers to commercial exchanges that are conducted across borders by individuals who operate as unregistered sole traders (Kahiya and Kadirov 2020). The sustainable development goals were born at the United Nations conference on sustainable development in Rio de Janeiro, in 2012 (Lu et al. 2015; Sachs 2012). Their main objective is to meet urgent environmental, political and economic challenges facing our world. Comprising 17 goals and 169 targets, the sustainable development goals aim at tackling complex and multiple challenges that are faced by humankind (Costanza et al. 2016; Nilsson et al. 2016). The sustainable development goals are successors of the millennium development goals which ran from 2000 to 2015 when the sustainable development goals took over. The millennium development goals and the sustainable development goals are interdependent, so by achieving one of them it would lead to the achievement of the other (Nilsson et al. 2016). The informal cross-border trade is an economic activity that can help to achieve sustainable development goals, particularly in dealing with poverty alleviation, which it tried to achieve during the implementation of the millennium development goals.

The informal cross-border tradein Africa holds fundamental economic and social development implications. It has tremendous potential from a sustainable development perspective in that it can help to tackle poverty reduction by generating substantial revenue for border communities, and its potential development benefits go much further (Brenton and Soprano 2018; Koroma et al. 2017; Sommer and Nshimbi 2018). Various experts have underlined that it significantly contributes to improve food security, promotes economic diversification and it even strengthens peace and stability by creating and sustaining social and economic ties across borders. Another key aspect is the strong gender dimension of such trade, as the majority of small-scale cross-border traders are women, meaning that it is a crucial factor for women's economic empowerment. The vast majority of Africa's informal cross-border traders are female, with up to 70–80% in some cases. Cross-border trade is often their only source of livelihood. According to an ILO study, about 60 percent of non-agricultural self-employment of women in sub-Saharan Africa comes from trade (Blankson et al. 2018; Peterson and Zehra 2018). Given this assertion, it is apparent that he informal cross-border trade encompasses important economic and social implications, making it an indispensable issue from a millennium development goals and sustainable development goals perspective.

The level of the informal cross-border trade in southern Africa remains high and many people are making a decent living out of this form of business. For many citizens of the Southern African Development Community (SADC), the informal cross-border trade has become a profitable enterprise to the extent that some of them have dared to leave formal employment in order to concentrate on informal cross-border trade (Africa Development Bank 2015; Babahet et al. 2019; Koroma et al. 2017; Manjokoto and Ranga 2017). As a result, the informal cross-border trade has enabled multitudes of cross-border traders to put food on the table, clothe their families, pay school fees for their children and purchase household furniture and appliances, among others (Dumbu 2018; Koroma et al. 2017; Manjokoto and Ranga 2017). Thus, the informal cross-border trade exemplifies a way of being and living for those who are engaged in it as they try to sustain themselves and earn a livelihood, thereby achieving the sustainable development goals in the process.

After independence, many African countries experienced a lot of challenges in economic development due to the economic structural adjustment programmes (ESAPs). In sub-Saharan Africa, the effects of the economic structural adjustment programmes led to the skyrocketing prices of goods, rising inflation, retrenchment of employees and a decline in standards of living (ILO 2018). The economic structural adjustment programmes were meant to alleviate poverty but they, however, actually worsened the poverty levels (Mwaniki 2005; Nkoroi 2016; Sandaca 2014). On the Zimbabwean economy, the economic adjustment programmes imposed austerity measures that led to the closure of many industries, increase in consumer prices, decline of real wages and massive retrenchments (Kanyenze 2011; Nkoroi 2016).

In the 1990s, Zimbabwe experienced a decline in the formal sector as a result of the economic structural adjustment programmes which led to the mushrooming of the informal sector as the majority of people started informal jobs due to the difficult economic situation (Chikanda and Tawodzera 2017; Kamete 2017). This was the period when the informal cross-border trade started to rise because many retrenched and unemployed people were forced to join this sector as a household livelihood strategy. Gaidzamwa (1998) and Kurebwa (2015)observed that the informal cross-border trade was not a new occurrence in Zimbabwe as it dates back to the 1990s when the majority of Zimbabweans started to experience severe economic hardships. The situation for Zimbabweans was further exacerbated by the effects of climate change which severely affected rain-fed agriculture, which was the anchor for survival for the rural population. This resulted in many women joining the informal cross-border trade as an alternative to earn a livelihood as they were the most vulnerable group of the knock effects of climate change (Madzwamuse 2010). This is the major reason why women dominate the informal cross-border tradein Zimbabwe (Chikanda and Tawodzera 2017; Muzvidziwa 2015; Rugube and Matshe 2011) and, in the process, empowering and improving the lives of women in fulfilment of the millennium development goal 3, and the sustainable development goal 5. The informal cross-border tradecontinued to rise from the 1990s to date, despite the various challenges faced in the sector (Chikanda and Tawodzera 2017; Dumbu 2018). This is because the informal cross-border trade improved the lives of many people who were engaged in it by increasing their income, reducing poverty and creating employment (Chikanda and Tawodzera 2017). From this discussion, it is apparent that the rise of the informal cross-border trade in Zimbabwe was mainly triggered by economic challenges which increased poverty levels, so the informal cross-border trade became an escape to alleviate this poverty, which between 2000 and 2015, was the first goal of the millennium development goals and is currently (2015 to 2030) the first goal of sustainable development.

Many researches that were carried out on the informal cross-border trade focused on various aspects other than the sustainable development goals. Although the researches that were carried out do acknowledge that the informal cross-border trade improved the household food security and improved the livelihoods of those engaged in it, they did not specifically clarify how the informal cross-border trade can achieve the sustainable development goals through its operation. There is a need to link the informal cross-border trade with the sustainable development goals that are derived from the millennium development goals, considering that the informal cross-border trade is a major economic activity that is taking place in Zimbabwe, particularly in the face of complex economic challenges being experienced in the country. Therefore, this research seeks to establish the link between the informal cross-border trade and the sustainable development goals; examine how the informal cross-border trade can achieve the sustainable development goals (from millennium development goals); and assess the extent of achieving the sustainable development goals through the informal cross-border trade in a challenging economic situation in Zimbabwe. The research follows four different stages of economic changes that were experienced by Zimbabweans, since the inception of the millennium development goals in order to reveal the challenges that must be dealt with so as to foster sustainable development in the country. These stages include the era of the Zimbabwean dollar, the multi-currency era, the bond note era and the cashless economy era.

2 Study Area

Gweru lies in agro-ecological region three, which experiences the tropical savannah climate, and it covers 16,700 km^2 (Jerie 2016). Its landscape is characterised by three main forms of soils, namely red loams, gravel and red basalt soils (Matsa 2007; Jerie 2016). The city lies on the watershed that stretches from Rusape to Bulawayo, with an average altitude of 1422 m (City of Gweru 2012; MSD 2014). It is 168 km from Bulawayo and 280 km from Harare, along the major Harare-Bulawayo road. There are numerous streams that dissect the municipal area, most of which drain into Gweru and Runde Rivers. The annual rainfall averages between 500 mm and 750 mm, and is characterised by mid-season dry spells. Although the normal rainy season is from October to April, the rainfall patterns have become increasingly poor, in some cases ending in March. Temperatures are low in winter (May to July), averaging 14° C and they are high in summer (September to April) when they can average 30° C (Jerie 2016).

Gweru comprises of high, medium and low density residential areas. The settlements in the city are divided into 18 wards, which are located in different directions from the central business district. The division of residential areas into wards was meant to ease service provision and administration by the local council. The total population of Gweru is 158,233 (73 768 males and 84 465 females), with a total of 41 149 households, according to the National Census of Zimbabwe (ZIMSTAT 2012). The largest low-income residential area is Mkoba high density suburb, which is located in the western section of the city and is divided into 21 sub-sections, referred to as 'villages'. The high-income low density residential areas are found mainly in the eastern and northern sections of the city.

Gweru is centrally located in the country and is the provincial capital of Midlands province, which makes it very accessible, and is the hub of traffic that traverses the country. The major industries include Bata Shoe Company and Zimbabwe Alloys, a chrome smelting company. Almost all the young and energetic generations are directly or indirectly involved in gold mining in the nearby Shurugwi District and other surrounding areas or are involved in informal economic activities Agricultural activities in the city include crop cultivation where maize, groundnuts, round nuts, beans, vegetables are grown and livestock is reared.

3 Theoretical Framework

The basic principles of the rational theory are derived from neo-classical economics. The rational choice theory focuses on actors who are seen as being purposive, that is, their actions are aimed at certain goals or ends (Rizter 1999). Actors also have preferences (values and utilities). The major assumption of the rational choice theory is that any individual action is purposive behaviour, which holds benefits for the actor in some ways (Vlaev 2018).

The individual's behaviour is seen as being motivated by goals, needs and wants. The theory also says that individuals act with specific given constraints that are based on the information that they have about the conditions under which they are acting (Weyl 2019). The rational choice theory holds that individuals usually anticipate the outcomes of alternative courses of action and they calculate that which would be the best for them. Rational individuals are believed to select the best alternative that is likely to grant them their goals at the minimal cost. Rational choice theorists identify two types of decision-making which are the event and involvement in decisions (Rizzo 2016; Vlaev 2018). Event decisions are those in which the strategies of carrying out an action are decided, while involvement decisions are those in which choices are made in order to become involved in a behaviour or an act. The retreat from or the continuity of such behaviour depends on the weight of the benefits or the costs of it.

The argument of this research is anchored on the notion that the economic activity of the informal cross-border trading is rooted in the rational choice theory, since any individual, before opting for cross-border trading, will have to weigh the pros and cons of such a decision. Since the cross-border trading does not require much capital, it is an easy option for many people as a means of achieving food security, reducing poverty, increasing income and employment creation. Along the line, as informal cross-border traders realise that the profits or benefits accrued from the trade far outweigh its costs, they develop ways of dealing with the challenges that are associated with the trade. This goes a long way to explain why there has been an increase in the informal cross-border trading in Zimbabwe since 2000. Coincidentally, it is the year when the millennium development goals were introduced, bringing to the fore the fact that an activity can only become attractive if the costs and the dangers associated with it are outweighed by its potential benefits. The study argues that the informal cross-border trade can help households and communities to achieve certain sustainable development goals in the difficult economic environment of Zimbabwe.

4 Methodology

4.1 Research Design

The research used the qualitative research design which allowed the researchers to explore and provide an in-depth understanding of how the informal cross-border trade can achieve the sustainable development goals (from millennium development goals). The design allowed the informal cross-border traders to elaborate on their views and it also enabled the researchers to probe for more explanations and in-depth understanding of their experiences. The major aim of the research was to link the achievements of the informal cross-border trade to the millennium development goals and the sustainable development goals. This was achieved by giving the research participants an opportunity to explain and elaborate on their experiences.

4.2 Target Population

The target population for this research was all the informal cross-border traders in Gweru, Zimbabwe. However, due to the trade's informality, the exact number of informal cross-border traders in Gweru could not be established. Therefore, all the informal cross-border traders who were accessible and reachable during the time of the data collection became part of the sample population, since the data was collected during the restrictive Covid-19 lockdown period.

4.3 Sampling Procedure and Sample Size

The research used snowball sampling. This was due to the lack of information about the total number of operating informal cross-border traders since they lacked records. More so, some of the informal cross-border traders at the stalls were mixed with other informal traders who were not practising cross-border trading. The informal cross-border traders were also scattered as some of them operated at various flea markets while others operated as street vendors. Snowball sampling was then used after the researchers were assisted to identify the first informal cross-border trader at one of the flea markets stall by the owner. This was then used as the first referral point upon which the participant referred the researchers to the next informal cross-border trader who, in turn, referred them to the next participant. This process went on until a total of 25 informal cross-border traders was arrived at. Twenty-five traders were chosen as the ideal number since the research was undertaken during the Covid-19 lockdown period when it was difficult to reach out to many people, considering that the total number of informal cross-border traders was not known due to the trade's informality.

4.4 Data Collection Instruments and Procedure

The research used semi-structured in-depth interviews to collect data from the informal cross-border traders. The purpose of the interview was explained to the participants before they participated in order to get their consent to be interviewed. The participants were given a consent form to sign as proof that they had agreed to participate in the research. In the consent form, they were assured of confidentiality in the interview, as well as the audio-recording that took place during the interview process. It was also made clear to them that they had the right to withdraw from the interview at any point. All the 25 in-depth interviews were carried out at various places and at different times, depending on the location of each trader. The information was tape-recorded, transcribed and then analysed through content analysis and ethnographic summaries. The interview guide followed the questions derived from the following key research questions:

1. What is the link between the informal cross-border trade and the sustainable development goals?
2. How does the informal cross-border trade help to achieve the sustainable development goals?
3. To what extent are the sustainable development goals achieved by the informal cross-border trade?

4.5 Data Analysis

Data from qualitative interviews was analysed using the thematic content analysis. This method was used because of its highly flexible nature as it could be modified to meet the needs of the research by providing a rich and detailed account of data. It was also used as a way of identifying, analysing and reporting themes and patterns from the data, being guided by the research objectives. The researchers had to first get familiar with the data during transcription by reading and re-reading it, searching for patterns and meanings. The researchers then generated the initial codes for the data. The sorting of the different codes was done in order to search for themes that were relevant and had meaningful data for the research. These themes were then reviewed before producing the report which was then presented under each theme that was formulated from the research objectives in order to make sure that all the research objectives were addressed.

5 Results

The major aim of the study is to examine how the informal cross-border trade ishelping households and communities in Zimbabwe to achieve the sustainable development goals that are derived from the millennium development goals. The study also focuses on the different phases that the country has gone through since the inception of the millennium development goals, bringing out how the millennium development goals and the sustainable development goals were either compromised or achieved during the process. A total of 25 respondents were interviewed (Table 1).

Out of the 25 respondents, 56% were females while 44% were males. The results of the research revealed that the major reason for this gender imbalance is that women were less educated, hence they were less likely to be able to secure employment in the formal sector. Subsistence farming, which they mainly used to depend on for survival, was no longer sustaining them due to recurrent droughts and the effects of climate

Table 1 Characteristics of the respondents

Bio data	Variable name	Number of people
Gender	Male	11
	Female	14
Age	Below 18 years	2
	18–25 years old	4
	26–35 years old	8
	36–45 years old	8
	Above 46 years old	4
Education level	Primary	4
	Secondary	13
	University/College	7
	None	0
Marital status	Married	8
	Widowed	4
	Divorced	6
	Single	9
Occupation	Both formal and informal sector	6
	Informal sector only	19

Source Compiled from the interviews, 2020

change. Hence, they were forced to look for an alternative, and this is in accordance with the rational choice theory, which they found in the form of informal cross-border trade, since it did not need much capital to start. These findings are in line with what was discovered by Chikanda and Tawodzera (2017). The findings of this research also revealed that 10 out of 14 women (71%) did not have spouses, hence they carried the family burden on their shoulders, which probably explains why a larger proportion of women than men was in the trade. The informal cross-border tradeis thus an economic activity that empowers women and promotes gender equality, thereby achieving both the millennium development goals 3 and the sustainable development goal 5.

The average age of people involved in the informal cross-border trade was around 26 years. From the findings, it was deduced that the reason for this age group could be that these people are mature and capable of making rational decisions that are associated with the challenges that are encountered in the informal cross-border trade. The informal cross-border trade is a business that operates illegally, therefore, it needs mature and strong individuals who can make rational decisions in their operation in order to maximise profits and reduce costs, according to the rational choice theory. The education level of the respondents shows that all of them had obtained some level of formal education. This assisted them to read and understand some business procedures and to make better decisions on business transactions that are aimed at profit-making. Most respondents were single and, amongst these, many had children and,therefore, had family responsibilities.

6 The Link Between the Informal Cross-Border Trade, Millennium Development Goals and Sustainable Development Goals

Reasons for Joining Informal the Cross-Border Trade

Recurrent Droughts: The major reason given for partaking in the informal cross-border trade, particularly by women, was that they could no longer depend on agriculture for survival due to frequent droughts. Those who had many years in the informal cross-border trade indicated that they used to spend much of their time in rural areas during the cropping season, focusing on agriculture. They recalled that Zimbabwe was struck by severe droughts in 2002, 2005 and 2007 that greatly affected their source of livelihood. It became difficult for them to continue relying on this source of livelihood, hence they had to seek an alternative for survival, and the informal cross-border trade came as the best escape for them and this is in fulfilment of the rational choice theory. This was the period when the millennium development goals had been introduced in 2000 and the informal cross-border trade played a major role in achieving the millennium development goals 1 and 3 as a source of livelihood in alleviating poverty and empowering women who were severely affected, since they could no longer depend on rain-fed agriculture for their survival in the rural areas.

Lack of Source of Income: The respondents indicated that the Zimbabwean economy has been retrogressive from 2000, at a time when the millennium development goals were introduced. Many people were retrenched from industries and some of the industries actually ceased operations. People were left with no other option but to get into the informal sector to earn income for survival. Many Zimbabweans were suffering from food insecurity and poverty, and the informal cross-border trade came as a solution to alleviate poverty and ensure food security amongst the poor households. As a result, the millennium development goals 1, 2 and 3 were achieved, thus alleviating poverty, empowering women and accessing education because of the increased income. So the major reasons behind the emergence of the informal cross-border trade were strongly related to achieving the millennium development goals and, later, the sustainable development goals.

7 Informal Cross-Border Trade as a Pathway to Achieving the Millennium Development Goals, Precursor to the Sustainable Development Goals

Zimbabwean Dollar Era (2000 to 2008) (Millennium Development Goals Era) Respondents indicated that, in general, the informal cross-border trade is a very profitable economic activity although the profits that can be realised from it differ in accordance with the different phases of economic development in the country. Those who ventured in the informal cross-border trade during the Zimbabwean dollar era, before the multi-currency system, had their own experiences. They indicated that before the hyperinflation of 2007, the informal cross-border trade was a very profitable business that helped them to sustain their families, despite the effects of climate change. The Zimbabwean dollar was very strong as a currency, hence the business of buying and selling was very lucrative and they could earn double the capital that they had invested. These profits meant that households involved in the informal cross-border trade could manage to feed their families, send their children to school, pay for all their social services and improve their livelihood. In the process, a number of millennium development goals were achieved. During that time, many industries were still functioning and this meant more customers for the informal cross-border trade as the majority of the people could afford purchasing products from the informal cross-border traders, since they had sources of income. During that period, informal traders were very few in the business competition and, in turn, they earned more income. A woman in the informal cross-border trade was quoted as saying:

I had no challenges at all in looking for customers during this period because I had established a strong relationship with workers from Bata Shoe Company, Zimbabwe Alloys Company and Zimbabwe Glass Company. They just made orders and I delivered the goods at the end of each month. However, the problem arose when the customers lost their jobs and could not afford my products anymore.

In 2007, the Zimbabwean dollar seriously lost value and, therefore, hyperinflation set in. Most industries downsized and many others closed as indicated by the respondents. This was the time when there was a surge in people joining the informal cross-border trade as it was considered to be the best alternative whose potential benefits outweighed costs, since it did not require much capital to start the business, and this is in accordance with the rational theory. Chikanda and Tawodzera (2017) argue that, by far, the largest percentage of the informal cross-border traders started operations during and after the economic crisis of 2008, with nearly three-quarters of them setting up businesses between 2006 and 2014. It was very difficult to find food stuffs in the formal markets and the situation was further worsened by the 2007/2008 hyperinflation. It was the informal cross-border trade that was mainly responsible for bringing in foodstuffs into Zimbabwe from the neighbouring countries and many people started sourcing food items from the informal market. Although the informal cross-border trade became the greatest source of goods and food items in particular, it started facing challenges of the exchange rate as the Zimbabwean dollar was losing value against the currencies of the product suppliers in the neighbouring countries. This negatively affected the profit margins of the informal cross-border trade which also, in turn, affected the possible achievement of the millennium development goals.

Multi-Currency Era From 2009–2015 (Millennium Development Goals Era)
In 2009, the government of Zimbabwe introduced the multi-currency system and phased out the Zimbabwean dollar. During this period, the United States dollar was the official currency that was used in the country although the South African rand and the Botswana pula were also used. The respondents indicated that, during this period, income from the informal cross-border trade increased because it was cheaper to buy goods outside the country and make more profit when selling the goods in American dollars, in Zimbabwe. One lady who was involved in selling second-hand clothing from Mozambique indicated that it was during this period that, in a single month, she managed to raise the profit that was more than the capital that she had invested and managed to expand her informal cross-border business. From the $500 that was invested, she earned around $1200 per month. With such huge incomes, it generally meant that several millennium development goals were achieved and this ensured the sustainability of the achievements during the period. All the respondents were of the same view that, during this period, their incomes from the informal cross-border trading were huge.

The following are some of the narrations of the informal cross-border traders:

1. *"I managed to buy a residential stand to build my own house and a private car for myself, with the profits that I raised from the informal cross-border trade."*
2. *"I managed to send my children to good boarding schools and had no challenge at all in raising their school fees. I also built a rural home for my parents."*
3. *"I managed to start another business of commuting people to and from town, using the profits that I earned from the informal cross-border trade. I bought two commuter omnibuses to start the business from the profits that I got."*
4. *"I built this house that I am staying induring the multi-currency period, usingthe money from my informal cross-border business."*

It is clear from the foregoing narrations that the informal cross-border traders accrued large profits from their businesses during this period. There was a lot of community development that took place as a result of profits that were obtained from the trade. This meant that poverty in the households that were involved in the informal cross-border trade was reduced as some of them revealed that they were earning far much more than those who were working in the formal sector. Since the millennium development goals were interdependent, just like the sustainable development goals, the reduction of poverty and hunger positively influenced better health, and all this was made possible by the increased income from the informal cross-border trade. The respondents also indicated that they managed to send their children to schools of their choice, with minimum challenges, thereby contributing to the achievement of the millennium development goals of universal education. Some of the cross-border traders managed to expand their businesses by acquiring two or more flea market stalls and this enabled them to employ more people as assistants in selling the products. The informal cross-border trade also provided employment to transporters who provided transport services to traders to and from the national borders, with some even crossing Zimbabwe's borders to inland towns like Messina in South Africa, Maputo in Mozambique and Francistown in Botswana.

Single mothers who participated in the study indicated that they managed to take care of their families without involving themselves in immoral activities such as prostitution. This meant that women were empowered, thereby contributing to the achievement of the millennium development goal 3 of gender equality by empowering women. This goal later became the sustainable development goal 5. The following are some of the statements from the single mothers:

1. *"During the multi-currency era, I had no challenges at all in looking after my siblings and my children."*
2. *"I actually joined the informal cross-border trade during the multi-currency period after seeing many women surviving on it. Initially, I was only involved in buying and selling locally, without crossing borders, because it was difficult for me since I did not have enough capital, so I lacked variety of merchandise which many customers preferred. Other traders got their merchandise from abroad. It was cheaper buying from abroad and reselling in the country because customers were transacting with the US dollar. I am a single mom and I managed to acquire a lot of things that include the car that I bought from the profits of the informal cross-border trade during that period."*

The Bond Note and the Cashless Economy Era From 2016 to 2019 (Sustainable Development Goals Era)

In 2016, the country ran fast out of the United States dollar, so the bond note was introduced as a surrogate currency. However, this bond note was not sufficient enough to service the whole country, which made the government of Zimbabwe introduce a virtual currency as an official method of transaction. From 2016, the US dollar became scarce on the formal market and this affected the income from the informal cross-border trade. However, during 2016 and 2017, the situation was better in the sense that the bond note was accorded the same value as the US dollar. The traders still achieved all the goals which they used to achieve during the multi-currency era, meaning, the sustainable development goals were being attained, since the millennium development goals were a carry-over of the sustainable development goals. From 2018, the situation worsened as the bond note was losing value at an alarming rate against the US dollar and that severely affected the incomes and profits from the informal cross-border trade. The worst scenario was experienced when the virtual currency was rated at lower than the bond note. The majority of customers in the country failed to access cash in banks, hence they bought the informal cross-border trade goods using the virtual currency. It was difficult for the informal cross-border traders to access foreign currency in order to import new stock using virtual money, hence they had to first buy bond notes at the parallel market rate so as to buy foreign currency. All these processes had premium rates which made traders lose a lot of money in the process. The reduction in profits meant that the traders could no longer afford to sustain their business as they used to, hence the achievement of the sustainable development goals was affected. The respondents indicated that they could no longer depend on the informal cross-border trade as they used to do. The business could not provide beyond mere subsistence needs. It was so because the buying rate by the customers had declined, since most of the customers earned their salaries in virtual currency, which could not sustain their basic necessities.

In an effort to survive, traders who used to employ people as assistants in selling their products had to retrench them as they could not manage to pay them. This meant that the sustainable development goal 8, which is about decent work and economic growth was negatively affected. Those who had expanded their informal cross-border business could only utilise one stall at the flea market and rented out the others. Some got to the extent of sharing stalls with other tenants as they could not manage to pay rentals for individual stalls. The traders preferred payment in cash from customers and they charged an extra amount on those who used plastic or mobile money to purchase their goods. All these issues, according to the rational choice theory, were rational decisions that were employed by the traders in the informal cross-border trade as alternatives for survival in order to reduce trading losses, given the state of the operating environment.

The informal cross-border trade no longer produced profits for them to survive. Some traders had to abandon flea markets stalls to sell in the streets or from their homes as they could not afford paying rentals, which were increasing every month, following the escalating exchange rate on the parallel market. Others had to remove their children from boarding schools as they could no longer afford paying the high school fees. Some of the sustainable development goals could no longer be sustained due to fiscal challenges that were faced in the country during this period.

The Zimbabwean Dollar and the Cashless Economy Era From 2019 to August 2020 (Sustainable Development Goals Era)

Respondents indicated that, in 2019, the bond note was converted to the Zimbabwean dollar, but still, cash was very scarce and it continued to maintain a much higher value against the virtual currency. The exchange rate of a US dollar against the Zimbabwean dollar continued to rise on the parallel market to the extent that it rose 17 times from

January 2019 to June 2020, a period of just one year and five months. In January 2019, 100 US dollar was equivalent to 200 Zimbabwean dollars (cash) and 300 Zimbabwean dollar (plastic money), but in August 2020 it was 8700 Zimbabwean dollars (cash) and 10 100 Zimbabwean dollars (plastic money). The traders indicated that it was difficult to survive in the business with such escalating and fluctuating rates and they had no option but to reject plastic money and mobile money as forms of payment from their customers, since they were running losses. The only forms of payment they accepted were foreign currency and the bond note after rating it against the prevailing exchange rate. This was another rational decision that they employed in order to reduce losses and maximise profits, according to the rational choice theory. This situation further repelled many customers since the majority of them were also earning salaries in plastic money. Although profits started to improve since they started charging products in foreign currency, they were not enough for one to survive like they used to do during the multi-currency era. The majority of those who were interviewed indicated that the income could only sustain them on some basic necessities.

Respondents indicated that, in 2020, during the Covid-19 lockdown period, the situation worsened as they could no longer generate anything from the informal cross-border trade. During the first 35 days of the lockdown period, there was hardly any business taking place because the movement of people was very restricted. Some indicated that although they had carried some stock home in order to sell during the period, it was difficult to do so as police officers were everywhere, monitoring the movement of people. This severely affected the sustainability of maintaining the sustainable development goals that had been achieved by the informal cross-border trade. After the 35 days, the lockdown was adjusted to a lower level, but the informal sector was not allowed to open and, for them, it meant there was no business taking place at all. Some indicated that they were not sure what was going to happen to their informal cross-border business as their areas of business operation had been destroyed by the city council during the lockdown period for reasons of cleaning up the city. They were not sure when and where they were going to be allocated new flea market stalls. The country's borders were closed and this meant that the importation of new stock for sale could not be done. The cost of living rose significantly during this period and many food products were in short supply in the country and yet this was actually a good opportunity for traders to import and supply the needed products.

To What Extent Does the Informal Cross-Border Trade Help Achieve Sustainable Development Goals?

The respondents indicated that the informal cross-border trade is an economic activity that has the potential to provide food security, alleviate poverty, offer employment, empower women and improve the living standards of many people in Zimbabwe, if the necessary support is given by the government. All the respondents had no intention of leaving the informal cross-border trade although they were facing challenges. They were hoping that the socio-economic situation would one day get back to its former state, when they used to earn huge profits from the business. They also pointed out that they always tried to find ways of surviving in the business whenever they are presented with challenges. The resilience that was exhibited by the traders shows that they have strong faith that the informal cross-border trade is capable of improving their livelihoods as it used to do. Figure 1 shows the millennium development goals and the sustainable development goals that were achieved through the informal cross-border trade in various phases, since the informal cross-border trade began. The difference in the shading depicts the impact of how the millennium development goals and the sustainable development goals were achieved by the informal cross-border trade with the darkest shaded portions indicating the greatest achievement.

Figure 2 shows the framework of the informal cross-border trade and the sustainable development goals. It is indicating that the informal cross-border trade can achieve the sustainable development goals 1,2,3,4,5,8 and 11, given a conducive operating environment. The income earned from the informal cross-border trade helped households to eradicate poverty, end hunger and, in the process, maintain good health and well-being. Poverty can be examined from various angles, which include capability deprivation, lack of income, little or no asset ownership, lack of access to basic goods and services. A critical analysis of the informal cross-border trade activities in this research suggests that the informal cross-border trade is an important income-earning activity for the small scale entrepreneurs, hence a mechanism of addressing poverty and the sustainable development goals. Quality education was achieved as the traders could afford to send their children to good schools and manage to pay for their education. The large number of women who were involved in the informal cross-border trade indicated that the informal cross-border trade was not

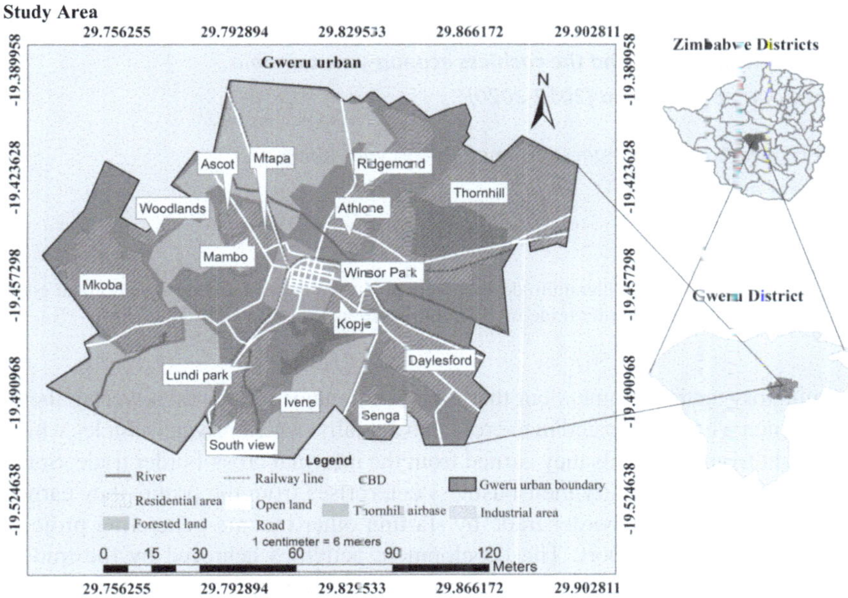

Fig. 1 Map of Gweru, Zimbabwe

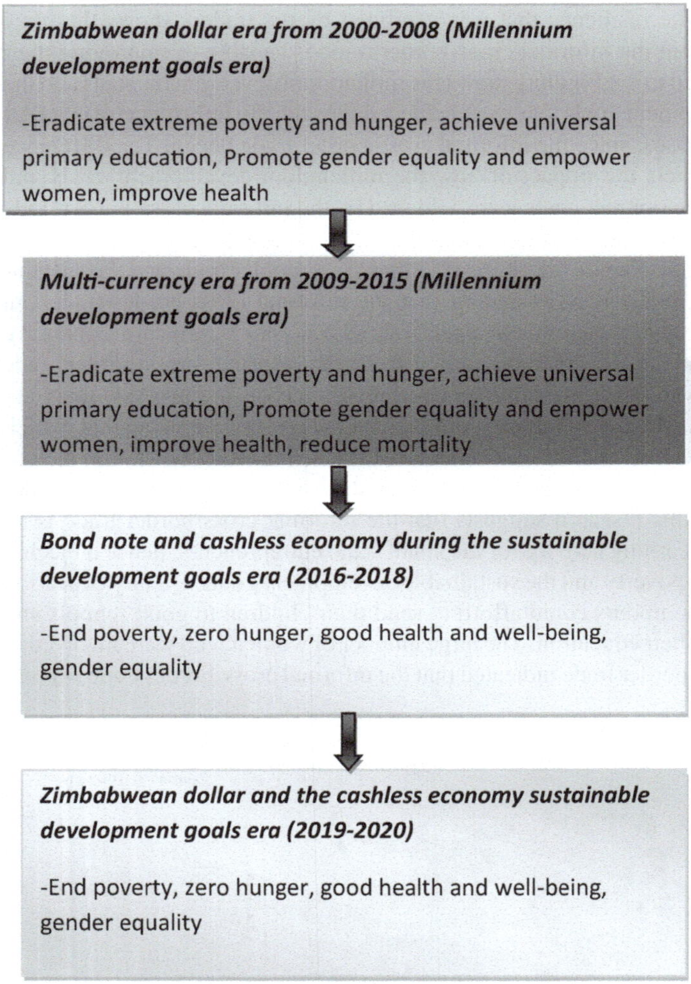

Fig. 2 Achievement of the millennium development goals and the sustainable development goals through the informal cross-border trade in different phases as experienced by Zimbabwe

discriminatory against women, but that it played a major role in empowering them. Some women who participated in the research actually owned private vehicles which they bought from the profits they earned from the informal cross-border trade. Some traders managed to diversify their business enterprises from the profits they earned from the informal cross-border trade by starting other income-generating projects like the commuter transport. The development activities achieved by the traders contributed in achieving sustainable communities. The major cause in the rise and increase of the informal cross-border trade was attributed to impacts of the climate change that significantly affected rain-fed agriculture, which they mainly depended

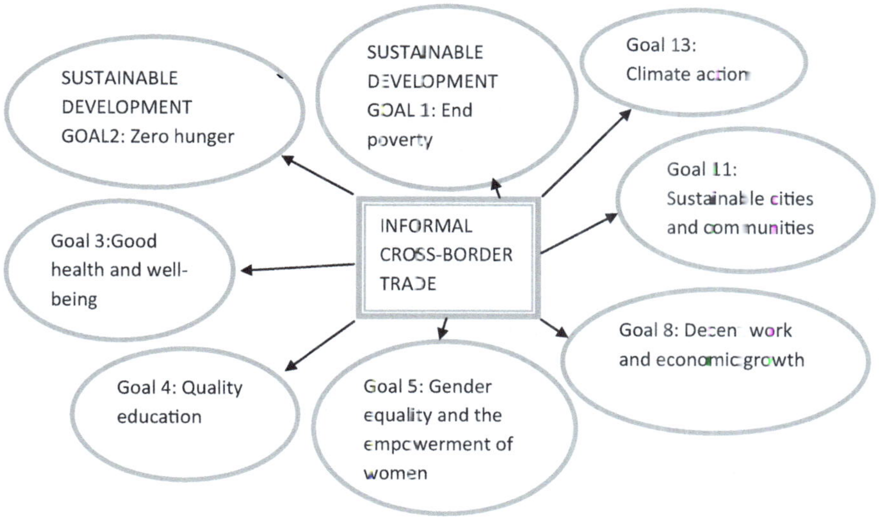

Fig. 3 Informal cross-border trade and sustainable development goals framework

on for their survival. Therefore, the informal cross-border trade became an adaptive strategy to survive in the face of the climate change impacts. From the findings of the study, it can be deduced that the informal cross-border trade is driven by the need to improve the welfare of the poor and that it can be used as a strategy for accumulating wealth. Within this context, it is apparent that the informal cross-border trade can achieve the sustainable development goals if the operating environment is improved (Fig. 3).

8 Conclusion

The findings from the study revealed, to a large extent, that the informal cross-border trade is induced by poverty as people engage in it for survival and to earn a living. This makes the informal cross-border trade a crucial economic activity that can help households to achieve some of the sustainable development goals, thereby ensuring sustainability. The income that is earned from the informal cross-border trade is able to sustain households in eradicating poverty, improving food security, empowering women and developing communities. However, the fiscal challenges and the unfavourable operating economic environment are the main obstacles that make it difficult for the informal cross-border trade to sustain the achievement of the sustainable development goals. This was the reason for the different levels of achievement in the attainment of the sustainable development goals and the millennium development goals in each phase in Zimbabwe.

The resilience of the traders to continue partaking in the informal cross-border tradeshows that the trade greatly improved their lives and they still hope that if the operating environment changes for the better, the informal cross-border trade will still continue improving their lives. For the traders, the informal cross-border trade remains the best alternative that gives them the greatest satisfaction to sustain their livelihoods and achieve the sustainable development goals, since it offers them opportunities to make rational choices to reduce losses and maximise profits, according to the rational choice theory.

Although the informal cross-border trade contributes considerably to the well-being of the Zimbabwean citizens, there is a need for planned steps to be taken in order to create an enabling policy and regulatory environment so that the sector can have a positive impact towards the attainment of the sustainable development goals.

9 Recommendations

The study recommends that the government of Zimbabwe should make it a priority to assist the informal cross-border tradein order to boost its economy and it should refrain from tightening the traders' conditions of operation. This can be achieved by allowing the informal cross-border traders to register without necessarily subjecting them to taxes until they have reached a stage where they can be able to pay such taxes, without threatening the viability of their businesses. This is because the sector has great potential to help the country in its journey towards achieving the sustainable development goals by 2030. Since the country is already struggling to resuscitate the formal sector, the government should take this opportunity to support this sector in which many people are involved in order to achieve the sustainable development goals by 2030. The government of Zimbabwe should also introduce policies that protect and govern the informal cross-border trade so that the traders are motivated to formalise their businesses.

References

African Development Bank (2015) Growth, Poverty, and Inequality Nexus: Overcoming Barriers to Sustainable Development. African Development Report, African Development Bank Group, number 2342, Abidjan, Ivory Coast

Babah D, Falylath PT, Ingenbleek M, Van Trijp HCM (2019) Living the African Dream: how subsistence entrepreneurs move to middle-class consumer markets in developing and emerging countries. J Public Policy Market 38(1):42–60

Blankson C, Cowan K, Darley WK (2018) Marketing practices of rural micro and small businesses in Ghana: the role of public policy. J Macro Market 38(1):29–56

Brenton P, Soprano C (2018) Small-scale border trade in Africa: why it matters and how it should be supported. Bridges Africa 7(4):3–6

Chikanda A, Tawodzera G (2017) Informal entrepreneurship and cross-border trade between Zimbabwe and South Africa (rep. i-41). Waterloo, ON: Southern African Migration Programme. SAMP Migration Policy Series No. 74

City of Gweru (2012) City of Gweru master plan written statement. Gweru, Gweru City Council

Costanza R, Fioramonti L, Kubiszewski I (2016) The UN sustainable development goals and the dynamics of well-being. Front Ecol Environ 14(2):59–79

Dumbu E (2018) Banking the unbanked: perceptions of the international cross-border women entrepreneurs in Beitbridge, Zimbabwe. London J Res Manage Bus 18(1):16–25

Gaidzanwa R (1998) Cross-border trade in southern Africa: a gendered perspective. In: Sachikonye L (ed) Labour markets and migration policy in Southern Africa. Harare, SAPES Trust

Jerie S (2016) Occupational risks associated with solid waste management in the informal sector of Gweru, Zimbabwe. J Environ Public Health 16(10):11–55

Kahiya E, Kadirov D (2020) Informal cross-border trade as a substratum marketing system: a review and conceptual framework. J Macro Market 40(1):88–109

Kamete A (2017) Governing enclaves of informality: unscrambling the logic of the camp in urban Zimbabwe. J Geoforum 81:76–86

Kanyenze G (2011) African migrant labour situation in Southern Africa. Paper Presented at the ICFTU-AFRO Conference on 'Migrant Labour,' Labour and Economic Development Research Institute of Zimbabwe (LEDRIZ), Nairobi

Koroma S, Nimarkoh J, You N, Ogalo V, Owino B (2017) Formalization of Informal Trade in Africa: Trends, Experiences and Socio-Economic Impacts. In: CUTS international–food and agriculture organizations of the United Nations, regional office for Africa, Ghana. http://www.fao.org/3/a-i7101e.pdf. Accessed 04 Aug 2020

Kurebwa J (2015) Thevulnerability of female cross-border traders to HIV and AIDS in Zimbabwe: the case of Beitbridge Border Post. IOSR J Humanit Soc Sci 20:65–69

Lu Y et al (2015) Policy: five priorities for the UN sustainable development goals. Nature 520(7548):432–433

Madzwamuse M (2010) Climate governance in Africa: Adaptation strategies and institutions. Heinrich Böll Stiftung (HBS), Berlin

Manjokoto C, Ranga D (2017) Opportunities and challenges faced by women involved in informal cross-bordertrade in the City of Mutareduring a prolonged economic crisis in Zimbabwe. J Indian Ocean Reg 13(1):25–39

Matsa M (2007) Environmental factors related to the spread of pollutants around Gweru dumpsite. Unpublished MA dissertation. Department of Geography and Environmental Science, University of Zimbabwe, Harare

MSD (2014) Meteorological service department: Zimbabwe weather report, June 2014. Harare, Zimbabwe

Muzvidziwa V (2015) Women without borders: informal cross-border trade among women in the Southern African development community region (SADC). OSSREA, Addis Ababa

Mwaniki J (2005) The impact of informal cross-border trade on regional integration in SADC and implications for wealth creation. IRED-ESA (The Corn Thematic Coordinator ForCross-Border Trade) Accessed 16 March 2013

Nilsson M, Griggs D, Visbeck M (2016) Policy: map the interactions between sustainable development goals. Nature 534(10):320–322

Nkoroi I (2016) Assessing the informal cross-border trade between Kenya and Uganda. Int Acad J Arts Humanit 1(1):54–90

Ritzer G (1999) Sociological theory, 4th edn. Sociology Series, McGraw Hill International Editions

Rizzo MJ (2016) Behavioural economics and deficient willpower: searching for Akrasia. Georgetown J Law Public Policy 14(1):789–806

Rugube L, Matshe I (2011) Economy livelihoods: Africa in informal cross-border traders and the creation of the SADC Common Market. In: Kondo K.W, Ejiogu C (eds) Africa in focus: Governance in the 21st century. Pretoria: Human Sciences Research Council

Peterson M, Zehra SS (2018) Putting African country development into macro marketing perspective. In: Agarwal J, Wu T (eds) Emerging Issues in Global Marketing. Springer, Cham, pp. 333–368. https://doi.org/10.1007/978-3-319-74129-1_13
Sachs JD (2012) From millennium development goals to sustainable development goals. Lancet 379(9832):2206–2211
Sandada M (2014) Transition from informality to formality perceptions of informal traders in the Harare metropolitan area. Int J Econ Commerce Manage 2(1):6–12
Sommer L, Nshimbi C (2018) The African continental free trade area: an opportunity for informal cross-border trade. Bridges Africa 7(4):7–11
Vlaev I (2018) Local choices: rationality and the contextuality of decision-making. J Brain Sci 8(1):1–22
Weyl GE (2019) Price theory. J Econ Lit 57(2):329–384
ZimStat (2012) Census preliminary report 2012. Harare, Zimbabwe

Framework for Coherent Formulation and Implementation of ICT Policies for Sustainable Development at the Bottom of the Pyramid Economy in Nigeria

Samuel C. Avemaria Utulu

Abstract Developing countries such as Nigeria adapt ICT policies formulated by UN agencies and other intergovernmental organizations to coordinate how they use ICT to reach the sustainable development goals. Different challenges usually impede ICT policy formulation and implementation at the bottom of the pyramid economy because the UN and its agencies deal mostly with national governments. Existing studies claim that factors such as geographical structure, political-administrative structure, weak institutions, and corruption are the main factors that impede the formulation and implementation of ICT policies for sustainable development at the bottom of the pyramid economy. The chapter, however, proposes that differences in the internalized views of those involved in the formulation and implementation of ICT policies and those operating at the bottom of the pyramid economy are the primary challenges to ICT policy formulation and implementation at the bottom of the pyramid economy. The chapter argues that problems connected to geographical structure, political-administrative structure, weak institutions, and corruption result from differences in the internalized views of those that formulate and implement ICT policies and those that operate at the bottom of the pyramid economy in Nigeria. The chapter draws from the extant literature and observation of experiences in Nigeria. It presents a useful framework that proposes how to align the internalized views of those that formulate and implement ICT policies and those that operate at the bottom of the pyramid economy. The framework will promote the coherent formulation and implementation of ICT policies at the bottom of the pyramid economy.

1 Introduction

Prahalad's (2005) idea of the importance of the bottom of the pyramid economy to promote sustainable development is welcoming. He argues that the bottom of the

---

S. C. A. Utulu (✉)
Department of Information Systems, School of Information Technology and Computing, American University of Nigeria, Yola By-Pass, Lamido Zubairu Way, PMB 2250, Yola, Adamawa State, Northeast Nigeria, Nigeria
e-mail: samuel.utulu@aun.edu.ng

W. Leal Filho et al. (eds.), *Sustainable Development in Africa*, World Sustainability Series,
https://doi.org/10.1007/978-3-030-74693-3_35

pyramid induced sustainable development ensures that the development needs of the future generation and the ecosystem are not compromised in developing countries. Prahalad argues that the bottom of the pyramid economy consists of the most impoverished, which are about half the world population and who mainly reside in countries in the Southern hemisphere. Therefore, scholars propose that if multinational corporations invest in the bottom of the pyramid, poverty will be eradicated (Kramer and Beltz 2017). Harnessing the fortunes at the bottom of the pyramid economy requires an economic development arrangement that makes investments that are not driven by profit-making. The investments are made by multinational corporations in poor people to eradicate poverty in regions where poverty is endemic (Prahalad 2005). The notion has been justified by practical examples that reveal how efforts made to implement the bottom of the pyramid economic development programs in Nigeria required sacrifice and not the desire to profit (Aideloje et al. 2018). The limitation of this economic development arrangements is that it depends entirely on multinational companies and how they are open to investing in poor people without making financial profits. Conversely, the good thing about the ideology is that it opens up to governments and development experts a unique philosophy about how developing countries in the South can easily reach critical sustainable development goals. The goal includes poverty eradication, zero hunger, economic growth, and sustainability (Pradhan et al. 2017; Sachs 2012). The philosophy that drives the bottom of the pyramid's economic development promotes the notion that devoting significant investments to poor people is the best way to promote sustainable development (Yaacoub and Alouini 2020).

Not all studies dealing with financial inclusion, digital inclusion, knowledge and information inclusion, etc., are explicitly devoted to assessing situations at the bottom of the pyramid economy. However, these studies deal with vital issues that could help stakeholders understand factors that determine sustainable development at the bottom of the pyramid economy. It is a known fact that those living in rural areas in many parts of the globe are usually excluded from benefiting from available social and economic resources. Therefore, inclusive theories propose that sustainable development can be hastened in developing countries if attention is paid to making those excluded from participating actively in social, economic, and technological activities to be included. Development frameworks based on inclusive theories enable government and development agents to focus on those operating at the bottom of the pyramid economy. Inclusive approaches argue that it is essential to take them as active players in sustainable development policy formulation and implementation processes (Asongu et al. 2020; Kramer and Belz 2017). Government and other stakeholders have implemented development programs focused on using information and communication technologies (ICT) to promote economic development among people in rural and sub-rural areas in sub-Saharan Africa in the past four or five decades. The development programs were the bottom of the pyramid economic development programs (Yaacoub and Alouini 2020; Boateng et al. 2012). Therefore, the bottom of the pyramid economic development does not have to be initiated by multinational corporations. In this chapter, it is taken that individuals, organizations, and governments can also trigger the bottom of the pyramid economic development in sub-Saharan Africa to promote sustainable development.

Given this, scholars further exposed the importance of ICT policies for sustainable development. The result is the growth in the number of studies carried out on the role of ICT policies in ICT-induced sustainable development in sub-Saharan Africa (Chigona 2019; Krauss 2013). An increase in the number of studies carried out on ICT policy formulation and implementation shows that stakeholders have started to value the notion that using appropriate policies to coordinate ICT adoption is likely to promote sustainable development. Therefore, the evolution of ICT policies in Africa can be traced to the early 2000s when the African Information Society Initiative (AISI) was launched (Hafkin 2002). The epoch marked when governments, inter-governmental organizations, and policymakers started to identify the importance of ICT to sustainable development programs of sub-Saharan African countries. Consequently, several ICT policies have been formulated at different levels to address the digital divide and knowledge divide and how they impact development in sub-Saharan Africa (Bornman 2016). Intergovernmental organizations such as the United Nations (UN), United Nations Economic Commission for Africa (UNECA), World Bank, African Union (AU), African Union Economic Community of West African States (ECOWAS), African Development Bank (ADB), etc. provided the basis for ICT policies formulation and implementation in sub-Sahara Africa.

Policies on ICT formulated with intergovernmental organizations' help were to be adapted and contextualized to suit the bottom of the pyramid economy's needs. In Nigeria, for example, due to the federal constitution in operation, states and local governments also develop their ICT policies following frames of references proposed by the federal government. An excellent example is the development plan tagged National Economic Empowerment and Development Strategy (NEEDS). The Federal Government of Nigeria prepared the plan. It was, however, condensed to State Economic Empowerment and Development Strategy (SEEDS) for state governments and Local Economic Empowerment and Development Strategy (LEEDS) for local governments (National Planning Commission 2004). The Federal Government of Nigerian derived important aspects of NEEDS, SEEDS, and LEEDS from the UN's millennium development goals (MDGs). Despite this, critics argue that the plans were not focused on real, sustainable development needs. For instance, Uche (2019) argues that they focused more on global economic growth parameters than local parameters to bring about real development at local levels. Some scholars say that implementing ICT policies embedded in the plans at local levels is the main reason for the development plans' failure (Ogundele and Hassan 2011). Issues such as this are persisting in Nigeria and other sub-Saharan African countries. Efforts made to assess the factors responsible for the persisting non-performance of ICT policies at the bottom of the pyramid economy usually reveal geographical structure, political-administrative structure, weak institution, and corruption. The persistence of these problems shows that the revelations have limitations. The chapter draws stakeholders' attention to these issues and addresses them. This conceptual research followed a literature review and observations in four rural settlements in Oyo (Moniya and Ide-Ose) and Osun (Ede and Akoda) States in Nigeria. A multi-theoretical approach has been used, similarly by scholars that deal with ICT policy.

2 ICT and Bottom of the Pyramid Economy Development

The term ICT encompasses three concepts, namely, information, communication, and technology. The concept, information is generally defined as processed data; that is, contextualized data modeled into some form of meaning. Information is new and questions or improves on existing knowledge. It is also relative and depends on the person(s) who did the processing (Tchamyou 2020; McNair 2017). The age-long belief that knowledge is not static came about because information produces new grounds to question or improve existing knowledge. The evolution of the information society resulted from the role information plays in achieving sustainable development goals. The crucial role information plays in the formulation and implementation of sustainable development programs led to coining the term, information society. Consequently, attaining sustainable development at the bottom of the pyramid economy is expected to eradicate ignorance through access to functional education, enlightenment programs, critical thinking skills, and culture change (Tchamyou 2020; Chandra et al. 2014).

The second concept used to derive the term ICT is communication. Communication has to do with transferring information from one person or entity, commonly referred to as the sender, to another person or entity, referred to as the receiver. The goal of communication is to transfer information through a medium that can be electronic, natural, or oral. People communicate to facilitate the transfer of new information between them (McNair 2017). The primary challenge those within the bottom of the pyramid economy face is the inability to communicate information as quickly as required to spur sustainable development. The third concept used to derive the term ICT is technology. Technology refers to the ways things are done that are coded into machines, models, processes, cultures, beliefs, and products (Utulu and Sote 2009; Heidegger 1954). Therefore, it follows that ICT means the way(s) information is created and communicated among people. Consequently, there are arguments that the bottom of the pyramid economy cannot achieve sustainable development goals without meaningful ICT investment (Bornman 2016; Boateng et al. 2012).

Given this, there are insights into the extant literature on how ICT has been used to promote survival in rural areas in sub-Saharan Africa. These insights show how ICT has been used to facilitate access to healthcare and health information (Adepoju et al. 2017). There are pieces of evidence of how pregnant women in rural areas who often suffer from lack of access to healthcare and health information use ICT to solve problems related to access to healthcare and health information (Obasola et al. 2015). Apart from facilitating sustainable development through healthcare, ICT has also promoted survival among those operating petty trades. Petty trading is the most common form of business activity within the bottom of the pyramid economy.

It provides funds required to support livelihood and has benefited from mobile phones (Boateng et al. 2012). Most petty traders use mobile phones to facilitate communication between them and buyers and suppliers. A few of them also use mobile phones to get financial information in bank transaction alerts.

There are also reports in the extant literature on how farmers, educators, and governments in rural areas use ICT to achieve sustainable development goals. Practical examples of these include the ways rural farmers use ICT to access information on subsidies and other forms of support provided by governments and non-governmental organizations (Oladele 2015). Reports on how farmers in rural areas get weather forecasts from the Internet, radio, and television stations are also available in the extant literature (Kiiza and Pederson 2012). There is also information on how ICT facilitates the sales of farm produce, leading to a loss of investments if not sold. Concerning education in rural areas, both students and teachers have recounted how ICT promotes teaching and learning in different fascinating ways (Chandra et al. 2014; Mthethwa and Munyoka 2013). With regard to government and governance in rural areas, there is a considerable number of reports in the extant literature that reveal how ICT provides opportunities to improve on existing government services on the one hand and to create access to government services on the other hand (Alzouma 2015). Electronic government and governance have been implemented in many sub-Saharan African countries for about three or more decades. Electronic government and governance have become popular because experts argue that access to government services is essential to attaining sustainable development in rural areas (Utulu and Sewchurran 2014). The chapter adds a critical dimension to existing narratives. The dimension is how the internalized views of those involved in the formulation and implementation of ICT policies and those operating in the bottom of the pyramid economy constitute a critical challenge to ICT policy formulation and implementation and sustainable development. The ways the challenges can be resolved were also addressed in the chapter.

3 Development Plans and ICT Policies in Sub-Saharan Africa

Indigenous development policies started to evolve in sub-Saharan Africa during the postcolonial and post-apartheid periods. The periods began in the 1950s through the 1960s and 1970s to early 1990s. Before this era, there were a couple of development policies initiated by colonial and apartheid governments. They developed the development policies as parts of constitutions and laws used for governing colonial and apartheid territories. There were, however, exceptions like the Ashby Commission and Elliot Commission, which were established in 1959 and 1943 respectively in British Colonies in West Africa. The commissions were established to deliberate and develop a policy to guide universities' establishment to fast track sustainable development (Dike 1983). However, the Lagos Plan of Action was a notable indigenous development policy formulated by African governments in the late 1970s to coordinate the region's development programs. Another noteworthy development policy, namely the New Partnership for Africa's Development (NEPAD), was introduced in the early stages of the 21st Century. Nigeria, South Africa, Ivory Coast, Kenya, and Egypt championed NEPAD. The African Union (AU) introduced the African Information Society Initiative (AISI) in 2005. The AU focused AISI on how ICT can promote sustainable development in African countries. Few sub-Saharan African countries had commitments to using ICT to spur sustainable development before the AU initiated AISI (Sturges and Neill 1990; Aiyepeku 1982).

The number of countries in sub-Saharan Africa that use formal ICT policies to coordinate their sustainable development plans are increasing. Given this, major global, regional, and national development plans used in the region now revolve around how to adopt ICT to promote sustainable development. The information and communication policy developed by the Ministry of Communication Technology (2012) used in Nigeria to coordinate sustainable development is a noteworthy example. Development plans meant to coordinate education, agriculture, health, commerce and trade, tourism, etc., all incorporate ICT adoption issues. In Nigeria, every civilian regime that has been voted into power since 1999 comes up with different development plans. One thing common to these development plans is that they strongly claim that ICT adoption offers the cheapest way to actualize Nigeria's sustainable development goals. The plans are designated with different names that reflect the philosophies behind them and cover short, medium, and long-term sustainable development plans. According to Uche (2019), "Since gaining independence in 1960, Nigeria had explored four 5-year development plans, one structural adjustment program, four visions and strategies including the recently launched Economic Recovery and Growth Plan (ERGP) (p. 27)." Between 2003 and 2007, the civilian regime headed by Chief Olusegun Obasanjo developed a national development plan that served as a framework for sectoral sustainable development plans. The development plan was named the National Economic Empowerment and Development Strategy (NEEDS).

The federal government encouraged and helped state governments to come up with a State Economic Empowerment and Development Strategy (SEEDS) and local governments to develop a Local Economic Empowerment and Development Strategy (LEEDS). NEEDS identified critical sustainable development goals and articulated a 4-point policy focus to promote ICT use to drive sustainable development in Nigeria (National Planning Commission 2004). The 4-point policy focus includes developing and sustaining ICT capacity, raising the level of computer usage and literacy, providing appropriate incentives for private sector involvement, and promoting ICT as an instrument of mass education, growth, and development. Another notable development plan in Nigeria that was championed by a democratic regime is Vision 20–2020. The development plan had an ambitious goal of making Nigeria one of the World's twenty largest economies by 2020. President Musa Yaradua introduced the plan, but it was implemented by President Goodluck Ebele Jonathan and helped Nigeria to rank 41st largest economy in the World and the largest economy in Africa in 2014. Like the NEEDs, Vision 20–2020 gave credence to the role of ICT. It was declared in the plan that "The complete integration of ICT into all areas and processes needed to conduct foreign trade will be enforced…[by] building capacity for complete digitization…(National Planning Commission 2009, p. 117)." The latest development plan adopted in Nigeria is the Economic Recovery and Growth Plan, which is expected to be used from 2017 to 2020. The plan states that making "…Nigerian economy more competitive in the 21st Century global economy, its industrial policy must be linked to a digital-led strategy for growth…[and] The Smart Nigeria Digital Economy Project to increase the contribution from ICT and ICT-enabled activity to GDP (Ministry of Budget and National Planning 2017, p. 13)".

It follows that efforts to formulate and implement ICT policies to promote sustainable development in sub-Saharan Africa naturally emanate from the national level and trickle down to state and local government levels. Several intergovernmental organizations are also involved in the formulation of ICT policies in the region. The development is welcoming because the processes required to be completed when formulating ICT policies are complex (Chigona 2019; Tamukong 2007). Experience shows that most sub-Saharan African countries may not be able to manage these complex processes without support. The impact of globalization on development planning and the obligation to consider needs related to regional integration also makes it reasonable to seek help from intergovernmental organizations. Being a member of the UN and its agencies, e.g., UNECA, UNDP, UNESCO, etc., Nigeria is expected to oblige to the UN and its agencies' leadership role in the formulation and implementation of ICT policies meant to promote sustainable development. Aside from this, Nigeria is also obliged to collaborate with regional and sub-regional intergovernmental organizations such as the African Union (AU), Economic Community of West Africa States (ECOWAS), ADB, etc. and implement formidable ICT policies. Apart from these external supports, Nigeria is also involved in in-country ICT policy formulation and implementation. As dictated by the Nigerian constitution, ICT policy formulation and implementation must be done at three governmental levels: federal, state, and local government levels. It is important to note that situations in which countries deal with diverse ICT policy formulation and implementation levels are peculiar to every country across the globe. Consequently, ICT policies used in Nigeria and most other sub-Saharan African countries are derived from the UN and its agencies' millennium development goals/sustainable development goals; AU's NEPAD, AISI, and related development policies of ADB.

The problem with the policies is the inability to align the internalized views of those involved in the formulation and implementation of ICT policies with those working at the pyramid economy resulting in contradictory and incoherent sustainable development programs. Given the geographical location and the nature of the bottom of the pyramid economy, the challenge of differences in internalized views negatively impacts it more than other economy levels. The scenario hinders sustainable development initiatives that rely on ICT adoption at the bottom of the pyramid in Nigeria and many other sub-Saharan African countries. A critical look at how these successive development plans were formulated and implemented shows that those involved did not consider the role uniform internalized views play in the extent to which the plans can spur ICT-induced sustainable development. The scenario has become endemic because those involved seem not to be familiar with the approaches required to use ICT to induce sustainable development at the bottom of the pyramid economy. The implementers of the plans seem to assume that many of the general statements outlined in the plans regarding using ICT to promote sustainable development would be easily translated to practical actions at the bottom of the pyramid economy. The role internalized views could play in ICT-induced sustainable development at the bottom of the pyramid economy was reinforced. It follows that in the past, factors such as geographical structure, political-administrative structures, weak institutions, and corruption were the factors identified as those that significantly impact the extent to which ICT has been used to promote sustainable development at the bottom of the pyramid economy (Alzouma 2015). Assessment done in the cause of the study reported in this chapter shows that the ways these factors impact ICT-induced sustainable development at the bottom of the pyramid were determined by the extent to which the internalized views of ICT policy formulators and implementers and those operating at the bottom of the pyramid economy aligned. In other words, the two groups' internalized views determine the ways they interpret and manage the alignment of these factors. The Figure below illustrates the gap.

Figure 1 shows that a wide gap exists in the experiences of those who participate in ICT policy formulation and implementation and those with everyday life knowledge of ICT needs at the bottom of the pyramid economy. While those who join in ICT policy formulation and implementation derive their internalized views from global practices and formal theories, those at the bottom of the pyramid economy derive their internalized views from their everyday life experiences. The scenario impedes ICT-induced sustainable development at the bottom of the pyramid economy.

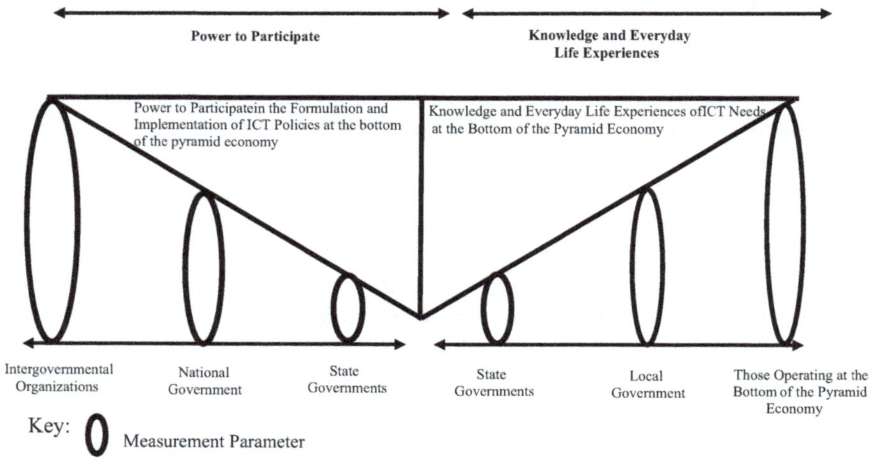

Fig. 1 Dynamics of ICT policy formulation and implementation at the bottom of the Pyramid in Nigeria

4 Internalized Views as Challenges to Sustainable Development at the Bottom of the Pyramid Economy

The chapter proposes that an internalized view is the primary factor that poses a challenge to using ICT to promote sustainable development at the bottom of the pyramid economy. Internalized view as a concept appears in the extant literature of different academic disciplines including, psychology, sociology, and ICT-related academic disciplines. The internalized view is used sometimes synonymously with terms such as mental frame, cognitive frame, frame of reference, espoused theory, etc., in management and ICT-related disciplines (Ifinedo 2014; Crossan 2003). There is a central idea common to how these terminologies are defined, irrespective of differences in the disciplines' academic traditions. The terminologies explain how cognition determines how people conceptualize things, people's beliefs, actions, and assumptions about what constitutes the best procedure(s) to achieve set goals (Ngwenyama and Nielsen 2014; Bjorn and Ngwenyama 2009; Argyris 1994). Internalized views provide grounds for the interpretations people give to issues and the actions they take. In the context of this chapter, such a case may include how they define geographical structures, political-administrative structures, institutions, and corruption. Internalized views produce cultures, beliefs, and norms critical to ICT policy formulation and implementation. It is, however, essential to note that internalized views are products of socialization. People are likely to have different internalized views about a particular issue because they are socialized differently.

It follows that globalization and internalized views impact how the geographical structures in Nigeria are interpreted and factored into ICT policy formulation and implementation. Nigeria evolved from the amalgamation of two British Colonial Protectorates, namely, Southern and Northern Protectorates. Lord Fredrick Luggard initiated Nigeria's formation, and the name Nigeria was proposed by Flora Shaw, his girlfriend (Akintola 2004; Kirk-Greene 1956). Due to its geographical landscape and the diversity of the tribes that make up Nigeria, it was further divided into three regions in 1939 to the Southern Region, Northern Region, and Western Region. In 1963, another region, namely, the Mid-Western Region, was created. It follows that the inclination to establish Nigeria as a federal state was the dream of the British Colonial Government, given that it believed in the principles of federalism, a system of government-aligned to Nigeria's diversity. The colonial-era laid the foundation of the ideologies used to determine and interpret Nigeria's geographical structure (Afigbo 1991). The further division of Nigeria into twelve states in 1967, nineteen states in 1976, twenty-one states in 1987, thirty states in 1991, and thirty-six states in 1996 were derived from the views used to determine its division in the colonial era. At the center of these decisions is the need to avoid Nigeria's disintegration and not take development initiatives further into the grassroots. The ways the Colonial governments treated the traditional monarchical political system in Nigeria became a blueprint for determining the role traditional rulers are to play in grass root development and effect bottom of the pyramid economic development. The much-needed local government system and the traditional political system headed by kings and queens, which offer a promising structure for reaching the bottom of the pyramid economy, were jettisoned (Naanen 1995). By this, successive Nigerian governments discard the bottom-up approach to attaining sustainable development for a top-down approach. Policymakers found it difficult to give more political and economic power to grass-root governments. The scenario remains one of the significant challenges to developing the bottom of the pyramid economy in Nigeria.

Another challenge to the development of the bottom of the pyramid economy promoted by globalization and internalized views is Nigeria's political-administrative structure. When the decision to take the capital city of Nigeria from Lagos to Abuja was taken, the then military ruler, General Ibrahim Babaginda, explained that it was expedient to do so to bring the political administration of Nigeria to a place geographically central to all the states in Nigeria (Thomas-Emeagwali 1989). Interestingly, the headquarters of all federal government ministries were moved to Abuja, including those that dealt with marine and crude oil issues playing out in the East, West, and South-South Region of the country. Although this anomaly was corrected in 1999, much is still left to be done to correct the adverse effects of Nigeria's political-administrative structures on the development of the country's bottom of the pyramid economy. Aside from ill-structured and ill-implemented federalism, lawmakers in Nigeria domiciled in Abuja are geographically too far from those they represent at the bottom of the pyramid. It negatively impacts the extent to which lawmakers get firsthand information and experiences of the people's events and needs at the bottom of the pyramid economy. The reality on the ground is that successive governments and lawmakers adopt the indirect rule, a replica of the style of governance used by the British Colonial Government (Mamdani 1999) given that they mostly depend on "middle-men" constituted by government agencies when dealing with those at the bottom of the pyramid. An excellent example of the agencies is the Niger Delta Development Commission (NDDC), which was established to see the Niger Delta Region's development needs. The inclination to set federal government commissions as mediators between those at the bottom of the pyramid economy and the federal government is persistent because it has become part of political office holders' internalized views.

Weak institutions are another challenge to the formulation and implementation of ICT policies to promote sustainable development at the bottom of Nigeria's pyramid economy. Most of the institutions established by the Nigerian constitution to encourage development at the bottom of the pyramid economy run with logics proposed by global intergovernmental organizations. A good example is the ongoing digitization of financial systems worldwide, which has become mandatory for countries like Nigeria. Financial system digitization has to do with computerizing processes and procedures to provide customers' financial services and provide necessary information to regulators (Lauer and Lyman 2015). Financial systems digitization affects microfinance banks in Nigeria. It is assumed that it will positively impact the drive to implement people and businesses' financial inclusion at the bottom of the pyramid. Presently, particularly as indicated by the Central Bank of Nigeria (CBN), the weakness of microfinance banks in Nigeria is primarily assessed based on the extent to which they have digitized and, in effect, make their transactions accessible to the CBN as a government regulatory agency (Hulikal Muralidhar 2018; Gammage et al. 2017). Essential questions regarding the extent to which microfinance banks in Nigeria can provide financial services and aid to businesses at the bottom of the pyramid are persistently being pushed off the table. No one is seriously asking questions regarding how the digitization of microfinance banks in Nigeria has promoted businesses at the bottom of the pyramid. There is no serious question on the extent to which people and businesses at the bottom of the pyramid economy can access financial services and support from microfinance banks. Experience shows that the gap between businesses operating at the bottom of the pyramid and available financial services and supports have further widened due to digitization exercises. The challenge to digitizing financial markets in Nigeria is promoted by globalization induced internalized views.

One major challenge to sustainable development initiatives in Nigeria that is known to everybody is corruption. Nigerians' inability to reach a consensus on what corruption is, further escalates the problem of corruption. The statement "stealing is not corruption," made by a former Nigerian president, received widespread condemnation from international and local commentators. The politicization of commentators' responses truncated the much-needed reflection that the statement could have spurred. Up till today, no one knows what the president in question had in mind when he made the statement. No one also knows if the press and misreported the statement. Common sense would have people assume that the intended message in the statement is "corruption is not only stealing." Transparency, due process, democratic principles, inclusion, and dialogue are always not put in the equation when defining corruption in Nigeria. Most politicians, law enforcement agencies, and some public commentators still see corruption only within the ambit of stealing money (Ibietan 2013; Osuagwu 2012). Nigerians still see misappropriation and mismanagement of funds as stealing. The erroneous conceptualization of corruption makes most public procurement processes and practices, budget implementation assessment parameters, and legal proceedings to focus more on the amount stolen than the procedures and practices that promote stealing in Nigeria. In Nigeria, legislations required to check corruption are slowed down and, in many cases, stopped due to the problem of lack of consensus on what corruption means. The group that suffers most from the effects of corruption on development comprises businesses and people operating at the bottom of the pyramid economy. Rural electrification, telephone wire, and cable laying, and broadband projects all suffer from corrupt practices. Corrupt practices impede productive ICT use at the bottom of the pyramid economy (Ukwayi and Okpa 2017; Omadjohwoefe 2014).

5 Framework for Implementing ICT Policy at the Bottom of the Pyramid Economy

The challenge identified in this chapter is the inability of ICT policy formulators and implementers to align their internalized views with the internalized views of those operating at the bottom of the pyramid economy. Policy formulators and implementers derive their internalized views mainly from access to a priori knowledge and formal theories. Those operating at the bottom of the pyramid economy derive their internalized views about how ICT can be used to reach sustainable development goals through real-time experiences gained within their local socio-economic contexts. It leads to a gap in the internalized views of the two groups. Therefore, the solution is to create ICT formulation and implementation processes that can enable both groups to understand and align the differences in their internalized views about how ICT policy can be formulated and implemented to promote sustainable development at the bottom of the pyramid economy. The framework shown below indicates that participatory observation, communication, interaction, and dialogue are germane to spurring coherent internalized views between ICT policy formulators and implementers and those operating at the bottom of the pyramid whose life is to be impacted by the policies (Fig. 2).

Participatory observation requires that ICT policy framework formulators and implementers do an extensive observation of the socio-technical situations prevalent in the bottom of the pyramid economy. Participatory observation will enable them to see and experience events the same way they are seen and experienced by those operating at the bottom of the pyramid. It allows them to see the subjective interpretations of events at the bottom of the pyramid economy and how this comes to play in their

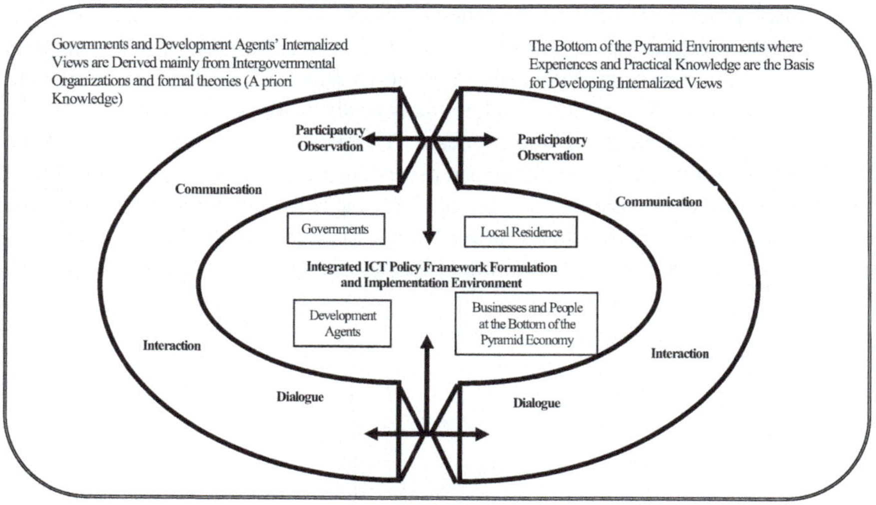

Fig. 2 Coherent ICT policy formulation and implementation framework

internalized views about how ICT can promote sustainable development. The importance of participatory observation also includes the opportunity to share knowledge and experiences. Policy formulators and implementers will share their knowledge and experiences and the more immense challenges that emanate outside the bottom of the pyramid economy that impact the formulation and implementation of ICT policies. It is important to note that participatory observation is a complex process. Training is required if ICT policy formulators and implementers are to adequately and appropriately implement participatory observation. The other three factors, namely, communication, interaction, and dialogue, are equally complex and important. They are to be implemented during and after participatory observation. Communication involves sharing information. The alignment of the internalized views of ICT policy formulators and implementers and internalized views of those operating at the bottom of the pyramid economy requires effective communication.

Effective communication promotes the productive transfer of (new) information that challenges existing internalized views and brings about interaction, a profound and informal way of communicating. Communication breeds trust and trust breeds interaction. Interaction is the informal sharing of new information, most of which may be difficult to share without trust. Conflicting internalized views detected during interactions are resolved through dialogue. Dialogue is formal and leads to agreeing on the way forward. It is used to resolve conflicting issues and enables the two groups to take responsibility for the decisions taken and their consequences. However, implementing this framework requires sacrifice, selfless commitment, and putting the comfort of those representing the government at risk. Issues connected to safety in rural areas in Nigeria make these factors necessary. The killings and kidnapping taking place in villages and local settings in most parts of Nigeria are significant challenges to implementing this framework. African countries with relatively peaceful local areas are likely to find it easier to implement.

6 Conclusion

The global community cannot overemphasize the importance of the bottom of the pyramid economy to sustainable development. The UN and its agencies and other regional agencies such as the ADB are well aware of this. Therefore, they help countries across the globe, particularly those in sub-Saharan Africa, where poverty is endemic, formulate ICT policies used to promote sustainable development at the bottom of the pyramid. While this is welcoming and has helped countries to reach set sustainable development goals, it also has unintended consequences. These unintended consequences have negatively affected how ICT policies have been appropriately formulated and implemented to spur sustainable development at the bottom of the pyramid economy. The challenge results because of the gap between the internalized views held by those in government about how to use ICT to achieve sustainable development and the internalized views held by those operating at the bottom of the pyramid economy about using ICT to achieve sustainable development. It limits the extent to which adequate and appropriate ICT policies have been formulated and implemented to promote sustainable development. The chapter proposes a framework that encompasses inclusion principles as a way out of this problem. It suggests that stakeholders can use participatory observation, communication, interaction, and dialogue to bridge the gaps in internalized views of groups involved in ICT policy formulation and implementation. The chapter argues that governments and development agents usually use assumptions and global ideologies that promote the use of metrics used for calculating economic growth as a basis for determining how to use ICT to attain sustainable development.

On the other hand, real-time experiences and knowledge of local contexts are used by those operating at the bottom of the pyramid as a basis for determining how to use ICT to attain sustainable development. It indicates that conflicting internalized views is the primary challenge to formulating and implementing ICT policies for sustainable development. The chapter proposes participatory observation, communication, interaction, and dialogue as crucial factors that can provide grounds for aligning the conflicting internalized views of both groups and, in effect, help to promote the coherent formulation and implementation of ICT policies at the bottom of the pyramid economy.

References

Adepoju I, Albertsen B, De Brouwere V, van Roosmalen J, Zweekhorst M (2017) mHealth for clinical decision making sub-Saharan Africa: a scoping review. JMIR Mhealth Uhealth, 5(3)

Afigbo A (1991) Background to Nigerian federalism: federal features in the colonial state. Publius J Federalism, 21(4), 13–29

Aideloje V, Okwudibe H, Jimoh A, Olawepo B (2018) Development and utilization of mobile cassava grating machines in Nigeria. J Environ Issues Agric Developing Countries, 10(3)

Aiyepeku W (1982) Information utilization by policy makers in Nigeria, Part I: assessing degree of information consciousness. J Inf Sci 4(5):203–210

Akintola E (2004) The Nigerian nation: a historic-futuristic perspective. Niger Educ Digest 8(1):33

Alzouma G (2015) The opportunity exists. Why don't they seize it? Political (in) competence and the potential of ICT for good governance in Niger Republic. Stab Int J Secur Dev 4(1)

Argyris C (1994) Good communication that blocks learning. Harvard Bus Rev 72(4):77–85

Asongu S, Nnanna J, Acha-Anyi P (2020) Inequality and gender economic inclusion: the moderating role of financial access in sub-Saharan Africa. Econ Anal Policy 65:173–185

Bjorn P, Ngwenyama O (2009) Virtual team collaboration: building shared meaning, resolving breakdowns and creating translucence. Inf Syst J 19(3):227–253

Boateng R, Hinson R, Galadima R, Olumide L (2012) Preliminary insights into the influence of mobile phones in micro-trading activities of market women in Nigeria. Inf Dev 30(1):32–50

Bornman E (2016) Information society and digital divide in South Africa: results of longitudinal surveys. Inf Commun Soc 19(2):264–278

Chandra V, Chandra R, Nutchey D (2014) Implementing ICT in schools in a developing country: a Fijian experience. In: Kaur H, Tao X (eds) ICTs and the millennium development goals: a United Nations perspective, pp. 139–160. Springer, New York

Chigona W (2019) External actors' role in forming national information policy in Malawi: a cause for concern in low income countries? Afr J Inf Syst 11(1)

Crossan M (2003) Altering theories of learning and action: an interview with Chris Argyris. Acad Manage Perspect 17(2):40–46
Dike K (1983) The Ashby commission and its report. In: Chizea C (ed) 20 Years of university education in Nigeria, pp. 1–9, National Universities Commission, Lagos
Federal Government of Nigeria (2009) Nigeria Vision 20: 2020: Economic Transformation Blueprint. www.nairametric.com. Accessed 1 Sept 2020
Gammage S, Kes A, Winograd L, Sultana N, Hiller S, Bourgault S (2017) Gender and digital financial inclusion: what do we need to know?. International Center for Research on Women, Washington
Hafkin N (2002) The African information society initiative: a seven year assessment (1996–2002). Perspect Global Dev Technol 1(2):101–142
Heidegger M (1954) The question concerning technology. Technol Values Essent Readings 99:113
Hulikal Muralidhar S, Bosen C, Mehra A, O'Neill J (2018) Digitization of monetary ecologies: intended and unintended consequences of introducing a financial management app in a low-resource setting. Proc ACM Hum Comput Interact 2(CSCW), 1–17
Ibietan J (2013) Corruption and public accountability in the Nigerian public sector: interrogating the omission. Eur J Bus Manage 5(15):41–48
Ifinedo P (2014) Information systems security policy compliance: an empirical study of the effects of socialization, influence, and cognition. Inf Manage 51(1):69–79
Kiiza B, Pederson G (2012) ICT-based market information and adoption of agricultural seed technologies: insights from Uganda. Telecommun Policy 36(4):253–259
Kirk-Green A (1956) Who coined the name 'Nigeria'? West Africa, (2071)
Kramer A, Belz F (2017) Consumer integration into innovation process: a new approach for creating and enhancing innovation for the bottom of the pyramid. In: sustainability challenges and solutions at the bottom of the pyramid, New York, Routledge, pp. 214–241
Krauss K (2013) Collisions between the world views of international ICT policy makers and a deep rural community in South Africa: assumptions, interpretations, implementation and reality. Inf Technol Dev 19(4):296–318
Mamdani M (1999) Historicizing power and responses to power: indirect rule and its reform. Soc Res, 859–886
McNair B (2017) An introduction to political communication. Taylor and Francis New York
Ministry of Budget and National Planning (2017) Federal Government of Nigeria economic recovery and growth plan, 2017–2020. MBNP, Abuja
Ministry of Communication Technology (2012) Federal Government of Nigeria national information and communication technology (ICT) policy. MCT, Abuja
Mthethwa P, Munyoka W (2013) Utility of information and communication technologies for educational changes in Africa. www.researchgate.net. Accessed 25 Sept 2020
Naanen B (1995) Oil producing minorities and the restructuring of Nigeria federalism: the case of the Ogoni people. J Commonwealth Comp Polit 33(1):46–78
National Planning Commision (2009) Vision 20:2020: economic transformation blueprint. Abuja: NPC. https://nairametrics.com/wp-content/uploads/2013/06/nigeria-vision-20_2020_draftetb.pdf. Accesed 24 Nov 2020
National Planning Commission (2004) Nigeria: nNational economic empowerment and development strategy. The Needs Secretariat, Abuja
Ngwenyama O, Nielsen P (2014) Using organizational influence processes to overcome IS implementation barriers: lessons from a longitudinal case study of SPI implementation. Eur J Inf Syst 23(2):205–222
Obasola O, Mabawonku I, Lagunju I (2015) A review of e-health interventions for maternal and child health in sub-Saharan Africa. Matern Child Health J 19(8):1813–1824
Ogundele O, Hassan A (2011) Projecting the possible impact of the national economic empowerment and development strategy (NEEDS) on human development in Nigeria. Int J Econ Financ 3(4):162–169

Oladele O (2015) Effect of ICT on agricultural information access among extension officers in North West Province, South Africa. S Afr J Agric Extension 43(2):30–41
Omadjohwoefe S (2014) Corruption and sustainable development in Nigeria: the imperative of a trado religious and anti-corruption strategy. Int J Soc Sci 5(5)
Osuagwu L (2012) Conceptualization of corruption in business organizations. Am Int J Contemp Res 2(5):18
Pieterse J (1995) Globalization as hybridization. Media Cult Stud, 658
Pradhan P, Costa L, Rybski D, Lucht W, Kropp J (2017) A systematic study of Sustainable Development Goal (SDG) interactions. Earth's Future 5(11):1169–1179
Prahalad C (2005) The fortune at the bottom of the pyramid: eradicating poverty through profits. Wharton Publishing, Upper Saddle River, NJ
Sachs J (2012) From millennium development goals to sustainable development goals. Lancet 379(9832):2206–2211
Sturges P, Neill R (1990) The quiet struggle: libraries and information for Africa. Mansell, London
Tchamyou V (2020) Education, lifelong learning, inequality and financial access: evidence from African countries. Contemp Soc Sci 15(1)
Thomas-Emeagwali G (1989) Notes on the history of Abuja Central Nigeria. www.repository.ku.lib.kyoto.u.ac.jp. Accessed 26 Sept 2020
Uche E (2019) Development plans and policies in Nigeria: observed impediments and practical best alternative. Int J Res Pract Innov 6(7):27–35
Ukwayi J, Okpa J (2017) The effect of electoral and economic crimes on sustainable development in Cross River State, Nigeria. Int J Soc Sci Res 5(2):32–42
Utulu S, Sewchurran K (2014) Designing e-government platform from the perspectives of systems thinking and performance measurement frameworks: how good thinking and good questions aid information systems design. In: Bwalya J (ed) Technology Development and Platforms Enhancement for Successful Global E-Government, pp. 97–117. IGI, Hershey
Utulu S, Sote T (2009) Africa information society initiative, technology transfer and the evolution of digital information and knowledge management systems in Africa
Yaacoub E, Alouini MS (2020) A key 6G challenge and opportunity—connecting the base of the pyramid: a survey on rural connectivity. Proc IEEE 108(4):533–582

How Disaster Risk Reduction Can Contribute to Sustainable Development: The EAGER Project

K. Fourie and A. C. Terblanché-Greeff

Abstract Disaster risk reduction [DRR] is an essential aspect of socio-economic development - if time and attention are allocated to DRR, development gains will be more. With this in mind, the African Centre for Disaster Studies conducted the Engaging African GIRRLs in Gender Enriched Disaster Reduction [EAGER] project from 2015–2018. The main target group of the project was adolescents in cross-border communities of Botswana and Namibia, as well as Swaziland and Mozambique. The goal of the project was to lessen the challenges that the participants face in pre- and post-disaster contexts by integrating education and participation in community-based disaster risk reduction and resilience. Additionally, the project intended to build local capacity and promote interaction between youth in bordering communities. By implication, activities aimed initially at DRR and capacity-building addressed aspects relating to various United Nations Sustainable Development Goals [SDGs]. It will be of value to explore the link between DRR and sustainable development by presenting the EAGER project. Additionally, focus will be allocated to the project in the context of some SDGs and how DRR contributes to sustainable development.

Keywords Sustainable development goals · EAGER project · Disaster risk reduction · Sustainable development

1 Introduction

The world is experiencing an increase in hazardous events (*e.g.* cyclones, droughts, floods, landslides, wildfires, and epidemics) that can and often do result in disasters (Vyas-Doorgapersad and Lukamba 2012). Without exception, the African continent is increasingly threatened by various disasters, as well as the risks posed by climate

K. Fourie (✉) · A. C. Terblanché-Greeff
African Centre for Disaster Studies, North-West University, 11 Hoffman Street, Potchefstroom, South Africa
e-mail: kristel.fourie@nwu.ac.za

School of Philosophy, North-West University, Potchefstroom, South Africa

W. Leal Filho et al. (eds.), *Sustainable Development in Africa*, World Sustainability Series,
https://doi.org/10.1007/978-3-030-74693-3_36

change which influences the frequency and intensity of natural phenomena (International Federation of Red Cross and Red Crescent Societies [IFRC] 2016). The effects of disasters are diverse, *viz.* disease, injury, and loss of life; destruction of assets; loss of infrastructure and essential services; socio-economic disruptions; as well as environmental destruction. These effects of disasters are exemplified when disasters impact development opportunities in cases where resources designated for development are redirected toward post-disaster recovery efforts. Consequently, disasters have been identified as a significant threat to sustainable development [SD] (Twigg 2015). More specifically, as disasters can have an immense impact on the social equity, environmental protection, and economic growth of African countries, it becomes clear that disasters negatively influence SD on the continent.

Extreme poverty contributes to the substantial negative impacts resulting from disasters that Africa experiences (van Niekerk 2015; van Niekerk et al. 2020). Additionally, underdevelopment and inadequate development decisions create the opportunity for hazardous events to have disastrous effects on humans, the environment, and economies. As a result, the dire consequences of disasters will continue to impede attempts aimed at alleviating poverty and securing SD unless disaster risks are sufficiently managed (United Nations International Strategy for Disaster Reduction [UNISDR] 2015b).

One way to manage risk is through disaster risk reduction [DRR], which refers to "[t]he concept and practice of reducing disaster risks through systematic efforts to analyse and manage the causal factors of disasters, including through reduced exposure to hazards, lessened vulnerability of people and property, wise management of land and the environment, and improved preparedness for adverse events" (UNISDR 2009). Noteworthy, DRR is more than mere defensive measures as it has the potential to promote positive change (Twigg 2015). In fact, "[i]mproved security and safety provide vital support and opportunity to households, communities, societies and governments so that they can undertake development initiatives that improve well-being, strengthen livelihoods and contribute to sustainable development" (Twigg 2015).

Overall, this chapter (through participant observation and authors reflections) aims to provide an account of how DRR is instrumental in SD and the achievement of the SDGs in Africa. Attention will be allocated to the aspects of SD and disaster risk reduction; also, the EAGER project will be introduced as an example of a community-based disaster risk reduction practice that can contribute to SD.

2 Sustainable Development, Disasters, and Disaster Risk Reduction

In the 1980s, the concept of sustainable development emerged in the global arena as an effort to "bridge the gap between environmental concerns about the increasingly evident ecological consequences of human activities and socio-political concerns about human development issues" (Robinson 2004:370). One of the most popular definitions of sustainable development is formulated by the World Commission on Environment and Development [WCED] in the document titled *Our Common Future* (Brundtland 1987). Accordingly, "[h]umanity has the ability to make development sustainable - to ensure that it meets the needs of the present [generation] without compromising the ability of future generations to meet their own needs" (Brundtland 1987:43).

The WCED focused on socio-economic conditions in developing countries, and the connection between these conditions and environmental degradation (Robinson 2004). It was acknowledged that anthropocentric behaviour associated with both under- and over development negatively influences the environment, but at the same time, poor communities are in need of more development (WCED 1987). SD can thus be described as development that "protects the environment while enhancing the quality of life for all, more particularly for those who are most affected by poverty and inequality" (Church 2012:512). Here, the focus is not on less development and growth, but on a different type of development, *i.e.* development that can be sustained.

Key motivations for the global call to SD by the WCED are based on concerns for social equity and environmental protection, as well as a focus on endless growth (Smythe 2014). From this the three interdependent dimensions (or pillars) of SD can be identified: the realisation of social equity (People), the protection of the environment (Planet), and the development of the economy (Prosperity). Based on the three pillars of SD, we can thus say that "growth must be both inclusive and environmentally sound to reduce poverty and build shared prosperity for today's population and to continue to meet the needs of future generations" (Muralikrishna and Manickam 2017:5).

2.1 Development and Disasters

One aspect that will influence SD is the occurrence of disasters. Here, the *United Nations Office for Disaster Risk Reduction* [UNDRR] (2020a) defines a disaster as a "serious disruption of the functioning of a community or a society at any scale due to hazardous events interacting with conditions of exposure, vulnerability and capacity, leading to one or more of the following: human, material, economic and environmental losses and impacts". As suggested by evidence, developed- and developing countries' hard-earned developmental gains are threatened by these impacts of disasters (UNISDR 2015b). Additionally, the consequences of a disaster often exacerbates poverty and can negatively affect social equity, environmental protection, and economic development, which are key aspects of SD.

It must, however, be noted that development also contributes to the occurrence of disasters. Since the 1970s, it has been acknowledged that disasters are often indicators of existing development problems or development that failed as this negatively influences vulnerability to hazards and it creates and exposes people to risk caused by investment decisions (O'Keefe et al. 1976; Cuny et al. 1983). Clearly, there is a relationship between unsustainable development and disasters, as such practices often create and contribute to the underlying cause of disasters (UNDP 2004).

Based on the reciprocal relationship between development and disasters, there is a need for mutually supportive undertakings to advance SD and increase resilience to disasters (Benson 2016:18). Per example, a positive way in which post-disaster situations can contribute to SD is through a "build-back-better" approach. Supportively, Dewald van Niekerk (2004) argues that disasters pose an opportunity for SD to be implemented and advanced as part of risk reduction to ensure systems are more resilient to future hazardous events.

2.2 Disaster Risk Reduction and the 2030 Agenda

In 2012, the *United Nations Conference on Sustainable Development*, was held in Rio de Janeiro, Brazil. A key outcome of the conference was the proposal to create goals as useful tools to attain SD. As such, the United Nations Development Programme [UNDP] (2015) published the document titled *Transforming Our World: The 2030 Agenda for Sustainable Development* [2030 Agenda] in 2015, which presents 17 specific sustainable development goals [SDGs], *e.g.* gender equality, universal education and health coverage, sustainable cities, food security, etc.

Twenty-five targets in ten of the SDGs are related to DRR and "[a] key part of the success of the 2030 Agenda…[will, to a degree], be measured by progress in [the] implementation of disaster risk reduction" (UNISDR 2015b:2). Thus, when considering the interlinked relationship between development and disasters, it can be argued that DRR is an important development strategy (UNISDR 2015b:2). Moreover, DRR strives toward the prevention of new risk, the reduction of existing risk, and the management of residual risk, which are all instrumental in improving resilience and the attainment of SD (UNDRR 2020a).

It is essential to grasp the prevalence and complexity of the relationship between SD and DRR. As such, the components of disaster risk must be understood.

3 Disaster Risk

Disaster risk is the "potential loss of life, injury, or destroyed or damaged assets which could occur to a system, society or a community in a specific period of time, determined probabilistically as a function of hazard, exposure, vulnerability and capacity" (UNISDR 2017b:98). Various components interact to create disaster risk. Scientific literature and international policies identify the components as *hazard*, *vulnerability*, *capacity*, and (more recently) *resilience* (cf. Twigg 2015; UNISDR 2015a, 2017a). More specifically, disaster risk is determined by the scope and loss caused by an identified hazard; the community's vulnerability (level of exposure to the hazard); and the capacity to lessen the adverse effects of the hazard as well as the vulnerability of the community.

Aimed at alleviating risk, DRR practices aim to facilitate 1) a reduction of the hazard-level by either changing the probability of the hazard occurring or the magnitude of the hazard; 2) a decrease in the systems' vulnerability by changing social, environmental, economic, or physical aspects of the system; and 3) an increase in the capacity of the system by strengthening the social, economic, physical, or institutional means, as well as improving the skilfulness of personnel or collective attributes (*e.g.* leadership and management).

Based on the descriptions of disaster risk and DRR, it becomes imperative to discuss the components (*viz.* hazard, vulnerability, capacity, and resilience) that interact and contribute to disaster risk. Additionally, cognisance of these components can contribute to sustainable DRR practices.

3.1 Hazard and Vulnerability

Hazards are considered as anything that poses a potential danger or threat to a system, community, or society (hereafter referred to as a *system*, which encompasses the concepts of *community* and *society*) (UNISDR 2009; Kabubi 2011). Hazards can be classified as natural or human-made. Natural hazards can be described as the elements in the physical environment that are harmful to humans and are caused by forces external to them (Burton and Kates 1964). Examples of natural hazards are floods (hydro-meteorological hazard), sinkholes (geological hazard), and epidemics (biological hazard). In contrast, human-made hazards are unwanted harmful phenomena caused by the actions of humans. Examples of human-made hazards are power outages (technological hazard), climate change (socio-environmental hazard), and drug addiction (sociological hazard).

The concept of vulnerability refers to the harm predisposition of a system due to the "exposure to stresses associated with environmental and social change and from the absence of capacity to adapt to that change" (Adger 2006:268). More specifically, vulnerability is the set of characteristics or circumstances that allows the opportunity for a hazard to cause adverse impacts in a system (Adger 2006). To what extent the impact on the system allows it to be overwhelmed depends on the extent to which the system is exposed to the hazard.

Disasters can be described as the result of the instance where a hazard exploits an existing vulnerability in a system; consequently, overwhelming that system to such a degree that it cannot subsist with the impact (UNISDR 2009; Twigg 2015). It should be noted that hazards will always exist in some form, intensity, or frequency (Kabubi 2011). However, a system's vulnerability and level of exposure toward the presented hazard determines the level of impact and whether that impact overwhelms the system or not. Supportively, a disaster will not present "if there are hazards but vulnerability to that hazard is (theoretically) nil, or if there is a vulnerable population but no hazard event" (Wisner et al. 2004:49). Consequently, if disasters are created by the exploitation of vulnerability (by hazards) to such an extent that the system cannot cope, then a closer look at vulnerability is necessary.

3.1.1 The Pressure and Release Model

The Pressure and Release [PAR] model (as developed by Wisner et al. 2004), represents how vulnerability is created and how it progresses to expose a system to a hazard.

In the PAR model, three components contribute to exposing a system to a hazard as illustrated by Fig. 1. Additionally, the three components work together to mark a progression of vulnerability. The first component is *root causes,* which pertain to "access" and "ideologies" as systemic circumstances created and maintained throughout history. The second component is *dynamic pressures,* which include macro-forces and trends expressed in social-, political-, and economic systems. Here, dynamic pressures can amplify the effect of root causes, contribute to the progression of vulnerability, and create conditions that are unsafe. The *unsafe conditions*, as the third component, might only become apparent when a hazard is introduced to the circumstances.

The vulnerability that developed through root causes and that is amplified by present dynamic pressures, allow the hazard to exploit the said vulnerability. It is this interaction between vulnerability and the hazard that contributes to the occurrence of a disaster event. Consequently, human behaviour influences, creates, and

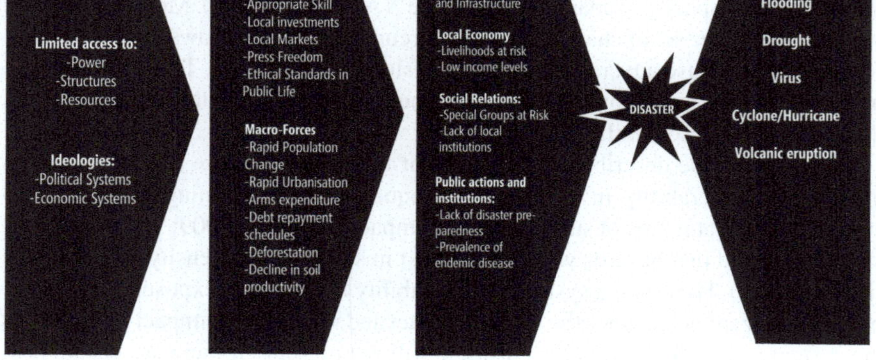

Fig. 1 Pressure and Release Model - adapted from Wisner et al. (2004)

maintains the root causes and dynamic pressures that result in unsafe conditions (all of which contributes to the progression of vulnerability). As such, human behaviour will determine if a disaster will occur or whether a system will be able to cope with the impact of the hazard.[1]

In DRR, it is imperative to recognise the vulnerability of a system to identify better the available resources that can be utilised in capacity-building activities to create a more resilient system. Furthermore, vulnerability and capacity are proportionately related. *I.e.* Increased capacity in a system means lowered vulnerability of that system, and *vice versa.*

3.2 Capacity and Resilience

The two other components of disaster risk that require attention are *capacity* and *resilience*. *Capacity* refers to the available assets or resources that enable the system to adequately cope when faced with a hazard. Additionally, capacity can ease the level of societal risk and mitigate the consequences of a disastrous event (UNISDR 2004). Examples of capacities are infrastructure, institutions, knowledge and skills, as well as collective attributes such as leadership, management, and social relationships (UNISDR 2017a).

The concept of *resilience* is related to that of capacity, however, resilience is more encompassing than the specific actions, plans, and resources that serve as a system's capacities (Twigg 2015). Furthermore, resilience strives to reinforce capacities used to cope with un/anticipated threats (Twigg 2015). In reference to DRR, resilience is defined as "the ability of a system… exposed to hazards to resist, absorb, accommodate, adapt to, transform and recover from the effects of a hazard in a timely and efficient manner, including through the preservation and restoration of its essential basic structures and functions through risk management" (UNGA 2016:22).

Based on the aforementioned, it is essential to remember that DRR is aimed at lowering the impact of a hazard to such a degree that a system is not overwhelmed. To avoid being overwhelmed, a system must have the capacity to deal with the resulting aftermath of the hazard without external assistance. One way in which a system's capacity can be developed and increased is through the implementation of the participatory approach in DRR practices (e.g. community-based disaster risk reduction).

[1]For in-depth discussions on the application of the PAR model in specific circumstances refer to studies done by Jakarta and Mumbai by Texier-Teixeira and Edelblutte (2017), as well as the study conducted by Awal (2015) in Bangladesh.

4 Community-Based Disaster Risk Reduction

Community-based disaster risk reduction [CBDRR] uses an approach that is based on the community's participation in the DRR process. The long-term goal of CBDRR is twofold as it aims to facilitate community empowerment as well as SD - for both of these aims risk reduction and -management is imperative (Delica-Willison 2005). To illustrate, local people are activated as the "prime movers in reducing disaster risks in their community" through empowerment (Delica-Willison 2005:68). Additionally, SD is grounded in socio-cultural development, environmental protection, economic growth, and political stability - all of which are subject to the reduction of disaster risks.

To implement CBDRR successfully, some principles must be adhered to (as identified by Delica-Willison and Gaillard 2012). Here, the primary focus is on the community, and their understanding of the risks they face forms the foundation for the objectives and activities of CBDRR (Delica-Willison and Gaillard 2012).

The first principle pertains to *community participation* and *empowerment*. Implementing a genuine participatory approach will allow community members to contribute to the accurate identification of vulnerabilities and priorities, as well as the development of appropriate interventions (Twigg 2015). Through participation, the community members can identify their context-specific disaster risks and they will be able to take ownership of the activities as well as outcomes of CBDRR (van Aalst et al. 2008; Delica-Willison and Gaillard 2012). Consequently, they can be empowered without necessarily creating a cycle of dependency. Participation and empowerment enable community members to partake in the decision-making process actively and to "build their own disaster-resistant future through a process of assessment, analysis and action" (Delica-Willison 2005:68). CBDRR can be sustainable because initiatives 1) build on local capacity, 2) can be developed based on community feedback before adoption, and 3) are consistent with long-term SD strategies (Twigg 2015).

The second principle relates to *development-orientated activities*. The root causes of a community's vulnerability (cf. PAR model) can be identified through the implementation of development-orientated activities, which can lead to an increase in the community's capacity and resilience (Delica-Willison and Gaillard 2012). Furthermore, CBDRR activities are development-orientated as illustrated by the following short- and medium-term goals: minimising human-, environmental-, and property losses; limiting economic and social disruption; as well as to enhance the benefits of a safe and secure environment (Delica-Willison 2005).

The third principle refers to a *multi-stakeholder approach*. The support of multiple stakeholders (*e.g.* local- and national government) is essential to empower the community in a sustainable way. Support can be provided by investing and contributing to social capital, local DRR planning, relevant management structures, as well as implementation and coordination mechanisms (Delica-Willison and Gaillard 2012). Noteworthy, the multi-stakeholders should not intervene in the CBDRR process in such a way that would result in ownership being taken away from the community.

To elucidate how DRR links with and contributes to empowerment and SD, it will be of value to discuss an example of CBDRR. With this in mind, the EAGER project will be presented.

5 The EAGER Project

The *Engaging African GIRRLs in Gender Enriched Disaster Reduction* [EAGER] project was conducted by the African Centre for Disaster Studies [ACDS] at the North-West University (South Africa) from 2015–2018. The project aimed to lessen the hardships that youths face in pre- and post-disaster contexts. This aim was to be achieved by integrating active participation as well as an action-based learning approach into CBDRR. Furthermore, the project intended to build local capacity and promote interaction between youths in bordering communities.

5.1 Background on the EAGER Project

The EAGER project draws from the success of the *Girls in Risk Reduction Leadership* [GIRRL] *Programme* of the ACDS and the various projects that have implemented the GIRRL approach (cf. UNISDR 2008, 2015c). The GIRRL programme is a research intervention that is aimed at engaging marginalised girl participants in disadvantaged communities (Forbes-Biggs and Maartens 2012). Furthermore, the programme is a CBDRR initiative aimed to help the adolescent girl participants understand the relationships between health, personal welfare, as well as individual vulnerability; and to challenge disaster-related risk (Forbes-Biggs and Martens 2012). The programme also aspires to minimise the social vulnerability of these marginalised girls through capacity-building activities (cf. Forbes-Biggs and Maartens 2012; Forbes-Genade and van Niekerk 2017, 2019).

These capacity-building interventions focuses on identifying factors that contribute to the vulnerability of adolescent girls and works to develop community-level risk reduction activities (Forbes-Genade and van Niekerk 2019). More specifically, participatory capacity-building is facilitated through the presentation of training and information in areas such as community-based disaster risk assessment, decision-making, effective communication, environmental awareness, fire safety, first-aid, peer education, personal (*viz.* mental-, physical- and sexual-) health, as well as self-defence (Forbes-Biggs 2016).

The implications and benefits of the GIRRL programme are multitudinous as well as significant, and the programme holds great potential for replication as illustrated by projects implementing the GIRRL approach (cf. UNISDR 2008, 2015c; NWU-ACDS 2016b). With this in mind, the EAGER project was developed based on the GIRRL approach.

5.2 Specifics of the EAGER Project

The EAGER project relied on the overall structure of the GIRRL approach (cf. Forbes-Biggs 2016b). Nonetheless, some differences and similarities need to be highlighted to better differentiate the unique aspects of the EAGER project from that of the GIRRL programme.

A notable difference was that the capacity-building interventions of the EAGER project included girls and boys as participants. The inclusion of male participants was motivated by lessons learned from previous DRR initiatives that established that dialogue and engagement from both genders are necessary for a holistic approach to integrating gender in DRR. Additionally, the age range stipulated by the GIRRL programme fell within the 13- 18 age bracket, but this was given as a mere guideline for the EAGER project trainers as they ultimately had to identify country-specific participants based on needs. For the EAGER project, forty participants from local primary- and secondary schools were identified in each targeted country.

The countries targeted to be part of the EAGER project were Botswana and Namibia, as well as Mozambique and Swaziland (now known as Eswatini). A prerequisite for the identification of the project sites was that they had to be cross-border communities in close proximity to each other. Cross-border communities most likely have similar and unique aspects that contribute to disaster risk. The rationale for community identification was that the EAGER project would establish a platform that facilitates cross-border networking. More specifically, such a network would aim to enable participants and trainers to share their experience and knowledge regarding hazards, vulnerabilities, and capacities in their unique environments.

Similar to the objectives of the GIRRL programme, the EAGER project also focused on addressing social vulnerability, as well as creating community safety and awareness of DRR through participation and empowerment. In the process of initiating the project, multiple stakeholders were engaged, *viz.* traditional leadership, disaster management entities, fire services, social workers, police, immigration services, school management structures and directorates of the district education, as well as private companies. With this multi-stakeholder approach, the EAGER project ensured the establishment of a support system network that is local and DRR-orientated.

Through the EAGER project, teachers from the participating schools were identified as in-country trainers to form part of the project teams. In line with the technical design of the GIRRL approach, the EAGER project started by training the teachers in a three-day Train-the-Trainer workshop. During the workshop, trainers were given the opportunity to decide how they wanted to structure the implementation of the EAGER project based on their country-specific contexts. Additionally, some sessions of the workshop were used to work through the *EAGER: Manual for Guiding Project Implementation* (NWU-ACDS 2016b), and the trainers were introduced to DRR principles.

In addition to the trainers, various other role players were involved in implementing the EAGER project, *e.g.* country coordinators, country facilitators, stakeholder body, trainers, parents and guardians, as well as participants. Each role player had a specific function to fulfil in the project implementation (cf. NWU-ACDS 2016a).

5.3 Implementation of the EAGER Project

The pre-implementation phase aimed at identifying all the relevant stakeholders in each of the four countries. To start off with, each country had a coordinator and a facilitator. The country coordinators consisted of the staff members of the ACDS, who had to identify and link up with the local facilitators and provide support to in-country activities and project implementation. The country facilitators were local community members from the project sites that have ample knowledge of the area and who could assist in setting up the necessary structures for the project implementation. However, the implementation was the responsibility of the trainers. More specifically, the trainers were responsible for the identification of the participants, the coordination of capacity-building sessions with the participants, the guidance and support of participants, as well as the organisation of stakeholder meetings.

The first phase in the implementation of the project was to bring the country teams (consisting of the country coordinator, -facilitator, and -trainers) together for the Train-the-Trainer workshop. Based on the participation of the country team, each attendee had the needed information regarding the planning, implementation, and evaluation of the project. In the second phase, the project team worked together to adapt the GIRRL approach based on their context. Additionally, disaster risk assessment activities were facilitated by the country coordinator(s) to enable the trainers to identify the context-specific hazards and vulnerabilities, which served as motivation for the topics selected for the capacity-building interventions. After the disaster risk assessment activities were concluded, the trainers were able to develop a project plan for the implementation of the project. The third phase pertained to the identification of the participants by trainers based on the criteria identified during the Train-the-Trainer workshop. As a fourth phase, the capacity-building interventions were implemented based on the project plan. These sessions were presented to the participants and focused on various themes covering hazards, vulnerabilities, and capacities.

5.4 Capacity-Building Interventions

In each of the four countries, 12 to 14 capacity building sessions were presented. Some of these sessions covered the proposed topics in the *EAGER: Manual for Guiding Project Implementation* (NWU-ACDS 2016b), while others focused on context-specific hazards, vulnerabilities, and capacities. Although it is not viable to discuss all session topics in detail, it is of value to mention some topics and how these relate to country-specific contexts.

Various sessions were related to the hazards identified. Examples of hazard-orientated topics are environmental awareness, drought, floods, structural- and wildfires, fire safety, wildlife and nature, human-wildlife conflict, water safety and pollution, traffic rules, as well as HIV/AIDS.

The sessions focusing on drought are an example of how various areas in the same country experience different hazards based on their respective contexts. In Mozambique, only one of the secondary schools (situated in *Escola Secundária Marcelino dos Santos)* covered the topic of drought - this specific school is situated in a context where the residents heavily rely on subsistence farming and natural resources (Fig. 2). Other schools in different locations that were part of EAGER Mozambique (*viz. Namaacha*) did not have sessions on the topic of drought.

In Namibia and Botswana, sessions covered topics pertaining to human-wildlife interaction and -conflict. All the project sites of EAGER Namibia (Fig. 3) are located in conservation areas and learners often miss school as they are unable to walk to school because a herd of buffalos or elephants are blocking their route. Additionally, the EAGER Botswana project sites (Fig. 3) were located in close proximity to the

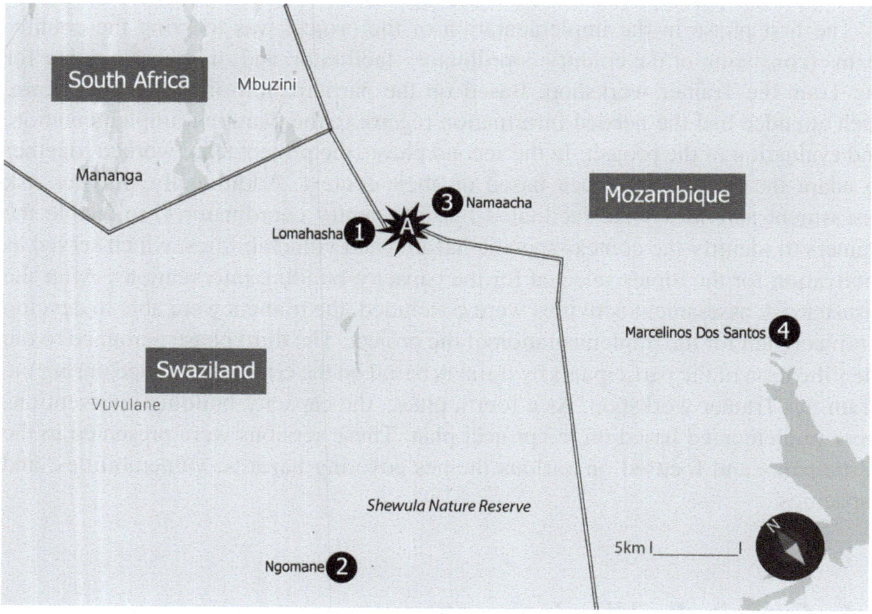

Fig. 2 Map indicating Swaziland (1. Lomahasha; 2. Ngomane) and Mozambique (3. Namaacha; 4. Escola Secundária Marcelino dos Santos), as well as the shared Namaacha border post (A) (Adapted from Google Maps 2020). (In Swaziland, Lomahasha and Ngomane are 20 km apart. In Mozambique, Namaacha and Escola Secundária Marcelino dos Santos are 28 km apart.)

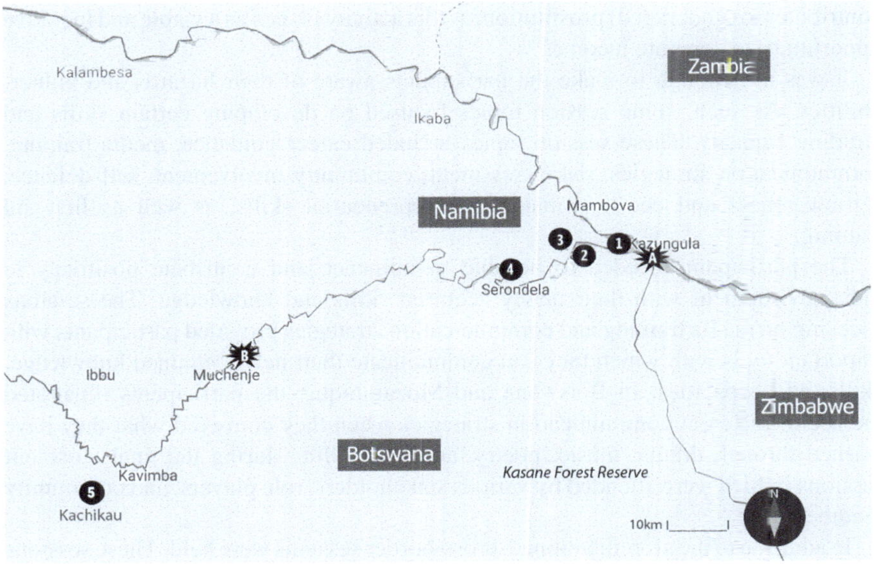

Fig. 3 Map indicating Namibia (B. Ngoma border post; 1. Impalila Island; 3. Kasika; 4. Mbalasinte) and Botswana (A. Kazungula border post; 2. Kasane; 5. Kachikau) (adapted from Google Maps 2020). (In Namibia, Impalila Island and Kasika are 14 km apart (via the Chobe River), while Kasika and Mbalasinte are 13 km apart. In Botswana, Kasane and Kachikau are 92 km)

Chobe National Park. Consequently, wild animals roam freely in the towns and between villages throughout the whole district to reach water sources (*e.g.* the Chobe River). Human-wildlife interaction and -conflict have created numerous dangerous situations between (urban-, peri-urban-, as well as rural) residents and animals.

Some session topics were characteristically socio-economic, and were shown to contribute to context-specific vulnerabilities. Such topics were illegal immigration; human trafficking; domestic violence; substance abuse; child abuse; child marriage; early parenthood; teenage pregnancy; prostitution; as well as personal (*viz.* mental-, physical-, and sexual-) health. The origin and root causes contributing to these vulnerabilities are particularly difficult to determine without further investigation. The progression of vulnerabilities is, however, influenced by the complex interaction between the components of disaster risk.

To illustrate the progression of vulnerability, consider the following case in point. In Kasane (Botswana) access can be gained to three other countries (*viz.* Namibia, Zambia, and Zimbabwe) via the local Kazungula border post (Fig. 3). Until fairly recent, a ferry was used to convey trucks on route to other African countries via Botswana to Zambia. The lengthy turnover time of the ferry created long queues of trucks, with drivers spending up to five days waiting to cross the border. Additionally, the Kasane area relies heavily on tourism with little other economic opportunity present, which results in high levels of poverty. The combination of these situations contributes to (underaged) prostitution, as this activity is seen as a viable and lucrative opportunity to generate income.

It was not enough to make the participants aware of their hazards and vulnerabilities. As such, some session topics focused on developing certain skills and building capacity. These session topics included career guidance, media training, communication strategies, risk assessment, community involvement, self-defence, self-awareness and decision making, entrepreneurial skills, as well as first aid training.

The participants needed to be able to influence and contribute positively to their environment with their newly acquired skills and knowledge. The sessions focusing on media training and communication strategies provided participants with important tools with which they can communicate their newly attained knowledge, skills, and capacities. In Botswana and Mozambique, the participants illustrated the use of different communication strategies when they conveyed what they have learned through drama, music, poetry, and storytelling during the final close-out sessions (which were attended by various stakeholders, role players, and community members).

In addition to the aforementioned, cross-border sessions were held. These sessions created the opportunity for the cross-border participants to interact and network with each other. Specifically, the networking session of the participants from Namibia and Botswana consisted of a sports-day. Here, the participants found common ground through shared language - this broke down some stereotyping as well as discrimination and facilitated the realisation that similarities exist between themselves and "the others" from across the border. In Swaziland and Mozambique, two cross-border sessions were held. The first session focused on fire safety training as this addressed a common hazard; the second session was a fun-day that included some media training. The participants experienced difficulty in communicating, which was ultimately overcame through the use of games and play.

Based on the experience by the authors and participant observation, it can be argued that the sessions presented as part of the CBDRR approach contribute to community empowerment and SD. As such, a reflection is presented in the next section on how the EAGER project links and contributes to SD and the attainment of the SDGs.

6 The EAGER Project and the Sustainable Development Goals

In disaster risk literature, the relation between DRR and SD has been discussed for some time. In 2015, the *United Nations Office for Disaster Risk Reduction* (formally known as the *United Nations International Strategy for Disaster Reduction* [UNISDR]) compiled a paper reflecting on the relationship between the *Sendai Framework for Disaster Risk Reduction 2015–2030* [Sendai Framework] and the

United Nations Sustainable Development Goals [SDGs] (cf. UNISDR 2015b). In various SDGs, reference is made to the Sendai Framework. Here, the link between the fields of DRR and SD is being recognised as indicated by the SDGs' explicit use of DRR-related terms. Examples of such terms are: hazards, vulnerability, adaptive capacity, resilience, exposure, mitigation, disasters, climate-related hazards and -extreme events, adaptation to climate change, holistic disaster risk management, water-related disasters, conservation, drought, floods, and access to decision-making.

With this link in mind, the EAGER project was created as a CBDRR intervention, which focuses on empowerment and SD. The EAGER project intended to address vulnerabilities and build capacity to combat the impact hazards might have in participating communities. Additionally, the project activities, and ripple effects thereof, indirectly contribute to SD and the objectives of the SDGs by strengthening resilience and contributing to the realisation of social equity (Pillar: People), environmental protection (Pillar: Planet), and economic growth (Pillar: Prosperity). To illustrate this reflective assertion, the project will be discussed in light of SD in general and some specific SDGs, viz. SDG 1 (No poverty), SDG 3 (Good health and well-being), SDG 5 (Gender equality), SDG 15 (Life on land), and SDG 16 (Peace, justice, and strong institutions).

6.1 Goal 1 - No Poverty

In the context of disaster risk, poverty is identified as one of the main drivers of vulnerability (UNISDR 2015b). Poverty is so prevalent in the progression of vulnerability that impoverishment can contribute to root causes, dynamic pressures, as well as unsafe conditions (as per the PAR model). Similarly, poverty also contributes to unsustainable development and negatively influences developmental gains. This is illustrated by some of the aims of SD: to alleviate poverty and ensure prosperity for all.

The EAGER Namibia sites were extremely isolated from the rest of the country, with isolation increasing during flood-season when mobility is severely limited. Minimal economic opportunities exist in the areas, and often employment options revolve around the high-end tourism lodges (with its limited employment capacity). With this in mind, the importance of the sessions focusing on economic capacity and skills development cannot be over-emphasised. These sessions covered topics such as entrepreneurship, as well as business and career guidance. Here, participants were introduced to a variety of professions, which provided more options than working for a tourism lodge, or fishing and subsistence farming. The sessions focusing on developing economic capacity and relevant skills link with SDG 1 (No poverty). More specifically, engaging with participants regarding possible careers, economic activities, and how to sustain oneself is a crucial step in addressing poverty. Such engagement can contribute to SD (Pillar: Prosperity) through DRR and the resultant resilience.

6.2 Goal 3 - Good Health and Well-Being

One way in which good health and well-being (SDG 3) will influence SD is through the realisation of social equity (Pillar: People). With that being said, various EAGER sessions related to topics that pertain to SDG 3. Aside from the sessions focusing specifically on personal health, the other sessions presented have some underlying cause of which the topic (covered in the sessions) is just a symptom. I.e. topics focusing on HIV/AIDS, substance abuse, early parenthood, teenage pregnancy, prostitution, traffic rules, and pollution are all symptoms of other issues that present in the community.

As illustration, a community elder in EAGER Namibia reflected that due to the lack of stimulating activities in the area (which is very isolated from the rest of the country), the youth often resort to alcohol consumption (*i.e.* substance abuse) (NWU-ACDS 2018a). Such behaviour might result in impaired judgement and early unsafe sexual behaviour, which can negatively influence the attainment of SDG 3 and the realisation of social equity. Additionally, teenage pregnancy and early parenthood can lead to more pressure on the youth's entire family - pregnant girls and adolescent mothers often drop out of school, and the family now has an additional dependant in already challenging economic circumstances.

CBDRR - with its primary focus on empowerment and SD - can be used to assist development-orientated activities in achieving some of the aims of SD and the SDGs objectives. More specifically, investigating session topics identified through the EAGER project with a tool like the PAR model can expose the root causes of the aforementioned "symptoms". As a result, vulnerability can be reduced and resilience strengthened, which will also contribute to SD.

6.3 Goal 5 - Gender Equality

All over the world, females of all ages often experience discrimination, gender-based violence, as well as an exclusion to education systems, power structures, and decision-making processes. These exclusions are root causes that contribute to the progression of vulnerability and, therefore, it negatively impacts the realisation of social equity needed for SD. With the aforementioned in mind, SDG 5 pertains to achieving gender equality and empowering all women and girls.

In the DRR context, gender is often (mis)understood as referring only to women and girls, with the exclusion of men and boys in DRR activities (Ngosi 2019). However, focusing only on females is too narrow a perspective and can limit empowerment and SD. Instead, focusing on all genders can help to address the root causes of gender inequality from all sides. Noteworthy, although the EAGER project did not exclude any gender in the sessions, gender sensitivity (that acknowledges the increased vulnerability of girls based on their gender) was built into the project intention and design. As such, some session topics contributed to SDG 5, *viz.* child marriage, early parenthood, teenage pregnancy, prostitution, sexual health, domestic violence, and human trafficking. Through inclusion and empowerment of female participants, social equity (Pillar: People) can be facilitated.

6.4 *Goal 15 - Life on Land*

The EAGER Namibia communities are heavily reliant on the Chobe River for transportation, as a food source, and for economic activity (e.g. small-scale fishing). Most of the Namibia project sites do not receive clean treated water and communities use water from the river (which is inhabited by crocodiles, hippopotamuses, etc.) on a daily basis. Furthermore, due to the isolation from the rest of the country, essential public services (e.g. refuse- and waste management) are either lacking or inadequately provided.

With the aforementioned vulnerabilities in mind, some of the Namibia sessions dealt with issues pertaining to water pollution, as well as the sustainable use and protection of natural resources - all of which are cardinal for the attainment of SD (Pillar: Planet). In the other EAGER countries, the importance and management of natural resources were discussed through topics such as environmental awareness and -health. These topics also contribute to the attainment of SD and the targets of SDG 15 with its focus on the protection, conservation, and sustainable use of terrestrial- and inland freshwater ecosystems.

6.5 *Goal 16 - Peace, Justice, and Strong Institutions*

SDG 16 focuses on promoting inclusive societies for SD, providing access to justice, and building strong institutions. The sessions with topics such as domestic violence and human trafficking support SDG16 and the attainment of SD. With this in mind, examples of the empowerment of the participants through CBDRR can be presented.

In Mozambique, time was allocated in the project schedules for sessions on topics that the participants selected themselves. I.e. the trainers gave freedom to and facilitated the agency of the participants by allowing them to participate in the identification of the sessions' content. Giving participants some power in the decision-making process contributes to the promotion of an inclusive society which is imperative for SD (Pillar: People). In addition to the sessions, the EAGER project also included a photo-voice component as part of the media aspect of the project (cf. University of Kansas - Centre for Community Health And Development [UK-CCHD] 2020). The participants were provided with digital cameras, and they were given agency through the opportunity to share their concerns and opinions (as it pertains to DRR) with the tool of photography (cf. NWU-ACDS 2018b).

These activities gave the participants confirmation that their opinions are important regardless of their age or gender. By empowering themselves and others, the adolescents can strengthen resilience and contribute to positive change in their society, which can play a part in SD and the attainment of SDG 16.

7 Conclusion

It was mentioned that disasters can pose a threat to SD by exacerbating poverty and negatively impacting social equity (Pillar: People), environmental protection (Pillar: Planet), and economic growth (Pillar: Prosperity). It is imperative to reduce a system's risk to ensure that exposure to a hazard does not result in the occurrence of a disaster. This can be done by decreasing a system's vulnerability and increasing a system's resilience through DRR practices.

The implementation of DRR practices is incredibly valuable in Africa when considering that the vulnerable communities on the continent experience a diverse range of disasters (cf. EM-DAT 2019), which have dire consequences for development gains and sustainable development. To strengthen the resilience of a system, the goals of DRR are to 1) reduce existing disaster risk, 2) prevent new disaster risk, and 3) manage residual risk - all of which will ultimately contribute to SD. The aforementioned is indicative of a link between DRR and SD.

In this chapter, the relationship between DRR and SD was discussed. CBDRR - with its aims of empowerment and SD - highlights this relationality and the EAGER project (as a practical example of CBDRR) was presented. The EAGER project was implemented in four African countries, and the approach utilised was not that of "big brother, big stick". Instead, the project implemented a participatory approach under CBDRR. In future it might be insightful to do follow-up research in the EAGER project sites to establish what participants and stakeholders have held over from the project years after the completion of the project.

The goal of the project was to reduce the challenges that the participants face in pre- and post-disaster contexts through capacity-building. The resultant resilience was strengthened through the integration of education and participation in the CBDRR approach. When considering the aforementioned three goals of DRR, it comes as little surprise that the capacity-building activities relate to SD and the attainment of the SDGs. That being said, the EAGER project is an example of community empowerment through DRR that can lessen vulnerability as well as strengthen resilience and contribute to the realisation of social equity, environmental protection, and economic growth as captured under SD.

References

Adger WN (2006) Vulnerability. Glob Environ Change 16(3):268–281

Awal MA (2015) Vulnerability to disaster: pressure and release model for climate change hazards in Bangladesh. Int J Environ Monit Prot 2(2):15–21

Benson C (2016) Promoting sustainable development through disaster risk management. Asian Development Bank Sustainable Development Working Paper Series No 41 Asian Development Bank

Brundtland G (1987) Report of the World Commission on Environment and Development: Our Common Future. United Nations General Assembly document A/42/427

Burton I, Kates RW (1964) The perception of natural hazards in resource management. Nat Res J 3:412–441

Church J (2012) Sustainable development and the culture of Ubuntu. De Jure Law J. 45(3):511–531

Cuny FC, Abrams S, America Oxfam (1983) Disaster and Development. Oxford University Press, New York

Delica-Willison Z (2005) Community based disaster risk management: local level solutions to disaster risk. Tropical Coast 12(1):66–73

Delica-Willison Z, Gaillard JC (2012) Community action and disaster. In: Ben W, Gaillard JC, Illan K (eds). The Routledge Handbook of Hazards and Disaster Risk Reduction Routledge, pp 711–722

Forbes-Biggs K, Maartens Y (2012) Adolescent girls at risk. Children, Youth and Environments 22(2):234–248

Forbes-Biggs KM (2016) A new protocol in Disaster Risk Reduction policy and praxis for the Southern Africa region: Gender-age socio-behavioural intervention and the GIRRL Programme Model. Doctoral thesis North-West University South Africa

Forbes-Genade K, van Niekerk D (2017) The GIRRL program: a human rights based approach to disaster risk reduction intervention in Southern Africa. Int J Disaster Risk Reduct 24:507–514

Forbes-Genade K, van Niekerk D (2019) GIRRL power! participatory action research for building girl-led community resilience in South Africa. Action Res 17(2):237–257

Google Maps (2020) Google Maps https://www.google.co.za/maps/@-22.3331071,24.21026,6.34z. Accessed Nov 2020

Kabubi J (2011) Varieties and characteristics of hazards. DRR Knowledge Shop. http://www.drrknowledgeshop.co.za/browse/?user_id=591&fids=6. Accessed 23 Nov 2020

Marulikrishna IV, Manickam V (2017) Environmental Management. Science and Engineering for Industry. Sustainable Development. Butterworth-Heinemann, p 5–21

Ngosi VJ (2019) Gendered approach toward Disaster Risk Reduction: case study of Chobe District, Botswana. Masters dissertation North-West University

NWU-ACDS EAGER Atlas. Project Documentation. North-West University – African Centre for Disaster Studies (2016a). http://natural-sciences.nwu.ac.za/sites/naturalsciences.nwu.ac.za/files/files/ACDS/50)%20EAGER%20ATLAS%20ENG.pdf. Accessed 23 Nov 2020

NWU-ACDS: Engaging African G.I.R.R.Ls in Gender Enriched Risk Reduction [EAGER]: Manual for Guiding Project Implementation. North-West University – African Centre for Disaster Studies (2016a). http://natural-sciences.nwu.ac.za/sites/naturalsciences.nwu.ac.za/files/files/ACDS/51)%20EAGER%20Manual%20English.pdf. Accessed 23 Nov 2020

NWU-ACDS EAGER Profile - Robert Mafila [Video]. North-West University - African Centre for Disaster Studies (2018a). https://www.youtube.com/watch?v=OZw6y5MuTHU&t. Accessed 23 Nov 2020

NWU-ACDS: EAGER Media Playlist [Video]. North-West University - African Centre for Disaster Studies (2018b). https://www.youtube.com/playlist?list=PL6doaQTA4_AMLOYSjpWrADPAlh3mXz1zr. Accessed 23 Nov 2018

O'Keefe P, Westgate K, Wisner B (1976) Taking the naturalness out of natural disasters. Nature 260(5552):566–567

Robinson J (2004) Squaring the circle? Some thoughts on the idea of sustainable development. Ecol Econ 48(4):369–384

Smythe KR (2014) An historian's critique of sustainability. Culture Unbound 6(5):913–929

Texier-Teixeira P, Edelblutte E (2017) Jakarta: mumbai—two megacities facing floods engaged in a marginalization process of slum areas. In: Sudmeier-Rieux K, Fernández M, Penna I, Jaboyedoff M, Gaillard JC (eds). Identifying Emerging Issues in Disaster Risk Reduction, Migration, Climate Change and Sustainable Development. Springer

Twigg J (2015) Good Practices Review 9: Disaster Risk Reduction. Overseas Development Institute London, United Kingdom

UNDP: Reducing Disaster Risk: A Challenge for Development. United Nations Development Programme. UNDP, New York (2004)

UNDP: Transforming our world: the 2030 Agenda for Sustainable Development. United Nations Development Programme. Division for Sustainable Development Goals: New York, NY, USA (2015)

UNGA: Report of the open-ended intergovernmental expert working group on indicators and terminology relating to disaster risk reduction (No. A/71/644). Seventy-first session. Agenda item 19(c) Sustainable development: Disaster risk reduction. United Nations General Assembly (2016). https://www.jcomm.info/components/com_oe/oe.php?task=download&id=33524&version=1.0&lang=1&format=1. Accessed 23 Nov 2020

UNISDR: Living with risk: A global review of disaster reduction initiatives. United Nations International Strategy for Disaster Reduction. Kobe: UNISDR (2004). https://www.undrr.org/publication/living-risk-global-review-disaster-reduction-initiatives. Accessed 23 Nov 2020

UNISDR (2008) Gender Perspectives: Integrating Disaster Risk Reduction Into Climate Change Adaptation. UNISDR, United Nations International Strategy for Disaster Reduction Geneva

UNISDR (2009) UNISDR Terminology on Disaster Risk Reduction. UNISDR, United Nations International Strategy for Disaster Reduction Geneva

UNISDR (2015a) Sendai Framework for Disaster Risk Reduction 2015–2030. UNISDR, United Nations International Strategy for Disaster Reduction Geneva

UNISDR: Disaster Risk Reduction and Resilience in the 2030 Agenda for Sustainable Development. United Nations International Strategy for Disaster Reduction Geneva: UNISDR (2015b). https://www.unisdr.org/files/46052_disasterriskreductioninthe2030agend.pdf. Accessed 23 Nov 2020

UNISDR (2015b) Women's Leadership in Risk-Resilient Development: Good Practices and Lessons Learned. UNISDR, United Nations International Strategy for Disaster Reduction Bangkok

UNISDR: Annual Report:2016–17 Biennium Work Programme Final Report. United Nations International Strategy for Disaster Reduction Geneva: UNISDR (2017a)

UNISDR: UNISDR terminology on disaster risk reduction. United Nations International Strategy for Disaster Reduction Geneva: UNISDR (2017b). https://www.unisdr.org/files/. Accessed 23 Nov 2020

UNDRR: Home: Terminology. United Nations Office for Disaster Risk Reduction (2020a). https://www.undrr.org/terminology/disaster. Accessed 23 Nov 2020

UK-CCHD: Section 20. Implementing Photovoice in Your Community University of Kansas - Centre for Community Health and Development (2020). https://ctb.ku.edu/en/table-of-contents/assessment/assessing-community-needs-and-resources/photovoice/main. Accessed 23 Nov 2020

van Aalst MK, Cannon T, Burton I (2008) Community level adaptation to climate change:the potential role of participatory community risk assessment. Global Environ. Change 18(1):165–179

van Niekerk D (2004) Disaster Reduction and Sustainable Development. Paper presented at Heads and Directors of Disaster Management Organisations Conference Abuja:Nigeria

van Niekerk D, Coetzee C, Nemakonde, L.: Implementing the Sendai Framework in Africa: Progress Against the Targets (2015–2018). Int. J. Disaster Risk Sci. **52**, 1–11 (2020)

van Niekerk D (2015) Disaster risk governance in Africa: A retrospective assessment of progress against the Hyogo Framework for Action (2000–2012). Disaster Prevention and Management 24(3):397–416

Vyas-Doorgapersad S, Lukamba TM (2012) Disaster risk reduction policy for sustainable development in the Southern African development community: a policy perspective. J Publ Adm 47(2):774–784

Wisner B, Blaikie P, Cannon T, Davis I (2004) At risk: natural hazards, people's vulnerability and disaster. Routledge, pp 3–87

Kristel Fourie is a researcher with the African Centre for Disaster Studies (ACDS) in the Unit for Environmental Science and Management at the North-West University. She has completed her MA in Development and Management with the main focus on the application of participatory development communication principles in the context of Disaster Risk Reduction. Kristel has been part of various research and community engagement projects in the past ten years at the ACDS. Her research interests include communication specifically in disaster risk reduction, community engagement and participatory decision-making as well as disaster response and the incident management system.

Aïda C. Terblanché-Greeff holds a Master's degree in Development and Management, with specialisation in disaster studies, which she obtained at the North-West University (NWU), South Africa. As of 2020, she is appointed as an Extraordinary Researcher in the Research Focus Area: Social Transformation, at the NWU. Her academic focus areas are inter- and transdisciplinary in nature with research topics such as disaster risk reduction, cross-cultural studies, and environmental ethics. As a PhD in Philosophy student at the University of Johannesburg, she plans to complete her degree in 2021.

Sailing Towards Prosperity: Africa's Ocean Governance Agenda in the 21st Century

Siqhamo Yamkela Ntola

Abstract African development policy in the 21st century has, in contrast to that of the 20th century, taken clear cognisance of the potential role of its vast maritime domain in aiding socio-economic development whilst mitigating environmental degradation. Mindful of the former, the United Nations Economic Commission for Africa noted in 2016 that despite the continent's rich resource endowments, it continues to suffer from large-scale poverty. The Commission postulated that the continent will account for much of the world's exponential growth in this century. Doing so requires, in cases of conflict, that African states prioritise the social and economic aspects of sustainable development over the physical environment. With less than ten years left to achieve the goals set under the United Nations (UN) Sustainable Development Goals (SDGs), the aim of this chapter is to shed light on Africa's ocean governance agenda by analysing key governance frameworks. Following this analysis, it is concluded that Africa's ocean governance agenda aligns with the continent's anthropocentric developmental outlook and contributes to the principle of intragenerational equity as outlined in the African Union's Agenda 2063: The Africa We Want. More so, this agenda is advanced by continental and regional legal frameworks. It is recommended that although intragenerational equity serves the continent's immediate development needs, African leaders ought to consider developing development thresholds that will enable the continent to later transition from adhering to the principle of intragenerational equity to intergenerational equity, so as to ensure that future generations also benefit from the continent's vast ocean related wealth. In this chapter, this issue is examined from Africa's ocean economy potential.

S. Y. Ntola (✉)
Department of Public, Constitutional and International Law, University of South Africa's College of Law, Room 92, Floor 7 at Cas van Vuuren Building, 1 Preller Street, Muckleneck, Pretoria, South Africa
e-mail: ntolasy@unisa.ac.za

W. Leal Filho et al. (eds.), *Sustainable Development in Africa*, World Sustainability Series,
https://doi.org/10.1007/978-3-030-74693-3_37

1 Introduction

From a governance perspective, developments in the 20th century have been pivotal in the struggle towards, inter alia, equality and freedom. For African states, pursuit of the latter now finds legal basis owing to the right to self-determination being entrenched in key international legal instruments such as the United Nations (UN) Charter (1945: art 1(2)). The UN Charter's adoption contributed towards the wave of independence that swept throughout the African continent in the mid 1900s. Integral to self-determination was territorial sovereignty, which for African coastal states, included regions in their adjacent seas yet to be determined at the time. With their detachment from colonial powers occurring around the same period when the international legal regime for the oceans was being re-negotiated (Rembe 1980; Akintoba 1996), the increase of states gaining independence translated to an increase in their participation from the first (commenced in 1958) to the third (commenced in 1974) UN Conference on the Law of the Sea (UNCLOS I – UNCLOS III), which culminated to the adoption of the 1982 United Nations Convention on the Law of the Sea (LOSC) (Tsamenyi and Ali 2012; Vrancken 2013).

African states made significant contributions in the negotiation, adoption and coming into effect of the LOSC (Tsamenyi and Ali 2012; Nmehielle and Pasipanodoya 2017). In fact, within the 20th century, their participation from UNCLOS I to UNCLOS III definitively portrayed the intended role of oceans in the continent's development agenda. For instance, the Organisation of African Unity's (OAU) 1980 Lagos Plan of Action for the Economic Development of Africa made constant reference to seabed exploitation as a contributor to the continent's development agenda. The potential impact of Africa's developing ocean economy has since been echoed, in the 21st century, in continental governance instruments such as the African Union's (AU) Agenda 2063: The Africa We Want ("Agenda 2063"); the 2050 Africa's Integrated Maritime Strategy ("2050 AIMS"); and the African Charter on Maritime Security and Safety and Development in Africa ("Lomé Charter").

At an international level, the LOSC is a key international legal instrument for ocean governance. It outlines, among other things, the rights and duties of coastal states in areas within and beyond their national jurisdiction. With a combined coastline of 41 162 km unevenly divided among Africa's 39 coastal states and with 47 of its 55 (or 85%) (Table 1) states having ratified the instrument, African states (coastal and landlocked) are well positioned, as far as the LOSC is concerned, to embark on the new international economic order which formed part of their negotiating rationale during the UNCLOS negotiations. Schrijer refers to this period in time as the "Noord-Zuidrecht" (North-South Law) (2003), which was when developing countries sought to alter international economic relations from those favouring developed states, to those that took into account the needs of the poor (developing states) and were more equitable (Tladi 2007). The concept of Noord-Zuidrecht generally permeated throughout international conferences during the late 1900s, contributing towards ideas fundamental to the concept of sustainable development.

Table 1 Ratification of key international and continental legal instruments essential for ocean governance

States (landlocked states highlighted)	International legal instrument	Continental legal instruments				Regional legal instruments					
	LOSC	AECT	AUCA	AMTC	Lomé Charter	Abidjan Convention	Barca Convention	Nairobi Convention	Abuja MOU	Indian Ocean MOU	Mediterranean MOU
1. Algeria	1996	1995	2001	X	X	X	1981	X	X	X	Ratified
2. Angola	1990	1992	2001	X	X	X	X	X	Ratified	X	X
3. Benin	1997	1999	2001	2012	2017	1997	X	X	Ratified	X	X
4. Botswana	1990	1996	2001	X	X	X	X	X	X	X	X
5. Burkina Faso	2005	1992	2001	2016	X	X	X	X	X	X	X
6. Burundi	X	1992	2001	X	X	X	X	X	X	X	X
7. Cameroon	1982	1995	2001	X	X	1983	X	X	X	X	X
8. Cape Verde	1987	1993	2001	X	X	X	X	X	Ratified	X	X
9. Central African Republic	X	1993	2001	X	X	X	X	X	X	X	X
10. Chad	2009	1993	2001	X	X	X	X	X	X	X	X
11. Comoros	1994	1994	2001	X	X	X	X	1994	X	Ratified	X
12. Congo	2008	1996	2002	2014	X	1987	X	X	Ratified	X	X
13. Côte d'Ivoire	1984	1993	2001	X	X	1982	X	X	X	X	X
14. Djibouti	1991	X	2000	X	X	X	X	X	X	X	X
15. DRC	1989	1993	2002	X	X	2014	X	X	Ratified	X	X
16. Egypt	1983	1992	2001	X	X	X	1978	X	X	X	Ratified
17. Equatorial Guinea	1997	2002	2000	X	X	X	X	X	X	X	X
18. Eritrea	X	X	2001	X	X	X	X	X	X	X	X
19. Ethiopia	X	1992	2001	2012	X	X	X	X	X	X	X
20. Gabon	1998	2007	2001	2013	X	1988	X	X	Ratified	X	X
21. Gambia	1984	1993	2001	2018	X	1984	X	X	Ratified	X	X
22. Ghana	1983	1991	2001	X	X	1989	X	X	Ratified	X	X
23. Guinea	1985	1992	2002	X	X	1982	X	X	Ratified	X	X
24. Guinea-Bissau	1986	1992	2001	X	X	2012	X	X	Ratified	X	X
25. Kenya	1989	1993	2001	2014	X	X	X	1990	X	Ratified	X
26. Lesotho	2007	1997	2001	X	X	X	X	X	X	X	X
27. Liberia	2008	1993	2001	X	X	2005	X	X	Ratified	X	X
28. Libya	X	1992	2000	X	X	X	1979	X	X	X	X
29. Madagascar	2001	X	2003	X	X	X	X	1990	X	X	X
30. Malawi	2010	1993	2001	X	X	X	X	X	X	X	X
31. Mali	1985	1992	2000	X	X	X	X	X	X	X	X
32. Mauritania	1996	2001	2001	X	X	2012	X	X	X	X	X
33. Mauritius	1994	1992	2001	2012	X	X	X	2000	X	Ratified	X
34. Morocco	2007	2017	2017	X	X	X	1980	X	X	X	Ratified
35. Mozambique	1997	1992	2001	X	X	X	X	1999	X	Ratified	X
36. Namibia	1983	1992	2001	X	X	X	X	X	X	X	X
37. Niger	2013	1992	2001	X	X	X	X	X	X	X	X
38. Nigeria	1986	1991	2001	X	X	1984	X	X	Ratified	X	X
39. Rwanda	X	1993	2001	X	X	X	X	X	X	X	X
40. Sahrawi Arab	X	1992	2000	X	X	X	X	X	X	X	X

(continued)

Table 1 (continued)

States (landlocked states highlighted)	International legal instrument	Continental legal instruments				Regional legal instruments					
	LOSC	AECT	AUCA	AMTC	Lomé Charter	Abidjan Convention	Barca Convention	Nairobi Convention	Abuja MOU	Indian Ocean MOU	Mediterranean MOU
Democratic Republic (SADR)											
41. São Tomé and Príncipe	1987	1993	2001	X	X	X	X	X	Ratified	X	X
42. Senegal	1984	1992	2000	X	X	1983	X	X	Ratified	X	X
43. Seychelles	1991	1991	2001	X	X	X	X	1990	X	X	X
44. Sierra Leone	1994	1994	2001	X	X	2005	X	X	Ratified	X	X
45. Somalia	1989	X	2001	X	X	X	X	1988	X	X	X
46. South Africa	1997	2001	2001	2016	X	2002	X	2003	Ratified	Ratified	X
47. South Sudan	X	X	2011	X	X	X	X	X	X	X	X
48. Sudan	1985	1993	2000	X	X	X	X	X	X	X	X
49. Swaziland	2009	2001	2001	X	X	X	X	X	X	X	X
50. Tanzania	1985	1992	2001	X	X	X	X	1996	X	Ratified	X
51. Togo	1985	1998	2000	2012	2017	1983	X	X	Ratified	X	X
52. Tunisia	1985	1994	2001	X	X	X	1977	X	X	X	Ratified
53. Uganda	1990	1991	2001	X	X	X	X	X	X	X	X
54. Zambia	1983	1992	2001	X	X	X	X	X	X	X	X
55. Zimbabwe	1993	1991	2001	X	X	X	X	X	X	X	X
Total	**47**	**50**	**55**	**10**	**2**	**17**	**5**	**9**	**17**	**6**	**4**

The 1987 Report of the World Commission on Environment and Development: Our Common Future ("the Brundtland Commission Report") commissioned by the UN General Assembly (UNGA) in 1983 (UN Res 38/11/61) is credited for its influence in mainstreaming sustainable development. Notorious for its definition of sustainable development, which is "development that meets the needs of the present generation without compromising the ability of future generations to meet their own needs" (the Brundtland Commission Report 1987: IV Conclusion), the Report's value extends farther than providing this meaning. Its general thrust of poverty alleviation through an integrated approach owing to how social, economic and environmental issues are interlocked, has, as far as the sea is concerned, arguably moulded governance frameworks and their implementation. An apt legal example demonstrating how these elements are interlocked is the International Tribunal for the Law of the Sea's (ITLOS) advisory opinion on the *Responsibilities and Obligations of States Sponsoring Persons and Entities with Respect to Activities in the Area (Case no.17).* Here, the ITLOS was seized of the matter by states desiring to undertake mineral exploitation (economic activity) in the Area, whose resources are for the common heritage of mankind (social). Among other things, included in the ITLOS' opinion is environmental considerations flowing from activities in the region.

Additionally, of value from the Brundland Report's definition of sustainable development, are the imbedded concepts of inter and intragenerational equity. The former recognises the poor's essential needs, "to which overriding priority should be given", with the latter accepting that the environment is limited in its ability to meet present and future needs. Intergenerational equity therefore infers that the "present generation owes a duty to future generations to leave the earth and its environment in no worse condition than they received it" (Tladi 2007: 41). Intragenerational equity requires that there be equity within generations, therefore concerning itself "with the distribution of the benefits of development activities and the distribution of costs for environmental protection" (Tladi 2007: 48). The concept's main concern is distributive justice. In this sense, considering the economic divide between developed and developing states, approaches to development and environmental protection thereto - vary. On the one hand, developed states having amassed their wealth through technological advancements employ stricter environmental protection measures, the finance for which is readily available. On the other hand, developing states are more liberal with environmental protection owing to constraints that conservative measures pose on immediate development goals. The latter scenario is reflected in the LOSC. For example, article 193 of the LOSC provides that states "have the sovereign right to exploit their natural resources pursuant to their environmental policies and in accordance with their duty to protect and preserve the marine environment". The provision demonstrates cognisance of the abovementioned integrated approach as it relates to ocean related development, as well as the varying approaches and degrees therein that developed and developing states may approach environmental protection flowing from natural resource exploitation. In no way is this an attempt to attribute influence of the Brundtland's Report on the LOSC's provisions. Indeed, this would be a far reach, especially considering the former's release date (1987, commissioned in 1983) versus the latter's adoption date (1982). However, it is reasonable to assert that the

LOSC, and the negotiations leading up to its adoption, contributed in sustainable development discourse and the principles underpinning it.

At this juncture, it is important to note that the above conceptualisation of sustainable development has been held to be too simplified, lending itself "to attack from different quarters" (Tladi 2007: 74). Desiring to formulate a better understanding of the normative implications of balancing environmental and development needs, Tladi (2007), having indicated that sustainable development requires the integration of its elements and discussed how such elements are to be integrated, provides a more nuanced conceptualisation of sustainable development by providing three variations. The first is an *economic growth-centred variation*, where economic growth and related values are favoured (anthropocentrism). The second is an *environment-centred variation*, where the natural environment is regarded more important and where there is a conflict with the latter variation, it triumphs (leaning towards biocentrism). The third is a *human-needs centred variation*, which places the social needs and general well-being of humanity at the centre of its concerns (anthropocentrism) (Tladi 2007: 80). Africa's governance frameworks on and relevant to ocean governance, as will be evident from discussions below, speak largely to the first and last variation.

To this end, in the context of sustainable development on the continent, the aim of this chapter is to shed light on Africa's ocean governance agenda through an analysis of key governance frameworks. Prior to doing so, the chapter aligns Africa's existing policy frameworks with the dominant ocean related development objectives, namely – goal 14 of the UN Sustainable Development Goals (SDGs), and ahead of concluding, highlights notable ocean governance strides made by African states at a domestic level.

2 A Key International Governance Framework: SDG 14

The UNGA's adoption of "Transforming our World: The 2030 Agenda for Sustainable Development" ("Agenda 2030") (UN Res 70/1) followed its adoption of *The Future We Want* (UN Res 66/288), an outcome document to the 2012 UN Conference on Sustainable Development held in Rio de Janeiro ("Rio + 20"). Agenda 2030 provides for the 17 SDGs, with 169 targets aimed at eradicating all forms and dimensions of poverty owing to the significant challenge it poses to sustainable development (Agenda 2030, 2015: Preamble). Goal 14 of the SDGs pertains to the conservation and sustainable use of "oceans, seas and marine resources for sustainable development" (Agenda 2030, 2015: goal 14.c). Of its ten targets, target 14.c best captures the goal's general ambition, which is to "[e]nhance the conservation and sustainable use of oceans and their resources by implementing international law as reflected in the United Nations Convention on the Law of the Sea, which provides the legal framework for the conservation and sustainable use of oceans and their resources, as recalled in paragraph 158 of the '[The Future We Want]'".

Goal 14 expressly calls for the implementation of the LOSC, underscoring, as previously mentioned, its key status in ocean governance. However, the LOSC is not the only instrument relevant to sustainable ocean governance. In fact, despite widespread ratification of the LOSC by African states, their participation in other international maritime legal instruments lends itself to an ocean governance deficit which accordingly has the potential to adversely affect its sustainable development objectives at sea. Of a sample of 70 multilateral agreements taken by Vrancken (2013) in 2013, approximately 29% of all African states participate in only 10% or less of the instruments. Seven years onwards, not much has changed. Reasons behind the stark contrast of ratifications of the LOSC versus the ratifications of other multilateral legal instruments which have a direct or indirect bearing on the marine environment is submitted to be unclear. What is however clear is that the Agenda 2030, as euphemised in its preamble, seeks to address failures with the Millennium Development Goals (MDGs).

The MDGs culminated from a process that departed from a UNGA resolution (A/Res/53/202) on 17 December 1998. The MDGs provided the international framework for development and according to Agenda 2030, progress on attaining the goals "has been uneven, particularly in Africa […]" (Agenda 2030, 2015: para 16–17). Goal 7 of the MDGs aimed at ensuring environmental sustainability. As far as the sea is concerned, world leaders undertook to reduce biodiversity loss at sea by 8.4% by 2015. In this respect, according to the "2015 Millennium Development Goals Report", there has generally been a substantial increase in marine protected areas globally (Millennium Development Goals Report 2015: 7 and 56). The increase is likely owing to the international community of states failing to meet the marine protected area targets set for 2012 at the 2002 Johannesburg World Summit on Sustainable Development (Santo 2013). Nonetheless, this does not clearly establish a link between an increase in marine protected areas and a reduction to the loss of biodiversity. The reason for this is because although establishing marine protected areas is a critical first step, putting in place measures (policy and/or laws) and ensuring their enforcement in those regions will speak to the extent to which the loss of biodiversity at sea has been reduced.

Unlike goal 14 of the SDGs, goal 7 of the MDGs acknowledges no existing legal instruments on the sea. Its focus on biodiversity conservation arguably acknowledges, albeit tacitly, those economic activities that are harmful to marine ecosystems. What is more, the MDGs limited language on the role of oceans as an economic enabler and ultimately a contributor to poverty reduction arguably reveals the low-level consideration afforded to oceans for socio-economic development. It is therefore not surprising that, among other things, goal 14 of the SDGs makes no reference to the MDGs but rather paragraph 158 of *The Future We Want* (adopted 2012) with regard to conserving and sustainably using the oceans in terms of the LOSC for sustainable development and poverty reduction. The reason for this is that world leaders, particularly those from least developed and small island developing states, appreciated the role of oceans in achieving prosperity. In fact, it is owing to the advocacy of the latter category of states that the concept of a "blue economy" was included in "green economy" negotiations during Rio + 20 (Hardy 2018).

The notion of a blue economy diagnoses that sustainable economic growth for aquatic and maritime-based economies is through productive, healthy freshwater and ocean ecosystems (UNECA 2016). The blue economy is "envisaged as the integration of ocean economy development with the principles of social inclusion, environmental sustainability and innovative, dynamic business models" (IORA 2016). Evidently, the notion of the blue economy resides within the purpose of sustainable development, that being a "vehicle for a paradigm shift from a situation where economic growth concerns are paramount, to one where social and environmental considerations are [also] of paramount importance" (Tladi 2007: 37). Such purpose is echoed in Africa's development governance frameworks relevant to the blue economy like the Agenda 2063.

3 Africa's Governance Frameworks

3.1 Continental Policy Frameworks

The Agenda 2063 was adopted for, among other reasons, to provide "a robust framework for addressing past injustices and the realisation of the 21st Century as the African Century" (Agenda 2063, 2015: para 1). Among the aspirations provided, the AU aims, by 2063, to have a "prosperous Africa based on inclusive growth and sustainable development" (aspiration one). Agenda 2063 in general, and the referred to aspiration in particular, arguably present a conflict between the inter and intra-generational equity principles of sustainable development on the continent. On the one hand, in a general sense, considering that 2063 is the target year for achieving all aspirations, there is an implied duty on present generations to undertake development in a manner that leaves the continent and its natural environment in no worse condition than it received it (intergenerational equity). Implied are stricter environmental protection measures, which is challenging considering the continent's desired development trajectory and the relatively liberal environmental measures required to aid it. Specifically, and on the other hand, the Agenda's first aspiration which indicates inclusive growth and sustainable development as elements of its prosperity, draws attention to intragenerational equity - the just distribution of developmental benefits within generations. As earlier indicated, this principle does not invite strict environmental protection measures. In fact, as far as the sea is concerned, its role in contributing towards the first aspiration appears more embracing of intragenerational equity owing to its relative silence on marine environmental protection. In this regard, African leaders have expressly noted that the continent's "blue/ocean economy, which is three times the size of its landmass, shall be a major contributor to continental transformation and growth, through knowledge on marine and aquatic biotechnology, the growth of an Africa-wide shipping industry, the development of sea, river and lake transport and fishing; and exploitation and beneficiation of deep sea mineral and other resources" (Agenda 2063, 2015: para 15). The latter view is further cemented in that "the goal" as far as oceans are concerned is for the blue/ocean economy to accelerate economic growth (https://au.int/en/agenda2063/goals), with its priority areas aligned accordingly. Such a goal is not surprising when considering that the value of the continent's maritime industry is estimated at USD 1 trillion a year (Ruppel and Biam 2016). Africa's strategy contained in the 2050 AIMS follows the same trajectory.

The 2050 AIMS is the roadmap produced by African leaders to effectively utilise the continent's ocean spaces to contribute towards meeting its development ambitions. The strategy "is an African-driven long-term and reasonably comprehensive vision crafted to better harness Africa's so-called 'blue economy', with the vision of using this to promote development in the continent" (Egede 2016). "AIMS reflects a turn of the page in Africa's history: a change from society's gaze towards the interior (the terrestrial part of the continent) to a view towards the exterior (the maritime part of the continent); a new understanding of the latter as a vehicle for communication with the world and inclusion in the global economy, rather than a cause of separation and a barrier towards development" (Nmehielle and Pasipanodya 2017). 2050 AIMS' strategic objectives include, inter alia, establishing a Combined Exclusive Maritime Zone (CEMZA); enhancing political will at a community, national, regional and continental level; and minimising damage to the environment and expediting recovery from catastrophic events (2050 AIMS 2012: para 21). Under its framework for strategic action, which are the activities necessary for its implementation, the Strategy provides, among others, the following:

(a) establishing the CEMZA with the objective of increasing intra-African trade;
(b) once established, sustainably managing fish stocks within the CEMZA;
(c) utilising the AU Border Programme (AUBP), in accordance with the LOSC, to peacefully resolve existing maritime boundary disputes and encourage states to claim maritime zones; and
(d) improving maritime governance in areas such as environmental crime, environmental and biodiversity monitory, flag and port state control, marine spatial planning, piracy and armed robbery at sea (2050 AIMS 2012: para XII).

3.2 *Continental Legal Frameworks*

Relevant to advancing a blue economy as envisaged in Agenda 2063, the 1991 Treaty Establishing the African Economic Community (AECT), 2000 AU Constitutive Act (AUCA), 2010 African Maritime Transport Charter (AMTC) and the Lomé Charter are key continental legal instruments discussed in this section.

The AECT, which has been ratified by 50 of Africa's 55 states (91%), captures the desire of African states and their leaders to economically integrate the continent. Broadly, the objectives outlined in the AECT capture the concept of a blue economy. These objectives include, inter alia,

(a) promoting economic, social and cultural development and integrating African economies "in order to increase economic self-reliance and promote an endogenous and self-sustained development";
(b) "promot[ing] cooperation in all fields of human endeavour in order to raise the standard of living of African peoples, and maintain and enhance economic stability, foster close and peaceful relations among member states and contribute to the progress, development and the economic integration of the continent"; and
(c) "coordinat[ing] and harmoniz[ing] policies among existing and future economic communities in order to foster the gradual establishment of the community" (AECT 1991: art 1(a)–(d)).

Regional Economic Communities (RECs) are identified as important vehicles for realising the above objectives. In terms of article 28(2) of the AECT, member states are directed to take "all necessary measures aimed at progressively promoting increasingly close co-operation among the communities, particularly through co-ordination and harmonisation of their activities in all fields or sectors in order to ensure the realisation of the objectives of the Community". Food and agriculture, which includes the development of fisheries to ensure food security; science, technology, energy, natural resources and the environment; as well as tourism and transport, are areas identified by the AECT for coordination and harmonisation on the continent (AECT 1991: art 46(1), chaps VIII, IX and X). To ensure alignment with the Agenda 2063, RECs development priorities were among those reviewed at a continental level and included in the Agenda's first ten-year implementation plan (https://au.int/en/agenda2063/priorities). Again, the blue economy concept is broadly captured in the identified priority areas.

So significant are the provisions of the AECT that they are echoed in the founding instrument of the AUCA. The AUCA's adoption was preceded by the Sirte Declaration following the Forth Extraordinary Session of the OAU in 1999. The session's primary agenda item was a discussion on "ways and means of making the [OAU] more effective so as to keep pace with the political, economic and social developments taking place within and outside [the] continent" (Sirte Declaration 1999: para 2). In fact, establishing the AU was deemed as "Africa's collective response to the phenomenon of globalisation" and thus the need to "reposition the OAU in the international scheme of things, reorient its objectives, and put in its place a new mechanism in order to reinvigorate the project of African integration [...]" (Maluwa 2003: 160). The AECT's principles and objectives informed the founding instrument of the AU. Concerning economic integration, the AECT's objectives are echoed in the existing AU framework. In this sense, the AUCA aims to integrate African economies through harmonised policies between existing and future RECs. Blue economy development, as envisaged in Agenda 2063 and the 2050 AIMS, will contribute towards meeting AECT and AUCA objectives. On the latter instrument, support for its provisions is evidenced by all African states ratifying the instrument.

The AMTC, a revision of the 1994 instrument, intends to provide a cooperative framework among African states and between African and non-African states to improve the poor state of the maritime sector on the continent. Among other things, it directs AU member states to cooperate in implementing the necessary legislation for achieving its objectives in the fields of maritime transport, inland waterways and port operations (AMTC 2009: art 23). To date, only ten (or 18%) of Africa's 55 states have ratified the instrument, leaving its provisions largely inactive on the continent. The ratifying states include Benin, Burkina Faso, Congo, Ethiopia, Gabon, Gambia, Kenya, Mauritius, South Africa and Togo.

The pace of ratification of the Lomé Charter is far worse, with only Benin and Togo having ratified it to date. The slow pace of ratifications of the Lomé Charter is concerning when considering the Agenda 2063's first aspiration and the intended contribution of Africa's blue economy. The Charter aims to promote "a flourishing and sustainable Blue/Ocean Economy" (Lomé Charter 2016: art 3(3)). Its provisions cover, among other things, measures for sustainably exploiting marine resources and optimising development opportunities of ocean related sectors (Lomé Charter 2016: art 4).

3.3 Regional Policy Frameworks

Within continental governance structures, blue economy ambitions are progressively seeping into regional precincts. In RECs, notable examples include the East Africa Community's (EAC) "Vision 2050" where infrastructure development is included among its areas of cooperation. Infrastructure development in this sense includes maritime and port development, food security, as well as environmental and natural resource management (chapter 3 of EAC Vision). Latest developments under the EAC's auspices include the adoption of the "ECOFISH Programme" aimed at enhancing equitable "economic growth by promoting sustainable fisheries in the East African-South African-Indian Ocean (EA-SA-IO)" (EAC 2020). With the support of the European Union, the programme will also be implemented in the Common Market for Eastern and Southern Africa (COMESA), Intergovernmental Authority for Development (IGAD) and the Southern African Development Community (SADC).

In southern Africa, SADC officials have confirmed that blue economy objectives are aligned with the regions development priorities and have gone as far as exploring the concept for regional purposes (Blue Economy Concept Paper for SADC). Considering that unlike other RECs, SADC has the largest maritime zone when considering the exclusive economic zones declared by its member states (Forrest 2017), the above confirmation by SADC officials is indeed an important realisation.

Mindful that that the functions of RECs are instrumental in achieving AECT objectives, provisions of their legal instruments generally align with it. An exhaustive demonstration is not necessary for present purposes. Suffice it to say that similar to the provisions of the AECT generally encompassing the blue economy concept, the legal instruments adopted by RECs generally do the same (Forrest 2017).

Outside AU auspices, Comoros, Kenya, Madagascar, Mauritius, Mozambique, Seychelles, Somalia, South Africa and Tanzania are among the 22 member states of the Indian Ocean Rim Association (IORA), an inter-government regional organisation that has taken upon itself the duty of ensuring cooperation in the Indian Ocean region. The blue economy is an adopted focus area by the IORA with two declarations revealing the organisation's thinking on the concept. These declarations are the 2015 Declaration of the Indian Ocean Rim Association on enhancing Blue Economy Cooperation for Sustainable Development in the Indian Ocean Region and the 2017 Declaration of the Indian Ocean Rim Association on the Blue Economy in the Indian Ocean Region. With the 2015 instrument recognising blue economy priority pillars such as fisheries, renewable energy and exploiting offshore hydrocarbon seabed minerals, the 2017 instrument outlines the principles for developing a blue economy in the region. Noteworthy is that these principles are derived from the LOSC and other international legal instruments on activities in the oceans and seas, goal 14 of the SGDs and relevant UNGA resolutions. The participation of the abovementioned east African coastal states to the IORA is symbiotic with the AU's development objectives concerning oceans, albeit more focussed. To this end, the regional legal instruments outlined and discussed below are those adopted under the auspices of the UN Environmental Programme – Regional Seas Programme (UNEP – RSP) and the International Maritime Organisation (IMO).

3.4 Regional Legal Frameworks

Key regional legal instruments on the continent are not only limited to those adopted under AU auspices. In fact, instruments adopted under the umbrella of other organisations, relevant to Africa's blue economy objectives have been ratified by African states. For instance, African coastal states have ratified legal instruments adopted under the UNEP – RSP, whose objectives include protecting the health of the world's ocean spaces through the development of treaties as well as rules and standards at a regional level. The UNEP – RSP was "conceived as an action-oriented programme encompassing a comprehensive, trans-sectoral approach to marine and coastal areas and to environmental problems, considering not only the consequences, but also the causes of environmental degradation" (Akiwumi and Melvasalo 1998: 229). UNEP – RSP action plans and conventions generally contain similar provisions on their objectives, with protocols differing on aspects relevant to a region. Concerning the former, the conventions provide that parties to it, both individually and jointly, take measures in accordance with international law and their protocols to prevent, reduce and combat marine pollution from its various sources of pollution which includes, inter alia, the atmosphere, dumping, land-based sources, seabed activities within national jurisdiction and vessels. There are currently 18 regions, three of which involve African states. These are the *east African region*, where the 1985 Nairobi Convention for the Protection, Management and Development of the Marine and Coastal Environment of the Eastern African Region applies ("Nairobi Convention"); the *west African region*, where the 1981 Convention for Cooperation in the Protection, Management and Development of the Marine and Coastal Environment of the Atlantic Coast of the West and Central and Southern Africa Region applies ("Abidjan Convention"); and the *Mediterranean region*, where the 1976 Convention for the Protection of the Mediterranean Sea against Pollution applies ("Barcelona Convention").

On Africa's eastern seaboard, Comoros, Kenya, Madagascar, Mauritius, Mozambique, Seychelles, Somalia, South Africa and Tanzania are the African states to have ratified the Nairobi Convention. Considering that the Red Sea is a seawater inlet of the Indian Ocean, the non-ratification by Africa's Red Sea states is likely owing to the Nairobi Convention's defined operating ambit being limited to the "western Indian Ocean", which excludes, in part, the Gulf of Aden and the Red Sea. The 1985 Protocol Concerning Co-operation in Combating Marine Pollution in Cases of Emergency in the Eastern Africa Region and the 1985 Protocol Concerning Protected Areas and Wild Fauna and Flora in the Eastern Africa Region are the only protocols adopted under the instrument.

On Africa's western seaboard, with the exception of Angola, Namibia, Morocco, and São Tomé and Príncipe, the Abidjan Convention enjoys ratifications from Benin, Cameroon, Congo, DRC, Côte d'Ivoire, Gabon, Gambia, Ghana, Guinea, Guinea-Bissau, Liberia, Mauritania, Nigeria, Senegal, Sierra Leone, South Africa and Togo. Presently, only the 1981 Protocol concerning Co-operation in Combating Pollution in Cases of Emergency has been adopted.

On the northern seaboard, Algeria, Egypt, Libya, Morocco and Tunisia have ratified the Barcelona Convention, which, under its auspices, has seen to the adoption of seven protocols pertaining to pollution from dumping from ships and aircraft, land-based sources, oil and other harmful substances in cases of emergency, transboundary hazardous waste, seabed activities within national jurisdiction and the protection of protected areas and biological diversity in the region.

As earlier indicated, the IMO is also an international organisation under which instruments relevant to Africa's blue economy have been adopted. These are in the form Memorandum's of Understanding (MoU) between the maritime authorities located in or involved in a particular region. In Africa, these regions include,

(a) in *east Africa*, the 1998 Memorandum of Understanding on Port State Control for the Indian Ocean Region ("Indian Ocean MoU"), to which Comoros, Kenya, Mauritius, Mozambique, South Africa and Tanzania's maritime authorities are party;
(b) in *west Africa*, the 1991 Memorandum of Understanding on Port State Control for West and Central African Region ("Abuja MoU"), to which Angola, Benin, Cameroon, Cape Verde, Congo, Côte d'Ivoire, Gabon, Ghana, Guinea, Guinea-Bissau, Liberia, Nigeria, São Tomé and Príncipe, Sierra Leone, Senegal, South Africa and Togo's maritime authorities are party; and

(c) in *north Africa*, the 1997 Memorandum of Understanding on Port State Control in the Mediterranean Region ("Mediterranean MoU"), to which Algeria, Egypt, Morocco and Tunisia's maritime authorities are party.

Generally, these regimes speak to safety at sea with regard to navigation, marine environmental protection and safe labour practices on board vessels. As such, they provide for the inspection of ships in other national ports by port state inspectors. Inspections are undertaken to ensure compliance with international instruments such as the 1974 International Convention for the Safety of Life at Sea (SOLAS), the 1973 International Convention for the Prevention of Pollution from Ships (MARPOL), the 1978 International Convention on the Standards of Training, Certification and Watch-keeping for Seafarer (STCW) and the 2006 Maritime Labour Convention (MLC). Of Africa's 39 coastal states, 27 (or 69%) maritime authorities are party to port state control MoU's, indicating, among other things, that a majority of the continent's coastal states agree with the objectives.

4 State Practice

In contributing to global and continental blue economy objectives, Mauritius, Seychelles and South Africa have taken notable governance strides. For instance, Mauritius' "Government Programme 2020–2024" aims to fast-track ocean economy development which the 2015–2019 programme indicated its intended contributions towards wealth generation, job creation and sustainable economic diversification. Mauritius' Ministry of Blue Economy, Marine Resources, Fishing and Shipping spearheads this programme (Govt of Mauritius 2020).

Seychelles adopted its "Blue Economy Strategic Policy Framework and Roadmap: Charting the Future (2018–2030)" in 2018. The Roadmap "is an integrated approach to ocean based sustainable development which brings together economy, environment and society, consistent with the Sustainable Development Agenda 2030 (SDG's), Aichi Target 11 of the Convention on Biological Diversity (CBD) and the Paris Agreement on Climate Change" (Govt of Seychelles 2017a, b: 3). The Roadmap includes four key pillars for implementation, which are (i) *economic diversification and resilience*, "to reduce economic vulnerability and reliance on a small number of sectors and to increase the % GDP derived from marine sectors"; (ii) *shared prosperity*, through the creation of high value jobs and local investment opportunities; (iii) *food security and wellbeing*; and (iv) *integrity of habitats and ecosystem services* for sustainable use and climate resilience (Government of Seychelles 2017a, b: 4). Initiatives under this strategy are championed by Seychelles' Blue Economy Department which sits under the office of the Vice-President (Govt of Seychelles 2017a, b).

South Africa adopted "Operation Phakisa: Oceans Economy" ("Phakisa") in 2014. Phakisa is South Africa's initiative aimed at developing its ocean economy to aid fast-tracking the implementation of its 2030 National Development Plan. Phakisa desires to unlock the country's economic potential by growing areas such as aquaculture, transport and manufacturing as well as offshore oil and gas. South Africa has identified marine protection services and ocean governance as important pillars for ensuring environmental and economic sustainability (DEAFF 2019). Significant legal developments made to date include the introduction of the Aquaculture Development Bill to Parliament in 2018 and the adoption and promulgation of the Marine Spatial Planning Act, 16 of 2018 (MSPA) (GN 641 in *GG* 42444 on 6 May 2019).

5 Conclusion

As far as Africa's oceans are concerned, their contribution to the continent's development objectives for the 21st century can best be seen in light of the Agenda 2063's first aspiration. That being, to foster prosperity through inclusive growth and sustainable development. In this respect, Africa's ocean governance agenda contributes to this anthropocentric development outlook, which is to say, it follows the *economic growth* and *human-needs* centred variations of sustainable development, with an emphasis on intragenerational equity. The agenda is consistent with Africa's development outlook since its states gained independence from colonial powers. Examples of such ambitions are replete in the UNCLOS' negotiation history.

It is not surprising therefore, that the above governance frameworks, all of which adopted prior to the Agenda 2063, complement the Agenda's development philosophy. With too few notable ocean governance advances at state level, concerns of meeting the continent's 2063 target and the even closer 2030 targets under goal 14 of the SDGs, are justified. Compounding this is the fact that far too few African states have ratified other multilateral legal instruments having a direct or indirect bearing on the marine environment. Considering the continent's developmental outlook, ratification of such instruments may impose onerous marine environmental protection obligations that limit desired development. While this may be the case, non-ratification may be to the detriment of future generations. It is advised that Africa's leaders consider developing development thresholds that will enable the continent to later transition from adhering to the principle of intragenerational equity to intergenerational equity. Considering the gravity of Africa's economic challenges, the implementation of the latter recommendation requires an in-depth analysis of the present status of the continent's socio-economic standing in relation to its aspirations. Such an analysis may aid in determining the said thresholds and hopefully prod Africa's leaders in crafting attainable targets in this respect. The present weight of political will towards developing Africa's blue economy is indeed necessary but insufficient. Moving forward, more in terms of political and financial capital will have to be invested, especially towards the strategic actions outlined in the 2050 AIMS. The state level blue economy initiatives are encouraging. With an increase in participation from RECs, it is likely that more states will adopt blue economy measures in the next decade.

References

Akintoba T (1996) African states and contemporary international law: a case study of the 1982 law of the sea convention and the exclusive economic zone. Martinus Nijhoff, Leiden

Akiwumi P, Melvasalo T (1998) The UNEP's regional seas programme: approach, experience and future plans. Marine Policy 22(3):229–234

African Union (n.d.) Flagship Projects of Agenda 2063. https://au.int/agenda2063/flagship-projects

African Union (2012) 2050 Africa's Integrated Maritime Strategy (2050 AIM Strategy). https://au.int/sites/default/files/documents/30928-doc-2050_aim_strategy_eng.doc

African Union (n.d.) The First-Ten Year Implementation Plan. https://au.int/en/agenda2063/ftyip

African Union. (n.d.) Goals & Priority Areas of Agenda 2063. https://au.int/en/agenda2063/goals

African Union (2016) African Charter on Maritime Security and Safety and Development in Africa (Lomé Charter). https://au.int/sites/default/files/treaties/37286-treaty-african_charter_on_maritime_security.pdf

African Union (2015) Agenda 2063: The Africa We Want. https://www.un.org/en/africa/osaa/pdf/au/agenda2063.pdf

African Union (1999) Sirte Declaration. https://archives.au.int/bitstream/handle/123456789/2475/1999_Sirte%20_Decl_%20E.pdf?sequence=1&isAllowed=y

Centre for Environmental Rights (2018) Aquaculture Development Bill. https://cer.org.za/wp-content/uploads/2018/05/Aquaculture-Development-Bill.pdf

Convention for Cooperation in the Protection, Management and Development of the Marine and Coastal Environment of the Atlantic Coast of the West and Central and Southern Africa Region, 20 ILM 746

Convention for the Protection of the Mediterranean Sea against Pollution, 1102 UNTS 27

Department of Environmental Affairs, Forestry and Fisheries (2019) Operation Phakisa: Oceans Economy. https://www.environment.gov.za/projectsprogrammes/operationphakisa/oceanseconomy

East African Community (2020) Ecofish Programme set to contribute to sustainable fisheries for the blue economy of the Eastern and Southern Africa and Indian Ocean regions. https://www.eac.int/press-releases/141-agriculture-food-security/1658-e€ofish-programme-set-to-contribute-to-sustainable-fisheries-for-the-blue-economy-of-the-eastern-and-southern-africa-and-indian-ocean-regions

Egede E (2016) Institutional gaps in the 2050 africa's integrated maritime strategy. J. Ocean Law Govern. Africa 1(1):1–27

Forrest C (2017) Eastern seaboard. In: Vrancken P, Tsamenyi M (eds) The Law of the Sea: The African Union and its Member States. Juta and Company, Cape Town, pp 63–110

Government of Mauritius (2020) Government Programme 2020–2024: Towards and Inclusive, High Income and Green Mauritius, Forging Ahead Together. https://foreign.govmu.org/Documents/2020%20-%20migrated%20data/Gov%20Program%2020-24/govt-programme-2020-2024.pdf

Government of Seychelles (2017) Seychelles' Blue Economy Strategic Policy Framework and Roadmap: Charting the Future 2018–2030. https://foreign.govmu.org/Documents/2020%20-%20migrated%20data/Gov%20Program%2020-24/govt-programme-2020-2024.pdf

Hardy D (2018) Challenges of indian ocean governance: the context of the seychelles. In: Attri V, Bohler-Muller N (eds) The Blue Economy Handbook of the Indian Ocean Region. AISA, Pretoria, pp 170–189

International Tribunal for the Law of the Sea (2011) Responsibilities and Obligations of States Sponsoring Persons and Entities with Respect to Activities in the Area (Request for Advisory Opinion submitted to the Seabed Disputes Chamber). https://www.itlos.org/fileadmin/itlos/documents/cases/case_no_17/17_adv_op_010211_en.pdf

Maluwa T (2003) The constitutive act of the african union and institution-building in postcolonial Africa. Leiden J. Int. Law 16:157–170

Nairobi Convention for the Protection, Management and Development of the Marine and Coastal Environment of the Eastern African Region, 91 RGIP 1122

Nmehielle V, Pasipanodya T (2017) African union. In: Vrancken P, Tsamenyi M (eds) The Law of the Sea: The African Union and its Member States. Juta and Company, Cape Town, pp 36–59
Organization of African Unity (1980) Lagos Plan of Action for the Economic Development of Africa 1980–2000. https://www.merit.unu.edu/wp-content/uploads/2015/01/Lagos-Plan-of-Action.pdf
Rembe N (1980) Africa and the international law of the Sea: A Study of the Contribution of African States to the Third United Nations Conference on the Law of the Sea. Sijthoff, Germantown
Southern African Development Community (n.d.) Blue Economy Concept Paper for SADC. https://www.sadc.int/files/5414/7306/3256/SADC_Blue_Economy_Concept_Paper.pdf
Government of Seychelles (2017) Seychelles' Blue Economy-Strategic Policy Framework and Roadmap: Charting the Future (2018–2030). https://seymsp.com/wp-content/uploads/2018/05/CommonwealthSecretariat-12pp-RoadMap-Brochure.pdf
Tladi D (2007) Sustainable Development in international Law: An Analysis of Key Enviro-Economic Instruments. Pretoria University Law Press, Pretoria
Treaty Establishing the African Economic Community, 2158 UNTS 3
Tsamenyi M, Ali K (2012) African states and the law of the sea convention: have the benefits been realized? Ocean Yearbook 26:113–160
United Nations Economic Commission for Africa (2016) Africa's Blue Economy: A Policy Handbook. https://www.uneca.org/sites/default/files/PublicationFiles/blueeco-policy-handbook_en.pdf
United Nations (1987) Report of the World Commission on Environment and Development: Our Common Future. https://sustainabledevelopment.un.org/content/documents/5987our-common-future.pdf
United Nations (2015) The Millennium Development Goals Report. https://www.un.org/millenniumgoals/2015_MDG_Report/pdf/MDG%202015%20rev%20(July%201).pdf
United Nations (2015) Transforming our World: The 2030 Agenda for Sustainable Development. https://sustainabledevelopment.un.org/content/documents/21252030%20Agenda%20for%20Sustainable%20Development%20web.pdf
United Nations Charter, 1 UNTS 16
United Nations Convention on the Law of the Sea, 1833 UNTS 3
Vrancken P (2013) UNCLOS at 30: Africa at 50. In: Xue G, White A (eds) 30 Years of UNCLOS (1982–2012): Progress and Prospects. Beijing University Press, pp 96–114

Sustaining Deliverables of Development Interventions in Africa: Lessons from the Implementation of National Fadama Development Project III in South–Western, Nigeria

Lukman Abiodun Oyebode and Lukman Abiodun Akinbile

Abstract Sustaining the project development objectives of government sponsored interventions beyond their official span have been a recurring challenge in Africa. Post official span assessment of development interventions establishes poor/low service delivery of deliverables. A close consideration of the approach used (Community Driven Development (CDD) in the implementation of the National Fadama Development Project (NFDP) III vis a vis other components raises the hope that the project will not tow the path of other development interventions whose deliverables were short lived. The Project Development Objective of the NFDP III was to increase the income of its beneficiaries on a sustainable basis. In view of the above, the paper identified key variables from field experiences that served as indicators in sustaining and achieving the objectives of NFDP III. By extension these indicators are necessary for sustaining deliverables of development projects in Africa. The paper deployed qualitative research methodology (Focus Group Discussion (FGD) and Key Informant Interview (KII)) to elicit information from its sample subjects. The respondents engaged in off farm/value addition enterprises (cassava, oil palm and palm kernel) served as sample subjects. Field data identified pre empowerment (group mobilization process and group cohesion) and post empowerment (capacity building support service, advisory service and adherence to the sustainable component of project) indices as necessary conditions for sustaining deliverables of development interventions in Africa beyond their official span.

Keywords National Fadama Development Project III · Sustaining deliverables and development interventions

L. A. Oyebode (✉)
Department of Agriculture, College of Agriculture Food Science and Technology, Wesley University Ondo, P.M.B. 507, Ondo, Ondo State, Nigeria
e-mail: lukman.abiodun@wesleyuni.edu.ng

L. A. Akinbile
Department of Agricultural Extension and Rural Development, Faculty of Agriculture, University of Ibadan, P.O. Box 22133, Ibadan, Oyo State, Nigeria

W. Leal Filho et al. (eds.), *Sustainable Development in Africa*, World Sustainability Series,
https://doi.org/10.1007/978-3-030-74693-3_38

1 Background to the Study

In recent times, there has been a paradigm shift in defining development and the way it is approached. This paradigm shift resulted in the adoption of Sustainable Development (SD), a concept which was highlighted in the Brunt land Report of 1987 and at the Stockholm Conference of 1972 (UNECA 2012). The shift was borne out of the global links between environmental problems and socio-economic concerns (Hopwood et al. 2005), and also as a result of the fact that earlier conceptions and approaches to development appeared to focus largely on economics and physical wealth despite the multi-dimensional and complex nature of development (Bellu 2011). Focusing on economics and physical wealth in the pursuit of development has proven to be unsustainable not only from the environmental perspective but also from the economic, social and employment perspectives because it is resource-intensive, hence the need to pursue SD (ILO 2012).

Numerous national agricultural programmes have been executed in the past in Nigeria. With some ongoing, it is also expected that more will still be implemented. However, very little tangible result has emerged from the implementation of these programmes, as changes in Nigerian agricultural policies and programmes vary only in nomenclature and organizational network (Iwuchukwu and Igbokwe 2012). Past agricultural programmes emphasize almost same objectives like: to provide food for the inhabitants of the nation (food security and sufficiency), export excess to other countries and to provide rural dwellers and farmers with extension services, agricultural support and rural development services etc. Recent programmes have tried approaches and strategies aimed at improving the conditions of the rural poor, with some of these efforts still on course, many have since gone moribund. Despite all the policies and laudable programmes with challenging themes, Nigeria is yet to achieve food security.

Ikotun (2002) noted that in spite of colossal amount of money that have gone into implementing rural development programmes and the proliferation of rural development agencies, not much impact has been made. He went further to state that in spite of pious official pronouncements and declaration of intentions as contained in their development plans, at the end of each programme, rural life remains significantly unchanged. Ugoh and Ukpere (2009) appropriately appraised the trend of policies on poverty alleviation programmes in Nigeria, highlighting both the successes and weaknesses. The authors mentioned poor targeting mechanisms, failure to focus on the poor, programme inconsistencies, poor implementation and corruption as major lapses of past programmes and concluded that the entire programmes were unable to involve the beneficiaries in their planning and implementation stages hence threatening their sustainability.

The last two decades ushered in new approaches or mechanisms to tackling rural problems with the development of different theories and frameworks that guide the way development programmes are being carried out. More importantly, to ensure that the programmes achieve its objective both in the short and long term. The National

Fadama Development Project (NFDP) III made use of the Community Driven Development (CDD) approach. The CDD approach has become a key strategy used by both government and development assistance programmes (Gillespie 2004; Manuri and Rao 2004; Platteau 2004). The popularity of the approach has been propelled by its potential to develop programmes and projects that are sustainable and responsive to local priorities, empower local communities to manage and govern their own development programs and more effectively target poor and vulnerable groups (Dongier et al. 2001; Gillespie 2004).

Sustainability is seen as the continued existence of programme outcomes long after their establishment. With huge investment in intervention programmes, the non/low sustained acquisition of assets, non-sustenance of project deliverables and the abandoning of project sites and assets acquired in the course of the developmental programmes after the official span of developmental efforts calls for worry taking into cognizance the huge investment of these developmental efforts.

1.1 Brief History of National Fadama Development Projects in Nigeria

The word "Fadama" is a Hausa word for low–lying flood plains usually with easily accessible shallow groundwater. The use of Fadama resource will ensure all year round farming of some staple food and vegetables especially where Fadama resources are more pronounce. The desire to harness the verse potentials of Fadama in Nigeria culminated in the design of National Fadama Development Project I which was implemented between 1993 to 1999. While Fadama I focused mainly on crop production, downstream activities such as processing, preservation and marketing were largely neglected. The design did not take into cognizance the need for spatial integration of the markets (creation of physical and market infrastructure). It also failed to take into consideration other Fadama resource users such as livestock producers, fishing folks, pastoralists, hunters etc. The project did not also support post-harvest technology, which manifested in reduced crop prices and increased storage losses during the period (Momoh et al. 2008).

Some of the lessons learnt in Fadama I informed the birth of Fadama II which was implemented from 2004 to 2009 and targeted at dry season farming, agro-processing, preservation and marketing. It also allowed for acquisition of productive assets, provision of rural infrastructure to ensure the efficient transportation of farm output to markets as well as marketing activities. The project development objective was to sustainably increase the incomes of the beneficiaries by empowering communities to take charge of their own development agenda through Community Driven Development (CDD) approach of project implementation in a socially inclusive manner. Fadama II also provided special preferences to groups of youths, women (especially widows), physically challenged, the elderly and people with HIV/AIDs (AfDB 2003).

Table 1 Indicative Project Cost of NFDP III by Components (in US$ m).

Component of NFDP III	Indicative costs	% of total	Bank financing	% of Bank financing
1. Capacity Building, Local Govt., Communication and Information Support	87.50	19.44	28.20	11.28
2. Small-Scale Community-Owned Infrastructure	73.57	16.67	66.07	27.00
3. Advisory Services and Input Support	39.50	8.78	24.60	9.84
4. Support to ADPs, Sponsored Research, and On-Farm Demonstration	37.43	8.11	6.43	2.20
5. Asset Acquisition for Individual FUGs/EIGs	150.0	33.33	105.00	42.00
6. Project Management, Monitoring and Evaluation	59.30	13.07	17.00	6.60
Project Preparation Facility	2.70	0.60	2.70	1.08
Total	**450.0**	**100.0%**	**250.0**	**100.0%**

Source NFDP III Project Implementation Manual (2008)

Fadama III project spanned from 2008 to 2013 was a follow-up to the Fadama II project which was asserted to have impacted the lives of rural farmers, raising their incomes by 63% (National Fadama Development Project 2009). It has a development objective of increasing the income of the users, hopes to reduce the prevalence of rural poverty and increase food security. The project like Fadama II also made use of the CDD approach, which places beneficiaries in the driver's seat thereby enabling them have primary stake in deciding what is best for them, thus fulfilling the bottom-top approach. With the local community members under the umbrella of Fadama Community Associations (FCAs) and Fadama User Groups (FUGs) oversee the design and implementation of the project and are empowered through skills and capacity building to improve their livelihoods by increasing income generating activities.

1.2 Description of National Fadama Development Project III According to Project Cost

Table 1 reveals that the total project cost is US$450 m, of which, Bank will finance with an IDA Credit of US$250 m, Borrower will contribute US$200 m (Federal

Government US$23 m, State Government US$77 m, State Government US$77 m, Local Governments US$40 m, and the Communities, including the private sector and civil society, US$60 m). Twenty percent of beneficiary contributions will be in kind (materials and/or labour). GEF aims to provide a grant of US$7 m under the Bank-led SIP umbrella (to be presented to the Board separately in December 2008).

1.3 Operational Definition of Terms as it Relates to NFDP III

Capacity Building Support Service: This component supports measures to build the capacity of Fadama Community Associations (FCAs) to carry out participatory planning, to implement, operate and maintain projects, to administer funds and to negotiate conflicts.

Advisory Service: An advisory service is one that provides Fadama users with advice on the how, when and why of an agro-technology with its associated input and markets. It is designed to guide and persuade Fadama beneficiaries to adopt more productive and profitable practices in their income generation activities using educational means.

Fadama Users' Equity Fund (FUEF): Is a saving scheme designed to ensure the sustainability of the investment activity, it is a savings scheme meant specifically for the replication of assets.

Fadama Community Associations (FCAs): This is regarded as an apex organization of Fadama User groups. It comprises a set of Fadama User Groups (FUGs)/Economic Interest Group (EIGs).

Fadama User Groups (FUGs)/Economic Interest Group (EIGs): This comprise a set of individuals who are engaged in the same enterprise activity under the NFDP III.

2 Methodology

2.1 The Study Area

The study was carried out in the South-Western, Nigeria. This area is one of the six geo-political zones of the country which comprises six states namely: Lagos,

Ogun, Oyo, Osun, Ekiti, and Ondo. The zone lies between latitude 60 and 90 N and longitudes 20E and 70E occupying an area of 76,851 Sq. kms. The 2006 census puts the population of the zone at 27,581,993. National Population Commission (NPC 2006). South-Western geo political zone is bounded in the south by the Atlantic Ocean, in the west by republic of Benin, in the East by Edo and Kogi states. From the South of the area towards the North, the ecology is characterized by fresh water swamp, tropical rainforest and derived savannah. The climate is predominantly humid with rainfall ranging between 2000 m in the coastal area to 1200 mm in the extreme northern parts. The rainfall is distributed over April to October followed by a short break between July and early August. The rainfall is bimodal with peaks in June and September. From November to February there is a dry season period characterized by harmattan, which the North Eastern trade winds brings from the sahara region, The mean maximum temperature ranges from 28° to 34 °C while the mean minimum temperature ranges from 20 °C to 24 °C. Agriculture is the main source of livelihood of the inhabitants of the zone with not less than 65% of the people depending on agriculture as their main source of livelihood (Oladeji and Thomas 2010). Common tree crops that thrive in this zone include cocoa, oil palm and cashew while arable crops such as yam, cassava, maize and rice also thrive well in this zone. Leafy and fruit vegetables such as amaranthus, jute malo, fluted pumpkin, tomato, pepper, Okra, egg-plant melon etc. are also crops of economic value in the area.

2.2 Study Population

The population of this study comprised of all processors (cassava, oil palm and palm kernel) that participated in the NFDP III in the study area.

2.3 Sampling Procedure

Multistage sampling procedure was employed in the sampling of respondents for the Qualitative survey. The first stage involved the random sampling of two states out of the six states in South-Western, Nigeria, The states sampled were Osun and Ogun. The second stage involved the purposive sampling of Fadama User Groups (cassava processing, palm oil processing and palm kernel processing) that has the highest combination of at least two of these enterprises in each of their Fadama Community Associations' (FCAs). The third stage involved the random sampling of 50% of the FCAs from the FCAs earlier purposively sampled. The fourth stage involved the random sampling of 50% of the FUGs for each of the enterprises under the FCAs that were initially sampled. The final stage involved the random sampling of group members from their respective FUGs for each of the enterprises (cassava, Oil palm and Palm kernel) for Focus Group Discussion. This was however disaggregated along gender. A total of 12 Key Informant Interviews (KII) was also conducted with identified stakeholders, four each for the respective enterprises with two conducted for each of the sampled states.

2.4 *Data Collection and Reporting*

Data were collected using an interview protocol for the key informant interviews and discussion guide for the focus group discussions. The instruments were back-stopped with audio-tape recording and notes taken by researchers which were later transcribed with particular attention to important quotes from respondents. Some of these quotes were used to support findings in this paper. The researchers also relied on the testimonies of the group members and the key informants who are primary stakeholders of the project. The data collected was processed and a trend, pattern and chain of activity was established. The questions covered group mobilization, group cohesion, effectiveness and efficiency of capacity building and advisory service and processors adherence to the sustainability component (FUEF) of the project. Their assertions served as first-hand information. Field visits was also relied on to make inferences after making physical observation at the centres where processing and livelihood activities are taking placed. Therefore mentioned were relied on to develop a schematic diagram to convey the indices needed to ensure sustainability of deliverables of development interventions (Fig. 1).

3 Literature Review

3.1 *Definition of Sustainability*

Over the years, the definition of sustainability in development literature has varied widely and broadened in scope. The concept arose in response to economic growth models that characterized development approaches over the last half century. It was eventually recognized that such models did not adequately address social inequalities and led to environmental degradation. The International Fund for Agricultural Development (IFAD) Strategic Framework 2007–2010 (IFAD 2007j) gave the following definition of sustainability:

Ensuring that the institutions supported through projects and the benefits realized are maintained and continue after the end of the project....

IFAD's Office of Evaluation adds to this definition by considering resource flows. It acknowledges that assessment of sustainability entails determining "whether the results of the project will be sustained in the medium or even longer term without continued external assistance". It further expands on the concept of programme

Fig. 1 The researcher with a stakeholder during key Informant Interview session in Abeokuta North Local Government Area of Ogun State

sustainability by distinguishing among several factors that either contribute to or detract from the long-term impact of IFAD interventions (IFAD, 2006a):

- Political sustainability – government commitment, an enabling policy environment, stakeholder interests, strong lobby groups and political influence/pressure;
- Social sustainability – social support and acceptability, community commitment, social cohesion; ownership – whether or not communities, local government and households accept and own the outcomes of the project in ways that are sustainable;
- Institutional sustainability – institutional support, policy implementation, staffing, recurrent budgets;
- Economic and financial sustainability – resilience to economic shocks, financial viability, reduced household vulnerability and increased capacity to cope with risk/shocks;
- Technical sustainability – technical soundness, appropriate solutions, technical training for operations and maintenance, access to and cost of spare parts and repairs;
- Environmental sustainability – projects' positive/negative contributions to soil and water preservation and management, resilience to external environmental shocks.

3.2 An Overview of Sustaining Development Projects in Africa and Asia

Comparing the huge investments sunk into developmental interventions in Africa and the deliverables it confers on a long term, it is observed that it often falls short of the expectations of the sponsors of such interventions and other stakeholders engaged in the. Issues of sustainability in development interventions became very important to donors in the 1980s (Scoons 2007). Sustainability became a very important criterion for project evaluation and represented the main purpose for donors providing aid in achieving or promoting development (Komalawati 2008). Sustainability implies the ability of a project to continue in operation to achieve its purpose for the longest time possible after the donor withdraws support Sally et al. (2013) and Martha (2013). In the context of this research, it represents the capacity of project beneficiaries to continue operating their farm activities/businesses and securing livelihoods after the financial support phase from donors has ended. A project is sustainable if it captured sustainability strategies from the design phase.

However, sustainability could be influenced by how the project is managed, monitored and evaluated IFAD (2009) and UNECO (2016). The UNESCO Science Report for 2016 reveals that many African countries have not waited for the adoption of the Sustainable Development Goals to embark on a more sustainable development path. Many of the long-term planning 'Vision' documents to 2020 or 2030 adopted by African countries over the past decade have identified sustainable development as being a pillar of their national development strategy, along with more inclusive growth and better governance. Regional economic communities are also conscious that developing an African Economic Community by 2028 will need to go hand in hand with greater scientific integration and more sustainable development. For instance, the Vision 2020 document adopted by the Economic Community of West African States (ECOWAS) in 2011 aspires to "create a borderless, prosperous and cohesive region built on good governance, where people have the capacity to access and harness its enormous resources through the creation of opportunities for sustainable development" UNECO (2016).

A report by Carbonnier et al. (2010) revealed that there is a difference between Africa and Asia in their transition from extensive to intensive agriculture. As shown by data on cereal yield and fertilizer consumption per hectare, South East Asian countries have made greater progress than sub-Saharan countries in modernizing their agricultural sector, thereby substantially increasing productivity per hectare. This has led to a higher GDP in Asian than African countries. According to them, African countries must be granted the policy space to adopt industrialization policies that are adapted to the reality of the twenty-first century globalized economy including effective measures to temporarily protect its infant industry (Harrison and Andres (2010).

On the other hand, income inequality tends to be higher in some sub-Saharan African countries than in South East Asia. While the high level of income concentration in many sub-Saharan countries may appear as counterintuitive according to

Mkandawire (2002), being relatively poor and with a large share of the population concentrated in agriculture, Africa should be characterized by a relatively lower degree of social inequality than richer, more industrialized countries. Since sustainability seeks to actively involve the people in the environment in order to attain sustainable development, the existence of poverty and environmental degradation pushes sustainability to fail. This is probably because sustainable development works with all the tools and human capital that can be found within a given project environment. This explains why development projects in Africa are mostly less sustainable than development projects in Asia (Carbonnier et al. 2010).

3.3 Guidelines for ensuring project sustainability

While this may not be a quick fix the template can serve as a guide to ensuring project sustainability, key variables include:

a) Effective linkages between project components: A key element in sustainable project outcomes is a design based on a holistic consideration of livelihood systems, needs and opportunities. Narrow, sector-focused interventions can be a risk to sustainability in various ways. For example, gains made in household food security can easily be lost due to disease outbreaks or adult mortality. Similarly, improved economic status can be compromised by shocks – natural or man- made – that deplete or destroy household and community assets. In short, if households and communities lack resilience in the face of natural, social or economic shocks, project impacts can quickly be lost.
b) Community participation: While many development programmes include participatory measures in project design, programmes that obtain sustainable results take the commitment seriously and put it into practice with sound concepts, focused dedication, careful monitoring, and appropriate adaptive measures when necessary. Successful programmes use bottom-up planning to determine priorities and then accurately reflect community needs in project design. Designs with promising sustainability results include plans for communities to manage both external and internal resources, which in turn promotes a greater sense of ownership.
c) Flexible design: Flexibility in design expands the ability of a project to be demand-driven, with home-grown initiatives that take advantage of local knowledge and practices and promote community ownership and sustainability. Moreover, flexibility facilitates quick adaptation to changing circumstances, as well as response to new opportunities. The flexible lending mechanisms currently in use by IFAD have improved the quality of implementation and have helped strengthen institutional capacity.
d) Institutional analysis: The design phase must undertake a comprehensive assessment of a project's institutional context. The ability and motivation of the country policy and institutional framework to promote sustainable

results greatly impact project success. Achievement occurs in projects with objectives linked to existing national policies and programmes, government strategies geared towards poverty reduction and supportive of participatory design, and activities mainstreamed into existing agencies. Determining the strengths of existing village institutions and other community organizations prior to 'Stakeholder and institutional analysis' simply means investigating what types of efforts are supported by Community Based Organizations (CBOs), Non-Governmental Organizations or the government.

e) Longer project cycle: Participatory problem analysis is an essential component in designing projects capable of achieving long-term development impact – and it requires a lengthy project planning cycle. Moreover, institutional strengthening and capacity-building, NRM practices aimed at achieving environmental sustainability and efforts to achieve structural equity are likely to require more than one project life cycle to demonstrate meaningful impact.

f) Risk assessment: Enhancing the risk management capacity of project participants is essential to long-term sustainability. This entails identifying risks that threaten the stream of benefits to households and communities and including risk mitigation in project design. Risk mitigation methods must be credible and implementable, particularly regarding responsiveness to social risks such as exclusion of key groups or lack of socio-political support by authorities or communities. Emergency savings accounts and weather and livestock insurance are methods of risk mitigation currently used in IFAD-supported projects, and which are likely to contribute to the sustainability of project activities and objectives. These types of analyses will become more important to project design as climate changes become more significant.

g) Consideration of environmental appropriateness: The establishment of environmentally productive systems is critical to sustainable income and asset streams. Beneficial systemic changes in the environment can occur when project design emphasizes the linkage between poverty reduction and environmental protection. Project design that focuses on environmental fortification (local and global) and incorporates measures to mainstream environmental considerations into economic growth objectives is more likely to reap long-term benefits. Building partnerships with functioning, sustainable institutions that are involved in collective action is vital as well. Moreover, successful programmes include environmental risk mitigation and methods of measuring environmental impact over several project cycles.

h) Consideration of environmental appropriateness: The establishment of environmentally productive systems is critical to sustainable income and asset streams. Beneficial systemic changes in the environment can occur when project design emphasizes the linkage between poverty reduction and environmental protection. Project design that focuses on environmental fortification (local and global) and incorporates measures to mainstream environmental considerations into economic growth objectives is more likely to reap long-term benefits. Building partnerships with functioning, sustainable institutions that are involved in collective action is vital as well. Moreover, successful programmes include environmental risk mitigation and methods of measuring environmental impact over several project cycles.

i) Building on existing assets and knowledge: Community participation in and ownership of initiatives – and thus their sustainability – is much greater when implementing agencies draw on existing practices and engage established community institutions, as opposed to creating new structures and mechanisms. By building on existing community assets and knowledge, development agencies can promote positive community attitudes towards collaboration and collective decision-making, as well as support social cohesion by strengthening relationships between internal and external organizations.

j) Consistent and objective monitoring and evaluation: A consistent and objective approach to project monitoring and evaluation is essential in achieving sustainability. Without reliable information on both the intended and unintended changes resulting from project interventions, project managers will be unable to make adjustments in the design and implementation of activities in response to contextual factors. Nor will they be in a position to formulate or carry out appropriate exit strategies – a critical component for ensuring sustainability. Project monitoring and evaluation plans should also include a set of standard sustainability indicators against which progress towards intended goals can be measured (IFAD 2007b).

4 Results and Discussion

4.1 Sustaining Deliverables of Development Interventions in Africa: Lessons from the National Fadama Development Project (NFDP) III

This section looks at the salient lessons learnt from field experience in the course of the research. It is hoped that these revealed truths from field experience will further guide and guard the operations and implementation off intervention projects. It is hoped that once these lessons are considered, the rate of sustaining the deliverables of interventions will increase far beyond the official span of development intervention. To further understand and appropriately convey the experience and thoughts of the beneficiaries the lessons will be discussed under identified sub headings.

4.1.1 Pre Empowerment Indices for Sustaining Deliverables of Development Interventions

This relates to necessary conditions that must be considered before beneficiaries are selected for the intervention. It is important to note that stakeholders engaged in the selection of beneficiaries of the intervention need to put necessary activities and actions in place to ensure that only the right set of people are picked as beneficiaries. With this in view, they will dispel the "elite capture syndrome" that has been a sharp practice bedeviling the sustainability of both government and privately sponsored interventions in Nigeria and by extension Africa.

4.1.2 Beneficiary Mobilization Process/Group Cohesion

This stage is perhaps the bed rock of the intervention. It is perhaps the most critical of all other activities that are to be carried out by the stakeholders. In this stage, the beneficiaries of the proposed projects are identified, profiled and eventually debriefed before they are eventually selected as beneficiaries of the proposed intervention. Apart from fear of the infiltration of fake/political beneficiaries otherwise referred to as "elite capture" there is fear that core beneficiaries may be few. Hence, they will eventually leave the future of the intervention in inexperienced hands. In view of the above, when beneficiaries are to be mobilized there is need to look out for and identify those that are already engaged in the activity/ line of business that the intervention is expected to offer. With this, the intervention will only come as an up scaling of their direct income generating activity/ engagement. Hence, proposed beneficiaries will not be seen as novices in the scheme of things and will better appreciate and will be better positioned to sustain the project development objectives of such interventions after its official span.

Apart from these, there is need to identify and isolate groups that are cohesive. The use of groups has been identified as a key component in the discharge of deliverables of interventions. Indices to identify as key precursors depicting cohesion includes the following but are not limited to the group bearing the following characteristics group norm (consensus of what is an acceptable behavior/group rule), group binding (desire of members to stay in group or stick together), goal attainment (achieving what the group set out as goals e.g. output quotas, purchase of equipment etc.), involvement of members in group activities, involvement of members in decision making, spontaneous cooperation (readiness of group members to assist each other when the need arises) and pressure to conform (ability of group to make members conform to established group rules e.g. sanctioning of erring members). It is observed that for prospective beneficiaries to share these characteristics they must have existed as a group long enough before their selection as beneficiaries of the intervention at hand. With cohesive groups used to achieve the project development objectives of interventions, beneficiaries will remain in perfect frame to withstand all shocks and threats aimed at hindering the sustainability of the intervention which may arise after the official span of the intervention.

4.1.3 Capacity Building Support Service for Beneficiaries before Empowerment

As earlier explained in the operational definition of terms, activities under this section covers training of the beneficiaries on a set of activities that are not directly related to their enterprise activities but indirectly has a role in ensuring that all activities on the external work in tandem with the project development objectives of the intervention. It is hoped that once there is effectiveness in training, beneficiaries of this intervention are all equipped for the smooth running of their enterprise. Among the services that can be rendered includes the following but are not limited to training on group management, training on book/record keeping, training on credit/savings management (community capitalization), training on maintaining viable investment in asset acquisition,, training on building a culture of social inclusion, training on the preparation of Local Development Plans (LDPs), training on adopting friendly environmental practices and training on enterprise expansion.

The following are some responses from stakeholders during the qualitative survey.

"We observed that after we were mobilized, some group members were never engaged in the primary occupation (cassava processing) we were empowered for. We observed that after they were empowerment with productive assets, they sublet /rent/ lease these assets to private individuals who later make returns/payments to them".

Key Informant Interview (KII) with a Male Leader of a Cassava Processor Group at Oyan Sadereku Community, in Odo-Otin Local Area of Osun State.

"Because we were not together as members of the same group before empowerment, it was very difficult for us to relate together and understand each other. This really led to friction and crisis. Later some members pulled out of the group and this affected our numerical strength, progress and ability to sustain the goals of the intervention".

A female discussant during a Focus Group Discussion (FGD) session at Ilewo orile at Abeokuta North, Local government area of Ogun State.

"We appreciate the training that was given to us at the inception of the intervention. We found it helpful because when we were eventually empowered there were threats of discontinuation coming from so many angles, but we were able to resist them. All these were possible as a result of the training that has prepared us ahead of time".

Key Informant Interview of a male executive member of an oil palm processing group at Erefe Community, Ife –South, Local Government Area of Osun State.

The use of groups in delivering extension services has been considered a veritable tool in sustaining development effort; hence its use has been deployed by recent agricultural interventions. However, the cohesion of this groups is considered more vital than its mere existence. Group cohesion is often viewed from an affective perspective as interpersonal attraction among members of a group. However, cohesion can also be considered as "attraction to collectivity" as opposed to an attraction to the individuals who make up that grouping. Thus, cohesion moves beyond simple interpersonal liking (Ofuoku et al. 2008). Cohesion is regarded as the degree to which members of a group desire to remain in the group, that is, how closely do the members interact or to what extent are the resultant forces acting on the members to remain in the group. Cohesiveness is central to groups as it is considered vital in a group decision making, goal attainment and member satisfaction (Ofuoku and Urang 2009). Hence group cohesiveness is vital to sustaining programme objectives.

In recent years, capacity building has been recognized as central to the quest for sustainable development and has emerged as a major focus for development institutions. It has been increasingly recognized that in addition to productivity, technological developments, infrastructure and good economic policies, human resource development plays an important role in poverty alleviation (Africa Development Bank 2003). Capacity building is a holistic enterprise encompassing a multitude of activities, it means building abilities, relationships and values that will enable organizations, groups and individuals to improve their performance and achieve their development objectives. It includes strengthening the processes, systems and rules that influence collective, individual behavior and performance in all development endeavors. It also entails enhancing people's technical ability and willingness to play new developmental roles and adapt to new demands and situations (Fig. 2).

Fig. 2 Processors in a NFDP III cassava processing hub at Iwo Local Government Area of Osun State

4.2 *Post Empowerment Indices for Sustaining the Deliverables of Development Interventions*

4.2.1 Advisory Service for Beneficiaries after Empowerment

These services are directly related to the chain and sequence of events engaged in during processing. It is expected that after beneficiaries would have been empowered and are already in the line of business, challenges would set in. However, with advisory services rendered by development interventions during the official span things can be placed in proper perspectives. Challenges that were not envisaged before empowerment and have risen will be carefully addressed in a bid to sustain their livelihood activities. The following are range of services that were accessed by the processors under NDFD III, training on enterprise/business management, training on enhanced marketing of products, training on efficient use of processing equipment for quality output, training on deploying economics of scale to purchasing of raw materials for processing, training on equipment maintenance, training on equipment installation, training on enhanced processing of produce/value addition and training on effective and efficient packaging of produce. These services can be adapted and up scaled in relation to the development intervention that is been carried out.

4.2.2 Adherence to the Sustainable Component of the Development Intervention

In the case of NFDP III the sustainability component was Fadama Users' Equity Fund (FUEF). Under this sustainability component design the FUGs will set aside funds as indicated above which will constitute a capitalization/revolving fund called the FUEF, which will be limited to only the economically taken active FUGs, will operate on guidelines outlined in its Operational Manual and integrated into the PIM. The facilitators will train the FUGs in financial literacy and will help in linking them with markets, where they can sell their produce/products at prevailing market prices, while at the same time ensuring that the agreed funds to be saved in the FUEF are deposited in the accounts as required.The FUEFs (as second generation funds) will be invested by the FUGs at the community level according to priorities and mechanism defined by them as the sole owners of the funds. The recovered amounts will thus remain as financial capital circulating in the community, or may take the form of physical capital depending on the investment decision made by the FUGs.

This will constitute the basis for the development of a sustainable savings and loan scheme. As good as the design is, most beneficiaries of development interventions do not adhere to the stipulated guidelines and the design of running the sustainability scheme/component. Under the NFDP III some of the component under the operations of FUEF includes payment of the replacement value of the productive asset (exact

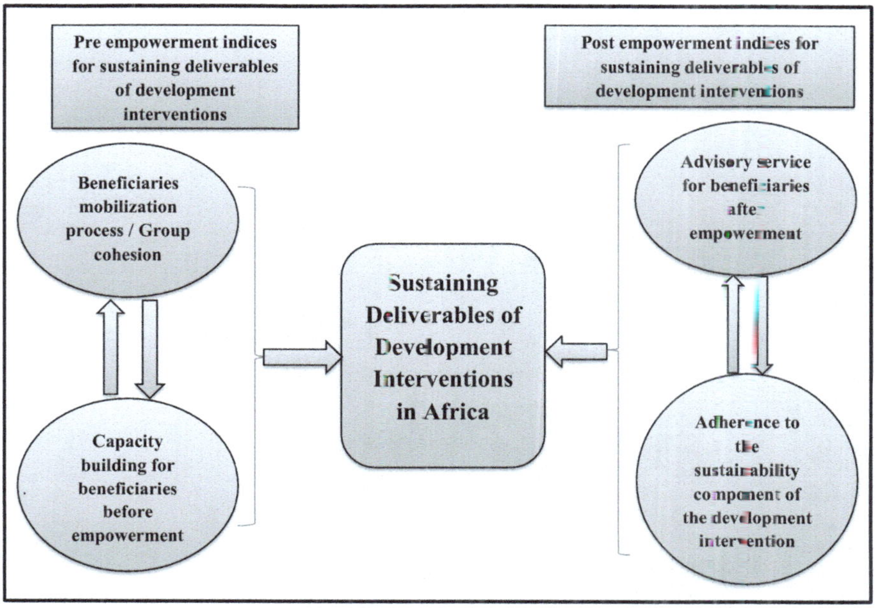

Fig. 3 Schematic diagram showing the interrelationship between pre and post empowerment indices for sustaining deliverables of development interventions in Africa

amount), regular payment of the replacement value of productive assets, meeting the timing of repayment agreed on (monthly/quarterly) and periodic reporting of the FUEF account to the FCA. It is observed that no two development interventions are alike. It is hoped that development interventions will develop and design practical sustainability scheme/component that will take enterprise and locality specific, it must also be cost effective, cost reflective and cost efficient to run for to enable the beneficiaries adhere to its operations all things been equal.

The schematic representation in Fig. 3 conveys the thoughts of stakeholders whose thoughts, opinions and experiences were sought in relation to how deliverables of development interventions can be sustained. This was also carried out alongside physical observation during field visits to the processing hubs of the beneficiaries (Fig. 4).

Fig. 4 Oil palm processing on going at one of the NFDP III processing hub at Irepodun local government area of Osun State

5 Conclusion

Sustaining the deliverables of development interventions beyond the official span of such interventions is indeed a huge task. Carefully looking at the implementation of the National Fadama Developmet project III vis a vis its components. The study has identified that both pre empowerment indices which include group mobilization process/group cohesion and capacity building and post empowerment indices which includes advisory service and adherence to the sustainable component of project indices are veritable indices that are necessary conditions for sustaining deliverables of development interventions in Africa. The study clearly identifies that the dynamics of recruiting beneficiaries is the main bedrock for ensuring the success of development interventions, this was carefully expressed under group mobilization process. In view of the above adequate time and attention should be given to these indices during project development and management. The study however identifies that development intervention are individually unique. Hence, indices beyond therefore mentioned can also be coopted to ensure the sustainability of the intervention considering local resources (human, material and financial) available at the disposal of the beneficiaries of such projects.

References

African Development Bank (AfDB) (2003) Appraisal report on community based agriculture and rural development project of the federal republic of Nigeria (2003). www.Afdb.org/project./NG-2003-En-Adf-Bd-Wp-NigeriaPDF. Accessed Nov 2011

Bellu LG (2011) Development and development paradigms: a (reasoned) review of prevailing visions, FAO, Italy

Carbonnier G, Chakraborty P, Mulle E, Cartografare I (2010) Asian and African development trajectories revisiting facts and figures. In: Proceedings of the Framework of the Traverse Conference Organized by SDC, Bern, Switzerland, 16 June 2010

Dongier P, Van Domelen, J, Ostrom E, Rizvi A, Wakeman W, Bebbington A, Alkire S, Esmail T, Polski M (2001) Community Driven Development. In: Klugman J (ed) A sourcebook for poverty reduction strategies, vol 1, no 1, pp 231–234. World Bank, Washington, DC

Fadama (2009) Third National Fadama Development Project (Fadama III) Advisory and Input Support. Manual No. 5:1– 29

Gillespie S (2004) Scaling up community-driven development: a synthesis of experience. Food and nutrition division discussion paper no 181. International Food Policy research Institute, Washington DC

Harrison A, Andres R (2010) From hard to soft industrial policies in developing countries, 27 June 2010. https://www.voxeu.org/index.php?q=node/5236. Accessed 16 Nov 2010

Hopwood B, Mellor M, O'Brien G (2005) Sustainable develoment: mapping different approaches. Sustain Dev 13:38–52. https://doi.org/10.1002/sd.244

Komalawati D (2008) Participatory and project sustainability: participatory integrated development in rain-fednareas (PIDRA) project in East Java-Indonesia. Master's Thesis, Massey University, Palmerstone North, New Zealand, p 96

IFAD (2006a) ARRI issues note: sustainability. Internal Document, Rome

IFAD (2007j) IFAD Strategic Framework 2007–2010, Rome

IFAD (2009) Sustainability of rural development projects, best practices and lesson learned by IFAD in Asia, North Eastern region community resource management project for upland areas, TANGO International, Tucson, AZ USA 2009, p 104

Ikotun A (2002) Strategies for promoting integrated rural development in Nigeria: theory and practice. Matram Press, Badagry, p 26

International Labor Organization (2012) Working towards sustainable development. www.ilo.org/publns

Iwuchukwu JC, Igbokwe EM (2012) Lessons from agricultural policies and programmes in Nigeria. J Law Policy Global 5:12–17

Manuri G, Rao V (2004) Community –based and driven development: a critical review. World Bank Res Observ 19(1):1–39

Martha C (2013) The sustainability of donor funded projects in Malawi. MCSER-CEMAS-Sapienza University of Rome, Rome, Italy, p 86

Mkandawire T (2002) The terrible toll of post-colonial rebel movements in Africa: towards an explanation of the violence against the peasantry. J Mod Afr Stud 2002(40):181–215

Momoh S, Fakoya EO, Gbadomi L, Okadapo T, Salawu RA, Onasanya IA (2008) Beneficiary Impact assessment of second Fadama development project in Ogun state – a technical report

Ofuoku AU, Enaikele M, Nnodim AU (2008) Cohesiveness of fish farmers group in Southern Nigeria. Agric Biol J 3(4):16–21

Ofuoku AU, Urang E (2009) Effecets of cohesion on loan repayment in farmers cooperative societies in delta state, Nigeria. Int J Sociol Anthroplol 1(4):070–076

Oladeji JO, Thomas KA (2010) Assessment of agricultural practical year training programmes of Nigeria universities in south western states. Niger J Rural Extension Dev 3:28–34

Platteau JP (2004) Monitoring elite capture in community–driven development. Dev Change 35(2):223–246

Sally Z, Gaskin SJ, Folifac F, Kometa SS (2013) The effect of urbanization on community-managed water supply: Case study of Buea Cameroon. Commun Dev J 2013(49):524–540
Scoons I (2007) Sustainability. Dev Pract 2007(17):589
Ugoh CC, Ukpere WI (2009) Appraising the trend of policy on poverty alleviation programmes in Nigeria with emphasis on the national poverty eradication programme (NAPEP) in African. J Bus Manage S Afr 3(12):847–854
United Nations Economic Commission for Africa (2012) Progress towards sustainable development report in Africa, UNECA, Addis Ababa
United Nations Educational, Scientific and Cultural Organization (UNESCO) (2016) It's in the numbers—how sustainability will support Africa's long-term growth and development. In: Proceedings of the world economic forum, Davos-Klosters, Switzerland, 20–23 January 2016. https://www.weforum.org/agenda/2019/09/why-sustainability-is-critical-for-the-long-termgrowth-and-development-of-africa/. Accessed 15 Nov 2019
National Population Commission (2006) National Population Census Figures. NPC Bulletin 4:12
National Fadama Development Project III. Project Implementation Manual (2009) pp 1–175

Power Taken, Power Given: Lessons for Collaborative Governance Praxis and Co-learning from Land Reform and Co-management in the Lekgalameetse Nature Reserve (South Africa)

Sharon Pollard, Charles Chikunda, Thabang Mohale, Lilian Goredema, and Karen Kotschy

Abstract Protected areas can meaningfully contribute to sustainable development in Africa and elsewhere – but only if well governed and managed. Among various models of shared governance of protected areas, co-management is a potentially innovative approach for addressing past injustices and meeting the Sustainable Development Goals through socio-economic redress, upliftment, and conservation, although there is mixed evidence of success. In establishing new governance arrangements, shifts in power between the state and new rightsholders can undermine good governance unless power dynamics are explicitly ackowledged. A case-study approach is used to describe and reflect on the transformative co-learning journey followed by land-reform beneficiaries and government partners at Lekgalameetse Nature Reserve in South Africa. Co-management was an entirely new practice unknown to both parties, and so *co-learning* was key. Recognising the need to support learning and adaptability, a strongly systemic, social learning approach was followed. The co-management support process covered all the aspects of *governance quality*. *Equitable governance* was promoted by enhancing the legitimacy and voice of the land-owner communities. *Fairness and rights* were promoted through legal support and co-developed principles used to guide the governance of benefits and park development; both highly contested issues that revealed hidden agendas. The work on equity laid the foundation for more effective governance, through a collaboratively developed vision and detailed understanding of roles and responsibilities needed to facilitate better performance and accountability. However, guidance on dealing with hidden agendas underlying power relations was identified as a critical gap. In concluding, the case study points to the importance of paying attention to power dynamics in development work. Secondly, the work emphasises that, within a land-reform context, co-management requires a shift from a contractual process to one that is socially and politically embedded and that is transformative. As such it has an important role to

S. Pollard (✉) · C. Chikunda · T. Mohale · L. Goredema · K. Kotschy
Association for Water and Rural Development, PO Box 1919, Hoedspruit 1380, Limpopo, South Africa
e-mail: sharon@award.org.za

W. Leal Filho et al. (eds.), *Sustainable Development in Africa*, World Sustainability Series,
https://doi.org/10.1007/978-3-030-74693-3_39

play in developing agency, dignity and respect whilst also contributing to people's livelihoods. Doing this means committing resources and capacities to the co-learning journey. Without appropriate support for both parties, and an orientation towards co-learning for new, largely uncharted governance arrangements, the co-management endeavour, and hence long-term sustainable development, will remain deeply fragile.

Keywords Protected areas · Co-management · Collaborative governance · Praxis · Social learning · Transfer of power · Transparency · Inclusive institutions

1 Introduction

Protected areas can meaningfully contribute to sustainable development in Africa and elsewhere through the provision of physical, recreational and spiritual benefits – but only if well governed and managed (CBD 2018). In establishing new and emerging governance arrangements, the need to address historical injustices is central since the establishment of protected areas throughout the twentieth century was associated with the forced removal and dislocation of indigenous communities from their land (Borrini-Feyerabend 2004; Jones and Murphree 2013). The impacts on communities were pervasive through the loss of access to resources, livelihood insecurity, increased risk of human-wildlife conflict, disruption of community cohesion and psychological suffering (Kepe et al. 2005; Cundill et al. 2017). The interests of indigenous communities were marginalised over those of the state or other powerful role players, leading to profound power imbalances, and ongoing clashes between these interests are often "thorny and eminently political" (Borrini-Feyerabend et al. 2004).

A century on, global moves towards decentralization of protected area management rights to communities and the public are seen as a way to address these wrongs, increase legitimacy, reduce costs and conflict and ensure greater long-term social-environmental sustainability (Borrini-Feyerabend 2004; Magome et al. 2012; Kothari et al. 2013). Among various models of shared governance of protected areas (Borrini-Feyerabend et al. 2014), co-management is a potentially innovative approach for addressing past injustices through socio-economic redress, upliftment, and conservation - although there is mixed evidence of success (Cundill et al. 2013; De Pourcq 2015; Chevallier 2016; Thondhlana et al. 2016). In establishing new governance arrangements, shifts in power between the state and new rightsholders – and within both of these groups - can undermine good governance unless attention is explicitly paid to supporting it. Various frameworks, based on extensive field experience, have been developed to support such processes (e.g. Barrow and Murphree 1998, Borrini-Feyerabend 2004; Cousins 2007) but applying these in practice can be immensely challenging since navigating the realities of context, existing power relations, resources and hidden agendas requires flexibility, adaptability and experience.

This chapter seeks to contribute to a growing body of knowledge regarding good governance and co-management of protected areas based on experiences from an unfolding land reform-co-management process in South Africa. Guided by Ostrom's seminal work in 1990, various definitions of good governance have been developed. Those sensitive to natural resources include the IUCN framework (Borrini-Feyerabend et al. 2014) which introduces the concepts of governance quality encompassing equity and effectiveness, and governance vitality, determined by factors such as integration, adaptability, wisdom, innovation and empowerment. These aspects were explored through an unfolding five-year facilitated process based on a systemic, social learning approach. In working out what governance means in practice – what it entails on a daily basis - multiple tensions were surfaced as the realities of the transference of power became apparent. We assert therefore that co-management is a transformative, co-learning journey where identities of authority and power are redefined. This requires a more careful, rigorous and yet flexible approach than those currently conceptualised and deployed in South Africa. The chapter explores what this framing of co-management means for the implementation of collaborative governance as part of sustainable development initiatives and biodiversity conservation in Africa.

Need to indicate your methodological approach to be followed, and suggest it is covered at that stage. It as requested by a reviewer and has not been attended to. It can be linked to the a case study, which you explore in a comprehensive way, but need to state how this was approached and how you did it.

A case study approach is used to describe and reflect on the way in which the transformative co-learning journey unfolded in this particular context. The combination of description and interpretation presented here is a product of the action learning approach followed by AWARD. A learning-centric monitoring and evaluation system prompted and captured regular reflections over the five years, in the form of post-fieldwork reports, monthly reflection reports and less frequent reflection events, which helped to make sense of what was happening 'on the ground' in terms of guiding theoretical frameworks (resilience, social-ecological systems, social learning and collective action). Three formative evaluation case studies were also conducted on the work at Lekgalameetse in order to guide decision-making and allocation of resources. These documents were used as primary data for the case study presented here.

After a short overview of co-management in South Africa and in Lekgalameetse Nature Reserve, the site of this work, co-management is framed within the context of transformation and co-learning before the governance learning journey is explored. The chapter concludes by reflecting on the process and providing recommendations for co-management and the implications of the work for sustainable development in South Africa and beyond.

2 Co-management in the South African Context

In South Africa, land dispossession was exacerbated by the racially-based Apartheid policies, resulting in the forced removal of more than 3.5 million people (Magome and Murombedzi 2012). Today South Africa's 'rural areas' are still characterised by vastly skewed land-ownership, extremely high densities of people on marginal land and widespread tenure insecurity (Pollard et al. 2014; Cousins and Pollard 2017).

South Africa's flagship land reform programme was central to the new democracy in 1994. The National Environmental Management: Protected Areas Act of 2003 recognised that protected areas can - and should - contribute to local livelihoods, and paved the way for co-management as a key mechanism for overcoming the highly contentious issue of land claims on protected areas (RSA 2003). While protected areas are legally required to remain under conservation, successful claimants are offered three options: equitable redress (e.g. alternative land); financial compensation, or return of original title with conservation in perpetuity and co-management. If the co-management option is selected, the claimants are reinstated land-ownership rights and afforded the opportunity to manage their land together with the conservation agency and to accrue benefits including job creation and resource use rights. In this way, the three national priorities of land reform, environmental conservation and socio-economic upliftment are reconciled in a so-called 'win–win' (RSA 2007).

However, the performance of co-management in South Africa has been unimpressive (Kepe et al. 2005; Kepe 2008; Cundill et al. 2013). The settlement of land claims on protected areas was extremely slow up until 2010 when the government published a National Co-Management Framework to clarify co-management strategies and fast-track the settling of claims. Three potential models were outlined, namely full co-management, full lease, and part-co-management/part-lease.

Despite this, the implementation of co-management agreements has remained slow and ineffective and the potential benefits are seldom realised (Cundill et al. 2013; de Koning and de Beer 2014). The reasons for these failures have been explored in several case studies over the years (e.g. Cousins and Kepe 2004; Thondhlana et al. 2011, 2016; Kepe 2012; Cundill et al. 2013, 2017; Mashale et al. 2014). Governance problems have consistently been identified, particularly the hugely unequal power balances between state conservation agencies and poor rural people, and capacity limitations leading to poor accountability and conflict.

Despite some understanding of why co-management initiatives have been disappointing in South Africa and the extensive international literature highlighting the importance of social capital and equity as enabling conditions for effective shared governance (Borrini-Feyerabend et al. 2004; Plummer and FitzGibbon 2006; Borrini-Feyerabend et al. 2013), support for good governance in the context of co-management in South Africa remains weak. Much of the effort from government is spent on drafting a co-management agreement as if this – in and of itself – will secure good governance. Co-management is treated as a technical process without due consideration of the need to support both parties in a new social process of co-managing. Additionally, the government entities responsible for supporting co-management are severely under-capacitated, with much of the required social, political and economic understanding falling outside their areas of expertise.

3 Background to the Lekgalameetse Co-management Support Work

The unique 19,200 ha Lekgalameetse Nature Reserve (LNR) lies in the foothills of the Drakensberg escarpment in Limpopo Province, South Africa (Fig. 1). Six land claimant communities and the provincial management authority (LEDET or Limpopo Economic Development, Environment and Tourism) had been struggling to revise a co-management agreement initially signed in 2007. Support for co-management of this reserve was initiated by AWARD, a research and development NGO, in 2015 under the auspices of a large, multi-disciplinary programme known as RESILIM-Olifants. This programme set out to reduce vulnerability and enhance the resilience of people and ecosystems based on systemic, social learning approaches.

The support was based on AWARD's experience of rural development, co-management and participatory processes, as well as on theories of social learning and adaptive management in complex and dynamic environments (see Pollard et al.

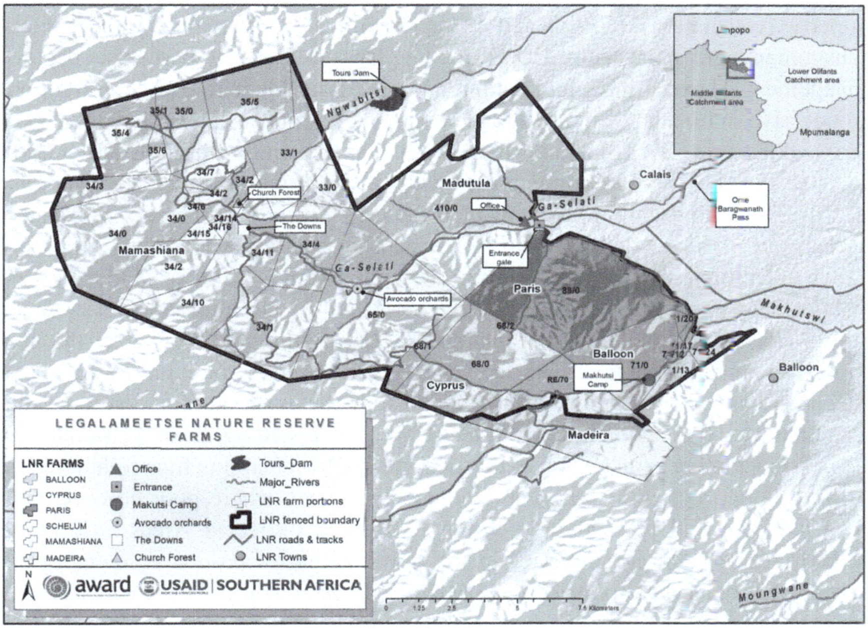

Fig. 1 Map of Lekgalameetse Nature Reserve indicating portions under claim by different communities

2020). Continued conservation of LNR was a key priority for the state, whilst expectations of financial benefits were all-encompassing for the poor, disenfranchised rural communities. These two seemingly opposing objectives needed to be integrated into a common vision for co-management, which was at an impasse in a context where history was central.

4 History of Forced Removals and the Move Towards Restitution Through Co-management

There are six communities with land claims in the LNR (see Fig. 1) representing an estimated 4200 claimants and 13 000–14 000 beneficiaries. Land-owners of Lekgalameetse Nature Reserve were forcibly removed between the late 1930s and the 1970s from a rich and diverse landscape to the hotter and more arid Lowveld area. Claimant communities lodged their claims between 1995 and 1998 and some won their claims nearly a decade later in 2005, whereupon they agreed to venture into co-management with the provincial management authority, LEDET.

Both institutionally and legally, the move towards land tenure security at Lekgalameetse has been complex. At the start of AWARD's engagement, only two of the six claimants had finalised their claims through settlement agreements (Mamashiana and Madutula). These two had formed legal entities called Communal Property Associations (CPAs) and were preparing to enter into a revised co-management agreement with LEDET, moving from full lease to part lease/ part co-management. The remaining four communities (Cyprus, Madeira, Paris and Balloon) did not have individual settlement agreements but fell under one land claim (Sekororo) and a single CPA (Makhutšwe) which had also claimed adjoining agricultural land (see Fig. 1). The four communities with claims in the protected area were not keen to remain under the Makhutšwe CPA due to allegations of poor governance. Furthermore, this CPA claimed that since it is an "agricultural CPA", the Cyprus and Madeira communities - without claims on agricultural land - were not valid members.

After nearly two decades, the lack of resolution for the four claimant communities was a major challenge that had stalled the drafting and signing of the new Co-Management Agreement with the state. Since only two communities had been receiving lease remuneration - and recognising the potential for conflict - the Lekgalameetse Management Committee (LMC), representing all six communities, had insisted that financial resources be put aside for the remaining four communities.

Institutional messiness was also evident in the state where poor co-ordination characterised LEDET's activities within LNR. The Protected Areas Directorate worked closely with communities in the development of co-management agreements, while other directorates such as Wildlife Resorts (formerly a separate entity) were responsible for economic development, tourism and associated infrastructure. The national department was also involved in funding development projects in LNR. Reserve staff such as rangers were uninformed despite their need to accommodate new arrangements. In general, other than three individuals, capacity, experience and understanding of co-management was limited in these government agencies at the start of the project.

The above is a poignant reminder that one seldom starts a journey into transforming natural resource management and governance with a "clean slate". Once the historical context and impacts on current relationships, power and institutional arrangements have been understood, an enabling environment needs to be built in order to promote equity and effectiveness as essential aspects of governance quality (Borrini-Feyerabend et al. 2014). An adaptive, learning approach is an important part of this enabling environment (CBD 2018).

5 The Praxis of Co-managing as a Complex, Transformative Learning Journey

Protected areas can be conceptualised as complex social-ecological systems (see Holling 2001) so that navigating co-management requires learning and adaptation because the process is influenced by, and influences, multiple socio-political, economic and environmental factors in a mutually influencing or 'co-evolving' system (Ison 2017). Accepting complexity means acknowledging that, despite the best plans, outcomes are not entirely predictable and cannot be known with certainty - which is a new, and often difficult reorientation for parties involved (Pollard et al. 2014). Moreover, co-management is an *entirely new practice* unknown to both parties, and so co-learning is key. Recognising the need to support learning and adaptability, AWARD adopted a strongly *systemic, social learning approach* to the praxis of co-management, acknowledging the connectedness between 'systems of interest' and co-management as a co-learning journey where actors develop new practices over time. Thus, co-management is a process and not an end in itself (Jones and Murphree 2004).

An important but sometimes ignored component is restorative justice and transformation (Zehr 2002). In the South African context, it is a *transformative* process which aims to address inequities of the past (Kepe 2008) and a failure to recognise this is unlikely to lead to sustainability in the long term. With forced removals, communal land-owners lost their rights and identity as both land-owners and custodians so that restitution is not just a legal process but also a socially and politically transformative one (Cousins 2007). It is not a favour; restitution is a right to regain a sense of identity and power over what was taken. In this regard, terminology is important. Referring to land-owners as 'communities', 'beneficiaries' or 'land claimants' does little to recognise this right; restituted communities should rather be called 'land-owners' as legitimate 'rightsholders'.

This framing moves co-management from a contractual process to one that has an important role to play in developing agency, dignity and respect whilst also contributing to people's livelihoods. Both the management authority and the new land-owners are embarking on a new, collaborative learning journey of co-managing. If successful, it will be a deeply transformative process in which power dynamics shift as each party negotiates new powers, interests and relationship in a landscape where the state was all-powerful (Borrini-Feyerabend et al. 2004; Cronkleton et al. 2012; de Koning and de Beer 2014). Moreover, because the redress of past injustices is central to this negotiated space, history-sensitive processes are required. At the same time a delicate balance is needed between the expectation that new land-owners take on a new identity as custodians over land that most of them have never known, and the relinquishment of certain governance powers and rights by the state (Kepe Kepe et al. 2005; Cundill et al. 2013).

6 Support for Co-management at LNR: Elements of the Co-learning Journey

One of the first steps was supporting both the land-owners and the state to clarify their concerns and challenges around the revised co-management agreement. This had been drawn up by LEDET in 2015 but was contested by the LMC on behalf of the wider community. In many respects this was not surprising since the move to a new model - part lease/part co-management - represented new governance arrangements with completely new roles and responsibilities, none of which had been detailed nor clarified. Most of the concerns raised by land-owners (Table 1) pointed to the complete lack of clarity on emerging governance arrangements which the state appeared to have no framework or process to meaningfully support.

6.1 Pre-conditions for a Joint Relationship

One of the assumptions behind any collaborative partnership is that the ‘marriage’ will work. However, partnerships cannot simply be established on a technical basis without working on the foundational elements of the relationship where the new land-owners and the state share values, capacities and vision for co-management. Based on AWARD’s own work facilitating arrangements between the state and communal land users and aligning with existing guidelines (e.g. Borrini-Feyerabend 2004), a number of factors or ‘pre-conditions’ for a tenable co-management relationship were assessed (Table 2). This collaborative exploration suggested that the foundations for entering into co-management were very weak. Not only was there no collaborative vision but, as emphasised before, no attention had been paid to governance arrangements, roles or responsibilities which suggested that the need to share power was poorly understood.

These starting conditions, together with the concerns listed in Table 1, guided much of the initial support given to both parties. By revealing challenges to good

Table 1 A summary of the key concerns of the land-owners to the draft, revised co-management agreement proposed by LEDET

Clause	Concerns raised (comments in italics)
1. Rental fee R12 ($0.6) per ha per year	i. Communities concerned how they arrived at R12 figure *This was explained by LEDET and resolved* ii. What was the plan for the four claimants whose settlement agreements have not been finalised and who could therefore not receive rental? *This was a major issue that had not been planned for by LEDET*
2. Commercial development to be approved by Management Authority (LEDET)	iii. This stood in direct contravention to statements in other documents (Cabinet Memo 2002, Inter-ministerial Memo 5; 2007) which require "All parties to define commercial activities within the agreement" iv. Where is an overarching commercialisation plan (According to the draft Co-Management Agreement: 7.1.3.3: *When commercial and development rights are issued to the third parties or LEDET it will be done within the Community Public Private Partnership Framework.*)? Why is there no involvement of land-owners over own land? *It was noted that this would not happen in private property arrangements. Moreover, there was no collaboratively-negotiated commercialisation or development plan, nor any framework to guide benefit-sharing*
3. Co-management Agreement may be reviewed anytime in line with the commercialization of LNR	v. Very unclear as to what this means. *Land-owners need to be fully part of any commercialization process. LEDET itself had not even developed a commercialization plan for LNR (see above)*
4. Moving from PPP (Public–Private-Partnership) to CPPP (Community – PPP)	vi. Land claimants felt that the (current) processes do not reflect this *This turned out to be a major source of tension as communities felt they were not being involved in a meaningful way especially in terms of financial benefits*
5. Park Management Plan	i. The Reserve Management Plan makes no mention of land-owners' involvement in the revision process *LEDET felt this was their responsibility alone. However, it was argued that land-owners should be involved in the process if they are expected to be custodians of the land*

(continued)

Table 1 (continued)

Clause	Concerns raised (comments in italics)
6. Employment and procurement- only Management Authority decides who gets appointed	ii. Major issue and strongly questioned by the community *This could be easily explained by the legislative procedures guiding government but was never adequately communicated. Nonetheless because of distrust around previous developments this remains a contentious issue and ultimately should be debated at a national level*
7. Training may be provided	iii. Very unclear: if training does not transpire how will capacity and skills be developed? *Without capacity it is impossible to co-manage as a new governance practice (see Fig. 3) for all involved including LEDET staff (see* de Koning and de Beer 2014*)*

governance, work could then start on the co-learning journey based on a shared vision and understanding of new and evolving identities in protected areas management. This was enriched by an understanding of challenges to practices that together comprise the co-management "activity system", discussed below.

6.2 Understanding Co-management as a Systemic, Expansive Learning Process

At the core of co-management is the *active practice* of co-managing and yet there was no guidance on process or content. To address this, the systemic, social learning framework focussed attention on *how* learning happens to enable growth and progress. Accordingly, co-management is a learning - or meaning-making - journey for all involved, with the 'practice' being produced over time by those who engage in it. This expansive learning process (Engeström 1987; Fig. 2), fosters the formation of an empowered collective or a new community of practice (Wenger 1998). According to Engeström, the collective mastery of societal problems (such as co-management) is achieved by resolving systemic challenges to the set of practices in question. Thus, co-management can be conceptualised as a set of interacting 'activity systems' that need to be understood, planned for, enacted and revised as learning happens.

Based on the activity system framework (and the CHAT tool[1]), co-management was explored collectively to understand the interaction of multiple activity systems, including those of national and provincial governments. This framing helped community representatives, the state (LEDET) and AWARD (as development facilitators) to conceptually reinterpret the challenges related to co-management more systemically

[1]Engeström's Cultural-Historical Activity Theory (CHAT).

Table 2 Analysis of 'pre-conditions' for collaborative co-management

Pre-condition	Land-owners (Lekgalameetse Management Committee)	The state
Potential for a collaborative vision based on shared principles	• No shared vision • Simplistic vision for their future in the context of LNR based on financial benefits only	• No shared vision • Unilateral, technicist vision of co-management in practice, to be concluded through signing of co-management agreement
Systems understanding	• Generally weak although some of the wider factors influencing and driving change were easily identified	• Poorly understood and an area needing support
Identity	• History of forced removals and alienation from the land undermined a sense of custodianship except in some elders Thus support aimed to build a sense of ownership and identity as land-owners and custodians	• See themselves as custodians and regulators - the guardians A shift in identity (and power) would be extremely difficult without a framework and strong leadership
Self-organisation	• Strong in some cases, extremely weak in others • "Some of us work together, not all of us"	• Very weak between directorates and park management and between state departments
Trust, transparency	• Complex relationships at different scales; within and between claimant communities, and between the LMC and individual communities • A major issue between owners and the state, especially around commercialization and settlement of claims	• Variable, with protected area facilitators eager to develop a relationship of trust with communities via the LMC Unclear within state directorates
Governance & institutional arrangements		
Joint	• No frameworks for co-management and no joint work on understanding collaborative arrangements • No platform established for joint co-management despite being required • Revised co-management agreement strongly contested	
Internal	• Some institutional arrangements but poorly codified (LMC) • Capacity a major issue (human, financial and administrative)	• Weak integration between various directorates • Treated as a technical exercise

(continued)

Table 2 (continued)

Pre-condition	Land-owners (Lekgalameetse Management Committee)	The state
Reflexivity / adaptive capacity	• Variable. Members of the LMC indicated clear commitment to reflect and adapt but needed support and mediation to do so	• Individually nuanced – not recognized as important
Facilitation – potential for facilitation within the group	• Chairperson of the LMC a committed and powerful force and supported by some • Greater pool of facilitators needed	• Individual within one directorate committed but without time, support or capacity to run ongoing processes
Potential to develop effective partnerships	• Poorly mapped out. In many respects a contentious issue particularly for development and commercialization • Some partnerships already existed (e.g. with NGOs and legal support)	

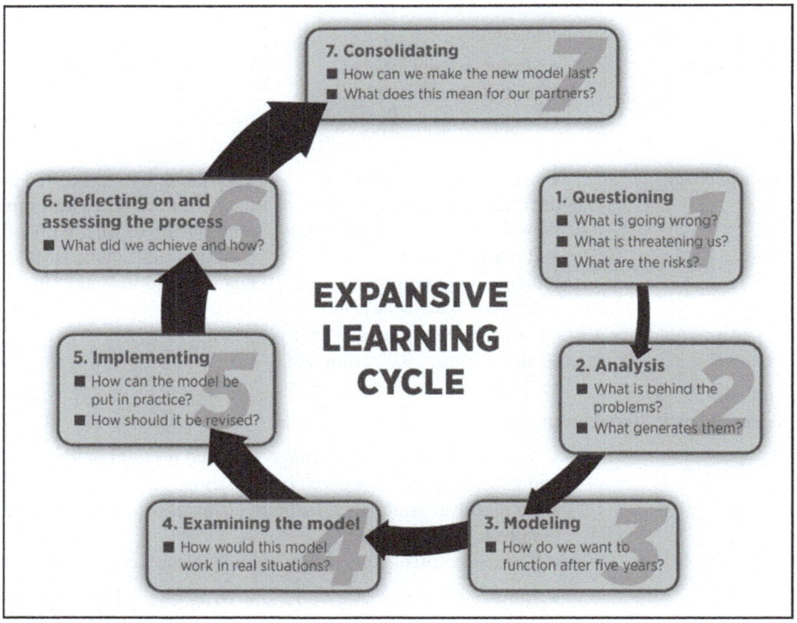

Fig. 2 The expansive learning cycle indicating specific learning actions (adapted from Engeström 1987)

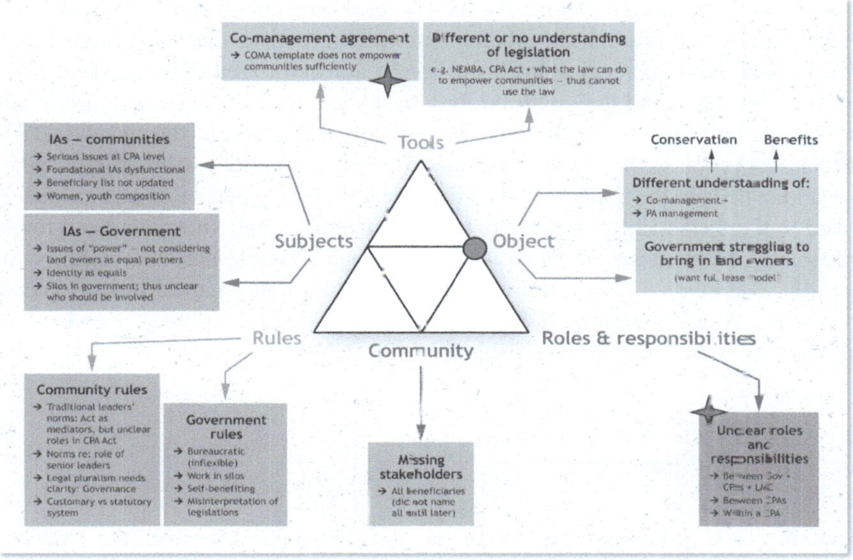

Fig. 3 Summary of results from a CHAT analysis of the co-management activity system between land-owners and LEDET

(Fig. 3). The process surfaced 'disturbances' related to each aspect of the activity system, from which a collaborative support plan was derived. The key challenges related to the co-management agreement, conflict between claimant communities internally and with the state, siloed roles of the state, the role of women and the youth, legal pluralism, different understandings and interpretation of conservation and benefits, and unclear roles and responsibilities. Many of the concerns related to assertions that the state was unable – or unwilling – to cede power and possible reasons behind this. AWARD was able to situate these discussions within guiding principles of good governance and specifically, to unearth concerns regarding perceived financial benefits from development and commercialisation. The lack of settlement agreements for four of the six claimant communities emerged as a major issue for co-management that, to date, remains unresolved. Most importantly it was clear that, without an understanding of co-management as a governing practice – and the roles and responsibilities flowing from this – tensions were likely to escalate. A number of processes were then agreed on to address these challenges.

6.3 *Eco-literacy, Historicity and Building a Sense of Identity*

At the start of the project many of the new land-owners had little knowledge of the reserve despite a long history of living on the land. The 50-year dislocation from

LNR was a long enough 'hiatus' that the younger generation knew very little of the reserve and of their forefathers' relationship with the land, so there was a real danger of losing long-term memory of the history and value of the area. A process of sharing between the elders and the youth through walks to sites of historical, spiritual and ecological importance revived a sense of identity in the elders and deepened trust and understanding across the group. It was also a deeply emotional process for the elders that acknowledged the impacts of forced dislocation from a place that was intimately linked to the 'self'. Elders were able to share memories and knowledge and reveal a distinctly different set of values around ecological assets than that held by the scientific community. The youth were also taken through a capacity development process regarding relevant policies and legislation and on eco-literacy, endemism, and the importance of rivers and wetlands.

6.4 Benefit-Sharing

Inevitably these processes led to conversations about benefits from the reserve - which was initially perceived to be of little value in its current form but which, it was suggested, could greatly benefit all through large-scale developments (such as hotels and golf-courses) and job creation. These naïve hopes were understandable given that developments by the state such as roads, camps and housing were going ahead without consultation or collaboration with the land-owners, in a continued flouting of co-management values. Adding to these perceptions and mounting anger, was that jobs (such as for clearing invasive plants and construction) were going to people outside the beneficiary community, raising questions about process. This led to protests and work stoppages and retaliatory accusations by the state of threatening behaviour, fostering greater distrust between the parties. Behind this lay perceptions of corruption, particularly with the sudden arrival of some two-million bricks which lay unused for the length of the project and about which the land-owners had little knowledge. Later, the promise of jobs based on very limited communication with the Lekgalameetse Management Committee further fuelled divisions as some members did not want to be seen as undermining potential jobs whilst others felt the process was flawed.

All of this pointed to the very muddy waters of co-management, the lack of any real process in place by the state and the urgent need to address governance and institutional arrangements to clarify where the power and authority lay for development and commercialisation (see de Koning and de Beer 2014; Knieter 2014). Whilst a detailed description of the processes related to benefits is beyond the scope of this chapter, some important learnings for co-governance emerged. Foremost was the need to understand which benefits and developments could be supported. Reframing benefits as both non-financial as well as financial enabled a broader conceptualisation of the real, seemingly 'hidden' value of LNR and therefore the role of custodianship. Land-owners acknowledged the value of the reserve for access to grazing, thatching

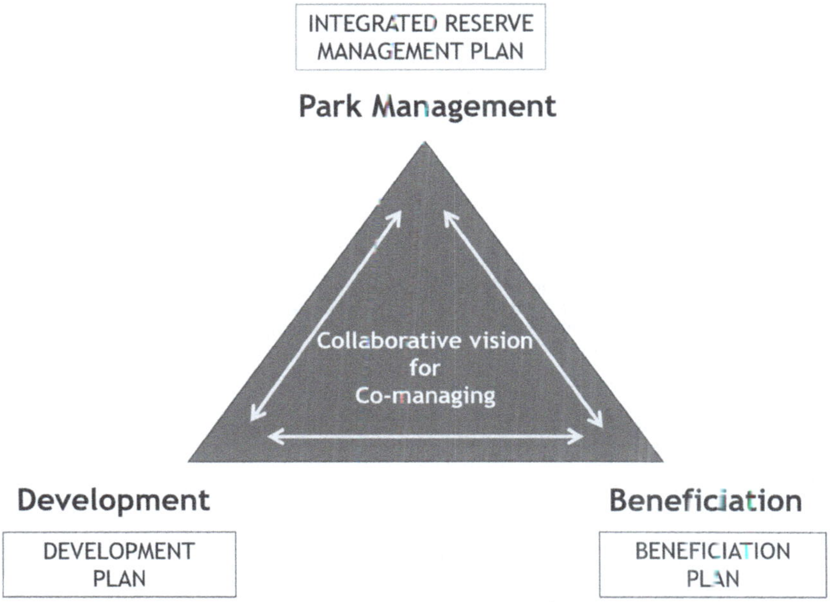

Fig. 4 Key strategic components of co-management under the model chosen for LNR; institutional arrangements and roles and responsibilities flow from these

and medicinal plants. As a contribution to the diversification of livelihoods from LNR, the concept of conservation-based enterprises was introduced and tested.

Through this process land-owners deepened their understanding of the ecological assets of the reserve and ultimately developed a Beneficiation Plan which included a strategic commitment to sustainable use and equity, as captured in the vision. It also illuminated the links between beneficiation and two other important strategic components of co-management that emerged through the process, namely the Reserve Management Plan which provides resource-use zones, and a Development Plan which sets the medium-term strategic framework for development and commercialisation (Fig. 4). Unpacking what each of these referred to and consisted of greatly advanced the capacity of both parties, as well as reducing tensions between them in many cases.

7 Explicating Co-management as Collaborative Governance Practice

At the heart of the process was the thorny issue of how power would shift and to whom. As noted, a fundamental question that had never been answered in the establishment of a relationship between the state and the new land-owners (prior to

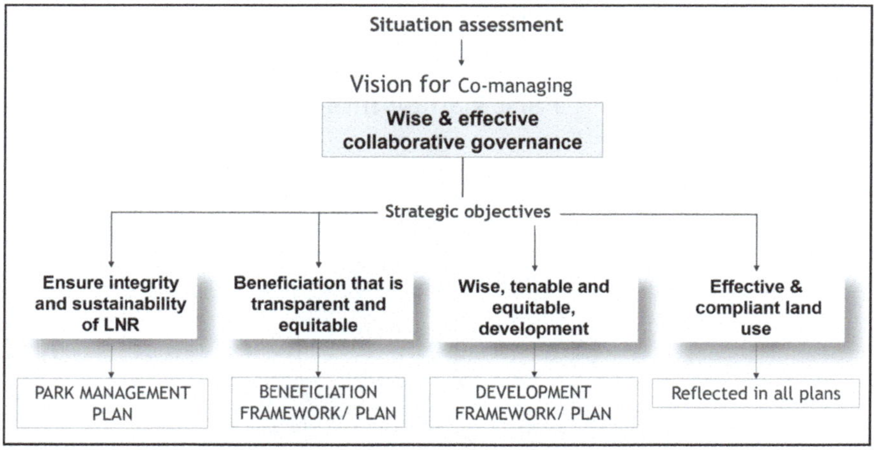

Fig. 5 A summary of the vision, values and strategic objectives set by the land-owners and state representatives for co-management of LNR

AWARD's support) was *what does co-managing of LNR under a part lease/ part co-management model mean in practice*? In other words, what are the governance and institutional arrangements and associated roles and responsibilities?

A major constraint was the lack of a Co-management Committee which should have been established[2] as the key mechanism for governance early in the process as part of the state's responsibility. It is through this committee that governance and co-management arrangements should have been negotiated and the co-learning journey undertaken. Despite repeated attempts by AWARD, this committee was only established in 2019 and even then, was led by LEDET (as Chair and Secretary) with minimal land-owner engagement. A final governance process resulted in a clearer understanding of the need for this committee and its position in terms of broader institutional arrangements.

With an institutional form finally in place, the development of a collaborative vision between land owners and the state was possible. This would normally be undertaken early in a co-management process so as to set the basis for co-managing. Despite its *critical importance* this did not happen until much later not only making it difficult to work together but also undermining the very basis for making explicit what co-management would entail.

The finalisation of a vision represented a major outcome for the parties involved. Emanating from the vision were four strategic objectives (Fig. 5) which are foundational for developing clear roles and responsibilities for both parties. Work on these is remains to be finalised but it is worthwhile highlighting that carefully working through the roles and responsibilities of each will go a long way towards producing a new co-management agreement that is acceptable to both parties.

[2]Required by the M.O.A (2007).

8 Discussion and Implications for Sustainable Development in Africa

The co-management support process described here covered all the aspects of governance quality as defined by the IUCN (Borrini-Feyerabend et al. 2014) *Equitable governance* was promoted by enhancing the *legitimacy and voice* of the land-owner communities through personal reconnection with their land, building an identity as custodians and land-owners, capacity development to increase their understanding of legislation and principles of good governance, and supporting them to resolve their own challenges as they arose. *Fairness and rights* were promoted through legal support and through the principles that were put in place to guide the governance of benefits through two important frameworks.

The Beneficiation Plan moderated expectations that LNR could provide 'everything for all' whilst addressing issues of equity and transparency through clarity of rules and protocols. Given the number of claimant households, the very nature of benefits had to be qualified to acknowledge those beyond financial alone and ensuring ways that financial gains could service the community more generally such as through schools, electrification and improved water supply, as has been done elsewhere (Cundill et al. 2017; Maluleke 2018). It also clarified how opportunities could be opened up for the land-owners through the establishment of a business entity that could bid for development work within the reserve, thereby going some way to satisfying the need for jobs. The complementary Development Framework also provided foundations for 'levelling the playing field' by moving the governance over development of the reserve into a collaborative space. Not only is this more transparent but, for the first time, sustainable development was prioritised through strategic integration with the Reserve Management Plan and Beneficiation Plan.

The work on equitable governance laid the foundation for more *effective governance*, the other important leg of governance quality (Borrini-Feyerabend et al. 2014), through the collaboratively developed direction (vision), and the work on roles and responsibilities which facilitates better *performance and accountability*. Nonetheless one cannot naively assume that issues of accountability will not continue as part of a dynamic landscape. In Lekgalameetse, simmering tensions are set against a backdrop of unilateral decision-making regarding developments by the state. Interpretations range from government continuing 'business-as-usual' by holding on to decision-making powers of the past (see Kepe et al. 2005) to perceptions of hidden interests driving the need for singular control. The latter talks to national dialogues of tender irregularities where service providers benefit through inflated costs along the value chain of service delivery (Daily Maverick 2020). Whatever the reality, clean governance and clarity on governance arrangements, together with open and transparent discussions as part of the ongoing co-management practices, is central to long-term sustainability.

Issues of power and social justice lie at the heart of sustainable development in Africa, given the historical injustices around land rights and access to natural resources on the continent (see Sect. 1). These injustices are in many cases perpetuated in the present by rent-seeking elites, land and natural resource "grabs" by foreign powers and the use of natural resources to fund conflict in conjunction with criminal syndicates and unscrupulous private companies (Bannon and Collier 2003; Amankor and Moyo 2008; Acemoglu and Robinson 2013). These factors substantially decrease the prospects for sustainable development, by disempowering large numbers of ordinary people and setting in motion vicious circles of unsustainable growth which only benefits a small minority and increases the likelihood of destructive conflicts - leading to "development in reverse" (Bannon and Collier 2003; Acemoglu and Robinson 2013).

The key to changing "vicious circles" to "virtuous circles" leading to broad-based benefit and sustainable development is inclusive institutions that allow and encourage broad participation by the great mass of people in governance, access to natural resources, economic activity, education, technology and innovation (Acemoglu and Robinson 2013). The IPBES[3] Regional Assessment for Africa also noted the value of inclusive governance for realising the potential syngergies between the Sustainable Development Goals on the continent and promoting benefits across several goals (IPBES (2018).

Athough the case study presented in this chapter is limited to a particular (local) scale and context, one of the lessons is that building inclusive institutions takes time, resources and support for participants as they travel on their co-learning journey. A failure to commit sufficient resources to this task is unlikely to lead to the type of transformative change outlined above.

Another lesson is the importance of paying attention to power in development work – not just in a static sense as part of a description of context, but in a relational, dynamic way as power relations change and are re-negotiated during the co-learning journey (Scoones 2009). Part of "paying attention to" power dynamics is describing and making them explicit. Tools and frameworks such as CHAT and visioning, as used here, can help to open the conversation by identifying conflicts and contradictions, and negotiate the practicalities of transfer and sharing of power.

However, there is also a need to find ways to surface and deal with hidden agendas and power relations which necessarily affect the actual outcomes of any collaborative governance process. Frameworks for collaborative governance of natural resources (e.g. Borrini-Feyerabend 2004, WIN 2020) tend to assume that the intentions and agendas of all parties can be understood through the initial context-scoping process at the start of negotiations, and provide little guidance on dealing with hidden agendas as they emerge. In this case study, such issues were partly revealed over time but remained largely unresolved. Practical tools and frameworks to address this limitation would be immensely valuable. These could build on existing guidelines for managing natural resource conflicts (e.g. Buckles 1999; Bannon and Collier 2003). Such tools would help to root the practice of sustainable development more firmly in a political economy framework as recommended by Amankor and Moyo (2008). This work emphasises that within a land-reform context, co-management requires a

[3]Intergovernmental Science-Policy Platform on Biodiversity and Ecosystem Services.

shift from a contractual process to one that is socially and politically embedded and that has an important role to play in developing agency, dignity and respect whilst also contributing to people's livelihoods. Doing this means committing resources and capacities to the co-learning journey. Without adequate and appropriate support for both parties, and an orientation towards co-learning for new, largely unchartered governance arrangements, the co-management endeavour, and hence long-term sustainable developmentity, will remain deeply fragile.

9 Conclusion and Recommendations for Co-management

Whilst co-management is guided by multiple policies in South Africa and beyond, much remains to be done on testing these in practice – particularly in the case of land reform and new governance arrangements that shift power away from the state to claimant land-owners with rights in decision-making. Recognition of this in the design of transformative processes can help to ensure that co-management contributes to sustainable development. Indeed, co-management has potential to contribute to achievment of many of the Sustainable Development Goals including conservation of natural resources, reduction in poverty and inequality, improved health, well-being and livelihoods, climate action, peace and justice, and strong partnerships (CBD 2018). However, the experience at LNR points to the need for a much more rigorous praxis of supporting collaborative governance of protected areas since 'the devil is in the detail'. Staff and facilitators alike often have only a rudimentary understanding of governance, shifting power dynamics and the importance of history, context and learning, and are therefore ill-equipped to mediate new governance arrangements whilst building governance quality and vitality (Borrini-Feyerabend et al. 2014).

This chapter emphasises a number of key points. First, co-management is a new, dynamic and on-going learning process for all involved; secondly, it is a practice embedded in a wider historical, socio-political, economic and environmental context making it, thirdly, a deeply transformative process requiring attention to socio-political dynamics. Fourth, working out the governance details – as difficult as this might be – must be an explicit focus of any facilitated process since this is where the seeds of discontent can grow. In conclusion therefore, we offer the following recommendations as a means to achieve a win–win in the 'marriage' between social justice, land reform and conservation so as to contribute to longer-term sustainable development.

1. Co-management is a new practice unknown to both parties, and so co-learning is key. Both the management authority and the new land-owners are embarking on a collaborative learning journey of *co-managing* – in essence forming a new community of practice. Thus, support for co-management as an evolving governance practice - not an endpoint – is needed. This continues to be a difficult framing for many involved, especially those that are engaged more infrequently such as the 'wider community' and sectors of government.

2. This requires acknowledging that it is a transformative process for all parties involved, including those supporting and facilitating it, and must thus be designed as a learning process with the humility and interest this requires. Whilst many gains were made by AWARD's engagement, resistance to relinquish power by some in government remains, pointing to how difficult but central this consideration is.
3. Whilst principles of transparency, trust and equity are central (Borrini-Feyerabend 2004), the process also requires reflection and adaptation since each situation is unique. It will catalyse and solidify new power relations as responsibilities are relinquished, constructed and taken up and therefore cannot be left to evolve 'naturally'. It is a process that requires mediation and support (Barrow and Murphree 1998; Borrini-Feyerabend 2004). Where a long history of unresolved land tenure, deeply skewed power relations and distrust is present, an extended phase of 'preparing for the partnership' (Borrini-Feyerabend 2004, p 139) is likely to be needed. Like others, AWARD's experiences have shown that this support role can be filled by NGOs in cases where government is unable or unwilling to do so (e.g. Magome et al. 2012).
4. Framing the praxis of co-management as a transformative, systemic social learning journey which recognises complexity, interconnectedness and the underlying history requires different approaches and tools to support a systemic approach of co-inquiry and co-learning (see also Pollard et al. 2014; 2020).
5. Clear and secure tenure rights are the basis for any co-management process (Hornby et al. 2017; Timko and Satterfield 2008). The tensions and uncertainties that surfaced at LNR from unresolved tenure arrangements continually eroded the goodwill or gains that were being made. At the time of writing, the communities with unfinalized land claims were considering legal action emphasising the importance of legal support as a key part of ongoing negotiations.
6. Benefit-sharing and protected area development lies at the heart of many potential areas of conflict. This work has made the case that at Lekgalameetse much of this could be tracked back to the lack of guidance as to what co-managing actually entails as a governance practice and what this means for power relations, as well as the 'hidden' dynamics of perceived or real corruption.

This work emphasises that within a land-reform context, co-management requires a shift from a contractual process to one that is socially and politically embedded and that has an important role to play in developing agency, dignity and respect whilst also contributing to people's livelihoods. Doing this means committing resources and capacities to the co-learning journey. Without adequate and appropriate support for both parties, and an orientation towards co-learning for new, largely uncharted governance arrangements, the co-management endeavour, and hence long-term sustainable developmentity, will remain deeply fragile.

References

Acemoglu D, Robinson J (2013) Why nations fail: the origins of power, prosperity and poverty. Profile Books, London

Amankor K, Moyo S (eds) (2008) Land and sustainable development in Africa. Zed Books, London and New York

Bannon I, Collier P (eds) (2003) Natural resources and violent conflict: options and actions. The World Bank Washington DC

Barrow E, Murphree M (1998) Community conservation: from concept to practice: a practical framework. working paper no. 8 (2012). https://www.researchgate.net/publication/242251300_Community_conservation_from_concept_to_practice_a_practical_framework

Borrini-Feyerabend G, et al (2004) Sharing power: learning by doing in co-management throughout the world, Cenesta. IIED and IUCN/CEESP/CMWG, Tehran. https://www.iucn.org/about/union/commissions/ceesp/ceesp_publications/sharing_power.cfm#sp_contents

Borrini-Feyerabend G, Dudley N, Jaeger T, Lassen B, Pathak Broome N, Phillips A, Sandwith T (2013) Governance of protected areas: from understanding to action. Best practice protected area guideline series no. 20. IUCN, Gland (Switzerland)

Borrini-Feyerabend G, Bueno P, Hay-Edie T, Lang B, Rastogi A, Sandwith T (2014) A Primer on governance for protected and conserved areas. Stream on enhancing diversity and quality of governance, World parks conference. IUCN. Gland (Switzerland)

Buckles D (Ed) (1999) Cultivating peace: conflict and collaboration in natural resource management. International development research centre (IDRC), Ottawa and World Bank Institute, Washington DC

CBD (2018) Protected and conserved areas governance in the convention on biological diversity: a review of key concepts, experiences, and sources of guidance. Convention on biological diversity subsidiary body on scientific, technical and technological advice, UNEP

Chevallier R (2016) The State of Community-Based Natural Resource Management in Southern Africa. SAIIA Occasional Paper No 240, August 2016

Cousins B (2007) More than socially embedded: The distinctive character of 'communal tenure' regimes in South Africa and its implications for land policy. J Agrarian Change 7(3):281–315

Cousins B, Kepe T (2004) Decentralisation when land and resource rights are deeply contested: a case study of the Mkambati eco-tourism project on the Wild Coast of South Africa. Eur J Dev Res 16(1):41–54

Cousins T, Pollard S (2017) Land tenure and the governance of wetlands in complex and transforming environments: lessons for understanding legal pluralism from Craigieburn, South Africa. In: Hornby D, Kingwill R, Royston L, Cousins B (eds) Untitled: securing land tenure in urban and rural South Africa. University of KwaZulu-Natal Press, Pietermaritzburg. South Africa

Cronkleton P, Pulhin JM, Saigal S (2012) Co-management in community forestry: how the partial devolution of management rights creates challenges for forest communities. Conserv Soc 10(2):91–102

Cundill G, Thondhlana G, Sisitka L, Shackleton S, Blore M (2013) Land claims and the pursuit of co-management on four protected areas in South Africa. Land Use Policy 35:171–178

Cundill G, Bezerra J, de Vos A, Ntingana N (2017) Beyond benefit-sharing: place attachment and the importance of access to protected areas for surrounding communities. Ecosyst Serv 28(B):140–148

de Koning M, De Beer F (2014) Land restitution and income generation in protected areas in Mpumalanga South Africa. Dev Prac 24(1):118–128

de Pourcq K, Thomas E, Arts B, Vranckx A, Léon-Sicard T, Van Damme P (2015) Conflict in protected areas: who says co-management does not work? PLoS One 10(12):e0144943

Engeström Y (1987) Learning by expanding. Helsinki, Orienta-konsultit

Holling CS (2001) Understanding the complexity of economic, ecological and social systems. Ecosystems 4:390–405

Hornby H, Kingwill R, Royston L, Cousins B (eds) (2017) Untitled: securing land tenure in urban and rural South Africa. University of KwaZulu-Natal Press, Pietermaritzburg, South Africa
IPBES (2018) Regional assessment report on biodiversity and ecosystem services for africa: summary for policymakers. Intergovernmental science-policy platform on biodiversity and ecosystem services, IPBES secretariat, Bonn, Germany, p 49
Ison R (2017) Systems practice: how to act. In situations of uncertainty and complexity in a climate-change world, 2nd edn. Springer, London and The Open University
Jones BTB, Murphree MW (2013) Community-based natural resource management as a conservation mechanism: lessons and directions. In: Child B (ed) Parks in transition: biodiversity, rural development, and the bottom line. Earthscan, London, pp 63–103
Kepe T (2008) Land claims and comanagement of protected areas in South Africa: exploring the challenges. Environ Manage 41(3):311–321
Kepe T (2012) Land and justice in South Africa: exploring the ambiguous role of the state in the land claims process. Afr Asian Stud 11(4):391–409
Kepe T, Wynberg R, Ellis W (2005) Land reform and biodiversity conservation in South Africa: complementary or in conflict? Int J Biodivers Sci Ecosyst Serv Manag 1(1):3–16
Knieter D (2014) Decolonizing conservation? Co-management of natural resources in Bushbuckridge Nature Reserve (BBR NR). University of Missouri-Columbia, South Africa
Kothari A, Camill P, Brown J (2013) Conservation as if people also mattered: policy and practice of community-based conservation. Conserv Soc 11(1):1
Magome H, Murombedzi J (2012) Sharing South African national parks: community land and conservation in a democratic south Africa. In: Adams WM, Mulligan M (eds) Decolonizing nature: strategies for conservation in a post-colonial era. Earthscan.
Maluleke L (2018) Rethinking protected area co-management in the Makuleke region, South Africa. Ph.D. thesis, Stellenbosch University
Wenger E (1998) Communities of Practice: Learning, Meaning, and Identity. Cambridge University Press, Cambridge
Water Integrity Network (WIN) and the CEO Mandate (2015) Guide for Managing integrity in water stewardship initiatives: a framework for improving effectiveness and transparency. https://www.waterintegritynetwork.net/stakeholder-engagement-tools/integrity-framework-for-water-stewardship-initiatives/
Mashale CM, Moyo T, Mtapuri O (2014) An evaluation of the public-private partnership in the Lekgalameetse nature reserve in South Africa. Mediterr J Soc Sci 5(20):855–862
Daily Maverick (2020) How can South Africa fix its tender system? Accessed Aug 2020. https://www.dailymaverick.co.za/article/2020-08-27-how-can-south-africa-fix-its-tender-system/
Ostrom E (1990) Governing the commons: the evolution of institutions for collective action. Cambridge University Press, Cambridge
Plummer R, FitzGibbon J (2006) People matter: the importance of social capital in the co-management of natural resources. Nat Resour Forum 30(1):51–62
Pollard S, Biggs H, du Toit DR (2014) A systemic framework for context-based decision making in natural resource management: reflections on an integrative assessment of water and livelihood security outcomes following policy reform in South Africa. Ecol Soc 19(2):63
Pollard SR, Chikunda C, Mohale T, Goredema L (2020) Systemic, social learning support for the co-management of the Lekgalameetse nature reserve Award internal report. https://award.org.za/wp/wp-content/uploads/2020/05/AWARD-BOOKLET-Social-learning-for-co-management-of-Lekgalameetse-2020-v2.pdf
RSA (2003) National Environmental management: protected areas act no. 57 of 2003. Government Printer, South Africa. https://www.environment.gov.za/sites/default/files/gazetted_notices/nempaa_actno57of2003_protectedareas.pdf
RSA (2007) Memorandum of agreement between the Minister of Agriculture and land affairs and the minister of environmental affairs and tourism, Pretoria
Scoones I (2009) Livelihoods perspectives and rural development. J Peasant Stud 36(1):171–196

Thondhlana G, Shackleton S, Muchapondwa E (2011) Kgalagadi Transfrontier Park and its land claimants: a pre- and post-land claim conservation and development history. Environ Res Lett 6:1–12

Thondhlana G, Cundill G, Kepe T (2016) Co-management, land rights, and conflicts around South Africa's Silaka Nature Reserve. Soc Nat Resour 29(4):403–417

Timko JA, Satterfield T (2008) Seeking social equity in national parks: experiments with evaluation in Canada and South Africa. Conserv Soc 6(3):238–254

Zehr H (2002) The little book of restorative justice. Good Books, Intercourse, Pennsylvania. https://sites.unicef.org/tdad/littlebookrjpakaf.pdf

Dr Sharon Pollard is the Director and founder of the Association for Water and Rural Development (AWARD). She has a background in the water sector focusing on integrated catchment management planning and implementation. In addition to pioneering research and advocacy work around water resorces management and systemic, social learning processes with stakeholders in underprivileged areas in South Africa, she has managed many internationally recognised development projects aimed at sustainability and equity.

Dr Charles Chikunda is a senior researcher focusing on stakeholder engagement, professional and capacity development and training to support resilience building. He holds a PhD in Education from Rhodes University with a focus on Science Education and Education for Sustainable Development. He has over fifteen years' experience in higher education focusing on these fields. He is involved in merging conceptual frameworks for learning and interacting with and engaging stakeholders at various levels. He is currently based in Zimbabwe.

Thabang Mohale holds a BSc (Hons) degree in Environmental Management. She has an interest in social sciences and a passion for environmental conservation and community development. She has worked extensively with communities at Lekgalameetse on land claims in protected areas and supporting co-management processes.

Lilian Goredema holds a Master of Science, Environment and Development from the University of KwaZulu-Natal. She has experience in community-based natural resource management in various countries in Africa and is currently working on a PhD on CBNRM forums as social networks and influences on policy and practice in southern Africa. She is currently based in Zimbabwe.

Dr Karen Kotschy is a monitoring, evaluation and learning specialist with a background in ecology, geography and education. She holds a PhD in Ecology from the University of the Witwatersrand. She has been responsible for implementing AWARD's pioneering monitoring, evaluation, reporting and learning (MERL) system since 2015.

Future Prospects of Sustainable Development in Africa

Walter Leal Filho, Luiza Olim de Sousa, and Rudi Pretorius

Abstract This final chapter provides a brief outline of the future prospects for sustainable development in Africa. Some prominent challenges to be faced as well as opportunities that exist for sustainable development and the attainment of the sustainable development goals (SDGs) are briefly reviewed. The progress so far on the implementation of the SDGs and targets has been uneven across Africa. A huge challenge therefore lies ahead for Africa if it wants to stay on track to meet the SDGs. Africa needs to accelerate reforms and implementation. With its emphasis on self-reliance and self-determination, the African Union's Agenda 2063 provides a framework for the attainment of sustainable economic development, social prosperity and well-being, lasting political stability, security and protection, and regional integration in Africa. Achievement of success with sustainable development and the attainment of the SDGs in Africa requires linkage with the lived-experiences, indigenous knowledge systems, and contextual realities of the ethnic groups of Africa. Furthermore, increased recognition of development initiatives that consider local perceptions and ways of doing and thinking are more likely to be successful and relevant to people's needs, are needed to support sustainable development.

Keywords Sustainable development · Africa · Resources · Challenges · Opportunities

---

W. Leal Filho (✉)
Research and Transfer Centre "Sustainable Development and Climate Change Management", Hamburg University of Applied Sciences, Ulmenliet 20, 21033 Hamburg, Germany
e-mail: walter.leal2@haw-hamburg.de

L. O. de Sousa
Community-Based Educational Research (COMBER), Geography Education and Environmental Education, School of Mathematics, Science and Technology Education, Faculty of Education, North-West University, Potchefstroom Campus, Potchefstroom 2520, South Africa
e-mail: Luiza.desousa@nwu.ac.za

R. Pretorius
Department of Geography, Science Campus, University of South Africa, Private Bag X6, Florida 1710, South Africa
e-mail: pretorw@unisa.ac.za

W. Leal Filho et al. (eds.), *Sustainable Development in Africa*, World Sustainability Series,
https://doi.org/10.1007/978-3-030-74693-3_40

1 Introduction

Africa is a continent rich in landscapes, natural resources, and biodiversity. Poor populations in particular are directly and existentially dependent on intact ecosystems and their services, as natural resources often form their livelihood and sole economic capital. These ecosystem services include the provision of elementary raw materials such as drinking water, food, timber, climate regulation or soil fertility.

At present, many African countries face serious environmental problems that affect socio-economic development and the livelihoods of the population. These include deforestation, land and soil degradation and loss of biodiversity. The over-exploitation of natural resources due to high population growth with rising living standards, is often a common problem. A further one, climate change, can contribute to a disruption of fragile ecosystems, placing high demands on the adaptive capacity of local populations who are particularly vulnerable to environmental change (Leal Filho et al. 2021).

Many studies have focused on sustainability practices in Africa. However, more recent research has highlighted that in order to achieve sustainability in the future, innovative practices need to be developed (Youssef et al. 2018).

A study undertaken in a sample of African countries identified the fact that innovation, improvement of institutional quality and entrepreneurship are vital in achieving sustainability in the future. More specifically, sustainability-orientated entrepreneurship is encouraged over conventional types of entrepreneurship. The latter contributes to environmental degradation in both the formal and informal sector. However, promoting innovation and adoption of new technologies by entrepreneurs allows for more sustainable practices and thus progression toward sustainability (Youssef et al. 2018). In terms of green-entrepreneurships, there are some evidences that Africa is moving towards sustainable urban agriculture that will promote both, economic transformation and growth in the future (Akinsemolu 2020).

Research in 13 Sub-Saharan African countries indicates that urbanisation, economic development, use of fossil fuels and electricity has contributed greatly to adverse environmental outcomes over recent years. Therefore, it is suggested that problems related to these sectors should be addressed to achieve sustainability. Emphasis is placed on switching to cleaner, environmentally friendly energy sources in Africa (Asongu et al. 2020). An article on sustainability futures suggests that Africa should use opportunities provided by the green economy to their advantage. An example is to address the pressures caused by climate change by developing new business and industries to ameliorate the problem. Researchers believe that this change will lead to a growth in the solar and wind energy industries, as seen in Kenya (Akinsemolu 2020).

Additionally, it is expected that lower rainfall may become more widespread in Sub-Saharan Africa, which may cause lower agricultural production in future. Using this information, innovative practices and adaptive measures to maximise productivity are being designed to ensure sustainable development in the future.

In this case, addressing water deficits is being seen as an opportunity to ensure sustainability (Nhemachena et al. 2020).

In other instances, an emphasis is being placed on the livestock industry, as current practices threaten sustainable development. New livestock policies are being designed to balance the cost and benefits of the livestock industry while ensuring a reduction in the impact on the environment in the future. More emphasis is also being placed on local solutions that can help to address the nexus environmental protection and nutrition (Mehrabi et al. 2020).

2 Challenges for Sustainable Development in Africa

Africa's population is growing at an exponential rate. The UN projects that 40% of the world's population will live on the continent by the end of the century (Coulibaly 2020). A top challenge that faces the continent is the continued discord between demographic trends and job creation. Africa's young and fast-growing population that is expected to reach over 2.5 billion by 2030 will require jobs, infrastructure and services (Begashaw 2020). Governments must attend to new policies that encourage job creation and maintain service delivery as the population grows rapidly and settlements become urbanized (Mbaku 2020).

Seventy-nine African cities, associated with 48% of the African gross domestic production, are amongst the world's fastest-growing cities and severely threatened by climate change and natural disasters (Coulibaly 2020; World Economic Forum 2019). The continent is coping with challenges that include fierce population growth, economic difficulties, corruption, authoritarianism, food shortages, health crises, extreme poverty, and extremism. Issues regarding service delivery and infrastructure are also holding the continent back from moving forward (Coulibaly 2020).

The progress so far on the implementation of the sustainable development goals (SDGs) and targets has been uneven across Africa. A lack of coordination across national and local governments is setting Africa back (Gilpin 2020). The Africa SDG Index and Dashboard (2019) shows that Mauritius, Botswana, Ghana, and Rwanda are the top performers. Of concern is that 18 out of a total of 46 countries in sub-Saharan Africa are, on average, less than 50% on the way to meeting the best possible outcome on all the SDGs. A huge challenge lies ahead for Africa if it wants to stay on track to meet the SDGs and it needs to accelerate reforms and implementation (Begashaw 2020). For Africa to accomplish the SDGs it must invest in good governance that is inclusive and builds on empowering women. Furthermore, Africa must build and sustain peace, and end corruption (Mbaku 2020). Despite visible improvement in education, progress still remains a concern. Africa is likely to meet the 2030 target of 100% education since current endeavours have achieved a primary school enrolment rate of over 90% in more than half of the countries in Africa. The challenge lies in increasing the net enrolment rate for lower secondary education since it stood at 28.3% in 2017 (Begashaw 2020)

Africa as a continent is also faced with the unique challenge of climate change (IPCC, 2018). Water, energy, and food are crucial resources that are vulnerable to the extreme weather events associated with climate change, putting vulnerable communities in Africa at risk of human migration (De Amorim et al. 2018). Water, energy, and food are interdependent, and these essential resources demand sustainable, integrated, and informed management (Ibid). In developing countries, 70% of industrial waste is untreated. 15 to 18 billion m^3 of freshwater resources are polluted by fossil fuel production annually (UNESCO 2020). Securing freshwater for growing populations is a challenge and a problem with multiple dimensions that presents unexpected consequences. Water challenges underscore the compromises that must be made between the environment and development (Sanya 2020). In Africa and the rest of the resource challenged Global South the rate of urbanisation is accelerating, with 43% of its population living in urban areas. Associated with rising urbanisation is the challenge to provide adequate safe water and sanitation for those who lack access (United Nations, Department of Economic and Social Affairs, Population Division 2019). These challenges are acknowledged in SDG 6 (Clean Water and Sanitation) (United Nations 2017).

Population and economic growth are primary driving forces of the continent's growth in CO_2 emissions (Habimana Simbi et al. 2021). African governments are faced with the challenge of supplying a growing demand for energy sources by business for economic growth without increasing CO_2 emissions and climate risks (Maji et al. 2019; Ssali et al. 2019). Ultimately, the challenge that lies ahead for Africa's future efforts in reducing CO_2 emissions should be to focus on scaling up energy-efficient technologies, renewable energy updates, emission pricing and long-term green development towards meeting the SDGs (Habimana Simbi et al. 2021). In sub-Saharan Africa coal (56%) is still the preferred choice for electricity generation. Other sources include hydro (22%), gas (9%), petroleum (9%), and nuclear (3%) (Maji et al. 2019). Renewable energy only accounts for 1% of electricity generation from wind, solar, biomass (unclean wood fuel) and geothermal sources (AEO 2014). The challenge for African countries is to move away from the heavy reliance on combustible biomass as a main source of energy for cooking and heating and include other cleaner energy sources like wind, solar, geothermal and clean biomass sources (Maji et al. 2019).

A key priority for Africa is to address the worsening food security situation that is aggravated by the effects of climate change, declining agricultural productivity, rapid population growth, and urbanization (Ehui et al. 2020). Millions of Africans living in low elevation coastal zones are also at risk of natural disasters and associated food insecurity (Mbaye 2020). Thirty-one African countries need food aid, and more than 30% of the 830 million people worldwide suffering from inadequate food supply live in Africa (Begashaw 2020).

3 Opportunities for Sustainable Development in Africa

Africa's potential lies in the collective positive action among its leaders and global policymakers to improve the livelihoods of all Africans guided by the SDGs and the African Union's Agenda 2063 (AUC 2014). Achievement of this Agenda is closely related to Africa's huge potential for growth, which indeed makes it one of the world's most promising continents, but not at all costs, and with cognisance of the full complement of sustainability constraints and concerns. However, it has to be pointed out that while several factors support sustainable development and the attainment of the SDGs in Africa, the persistence of some antitheses in this regard have to be noted. An example is that although Africa is blessed with an extensive natural resource base that can be used for development initiatives, these resources are often not evenly distributed between countries and/or segments of society. The latter gives rise to an associated occurrence of negative environmental impacts in contexts where environmental management practices are not in place (Gasparatos et al. 2017). Yet another antithesis is presented by the fact that although Africa is characterized by large tracts of agricultural land which have the potential to stimulate economic development (Schoneveld 2014), high levels of under-nutrition and food insecurity are rampant (EIU 2015) while agricultural practices are frequently linked to the occurrence of environmental degradation (Reynolds et al. 2015). In order to address antitheses as these, and to capitalize on the opportunities for sustainable development, it is crucial to optimally implement Agenda 2063. Agenda 2063 is the continent's strategic development program that spells out Africa's long term economic and social aspirations. Agenda 2063 is an endogenous transformation plan rooted in African values and principles, such as solidarity and *Ubuntu*. The Agenda is focused on the strategic and competitive repositioning of Africa to ensure the eradication of poverty, in addition to socio-economic and technological transformation (Mhangara et al. 2019).

With its emphasis on self-reliance and self-determination, Agenda 2063 provides a framework for the realization of sustainable development and the SDGs in Africa over medium to longer time horizons. It is generally accepted that successful implementation of Agenda 2063 is critical for the attainment of sustainable economic development, social prosperity and well-being, lasting political stability, security and protection, and regional integration in Africa (Mhangara et al. 2019). Agenda 2063 furthermore embraces the notions of Pan Africanism and the African Renaissance, with a focus on liberation, political and economic emancipation, and which Nahmo (2017) regards as essential prerequisites for sustainable development in Africa. The fact that up to two thirds of Africa consists of vast stretches of unused arable land provides the continent with the potential to not only sustainably grow its own food, but to provide food to the rest of the world as well (Muswere 2020). Although Africa therefore indeed has the resource base to support its sustainable development, accompanied by significant growth potential (Nahmo 2017), this potential will not realize if the practice persists to look for resources elsewhere instead of raising development finance from African markets (AUC 2014). Expanding on the portrayal of

Africa by August (2013) as ‘A hopeful continent’, Nahmo (2017) maintains that Africa will indeed ‘rise’ during the post-2015 Development Agenda, and supports his observation as follows:

- Africa now has its own vision, contained in Agenda 2063.
- Africa now realizes that domestic funding is crucial, including the need to mobilize these financial, as well as other resources.
- The ideal of a United Africa is closer than ever before.
- Africa is positioned strongly as part of the post-2015 Development Agenda.
- The required ingredients for development growth are slowly but surely realized.
- The view held by the world on Africa is changing positively.
- If turned to a real opportunity, the involvement of China can be of benefit to the continent.

In line with the remarks by President Obama on 26 July 2015 during his state visit to Kenya that the ‘future of Africa is up to Africans’ (Holmes and Scott 2015), the achievement of success with sustainable development and the attainment of the SDGs in Africa requires the linkage of sustainable development to the lived-experiences, indigenous knowledge systems, and contextual realities of the different peoples of Africa (Manteaw 2012). The latter is in alignment with the increased recognition that development initiatives that consider local perceptions and ways of doing and thinking are more likely to be successful and relevant to people’s needs and to support sustainable development, as alluded to by Sillitoe (1998) more than twenty years ago. Manteaw (2012) points out that the privileging as well as dominance of Western approaches to sustainable development over local approaches has not been helpful in advancing the SDGs in some regions of the world, and particularly in Africa. Template transfers of sustainability solutions between regions and between cultures have proven to be less successful. A spectrum of opportunities is provided for Africans to drive the transformative shifts required for the realization of the SDGs in a framework grounded in their knowledge and value systems, context and aspirations. The significant number of chapters in this book which deals with the nexus between indigenous knowledge systems, sustainable development and the African continent confirm the importance attached to the so-called ‘localisation’ (Jones and Comfort 2020) of the SDGs, thus making it relevant for Africa, by Africa. In this regard Chilisa (2017) holds the viewpoint that “The unequal power relations between indigenous and western academic knowledge are the greatest threat to any form of collaborative research that seek to address Africa’s sustainability challenges”. An approach relying only on contextualization, but without consideration of decolonization and a reflection on associated research philosophies, is not sufficient.

4 Conclusion

Africa's sustainable development requires stable living and economic spaces and the careful use of natural resources. This book has provided various examples of such research being conducted in Africa.

Governments in Africa need to adopt more pragmatic policies and treaties that address sustainability in the future (Asongu et al. 2020). Goal 1 of the African Union's Agenda 2063 is for Africa to be "a prosperous continent with the means and resources to drive its own development, with sustainable and long-term stewardship of its resources."

Such international goals take up the importance of the issue for sustainable development, poverty reduction and Africa's competitiveness: protection of ecosystems and their sustainable use are anchored in international environment and development goals and regulations on climate protection, biodiversity conservation, combating desertification and forest protection, among others.

Even though the goals set by Agenda 2063 may be feasible, it is noticeable that there is often an inverse correlation between government initiatives and environmentally sustainability. Therefore, it is vital that, in the future, governments work together with other relevant stakeholders such as higher education institutions, NGOs, companies and citizens, so as to maximise synergies and lead to more -and better- initiatives, which may support the efforts to pursue and ensure sustainability in their countries. In Africa, inter-governmental collaborations should be encouraged (Adekunle 2020), since they provide a good basis to gather capital and technical support, both within Africa and in cooperation with countries outside it.

The prospects of a sustainable Africa are bright. If carefully pursued and implemented. Taking into account aspects such as structural change and structural policy, agricultural investment, supply chains, digitalisation, trade policy, direct investment and job creation, and financial market development, sustainability policies in African countries will enable them to embark on a journey, where development may be socially acceptable, economically sound, ethically correct and socially just. Taken together, these provide suitable conditions for inclusive development, pursued along the lines of structural change and social cohesion in the region.

References

Adekunle IA (2020) On the search for environmental sustainability in Africa: the role of governance. Environ Sci Pollut Res 1–14.https://doi.org/10.1007/s11356-020-11432-5

African Union Commission (AUC) (2014). AGENDA 2063: The Africa we want. African Union Commission, Addis Ababa

AEO (African Energy Outlook) (2014) A focus on energy prospects on SubSaharan Africa. World energy outlook special report. www.iea.org. Accessed 10 Mar 2021

Akinsemolu AA (2020) The future: achieving sustainability. In: The principles of green and sustainability science. Springer, Heidelberg, pp 361–380

Asongu SA, Agboola MO, Alola AA, Bekun FV (2020) The criticality of growth, urbanization, electricity and fossil fuel consumption to environment sustainability in Africa. Sci Total Environ 712:136–376. https://doi.org/10.1016/j.scitotenv.2019.136376

August O (2013) Africa rising: a hopeful continent. The economist, 2 Mar 2013. https://www.economist.com/special-report/2013/03/02/a-hopeful-continent. Accessed 11 Mar 2021

Begashaw B (2020) Strategies to deliver on the sustainable development goals in Africa. In: Coulibaly BS (ed) Foresight Africa: top priorities for the continent 2020 to 2030, pp 9–13. Africa Growth Initiatives at Brookings

Chilisa B (2017) Decolonising transdisciplinary research approaches: an African perspective for enhancing knowledge integration in sustainability science. Sustainability Science 12(5):813–827

Coulibaly BS (2020) Foresight Africa: top priorities for the continent 2020 to 2030. https://www.brookings.edu/multi-chapter-report/foresight-africa-top-priorities-for-the-continent-in-2020/. Accessed 10 Mar 2021

De Amorim WS, Valduga IB, Ribeiro JMP, Williamson VG, Krauser GE, Magtoto MK, de Andrade JBSO (2018) The nexus between water, energy, and food in the context of the global risks: an analysis of the interactions between food, water, and energy security. Environ Impact Assess Rev 72:1–11

Ehui S, Kray H, Mghenyi E (2020) Policy priorities for achieving food and nutrition security by 2030. In: Coulibaly BS (ed) Foresight Africa: top priorities for the continent 2020 to 2030. Africa Growth Initiatives at Brookings, pp 56–59

EIU (The Economist Intelligence Unit) (2015) Global food security index 2015. An annual measure of the state of global food security. The Economist Intelligence Unit Limited, London

Gasparatos A, Takeuchi K, Elmqvist T, Fukushi K, Nagao M, Swanepoel F, Swilling M, Trotter D, von Blottnitz H (2017) Sustainability science for meeting Africa's challenges: setting the stage. Sustain Sci 12(5):635–640

Gilpin R (2020) Unpacking the implications of future trends for security in Africa. In: Coulibaly BS (ed) Foresight Africa: top priorities for the continent 2020 to 2030. Africa Growth Initiatives at Brookings, , pp. 31–33

Habimana Simbi C, Lin J, Yang D, Ndayishimiye JC, Liu Y, Li H, Xu L, Ma W (2021) Decomposition and decoupling analysis of carbon dioxide emissions in African countries during 1984–2014. J Environ Sci 102, 85–98

Holmes K, Scott E (2015) Obama lectures Kenyan president on gay rights. https://edition.cnn.com/2015/07/25/politics/obama-kenya-kenyatta/. Accessed 11 Mar 2021

Jones P, Comfort D (2020) A commentary on the localisation of the sustainable development goals. J Publ Aff 20(1):e1943

IPCC (Intergovernmental Panel on Climate Change) (2018) Global warming of 1.5 °C. ISBN 978-92-9169-151-7. www.ipcc.ch. Accessed 10 Mar 2021

Leal Filho W, Oguge N, Ayal D, Adeleke L, da Silva I (2021) African handbook of climate change adaptation. Springer, Cham

Mhangara P, Lamba A, Mapurisa W, Mudau N (2019) Towards the development of agenda 2063 geo-portal to support sustainable development in Africa. ISPRS Int J Geo-Inf 8(9):399

Maji IK, Sulaiman C, Abdul-Rahim AS (2019) Renewable energy consumption and economic growth nexus: a fresh evidence from West Africa Energy Rep 5:384–392

Manteaw OO (2012) Education for sustainable development in Africa: the search for pedagogical logic. Int J Educ Dev 32(3):376–383

Mbaku JM (2020) Good and inclusive governance is imperative for Africa's future in Coulibaly. In: Coulibaly BS (ed) Foresight Africa: top priorities for the continent 2020 to 2030. Africa Growth Initiatives at Brookings, pp. 23–30

Mbaye AA (2020) Confronting the challenges of climate change on Africa's coastal areas. In: Coulibaly BS (ed) Foresight Africa: top priorities for the continent 2020 to 2030, pp. 52–56. Africa Growth Initiatives at Brookings

Mehrabi Z, Gill M, van Wijk M, Herrero M, Ramankutty N (2020) Livestock policy for sustainable development. Nat Food 1(3):160–165. https://doi.org/10.1038/s43016-020-0042-9

Muswere A (2020) Is Africa capable of achieving the sustainable development goals in the 2020–2030 decade of programming? J Publ Admin Gov 10(4):5372–5372

Nhemachena C, Nhamo L, Matchaya G, Nhemachena CR, Muchara B, Karuaihe ST, Mpandeli S (2020) Climate change impacts on water and agriculture sectors in Southern Africa: threats and opportunities for sustainable development. Water 12(10):2673. https://doi.org/10.3390/w12102673

Nhamo G (2017) New global sustainable development agenda: a focus on Africa. Sustain Dev 25(3):227–241

Reynolds TW, Waddington SR, Anderson CL, Chew A, True Z, Cullen A (2015) Environmental impacts and constraints associated with the production of major food crops in Sub-Saharan Africa and South Asia. Food Secur 7(4):795–822

Sanya T (2020) Freshwater: towards a better understanding of a wicked problem. Environ Sci Sustain Dev 48. https://doi.org/10.21625/essd.v5i2.759

Schoneveld GC (2014) The geographic and sectoral patterns of large-scale farmland investments in sub-Saharan Africa. Food Policy 48:34–50

Sillitoe P (1998) The development of indigenous knowledge: a new applied anthropology. Curr Anthropol 39(2):223–235

Ssali MW, Du J, Mensah IA, Hongo DO (2019) Investigating the nexus among environmental pollution, economic growth, energy use, and foreign direct investment in 6 selected sub-Saharan African countries. Environ Sci Pollut Res Int 26:11245–11260

United Nations (2017) Sustainable development goals-17 goals to transform our world. United Nations. https://www.un.org/sustainabledevelopment/water-and-sanitation/

United Nations, Department of Economic and Social Affairs, Population Division (2019) World urbanization prospects: the 2018 revision (ST/ESA/SER.A/420). United Nations, New York

United Nations Educational Scientific and Cultural Organisation (UNESCO) (2020). The United Nations world water development report 2017: wastewater the untapped resource. United Nations Educational, Scientific and Cultural Organization (UNESCO), Paris. https://unesdoc.unesco.org/images/0024/002471/247153e.pdf

World Economic Forum (2019) The global risks report 2019 14th edn. In: African fastest-growing cities are the most vulnerable to climate change globally. https://www.weforum.org/agenda/2018/12/africa-s-fastest-growing-cities-are-the-most-vulnerable-to-climate-change-globally. Accessed 10 Mar 2021

Youssef AB, Boubaker S, Omri A (2018) Entrepreneurship and sustainability: the need for innovative and institutional solutions. Technol Forecast Soc Chang 129:232–241. https://doi.org/10.1016/j.techfore.2017.11.003

The manufacturer's authorised representative in the EU is Springer
Nature Customer Service Centre GmbH, Europaplatz 3, 69115 Heidelberg,
Germany. If you have any concerns regarding our products, please
contact ProductSafety@springernature.com

Printed and bound by CPI Group (UK) Ltd, Croydon, CR0 4YY
30/09/2026
02225690-0020